Applied Multidimensional Geological Modeling

Informing Sustainable Human Interactions with the Shallow Subsurface

Applied Multidimensional Geological Modeling

Informing Sustainable Human Interactions with the Shallow Subsurface

Edited by

Alan Keith Turner
Department of Geology & Geological Engineering
Colorado School of Mines
Golden, CO 80401
USA

British Geological Survey
Environmental Science Centre
Keyworth, Nottingham NG12 5GG
UK
Email: kturner@mines.edu

Holger Kessler
British Geological Survey
Environmental Science Centre
Keyworth, Nottingham NG12 5GG
UK
Email: hke@bgs.ac.uk

Michiel J. van der Meulen
TNO, Geological Survey of the Netherlands
PO Box 80015
NL-3508 TA Utrecht
The Netherlands
Email: michiel.vandermeulen@tno.nl

 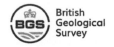 

This edition first published 2021
© 2021 John Wiley & Sons Ltd

The right of Alan Keith Turner, Holger Kessler, and Michiel J. van der Meulen to be identified as the authors of the editorial material in this work has been asserted in accordance with law.

Registered Offices
John Wiley & Sons, Inc., 111 River Street, Hoboken, NJ 07030, USA
John Wiley & Sons Ltd, The Atrium, Southern Gate, Chichester, West Sussex, PO19 8SQ, UK

Editorial Office
The Atrium, Southern Gate, Chichester, West Sussex, PO19 8SQ, UK

For details of our global editorial offices, customer services, and more information about Wiley products visit us at www.wiley.com.

Wiley also publishes its books in a variety of electronic formats and by print-on-demand. Some content that appears in standard print versions of this book may not be available in other formats.

Library of Congress Cataloging-in-Publication Data applied for
9781119163121

Cover Design: Wiley
Cover Graphics: © TNO

Set in 9.5/12.5pt STIXTwoText by SPi Global, Chennai, India

Printed and bound by CPI Group (UK) Ltd, Croydon, CR0 4YY
C9781119163121_010621

Contents

List of Contributors

The Editors

Alan Keith Turner
Department of Geology & Geological Engineering
Colorado School of Mines
Golden, CO
USA

British Geological Survey
Keyworth, Nottingham
UK

Holger Kessler
British Geological Survey
Keyworth, Nottingham
UK

Michiel J. van der Meulen
TNO, Geological Survey of the Netherlands
Utrecht
The Netherlands

The Authors

Mar Alcaraz
Institute of Environmental Assessment and Water Research
(IDAEA)
Spanish Council of Scientific Research (CSIC)
Barcelona
Spain

Cécile Allanic
Bureau de Recherches Géologiques et Minières (BRGM)
Orléans Cedex 02
France

Dirk Arndt
Geo-Explorers AG
Liestal
Switzerland

Kristian Bär
Institute of Applied Geosciences
Technische Universität Darmstadt
Darmstadt
Germany

Garry Baker
British Geological Survey
Keyworth, Nottingham
UK

Hugh F. Barron
British Geological Survey
Edinburgh
UK

Vivien Baudouin
Bureau de Recherches Géologiques et
Minières (BRGM)
Nantes, France

Johannes de Beer
Geological Survey of Norway (NGU)
Trondheim
Norway

John G. Begg
GNS Science
Lower Hutt
New Zealand

Wilbert Berendrecht
TNO, Geological Survey of the Netherlands
Utrecht
The Netherlands

Richard Berg
Illinois State Geological Survey
University of Illinois at Urbana-Champaign
Champaign, IL
USA

Marco Bianchi
British Geological Survey
Keyworth, Nottingham
UK

T. P. Bide
British Geological Survey
Keyworth, Nottingham
UK

Helen C. Bonsor
British Geological Survey
Edinburgh
UK

Bernard Bourgine
Bureau de Recherches Géologiques et Minières (BRGM)
Orléans Cedex 02
France

Helen Burke
British Geological Survey
Keyworth, Nottingham
UK

Abigail Burt
Ontario Geological Survey
Sudbury, ON
Canada

Freek S. Busschers
TNO, Geological Survey of the Netherlands
Utrecht
The Netherlands

Paula Cabrero
Formerly Dr. Sauer & Partners Ltd
London, UK

Philippe Calcagno
Bureau de Recherches Géologiques et Minières (BRGM)
Orléans Cedex 02
France

S. Diarmad G. Campbell
British Geological Survey
Edinburgh
UK

Ignace P.A.M. van Campenhout
Gemeentewerken Rotterdam
Rotterdam
the Netherlands

Séverine Caritg
Bureau de Recherches Géologiques et Minières (BRGM)
Orléans Cedex 02
France

Jerome Chamfray
Jacobs UK Ltd
London
UK

Anthony H. Cooper
Honorary Research Associate
Environmental Science Centre
Keyworth, Nottingham
UK

Pierre Conil
Bureau de Recherches Géologiques et
Minières (BRGM)
Nantes, France

Albert Corbera
Institute of Environmental Assessment and Water Research
(IDAEA)
Spanish Council of Scientific Research (CSIC)
Barcelona
Spain

Gabriel Courrioux
Bureau de Recherches Géologiques et Minières (BRGM)
Orléans Cedex 02
France

Catherine Cripps
British Geological Survey
Keyworth, Nottingham
UK

Martin Culshaw
Honorary Research Associate
British Geological Survey
Keyworth, Nottingham
UK

Willem Dabekaussen
TNO, Geological Survey of the Netherlands
Utrecht
The Netherlands

Oliver J. N. Dabson
Jacobs UK Ltd
London
UK

Roula M. Dambrink
Wageningen Economic Research
The Hague
The Netherlands

Maxime Delayre
Bureau de Recherches Géologiques et
Minières (BRGM)
Nantes, France

Michael A. Ellis
British Geological Survey
Keyworth, Nottingham
UK

David Entwisle
British Geological Survey
Keyworth, Nottingham
UK

Ross J. Fitzgerald
Jacobs UK Ltd
London
UK

Jonathan R. Ford
British Geological Survey
Keyworth, Nottingham
UK

Matthew Free
ARUP
London
UK

Johann-Gerhard Fritsche
Hessisches Landesamt für Umwelt, Naturschutz
und Geologie
Wiesbaden, Germany

Sunseare Gabalda
Bureau de Recherches Géologiques et Minières (BRGM)
Orléans Cedex 02
France

Angelos Gakis
Dr. Sauer & Partners Ltd
London
UK

Marieta Garcia-Bajo
British Geological Survey
Keyworth, Nottingham
UK

Alejandro García-Gil
Institute of Environmental Assessment and Water Research
(IDAEA)
Spanish Council of Scientific Research (CSIC)
Barcelona
Spain

Jennifer Gates
ARUP
London
UK

Frans van Geer
Formerly TNO, Geological Survey of the Netherlands
Utrecht, The Netherlands

Ben Gilson
ARUP
London
UK

Sascha Görne
Department of Geology
Saxon State Office for Environment, Agriculture and
Geology
Freiberg
Germany

Rubén C. Lois González
Department of Geography
Universidade de Santiago de Compostela
Santiago de Compostela
Spain

Andreas Guenther
Bundesanstalt für Geowissenschaften und Rohstoffe (BGR)
Hannover
Germany

Cécile Le Guern
Bureau de Recherches Géologiques et
Minières (BRGM)
Nantes, France

Jan L. Gunnink
TNO, Geological Survey of the Netherlands
Utrecht
The Netherlands

Christina Habenberger
Hessisches Landesamt für Umwelt, Naturschutz
und Geologie
Wiesbaden, Germany

Elizabeth D. Hannon
School of Engineering
Newcastle University
Newcastle upon Tyne
UK

Robin Harrap
Department of Geological Sciences and Geological
Engineering
Queen's University
Kingston, ON
Canada

Ronald Harting
TNO, Geological Survey of the Netherlands
Utrecht
The Netherlands

Eppie de Heer
TNO, Geological Survey of the Netherlands
Utrecht
The Netherlands

Heiner Heggemann
Hessisches Landesamt für Umwelt, Naturschutz
und Geologie
Wiesbaden, Germany

Richard Hosker
ARUP
London
UK

James Howard
JBA Consulting
Newcastle upon Tyne
UK

Andy Hulbert
British Geological Survey
Keyworth, Nottingham
UK

Jan H. Hummelman
TNO, Geological Survey of the Netherlands
Utrecht
The Netherlands

Ian Jackson
Formerly British Geological Survey
Seatsides, Bardon Mill
Northumberland
UK

Mitja Janža
Geological Survey of Slovenia
Ljubljana
Slovenia

Eva Jirner
Sveriges Geologiska Undersökning (Swedish Geological
Survey)
Uppsala
Sweden

P-O. Johansson
Artesia Grundvattenkonsult AB
Täby
Sweden

Katie E. Jones
GNS Science
Lower Hutt
New Zealand

Lee Jones
British Geological Survey
Keyworth, Nottingham
UK

Donald A. Keefer
Illinois State Geological Survey
University of Illinois at Urbana-Champaign
Champaign, IL
USA

Michael Kehinde
Environment Agency
Welwyn Garden City
UK

Samuel Kelley
Department of Earth & Environmental Sciences
University of Waterloo
Waterloo, Ontario
Canada

Matthias Kracht
Hessisches Landesamt für Umwelt, Naturschutz
und Geologie
Wiesbaden, Germany

Ottomar Krentz
Department of Geology
Saxon State Office for Environment, Agriculture and
Geology
Freiberg
Germany

Murray Lark
School of Biosciences
University of Nottingham
Sutton Bonington
Nottingham
UK

Matthew Lato
BGC Engineering Inc.
Ottawa, ON
Canada

Amanda K. Lawson
Delaware Geological Survey
University of Delaware
Newark, DE
USA

Rouwen Lehné
Hessisches Landesamt für Umwelt, Naturschutz
und Geologie
Wiesbaden, Germany

Elizabeth Lewis
School of Engineering
Newcastle University
Newcastle upon Tyne
UK

Melinda Lewis
British Geological Survey
Keyworth, Nottingham
UK

Simon Lopez
Bureau de Recherches Géologiques et Minières (BRGM)
Orléans Cedex 02
France

Aris Lourens
TNO, Geological Survey of the Netherlands
Utrecht
The Netherlands

Rolf R. Ludwig
Bundesanstalt für Geowissenschaften und Rohstoffe (BGR)
Hannover
Germany

Gerard McArdle
TSP Projects
York
UK

Duncan McConnachie
WSP Sverige AB
Stockholm-Globen
Sweden

Kelsey MacCormack
Alberta Geological Survey, Alberta Energy Regulator
Edmonton, AL
Canada

Peter P. McLaughlin, Jr.
Delaware Geological Survey
University of Delaware
Newark, DE
USA

D. Maljers
TNO, Geological Survey of the Netherlands
Utrecht
The Netherlands

J. M. Mankelow
British Geological Survey
Keyworth, Nottingham
UK

Jason Manning
ARUP
London
UK

Miguel Á. Marazuela
Institute of Environmental Assessment and Water Research
(IDAEA)
Spanish Council of Scientific Research (CSIC)
Barcelona
Spain

B. P. Marchant
British Geological Survey
Keyworth, Nottingham
UK

Brian Marker
Independent Consultant
London
UK

K. Mee
British Geological Survey
Keyworth, Nottingham
UK

Armin Menkovic
TNO, Geological Survey of the Netherlands
Utrecht
The Netherlands

Simon R. Miles
Atkins
Epsom
UK

Gary Morin
Previously Keynetix Ltd.
Redditch
UK

Martin L. Nayembil
British Geological Survey
Keyworth, Nottingham
UK

Miguel Pazos Otón
Department of Geography
Universidade de Santiago de Compostela
Santiago de Compostela
Spain

Björn Panteleit
Geological Survey for Bremen
Bremen
Germany

G. Parkin
School of Engineering
Newcastle University
Newcastle upon Tyne
UK

Andrés Payo
British Geological Survey
Keyworth, Nottingham
UK

Denis Peach
Honorary Research Associate
British Geological Survey
Keyworth, Nottingham
UK

Martin R.H.E Peersmann
Ministry of the Interior and Kingdom Relations
The Hague
The Netherlands

Mark S. Rattenbury
GNS Science
Lower Hutt
New Zealand

Reinder N. Reindersma
TNO, Geological Survey of the Netherlands
Utrecht
The Netherlands

Martin Ross
Department of Earth & Environmental Sciences
University of Waterloo
Waterloo, Ontario
Canada

Ingo Sass
Institute of Applied Geosciences
Technische Universität Darmstadt
Darmstadt, Germany

Baptiste Sauvaget
Bureau de Recherches Géologiques et
Minières (BRGM)
Nantes, France

Lars Schimpf
State Office for Geology and Mining (LAGB) Saxony-Anhalt
Halle (Saale)
Germany

Jeroen Schokker
TNO, Geological Survey of the Netherlands
Utrecht
The Netherlands

Katherina Seiter
Geological Survey for Bremen
Bremen
Germany

Philip Sirles
Collier Consulting
Lakewood CO
USA

Jan Stafleu
TNO, Geological Survey of the Netherlands
Utrecht
The Netherlands

Andrew J. Stumpf
Illinois State Geological Survey
University of Illinois at Urbana-Champaign
Champaign, IL
USA

Enric Vázquez-Suñé
Institute of Environmental Assessment and Water Research
(IDAEA)
Spanish Council of Scientific Research (CSIC)
Barcelona
Spain

Amanda Taylor
Department of Earth & Environmental Sciences
University of Waterloo
Waterloo, Ontario
Canada

Ricky Terrington
British Geological Survey
Keyworth, Nottingham
UK

Jason F. Thomason
Illinois State Geological Survey
University of Illinois at Urbana-Champaign
Champaign, IL
USA

Harvey Thorleifson
Minnesota Geological Survey
University of Minnesota
St. Paul, MN
USA

Jaime Tomlinson
Delaware Geological Survey
University of Delaware
DGS Building, 257 Academy Street
Newark, DE
USA

Violeta Velasco
Institute of Environmental Assessment and Water Research
(IDAEA)
Spanish Council of Scientific Research (CSIC)
Barcelona
Spain

Tamara J.M. van de Ven
TNO, Geological Survey of the Netherlands
Utrecht
The Netherlands

Ronald W. Vernes
TNO, Geological Survey of the Netherlands
Utrecht
The Netherlands

Nicholas Vlachopoulos
Royal Military College of Canada
Kingston, ON
Canada

Jacob Wächter
Hessisches Landesamt für Umwelt, Naturschutz
und Geologie
Wiesbaden, Germany

Colin Waters
Department of Geology
University of Leicester, Leicester
UK

Carl Watson
British Geological Survey
Keyworth, Nottingham
UK

Benjamin Wood
British Geological Survey
Keyworth, Nottingham
UK

Peter Wycisk
Institute of Geosciences and Geography
Faculty of Geosciences, Martin-Luther University
Halle (Saale)
Germany

Acknowledgments

Developing this book has been a substantial task that has taken about five years, and could not have been accomplished without the help of many people. First and foremost, we would like to acknowledge the over 100 authors who contributed to one or more of the 26 chapters. The majority of the authors work at geological surveys where their primary task is to provide society with information about geological resources and potential hazards in their jurisdictions. While they produce scientific publications to describe their activities, producing publications suitable for this book are not as central to their work as publications are for academics. The same is true, perhaps even more so, for the authors who work in the private sector. So, the editors offer a big "thank you" to the authors for making the effort, undoubtedly partly or even largely in their free time, to share their knowledge that would otherwise largely go unpublished.

Most of the geological survey authors are members of two overlapping communities of practice that have been instrumental in advancing the state-of-the-art of geological modeling. Since 2001, North American and European geologists have gathered in a series of biennial workshops, organized as fringe-events at Geological Society of America Annual Meetings and also as regular sessions. In 2011, a group of European participants decided to establish an equivalent *European 3D Geological Modeling Community* to help coordinate and exchange information among the geological surveys of Europe. By exchanging best practices, organizing modeling courses, formulating research questions, and interacting with model users and modeling software developers, these two communities have fostered a congruence of thinking and a considerable momentum which underpins this book and the science behind it.

During the work, the editorial team developed a *modus operandi* in which each took up a distinctive role and approach, gradually picking up speed and efficiency, and ultimately becoming a fully functional editorial board. The co-editors Michiel J. van der Meulen and Holger Kessler offer their sincere thanks to Editor-in-chief Alan Keith Turner for his dedication, energy, tenacity, and wisdom during the creation of this publication. He solicited and processed any type of input that authors provided, ranging from PowerPoint presentations and conference abstracts to full-blown journal-grade manuscripts, and he managed to morph all of those into a book that reads as one.

The editors thank the British Geological Survey (BGS) and TNO/Geological Survey of the Netherlands for supporting this publication financially and logistically, as well as granting the time for the co-editors to work on it. This book was produced with the back office support of Ina Vissinga (TNO/Geological Survey of the Netherlands) and Steve Thorpe (BGS); the editors thank them for their assistance. Securing copyright approvals was a major task with so many contributors providing material from numerous sources. The editors are deeply grateful to Bryony Chambers-Towers (BGS), Susan Krusemark (Illinois State Geological Survey), Marg Rutka (Ontario Geological Survey), and all the individual contributors who assisted with this task. N.N. Trabucho (TNO/Geological Survey of the Netherlands) created the book cover design. Finally, a "thank you" is due to the families of the editors and authors for putting up with many hours of time spent working on the book.

We hope that this book will be a valuable resource for those who want to begin geological modeling, to further develop geological models, or to apply geological models to resolving societal issues.

The Editors

Part I

Introduction and Background

1

Introduction to Modeling Terminology and Concepts

Alan Keith Turner[1,2], Holger Kessler[2], and Michiel J. van der Meulen[3]

[1]*Colorado School of Mines, Golden, CO 80401, USA*
[2]*British Geological Survey, Keyworth, Nottingham NG12 5GG, UK*
[3]*TNO, Geological Survey of the Netherlands, 3584 CB Utrecht, The Netherlands*

In response to various technological and economic trends, geological survey organizations (GSOs) have increasingly relied on digital solutions to support the collection, maintenance, and distribution of subsurface geological data and information for use by increasingly diverse stakeholder communities. Raper (1989, 1991) provided some initial discussions of the applications and key concepts governing the development and applications of 3-D geological models. These efforts have been guided by the definition of the basic requirements for subsurface characterization and modeling first stated by Kelk (1991): "*The industry requires a system for interactive creation of spatial and spatio-temporal models of the physical nature of portions of the Earth's crust.*" Kelk elaborated further by stating: "*these models must have the capability to effectively* **model & visualise the geometry** *of rock- and time-stratigraphic units, the spatial and temporal* **relationships** *between geo-objects, the variation in internal* **composition** *of geo-objects, the* **displacements** *or distortions by tectonic forces, and the* **fluid flow** *through rock units.*" These capabilities involve time-varying as well as spatially-varying phenomena; thus, the modeling process is multi-dimensional.

While computer-based subsurface geological models were initially created within individual projects and for specific applications, over the past two decades several GSOs have produced models at national, regional, and local scales suitable for use by multiple applications and have developed corporate data management protocols to manage both the models and the supporting databases. Many other GSOs and, increasingly, private-sector firms are creating and using models and are looking to the "early-adopters" among the GSOs for advice in model development and management.

However, there has so far been no comprehensive review of these subsurface geological modeling experiences, only a series of peer-reviewed articles in geological journals, and a small number of conference proceedings volumes. The first peer-reviewed papers written by end-users of shallow subsurface geological models have been published only recently. These demonstrate that the timely availability of appropriate geological information produces economic savings and reduces geotechnical risk to project developers (Chapter 4).

This book has been developed to provide a citable central source that documents the current capabilities and contributions of several GSOs and other practitioners in industry and academia that are already producing multidimensional geological models. Each of these groups has its own agenda, aims, and remit. The groups employ a variety of modeling approaches developed in response to local geological characteristics, historical data collections, and economic, societal, and regulatory requirements. The book's contents are arranged to provide a "shop window" describing what multidimensional geological modeling can do and the value of this information to a large and varied potential stakeholder community.

The subtitle emphasizes that this book focuses on applications related to human interactions with subsurface conditions; these interactions mostly occur in the shallow subsurface, typically considered to be within 100–200 m (300–600 ft) of the surface. This has been defined as the zone of human interaction. Several books address the applications of geological models of the deeper subsurface for petroleum and minerals exploration or recovery. These applications utilize data sources, data types, interpretation procedures, and software solutions that are not especially suitable, and are often unaffordable for extensive adoption, by many GSOs or by those involved with design and construction of civil engineering infrastructure, urban planning, or environmental evaluations. The economic and administrative realities of these user communities are quite different from the petroleum and minerals industries. Therefore, this book does not discuss methods used by the petroleum industry for deep regional or offshore explorations, nor does it extensively explore integrated environmental modeling. While such integrated modeling applications are logical extensions to the application of multidimensional geological models, they are still in early stages of development.

Initial development of shallow subsurface geological models was mostly related to assessments of groundwater resources or mitigation of sources of subsurface pollution. More recently, the applications of geological models have diversified considerably with the realization that geotechnical risk is inversely proportional to the level of detail and accuracy of the geological information. As discussed in Chapter 4, geological models have become an integral part of major civil engineering design, construction, and management projects (Turner et al. 2008, 2009). These applications have been enhanced as projects encounter the reality that the shallow underground space is increasingly the subject of multiple competing potential uses and the very shallow portions are increasingly congested with existing underground structures, transportation tunnels, and utility networks. In the last few years, geological models have been used to assess the potential of geothermal energy to provide a renewable and sustainable source of heating and cooling for buildings, and even for electrical power generation (Chapter 20). This is an important and rapidly growing application of geological models.

The need to ensure that new and existing developments are sustainable under a variety of external conditions has led to an awareness that geological subsurface information is important and must be considered by public sector initiatives to achieve greater efficiency, resilience, and cost savings when managing state-owned infrastructure. The general sustainability of urban areas under climate change scenarios has provided an additional impetus toward the use of multidimensional geological subsurface models by various planning studies. In the Netherlands, shallow subsurface information is now stored in an official register administered by the Geological Survey of the Netherlands, with legal requirements for its use by any projects involving the subsurface. Several other countries, including Norway, Australia, Singapore, and the United Kingdom, are considering the establishment of national geospatial strategies or registers. National subsurface geological models are increasingly recognized as an important and integral part of an official data repository.

Many of the stakeholders responsible for regulating or utilizing the shallow subsurface to resolve societal needs in environmental protection, civil infrastructure works, and urban planning are not trained in the geosciences. Support

for public debates and decision-making requires the ability to provide basic geological data in open and understandable formats – often by using only commonly accepted web-based graphical displays. Thus, the evolving applications of subsurface geological models are raising issues of communication as well as technology (Branscombe and MacCormack 2018). In fact, communication is becoming appreciated as the critical component in assuring the models are accepted and utilized to their full potential by diverse user communities. Unlike the resource-industry based user-community, many of today's potential users of geological subsurface models and visualizations cannot interpret basic geoscience data or evaluate the merits of alternative interpretations. They may be unable to distinguish between theories and facts. Turner (2006) states that these communities clearly desire *"solutions, not data"* and *"information in understandable form"* and proposed that model users be classified in terms of their information acceptance capabilities into two groups: users who can accept and interpret or evaluate a great deal of raw data, and relatively unsophisticated users desiring relatively simple and concise answers.

1.1 Mapping or Modeling – Which Is Correct?

GSOs exist as both national geological surveys and regional state or provincial geological surveys. Many GSOs are now heavily committed to the creation and dissemination of 3-D computer-generated geological models. These efforts are a logical modern version of their traditional role of providing geological maps. The development of digital technology made it economically justified for the GSOs to divert resources from printing and storing traditional maps in order to rely on digital media to store and disseminate their information. A considerable debate arose in some organizations as to whether this new digital approach was "3-D Geological Mapping" or "3-D Geological Modeling."

The transition from a paper-map publication environment to a fully digital computational approach adds considerable new flexibility and capability to the assessment and interpretation of subsurface conditions. It also represents a considerable change in operation and management functions within the GSO. Hence, it is more than a modern version of the traditional mapping function – although the importance of making geological observations remains at the heart of the process. Accordingly, this book will use the term "3-D geological modeling" and "3-D geological models" to describe these digital products and will refer to "3-D geological mapping" only when describing historical efforts.

In fact, the concept that 3-D geological models of the shallow subsurface should be used to support the planning and design of infrastructure projects, especially by supporting more effective site investigation, has been considered within the engineering geology community for several decades and predates the development of effective digital 3-D geological framework models (Parry et al. 2014). These conclusions were reached by Commission C25 of the International Association for Engineering Geology and the Environment (IAEG) which was charged with evaluating the "Use of Engineering Geological Models." It thus reflects the consensus of a respected group of noted engineering geologists who represent a broad range of international experience. It also documents the current position of the IAEG on the topic of geological models.

1.1.1 Definition of the Term "Model"

Parry et al. (2014) note that the term "model" is broadly used by scientists and engineers to define a variety of products, including scaled physical replicas, drawings, governing equations, and computer simulations. In fact, this variation in the use of the term is partly the reason for the controversy between "3-D geological mapping" and "3-D geological modeling." In response to this dilemma, Parry et al. (2014) proposed the following definition: *"A model is an approximation of reality created for the purpose of solving a problem."* While this definition is elegant, there is a caveat when applying it to the geological models issued by GSOs, which are, by definition – just like their map predecessors – not created for a single, known purpose. For example, the systematically produced, freely available, Dutch model GeoTOP (Chapter 11) was originally conceived as a refinement of a pre-existing hydrogeological model, but is now rapidly developing a geotechnical user community, and has even been used to predict ground motion amplification of induced seismicity in a major gas field. This is why adequate metadata of and communication about the model must be made available, in terms that can be understood outside the community of geomodelers, to inform potential users about its purpose and how this translates into its specifications and its uncertainties.

The definition of Parry et al. (2014) does not restrict "modeling" to only activities related to the use of computers and digital technologies; it refers to a much broader set of activities that approximate geological conditions, at varying scales, to solve a problem. The model is a hypothesis that must be tested, usually by some form of investigation. Usually, this investigation follows the classic scientific method, which typically involves a sequence of steps: (i) observation; (ii) defining a question or problem; (iii) research (planning, evaluating current evidence); (iv) forming a hypothesis; (v) prediction from the hypothesis (deductive reasoning); (vi) experimentation (testing the hypothesis); (vii) evaluation and analysis; (viii) peer review and evaluation; and (ix) publication. This is not a linear process but an iterative one because the results of the predictions, hypothesis testing, and evaluations (steps (v)–(vii)) frequently require refinement and revision of the problem definition (the conceptual model) and collection of additional observations (steps (ii) and (iii)).

1.1.2 Evolution of the Geological Model Concept

The concept of geological models is fundamental to geology. In the late 1790s, while working in the Somerset coalfields, William Smith realized that the rock layers were arranged in a predictable pattern and that the various strata could always be found in the same relative positions. Additionally, Smith observed that each stratum contained distinctive fossils and the same succession of fossils could be found in many parts of England. This gave Smith a testable hypothesis (a "geological model"), which he termed "The Principle of Faunal Succession." Initially he drew sketches showing the vertical extent of strata at various locations and developed some cross-sections and tables of what he saw. In 1799, Smith produced his first geological map for the area around Bath, Somerset, after seeing a colored map showing the types of soils and vegetation around Bath. Smith subsequently produced a series of maps showing the geology of England, Wales, and southern Scotland – with the most famous being the 1815 and subsequent 1828 editions. During the nineteenth century, geological maps were sometimes supplemented by 3-D physical models. Sets of small wooden models were developed to explain structural geology concepts (Turner and Dearman 1979) and museums included displays of larger 3-D models or dioramas to explain the geological subsurface of interesting or important locations.

The use of geological models, and maps, to support multiple applications, including engineering projects, is a more recent topic of discussion. Varnes (1974) wrote a carefully reasoned explanation of the "logic of geological maps" and especially how their interpretation can benefit engineering works. Varnes discussed the fundamentals of map design and purpose, their analytical capabilities, their limitations, and potential improvements. While this report predates any use of computer technology, it clearly addresses several aspects of geological modeling, which are illustrated by examples of several mapping approaches in Europe and North America that attempt to display 3-D geological conditions in the shallow (few tens of meters) subsurface. Beginning in the mid-1960s, prior to any computer-based methods, the Illinois State

Geological Survey began to develop an alternative unitary mapping system, which is termed "stack-unit mapping" (Kempton 1981). Chapter 5 contains additional details about these mapping approaches.

The Geological Survey of the Netherlands developed a rather complex version of unitary mapping, which Varnes (1974) referred to as "profile-legend maps," to show the great variety of relationships among the Dutch Holocene and Pleistocene deposits (Hageman 1963). In these maps, each color, supplemented where necessary with patterns, represents a particular succession of as many as five deposits, as well as the interfingering and erosional relations between them. Thus, these maps were a forerunner of the very extensive development of 3-D subsurface geological models by the Geological Survey of the Netherlands (TNO) beginning in the late 1980s and discussed in Chapter 11.

1.2 Why Use "Multidimensional"?

Some readers may wonder why the title of this book includes the word "multidimensional" rather than a specific number of dimensions. After all, aren't the subsurface models we are building three-dimensional (3-D)? The decision to use "multidimensional" reflects on some important research concepts in spatial analysis and the needs of geoscientists (Raper 2000).

Geological modeling can be more than 3-D volumetric representation. Geology involves time; time can be considered another dimension. So volumetric geological models that incorporate dynamic aspects, i.e. changes over time, are 4-D models (three spatial coordinates plus time). Kinematic modeling approaches have proven valuable in modeling complexly deformed geological environments (Wellmann et al. 2016). They use algorithms to evaluate and restore 3-D geological models through time while maintaining line length, area, and volume balancing principles. Because these approaches have only limited applications to most shallow subsurface investigations, they are not discussed further in this book. However, many applications of 3-D geological framework models discussed in Chapter 14 deal with time-varying phenomena and thus are 4-D models. Groundwater resource evaluations are 4-D, as are subsurface pollution studies using contaminant transport models. Analyses of geological hazards, the stability of underground excavations, and groundwater and surface water interactions also involve complex time-varying processes, and so are 4-D.

Van Oosterom and Stoter (2010) argue that scale should be considered as a dimension, making 5-D models a possibility. Such an implementation would potentially address the current problems of upscaling (or downscaling) models. In a 5-D view, 3-D model geometries could not only change over time but could also be represented at different resolutions, or levels of detail, on the scale dimension. Then it might be possible to represent the geometries at intermediate scales, creating truly scalable geological information. Research continues at Delft University of Technology by a research team led by Professor van Oosterom on 5-D applications to geoscience topics. Considerable additional 5-D research is underway within the Building Information Management (BIM) community, where 5-D BIM refers to the virtual modeling of buildings (3-D spatial objects + time + cost), and a 6-D BIM has been proposed that includes all aspects of life-cycle facility management.

Finally, there remains the issue of describing uncertainty (Chapter 15). The uncertainty (or validity) of the geological subsurface models results from several contributors. There are alternative methods of assessing uncertainty, ranging from qualitative to quantitative. These seem to always require separate discussions and displays; making the clear communication of uncertainty from model developer to model user rather difficult (Chapter 15). It seems that the communication of uncertainty would be enhanced if it was presented in parallel with the model, rather than as a separate series of texts and illustrations. Several years ago, it was suggested that the audio channel on a computer workstation could be used to produce appropriate sounds of approval or disapproval as the spatial locations within a model were interrogated. Nothing has resulted from this suggestion, and it may be impractical, or even annoying, but the idea shows that uncertainty has the characteristics of another dimension. So maybe in the future we will have 6-D geological models; three spatial coordinates + time + scale + uncertainty!

1.3 Evolution of Digital Geological Modeling

Digital geological maps were first produced in the early 1970s using computer-aided drafting methods when the Experimental Cartography Unit (ECU), established at the Royal College of Art in London, experimented with the production of topographic, geological, and geochemical maps in cooperation with the Institute of Geological Sciences (Rhind 1988). The production by the ECU of the 1:50 000 scale geologic map of Abingdon revealed the difficulties in matching geologic and topographic features (Rhind 1971). Throughout the 1970s, no readily available commercial software products provided the generic tools necessary for handling spatially distributed data (Rhind 1981). This changed rapidly within a decade, and the early 1980s saw the development of commercial software products that provided the generic tools necessary for handling spatially distributed data. "Geographic Information Systems" with the acronym "GIS" became widely adopted by government regulators and private-sector investors to support a tremendous variety of applications, including social and economic planning, marketing analysis, facilities management by utility and transportation interests, and environmental and resource studies (Burrough 1986; Maguire et al. 1991). Most of these applications relied on two-dimensional (2-D) spatial relationships. Accordingly, GIS developed around a 2-D data model. GIS evolved to provide limited 3-D display capabilities, such as contoured maps or isometric views by treating elevation (the z-axis) as a dependent variable, or attribute, of spatial location (defined by x and y coordinates). This approach was described as "2.5 D" capability because it is more than 2-D, but not a full 3-D representation.

At about the same time, Computer Aided Design (CAD) systems were developed for engineering applications. Based on a true 3-D data model, CAD systems can efficiently manage and display engineered objects, such as machine components, which are described by precise dimensions and well-defined boundaries. Several mine modeling software products were developed based on a CAD, rather than a GIS, approach (Houlding 1991, 1994). These systems readily supported mine design and operation functions, but had difficulty representing typical subsurface geological features described by relatively few scattered and often imprecise measurements representing complex shapes, often with transitional boundaries. Thus, it is not surprising that problems arose when CAD techniques were applied to geological modeling (Fisher and Wales 1992; Kelk and Challen 1992). Recognizing these problems, Mallet (1991, 1992) developed the Discrete Smooth Interpolation (DSI) algorithm as a core capability of the GOCAD software, which became a widely used geological modeling tool.

By the late 1980s it was apparent that the conventional GIS products, with their predominantly 2-D spatial viewpoint, could not entirely satisfy geological exploration requirements. Definition of subsurface conditions requires volumetric representation referenced to three orthogonal axes. A number of research initiatives began to explore possible extensions of conventional GIS to better serve the geoscience community. As far as can be determined, German geoscientists first used the term "geoscientific information system" to describe these new research systems (Vinken 1986, 1988, 1992a,b). At about the same time, Dutch geoscientists developing computer-based comprehensive data handling and modeling applications for oil and gas exploration used the terms "Geoscientific Information System" and the acronym "GSIS" to describe their applications (Ritsema et al. 1988). In contrast, Raper (1989) and many British researchers were referring to similar systems as "Geoscientific Mapping and Modelling Systems" with the acronym "GMMS." Turner (1991) and Bonham-Carter (1994) used the "GSIS" acronym to describe computer-based three-dimensional systems designed to handle the variety of geoscience data and analysis needs. However, many questioned the use of the acronym "GSIS," preferring the term "3-D GIS" because they consider these systems merely support standard GIS functions in a three-dimensional context.

However, the GSIS concept included more than support for a three-dimensional geometry. When attempting to define, model, or visualize subsurface conditions, the geoscientist is faced with extremely sparse data, considerable complexity in subsurface conditions, and various types of discontinuities. Thus, the GSIS concept specified additional capabilities and functions in order to be able to support subsurface conceptualization. Generally, these capabilities must include rapid and complex data interpolation and extrapolation procedures, provide for interactive data manipulation, and support rapid visualization. Also, they must support the range of distinctive data formats commonly encountered within a project and should provide ways to illustrate the uncertainty inherent in the conceptualization. Interpolation between widely spaced field observations requires geological knowledge to successfully replicate actual geological environments. Iterative methods involving assessments and progressive refinements add considerable time and cost to the creation of subsurface models (Turner 2006).

Initially, computers were of little assistance to 3-D data handling and representation problems. Memory was too expensive to handle the huge amounts of data required by 3-D assessments; computational speeds were too slow to perform the necessary calculations within a reasonable time; and graphical displays had too low a resolution or were much too expensive to produce useful visualizations. The advent of the modern computer workstation, with its enhanced memory and graphical capabilities, largely overcame these earlier constraints. During 1988 and 1989, critical performance/cost thresholds were crossed, and the initial commercial 3-D geological modeling software products were announced and demonstrated. These systems were still too expensive to be widely adopted outside the petroleum and mineral industries.

In the late 1990s the British Geological Survey (BGS) undertook a major effort to convert all their records and maps to digital formats. In the early 2000s, trends in computer hardware development finally produced affordable, more powerful, faster platforms incorporating more memory, greater computing speed, and even greater display capabilities. In 2000, the BGS began the Digital Geoscience Spatial Model (DGSM) research and development project to establish the modeling methodology and cyberinfrastructure required to manage and utilize the recently developed digital data resources. DGSM was first deployed in 2005 with seven main components (Smith 2005). DGSM combined the digital data sources with the recently developed geological modeling systems that explicitly supported the geological knowledge of the geologist and the traditional mapping methods of geologists (Chapter 10). Since this modeling approach is easier for some geologists to use than other methods, the DGSM workflow and its various components have been used routinely since then on all BGS 3-D modeling projects for a variety of clients (Rosenbaum and Turner 2003; Culshaw 2005).

While geological modeling was initially adopted as a central function at only a few GSOs, the subsequent availability of several additional affordable modeling systems has resulted in many additional GSOs applying a variety of modeling approaches to a more diverse set of applications of geological subsurface models (Berg et al. 2011). This is a fast-moving field and therefore this book provides a snapshot in time of current best practices along with a discussion of anticipated technological advances in Chapter 26.

1.4   Overview of the Book

The book has 26 chapters organized into five parts. This arrangement was adopted to group chapters into related multidimensional geological modeling topics. Although considerable effort was expended to eliminate repetition of material in different chapters, some reiteration was necessary to provide continuity of thought and to allow adequate explanation of specific topics. Such repetition was judged more acceptable than excessive cross-referencing within the text to other parts and chapters, as a result of that, all chapters read like individual papers.

1.4.1   Intended Audience

This volume has been written and organized to be a useful information source for any geologist, scientist, engineer, urban planner, or decision maker whose practice includes evaluation of underground space. The text has been designed to appeal to a diverse audience, including:

- Those responsible for the creation and management of multidimensional geological models, especially staff members working at GSOs throughout the world;

- Individuals and organizations wishing to use multidimensional geological models to design new underground facilities or to better evaluate, utilize, regulate, and manage the Earth's subsurface, especially in urban centers;
- Students in geoscience, environmental, and geotechnical fields with an interest in using multidimensional geological models to define subsurface conditions;
- Researchers needing a definitive information source for multidimensional geological modeling procedures.

Each of these audiences has different information needs and expectations; this book attempts to address them while maintaining a balance and some brevity within a reasonably complete single source of information concerning the construction, management, and utilization of multidimensional geological models. The authors hope this book will assist all of its readers in reaching the same goal – a better understanding of the subsurface to support evidence-based decision making.

The initial chapters discuss the organizational, legal, and economic contexts that influence the development and application of geological models. Comprehensive discussions of how the multidimensional geological models can be developed are linked to explanations of user needs and how models may be distributed to, and understood by, diverse user communities. Both topics are illustrated by case history examples. The book covers all stages of modeling capability development, with the aim of being as useful to organizations or individuals that aspire to develop such capability as it is to those who want to further or systemize existing capabilities. A final chapter concludes with an assessment of future possibilities and challenges.

Many students and researchers desire comprehensive references to the literature and discussions of case studies, state-of-the-art techniques, and research directions. Accordingly, considerable effort was expended in identifying suitable literature citations and in providing some representative case studies that illustrate current capabilities of multidimensional geological models. References to specialized and hard-to-obtain sources were avoided as much as possible; most of the cited references will be readily available through university and special libraries.

1.4.2   Part I: Introduction and Background

The first four chapters cover topics related to the relevance and importance of multidimensional geological models to broader societal needs. Chapter 1 introduces basic modeling terminology and concepts, explains the importance of the term "multidimensional," surveys the evolution of digital geological modeling, and provides an overview of the book's focus, audience, and chapter topics. GSOs are the primary developers of multidimensional geological models; Chapter 2 describes their role in model creation, management, and distribution. Chapter 3 introduces the importance of subsurface regulation and management and the role of multidimensional geological models in implementing appropriate procedures. Chapter 4 explores the economic case for understanding subsurface ground conditions with emphasis on the reduction of geotechnical risk during the site investigation process.

1.4.3 Part II: Building and Managing Models

Eleven chapters discuss, in some detail, the entire spectrum of scientific, technical, and economic issues that influence the creation of models. These chapters are most easily summarized as three subareas – technical considerations, alternative model building approaches, and model application and evaluation.

1.4.3.1 Technical Considerations – Chapters 5–8

Chapter 5 provides an overview of the entire multidimensional geological modeling process, which is defined in greater detail in the subsequent chapters. While the creation of multidimensional geological models may be achieved by any one of several distinctly different approaches, efficient model construction often requires the use of several methods in combination. Chapter 6 discusses the importance of establishing an efficient modeling process and addresses two major topics – the establishment of appropriate data and information management procedures to support the creation, storage and distribution of models, and the creation and management of suitable workflows to support the entire modeling process. Chapter 7 describes the various data sources used to create multidimensional geological models and the issues influencing the selection and utilization of these data. Chapter 8 assesses data management considerations associated with the organization and archiving of source data, previously developed models, and material properties data using examples from the BGS experience. It also discusses the important data management issues when integrated multiple models are used to assess complex situations, or when models extend across political boundaries and encounter diverse data classifications and standards.

1.4.3.2 Alternative Model Building Approaches – Chapters 9–12

Chapters 9–12 describe the four principal alternative approaches to model development. Chapter 9 describes how a 3-D model can be created by stacking a series of surfaces defining geological unit boundaries. This method allows the incorporation of legacy gridded surfaces developed by GIS procedures and is an attractive and efficient model generation procedure used by several geological organizations when modeling areas of relatively undeformed sedimentary strata. Chapter 10 discusses the creation of models based on digital borehole records and interpreted geological cross-sections, while Chapter 11 describes the creation and application of models formed by 3-D cellular voxel arrays using the example of the GeoTOP model in the Netherlands. Chapter 12 explores explicit and implicit approaches to integrated rule-based geomodeling.

1.4.3.3 Model Application and Evaluation – Chapters 13–15

Chapter 13 describes how geological models may be discretized (further subdivided) to allow for assignment of variable ground-property values required when analyzing groundwater flow or ground conditions for geotechnical design by finite-difference, finite-element, or discrete-element methods. Chapter 13 also briefly describes how stochastic modeling using discretized cellular models can assess heterogeneity and uncertainty of material properties estimates. Chapter 14 then expands upon the topics in Chapter 13, by describing how discretized 3-D geological framework models are used to evaluate time-varying (4-D) phenomena for groundwater, geothermal, and geotechnical applications, and to assess geohazards. It also briefly describes the use of integrated surface–subsurface models. Chapter 15 evaluates the causes of uncertainty in multidimensional geological models, how this uncertainty can be assessed, and how it can be communicated to users of the models.

1.4.4 Part III: Using and Disseminating Models

While the preceding chapters on approaches to modeling discuss conventional model applications, Chapter 16 focuses on emerging user needs in the field of urban planning and provides an overview of the importance of models as communication tools within existing decision-making processes. It discusses how, especially in urban centers, the provision of sustainable environments that are resilient to hazards posed by anticipated climate changes is leading to an increased reliance on information provided by multidimensional geological models.

Chapter 17 discusses alternative and novel technologies that can be used to deliver completed multidimensional geological models to potential users. Until recently, dissemination mostly involved peer-to-peer delivery of geological information and models from their producer (usually a GSO) to a primary model-user organization that was able to use the geological framework information to support further technical analysis and, ultimately, for decision making. The evolving demands for model distribution to much more diverse user communities, including the general public, local authorities, and public interest groups, require a simpler access to the models of interest; one that uses commonly available technologies that are free, or have a very low cost, and are easy to operate. Chapter 17 reviews a variety of distribution mechanisms including direct distribution of digital data files, use of specialty digital "viewers" and animations, and production of interactive documents, digital models, and new physical models by 3-D Printing or laser-etching of glass blocks.

1.4.5 Part IV: Case Studies

Chapters 18–25 contain 25 case studies that have been grouped into eight model application themes; each theme chapter includes a general explanation of the use of multidimensional geological models for the theme followed by several example case studies. Chapter 18 addresses the urban planning application with three case studies (Darmstadt, Germany, Glasgow, Scotland, and Dahka, Bangladesh). Chapter 19 illustrates groundwater evaluations using geological models with case studies from Sweden, Southern Ontario (Canada), the State of Delaware, and the Netherlands (REGIS II model).

Chapter 20 demonstrates a relatively new, but rapidly growing, application – the assessment of geothermal resources for heating/cooling – with three examples: Zaragosa, Spain; the German-Polish TransGeoTherm Project, and the Rhine graben geothermal study in the German State of Hesse.

Chapter 21 provides two examples of the use of models to support legislative and regulatory administration of urban area planning with an example from the Environment Agency of England and for the German State of Bremen. Chapter 22 explores the use of models to identify and analyze several geohazards with five case studies including utilization of a subsurface geological model for post-earthquake reconstruction planning in Christchurch, New Zealand, coastal cliff retreat at two locations in England, and assessment of contaminated sites in France and Slovenia.

Chapter 23 provides examples of geological models supporting the design and construction of transportation infrastructure with three United Kingdom examples: the new Crossrail Farringdon Station in London, several railway infrastructure renewal projects, and preliminary design for the Silvertown Tunnel under the Thames in east London. Chapter 24 shows how geological models can improve construction resource evaluations with examples from the United Kingdom and the Netherlands. Three case studies in Chapter 25 examine the geological modeling methods used in locations where historical settlement has produced anthropogenic deposits (often referred to as artificially modified ground) and how geological modeling assists in archeological studies.

1.4.6 Part V: Future Possibilities and Challenges

The final chapter, Chapter 26, discusses anticipated technological advances and their impacts on future multidimensional geological models. While predicting the future is imprecise, and actual developments often vary considerably from expectations, continued evolution of methods to convert data to information, information to knowledge, and knowledge to action seem likely to create new pathways to better understanding of geological systems. Future developments are likely to include improved data visualization options and better integration of data derived from and external to the geosciences, while open source portals and modeling systems will increase the range of application of geological models.

References

Berg, R.C., Mathers, S.J., Kessler, H. and D.A. Keefer [Eds.] (2011). *Synopsis of Current Three-dimensional Geological Mapping and Modeling in Geological Survey Organizations.* Champaign, IL: Illinois State Geological Survey, Circular 578. 92 pp. [Online: Available at: https://isgs.illinois.edu/publications/c578] (Accessed December 2017)

Bonham-Carter, G.F. (1994). *Geographic Information Systems for Geoscientists – Modelling with GIS.* Oxford, UK: Pergamon/Elsevier Science Publications. 398 pp. https://doi.org/10.1016/C2013-0-03864-9.

Branscombe, P. and MacCormack, K.E. (2018). Delivering to the client—communication and delivery for successful application of 3D models. In: *Three-Dimensional Geological Mapping: Workshop Extended Abstracts* (eds. R.C. Berg, K.E. MacCormack, H.A.J. Russell and L.H. Thorleifson), 24–27. Champaign, IL: Illinois State Geological Survey,

Open File Series 2018-1. [Online: Available at: https://isgs.illinois.edu/publications/ofs2018-1] (Accessed December 2017).

Burrough, P.A. (1986). *Principles of Geographical Information Systems for Land Resources Assessment.* Oxford, UK: Oxford University Press. 194 pp.

Culshaw, M.G. (2005). From concept towards reality: developing the attributed 3D geological model of the shallow subsurface (the seventh Glossop lecture). *Quarterly Journal of Engineering Geology and Hydrogeology* 38: 231–284. https://doi.org/10.1144/1470-9236/04-072.

Fisher, T.R. and Wales, R.Q. (1992). Rational splines and multi-dimensional geologic modelling. In: Computer Graphics in Modeling (eds. R. Pflug and J.W. Harbaugh), 17–28. Heidelberg, Germany: Springer-Verlag, Lecture Notes in Earth Sciences, vol. 41. https://doi.org/10.1007/BFb0117782.

Hageman, B.P. (1963). A new method of representation in mapping alluvial areas. *Verhandelingen van het Koninklijk Nederlands Geologisch Mijnbouwkundig Genootschap, Geologische Serie* 21–22, (2): 211–219.

Houlding, S.W. (1991). The application of new 3-D computer modelling techniques to mining. In: *Three-Dimensional Modeling with Geoscientific Information Systems* (ed. A.K. Turner), 303–325. Dordrecht, Netherlands: Kluwer Academic Publishers, NATO ASI Series C: Mathematical and Physical Sciences, vol. 354. https://doi.org/10.1007/978-94-011-2556-7_20.

Houlding, S.W. (1994). *3D Geoscience Modeling – Computer Techniques for Geological Characterization.* Berlin: Springer. 309 pp. https://doi.org/10.1007/978-3-642-79012-6.

Kelk, B. (1991). 3D modelling with geoscientific information systems: the problem. In: *Three-Dimensional Modeling with Geoscientific Information Systems* (ed. A.K. Turner), 29–37. Dordrecht, Netherlands: Kluwer Academic Publishers, NATO ASI Series C: Mathematical and Physical Sciences, vol. 354. https://doi.org/10.1007/978-94-011-2556-7_4.

Kelk, B. and Challen, K. (1992). Experiments with a CAD system for spatial modelling of geoscientific data. *Geologisches Jahrbuch A* 122: 145–153.

Kempton, J.P. (1981). *Three-Dimensional Geologic Mapping for Environmental Studies in Illinois.* Champaign, IL: Illinois State Geological Survey, Environmental Geology Notes 100. 43 pp. [Online: Available at: https://isgs.illinois.edu/publications/eg100] (Accessed December 2020).

Kessler, H., Turner, A.K., Culshaw, M.G. and Royse, K.R. (2008). Unlocking the potential of digital 3D geological subsurface models for geotechnical engineers. In: *Proceedings, European Conference of the International Association for Engineering geology, Madrid, Spain, 15–20 Sept 2008.* Madrid, Spain: Asociación Española de Geología Aplicada a la Ingeniería. 8 pp. [Online: Available at: http://nora.nerc.ac.uk/id/eprint/3817/] (Accessed August 2018).

Maguire, D.J., Goodchild, M.F. and Rhind, D. (1991). *Geographical Information Systems: Principles and Applications, vol.* 2. London, UK: Longman Scientific and Technical Press. 1096 pp.

Mallet, J.L. (1991). GOCAD: a computer aided design program for geological applications. In: *Three-Dimensional Modeling with Geoscientific Information Systems* (ed. A.K. Turner), 123–142. Dordrecht, Netherlands: Kluwer Academic Publishers, NATO ASI Series C: Mathematical and Physical Sciences, vol. 354. https://doi.org/10.1007/978-94-011-2556-7_11.

Mallet, J.L. (1992). Discrete smooth interpolation in geometric modelling. *Computer-Aided Design* 24 (4): 178–191. https://doi.org/10.1016/0010-4485(92)90054-E.

Parry, S., Baynes, F.J., Culshaw, M.G. et al. (2014). Engineering geological models: an introduction: IAEG commission 25. *Bulletin of Engineering Geology and the Environment* 73: 689–706. https://doi.org/10.1007/s10064-014-0576-x.

Raper, J. [Ed.] (1989). *Three-Dimensional Applications in Geographical Information Systems.* London, UK: Taylor and Francis. 280 pp.

Raper, J. (1991). Key 3D modelling concepts for geoscientific analysis. In: *Three-Dimensional Modeling with Geoscientific Information Systems* (ed. A.K. Turner), 215–232. Dordrecht, Netherlands: Kluwer Academic Publishers, NATO ASI Series C: Mathematical and Physical Sciences, vol. *354.* https://doi.org/10.1007/978-94-011-2556-7_15.

Raper, J. (2000). *Multidimensional Geographic Information Science.* London, UK: Taylor & Francis. 300 pp. https://doi.org/10.4324/9780203301227.

Rhind, D.W. (1971). The production of a multi-colour geological map by automated means. *Nachrichten aus dem Karten- und Vermessungswesen* 52: 47–52.

Rhind, D.W. (1981). Geographical information systems in Britain. In: *Quantitative Geography* (eds. N. Wrigley and R.J. Bennett), 17–39. London, UK: Routledge and Kegan Paul.

Rhind, D.W. (1988). Personality as a factor in the development of a discipline – the example of computer-assisted cartography. *The American Cartographer* 15 (3): 277–289. https://doi.org/10.1559/152304088783886928.

Ritsema, I.L., Riepen, M., Ridder, J. and Paige, S.L. (1988). *Global System Definition for Version 0.x of the GeoScientific Information System, SWS Document GSIS 1.1*. Delft, Netherlands: TNO, Netherlands Institute for Applied Geosciences.

Rosenbaum, M.S. and Turner, A.K. (2003). *New Paradigms in Subsurface Prediction: Characterisation of the Shallow Subsurface: Implications for Urban Infrastructure and Environmental Assessment*. Berlin: Springer, Lecture Notes in the Earth Sciences, Vol. 99. 397 pp. https://doi.org/10.1007/3-540-48019-6.

Smith, I.F. (ed.) (2005). *Digital Geoscience Spatial Model Project Final Report*. Keyworth, UK: British Geological Survey, Occasional Publication no. 9. 56 pp. [Online: Available at: http://nora.nerc.ac.uk/id/eprint/2366/] (Accessed December 2019).

Turner, A.K. [Ed.] (1991). *Three-Dimensional Modeling with Geoscientific Information Systems*. Dordrecht, Netherlands: Kluwer Academic Publishers, NATO ASI Series C: Mathematical and Physical Sciences, vol. 354. 443 pp. https://doi.org/10.1007/978-94-011-2556-7.

Turner, A.K. (2006). Challenges and trends for geological modelling and visualisation. *Bulletin of Engineering Geology and the Environment* 65 (2): 109–127. https://doi.org/10.1007/s10064-005-0015-0.

Turner, S. and Dearman, W.R. (1979). Sopwith's geological models. *Bulletin of the International Association of Engineering Geology* 19: 331–345. https://doi.org/10.1007/BF02600498

Turner, A.K., Kessler, H., Ford, J. and Terrington, R. (2009). Geological models for engineers – finally overcoming the barriers. AEG News 52: 108–109.

Van Oosterom, P. and Stoter, J. (2010). 5D data modelling: full integration of 2D/3D space, time and scale dimensions.

Varnes, D.J. (1974). *The Logic of Geological Maps, with Reference to Their Interpretation and Use for Engineering Purposes*. Reston, VA: U.S. Geological Survey, Professional Paper 837. 48 pp. https://doi.org/10.3133/pp837.

Vinken, R. (1986). Digital geoscientific maps: a priority program of the German society for the advancement of scientific research. *Mathematical Geology* 18: 237–246. https://doi.org/10.1007/BF00898285.

Vinken, R. [Ed.] (1988). *Construction and display of geoscientific maps derived from databases*. Stuttgart, Germany: Scheizerbart, Geologisches Jahrbuch, Series A, Vol. *104*. 475 pp.

Vinken, R. [Ed.] (1992a). *From geoscientific map series to geo-information systems*. Stuttgart, Germany: Scheizerbart, Geologisches Jahrbuch, Series A, Vol. 122. 501 pp.

Vinken, R. (1992b). From digital map series in the geosciences to a geo-information system. Geologisches Jahrbuch A 122: 7–25.

Wellmann, J.F., Thiele, S.T., Lindsay, M.D. and Jessell, M.W. (2016). Pynoddy 1.0: an experimental platform for automated 3-D kinematic and potential field modelling. *Geoscientific Model Development* 9: 1019–1035. https://doi.org/10.5194/gmd-9-1019-2016.

2

Geological Survey Data and the Move from 2-D to 4-D

Martin Culshaw[1], Ian Jackson[1], Denis Peach[1], Michiel J. van der Meulen[2], Richard Berg[3], and Harvey Thorleifson[4]

[1] *Retired from British Geological Survey, Keyworth, Nottingham NG12 5GG, UK*
[2] *TNO, Geological Survey of the Netherlands, 3584 CB Utrecht, The Netherlands*
[3] *Illinois State Geological Survey, University of Illinois at Urbana-Champaign, Champaign, IL 61820, USA*
[4] *Minnesota Geological Survey, University of Minnesota, St Paul, MN 55114, USA*

2.1 Introduction

Geologists are involved in the acquisition and interpretation of geological data and information. This chapter discusses the role of geological surveys, the sorts of data they hold, and how these data holdings provide the public with information useful for multiple applications. Geological surveys have a specific interest in maintaining an accessible archive of geological information. This contrasts with the interests of geologists employed by commercial or academic entities where most projects are relatively short term (less than 10 years), and upon project completion the data are either destroyed or archived without adequate cataloguing.

The authors of this chapter are current or former senior executives of the British Geological Survey (BGS), the Geological Survey of the Netherlands (GDN), and the geological surveys of Illinois and Minnesota in the United States. These organizations have proactively developed and used digital geological models and have experienced the impacts of these new technologies. However, they also represent a much larger group of geological surveys which are increasingly relying on digital geological models to satisfy their overall missions and day-to-day operations.

2.2 The Role of Geological Survey Organizations

A geological survey may be distinguished from all other types of geological organization by its commitment to gather, archive, and interpret geological data. These activities are likely to be *national* or *regional* in coverage, *strategic* in their role for the country or region and *long-term* in that the geological data acquired are intended to be stored, catalogued, managed and used for posterity. *Long-term* also implies that data collection and storage should go on almost indefinitely. The general purpose of geological surveying is to explore natural capital and natural hazards.

2.2.1 Establishment of Geological Surveys

The BGS is the oldest such undertaking to have functioned continuously since its inception, in May 1832, and formal establishment on 11 July 1835 (Bate 2010). In common with several other countries, the British Government had previously recognized the importance of accurate topographical maps. The Ordnance Survey (then the Ordnance Trigonometrical Survey) had been formed in 1791 to produce topographical maps of Britain and (subsequently) Ireland. When Major General Thomas Colby took over as Superintendent of the Ordnance Survey in 1820, he intended that the topographical maps should serve as the base for geological survey.

In 1826, Henry De la Beche published a geological map of South Pembrokeshire (Figure 2.1) using recently completed one-inch to the mile Ordnance Survey maps (De la Beche 1826). Subsequently, De la Beche undertook geological surveys in Cornwall and Devon, again using Ordnance Survey topographical maps. In the early 1830s, several proposals to add geological information to Ordnance Survey maps led to the establishment in 1835 of the Geological Ordnance Survey, the precursor of the BGS, with De la Beche as its first director (Bate 2010). In 1827, Captain J.W. Pringle was asked to form a geological branch of the Ordnance Survey in Ireland, but the work lapsed when it became clear that the surveyors were not adequately qualified to supply the

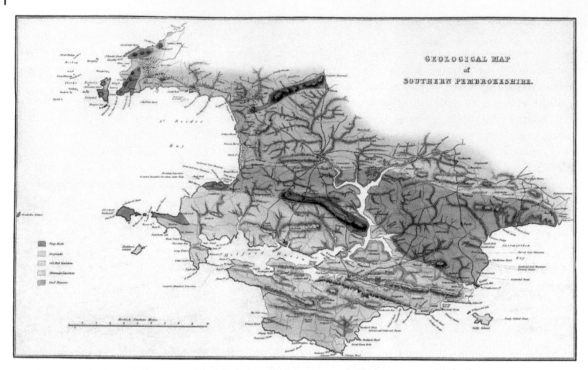

Figure 2.1 Geological map of southern Pembrokeshire, surveyed by De la Beche in 1822 and published in 1826. (De la Beche 1826).

geological information. Subsequently, Richard Griffith, a Fellow of the Geological Society, took over the work in Ireland and, in 1838, published a geological map of Ireland utilizing a topographical base specially prepared by the Ordnance Survey (Davis 1950; Bate 2010). In 1845, the Geological Survey of Great Britain and Ireland was formed as an independent organization (Owen and Pilbeam 1992).

Charles Lyell, in his 1836 Presidential Address to the Geological Society, described the establishment of the Geological Ordnance Survey and noted that a report (prepared by Lyell, William Buckland, and Adam Sedgwick) stated that it was

> … fully our opinion as to the great advantages which must accrue from such an undertaking not only as calculated to promote geological science, which would alone be sufficient object, but also as a work of great practical utility, bearing on agriculture, mining, road-making, the formation of canals and railroads and other branches of national industry.
>
> (Bailey 1962)

In other words, the Geological Survey was seen as having a strong practical emphasis. By the First World War (1914–1918), some 80 years later, Britain and Ireland had been mapped at 1:63 360 (1 in. to 1 mile) scale.

The first geological survey on the European mainland was established in 1849 by the Austrian Empire as the

Kaiserliche und Königliche Geologische Reichsanstalt; Spain also established a geological survey in 1849. In fact, most European countries had started or were about to start mapping their geology by the mid-nineteenth century, but this activity was not always immediately institutionalized. The French Geological Survey (*Service de la Carte Geologique Detailleé de la France*), for example, was not inaugurated until 1868, but engineers of the *Corps Royal des Mines* began a geological survey of France as early as 1825 (Eyles 1950). The survey was completed by 1835 (Brochant de Villiers 1835), but lack of a suitable topographical base-map delayed publication of the map until 1841 (Bate 2010). The French Geological Survey was established when more detailed geological mapping, funded by the regional *Départements* and under the supervision of local mining engineers, proved ineffectual (Savaton 2007).

These early European geological surveys were established in countries where geology is a manifestly important element of the landscape, and geological mapping was an obvious extension of topographic surveying. Developing industrialization needed new supplies of energy, metals, and minerals. But even though such practical use of geological surveying was advocated by pioneers such as Lyell, strong national geological communities often independently encouraged and undertook geological mapping to answer curiosity-driven questions as well as for economic reasons.

For example, in the Netherlands, which first appeared on a geological map almost entirely as "terrains with mammal bones" (*terrains mastozootiques*, d'Omalius d'Halloy, 1823), a geological community encouraged geological mapping efforts because the colonies and the metropole needed geological information to serve agricultural development. The first round of systematic geological mapping of the Netherlands at a scale of 1:200 000 – first proposed in 1844, initiated in 1852 and delivered in 1868 – was undertaken by W. Staring, who is now considered a pioneer by both the Dutch soil science and geological communities (Staring 1844, 1858–1869). His first maps won a prize at the World Exhibition of 1860 in London, for the revolutionary way in which Quaternary deposits were subdivided and mapped. After completion, however, geological surveying was discontinued until 1903 when a state agency was established to explore natural geological resources (primarily coal, Van Waterschoot van de Gracht 1918). The GDN was formally established in 1918, tasked with systematic geological mapping, which has continued until the present. By 1918, virtually all European countries had a geological survey.

In the USA during the early nineteenth century, scientific investigations and the construction of public works were considered the responsibility of the individual States or private institutions rather than the Federal Government. Farmland in the Eastern and Southern States was beginning to lose its fertility; this spurred westward migration. Between 1823 and 1836, several eastern States initiated geological surveys in support of agriculture. Identifying mineral resources was still of lesser economic importance and thus was of secondary interest. These early surveys had relatively short durations; they were provided with very small financial appropriations and were terminated when funds were exhausted. In 1836, New York established the first continuously functioning state survey (Hayes 1911; Socolow 1988; Rabbitt 1989; Bate 2010). Hendrickson (1961) provides a useful overview focusing on the role of government in funding these early geological surveys.

The Federal Government established a Corps of Topographical Engineers in 1838 to explore and map the entire USA. For the next 20 years the Topographical Engineers provided geologists with opportunities to study the western territories. The discovery of gold in California expanded interest in mining and mineral exploration throughout the country. Several southern and mid-western States established new State surveys in the 1850s to encourage and support mineral exploration. In 1853, Congress appropriated $150 000 for surveys to locate the most practical and economic route for a railroad between the Mississippi River and the Pacific. Following the Civil War, four major surveys of the western USA were authorized with geology the principal objective. Clarence King led a survey which explored along the fortieth parallel from northeastern California, through Nevada, to eastern Wyoming between 1867 and 1872; Ferdinand Hayden conducted several surveys in the Nebraska Territory between 1867 and 1872; John Wesley Powell conducted two trips along the Colorado River in 1869 and 1870, and subsequently was funded to make additional reports; and Lieutenant George Wheeler, Engineer Officer with the Army's Department of California (which covered California, Nevada, and Arizona), undertook surveys during 1869 and 1871 in southern and eastern Nevada and Arizona (Rabbitt 1989). Ultimately, in 1879, after several intermediate steps, Congress established the U.S. Geological Survey within the Department of Interior. Clarence King was its first director.

2.2.2 Systematic versus Strategic Mapping Approaches

In the nineteenth century, when many Geological Surveys were founded, they did not necessarily operate strategically. For example, the BGS began systematically mapping starting in the south and progressing northward; this was believed to be the best approach to acquiring geological information and knowledge in a form suitable for multiple applications. However, fears about the exhaustion of coal resources led to the adoption of a more strategic approach; geological mapping was prioritized in those areas where coal resources were suspected to exist at depth below Permo-Triassic strata. Mapping elsewhere was reduced or stopped; some areas mapped in the mid-nineteenth century were not remapped for 150 years, while other areas were never mapped.

2.2.3 Geological Mapping by Geological Surveys

Geological maps are a means to an end, not an end in themselves – however much geologists would like to think otherwise. Geological models are precisely the same. While a geological model may seem to be a far more dynamic and superior way to convey understanding of the subsurface within the geoscience community, it does not usually significantly aid the comprehension of many important user communities – even if presented via visualization and virtual reality systems.

Geological models produced by geological surveys exist to answer a fundamental question: "*What are the properties of that piece of the Earth at the coordinates x, y, z (and t)?*" That question may be posed in different ways by other disciplines, such as miners, civil engineers, or insurance brokers. The model should give geoscientists the essential, baseline knowledge to begin to answer that question in a way that best helps the person asking it. Depending on the user and the application, the answer can be lithologic or stratigraphic; it can be used to support resource assessment, define bearing strength, or evaluate probability of geohazard. Delivering a digital geological model does not obviate a geological survey from developing products and services that communicate to those who are, and should be, using geological knowledge outside the geological profession. However, the model should make any interpretation of the subsurface a lot more accurate, consistent, reliable, and responsive.

2.2.4 Difficulty in Maintaining Adequate Financial Support

Many geological surveys are facing the challenge of establishing appropriate budgetary support levels and determining financial sources to support the costs resulting from the transition from traditional geological mapping approaches to 3-D geological mapping programs. In spite of this challenge, national, state, and provincial geological surveys are embracing 3-D geological modeling because this provides appropriate information required by their diverse clients. The decision-making based on geological information ranges from energy and mineral resource assessments required by the government and private-sector companies, to public–private collaborations on infrastructure developments, and various government regulatory agencies.

Bhagwat and Berg (1991) suggest that the value of geological information is often augmented by its use as new information combined with existing knowledge. Häggquist and Söderholm (2015) provide an extensive review of the alternative methods for assessing the economic value of geological information and address the difficulties in accurately valuing the impact of geological information. Numerous assessments of the value of geological mapping have been conducted in many countries by both national and regional geological surveys, largely in support of their budgetary requests (Bernknopf et al. 1993, 1997, 2007; Bhagwat and Berg 1991, 1992; Bhagwat and Ipe 2000; Halsing et al. 2004; Ovadia 2007; Scott et al. 2002). While there are considerable variations in the assessment methods, the general conclusion is that geological mapping provides significant economic benefits (Häggquist and Söderholm 2015). Because 3-D geological modeling is a more recent development, there have been fewer assessments of its economic value, but these also identify positive economic values (Berg 2005; Diepolder and Lehné 2016; Gill et al. 2011; Raiber et al. 2012; Tame et al. 2013). Some additional

assessments have identified positive economic benefits derived from digital geoscience data holdings and the digital delivery of geological information (Castelein et al. 2010; de Mulder and Kooijman 2003; Frey et al. 2016; Giles 2006; Riddick et al. 2017).

National or regional governments are typically the ultimate owners or custodians of the data on behalf of their nation or region. Thus, it can be argued that the government should be responsible for bearing much of the cost of obtaining and looking after the geological data. However, while non-geological government authorities may see the benefit of holding the data and making it available to users, they frequently do not appreciate or understand why continued acquisition of geological data is critically important. In most developed countries where geological surveys have been established for decades, the diminishing return-on-investment of repeated geological re-mapping and revision of 2-D geological maps has been recognized. Systematic and strategic data collection has become the goal of these geological surveys. The move toward digitization of the data that took place at around the end of twentieth century was a major change; it emphasized the importance of spatial geological data. Developing and compiling models from the extensive data held within corporate information systems and re-purposing existing models is becoming the predominant national modeling function of these geological surveys. As discussed in Section 2.5, this approach depends on effective data management. There is a clear case for treating geoscience data as a national asset and managing it professionally.

In contrast, geological surveys in many countries where mineral or energy resource development forms a major component of their GDP (Gross Domestic Profit) continue to undertake major field data acquisition programs. While these include geological mapping, because it is required to provide the necessary basic information, many of these programs also involve 3-D geological modeling and digital data management (Berg 2005; Gill et al. 2011; Raiber et al. 2012; Tame et al. 2013).

The level of financial support provided to geological surveys is subject to the unique current political and economic climates and trends in each country. While some geological surveys have recently experienced level or reduced budget authorizations (e.g. Buchanan 2016), others have experienced a gradual long-term increase in funding as the availability of digital geoscience data has demonstrated value for resolving additional societal and economic goals. For the USA and Europe, this is shown by data from the American Association of State Geologists (www.stategeologists.org) and EuroGeoSurveys (www .eurogeosurveys.org), respectively.

2.3 Challenges Facing Geological Survey Organizations

The unique mission of any geological survey is to be the provider of modern geoscience spatial knowledge for *all* their nation or region, not merely localized and spasmodic information from research locations or projects – the domain of academia. In the context of 3-D modeling, this mission presents geological surveys with their most daunting challenge – how to develop and deliver a practical and competent strategy to model their territories. Many of the (approximately) 200+ geological survey organizations in the world are continuing to develop a strategy that meets their unique circumstances. While several geological surveys have produced sophisticated 3-D models, with very few exceptions the models are restricted to specific parts of their territory and typically address specific economic or scientific questions posed by paying clients. Delivering such models is no longer difficult; the software is available, as are the digital data and the workflows.

However, because these models support localized (or tactical) research projects rather than strategic survey projects, they generally have different resolutions and diverse parameters. A larger challenge for geological surveys is how to specify and resource the creation of a multi-purpose national model, or a series of compatible and integrated models for the entire nation or region. This is difficult for geological surveys; they know from their own research project experiences, and from the experiences of the front-running surveys, that five important capabilities must be in place to successfully implement such a vision: (i) comprehensive digital databases for the prime geological input parameters; (ii) cost-effective software that is straightforward to use; (iii) established appropriate workflows, standards, and protocols; (iv) systems to manage models and data and to allow discovery and re-purposing; and (v) personnel with the right skills and experience. Thus, setting out on such a venture will mean for many surveys not only a radical reorganization of their data management approaches and staff competencies; it will also require changes to their science priorities, structures, and cultures.

How these challenges are perceived by any survey depends on several things. Where is the survey on the digital evolution pathway? How much of its data remains analog? Does it manage its data as a corporate asset, or is it still a personal and project-based operation? Are there skilled personnel available, or do staff have the aptitude to re-skill as information system specialists and geological modelers? Is the survey prepared to devote sufficient financial resources and priority to achieving a sustainable

3-D model? This quite often boils down to a matter of prioritization between 2-D and 3-D mapping efforts, a rather radical example of which is presented by the Netherlands.

In the late 1990s, the Dutch geological survey, at that time a government agency embedded in the Ministry of Economic Affairs, became incorporated into the Nederlandse Organisatie voor Toegepast Natuurwetenschappelijk Onderzoek (TNO) (in English: Netherlands Organization for Applied Scientific Research) to achieve more coherency and efficiency in the applied earth sciences in the Netherlands. One of the first major decisions taken by the new organization was to discontinue traditional systematic geological mapping, as the ongoing second round of revised national geological maps, which was initiated in the 1950s, had lost all momentum and support and was considered to be too costly to pursue. This radical step, which most likely would not have been taken in the precursor organization, freed funds for a substantial investment in 3-D mapping, which eventually evolved into a suite of systematically produced models delivered by a program that has experienced no serious problems in maintaining funding since (further discussed in Section 2.8).

Thus, more than anything, the future depends on the chosen mission and vision of a geological survey, the priorities of its leadership team, and its ability to organize itself to achieve its goals. Some surveys have made this adjustment; they have revised their corporate structure and are delivering the best resolution geological models of their territories. Many other surveys are in transition, but recognize these adaptations are fundamental to their function and are required to meet societal needs in the twenty-first century. Geological modeling is increasingly recognized as the raison d'être and prime responsibility of geological surveys.

2.4 A Geological Map is Not a Piece of Paper

The geological map is a two-dimensional representation of a three-dimensional object – the ground. As such, it is a form of geological model. The use of the term 'map' to describe this model confuses it with topographical or hydrological maps that provide only a representation of a surface. This difference is fundamental. The makers of topographical maps can observe almost 100% of what they record, whereas geologists might see less than 1%. The geological map is, therefore, mostly an interpretation.

The earliest survey geologists, field geologists, developed 3-D models; they and their successors were aware of, and attempted to understand and interpret, the multiple dimensions of the rocks they surveyed. They held 3-D and

4-D models in their heads but the only way they could communicate these models was as 2-D maps and sections on a 2-D medium – paper. However, it was realized from the beginning that the information being presented was of a 3-D ground model. The problem was how to visualize it. Geological maps were accompanied by vertical and horizontal sections. However, these were quite hard for non-geologists to understand; thus, solid geological models and dioramas were constructed and displayed in the Museum of Practical Geology in London (now part of the Natural History Museum). In 1841 Thomas Sopwith began to produce small wooden geological models (Turner and Dearman 1979). Sets of twelve small 7.5 by 7.5 cm (3 by 3 in.) models were sold to the public and adopted by universities as instructional tools; some could be reconfigured, allowing subterranean structures to be viewed from different angles.

2.4.1 Early Geological Maps

In 1684, the British physician and naturalist Martin Lister suggested that the distribution of the different soil types of the British landscape could accurately be represented on a topographic map:

The Soil might either be coloured, by variety of Lines, or Etchings; but the great care must be, very exactly to note upon the Map, where such and such Soiles are bounded… Now if it were noted, how far these extended, and the limits of each Soil appeared on a Map, something more might be comprehended from the whole, and from every part, then I can possibly foresee, which would make such a labour very well worth the pains,

(Lister 1684)

While Lister may be justly credited with being the first to realize the importance of a geological survey, he never created a map based on this idea.

Early attempts at geological mapping were hampered by the lack of suitable topographical base maps; thus, as discussed in Section 2.2.1, much initial geological surveying and mapping was conducted by government-sponsored topographical surveys, often termed "trigonometric" or "ordnance" surveys. Nevertheless, in 1809, William Maclure published his "*Observations on the Geology of the United States*" which included a colored geological map of the regions east of the Mississippi River (Figure 2.2). This effort pre-dated any government support; Maclure personally financed the survey and publication.

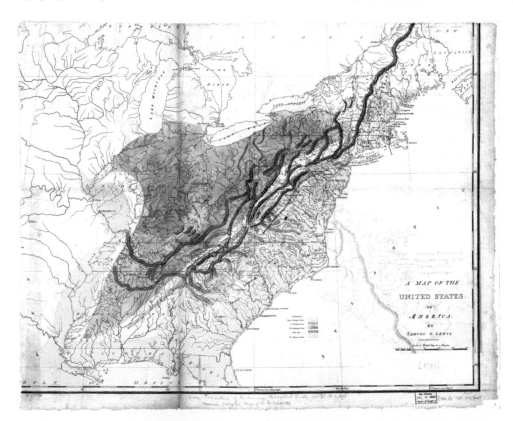

Figure 2.2 The first geological map of the United States. Source: Maclure et al. 1809; Library of Congress, Geography, and Map Division.

Even though the basic principles of stratigraphy were formulated as early as in the late seventeenth century (Steno 1669), they were not used to the fullest extent in many early geological maps a century later, even though they are a sine qua non for consistent mapping of rock sequences and their 3-D structure. The first modern geological maps, i.e. ones that did use stratigraphic principles, were produced by William Smith in Britain (Smith 1815), by Georges Cuvier and Alexandre Brongniart in France (Savaton 2007), and by Abraham Gottlob Werner in Germany. Smith was able to trace sequences of rocks because he recognized groups of fossils characterized rocks of a particular age. In 1799 he produced a small geological map of the area around Bath (in the west of England) based on the relative age of the strata (Figure 2.3). A similar approach was followed by Cuvier and Brongniart mapping the Paris basin (Cuvier and Brongniart 1811). Werner was commissioned to map the Kingdom of Saxony in 1791, the

map was completed and published posthumously by his successor Georg Naumann with the aid of Bernhard von Cotta in 1846. Werner's mapping protocol honored stratigraphic principles and was modern in that sense (Werner 1811), even though his approach was rooted in the now obsolete concept of Neptunism that defined crystalline rocks as being chemical precipitates formed in the early Earth's oceans (Werner 1787).

2.4.2 Early Digital Mapping and Modeling

Throughout much of the twentieth century the interpretation and depiction of rocks and sediments by geological surveys was largely, as it had been in the nineteenth century, shackled by the limitations of the two dimensions and inflexibility of paper. From the publication of William Smith's first geological maps until the 1980s, all but the simplest maps could only be visualized on paper or film;

Figure 2.3   William Smith's 1799 geological map of Bath, UK. Source: BGS digital image of original map by William. Smith, presented by Smith to Geological Society, London on 18 February 1831.

although some earlier experiments explored the potential of digitally producing geological maps (Rhind 1971) and some forms of digital geological contouring were evaluated (Coe and Cratchley 1979). As digital data and systems matured, and the expertise of those using them grew, new options became available at the turn of the millennium. Not surprisingly, the information technology revolution has had as big an effect on the development of geological maps as on other aspects of life.

Beginning in the 1980s geoscientists in several geological surveys instigated projects to explore the creation of geological models. In the vanguard were survey geophysicists, a more numerate and computer-literate discipline, who rapidly developed models using geophysical data and software. Conventional survey geologists lagged someway behind. The challenge of modeling "real" rocks was made significantly greater by the variability, irregular distribution, and the interpretive nature of their source data and information. The inadequacies of the available modeling software (both the algorithms and interfaces) and a lack of information technology expertise further hampered the implementation of geological models. Turner (1991) described the first international conference devoted to applications of 3-D geoscientific mapping and modeling. Orlić and Rösingh (1995) described the application of 3-D modeling to offshore aggregate assessment in Hong Kong.

Faced with software limitations, many geologists at that time (and still today) quickly exploited Geographic Information Systems (GIS) to produce models (using software such as ArcInfo, MapInfo, and MicroStation). These products were off the shelf, relatively intuitive to use and presented data in a familiar point, line, and polygon map-layer format. Much progress was, and is, being made through applying GIS technology, but it is generally accepted as being a 2-D or 2.5-D solution: the models are rock surfaces (flying carpets), not rock volumes; representation of real-world discontinuities and other geometrical complexities (typical of igneous and metamorphic terrains) continues to challenge software that was, and is, designed mainly for 2-D geographical data. Turner (2006) discussed the historical development of 3-D digital geological models.

Some geological surveys have the resources and capacity to use commercial software specifically designed to construct sophisticated 3-D and 4-D models. The outputs are impressive, although the models are generally geographically or geologically constrained, reflecting the application focus of the sponsoring projects. These systems can deal with complex geological structures as well as simple layer-cake stratigraphy, but they are complex software applications. Their capabilities may be extensive, but their interfaces and processes are usually too complicated for use by geologists whose prime expertise is rocks not

computing. They are also expensive with license fees well beyond the resources of smaller surveys. It did not take long for entrepreneurial individuals to realize there was a gap and a business opportunity. Beginning in 2000, more affordable software systems became available that provided more geoscience utility than GIS but required less digital expertise than the complex software.

Today, geological surveys are faced with choosing among: (i) simple, affordable, and widely employed GIS technology; (ii) expensive and complex specialized commercial software developed for petroleum or mining applications; or (iii) affordable software systems providing basic geological modeling capabilities. Surveys adopt one or several of these methodologies, usually reflecting their capabilities and resources, but also sometimes, less rationally, on the personal predilections of their scientists (Berg et al. 2011). The key step forward was not the ability to digitize a 2-D geological map but the technological ability to create digital 3-D (and 4-D) geological models.

2.4.3 Advantages of Digital Mapping

A digital map consists of *points*, *lines*, and *polygons*. GIS software allows other information such as text, data tables, photographs, diagrams etc. to be "attached" to a point, line, or polygon. These attributes can be evaluated when the digital geological map is used. On a paper-based map this additional information is better known as the "key" or "legend." However, the amount of additional information that can be provided is limited by the area of paper available to print the information. With a digital map, the amount of additional information that can be included is only limited by the time taken to input and check the data and the capacity of the computer hardware and software to store and process it.

This ability to hold additional information in a digital map system provides new possibilities, since maps are interpretations of the 3-D geology and, as such, are syntheses of many diverse data types. A digital map that is simply a display on a computer screen of what could also be presented on paper does not represent much of an advance. Paper-based maps, when updated, require reprinting, an expensive process that makes existing unsold maps redundant. Often many decades pass between updates; for some areas, the latest geological map may be over 100 years old! This situation changes when digital mapping procedures are adopted; maps can be updated and printed on-demand, so each user receives the latest version. Also, the map is no longer restricted to a predefined extent; any desired area can be specified, such as a local authority boundary, a mineral license area, or an engineering site.

Paper-based geological maps typically include a few geological cross sections, providing an interpretation of the geology with depth. However, because of the lack of space, only a few sections can be included, and these sections may not be in the places of interest to the map reader. The ability to hold additional information in a digital database allows the production of cross sections at any chosen location and along any desired direction. Cross sections can be created that link information recorded in multiple boreholes to help define the subsurface geology.

2.5 The Importance of Effective Data Management

Geological surveys have the unique responsibility to provide geoscience knowledge for the *whole* of their territories. Creating a competent 3-D model of an area, however small or large, needs a substantial amount of consistent, quality-controlled data. Thus, the modeling aspirations of geological surveys depend on the availability of such data for all their territories. Thus, effective management of data at a survey (corporate) level is essential.

Geological surveys have two major assets, or core competencies: the expertise of their staff and their long-term data holdings. While a significant percentage of these data holdings was produced by survey projects, the majority was commissioned for other diverse purposes, ranging from coal prospecting and mining to civil engineering and environmental monitoring.

Yet, for most of the twentieth century, it is a sad fact that these data (paper records, books, maps, logs, samples, core, digital content, etc.), which are held by virtually every survey across the world, were not highly valued and have never been accorded high priority. In 1998, a project funded by the European Commission – *"Geological Electronic Information Exchange System (GEIXS)"* – discovered that only one or two of the fifteen participating European geological surveys had systems in place that allowed discovery and access to the data many had been collecting at public expense for more than 150 years (Bonnefoy 2005).

If a geological survey is genuinely serious about becoming an organization whose mission is to deliver a national digital model, then it is axiomatic that it must completely revise its approach to the entire geological data lifecycle. Data must be collected with three spatial dimensions in the field, in 4-D where there is a time component, in 5-D if scale is considered a higher-dimensional geometrical and topological primitive (Van Oosterom and Stoter 2010) and in 6-D if the parameter of uncertainty is to be similarly assessed! The goal has to be to get as close as possible to capturing the multidimensional conceptual model of the geoscientist and to ensure that there is minimal regression to the 2-D paradigm at all stages of data processing and analysis. Equally challenging is how to store, integrate, and re-use existing models, constructed at different resolutions, by different authors and for different purposes, to produce national generic/multi-purpose national geological models.

2.6 The Challenges of Parameterization – Putting Numbers on the Geology

Combining geological models with process models permits the integration of existing scientific knowledge to define natural earth systems and how they work (see Chapter 14). Society would be foolish to ignore this knowledge when taking decisions that impact the natural environment. In addition to defining the geological framework, geological models developed to assist in problem-solving and effective environmental management must be "parameterized" – attributed with subsurface properties that are meaningful to the likely users (see Chapter 13).

2.6.1 Parameterization of Geological Models

Geological maps (2-D representations of the 3-D geology) are attributed with stratigraphy and lithology. Geological models, in the first instance, are often similarly attributed. However, the prerequisite of being mappable plays out differently in 2-D and 3-D mapping, so 3-D modeling typically leads to the reconsideration of stratigraphic units. Boundary conditions (formations, faults, facies changes, etc.) must be specified; from a geologist's perspective these specifications may depend on how recently the original mapping took place and the detail and scale at which the mapping was carried out. Engineering and groundwater management applications require estimates of geotechnical or hydrological parameters that are related to the lithologies or formations. Geological framework models normally define volumes; parameter attribution requires the filling of specific volumes with distributions of parameter values. This requires discretization of the geological model to create a grid-based model; then each cell in the grid may be attributed with appropriate parameter values (Chapter 13). Parameters calculated from observed data are normally relatively scarce, so statistical techniques are often employed to interpolate, extrapolate, or average observed values (Chapter 13).

Three factors influence the selection of appropriate parameter values and the definition of their spatial and temporal variation: (i) model scale, (ii) parameter heterogeneity, and (iii) model uncertainty. These factors interact; model scale and parameter heterogeneity influence the capability of a model to represent smaller- or larger-scaled variability or processes. Together, model scale and parameter heterogeneity influence model uncertainty.

2.6.2   Model Scale

Geological models can be developed at any appropriate scale; models may address national, regional, or local problems. For practical reasons, national models are more generalized and ignore minor local variations. They approximate "small-scale" maps. In contrast, local geological models cover relatively small areas or volumes and contain detailed descriptions of local variations of parameter values. Regional models occupy a middle position between the generalized national models and the detailed local models.

Over approximately the past two decades, geological surveys all over the world have created numerous local and regional models designed to address the requirements of various projects and users. Figure 2.4 illustrates the status of 3-D geological models in the United Kingdom; similar situations exist in other countries (Berg et al. 2011). It became generally apparent that a national model would not only be useful in addressing national problems, it would provide a "national framework" that allows for either the integration of local and regional models, or, conversely, as a starting point for local and regional exercises. Combining and "nesting" existing models at different scales promotes greater operational efficiency and potentially adds value to existing information, models, and associated databases.

A common problem facing geologists is the uneven distribution of available data. This leads to the requirement to "upscale" and "downscale." "Upscaling" is the process of generalizing from highly detailed local data to a more regional understanding, while "downscaling" is the reverse process, limited regional scale information is used to produce a more detailed local-scale understanding.

Model scale influences how process models describe heterogeneous natural environments. Critical parameters may occur at many scales. Heterogeneity of granular sedimentary sequences has a fractal nature through scales ranging from microscopic to centimetric, decametric, and kilometric (Weber 1986). Groundwater resource models must be able to handle the wide range of parameter scales found in nature. Groundwater flow processes may be partly constrained by the size of the pore spaces between grains but are also affected by changes in formation lithologies that may occur over kilometers. Groundwater flow is controlled by the hydraulic gradient, which may vary over meters to kilometers and may be controlled by base levels defined by springs, rivers, lakes, or sea level.

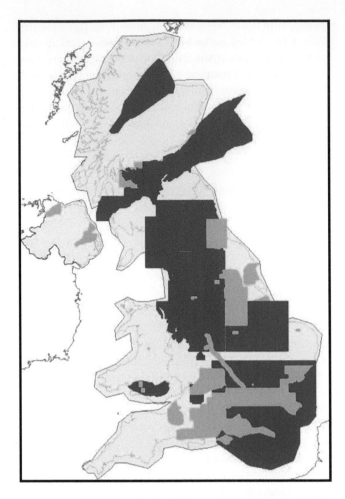

Figure 2.4   BGS 3-D geological framework models created prior to 2016. Red areas are regional and basin-scale models; Green areas are more detailed shallower models, including Quaternary models; Green areas enclosed within larger red polygons usually contain models of both types; Pale green area is coverage of the national UK3D_v2015 model. (Waters et al. 2016).

In applications in support of urban development, issues such as subsidence become important; thus, material strength at the building-site scale might be required for infrastructure renewal projects, while planning applications need information on variations in geotechnical properties within an urban area covering many kilometers (for examples, see Terrington et al. in Chapter 25).

2.6.3   Parameter Heterogeneity

Natural systems are inherently complex; models of these systems must recognize that natural systems are composed of numerous interacting or independent components and processes. These processes may have spatial or temporal variations.

All natural systems are spatially heterogeneous. Small-scale models usually generalize or average parameter values so that the complexity or heterogeneity cannot be seen in the model. Large-scale more detailed models must identify and contain the inherent parameter heterogeneity. The use of "upscaling" or "downscaling" procedures to partially resolve parameter heterogeneity introduces uncertainty in the model which should be recognized in the model outputs. This problem was articulated by Sivapalan et al. (2003).

Urban areas often have considerable volumes of existing site-investigation data. Kearsey et al. (2015) tested the use of stochastic methods to predict the lithology of complex glacial and fluvial deposits beneath Glasgow. They found a moderate improvement in the prediction of lithology when using a lithologically-derived stochastic model compared with a conventionally interpolated lithostratigraphic model. The GDN uses these approaches sequentially: first create a lithostratigraphic model, then parameterize the lithostratigraphic units individually (Van der Meulen et al. 2013). Stochastic methods also facilitate reporting uncertainty within the resulting models, either with probability maps or through a suite of plausible simulations of the lithologies across the study region (Chapters 13 and 15).

Temporal heterogeneity reflects natural cycles. Societally -induced changes may cause perturbations in time-series records and so increase the heterogeneity of these records. Natural cycles may occur over a short time-scale (for example, seasonal or diurnal) or over much longer time-scales (for example, due to orbital forcing and resulting climate change, or climate variation caused by shifts in the jet stream). Observed time-series data of component processes within Earth systems may be of inadequate duration to capture the whole natural complexity of longer cycles.

2.6.4 Model Uncertainty

Chapter 15 explores the sources of uncertainty in geological models and describes methods to assess this uncertainty. All models are uncertain to some degree, whether the uncertainties are known or ignored. Clearly the degree of parameter heterogeneity will make assessment of uncertainty more challenging. When practitioners who are not well versed in how the model was created and parameterized accept model predictions as absolutely valid, without any uncertainty, poor decisions may result. The problem of uncertainties has long been recognized by statisticians and scientists (Chatfield 1995; Beven and Lamb 2017).

Oreskes (2003) described the complexity paradox. As understanding increases, the natural reaction of any scientist is to add complexity to their models. In other words, as data are collected and understanding correspondingly improves, more and different processes can be added to

any model. However, as more processes are added, the model requires more parameters which have associated uncertainties. So, although a more complex model may better describe the natural system, it is more difficult to determine whether the model successfully reproduces the natural system. Whilst the overall system is better represented, more parameters have been introduced, uncertainty has increased, and model evaluation becomes a very important activity.

Evaluation of model uncertainty requires testing against observed data, but there are practical and economic constraints on investigation and monitoring of the environment. All observed data contain errors and are inherently uncertain. For example, boreholes cannot be drilled everywhere, so geophysical and remote sensing techniques may be used instead. These may introduce additional sources of uncertainty. Organizations such as the Environment Agency (in England) and the BGS have difficulty justifying recurrent expenditures to achieve the required numbers of observation records.

When process models are linked in a model cascade, that is, when the outputs of one become the inputs to another, there can be many kinds of uncertainty to be evaluated (Beven 2009, 2012). Beven and Lamb (2017) provides a comprehensive discussion of the uncertainty problem and clearly suggests that a precautionary approach is to be preferred when model outputs are used for decision-making. Visualizing and communicating uncertainty is difficult, especially by non-specialist users who are uncomfortable with the inherently large uncertainties associated with geological modeling and prefer to see 3-D models as the absolute truth. Wellman and Regenauer-Lieb (2012) propose the use of information entropy as an objective measure to compare and evaluate model and observational results. The advantage is that it is a catch-all parameter that is fairly easy to calculate, visualize and communicate: high entropy values occur where the model predicts accurately and vice versa. Chapter 15 provides more details on information entropy.

2.7 Use of 3-D Geological Models with Process Models

Numerical groundwater flow modeling began in the late 1960s and early 1970s with the development of sufficiently powerful computers (Chapter 6). Concurrently, the oil industry was exploring similar models to assess oil resources. The initial models were generally restricted to simulations of single aquifer layers, or simple oil reservoirs, but increasing computer processing power and speed permitted development of multilayer models. A conceptual understanding of the groundwater flow regime was required to design the domains, boundary conditions, and properties of these models. This was achieved by evaluating the geology from maps, borehole logs, and geological cross-sections but 3-D geological models increased the accuracy of this process, while visualization techniques improved the degree of understanding. Chapter 14 provides additional discussions of the interactions between geological and process models.

Robins et al. (2003) used 3-D visualization and modeling of the South Downs (a line of Chalk hills along the south coast of England) to improve the conceptual understanding of the groundwater flow regime. Subsequently, the area was modeled using a distributed finite difference groundwater flow model. Bricker et al. (2014) similarly found that 3-D geological modeling helped modify the conceptual understanding of the hydrogeology of the Jurassic aquifers of Cotswold aquifers in south-central England. This understanding provided the basis for the development of a lumped parameter, semi-distributed model (Peach et al. 2017).

Integrated modeling is a relatively recent discipline that combines appropriate data and models to facilitate the solution of complex problems by joining models from a range of different sources (Laniak et al. 2013). It involves selecting a series of model codes, making them linkable, and then allowing the scientist or other end-user to answer their question by joining together models in a bespoke combination (Barkwith et al. 2014). The aim is to develop methods capable of answering more complex questions than can currently be addressed using single instance or single discipline models. This results in a multidisciplinary approach to problem solving. Integration may be achieved by building one large model that simulates several processes, such as SHETRAN (Ewen et al. 2000), or by using a generic coupling system, such as the Open Modeling Interface (OpenMI) to dynamically link separate models during runtime (Gregersen et al. 2007). Payo et al. in Chapter 22 provides a more detailed example of these methods applied to modeling coastal change in Norfolk, eastern England.

The integrated model can be thought of as a unit operation with two facets: model codes, which solve the governing equations, and model instances, which apply those equations (coded in a model) to a particular area or problem. By considering a model as a unit operation and defining the input and output data such that the output from one model code can be passed to another, a linked series of models can be created with a workflow to address a specific question or problem. Several important requirements are required to achieve successful model integration, including:

- The model should be described with metadata catalogs along with documentation of model operation procedures; this includes definition of the model code development process and the underlying physics/mathematics and model assumptions (Harpham and Danovaro 2015);
- The models should be linked by either using well-described file formats or runtime coupling, such as OpenMI (Gregersen et al. 2007; Harpham et al. 2016) or CSDMS (Peckham et al. 2013);
- Linkages between models must be described with semantics/ontologies so that different disciplines can recognize the same variable, even if it is described differently (Nativi et al. 2013);
- Required computational resources must be known and available. Some models need to run in the Cloud and be made available via Web Services (Castronova et al. 2013);
- Uncertainty must be fully evaluated for the entire model chain with procedures such as UncertWeb (Bastin et al. 2013); and
- Results must be visualized appropriately so that decision-makers or other scientists and stakeholders can use them appropriately (Robins et al. 2005).

2.8 The Evolving Mission of the Geological Survey of the Netherlands

The GDN was the first, and to date remains the only such organization, to be engaged in national-scale systematic 3-D modeling, at various scales and depth ranges, serving the energy, water and built-environment domains. Van der Meulen et al. (2013) provides a comprehensive review of the experiences of the GDN, along with an assessment of its anticipated future role. Around 2000, the GDN replaced the systematic production of geological map sheets with 3-D subsurface models that are distributed electronically. Several factors contributed to this particular survey being an early adaptor. Most importantly, at just over 100 000 km^2 the country (including the North Sea Exclusive Economic Zone) is comparatively small, but it has a sizeable population and GDP, so the survey activities are supported by 100 times more tax payers per square km than for example in Canada. Second, the sedimentary basin setting makes for a level of geological complexity that is manageable for common 3-D modeling techniques. Third, the survey has an excellent data position, benefiting from a favorable mining law and being situated in a mature hydrocarbon province, as well as from a decision to undertake a major data digitization effort, made in the 1980s. Finally, the abandonment of superficial mapping freed funds for a substantial 3-D modeling effort of the shallow subsurface (upper hundreds of meters), which could draw from experience in deep subsurface mapping that had been going on since the mid-1980s.

This transition has been accelerated in 2015 by the enactment of a new law that governs the management and utilization of subsurface information (Anonymous 2015). This law commissions the GDN to build and operate an official government register, a "key register" (*basisregistratie*) for subsurface information called BRO *(Basisregistratie Ondergrond),* on behalf of the Ministry of the Interior and Kingdom Relations (Figure 2.5). A key register contains high-quality data that the government is obliged to use for its public tasks. The objective is to enhance the efficiency in data management and to avoid errors in using data. This is often defined by the phrase "Capture once, use many times." Chapter 4 (Section 4.5.6) describes in more detail how and why the BRO was established.

All Dutch government bodies are obligated to use BRO for storing their own subsurface data, and when making policies or decisions pertaining to, or affected by, the subsurface. In Figure 2.5, "Use" refers to all possible applications of subsurface data and information by government and industry, while "Advice" refers mainly, but not exclusively, to GDN advice on geological matters under the Mining Act.

In many ways, this new legislative environment represents a culmination of the ongoing technological or digital transition. New requirements and responsibilities have been placed on the GDN; it has had to rethink and redesign most of its operation, from data acquisition and interpretation to data delivery. Although the GDN is the first geological survey to have received official authorization for such a central information management role, it appears to offer a view of how other geological surveys may be required to operate in the future.

The GDN has an existing geological database, DINO (Data en Informatie van de Netherlandse Ondergrond), that has served the public since 2001 (Lance et al. 2010). Users access DINO with a web portal (WP), DINOloket (www.dinoloket.nl), assisted by a service desk. DINOloket has many thousands of registered users, ranging from individuals to institutions, and it annually handles requests for data from hundreds of thousands of boreholes and cone penetrations tests, hundreds of millions of groundwater levels, and many thousands of other data types. The responsibilities of a key register exceed the capabilities and capacities of DINO; so DINO is being developed toward the internal database, while the BRO is being built from scratch.

BRO will initially contain information currently stored in DINO and in BIS (Bodem Informatie Systeem), the Dutch national soil survey database, holding four major classes of data (Table 2.1). The BRO will have to handle and store substantially more data than DINO, while its user community will become much larger and more diverse. Current users will be joined by new users who are less familiar with subsurface conditions. For example, DINO currently serves only a few of the 380 Dutch municipalities; all will have to use BRO. Beyond a legal "umbrella," that supports long-term continuity, the BRO is important to GDN in being the only conceivable way to significantly

Figure 2.5 BRO data flow diagram showing the key activities and processes of GDN: data intake, database management, data delivery, modeling, and advice. Red arrows indicate QC and error reporting processes; WS indicates a web service; WP indicated a web portal. (Van der Meulen et al. 2013).

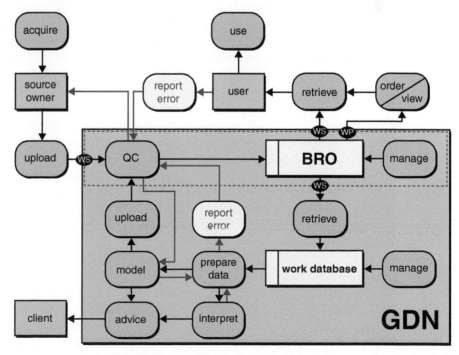

Table 2.1 Data types contained in BRO key register (Van der Meulen et al. 2013).

Data type	Description/examples
Survey Data (*verkenningen*)	Borehole (including sample analyses), cone penetration test, well log, seismic survey, geo-electric survey
Models (*modellen*)	Four national geological models maintained by GDN: DGM (Digital Geological Model), DGM-Deep, REGIS-II (hydrogeological model), and GeoTOP (geological model of top 30–50 m)
	National geomorphological and soils maps of Alterra (part of Wageningen University and Research – WUR)
Rights of use (*gebruiksrechten*)	Exploration and production licenses for hydrocarbons, minerals, and geothermal heat (required by Mining Act) and groundwater-abstraction permits (required by Water Act)
Constructions (*constructies*)	Infrastructure elements that extract substance from the subsurface (hydrocarbon or water wells), store substances in the subsurface, or monitor subsurface processes (provincial and national monitoring networks for soil quality, groundwater quality, and groundwater levels)

increase the flow of data to the survey. As opposed to a decade ago, computer power is no longer a limiting factor to 3-D modeling; however, getting hold of more and better data is crucial for improving model resolution and quality.

2.9 Experience With a Multiagency and Multijurisdictional Approach to 3-D Mapping in the Great Lakes Region

The Great Lakes region is characterized by multiple successive and very complex landscapes, resulting from multiple glacial advances and retreats during the Wisconsin, Illinois, and pre-Illinois glacial episodes, and deposits from two primary interglacial episodes. Glaciation has largely buried the bedrock topography and the current relatively flat landscape means exposures of geologic materials are rare. Therefore, subsurface information must be obtained using test drilling and geophysical techniques, supplemented with lower-quality data from validated water-well drilling logs. This is a challenging process to ascertain the geometries and distributions of the subsurface units and define their sedimentology and environments of deposition. Questions arise as to how 3-D geologic information, often to depths of several hundred meters, can be provided at a scale usable to local jurisdictions for land- and water-use planning, particularly in urban or rapidly expanding suburban settings. This information is needed within a reasonable period if these issues are to be addressed before they become serious problems (Berg et al. 2016).

Because no single agency had the financial, intellectual, or physical resources necessary to conduct a massive geologic mapping effort at a detailed scale over this region, the directors of the state geological surveys of Illinois, Indiana, and Ohio approached the U.S. Geological Survey (USGS) in 1997 with a strategy for generating support to conduct 3-D geologic mapping of the glacial sediments and shallow bedrock that cover their states. The strategy involved establishing a mapping coalition of geological surveys that would seek federal funds, pool physical and personnel resources, and share mapping expertise to characterize the thick cover of glacial sediments and shallow bedrock in three dimensions, particularly in areas of greatest societal need.

This initial meeting led to the formation of the *Central Great Lakes Geologic Mapping Coalition*, consisting of the state geological surveys from Illinois, Indiana, Michigan, and Ohio, and the USGS. The Coalition conducted regional mapping forums with stakeholders to establish and document the need for mapping among various public and private user groups. A planning document (Berg et al. 1999) and a promotional circular (Central Great Lakes Geologic Mapping Coalition 1999) were developed. The coalition remained as a five-member organization until 2008, when it expanded to include the state geological surveys of Minnesota, Wisconsin, Pennsylvania, and New York. Subsequently, the name changed to the *Great Lakes Geologic Mapping Coalition* (GLGMC). It expanded again in 2012 by adding the Ontario Geological Survey (OGS) as a non-funded partner. As a regional entity extending east–west from New York to Illinois and northward into Ontario, the GLGMC states and province are unique because they share (i) multiple continuous, thick, complex layers of glacial deposits containing groundwater resources used by ~50% of their residents; (ii) similar geomorphology; (iii) high populations, (iv) a long-standing tradition of developing light and heavy industry; (v) serious brownfield redevelopment issues; (vi) high agricultural productivity; and (vii) similar weather and climate. To meet its goals, the GLGMC has mapped the glacial geology of identified priority areas within the region and provided 3-D geological maps to depths of several hundred meters. These provide scientific interpretations of value to decision makers. Figure 2.6 shows the locations of priority areas for short- and long-range mapping actions within the GLGMC.

The GLGMC efforts in Minnesota have concentrated on Quaternary stratigraphy, 3-D modeling, and mapping sand bodies. The Quaternary stratigraphy atlas plate depicts three to five cross-sections, whereas the sand body thickness and extent models are based on cross-sections at 1 km spacing. Bedrock topography and depth to bedrock maps are important elements of each atlas, whereas the bedrock surface and its elevation are mapped mostly with data from

the records of wells and geotechnical drill holes, with some support from seismic surveys (Setterholm 2012).

In Illinois, 3-D geologic information has enabled the Illinois State Geological Survey (ISGS) to counsel local Lake County officials, and their hydrogeologic consultants, about the feasibility of continuing to develop groundwater resources rather than building infrastructure to withdraw, treat, and move water from Lake Michigan to communities inland from the lake at an estimated cost of more than US$200 million (Figure 2.7).

The Ohio Division of Geological Survey (ODGS) has used multiple methods to deliver geologic information, including new map formats, 3-D models, and public outreach. These have addressed multiple applications, including mapping the flood-prone sinkhole basins in a four-county karst region around Bellevue, Ohio. This area experienced major groundwater flooding in 2008 that inundated homes and roads and resulted in closure of a state highway. Using 3-D methods, the ODGS provided local policymakers with tools to help mitigate future risks (Figure 2.8).

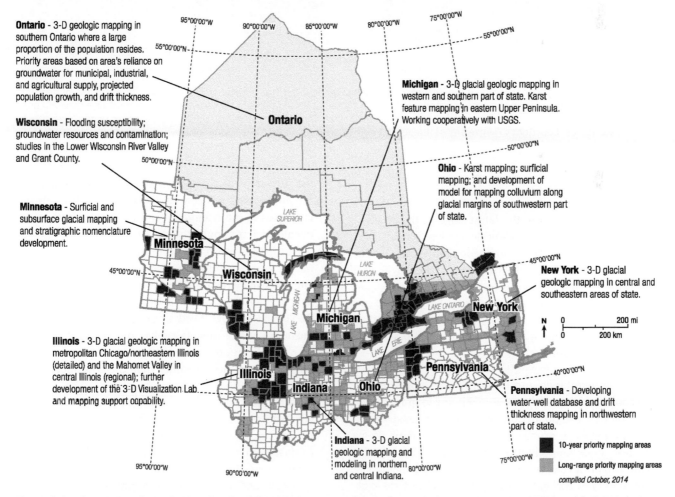

Figure 2.6 Great Lakes Geologic Mapping Coalition, 2014 short- and long-range priority mapping areas (USGS – U.S. Geological Survey). Source: Copyright © 2014 Great Lakes Geologic Mapping Coalition. Used with permission.

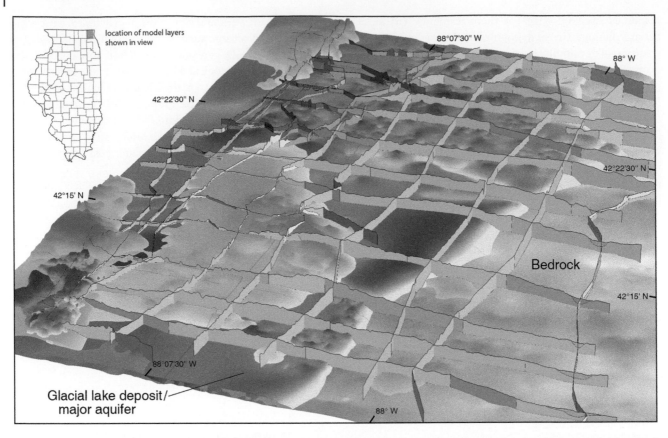

Figure 2.7 Three-dimensional image of the western part of Lake County, Illinois, as viewed from the southeast. A network of geologic cross-sections was derived from the geologic model layers. Not all geologic model layers are shown. The lowermost layer, shaded gray, is the bedrock surface. The model layer, shaded brown, is a glacial lake deposit and is a major aquifer that supplies water to northern communities. Source: Copyright © University of Illinois Board of Trustees. Used with permission of the Illinois State Geological Survey.

2.10 Conclusions

By way of introduction to the latest stage in the development of 3-D and 4-D geoscientific models intended for a range of potential (often non-geoscientific) users, this chapter has taken the reader from the start of geological modeling at around the beginning of the nineteenth century to some of the 4-D models that enable us to answer complex environmental and societal questions. However, regardless of this increasing complexity, some basic requirements remain:

1. The geoscientific data to input into the models has to be collected, managed, and stored in the long-term. The accuracy of the predictions and forecasts made using these models is crucial to solving societal problems caused by climate change, energy insecurity, sustainable urban development and water management. The complexity of geological and other natural systems leads to the need for an understanding of uncertainty caused by heterogeneity, scaling, and measurement error. Thus,

the fundamental question that has bedeviled geological surveys and environmental regulators through the latter part of the twentieth century and the start of the twenty-first is should the data be collected when the need arises or on a continuing basis, on the grounds that, if we wait until the need arises it will be too late and too expensive to obtain the types and quantity of data needed.

2. Geological surveys need to continually adapt as the means to create models evolve and improve (as they surely will). In other words, the emphasis is on development of the models rather than the collection of the data.

3. Only by blending traditional geoscientific method and advances in information technology can geoscientific models provide the answers that will help society to deal with the many environmental issues that lie ahead.

4. Linkages and collaborations between disciplines are essential to providing effective management of the environment for the benefit of society.

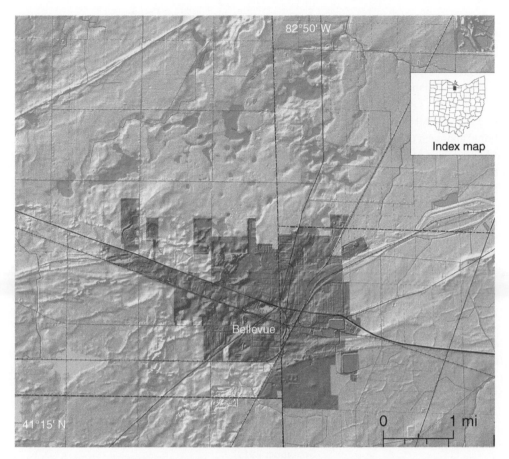

Figure 2.8 Portion of a karst-induced groundwater flooding map near Bellevue, Ohio. Groundwater flooding areas are shown in green. Pink area designates the municipality of Bellevue (Pavey et al. 2012). Source: Copyright © 2014 State of Ohio, Department of Natural Resources. Used with permission of the Ohio Division of Geological Survey.

References

Anonymous (2015). *Wet Basisregistratie Ondergrond* (in Dutch; Law Establishing the Key Register for the Subsurface BRO). Adopted in 2015, Amended in 2018. The Hague, Netherlands: Ministry of Infrastructure and Water Management. [Online: Available at: http://wetten.overheid.nl/jci1.3:c:BWBR0037095&z=2018-07-28&g=2018-07-28] (Accessed September 2018).

Bailey, E.B. (1962). *Charles Lyell*. London: Thomas Nelson and Sons Ltd. 214 pp.

Barkwith, A.K.A.P., Pachocka, M., Watson, C. and A.G. Hughes (2014). *Couplers for linking environmental models: scoping study and potential next steps*. Keyworth, UK: British Geological Survey, Report OR/14/022. 42 pp. [Online: Available at: http://nora.nerc.ac.uk/id/eprint/508423/] (Accessed November 2019).

Bastin, L., Cornford, D., Jones, R. et al. (2013). Managing uncertainty in integrated environmental modelling: the Uncertweb framework. *Environmental Modelling & Software* 39: 116–134.

Bate, D.G. (2010). Sir Henry Thomas De la Beche and the founding of the British Geological Survey. *Mercian Geologist* 17 (3): 149–165.

Berg, R.C. (2005). Societal and economic benefits of three-dimensional geological mapping for environmental protection at multiple scales: an overview perspective from Illinois, USA. In: *The Current Role of Geological Mapping in Geosciences* (ed. S. Ostaficzuk), 97–114. Dordrecht, Netherlands: Springer. https://doi.org/10.1007/1-4020-3551-9_10.

Berg, R.C., Bleuer, N.K., Jones, B.E. et al. (1999). *Mapping the glacial geology of the Central Great Lakes region in three dimensions – a model for state–federal cooperation*. Reston, VA: U.S. Geological Survey, Open-File Report 99–349. 41 pp. https://doi.org/10.3133/ofr99349.

Berg, R.C., Mathers, S.J., Kessler, H. and D.A. Keefer [Eds.] (2011). *Synopsis of current three-dimensional geological mapping and modeling in geological survey organizations.* Champaign, IL: Illinois State Geological Survey, Circular 578. 92 pp. [Online: Available at: http://hdl.handle.net/2142/43535] (Accessed November 2019).

Berg, R.C., Brown, S.E., Thomason, J.F. et al. (2016). A Multiagency and multijurisdictional approach to mapping the glacial deposits of the Great Lakes region in three dimensions. In: *Geoscience for the Public Good and Global Development: Toward a Sustainable Future* (eds. G.R. Wessel and J.K. Greenberg), 415–447. Geological Society of America, Special Paper 520. https://doi.org/10.1130/2016.2520(37).

Bernknopf, R.L., Brookshire, D.S., McKee, M. et al. (1993). *Societal value of geologic maps.* Reston, VA: U.S. Geological Survey, Circular 1111. vi + 53 pp. https://doi.org/10.3133/cir1111.

Bernknopf, R.L., Brookshire, D.S., McKee, M. and Soller, D.R. (1997). Estimating the social value of geologic map information: a regulatory application. *Journal Environmental Economic Management* 32 (2): 204–218. https://doi.org/10.1006/jeem.1996.0963.

Bernknopf, R.L., Wein, A.M., St-Onge, M.R. and Lucas, S.B. (2007). *Analysis of improved government geological map information for mineral exploration: incorporating efficiency, productivity, effectiveness, and risk consideration.* Reston, VA: U.S. Geological Survey, Professional Paper 1721. vi + 45 pp. https://doi.org/10.3133/pp1721.

Beven, K.J. (2009). *Environmental Modelling: An Uncertain Future?* London, UK: Routledge. 328 pp.

Beven, K.J. (2012). How much of your error is epistemic? Lessons from Japan and Italy. *Hydrological Processes* 27: 1677–1680. https://doi.org/10.1012/hyp.9648.

Beven, K. and Lamb, R. (2017). The uncertainty Cascade in model fusion. *Geological Society Special Publications* 408: 255–266. https://doi.org/10.1144/SP408.13.

Bhagwat, S.B. and Berg, R.C. (1991). *Benefits and costs of geologic mapping programs in Illinois: case study of Boone and Winnebago Counties and its statewide applicability.* Champaign, IL: Illinois State Geological Survey, Circular 549. 40 pp. [Online: Available at: https://archive.org/details/benefitscostsofg549bhag] (Accessed November 2019).

Bhagwat, S.B. and Berg, R.C. (1992). Environmental benefits vs. cost of geologic mapping. *Environmental Geology and Water Science* 19 (1): 33–40. https://doi.org/10.1007/BF01740575.

Bhagwat, S.B. and Ipe, V.C. (2000). *Economic benefits of detailed geologic mapping to Kentucky.* Champaign, IL: Illinois State Geological Survey, Special Report 3. 39 pp.

[Online: Available at: https://isgs.illinois.edu/publications/spr003] (Accessed November 2019).

Bonnefoy, D. (2005). EuroGeoSurveys: GEIXS: geological exchange information system – description – metadata. In: *International Cartographic Association, World Spatial Metadata Standards* (eds. H. Moellering, H.J.G.L. Aalders and A. Crane), 601–617. Amsterdam, Netherlands: Elsevier. https://doi.org/10.1016/B978-008043949-5/50034-6.

Bricker, S.H., Barron, A.J.M., Hughes, A.G. et al. (2014). From geological complexity to hydrogeological understanding using an integrated 3D conceptual modelling approach–insights from the Cotswolds, UK. In: *Fractured Rock Hydrogeology* (ed. J.M. Sharp), 99–114. Leiden: CRC Press/Balkema. https://doi.org/10.1201/b17016-7.

Brochant de Villiers, A. (1835). Notice sur la Carte Géologique Générale de la France. *Compte rendus hebdomadaires des séances de l'Académie des sciences* 1: 423–429.

Buchanan, R. (2016). State Budgets, Geological Surveys, and the New Reality. *EOS 97.* https://doi.org/10.1029/2016EO059893.

Castelein, T.W., Bregt, A. and Pluimers, Y. (2010). The economic value of the Dutch Geoinformation sector. *International Journal Spatial Data Infrastructure Research* 5: 58–76. [Online: Available at: https://ijsdir.sadl.kuleuven.be/index.php/ijsdir/article/view/148] (Accessed November 2019).

Castronova, A.M., Goodall, J.L. and Elag, M.M. (2013). Models as web services using the open geospatial consortium (OGC) web processing service (WPS) standard. *Environmental Modelling & Software* 41: 72–83. https://doi.org/10.1016/j.envsoft.2012.11.010.

Central Great Lakes Geologic Mapping Coalition (1999). *Sustainable growth in America's heartland – 3-D geologic maps as the foundation.* Reston, VA: U.S. Geological, Survey, Circular 1190. 17 pp. https://doi.org/10.3133/cir1190.

Chatfield, C. (1995). Model uncertainty, data mining and statistical inference. *Journal of the Royal Statistical Society, Series A* 158 (3): 419–466. https://doi.org/10.2307/2983440.

Coe, L. and Cratchley, C.R. (1979). The influence of data point distribution on automatic contouring. *Bulletin of the International Association of Engineering Geology* 19: 284–290. https://doi.org/10.1007/BF02600490.

Cuvier, G. and Brongniart, A. (1811). *Essai sur la géographie minéralogique des environs de Paris, avec une carte Géognostique, et des coupes de terrain.* Paris, France: Baudouin, Imprimeur de l'Institut Impérial de France. 211 pp.

Davis, A.G. (1950). Notes on Griffith's geological maps of Ireland. *Journal of the Society for the Bibliography of Natural History* 2 (6): 209–211.

De la Beche, H.T. (1826). On the geology of southern Pembrokeshire [read May 1823]. *Transactions of the Geological Society of London, for 1829, Ser. 2* 2 (1): 1–20.

De la Beche, H.T. (1839). *Report on the Geology of Cornwall, Devon and West Somerset*. London, UK: HM Stationary Office. 648 pp.

De Mulder, E.F.J. and Kooijman, J. (2003). Dissemination of geoscience data: societal implications. In: *New Paradigms in Subsurface Prediction: Characterization of the Shallow Subsurface: Implications for Urban Infrastructure and Environmental Assessment* (eds. M.S. Rosenbaum and A.K. Turner), 191–200. Berlin, Germany: Springer, Lecture Notes in Earth Sciences 99. https://doi.org/10.1007/3-540-48019-6_17.

Diepolder, G. and Lehné, R. [Eds.] (2016). *Applications in 3D geological modelling*. Stuttgart, Germany: Schweizerbart, Zeitschrift der Deutschen Gesellschaft für Geowissenschaften 167 (4). 88 pp.

de Mulder, E.F.J. and Kooijman, J. (2003). Dissemination of geoscience data: societal implications. In: *New Paradigms in Subsurface Prediction: Characterization of the Shallow Subsurface: Implications for Urban Infrastructure and Environmental Assessment* (eds. M.S. Rosenbaum and A.K. Turner) Lecture Notes in Earth Sciences 99, 191–200. Berlin: Springer.

D'Omalius d'Halloy, J.-B.-J. (1823). *Observations sur un essai de carte geologique de la France, des Pays-Bas, et de quelques contrées viosines*. Paris, France: Imprimerie de Madame Huzard (née Vallat la Chappelle) 26 pp + 1 map sheet.

Ewen, J., Parkin, G. and O'Connell, P.E. (2000). SHETRAN: distributed River Basin flow and transport modeling system. *Journal Hydrological Engineering* 5: 250–258. https://doi.org/10.1061/(ASCE)1084-0699(2000)5:3(250).

Eyles, V.A. (1950). The first National Geological Survey. *Geological Magazine* 87: 373–382. https://doi.org/10.1017/S0016756800077347.

Frey, S.K., Berg, S.J. and Sudicky, E.A. (2016). *A feasibility study of merits and development strategies for a regional water resources modelling platform for southern Ontario–Great Lakes Basin*. Ottawa, ON: Geological Survey of Canada, Open File 8021. 46 pp. https://doi.org/10.4095/298816.

Giles, J. (2006). Geological map database—a Practitioner's guide to delivering the information. In: *Digital Mapping Techniques '06 – Workshop Proceedings* (ed. D.R. Soller), 77–84. Reston, VA: U.S. Geological Survey, Open-File Report 2007-1285. [Online: Available at: https://pubs.usgs.gov/of/2007/1285/pdf/Giles.pdf] (Accessed November 2019).

Gill, B., Cherry, D., Adelana, M. et al. (2011). Using three-dimensional geological mapping methods to inform sustainable groundwater development in a volcanic landscape, Victoria, Australia. *Hydrogeology Journal* 19 (7): 1349–1365. https://doi.org/10.1007/s10040-011-0757-7.

Gregersen, J.B., Gijsbers, P.J.A. and Westen, S.J.P. (2007). OpenMI: open modelling interface. *Journal of Hydroinformatics* 9: 175–191. https://doi.org/10.2166/hydro.2007.023.

Häggquist, E. and Söderholm, P. (2015). The economic value of geological information: synthesis and directions for future research. *Resources Policy* 43: 91–100. https://doi.org/10.1016/j.resourpol.2014.11.001.

Halsing, D., Theissen, K. and Bernknopf, R.L. (2004). *A Cost-Benefit Analysis of The National Map*. Reston, VA: U.S. Geological Survey, Circular 1271. VI + 41 pp. [Online: Available at: https://pubs.usgs.gov/circ/2004/1271/] (Accessed November 2019).

Harpham, Q. and Danovaro, E. (2015). Towards standard metadata to support models and interfaces in a hydro-meteorological model chain. *Journal of Hydroinformatics* 17 (2): 260–274. https://doi.org/10.2166/hydro.2014.061.

Harpham, Q., L'homme, J., Parodi, A. et al. (2016). Using OpenMI and a model MAP to integrate WaterML2 and NetCDF data sources into flood modeling of Genoa, Italy. *JAWRA Journal of the American Water Resources Association* 52 (4): 933–949. https://doi.org/10.1111/1752-1688.12418.

Hayes, C.W. (1911). *The State Geological Surveys of the United States*. Reston, VA: U.S. Geological Survey, Bulletin 465. 177 pp. https://doi.org/10.3133/b465.

Hendrickson, W.B. (1961). Nineteenth-century state geological surveys: early government support of science. *Isis* 52: 357–371.

Kearsey, T., Williams, J., Finlayson, A. et al. (2015). Testing the application and limitation of stochastic simulations to predict the lithology of glacial and fluvial deposits in Central Glasgow, UK. *Engineering Geology* 187: 98–112. https://doi.org/10.1016/j.enggeo.2014.12.017.

Lance, K.T., Geogiadou, Y.P. and Bregt, A.K. (2010). Evaluation of the Dutch subsurface geoportal: what lies beneath? *Computers, Environment and Urban Systems* 35: 150–158. https://doi.org/10.1016/j.compenvurbsys.2010.09.002.

Laniak, G.F., Olchin, G., Goodall, J. et al. (2013). Integrated environmental modeling: a vision and roadmap for the future. *Environmental Modelling & Software* 39: 3–23. https://doi.org/10.1016/j.envsoft.2012.09.006.

Lister, M. (1684). An ingenious proposal for a new sort of maps of countries, together with tables of sands and clays, such chiefly as are found in the north parts of England, drawn up about 10 years since, and delivered to the Royal Society Mar. 12. 1683. by the Learned Martin Lister M.D. *Philosophical Transactions of the Royal Society* 14: 739–746.

[Online: Available at: https://royalsocietypublishing.org/doi/pdf/10.1098/rstl.1684.0067] (Accessed December 2019).

Maclure, W., Tanner, H.S., Lewis, S. and American Philosophical Society (1809). *Maclure's Geological Map of the United States*. Philadelphia, PA: American Philosophical Society. 1 map sheet. [Online: Available at: https://www.loc.gov/item/85694780] (Accessed April 2018).

Nativi, S., Mazzetti, P. and Geller, G.N. (2013). Environmental model access and interoperability: the GEO model web initiative. *Environmental Modelling & Software* 39: 214–228. https://doi.org/10.1016/j.envsoft.2012.03.007.

Oreskes, N. (2003). The role of quantitative models in science. In: *Models in Ecosystem Science* (eds. C.D. Canham, J.J. Cole and W.K. Lauenroth), 13–31. Princeton, NJ: Princeton University Press.

Orlić, B. and Rösingh, J.W. (1995). Three-dimensional geomodelling for offshore aggregate resources assessment. *Quarterly Journal of Engineering Geology* 28: 385–391. https://doi.org/10.1144/GSL.QJEGH.1995.028.P4.08.

Ovadia, D. (2007). Geology as a Contributor to National Economies and their Development. In: *Proceedings, CCOP Annual Session, Cebu, Phillipines 21–26 Oct 2007* (eds. A.J. Reedman, N.H. Minh and N. Chaimanee), 41–48. [Online: Available at: http://www.ccop.or.th/download/pub/44as_ii.pdf] (Accessed November 2019).

Owen, T. and Pilbeam, E. (1992). *Ordnance Survey – Map Makers to Britain Since 1791*. London, UK: HM Stationary Office. 196 pp.

Pavey, R.R., Angle, M.P., Powers, D.M. and Swinford, E.M. (2012). *Map EG-5: Karst Flooding in Bellevue, Ohio, and Vicinity – 2009.*. Columbus, OH: Ohio Department of Natural Resources, Division of Geological Survey. 1 map sheet, scale 1 : 24,000.

Peach, D.W., Riddick, A.T., Hughes, A. et al. (2017). Model Fusion at the British Geological Survey: experiences and future trends. *Geological Society Special Publications* 408: 7–16. https://doi.org/10.1144/SP408.13.

Peckham, S.D., Hutton, E.W. and Norris, B. (2013). A component-based approach to integrated modeling in the geosciences: the design of CSDMS. *Computers & Geosciences* 53: 3–12. https://doi.org/10.1016/j.cageo.2012.04.002.

Rabbitt, M.C. (1989). *The United States Geological Survey, 1879–1989*. Reston, VA: U.S. Geological Survey, Circular 1050. 52 pp. https://doi.org/10.3133/cir1050.

Raiber, M., White, P., Daughney, C. et al. (2012). Three-dimensional geological modelling and multivariate statistical analysis of water chemistry data to analyse and visualise aquifer structure and groundwater composition in the Wairau Plain, Marlborough District, New Zealand.

Journal of Hydrology 436–437: 13–34. https://doi.org/10.1016/j.jhydrol.2012.01.045.

Rhind, D.W. (1971). The production of a multi-colour geological map by automated means. *Nachrichten aus den Karten- und Vermessungeswesen* 52: 47–52.

Riddick, A.T., Kessler, H. and Giles, J.R.A. [Eds.] (2017). *Integrated Environmental Modelling to Solve Real World Problems: Methods, Vision and Challenges*. London, UK: Geological Society, Special Publication 408. 274 pp. https://doi.org/10.1144/SP408.

Robins, N.S., Dumpleton, S. and Packman, M.J. (2003). 3-D visualization as an aid to the hydrogeological conceptualization of the central south downs. *Quarterly Journal of Engineering Geology and Hydrogeology* 36 (1): 51–58. https://doi.org/10.1144/1470-9236/01051.

Robins, N.S., Rutter, H.K., Dumpleton, S. and Peach, D.W. (2005). The role of 3D visualisation as an analytical tool preparatory to numerical modelling. *Journal of Hydrology* 301 (1–4): 287–295. https://doi.org/10.1016/j.jhydrol.2004.05.004.

Savaton, P. (2007). The first detailed geological maps of France: the contributions of local scientists and mining engineers. *Earth Sciences History* 26: 55–73. https://doi.org/10.17704/eshi.26.1.028355877th55714.

Scott, M., Dimitrakopoulos, R. and Brown, R.P.C. (2002). Valuing regional Geoscientific data acquisition programmes: addressing issues of quantification, uncertainty and risk. *Natural Resources Forum* 26: 55–68. https://doi.org/10.1111/1477-8947.00006.

Setterholm, D. (2012). *Geologic atlas user's guide: using geologic maps and databases for resource management and planning*. St. Paul, MN: Minnesota Geological Survey, Open-File Report OFR-12-1. 24 pp. [Online: Available at: https://conservancy.umn.edu/handle/11299/166713] (Accessed November 2019).

Sivapalan, M., Takeuchi, K., Franks, S.W. et al. (2003). IAHS decade on predictions in Ungauged basins (PUB), 2003–2012: shaping an exciting future for the hydrological sciences. *Hydrological Sciences Journal* 48 (6): 857–880. https://doi.org/10.1623/hysj.48.6.857.51421.

Smith, W. (1815). *A Memoir to the Map and Delineation of the Strata of England and Wales, with Part of Scotland*. London, UK: John Cary. 52 pp.

Socolow, A.A. [Ed.] (1988). *The State Geological Surveys: A History*. Tuscaloosa, AL: Association of American State Geologists. 499 pp.

Staring, W.C.H. (1844). *Proef eener Geologische Kaart van de Nederlanden*. Groningen, Netherlands: J. Oomkens and Zoon. 1 map sheet.

Staring, W.C.H. (1858–1869). *Geologische kaart van Nederland : schaal van 1:200.000, uitgevoerd door het*

Topographisch Bureau van Oorlog, uitgegeven op last van Zijne Majesteit Den Koning. Haarlem, Netherlands: Kruseman. 24 map sheets.

Steno, N. (1669). *De solido intra solidum naturaliter contento dissertationis prodromus*. Florence, Italy: Typographia sub signo Stellae. 79 pp.

Tame, C., Cundy, A.B., Royse, K.R. et al. (2013). Three-dimensional geological modelling of anthropogenic deposits at small urban sites: a case study from Sheepcote Valley, Brighton, UK. *Journal of Environmental Management* 129: 628–634. https://doi.org/10.1016/j .jenvman.2013.08.030.

Turner, A.K. [Ed.] (1991). *Three Dimensional Modelling with Geoscientific Information Systems*. Dordrecht, Netherlands: Kluwer Academic Publishers, NATO ASI Series C: Mathematical and Physical Sciences, vol. 354. 443 pp.

Turner, A.K. (2006). Challenges and trends for geological modelling and visualisation. *Bulletin of Engineering Geology and the Environment* 65: 109–127. https://doi.org/ 10.1007/s10064-005-0015-0.

Turner, S. and Dearman, W.R. (1979). Sopwith's geological models. *Bulletin of the International Association of Engineering Geology* 19: 331–345. https://doi.org/10.1007/ BF02600498.

Van der Meulen, M.J., Doornenbal, J.C., Gunnink, J.L. et al. (2013). 3D geology in a 2D country: perspectives for geological surveying in the Netherlands. *Netherlands Journal of Geosciences* 92 (4): 217–241. https://doi.org/10 .1017/S0016774600000184.

Van Oosterom, P.J.M. and Stoter, J.E. (2010). 5D data modelling: full integration of 2D/3D space, time and scale dimensions. In: *Proceedings of the Sixth International Conference GIScience 2010* (eds. S.I. Fabrikant, T. Reichenbacher, M. Van Kreveld and M. Schlieder), 311–324. Berlin: Springer-Verlag. https://doi.org/10.1007/ 978-3-642-15300-6_22.

Van Waterschoot van de Gracht, W.A.J.M. (1918). *Eindverslag over de onderzoekingen en uitkomsten van den Dienst der Rijksopsporing van Delfstoffen in Nederland 1903–1916*. The Hague, Netherlands: Martinus Nijhoff. 664 pp.

Waters, C.N., Terrington, R.L., Cooper, M.R. et al. (2016). *The Construction of a Bedrock Geology Model for the UK: UK3D_v2015*. Keyworth, UK: British Geological Survey, Report OR/15/069. 22 pp. [Online: Available at: http://nora .nerc.ac.uk/id/eprint/512904/] (Last accessed February 2018).

Weber, K.J. (1986). How heterogeneity affects oil recovery. In: *Reservoir Characterisation* (eds. L.W. Lake and H.B. Carroll Jr.,), 487–544. Orlando, FL: Academic Press.

Wellman, J.F. and Regenauer-Lieb, K. (2012). Uncertainties have a meaning: information entropy as a quality measure for 3-D geological models. *Tectonophysics* 526–529: 207–216. https://doi.org/10.1016/j.tecto.2011.05.001.

Werner, A.G. (1787). *Kurze Klassifikation und Beschreibung der verschiedenen Gebirgsarten*. Dresden, Germany: Walther. 28 pp.

Werner A.G. (1811). *Illuminierte petrographische Karte*. Freiberg, Germany: Bergakademie Freiberg. 1 map sheet.

3

Legislation, Regulation, and Management

Brian Marker[1] and Alan Keith Turner[2,3]

[1] *Independent Consultant, London W13 9PT, UK*
[2] *Colorado School of Mines, Golden, CO 80401, USA*
[3] *British Geological Survey, Keyworth, Nottingham NG12 5GG, UK*

3.1 Introduction

Competition for space has stimulated widespread urban high-rise developments but, still, surface space is inadequate. The subsurface is extensively used for tunnels and underground car parks but has the potential for much greater use, particularly in congested urban areas. It can provide a sustainable option for social, environmental and economic reasons (Sterling and Godard 2000) including lack of surface space, need for infrastructure, secure storage, adaptation to climate change and geothermal heat resources. In places, building height restrictions and safeguarding of viewpoints have stimulated "building downwards," rather than upwards, to provide convention halls, commercial, retail, parking and storage spaces (Box 3.1). Constraints on subsurface development include cost, existing development, environmental problems, ground conditions, other natural resources, limited awareness of opportunities and lack of appropriate regulations (Table 3.1). Although subsurface development is expensive compared with "greenfield" sites, demand, surface land prices, and social and environmental advantages can make it a preferred option. However, legislation and regulation specifically designed for such development is currently uncommon.

The COST (European Cooperation in Sciences and Technology) Research Programme, part of the EU Framework Programme "Horizon 2020," included the COST Action TU1206 Sub-Urban Network, which was undertaken to improve understanding and use of the ground beneath European cities. It provided extensive state-of-the-art and state-of-practice reviews of the use of 3-D information for subsurface management in several European cities. This produced two major reports and numerous European case studies that are available online (http://sub-urban .squarespace.com/new-index/#publications). Van der Meulen et al. (2016a) evaluated the importance of knowledge about the urban subsurface and confirmed that the urban subsurface is still largely "out of sight, out of mind," not yet a daily concern of city planners and managers. Mielby et al. (2017) give a comprehensive overview of use of subsurface information in urban planning and development. Both reports have been drawn on in this chapter.

3.2 Layers of the Subsurface

In applied contexts, the subsurface is usually considered as two zones (Figure 3.1): "shallow" and "deep." The shallow subsurface is the zone of intense human interaction; it is used for, and congested by, basements, foundations, quarries, and, more recently, utilities and transport tunnels. Developing such facilities and functions depends on subsurface properties but also how changes to these properties affect stability and groundwater flow.

The deep subsurface is traditionally defined by its resource potential for groundwater, minerals and energy and, more recently, by other types of use, such as the disposal of hazardous wastes, geothermal energy production, and carbon capture and storage. The use of the shallow subsurface is generally managed to minimize costs while use of the deeper subsurface is guided by the goal of maximizing profits (Van der Meulen et al. 2016a). Depending on local geology and jargon, the shallow subsurface is defined as the upper tens of meters to a maximum depth of about 200 m (650 ft). The underlying deep subsurface is economically defined and is typically mapped and considered down to a couple of kilometers, which is still very shallow from

Applied Multidimensional Geological Modeling: Informing Sustainable Human Interactions with the Shallow Subsurface, First Edition.
Edited by Alan Keith Turner, Holger Kessler, and Michiel J. van der Meulen.
© 2021 John Wiley & Sons Ltd. Published 2021 by John Wiley & Sons Ltd.

Box 3.1 New Hotels in Washington DC and London Meet Building Height Restrictions by Using Underground Space

The Marriott Marquis Hotel, Washington DC
The Marriott Marquis Hotel, completed in 2014, was designed to conform to Washington's strict building height limits. The Height of Buildings Act passed by Congress in 1899 limited buildings to the height of the Capitol but was amended in 1910 to allow commercial buildings to have a maximum height of 40 m (130 ft), or to the width of the adjacent street plus 6.1 m (20 ft). Thus, a building facing a 27 m (90 ft) street cannot exceed 34 m (110 ft) (Congressional Record 1910). Additional restrictions control developments near important government buildings. Minor revisions in 2013 and 2016 have retained these height restrictions.

The large hotel is located on a prime site, which allows for an underground connection to the adjacent convention center. The hotel has 15-stories above ground and seven below-ground levels housing extensive meeting event spaces, including three double-height ballrooms, and parking for 400 cars. The hotel construction required innovative top-down subsurface excavation controls and sophisticated structural design.

Edwardian Hotel, Leicester Square, London
Planning consent was granted for a new hotel occupying an entire city block on the perimeter of London's famous Leicester Square (City of Westminster 2013). Local planning policy locates the site within a Landmark Viewing corridor, the Wider Setting Consultation Area of Protected Vistas at the Palace of Westminster, and for two aspects from Parliament Hill.

Thus, to maximize floor area and property value while meeting these regulations, the building was designed with 10 floors and five basement levels. The building rises 35.4 m above street level and extends 28.9 m below ground.

Table 3.1 Issues influencing subsurface development.

Driving forces promoting development	Constraints restricting development
1. Competition for space where existing and proposed development leave few other options; either build high or develop beneath the surface	1. Existing surface development, deep foundations, and previous subsurface structures
2. Secure adequate water supply, sewage and communications infrastructure and pipelines	2. Environmental problems associated with groundwater, gases, contaminants and pollution, often associated with previous human activities;
3. Develop underground road and rail systems due to existing surface development	3. Unfavorable properties of engineering soils and bedrock
4. Overcome topographical barriers to communication by securing more direct routes	4. Existing, or future, extraction of water, minerals, oil, or natural gas.
5. Reduce impacts on populations of climate extremes	
6. Secure domestic and geothermal heating and cooling systems to reduce emissions	

a more academic point of view, representing only about 0.03% of the Earth's radius.

Dutch planning authorities use a conceptual threefold division in three layers (Figure 3.2) that are defined according to residence time and changes of use: an *occupation layer* with buildings and foundations having lifetimes generally in the order of decades; a *network layer* with transport infrastructure and utilities networks having lifetimes in the order of centuries; and a *subsurface layer* with much slower geological processes such as deformation and groundwater flow (Van der Meulen et al. 2016a). These layers, even though they are more or less superimposed, do not coincide with actual subdivisions of the subsurface. It is, however, the first spatial planning concept that

Figure 3.1 Classification of the subsurface into two layers. (Van der Meulen et al. 2016a).

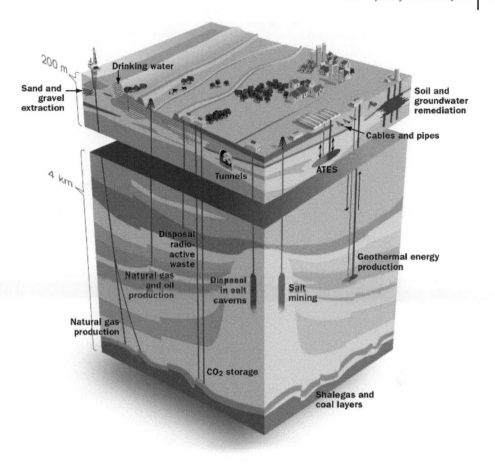

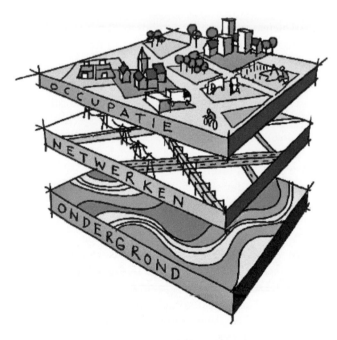

Figure 3.2 The Dutch spatial planning layer approach distinguishing three layers: the *occupation layer* ("Occupatie"); *network layer* ("Netwerken"); and *subsurface layer* ("Ondergrond"). (Van der Meulen et al. 2016a).

explicitly acknowledges the subsurface and serves as an important communication tool between planners and geoscientists.

In old cities, the uppermost 3 m (10 ft) of the ground is often congested with aging, poorly documented, water, gas, telecommunications, and electrical utilities (Figure 3.3) and is penetrated by foundations. Even in newer cities, such as Abu Dhabi and Dubai, deep piled foundations of high-rise developments, and utility and drainage networks, already occupy much of the uppermost 80 m (270 ft) of the subsurface (Price et al. 2016) and present potential problems for future subsurface uses.

Urban substrates consist of bedrock and poorly consolidated superficial deposits but, also, anthropogenic deposits that are often contaminated, polluted, or structurally weak. This "Artificial ground" is particularly important, as it is often associated with potentially contaminated material, unpredictable engineering conditions and unstable conditions (Rosenbaum et al. 2003).

Principles of legal, regulation and management issues have been developed mainly for the deeper subsurface. This chapter includes examples of these as reference material for the shallower applications that are the subject of this book.

Figure 3.3 Underground utilities exposed in a New York City street after removal of the road surface. Source: New York City Department of Transportation – Federal Highway Administration Webinar, September 24, 2012. See: http://ops.fhwa.dot.gov/wz/construction/webinar92412/nycdot/index.htm.

3.3 Legal Systems

Subsurface development often has less legislative oversight than surface activities. Traditionally, deep subsurface activities have been regulated through provisions for resource exploitation while the shallow subsurface has been subject to diverse provisions intended originally to control surface development.

Legislation is the framework for regulation and management that must be complied with. But legal and regulatory systems are diverse, reflecting history, geography and ideology (Zweigert and Kötz 1998). Many publications describe the role of legislation in developing subsurface spaces including: Bobylev (2009), De Mulder et al. (2012, 2013, 2015), Evans et al. (2009), Guo et al. (2013), ITA (2000), ITACUS (2016), Johnson (2010), Larsson (2006), Marker (2015b, 2016), and Vähäaho (2014a). These all broadly agree with three points summarized by a US National Research Council report (NRC 2013):

1. In many jurisdictions there is still little strategic coordination of underground infrastructure development. Rather, market forces encourage "ad hoc" development.
2. Coordinated local and regional planning are needed to manage the underground as part of an integrated urban system to encourage better management of investments, optimization of decisions, reduction of project risk and increase leveraging between government agencies and the private sector (e.g. public–private partnerships).

3. Development of urban underground space requires expanded and coordinated communication with stakeholders to have better regard to local circumstances, encourage greater flexibility, and accommodate long-term community and environmental needs. Measures that explicitly address shallow subsurface development remain unusual.

Recent publications (David 1950; Mattei 1997; Zweigert and Kötz 1998; Siems 2007; De Mulder et al. 2012) classify legal systems. Table 3.2 compares three widely referenced classifications of legal systems. Mattei (1997) identified three "patterns" of law (Table 3.3): traditional, professional, and political:

- *Professional Law*, also called the "Western Legal Tradition" is based on the separation between law and politics and between law and religious and/or philosophical tradition. It includes: *Civil Law* systems of western Europe; *Common Law* systems of England, North America and Oceania; and *"Mixed"* which exhibits components of both Civil and Common Law.
- *Political Law*, under which there is no formal law impinging on Government and the legal process is often determined by political relationships. Some Governments may make efforts to recognize aspects of "Western" international financing institutions, but societal norms, and the need to maintain stable authority, can bypass formal law.
- *Traditional Law*, also called the *"Eastern Legal Tradition"* in which formal legal institutions exist, but law is not separated from religious and/or philosophical traditions.

These were subdivided into 14 individual systems which, individually, rarely conform to a main pattern but, rather, mix characteristics of several patterns. The "families" are summarized in Figure 3.4 in which legal systems plotted close together tend to share a greater number of characteristics.

However, this taxonomy did not explicitly define "Customary Law," identified by De Mulder et al. (2012), under which land belongs to the community, individuals have cultivation rights but the land cannot be sold (Williamson 2001). This system exists in, for example: designated Native American (First Nation) lands of the Americas (USA, Canada, and South America); Australian Aboriginal Lands; and several countries in Africa (e.g. Malawi, Ghana, and Uganda) and Oceania (e.g. Papua New Guinea and the Solomon Islands). That requires careful interactions with whole communities.

Legal systems operate at various levels depending on governance structures: international, national, regional

Table 3.2 Comparison of three widely referenced classifications of legal systems.

David (1950)	Zweigert and Kötz (1998)	De Mulder et al. (2012)
1. Western laws, a) Romano-Germanic sub-group (civil law) b) Anglo-Saxon subgroup (common law)	1. Roman family	1. Civil law
	2. German family	
	3. Common law family	2. Common law
	4. Nordic family (mixed common and civil law)	
2. Soviet law	5. Prior to 1996: soviet or socialist law (dropped after fall of "soviet communism")	
3. Muslim law (sharia)	6. Religious family (Jewish, Muslim, and Hindu law)	3. Islamic law
4. Jewish law (halakha)		4. Customary law
5. Hindu law		
6. Chinese law	7. Far east laws family (China, and Japan)	

or local (Mattei 1997; De Mulder et al. 2012). It is important that geoscientists understand and comply with all legal provisions when designing and undertaking ground investigations, analysis and interpretation of materials and reporting and providing advice on development proposals, as well as engagement with stakeholders.

3.4 Land Ownership

Land ownership is a major issue. Legislation aims to protect legitimate rights of land owners while securing safe, environmentally acceptable development in the public interest. Underground infrastructure is usually owned and operated by many entities, with different, sometimes opposing, needs, interests, governing structures, and resources. This can lead to disputes.

Subsurface development inevitably affects owners' rights. The landowner's permission is often required for access to, and use of, the subsurface, even in the prospecting stage. Land can be purchased for access but, sometimes, provisions allow authorities to have statutory access for certain purposes, or to expropriate lands. Legal provisions are intended to help to resolve disputes during land acquisition. Before investments are made, permissions to proceed

are given; additional permissions control construction, occupation, use, and decommissioning.

In most "Western democracies", land rights are held by individual private or public legal entities. "*Public land*" is acquired occupied or used by the Government or other public authorities. "*Private land*" is owned, held, used, or occupied under a freehold title, a leasehold title, or a certificate of claim. Landowners' rights are generally registered and asserted under Land Registration Acts on a definitive cadastral map (De Mulder et al. 2012). In Communist countries, however, most land is owned by the State, although China and Vietnam recently introduced provisions allowing individuals to hold land-use rights for periods up to 50–70 years and, in October 2007, China legally reintroduced private land ownership (De Mulder et al. 2012).

Extensive literature on land ownership, land information systems and reforms over the last two centuries was summarized by Williamson (2001). Historically, extensive lands were held by few people but, over time, ownership has often become fragmented, causing increased costs and delays in assembling land for large projects. A particular instance was in Japan where traditional family inheritance rights caused significant barriers to underground development. In 2000, the Japanese Parliament adopted a "Special

Table 3.3 Mattei's taxonomy of world's legal systems.

Main pattern	Family	Individual system	Description
Rule of professional law	Civil law	French inspired	Both systems are based in Roman law. French system evolved from the Napoleonic Code; German system from the German Civil Code in 1900. Differences are important at local project scales.
		German inspired	
	Common law	English law	Although with a common root, currently English and American law exhibit some important differences. Canada and Australia combine significant characteristics of both systems.
		American	
	Mixed	Scandinavian	A mixture of civil and common law, that evolved independently in the seventeenth and eighteenth centuries.
		Technically mixed	Systems combining elements of both Civil and Common law are found in Scotland, Louisiana, Quebec, and South Africa due to their history.
Rule of political law	Transition	Socialist	Russia and most ex-socialist European countries, except for Poland, Hungary, and the Czech Republic where Socialist law faced a highly sophisticated civilian legal heritage, resulting in less impact. Excludes Asian socialist countries.
		Post-socialist	
	Development	Africa	Legal process influenced by political relationships. Western legal models from the colonial period combined with traditional (customary) law which often has strong influence.
		Latin America	Legal process is often determined by political relationships. Legal models from the colonial period result in some aspects of professional law combined with traditional (customary) law.
Rule of traditional law	Traditional	Chinese	Combines Confucian conceptions of with traditional philosophic-behavioral rules. Communist dogma provides characteristics of the rule of political law, but differs greatly from other Socialist systems.
		Japanese	Japanese social structures result in considerable influences from traditional law; these modified by western law models supporting industrial development.
	Religious	Islamic	In North Africa and Middle East, Islamic principles of jurisprudence are applied. It has some similarities with common law
		Hindu	Indian law is inherently complex. It combines characteristics of a western constitutional system (based on English Common Law) and a great deal of traditional law with special characteristics due to cohabitation of Hindu and Muslim communities.

Measures Act for Public Use of Deep Underground" (Box 3.2), which gives public organizations prior rights to develop deep underground space (Konda 2003).

Sandberg (2003) defined three legal options for the use and regulation of the subsurface, including multiple-uses:

(full) ownership; condominium law; and lease and easement.

Full (or sole) private ownership is the simplest, most widespread, option in which the landowner owns a theoretical segment of the Earth, extrapolated from the

Figure 3.4 Graphical representation of taxonomy of World's legal systems. Modified from Mattei (1997).

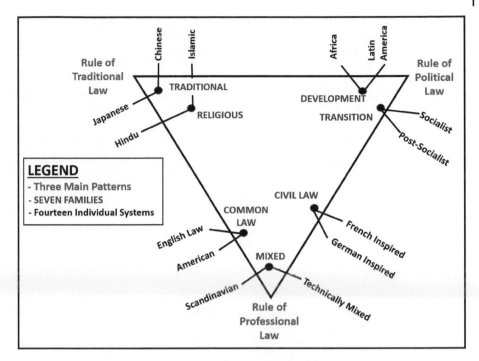

boundaries of the surface holding to the center of the planet. However, Nations increasingly limit landowners' rights to specific depths to safeguard direct interests in future subsurface uses, e.g. they may not claim rights to prevent the construction of a tunnel or underground structure below their property unless it would adversely affect their normal use of the land. Another example is state ownership of (deeper) minerals irrespective of land ownership, as defined originally in Napoleonic Code mining laws that exist in parts of Western Europe.

Condominium Law was developed in the USA within a long tradition of multi-level and multi-purpose building. It recognizes three-dimensional units (Sandberg 2003) in which owners share the property but have individual exclusive rights to use a certain part. Regulations define interrelationships between individual owners. This approach is often combined with transport routes, thus mixing private and public property rights.

Under *Leasehold provisions*, the owner grants a right, usually time limited, to use or build on land to a lessee. Easements allow the property owners to assign access to the property, conditionally, to third parties for specific purposes. Rights-of-way across private property are a specific type of easement so this provision is often used for underground construction of utilities. Leases are widely used in exploring and exploiting minerals, oil or gas. However, if underground space is created under terms of a lease, it is impossible to return that space to its original condition on decommissioning so the lease must define how any changed property value should be assigned.

Land ownership relates to surveying and geological modeling in several ways. There will always be an element of site selection involved before land is actually acquired and developed, whether the development is for land use or infrastructure planning by government, or private undertakings such as mining, or whether development occurs underground or above-ground. This implies that lands that are considered for development may not be accessible for surveying, and if 3-D modeling is to support the planning process, it will have to be undertaken with sparse data. The results are likely to become a matter of debate, or even dispute, between plan proponents and opponents. In the Netherlands, it has been argued that situations like these call for a strong and independent public information service, implemented as the Key Register for the Subsurface (see Chapter 2). This includes systematically produced 3-D geological models, since these are the natural successor of the geological map that would have been routinely consulted previously. When such models are produced independently of the planning and development process, or of the organizations involved in that process, they become more acceptable as a base for decision making.

Box 3.2 The Japanese Special Measures Act for Public Use of Deep Underground

In Japanese cities, extreme fragmentation of privately-owned lands, resulting from inheritance rights, poses significant legal barriers to underground development. Many tunnels have been constructed below public roads, which are under the jurisdiction of national, prefectural, or municipal agencies and can be used after acquiring permission from the relevant agency. This has resulted in sharp corners and steep gradients in some railway tunnels and serious cost inefficiencies due to technical difficulties. The shallow subsurface below roads is congested with buried underground utilities. However, the soft estuarine soils below Osaka and Tokyo are suitable for construction of bored tunnels located 50 m or more below the surface.

In 2000, the Japanese Parliament adopted the "Special Measures Act for Public Use of Deep Underground" (ACT No. 87 of 2000) to provide public organizations prior rights to develop 'deep underground' zones normally not used by landowners (Konda 2003). This legislation makes it possible to route infrastructure elements at depth beneath private land, significantly streamlines the process, and overcomes legal/economic barriers due to private property concerns. This results in more versatile and rapid development of deep underground space.

The Special Measures Act for Public Use of Deep Underground preserves traditional private-sector land-use rights (Figure 3.2.1). In many areas, basements of even the tallest buildings do not extend more than 25 m below the surface. In these locations, "deep underground" is defined as deeper than 40 m to allow for any additional construction works. In areas where buildings are supported by pile foundations, 'deep underground' is defined as depths 10 m below the local supporting layer used by the piles.

Currently, application of the act is limited to Tokyo, Nagoya, and Osaka – areas with dense land-use.

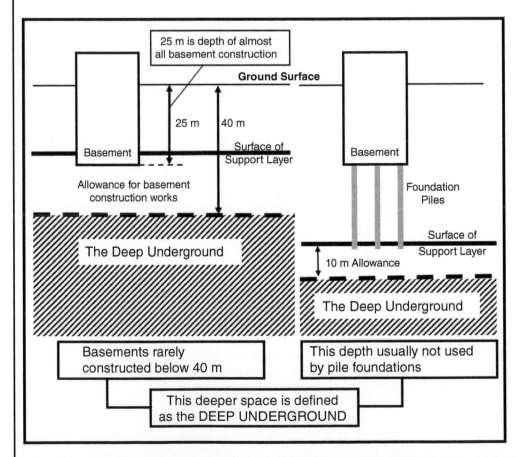

Figure 3.2.1 Definition of "Deep Underground" according to the Japanese Special Measures Act for Public Use of Deep Underground. Modified after Konda (2003).

3.5 Regulation and Management

Regulation provides detailed procedures for implementing international, national, regional, and/or local legislation involving permitting, licensing, inspection, monitoring, and reporting. Regulation requires compliance with legally-binding regulations and standards but is also supported by advisory guidance documents, codes of practice and voluntary legal agreements. Regulations that streamline permitting processes provide clarity, certainty and reduce delays and project costs (NRC 2013). Poorly designed regulations impose unnecessary burdens and delays. Government deregulation initiatives may eliminate seemingly redundant regulations, only to find that new regulations are needed if, for example, a hazardous event or many smaller events reaffirm the need. Regulations are only effective if they are properly enforced. But enforcement is not always properly resourced or effective. Improvements require new legislation (e.g. UK Regulatory Enforcement and Sanctions Act; Anonymous 2008). Appropriate measures are needed to improve planning, design, construction, operation and maintenance but, as yet, these are often weak and poorly adapted to shallow subsurface development.

Often, applications must be made to multiple authorities for subsurface developments within regulations that sometimes conflict or overlap (ITA 2000). However, coordinated permitting procedures have been implemented in some places, including Helsinki (City of Helsinki 2009), in Japan (Konda 2003), and in the Netherlands (De Mulder et al. 2012).

Regulation is shared by National, Regional and local Government Agencies according to the nature and depth of the activity. While activities in the deeper subsurface are largely regulated and managed by National governments, activities in the shallow subsurface are usually matters for City and local authorities.

Coordination of development above and below ground remains rare (NRC 2013) despite the dependence of urban areas on infrastructure, foundations and diminishing space (Sterling et al. 2015). Concern about this is not new; more than a century ago George Webster, chief engineer and surveyor of Philadelphia, USA, advocated that underground space be regulated to facilitate future installations and minimize future installation costs and delays stating:

> Although much careful study has been given by trained experts to the preparation of plans for the rebuilding and extension of large cities and the laying out of new towns, and to the development and improvement of street systems so as to provide for present and future surface traffic and to best serve the convenience, health and welfare of the people, but little thought has been given to the subterranean street. In only a very few of our large cities has any attempt been made to plan subterranean streets or to chart the structures which they contain.
>
> (Webster 1914)

Some aspects of regulation of subsurface development concern: ground investigations, spatial planning, exploitation of natural resources, subsurface construction, safety and health and environmental and cultural issues.

3.5.1 Ground Investigation

Ground investigations are essential precursors to development, contribute to environmental assessments and provide data used in 3-D modeling. High quality records of boreholes, trial pits and trenches are required. Adequacy and quality are regulated through various standards and codes of practice. Numerous provisions cover all aspects of site investigation, description of rocks and soils, sampling and testing and presentation of results. For instance, ground investigation procedures are defined by international and national standards and codes of practice. Examples of international standards provided by the International Organization for Standardization include ISO (2002) on the identification and classification of soil and ISO (2003) on the identification and description of rock, while Eurocode 7 provides guidelines on geotechnical design (European Committee for Standardization 2004), and ASTM has guidelines on the selection of drilling methods.

In addition, practitioners are widely required to be licensed or accredited members of professional bodies which require them to practice in an ethically responsible matter within the limits of their expertise and experience. This encourages the best possible investigation analysis, interpretation and reporting of results.

3.5.2 Spatial Planning

Spatial planning guides development and conservation, taking account of economic, social and environmental implications, to secure sustainable outcomes. Traditionally, many local governments directed acceptable growth by "*zoning*" surface spaces for development or conservation and allocated associated "*air rights*" above the ground. This "*land use planning*" concept broadened into "*spatial planning*," which considers all relevant implications for society. This process, in democratic systems, has four components.

- *Formulation of Policies (i.e. spatial strategies, planning guidance documents)*: assessment of the sustainability of strategic policy options, public consultation on a preferred option, amendment and, then, approval by elected members followed by periodic review and updating.
- *Evaluation of Planning Applications,* based partly on ground investigations or, for large projects, environmental assessment, and public consultation. (A planning application is defined here as the required documentation submitted to a planning authority by a developer or owner seeking permission to modify existing development, undertake new development and/or to change the use of land.)
- *Approval or Refusal of Applications,* which results in requirements for the control of construction and the use and monitoring of the facility.
- *Enforcement of Planning Conditions and Legal Agreements attached to planning consents*: these are monitored and reviewed by relevant authorities, but inadequate enforcement may cause social or environmental problems.

However, private operators or Government agencies commonly own, administer, and operate large underground infrastructure and utility networks for sewerage, water, energy, storage, transport and communications, overseen by multiple regulatory authorities, often outside planning systems. Integrated planning of surface and subsurface development within cities remains uncommon (NRC 2013) although some city authorities have begun to recognize the importance of underground space and to coordinate planning functions within a 3-D framework (Van der Meulen et al. 2016a; Mielby et al. 2017). A variety of approaches have been tried, ranging from non-binding promotion of integrated planning concepts, to more rigorously enforced regulations and adoption of legally binding urban master plans.

The Dutch three-layer conceptual spatial planning model (Figure 3.2) encourages underground development. Arnhem City Council promotes public and private use for all new developments because of shortages of surface space and the need to maintain environmental qualities of the city. A "Vision on the Underground of Zwolle," approved in 2007, comprehensively evaluated underground space beneath that City. The resulting plan, although not legally binding, provides a strategy for underground development (ITACUS 2016). A new Dutch legal system that regulates subsurface information with the aim of improving decision-making in infrastructure and land-use planning is discussed in Chapters 2 and 4.

The British Geological Survey (BGS) has long collaborated with planners of Glasgow City Council, initially through environmental geology mapping and later by 3-D modeling. The city presents many redevelopment challenges because of the geological setting and a legacy of industrially damaged land. In 2012, as models matured, the City Council and BGS launched a partnership "Accessing Subsurface Knowledge (ASK)" to facilitate coordination between the public and private sectors, academic research community, and the BGS (Barron and Campbell 2013; also Chapter 18). This coordination was established to help deliver successful construction and regeneration projects because poor understanding of ground conditions has been the main cause of project delays and cost overruns. Members of the ASK Network partnership can access the 3-D models of superficial deposits and bedrock of the central city without charge.

Following extended public debates, in 2010, Helsinki, Finland, adopted an Underground Master Plan (UMP) which guides current use of underground space and also safeguards spaces for future uses (Box 3.3). The Geotechnical Division of the Helsinki City Planning Department maintains a Geographic Information System (GIS) database that defines areas and depths suitable for the construction of large, hall-like spaces to safeguard these for possible future development. The Master Plan is a legally binding document by making it obligatory to *consider* the use of underground space in development plans (Vähäaho 2014a,b). However, so far, 3-D subsurface geological models have had limited use in Helsinki. Singapore also safeguards long-term availability of underground spaces through regulations (Chang et al. 2012).

Some developments regarded as too small to require planning permission are dealt with through building control, such as limited extensions to existing buildings. This can lead to problems. For example, high property costs and limited space in London, UK, stimulated demand for new basements. There are numerous reports about disputes over, and the extravagances of, "iceberg homes" (Dowling 2014). A few buildings have been damaged or even collapsed during the excavation of new basements (Mann 2015). Therefore, some local authorities have issued supplementary planning guidance to control these activities (City of Westminster 2014, 2016; Kensington and Chelsea 2015, 2016; Box 3.4).

Box 3.3 Regulation and Management of Underground Helsinki

Helsinki, the capital of Finland, occupies a relatively small area (214 km^2) on the Baltic shore. The city center is on a peninsula with a population density as high as 16 500 inhabitants/km^2, although the city lacks high-rise buildings. However, Finns expect to have extensive "green-spaces" even in urban areas. These social and economic reasons have encouraged the use of underground space, and this has been further encouraged by the local geology. Precambrian bedrock, found outcropping or at shallow depths under Helsinki, is mostly ideal for tunneling and building underground spaces. One third of Helsinki is underlain by glacial clay, with an average thickness of only 3–7 m. Thus, underground construction can be undertaken at multiple depths with minimal disturbance to surface uses.

In the 1980s, Helsinki instituted an underground space allocation plan to guide the efficient use of underground space. By the early 2000s, the growing number of competing underground installations required more comprehensive regulation and management. In 2010, after an extensive public review process (Vähäaho 2014a,b), the Helsinki UMP, which governs the planning and management of all of Helsinki's current and future underground facilities, was approved (City of Helsinki 2009).

The UMP covers the entire city; it is a juridical plan binding property owners and public officials. It also guides preparation of above-ground zoning plans; new development must not conflict with the main underground functions indicated in the UMP. Currently, it coordinates the management of more than 400 underground installations, 220 km of transportation tunnels, 24 km of raw water tunnels, and 60 km of utility tunnels. These uses total about 9 million m^3 of underground spaces, but these underlie only about 1% of the total surface area. Thus, there is considerable growth potential; the UMP balances demands for underground facilities in the city center by reserving areas defined as "unnamed rock resource," areas of bedrock suitable for future underground development. The UMP contains 40 new areas reserved as rock resources and 100 new space allocations for future underground construction. These are protected, even though no plans yet exist, nor are specific uses identified. Because property owners must include civil defense shelters in buildings larger than 1200 m^2, all large underground facilities are designed for use as civil defense shelters. As additional facilities serving more varied purposes are placed underground, they require interconnections to form coherent and interrelated complexes. The Helsinki City Planning Department administers the UMP; its Geotechnical Division provides technical expertise, such as defining areas and depths suitable for the construction of large, hall-like spaces.

The Helsinki UMP results from the convergence of several factors: the favorable characteristics of the bedrock and the severe winter climate conditions have often been suggested as the main drivers. But the UMP also results from the Finnish public desire for open spaces even in the city center, the excellent and long-term cooperation between technical departments and commercial enterprises, and the small size of Helsinki. Urban planning is made easier in Helsinki because the city owns about 62% of the real estate.

3.5.3 Natural Resources

Economically important natural resources, such as metals and energy sources, are mostly regulated at National/State levels, especially when the state owns these resources. In contrast, permits to manage the generally superficial extraction of construction materials, or new construction or closure and rehabilitation of existing facilities, are normally granted by local government (De Mulder et al. 2012). Mining legislation developed by the early twentieth century was widely updated toward the end of the century to address increasing environmental concerns.

Johnson (2010) identified three systems for regulating and managing exploration and exploitation of natural resources: (i) direct ownership, (ii) concession systems, and (iii) claim systems. In direct ownership, the owner explores and exploits the resource. In the claim system, prospectors make a "find" and register it with officials. Both were common in the past but concession systems, in which a National authority grants exploration and exploitation rights now dominate. There are three stages: exploration, exploitation, and closure/decommissioning.

Individual exploration rights are granted for various periods, often 12 months to 3 years or, in some Asian countries, for 6–8 years, while other countries leave periods

---

Box 3.4 Regulating Construction of Basement Extensions in London

In response to extremely high housing values, individual property owners in parts of London, have begun to excavate basement extensions to existing buildings. Over the past decade, the number of proposed extensions has dramatically increased because they provide much-needed extra accommodation where room to extend in other ways is limited. Several completed projects have more than doubled the living space and have included multiple levels below ground. In some cases, almost the entire rear garden area was excavated and, in a few cases, the subsurface beneath the street in front of the house was used.

Economic factors concentrated demands for new basement construction within two high-income and historical sections of London – the Royal Borough of Kensington and Chelsea and the City of Westminster. Development of basement extensions was not regulated by existing planning regulations, so both city councils initially produced planning policy documents (City of Westminster 2014; Kensington and Chelsea 2015). Subsequently, new regulations were introduced in official city planning documents (City of Westminster 2016; Kensington and Chelsea 2016).

Regulations encourage "appropriate and sensitively managed" basement extensions and respect the rights of individual property owners to undertake these developments.

However, because many projects caused objections from other residents, the regulations address: (i) limits to the size of basement extensions; (ii) duration of construction phases; (iii) mitigation of neighborhood disruption; and (iv) damage to adjoining properties. These new regulations are integrated with existing policies covering related relevant issues such as design, flood risk, trees, heritage assets, and residential amenity. As these regulations were being developed, numerous newspaper articles reported on planning disputes, digging disruption, and the sheer extravagance of "iceberg homes" (Dowling 2014). A few buildings collapsed during the excavation of a new basement (Mann 2015).

---

indefinite (Naito et al. 1998). As exploration continues, strict and increasingly limiting procedures are usually imposed. Exploitation normally requires government permits or licenses with parties holding exploration rights having priority to applications. Rights for actual mining are granted in most countries for various periods from 25 years to indefinite (Naito et al. 1998; Bastida 2002).

Groundwater is vital for drinking, agriculture, and industry and, through springs, supplies surface waters. Quality and quantity must be safeguarded leading to extensive regulations controlling abstraction, management and protection. Disputes over groundwater rights are common in areas where water resources are scarce and easily over-exploited, both nationally and internationally. In the arid western USA, water resources are regulated and managed through complex inter-State agreements and "water courts," while eastern, more humid, States have no such system.

Groundwater regulations initially concentrated on quantitative aspects but quality and ecosystems later became significant. Increasing use of groundwater for geothermal heating and cooling has raised concerns about "thermal pollution" where facilities are concentrated.

Groundwater abstraction may also cause extensive land subsidence and stability problems, for example in Mexico City, Bangkok, and Shanghai (Konikow & Kendy 2005; Rudolph 2001), in the Santa Clara and San Joaquin valleys of California, the Houston–Galveston area in Texas, the Las Vegas Valley, Nevada, and south-central Arizona in the

USA (Galloway et al. 2000), as well as in the Netherlands, where water tables have been managed for many centuries (Erkens 2011).

Subsurface space is also a non-renewable natural resource but specific regulations for shallow subsurface construction are still uncommon and mostly derive from building, mining and planning legislation. The shallowest subsurface is used for utilities such as cables, pipelines and sewerage-systems and their repair is typically by socially and environmentally disruptive emergency open-cut excavation while minimizing costs (Hunt et al. 2014). Consequently, some cities have invested in co-locating multiple utilities in Multi-Utility Tunnels (MUTs) to facilitate maintenance and renewal with less excavation and to reduce shallow underground congestion. However, management is complicated if the MUT is shared by multiple utility owners (Canto-Perello and Curiel-Esparza 2013).

In the USA, new safety regulations govern the underground storage of natural gas in depleted hydrocarbon or aquifer reservoirs or in brine-mined salt caverns (Inter-agency Task Force on Natural Gas Storage Safety 2016) conforming to recently recommended practices (American Petroleum Institute 2015a,b) but many countries lack such provisions.

National governments initiate, finance and regulate major new transport infrastructure investments such as underground or high-speed rail lines, such as the Crossrail project in London, UK (Aldiss et al. 2009; Gakis et al. 2016), or major road tunnels such as the St Gotthard Base Tunnel,

Switzerland. Development in cities such as Singapore (Chang et al. 2012) and Hong Kong (CEDD-ARUP 2011; Cervero & Murakami 2009; Chan 2011; Lam 2011; Ling 2011; Box 3.5) are often initiated and financed, at least in part, by national governments with regulatory functions often shared by national, regional, and local governments.

Liability for health and safety during construction, occupancy and use of developments is important to avoid adverse events, damage, injuries and other losses and is widely covered by legislation and regulations. For instance, the UK Health and Safety at Work etc. Act 1974 (Anonymous 1974) places a duty on all employers "to ensure, so far as is reasonably practicable, the health, safety and welfare at work" of all employees. Subsequent regulations and a code of practice clarified procedures (Health and Safety Executive 2013). Safety during deeper subsurface work, such as tunneling, is widely governed by mining legislation which is usually established at national, or state levels (Government of New South Wales 2014; British Standards Institution 2011). Regulations also address the health and safety of occupants of underground facilities, including ventilation, lighting, and directional signs. Long road tunnels are challenging,

---

Box 3.5 The Role of Underground Space in Hong Kong

The steep hilly terrain of Hong Kong limits the expansion of urban areas, but also provides an opportunity for placing urban facilities underground. The strong granitic and volcanic rocks underlying 80% of Hong Kong provide excellent conditions to develop underground space (Ho et al. 2016).

Underground space is used extensively in Hong Kong for mass transit stations, rail, utility, and highway tunnels, and property basements. Additional facilities, including a refuse transfer station, a sewage treatment plant, salt water service reservoirs, and an explosives magazine have already been housed in rock caverns (CEDD-Arup 2011). Many of these facilities can be considered "bad neighbor" facilities. Placing them underground can result in significant benefits to the community and the environment. Given the demand for surface land, the Hong Kong Government has concluded that greater use of underground space can increase land supply – especially if existing government facilities are transferred underground to release the land surface for other beneficial uses.

The Geotechnical Engineering Office of the Hong Kong Civil Engineering and Development Department (CEDD) commissioned a strategic planning and technical study on the "Enhanced Use of Underground Space in Hong Kong" (CEDD-Arup 2011). The study reviewed local and overseas experience with various facility types that could be transferred underground, identified suitable areas for underground developments, and recommended approaches for promoting future development of underground space in Hong Kong.

The benefits of going underground partly depend on the type of facility to be housed. Underground development in Hong Kong offers: (i) reduction of urban sprawl, (ii) preservation of the natural environment, (iii) energy efficiency of underground space, and (iv) favorable construction costs and conditions (Ng et al. 2013). During redevelopment of existing older urban areas, underground development can be constructed concurrently beneath new facilities with minimum disruption to the surface and public. Prime land areas that have become vertically constrained are likely to adopt underground options in the future, particularly if incentives are available to the private sector. Locating refuse transfer stations and sewage treatment works above-ground can be a contentious aspect within existing communities. Placing such facilities underground may reduce local opposition; underground developments have a relatively low visual impact with only shafts and portals visible at the surface. Use of underground spaces as data centers, document storage facilities, and for some manufacturing processes, can benefit from increased security and reduced energy requirements. With Hong Kong's climate, underground facilities offer superior uniform temperature control options and potentially require 50–80% less energy for heating and cooling than a surface building. The construction cost of underground facilities in Hong Kong may sometimes be cheaper than that of an above-ground alternative when the very high values of surface land are included. Innovative design schemes have been proposed, including development of multi-facility caverns and creation of underground space by underground quarrying, these offer numerous societal and economic benefits (Ross et al. 2014; Wallace and Ng 2016).

The CEDD-Arup study identified five strategic areas for future underground development; each area is greater than 20 ha, can accommodate multiple facilities, offers relatively easy access to the surrounding infrastructure network, offers some public benefit, and conforms to Hong Kong's strategic planning initiatives. Three developments were proposed to showcase the feasibility of transferring existing government facilities to underground rock caverns. Preliminary feasibility was explored by reviewing the technical challenges, constraints and possible solutions.

with responses such as European Union legislation for tunnels on the Trans-European Road Network (Council of Europe 2004a). In the USA, the National Fire Protection Association (NFPA) established fire protection standards for occupied underground facilities defining distances to and numbers of exits, ventilation, communication, and underground routes (NFPA 2016). Kansas City codes establish minimum safety requirements for underground space, including mine roof supports (City of Kansas City 2020).

Damage and consequential losses mostly concern land subsidence through groundwater or natural gas extraction, mine roof collapses, contamination of soils or groundwater, and leakage causing explosions or fires from underground fuel stores or ruptured pipelines. In the UK, mine instability was addressed by the Coal Mining (Subsidence) Act of 1957 (Anonymous 1957) and the Coal Industry Act (Anonymous 1975), yet insurance claims for subsidence damage to private properties remain as high as £100 million annually. However, destructive subsidence can also be caused by tunneling, as in Cologne (Box 3.6). In the Netherlands, annual costs of land subsidence due to natural gas extraction amount to about €1.65 billion (De Mulder et al.

Box 3.6 Collapse of Cologne Archive Building During Metro Line Construction

Subsurface construction is difficult in central Cologne; it is underlain by 20–30 m of alluvial sand and gravel, and groundwater is typically found within 12 m of the surface. Cologne transit authority Kölner VerkehrsBetriebe (KVB) proposed a new 4-km North-South Metro line with seven stations beneath the historic city center, relieving congestions and increasing public transport network capacity (Figure 3.6.1). Twin tunnels were constructed without major problems using tunnel boring machines between 2003 and 2008. Cut-and-cover excavation of the stations and a track crossover followed; diaphragm walls, dewatering by pumping, and limited ground freezing were used to stabilize these excavation pits (Haack 2009).

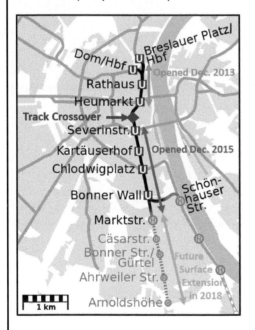

Figure 3.6.1 Map of Cologne new North-South Metro Line. Modified from Wikipedia (2017a).

On March 3, 2009, the track crossover excavation, located in front of the Cologne Historical Archive building, suffered a rapid catastrophic failure. Site monitoring systems did not provide any warning.

Water and soil from behind the diaphragm wall partially filled the excavation pit, damaged the dewatering system and the temporary steel ceiling above the excavation. The archive building was undermined, resulting in its collapse. Construction workers managed to warn and evacuate 45 people, but two individuals were asleep and died in the rubble (Haack 2009). The collapse sparked debates in other European cities, including Amsterdam and Copenhagen, where new subway lines are being built in historic areas with buildings that could be vulnerable to construction.

Investigation into the collapse continues; it has been surrounded by controversy (Rowson, 2009; Wallis 2009; Goebel, 2010; Rosie, 2010; Turner, 2010). In February 2015, six years after the collapse, the KVB reported to City Council members that the assumed cause of collapse was "a construction defect" in the underground diaphragm walls (Damm 2015). However, the project contractors have proposed an alternative cause; a sudden, explosive flooding of the excavation pit (Damm 2015).

The complete North-South Metro line will not be open for through traffic, and thus satisfy its original goal, until at least 2023 – about 12 years after its planned completion (Wikipedia 2017a). Limited service using the completed tunnels and stations was initiated to the north of the collapse site in December 2013, and to the south in December 2015 (Figure 3.6.1).

The collapse has large financial consequences. The archive contained around 65 000 historical documents dating back to the year 922. Document restoration is expected to take up to 30 years and cost €500 million. The KVB has budgeted an extra €4 million to mitigate negative perceptions of the project. Direct costs related to the collapse combined with inflationary increases due to delayed completion have increased the total project cost from its estimated €600 million to between €960 million and €1 billion.

2012). In Japan, the "Industrial Water Law and the Law for Groundwater Use in Buildings" forbids withdrawing groundwater in subsidence-prone areas without a permit (World Bank 2006).

Safety of utilities that supply electricity, gas, water, and telecommunications or drainage for waste and storm water is also a concern. There can be interactions between underground services, including exposure to buried high-voltage electric cables, fires/explosions from gas or oil pipelines, or flooding and collapses due to ruptured water supplies. Damage to telecommunication and television cables also may require disruptive and expensive repairs.

There is potential for damage during ground investigation and excavation of contaminated soils. Particular care is needed at former landfill sites to prevent pollution incidents (Council of Europe 2006). Potential problems require risk assessment (Rudland et al. 2001).

3.5.4 Environmental and Cultural Issues

Environmental protection measures minimize risks to people and ecosystems and require "clean-up" of previous damage (Rudland et al. 2001). For example, EU Directive 2004/35/CE (Council of Europe 2004b) addressed liability for remediation of environmental damage including that to the subsurface. In the USA, contaminated sites were identified as a liability for responsible corporations: in 1997, the total cleanup cost was estimated to be $250 billion. Initially, legal fees accounted for almost 90% of the insurance claims paid relating to USA "Superfund" legislation (De Mulder et al. 2012).

Regulations protecting ecosystems habitats, biodiversity and potentially threatened species are widespread (Brown et al. 2005) and apply to both surface and some underground environments (e.g. Goonan et al. 2015). For example, the EU Habitats Directive (Anonymous 1992a) has been transposed into the laws of each Member State alongside implementing national regulations and local guidance documents, e.g. guidance concerning protection of bat colonies inhabiting old limestone mines in Wiltshire, England (Natural England and Wiltshire Council 2015). But, as yet, other underground organisms/ecosystems have been little considered: for instance, in groundwater within pores and fissures (Gibert et al. 1994; Griebler and Lueders 2009) or bacteria in hydrocarbon reservoirs (Chapelle 2000). Steps are needed to secure appropriate protective designations (Kløve et al. 2011).

The shallow subsurface preserves a record of the geological past and of human cultures; development can damage or destroy archeological layers. Administrative provisions protect significant archeological remains "in situ" or require excavation and recovery prior to development

(Council of Europe 1992b). Damage can be reduced by using access points in the least sensitive locations and carefully developing spaces beneath archeologically important layers. Caution is needed because dewatering during construction and developments that changes groundwater flow can cause organic remains in previously waterlogged soils to dry and crumble (De Beer et al. 2012a,b; Holden et al. 2009). In addition, vibrations from construction works can fracture artifacts (Historic England 2007). However, subsurface development can provide good opportunities to protect archeological remains in place and to sometimes secure underground public viewing through suitable access arrangements.

3.6 Approaches to Subsurface Development

There are two main aspects of subsurface development: (i) re-use of existing spaces and (ii) excavation of new spaces.

3.6.1 Existing Spaces

Natural caves and disused mines have long been used for storage, such as wine cellars. Additional potential uses of existing underground spaces include military and national security purposes, including nuclear waste disposal and carbon sequestration. More recently, major developments have been undertaken in room-and-pillar mines in horizontal strata with sound roof rocks, limited jointing, no significant hydrogeological problems and good road or rail access. Commercially viable re-purposing of existing underground spaces includes storage (including archiving of valuable legal, banking and other archive documents), warehousing, computer back-up facilities, and manufacturing. The largest and most widely promoted commercial facility is "SubTropolis" in the Kansas City Metropolitan Area in the mid-western USA, although the area has several other underground facilities such as those in a limestone mine at Park University (Box 3.7) and a mine at Hutchinson, Kansas.

At Hutchinson, room-and-pillar mining of salt began in 1923. In 1959, a group of businessmen conceived the idea of underground records storage for security purposes and began to lease abandoned portions of the Carey Salt Mine, now owned by the Hutchinson Salt Company. Security is enhanced by limited access via a single vertical shaft shared with the operating mine. The cool and dry conditions of the mine have attracted both "Fortune 500" companies (those which, together, represent about two-thirds of the US economy) and small businesses to store sensitive information and assets. It is the largest single storage facility for

---

Box 3.7 Underground Commercial Centers in Former Mining Excavations, Kansas City Metropolitan Area, USA

About 10% of Kansas City's commercial real estate is underground, and with 90% of the world's subsurface office space, it is the world leader in the use of underground space for human occupancy. Over the past century, extensive limestone mining utilized "room and pillar" (also called pillar and stall) excavation methods to create underground spaces scattered throughout the metropolitan Kansas City area in Kansas and Missouri. Several mines are located below bluffs along the Missouri River; this allows for easy access. Although limestone mining continues, extensive mined areas were largely abandoned; beginning in the 1950s, these attracted the interest of investors as sites for warehouses, document storage, and other businesses (Clark 2015; EDCKC 2016).

"SubTropolis," the largest of these underground facilities, opened in 1964. The room-and-pillar limestone excavation plan produced a grid of 4.9 m (16 ft) high, 12 m (40 ft) wide tunnels separated by 7.6 m (25 ft) square limestone pillars. Located up to 49 m (160 ft) below the surface, the complex contains almost 11 km (7 mi) of illuminated paved roads and 3 km (2 mi) of railroad track (Nadis 2010). Currently 460 000 m² (5 million ft²) is occupied and 930 000 m² (10 million ft²) are "improved." Active mining adds about 13 000 m² of new space each year.

Relatively constant year-round temperatures ranging between 18–21°C (65–70°F) and low humidity eliminate the need for air conditioning or heating. Tenants have reported saving as much as 70% on their energy bills. Rents are about half the going rate for surface facilities (Midwest 2014; Taves 2014).

This climate control attracted a cloud computing company to house its servers. Several tenants use SubTropolis for secure records storage, including the U.S. Environmental Protection Agency (EPA), the National Archives and Records Administration (NARA), and the United States Postal Service for its collectible stamp operations. Today, over 55 diverse businesses lease space in SubTropolis, including: Ford Motor Company, a commercial secure-records-storage facility, a massive frozen-food storage complex, a paintball business, and EarthWorks – an educational program defining the Midwest's natural habitats for schools.

Within the Kansas City metropolitan area there are several other smaller underground facilities. Park University, in the Kansas City suburb of Parkville, has dozens of classrooms, offices, and a library in an abandoned limestone mine, and rents out other underground space to businesses.

In 1991, SubTropolis experienced a major and difficult-to-control fire that burned for weeks in spite of efforts to control it (Buzbee 2011). After similar problems were encountered at another underground facility located in Louisville, Kentucky, the National Fire Protection Association (NFPA) established fire protection standards defining distance to and numbers of exits, ventilation, communication, and underground way-finding (NFPA 2016). Kansas City has since revised its codes with new safety language that establishes minimum safety requirements governing the use of underground space, including roof supports (City of Kansas City 2020).

---

movie and television film in the World. A new shaft was sunk in 2004 and part of the extension became the Kansas Underground Salt Museum, recently renamed "Strataca" (Wikipedia 2017b). Smaller, storage facilities for similar purposes are found in Europe, such as the Winsford Salt mine in Cheshire and limestone mines at Box, Wiltshire, UK (British Geological Survey 2008).

Existing disused tunnels can also be used for imaginative purposes. For example, there is a prototype hydroponic farm in tunnels, originally excavated as air-raid shelters in London, UK that supplies restaurants and shops in the city with green leafy vegetables and herbs on a fully commercial scale (Anonymous 2017).

3.6.2 New Developments

New developments are usually stimulated by local requirements but have often been limited to underground car parks. But there are several examples of major subsurface developments driven by "entrepreneurs" and "market forces."

Because of the harsh winters, Montreal and Toronto, Canada, have large systems connecting basements of privately owned commercial and government buildings with tunnels underneath public streets and linking them to well-lit climate-controlled pedestrian shopping malls, restaurants, metro transit stations, theaters, and other facilities at several levels (Box 3.8). Management of these networks involves agreements between public and private entities concerning safety, operations, and liabilities (Wikipedia 2017c; RÉSO 2017).

Subsurface developments have grown rapidly in China in the last two decades, although conflicts with existing uses caused major difficulties. Therefore, in Shanghai and Beijing recent local regulations coordinate underground space and surface uses and reduce spatial conflicts by regulating

---

Box 3.8 Underground Urban Complexes in Montreal and Toronto

Winters are cold in eastern Canada; this has encouraged the creation of large underground pedestrian complexes beneath the urban centers of Montreal and Toronto.

In 2004, Montreal's underground network of pedestrian walkways, retail centers, residential areas, and public transportation was re-branded and given the name RÉSO, a combination of the French word "réseau" (network) and the letter "O," the logo of the Montreal Metro (RÉSO 2017). The RÉSO is promoted as an important tourist attraction by most Montreal travel guidebooks and is an impressive urban planning achievement (Wikipedia 2017d). Most residents consider it as a large mall complex linking Metro stations. Nearly 500 000 people use RÉSO each day. It stretches for 32 km (20 mi) and covers 4.5 square miles (4 million m²) of the downtown core. Construction of the network began in 1962; growth was encouraged when the Montreal transit commission offered aerial rights above Metro station entrances for construction of new buildings. When the Metro began operations in 1966, 10 buildings were already connected directly to Metro stations; today essentially all stations are connected to building complexes.

Toronto's underground network, termed PATH, consists of 28 km (17 mi) of walkways connecting more than 50 buildings/office towers, 20 parking garages, five subway stations, two major department stores, six major hotels, a railway terminal, and several major tourist and entertainment attractions (Wikipedia 2017c).

Limited underground connections in Toronto were developed prior to the mid-1960s, when the first major urban development in Toronto was designed to include an underground shopping complex, with the possibility of future expansion. The city helped fund this project, but subsequent city administrations did not encourage additional underground construction. However, developers responded to their tenants' wishes and continued to connect their buildings to the system. This converted low-valued basements into valuable retail spaces.

Toronto's PATH shares many characteristics with Montréal's RÉSO; but there are some differences. Montréal's RÉSO has the largest underground system overall, however Toronto's PATH is the longest continuous system and current expansion plans will make it even larger. Montreal civic authorities initially encouraged the growth of RÉSO; Toronto considered PATH as a private-sector development, but recently has realized its importance to the urban core and is now more actively managing and financing current expansion. Montreal's RÉSO is considered a tourist attraction; Toronto's PATH is mostly used by commuters and people who work within the downtown core of Toronto.

---

parking, commercial uses and how much underground space can be developed beneath high-rises. Almost 20 cities in China now have plans for the use of underground spaces that document size, layout, function, development depth, and timescale for planned projects (ITACUS 2016; Guo et al. 2013). A side-benefit of urban underground development has occurred in places such as Helsinki and Hong Kong; suitable waste rock from underground excavations is a source of construction aggregate or can be used as fill for waterfront land reclamation, thus reducing demands for surface quarrying (British Geological Survey 2013).

3.7 Involving Stakeholders

There are many future options for beneficial subsurface development but many stakeholders (including politicians, administrative officials, business and financial managers, commerce and the general public) are not equipped to evaluate these because of diverse educational and cultural backgrounds and a lack of geoscience knowledge or of awareness of advantages and disadvantages. This is exacerbated by the need for relevant legal and regulatory frameworks to avoid "sterilization" of future potential and facilitate appropriate development. Improvements are needed.

Specifically, advice is needed on positive and negative social, economic and environmental impacts of policy options, the time needed for implementation of options, how recommended actions might constrain future decisions and where additional information and advice can be obtained (Marker 2015a). Planning and decisions require (Van der Meulen et al. 2016a):

- An improved general level of *awareness* of subsurface and geological issues among the professional planning and geoscience communities, elected officials, and the general public, which will also lead to better cooperation between planning and geoscience professionals;
- An understanding among planning and geoscience professionals, so that: planners are able to *articulate* their information needs and *understand* what is available; while geoscientists are able to *understand* what is needed and *articulate* what is available or developable;
- Availability of systems to *capture* relevant subsurface data and information that meet baseline requirements at each urban planning and development stage; and

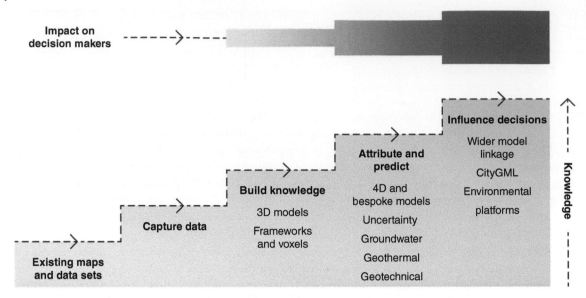

Figure 3.5 Important information levels in the knowledge building process and potential impact on decision makers. (Mielby et al. 2017; illustration developed by D. Campbell, British Geological Survey).

- A shared *joint vision* between planners and geoscientists on how support systems can evolve to address additional information needs.

But that, also, requires adequate budgeting for development, maintaining and updating of databases and 3-D models. Mielby et al. (2017) defined a five-stage improvement process (Figure 3.5):

- Investigate available knowledge from existing maps, reports, and data sets;
- Capture essential data;
- Create the knowledge base by organizing all the collected information into a knowledge framework, including databases and 3-D subsurface geological framework models;
- Attribute any 3-D geological model with physical parameters to predict future situations or explain processes; and
- Address the incorporation of subsurface information in decision-making.

3.8 Delivery of Information

A geoscientist requires several years of training and significant practical experience to interpret information concerning ground conditions and assess implications. A non-specialist cannot be expected be able to bypass that process. Some geoscientists might assume that these models will be readily appreciated by non-geoscientists, but the uses and uncertainties of 3-D models are not easily understood by many stakeholders. Geoscientists are used to thinking about the ground in 3-D, but many non-geologists are not. Careful explanation, illustration, simplification and communication are required to make geological information readily appreciated. Customizing interfaces to suit the specific requirements of stakeholder groups will aid planning, reduce bad decisions and save substantial amounts of money in the long run (Marker 1998). Mielby et al. (2017) also identified the knowledge and communication gap between subsurface experts and urban planners and decision makers, suggesting that bridging the gap requires:

- Provision of the right type of subsurface information, in the right format, at the right time; and
- Ensuring that people receiving the information can understand and use it when taking decisions.

Marker (1998) recommended three steps to improve the communication process: (i) making information quick and easy to find; (ii) customizing information in ways that are directly relevant to, and can be readily understood by, users; and (iii) undertaking capacity building and public awareness initiatives. However, professional geoscientists are not always trained to undertake outreach activities. Because stakeholders' interests and requirements are diverse, a single technical output typically will not meet the all their needs. Results need to be customized (Forster and Freeborough 2006) to reflect individual interests at the right levels of detail. 3-D subsurface models will help only if they are customized in the right ways. Most stakeholders focus on social and economic data organized by administrative or postal/zip code areas. These rarely correspond to physiographical or geological units. Optimum use of geoscience information is unlikely to occur until it is harmonized with other data sets (Marker 1998, 2008, 2015a). This requires scientific, economic and social skills within each team.

3.9 The Role of 3-D Subsurface Models

Technological advances have made the use of 3-D models feasible during the stages of planning, design, and construction, and they are already being used in support of development (Bridge et al. 2005). The COST SubUrban Project (Van der Meulen et al. 2016a; Mielby et al. 2017) provided case studies for several European cities. The shallow subsurface of most urban areas contains a complex mixture of materials produced over time by human activities. Information about the importance of 3-D geological models for regulation and management has been recognized by many various agencies. In the UK, the Environment Agency initially used models to support regional groundwater assessments, but recent work has focused on specific catchments in which abstraction can be regulated through high-resolution hydrogeological and geotechnical characterizations. The use of 3-D models for planning, design, and construction of new urban transit tunnels have reduced geotechnical risk by reducing hazards and saving time and costs in several projects in London, including the Crossrail Farringdon Station (Aldiss et al. 2009; Gakis et al. 2016; Chapter 23) and Silvertown road tunnel (see Chapter 23).

The current variety of 3-D models, ranging from National to regional, local and site specific, is illustrated in Table 3.4. The Netherlands and the United Kingdom are the present leaders in strategic national modeling, but other nations are catching up fast. The Geological Survey of the Netherlands builds and systematically maintains four 3-D national models; each has a distinct application for a specific subsurface depth range (Van der Meulen et al. 2013, 2016b). In 2012, the British Geological Survey (BGS) released a 3-D model of Great Britain (Mathers et al. 2014) consisting of digital geological cross-sections forming a "fence diagram" (Figure 3.6) initially for groundwater studies (Figure 3.7). A screen-shot of the interface display is illustrated in Figure 3.8. This model was extended to the whole of the United Kingdom in 2016 partly to support national geological screening of potential nuclear waste repositories (British Geological Survey 2017). Figure 3.9 illustrates the "lithoframe concept," which allows the "nesting" of local or regional models with higher levels of resolution within a national model framework.

Other countries are now planning national coverage, usually to support natural resource regulation or management functions. The Geological Survey of Denmark and Greenland (GEUS) had produced various 3-D geological models primarily to evaluate oil/gas, groundwater, geothermal, and raw materials/minerals resources, or to assess soil/groundwater contamination issues. Many are publicly available, especially those resulting from an intensive groundwater mapping campaign during the last 15 years (Pedersen et al. 2016). But these models, created by different methods for specific areas, purposes, and of variable quality, are difficult to integrate. Therefore, GEUS is building a coherent national model drawing on results of earlier mapping projects (Sandersen et al. 2016). Unusually, the Dutch National 3-D subsurface models are legally mandated (Van der Meulen et al. 2016b; also see Chapter 4). That is also intended in Denmark (Sandersen et al. 2016).

Internationally, most models (Table 3.4) address groundwater, energy resources and underground storage; these models have a bias toward deeper levels of the subsurface. A few models specifically focus on the shallow subsurface. Examples include the Dutch National GeoTOP model (Chapter 11), a Danish assessment of soil characteristics, and a model for Christchurch, New Zealand, post-earthquake redevelopment (see Chapter 22). These examples support the basic communication and dissemination of information to stakeholders and the general public. The Alberta Geological Survey is in the process of developing an extensive communication and dissemination plan to provide stakeholders with open access to consistent and reliable 3-D geological models and geospatial data to support science-based decision making related to land-use planning, environmental sustainability, economic diversification, and public safety (MacCormack 2016, 2018; Figure 3.10).

Some projects are addressing international trans-border groundwater and geothermal energy resource issues through international collaborations. Examples of this include the H3O project, which involves Belgium, the Netherlands, and Germany (see Chapter 8), and the GeoMol project (Diepolder 2015) involving a consortium of Alpine nations (Figure 3.11 and Chapter 8). The ongoing EU GeoERA program with a 2017–2021 budget of 30 million euro (see http://geoera.eu) is intended to stimulate such collaboration amongst 45 geological survey organizations in 33 European countries. This is important because Europe consists of 51 relatively small states, and the overall body of geological data and information is fragmented accordingly.

The Environment Agency of England and Wales has used 3-D models prepared by the BGS to manage groundwater abstraction from several major aquifers, initially for regional assessments (Figure 3.8), but later for work on abstraction sites that are particularly vulnerable to

Table 3.4 Examples of 3-D models at the time of writing.

Country	Extent	Purpose(s)	References	Comments
Bangladesh	Dhaka	Geological subsurface models for urban planning in mega-cities	See Chapter 18	Examples developed by German foreign assistance program
Canada	Alberta	Alberta geological survey – model of the province with more detailed models of target areas for specific purposes	MacCormack (2016, 2018) Figure 3.10	To communicate and disseminate information to stakeholders and the general public
	Southern Ontario	Ontario geological survey – several models of areas in southern Ontario	See Chapter 19	For groundwater resource management
Denmark	Regional and local	The geological survey of Denmark and Greenland – models for: evaluation of oil/gas, groundwater, mineral resources; and assessment of soil and groundwater characteristics	Pedersen et al. (2016), Sandersen et al. (2016).	Many, especially for groundwater prepared over the last 15 yr, are publicly available. Now building a coherent national model based on earlier mapping projects
France	City of Nantes	3-D geochemical modeling of contaminated "brownfield" site	See Chapter 22	Management of excavated materials linked to urban redevelopment
Germany	Hesse	Hessen – model for assessment of geothermal energy for the whole State.	See Chapter 20	Part of potential commercial geothermal energy development
	Bremen	Model for management of groundwater supply and pollution problems	See Chapter 19	Used by Bremen government for urban planning.
The Netherlands	National	Netherlands geological survey (TNO) GeoTOP model for civil infrastructure development	Van der Meulen et al. (2013, 2016b). Also, see Chapter 11	Lithological and stratigraphic attributes of the top 50 m of the ground with $100 \times 100 \times 5$ m voxels[a].
	National	REGIS- 11 – for groundwater management in the top 1000 ms.	Van der Meulen et al. (2013, 2016b). Also see Chapter 19	Model has 133 parameterized hydro-stratigraphic units[a].
	National	Digital Geological Model (DGM) – extends to 1500 m for energy resource management	Van der Meulen et al. (2013, 2016b).	Model has 34 Neogene and Quaternary lithostratigraphic units[a].
	National	DGM – Deep – extends to 7000 m for energy resource management	Van der Meulen et al. (2013, 2016b).	Model has 13 Carboniferous to Neogene seismo-stratigraphic units[a].
New Zealand	Christchurch	Models for evaluation of post-earthquake reconstruction plans	See Chapter 22	Models formed basis for evaluation of potential liquefaction.
Norway	Bergen	Geological modeling of anthropogenic deposits	See Chapter 25	Used to design protection of archeological remains
Poland	Regional	Model of the Lublin Basin for consideration of energy resources/geothermal potential and salt mobility	Małolepszy (2005), Małolepszy et al. (2016)	Systematic modeling of all sedimentary basins for regional decision making
Slovenia	Ljubljana	Models to aid management of groundwater supply	See Chapter 22	Also used to predict development of potential pollution plume.

(Continued)

Table 3.4 (Continued)

Country	Extent	Purpose(s)	References	Comments
Spain	Zaragosa	Assessment of shallow geothermal resources	See Chapter 20	Local authorities plan to use shallow geothermal for heating & cooling
Sweden	Uppsala	Groundwater modeling of an esker	See Chapter 19	To manage municipal water supplies
Trans-border models	Alpine foreland basins of the Alpine co-operation area	GeoMol project – mainly focused on geothermal energy issues.	See Chapter 8 and Figure 3.11	Fourteen institutions from six member states – Co-ordination by using data exchange standards to evaluate subsurface for multiple future development
	Netherlands Flemish and German trans-border areas	H30 project – evaluates the geothermal potential of the Roer Graben	See Chapter 8	An example of cross-border standardization of stratigraphic terminology and the resulting improved model development.
	Border of Germany and Poland	TransGeoTherm project – to assess trans-border geothermal energy issues	See Chapters 8 and 20	Cross-border evaluation of geothermal resources
United Kingdom	National	British geological survey (BGS) "GB3D" model of great Britain. Funded with the environment agencies of England/Wales and Scotland f	Mathers et al. (2014), Figures 3.6 and 3.8	To evaluate geothermal energy potential and manage groundwater abstraction. Digital geological cross-sections organized into a "fence diagram."
	National	BGS "UK3D" for the whole of the UK including selected sections of the shallow near-shore of England and Wales. Funded by Radioactive Waste Management Ltd. to screen for potential radioactive waste repositories.	British Geological Survey (2020), Figure 3.7	Downloadable in a number of formats including 3-D pdf, 3-D Shapefiles, and KMZ files for Google Earth either in a bespoke BGS view or as a file for use in specialist geological modeling packages.
United Kingdom (Cont'd)	Glasgow	Subsurface models of Glasgow and Clyde region	See Chapter 18	Officially part of urban planning process. ASK network distributes information to users.
	Herts and North London	Manage the groundwater resources of the lower greensand confined aquifer	See Chapter 21	Environment agency funded model for regulating groundwater abstraction.
	Hampshire	Evaluation of cliff instability at Barton-On-Sea	See Chapter 22	Assessment of value of 3-D models by geotechnical consultancy.
	London	Assist the design and construction of the new Crossrail Farringdon Station	See Chapter 23	Formed an integral part of the underground construction process.
	London	Geological modeling for the preliminary design of the Silvertown Tunnel	See Chapter 23	Model reduced required field exploration for preliminary design
	London	3-D models of volume change potential in the London Clay	See Chapter 24	A prototype of possible future 3-D geo-hazard assessments

(Continued)

Table 3.4 (Continued)

Country	Extent	Purpose(s)	References	Comments
	Multiple project locations	Using 3-D models to evaluate designs for railway infrastructure renewal	See Chapter 23	Provided geological data to integrated project designs.
	Newcastle	Characterizing the near-surface geology of Newcastle upon Tyne	See Chapter 25	Masters research in environmental geology.
	Norfolk	Evaluation of coastal cliff retreat at Trimingham	See Chapter 22	A proof-of-concept to show value of geological models.
	Thames Basin	Using 3-D models for evaluating sand and gravel resources.	See Chapter 24	Experimental use of 3-D models for aggregate resource assessment.
United States of America	Counties in State of Illinois	Illinois State Geological Survey – modeling for groundwater resource management in several areas including Chicago suburbs	See Chapters 6 and 9	Illinois is part of a Great Lakes Consortium (x States and Province of Ontario) that conduct 3-D modeling of glaciated terrains.
	Counties in State of Delaware	Delaware Geological Survey – models for groundwater resource assessment	See Chapter 19	Used to manage groundwater in multiple aquifer system in a coastal environment.

a) New versions are to be issued whenever significant improvements are made rather than waiting for full updates.

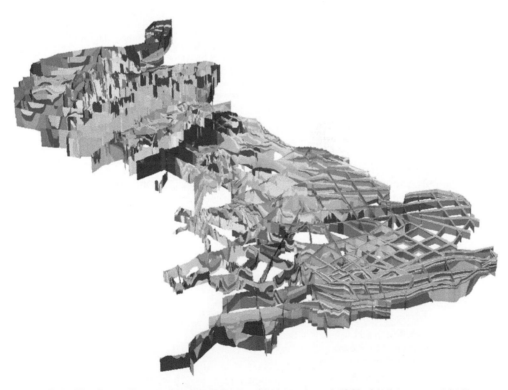

Figure 3.6 The fence diagram developed for the 2012 Version of GB3D. (Mathers et al. 2014).

Figure 3.7 Screenshot of UK3D cross sections for northern England, demonstrated in Google Earth. The sections are displayed with two times vertical exaggeration. British Geological Survey (2020).

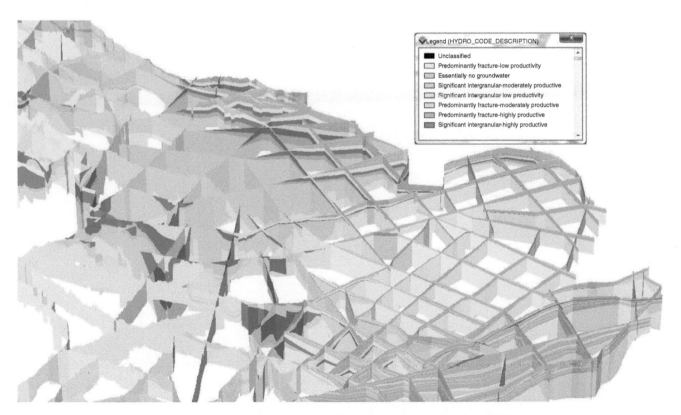

Figure 3.8 The GB3D national model showing an aquifer classification for south-east England. Blue shades are significant inter-granular aquifers, shades of green are fracture controlled aquifers, and brown units are aquicludes. (Mathers et al. 2014).

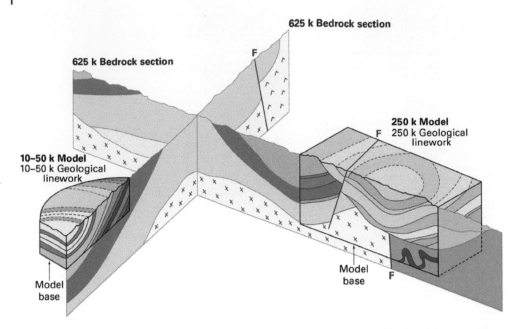

Figure 3.9 The LithoFrame concept showing varied detail at differing levels of resolution. (Mathers et al. 2014).

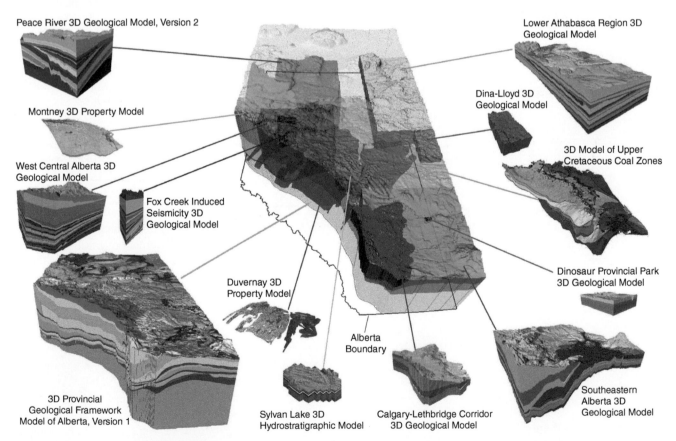

Figure 3.10 Variety of provincial and sub-model scale 3-D models that are contained within the Alberta geological survey's 3-D geological framework. (MacCormack 2018).

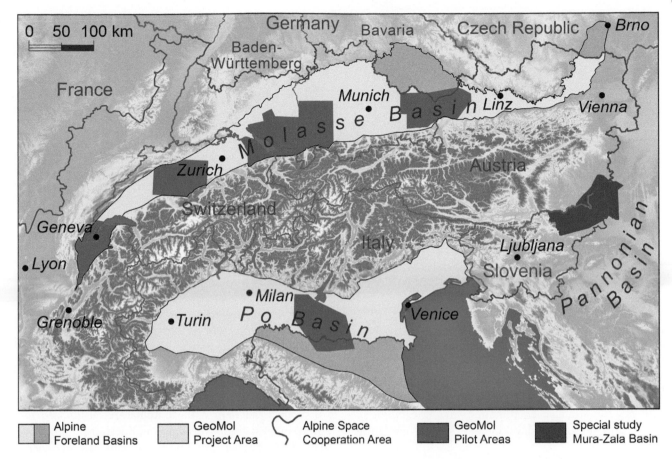

Figure 3.11 GeoMol project area, including five pilot areas and one special study area in Slovenia. (Diepolder 2015).

pollution. Many geotechnical companies now routinely use 3-D models in infrastructure planning, design and construction because these can increase efficiency and reduce geotechnical risk (Chapter 4). Models have also assisted in securing essential transport infrastructure, including underground mass-transit systems, and rail and road tunnels (Chapter 23). Subsurface 3-D models have also been used to evaluate the extent of contamination and pollution (Chapter 22). Archeological investigations are increasingly using 3-D geological models (Carey et al. 2018); de Beer et al. (2012a, Chapter 25) used a 3-D geological model to support archeological evaluations of the important World Heritage site of Bryggen at Bergen, Norway.

However, most 3-D geological models are designed for use by technical experts rather than the full spectrum of stakeholders. This situation is not new. In the late twentieth century, there were many attempts to make 2-D engineering and environmental geology mapping results accessible to various groups of users, particularly planners, through simplification in terms of their specific interests which were later facilitated further by the availability of GIS (Brook and Marker 1987). A recent example of 2-D simplification, in Rotterdam, is to use simple "traffic light" maps to communicate development constraints in terms of relative construction costs (Figure 3.12). The challenge now is to develop customized interfaces for 3-D models.

3.10 Conclusions

The subsurface could make a greater contribution to socially, economically and environmentally beneficial development, but that requires awareness of opportunities in both the public and private sectors and an appropriate legal and technical framework. Legislation and regulations

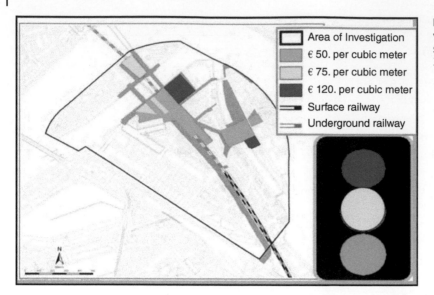

Figure 3.12 Rotterdam "traffic light" visualization of development constraints due to subsurface conditions. (Van Campenhout et al. 2016).

governing subsurface development are mainly drawn from mining and groundwater management for the deeper subsurface while the shallow subsurface is mainly governed through extrapolation of diverse measures originally designed for control of surface development. Currently, there is little strategic coordination and integration of surface and subsurface development, storing up future problems for both.

Legally binding regulations and standards, and advisory guidance documents and codes of practice to implement legislation, guide all stages of the development process, from planning to use. But existing measures are not always well adapted to subsurface activities and can be contradictory or overlapping at various levels (National, Regional, and Local), and between responsible agencies, leading to delays and additional costs. Coordinated permitting procedures are needed. Integrated and improved legislation, regulation and planning, supported by better delivery of technical information, are required to facilitate appropriate, cost-effective and safe subsurface development.

Spatial planning can play a key part but is still widely focused on the ground surface (2-D), but 3-D approaches are emerging (Biljecki et al. 2015; Stoter et al. 2017). Integration of strategies for development and conservation for the surface and subsurface together require adaptation to consider 3-D blocks of ground. Even so, there are now good examples to learn about market-led and planned subsurface initiatives that have re-used existing spaces or created new spaces for many purposes.

But many stakeholders remain poorly equipped to make that transition because of a lack of awareness of subsurface opportunities and constraints or limited geoscience information and knowledge. Effective communication between geoscientists and other stakeholders is essential. 3-D models can help if the outputs are customized for various interest groups through simplification, appropriate changes of emphasis and integration with relevant social and economic data sets, perhaps aided by decision support systems. That requires multi-disciplinary teams and close collaboration with users, but geoscientists are not always well equipped for communication with other professionals and the public. More training would be helpful.

In addition, there are now good examples of National, Regional and Local and site specific 3-D models, as well as some that address trans-border issues. However, other than site-specific models, few focus specifically on the shallow subsurface while many address a single theme, such as groundwater or geothermal energy. More emphasis is needed on models that can serve a wide range of issues at all levels. That requires standardization of approaches to allow coordination between models.

Some possible key steps for the future are summarized in Table 3.5. Geological surveys are uniquely equipped to provide essential geological information when it is needed, and their maps, databases and reports reflect long-term commitments to information gathering and delivery. Several surveys are well advanced in adapting their traditional approaches to 3-D.

Table 3.5 Planning a 3-D modeling initiative that delivers appropriate information to stakeholders.

Stage	Steps
Preparation of a 3-D model (modified from Mielby et al. 2017)	Secure adequate funding
	Undertake the study (with an initial pilot a pilot study, if necessary)
	Review existing 3-D models to see whether aspects can be used or adapted
	Investigate and capture existing data
	Organize data into databases
	Prepare a 3-D model
	Attribute the model with physical parameters that explain processes and predict future situations
	Prepare a quickly accessible web-based delivery system
	Undertake adequately funded continuing maintenance and updating
Steps toward customizing for selected groups of users	Discuss requirements with each selected user group
	Develop simplified user interfaces linked to the users' decision frameworks
	Harmonize with appropriate social and/or economic data if required, in liaison with other relevant experts
Incorporating evidence from 3-D modeling into spatial planning	Undertake an audit of existing underground space (used and unused)
	Identify potentially usable blocks of subsurface space in terms of suitability for various purposes and potential for access
	Assess the potential for safe and practical use of existing underground voids
	Examine potential advantages
	Examine constraints in terms of surface development, underground natural resources and safety.
	Set out policies for development and safeguarding of underground space for future use in the draft development plan, for amendment and approval after public consultation and periodic review and revision

References

Aldiss, D.T., Entwisle, D.C. and Terrington, R.L. (2009). *A 3D geological model for the proposed Farringdon underground railway station, central London*. Keyworth, UK: British Geological Survey, Commissioned Report, CR/09/014.

American Petroleum Institute (2015a). *Design and Operation of Solution-Mined Salt Caverns Used for Natural Gas Storage*. Washington, DC: American Petroleum Institute, Recommended Practice 1170. 52 pp.

American Petroleum Institute (2015b). *Functional Integrity of Natural Gas Storage in Depleted Hydrocarbon Reservoirs and Aquifer Reservoirs*. Washington, DC: American Petroleum Institute, Recommended Practice 1171. 52 pp.

Anonymous (1957). *Coal-Mining (Subsidence) Act 1957 c. 59*. London, UK: HM Government. [Online: Available at: http://www.legislation.gov.uk/ukpga/Eliz2/5-6/59/contents] (Accessed June, 2020).

Anonymous (1974). *Health and Safety at Work etc. Act 1974 c. 37*. London, UK: HM Government. [Online: Available at: http://www.legislation.gov.uk/ukpga/1974/37/contents] (Accessed June 26, 2020).

Anonymous (1975). London, UK: HM Government. *Coal Industry Act 1975 c. 56*. [Online: Available at: http://www.legislation.gov.uk/ukpga/1975/56/contents] (Accessed June 26, 2020).

Anonymous (2008). *Regulatory Enforcement and Sanctions Act 2008 c. 13*. London, UK: HM Government. [Online: Available at: http://www.legislation.gov.uk/ukpga/2008/13/contents] (Accessed June 26, 2020).

Anonymous (2017). *Inside London's First Underground Farm*. The Independent, 3 February 2017 [Online: Available at: www.independent.co.uk/Business/indyventure/growing-underground-london-farm-food-waste-first-food-miles-a7562151.html] (Accessed March 3, 2019).

Bastida, E. (2002). *Integrating Sustainability into Legal Frameworks for Mining in some Selected Latin American Countries*. London, UK: International Institute for

Environment and Development and World Business Council for Sustainable Development, Mining, Minerals and Sustainable Development Project, Report 120. 33 pp. [Online: Available at: https://pubs.iied.org/pdfs/G00577 .pdf] (Acessed June 22, 2020).

Biljecki, F., Stoter, J.E., Ledoux, H. et al. (2015). Applications of 3D city models: state of the art review. *ISPRS International Journal of Geo-Information* 4 (4): 2842–2889. https://doi.org/10.3390/ijgi4042842.

Bobylev, B. (2009). Mainstreaming sustainable development into a city's master plan: a case for urban underground space use. *Land Use Policy* 26: 1128–1137. https://doi.org/10.1016/j.landusepol.2009.02.003.

Bridge, D., Hough, E., Kessler, H. et al. (2005). Urban geology: Integrating surface and sub-surface geoscientific information for developing needs. In: *The Current Role of Geological Mapping in Geosciences* (ed. S.R. Ostaficzuk), 129–134. Dordrecht, the Netherlands: Springer.

British Geological Survey (2008). *Mineral Planning Fact Sheet: Underground storage.* Keyworth, UK: British Geological Survey. 22 pp. [Online: Available at: https://www2.bgs.ac .uk/mineralsuk/download/planning_factsheets/mpf_ storage.pdf] (Accessed March 4, 2019).

British Geological Survey (2013). *Mineral Planning Fact Sheet: Construction Aggregates.* Keyworth, UK: British Geological Survey. 31 pp. [Online: Available at: https://www2.bgs.ac .uk/mineralsuk/download/planning_factsheets/mpf_ aggregates.pdf] (Accessed August 9, 2017).

British Geological Survey (2020). *UK3D.* Keyworth, UK: British Geological Survey. [Website: https://www.bgs.ac .uk/datasets/uk3d/] (Accessed June 2020).

British Standards Institution (2011). *Code of Practice for Health and Safety in Tunnelling in the Construction Industry.* London, UK: British Standards Institution, BS 6164:2011. 168 pp.

Brook, D. and Marker, B.R. (1987). Thematic geological mapping as an essential tool in land-use planning. *Engineering Geology Special Publications* 4: 211–214. https://doi.org/10.1144/GSL.ENG.1987.004.01.26.

Brown, J., Mitchell, N. and Beresford, M. [Eds.] (2005). *The Protected Landscape Approach: Linking Nature, Culture and Community.* Gland, Switzerland: IUCN World Commission on Protected Areas. 270 pp. https://doi.org/10 .2305/IUCN.CH.2005.2.en.

Buzbee, J. (2011). Business Goes Underground. *Progressive Engineer, May/June 2011.* [Online: Available at: http:// progressiveengineer.com/feature-business-goes- underground/] (Accessed January 30, 2017).

Canto-Perello, J. and Curiel-Esparza, J. (2013). Assessing governance issues of urban utility tunnels. *Tunnelling and Underground Space Technology* 33: 82–87. https://doi.org/ 10.1016/j.tust.2012.08.007.

Carey, C., Howard, A.J., Knight, D., Corcoran, J. and Heathcote, J. [Eds.] (2018). *Deposit Modelling and Archaeology.* Brighton, UK: University of Brighton. 230 pp. [Online: Available at: www.brighton.ac.uk/_pdf/research/ set-groups/deposit-modelling-and-archaeology.pdf] (Accessed April 2018).

CEDD-Arup (2011). *Executive Summary on Enhanced Use of Underground Space in Hong Kong - Feasibility Study.* Hong Kong: Government of the Hong Kong Special Administrative Region and Arup. 30 pp.

Cervero, R. and Murakami, J. (2009). Rail and property development in Hong Kong: experiences and extensions. *Urban Studies* 46: 2019–2043. https://doi.org/10.1177/ 0042098009339431.

Chan, R.K. (2011). Planning Future Cavern Development in Hong Kong. In: *Proceedings, Joint HKIE-HKIP Conference on Planning and Development of Underground Space, Hong Kong, 23–24 September 2011.* Hong Kong: Hong Kong Institution of Engineers and Hong Kong Institute of Planners. 228 pp.

Chang, A., Ng, K.W., Schimid, H. et al. (2012). Feasibility study of underground Science City Singapore: Part 2 planning, architectural, engineering, fire safety and sustainability. In: *Proceedings of the ACUUS 2013 conference "Advances in Underground Space Development", Singapore.*

Chapelle, F.H. (2000). The significance of microbial processes in hydrogeology and geochemistry. *Hydrogeology Journal* 8: 41–46. https://doi.org/10.1007/PL00010973.

City of Helsinki (2009). *Underground Master Plan of Helsinki: A City Growing within Bedrock* (leaflet). Helsinki, Finland: Helsinki City Planning Department. 4 pp. [Online: Available at: https://www.hel.fi/hel2/ksv/julkaisut/ esitteet/esite_2009-8_en.pdf] (Accessed: August 2, 2017).

City of Kansas City (2020). *City of Kansas City Missouri, Current 2012 Building Rehabilitation Code, Chapter 18, Article XI: Underground Space.* [Online: Available at: https://library.municode.com/mo/kansas_city/codes/ code_of_ordinances?nodeId=PTIICOOR_CH18BUBURE_ ARTXIUNSP] (Accessed June 24, 2020).

City of Westminster (2013). *Leicester Square Hotel Project.* London, UK: City of Westminster, Planning Application Reference 13/07443/FULL [Online: Available at: http:// idoxpa.westminster.gov.uk/online-applications/ applicationDetails.do?activeTab=summary& keyVal=MQQYHARPZ5000] (Accessed February 20, 2017).

City of Westminster (2014). *Basement development in Westminster: supplementary planning document.* London, UK: City of Westminster. [Online: Available at: http://transact.westminster.gov.uk/docstores/ publications_store/adopted%20SPD%20publication %20version.pdf] (Accessed February 6, 2017).

City of Westminster (2016). Policy CM28.1 Basement Development. In: *Revision to Westminster's City Plan, November 2016*, 121–129. London, UK: City of Westminster. [Online: Available at: http://transact .westminster.gov.uk/docstores/publications_store/ cityplan/Westminster_City_Plan_consolidated_version_ Nov_2016.pdf] (Accessed February 11, 2017).

Clark P. (2015). Welcome to SubTropolis: The Massive Business Complex Buried Under Kansas City. *Bloomberg Businessweek, February 4, 2015*. [Online: Available at: https://www.bloomberg.com/news/features/2015-02-04/ welcome-to-subtropolis-the-business-complex-buried- under-kansas-city] (Last accessed December 16, 2016).

Congressional Record (1910). *An Act to regulate the Height of Buildings in the District of Columbia*. Sixty-First Congress, Session II, Chapter 263: 452–455. [Online: Available at: https://www.loc.gov/law/help/statutes-at-large/61st- congress/session-2/c61s2ch263.pdf] (Last accessed February 24, 2017).

Council of Europe (1992a). Council Directive 92/43/EEC of 21 May 1992 on the conservation of natural habitats and of wild fauna and flora. *Official Journal L* 206: 7–50. [Online: Available at: https://eur-lex.europa.eu/legal-content/EN/ TXT/?uri=CELEX:31992L0043] (Accessed June 26, 2020).

Council of Europe (1992b). *European Convention on the Protection of the Archaeological Heritage (Revised)*. Strasbourg, France: Council of Europe, European Treaty Series 143. 8 pp. [Online: Available at: https://rm.coe.int/ 168007bd25] (Accessed February 28, 2017).

Council of Europe (2004a). Directive 2004/54/EC of the European Parliament and of the Council of 29 April 2004 on minimum safety requirements for tunnels in the Trans-European Road Network. *Official Journal L* 167: 39–91. [Online: Available at: http://data.europa.eu/eli/dir/ 2004/54/oj] (Accessed June 24, 2020).

Council of Europe (2004b). Directive 2004/35/CE of the European Parliament and of the Council of 21 April 2004 on environmental liability with regard to the prevention and remedying of environmental damage. *Official Journal L* 143: 56–75. [Online: Available at: http://data.europa.eu/ eli/dir/2004/35/oj] (Accessed June 24, 2020).

Council of Europe (2006). Directive 2006/21/EC of the European Parliament and of the Council of 15 March 2006 on the management of waste from extractive industries and amending Directive 2004/35/EC – Statement by the European Parliament, the Council and the Commission. *Official Journal L* 102: 15–34. [Online: Available at: https:// eur-lex.europa.eu/eli/dir/2006/21/oj] (Accessed June 24, 2020).

Damm, A. (2015). KVB-Vorstand hält Baufehler für die Einsturzursache (in German). *Kölner Stadt-Anzeiger, March 2, 2015*. [Online: Available at: https://www.ksta.de/ koeln/koelner-stadtarchiv-kvb-vorstand-haelt-baufehler- fuer-die-einsturzursache-1513120] (Accessed July 16, 2017).

David, R. (1950). *Traité Élémentaire de Droit Civil Comparé: Introduction à l'Étude des Droits Étrangers et à la Méthode Comparative*. Paris, France: Librairie Générale de Droit et de Jurisprudence. 556 pp.

De Beer, J., Price, S.J. and Ford, J.R. (2012a). 3D modelling of geological and anthropogenic deposits at the world heritage site of Bryggen in Bergen, Norway. *Quaternary International* 251: 107–116. https://doi.org/10.1016/j .quaint.2011.06.015.

De Beer, J., Matthiesen, H. and Christensson, A. (2012b). Quantification and visualization of in situ degradation at the world heritage site Bryggen in Bergen, Norway. *Conservation and Management of Archeological Sites* 14 (1–4): 215–227. https://doi.org/10.1179/1350503312Z .00000000018.

De Mulder, E.F.J., Hack, H.R.G.K. and van Ree, C.C.D.F. (2012). Chapter 6: Legal aspects, policy and management. In: *Sustainable Development and Management of the Shallow Subsurface* (eds. E.F.J. de Mulder, H.R.G.K. Hack and C.C.D.F. van Ree), 133–167. London: Geological Society. https://doi.org/10.1144/MPSDM.6.

De Mulder, E.F.J., Besner, J. and Marker, B.R. (2013). Underground cities. In: *Megacities: Our Global Urban Future* (eds. F. Kraas, S. Aggarwal, M. Coy and G. Mertins), 25–32. Dordrecht, Netherlands: Springer. https://doi.org/ 10.1007/978-90-481-3417-5_3.

De Mulder, E.F.J., van Ree, C.C.D.F. and Wang, K. (2015). Underground urban development: an overview. *Engineering Geology for Society and Territory*, 5: 25–29. https://doi.org/10.1007/978-3-319-09048-1_4.

Diepolder, G.W. and GeoMol Team (2015). *GeoMol – Assessing Subsurface Potentials of the Alpine Foreland Basins for Sustainable Planning and Use of Natural Resources. Project Report*. Augsburg, Germany: Bayerisches Landesamt für Umwelt (Bavarian Environment Agency). 188 pp. [Online: Available at: http://www.geomol.eu/geomol/report/GeoMol_Report_ web_reduced.pdf] (Accessed October 2018).

Dowling, T. (2014). Deep Concerns: The Trouble with Basement Conversions. *The Guardian, August 18, 2014*. [Online: Available at: https://www.theguardian.com/ lifeandstyle/2014/aug/18/basement-conversions-disputes- digging-iceberg-homes] (Accessed August 30, 2016).

EDCKC (2016). *Historical Building Highlight: SubTropolis*. Kansas City, MO: Economic Development Corporation [Website: http://edckc.com/subtropolis] (Accessed January 30, 2017).

England, H. (2007). *Piling and Archaeology: Guidelines and Best Practice*. London, UK: Historic England. 52 pp.

Erkens, M. (2011). *Water Management in the Netherlands*. The Hague, Netherlands: Ministry of Infrastructure and the Environment. 22 pp. [Online: Available at: http://www.iiea .com/documents/09%20Nov%202011_Marco%20Erkens_ IIEA.pdf] (Accessed March 4, 2019).

European Committee for Standardization (2004). *European Standard EN 1997-1. Eurocode 7: Geotechnical Design – Part 1: General Rules*. Brussels, Belgium: European Committee for Standardization (CEN). 168 pp.

Evans, D., Stephenson, M. and Shaw, R. (2009). The present and future use of 'Land' below ground. *Land Use Policy* 26S: S302–S316. https://doi.org/10.1016/j.landusepol.2009 .09.015.

Forster, A. and Freeborough, K. (2006). *A Guide to the Communication of Geohazards Information to the Public*. Keyworth, UK: British Geological Survey, Urban Science and Geohazards Programme Internal Report, IR/06/009. 38 pp. [Online: Available at: http://nora.nerc.ac.uk/id/ eprint/7173/] (Accessed December 2020).

Gakis, A., Cabrero, P., Entwisle, D. and Kessler, H. (2016). 3D geological model of the completed Farringdon underground Railway Station. *Crossrail Project: Infrastructure Design and Construction* 3: 431–446. London, UK: Institution of Civil Engineers Publishing.

Galloway, D.L., Jones, D.R. and Ingebritsen, S.E. (2000). *Land Subsidence in the United States*. Reston, VA: U.S. Geological Survey, Fact Sheet-165-00. 4 pp. [Online: Available at: https://water.usgs.gov/ogw/pubs/fs00165/SubsidenceFS.v7 .PDF] (Accessed February 25, 2017).

Gibert, J., Danielopol, D.L. and Stanford, J.A. [Eds.] (1994). *Groundwater Ecology*. San Diego, CA: Academic Press. 571 pp.

Goebel, N. (2010). Probe into collapse of Cologne archive mired in scandal one year on. *Deutsche Welle, March 3, 2010*. [Online: Available at: http://www.dw.com/en/probe-into-collapse-of-cologne-archive-mired-in-scandal-one-yearon/a-5281285] (Accessed July 16, 2017).

Goonan, P., Jenkins, C., Hill, R. and Kleinig, T. (2015). *Subsurface Groundwater Ecosystems – A Briefing Report on the Current Knowledge, Monitoring Considerations and Future Plans for South Australia*. Adelaide, Australia: Environment Protection Agency South Australia. 20 pp.

Government of New South Wales (2014). *Work Health Safety (Mines) Regulation* 2014 (799). Sidney, Australia: Parliamentary Counsel's Office, No 799. 171 pp. [Online: Available at: http://www.legislation.nsw.gov.au/inforce/ 500011c6-2a99-440f-8799-8cbf39393930/2014-799.pdf] (Accessed August 9, 2016).

Griebler, C. and Lueders, T. (2009). Microbial biodiversity in groundwater ecosystems. *Freshwater Biology* 54: 649–677. https://doi.org/10.1111/j.1365-2427.2008.02013.x.

Guo, D., Nelson, P.P., Xie, J. and Chen, Z. (2013). Vertical planning of urban underground space use in China. In: *Proceedings, ACUUS 2013 "Advances in Underground Space Development"* (eds. Y. Zhou, J. Cai and R. Sterling). 190. Singapore: The Society for Rock Mechanics & Engineering Geology. https://doi.org/10.3850/978-981-07-3757-3_RP-039-P190.

Haack, A. (2009). *Construction of the North-South-Metro Line in Cologne and the Accident on March 3rd, 2009*. Cologne, Germany: STUVA (Research Association for Underground Transportation Facilities). 5 pp. [Online: Available at: http://ssms.jp/img/files/2019/04/sms10_194.pdf] (Accessed 28 December, 2016).

Health and Safety Executive (2013). *HSE Workplace (Health, Safety and Welfare) Regulations 1992 Approved Code of Practice and Guidance*. Norwich, UK: The Stationary Office. 60 pp. [Online: Available at: www.hse.gov.uk/ pUbns/priced/l24.pdf] (Accessed August 9, 2017).

Ho, Y.K., Shum, K.-W. and Wong, J.C.-F. (2016). Strategic use of rock caverns and underground space for sustainable urban development of Hong Kong. *Procedia Engineering* 165: 705–716. https://doi.org/10.1016/j.proeng.2016.11.768.

Holden, J., Howard, A.J., West, L.J. et al. (2009). A critical review of hydrological data collection for assessing preservation risk for urban waterlogged archaeology: a case study from the City of York, UK. *Journal of Environmental Management* 90 (11): 3197–3204. https://doi.org/10.1016/j .jenvman.2009.04.015.

Hunt, D.V.L., Nash, D. and Rogers, C.D.F. (2014). Sustainable utility placement via multi-utility tunnels. *Tunnelling and Underground Space Technology* 39: 15–26. https://doi.org/ 10.1016/j.tust.2012.02.001.

Hunt Midwest (2014). *SubTropolis Benefits*. Kansas City, MO: Hun Midwest. 1 pp. [Online: Available at: http://www .huntmidwest.com/pdfs/subtropolis_benefits.pdf] (Accessed January 30, 2017).

Interagency Task Force on Natural Gas Storage Safety (2016). *Ensuring Safe and Reliable Underground Natural Gas Storage - Final Report*. 92 pp. [Online: Available at: https:// www.energy.gov/sites/prod/files/2016/10/f33/Ensuring %20Safe%20and%20Reliable%20Underground%20Natural %20Gas%20Storage%20-%20Final%20Report.pdf] (Accessed August 9, 2017).

ITA (2000). Planning and mapping of underground space — an overview. *Tunnelling and Underground Space Technology* 15 (3): 271–286. https://doi.org/10.1016/S0886-7798(00)00056-0.

ITACUS (2016). *Planning the Use of Underground Space*. Châtelaine, Switzerland: International Tunnelling Association Committee on Underground Space (ITACUS) White Paper #2, 4 pp.

Johnson, E.L. (2010). *Mineral Rights. Legal Systems Governing Exploration and Exploitation (doctoral thesis)*. Stockholm, Sweden: Royal Institute of Technology (KTH). 285 pp. [Online: Available at: http://kth.diva-portal.org/smash/ get/diva2:300248/FULLTEXT01] (Accessed February 7, 2017).

Kensington and Chelsea (2015). *Basements Planning Policy. Policy CL7: Basements in Chapter 34 of the Consolidated Local Plan. Adopted 21 January 2015*. London, UK: Royal Borough of Kensington and Chelsea. 20 pp. [Online: Available at: www.rbkc.gov.uk/pdf/Final%20Basements %20Policy%20Jan%202015%20adopted%20web.pdf] (Accessed February 11, 2017).

Kensington and Chelsea (2016). *Basements Supplementary Planning Document. Adopted 14 April 2016*. London, UK: Royal Borough of Kensington and Chelsea. 20 pp. [Online: Available at: www.rbkc.gov.uk/sites/default/files/atoms/ files/01%20160414%20Final%20Basements%20SPD.pdf] (Accessed February 11, 2017).

Kløve, B., Allan, A., Bertrand, G. et al. (2011). Groundwater dependent ecosystems. Part II. ecosystem services and management in Europe under risk of climate change and land use intensification. *Environmental Science and Policy* 14 (7): 782–793. https://doi.org/10.1016/j.envsci.2011.04 .005.

Konda, T. (2003). Reclaiming the underground space of large cities in Japan. In: *(Re)Claiming the Underground Space: Proceedings of the ITA World Tunneling Congress, Amsterdam 2003* (ed. J. Saveaur), 13–23. Lisse, The Netherlands: Swets & Zeitlinger.

Konikow, L.F. and Kendy, E. (2005). Groundwater depletion: a global problem. *Hydrogeology Journal* 13: 317–320. https://doi.org/10.1007/s10040-004-0411-8.

Lam, C. (2011). Optimal use of underground space in Hong Kong – the new era. In: *Proceedings, Joint HKIE-HKIP Conference on Planning and Development of Underground Space, Hong Kong, 23–24 September 2011*. Hong Kong: Hong Kong Institution of Engineers and Hong Kong Institute of Planners. 228 pp.

Larsson, G. (2006). *Spatial Planning Systems in Western Europe: An Overview*. Delft, Netherlands: Delft University Press. 228 pp.

Ling, K-K. (2011). Towards an underground development strategy for Hong Kong. In: *Proceedings, Joint HKIE-HKIP Conference on Planning and Development of Underground Space, Hong Kong, 23–24 September 2011*. Hong Kong: Hong Kong Institution of Engineers and Hong Kong Institute of Planners. 228 pp.

MacCormack, K.E. (2016). Characterizing Alberta's subsurface in 3D – exploring innovative solutions to enhance communication of geoscience information to stakeholders and provide decision support. In: *Three-Dimensional Geological Mapping: Workshop Extended Abstracts* (eds. K.E. MacCormack, L.H. Thorleifson, R.C. Berg and H.A.J. Russell), 95–99. Edmonton, AL: AER/AGS, Special Report 101 [Online: Available at: https://ags.aer.ca/document/SPE/SPE_101 .pdf] (Accessed June 24, 2020).

MacCormack, K.E. (2018). Developing a 3D geological framework program at the Alberta geological survey; optimizing the integration of geologists, Geomodellers, and Geostatiticians to build multi-disciplinary, multi-scalar, Geostatistical 3D geological models of Alberta. In: *Three-Dimensional Geological Mapping: Workshop Extended Abstracts, Resources for Future Generations conference, Vancouver, Canada, June 16–17, 2018* (eds. R.C. Berg, K.E. MacCormack, H.A.J. Russell and L.H. Thorleifson), 61–67. Champaign, IL: Illinois State Geological Survey, Open File Series 2018-1. [Online: Available at: http://library.isgs.illinois.edu/Pubs/pdfs/ofs/ 2018/ofs2018_1.pdf] (Accessed June 24, 2020).

Małolepszy, Z. (2005). Three-Dimensional Geological Model of Poland and its Application to Geothermal Resource Assessment. In: *Three-dimensional geological mapping for groundwater applications: workshop extended abstracts* (eds. H.A.J. Russell, R.C. Berg and L.H. Thorleifson), 47–50. Ottawa, ON: Geological Survey of Canada, Open File 5048. https://doi.org/10.4095/221888.

Małolepszy, Z., Dąbrowski, M., Mydłowski, A. et al. (2016). Modelling the geological structure of Poland – approach, recent results and roadmap. In: *Three-Dimensional Geological Mapping: Workshop Extended Abstracts*: (eds. K.E. MacCormack, L.H. Thorleifson, R.C. Berg and H.A.J. Russell), 45–50. Edmonton, Canada: AER/AGS Special Report 101. [Online: Available at: https://ags.aer.ca/ document/SPE/SPE_101.pdf] (Accessed June 24, 2020).

Mann, T. (2015). £3.8million House in London Falls in on Itself during Basement Works. *Metro, November 26, 2015*. [Online: Available at: http://metro.co.uk/2015/11/26/3- 8million-house-in-london-falls-in-on-itself-during- basement-works-5528488/#ixzz4IopAaqaT] (Accessed August 30, 2016).

Marker, B.R. (1998). Incorporating information on Geohazards into the planning process. *Engineering Geology Special Publication* 15: 385–389. https://doi.org/10.1144/ GSL.ENG.1998.015.01.38.

Marker, B.R. (2008). Communication of geoscience information in public administration: UK experiences. *Geological Society Special Publications* 305: 185–196. https://doi.org/10.1144/SP305.16.

Marker, B.R. (2015a). Communication of geological information in planning of urban areas. In: *Engineering Geology for Society and Territory – Volume 5* (eds. G. Lollino, A. Manconi A., F. Guzzetti F. et al.), 335–338.

Cham, Switzerland: Springer. https://doi.org/10.1007/978-3-319-09048-1_64.

Marker, B.R. (2015b). Planning for underground development: principles and problems. In: *Engineering Geology for Society and Territory – Volume 5* (eds. G. Lollino, A. Manconi A., F. Guzzetti F. et al.), 1205–1208. Cham, Switzerland: Springer. https://doi.org/10.1007/978-3-319-09048-1_230.

Marker, B.R. (2016). Urban Planning: The Geoscience Input. *Engineering Geology Special Publications* 27: 35–43. https://doi.org/10.1144/EGSP27.3.

Mathers, S.J., Terrington, R.L., Waters, C.N. and Leslie, A.G. (2014). GB3D – a framework for the bedrock geology of Great Britain. *Geoscience Data Journal* 1: 30–42. https://doi.org/10.1002/gdj3.9.

Mattei, U. (1997). Three patterns of law: taxonomy and change in the world's legal systems. *American Journal of Comparative Law* 45 (1): 5–44. https://doi.org/10.2307/840958.

Mielby, S., Eriksson, I., Campbell, D. et al. (2017). *Opening up the Subsurface for the Cities of Tomorrow: Considering Access to Subsurface Knowledge – Evaluation of Practices and Techniques*. COST TU1206 Sub-Urban Report TU1206-WG2.0–001. 111 pp. [Online: Available at: http://sub-urban.squarespace.com/s/TU1206-WG2-001-Opening-up-the-subsurface-for-the-cities-of-tomorrow_Summary-Report.pdf] (Accessed July 16, 2017).

Nadis, S. (2010). SubTropolis, USA. *Atlantic Magazine, May 2010*. [Online: Available at: http://www.theatlantic.com/magazine/archive/2010/05/subtropolis-usa/8033] (Accessed January 30, 2017).

Naito, K., Myoi, H., Ono, J. et al. (1998). Mineral projects in Asian countries. Geology, regulation, fiscal regimes and the environment. *Resources Policy* 24: 87–93. https://doi.org/10.1016/S0301-4207(98)00012-9.

Natural England & Wiltshire Council (2015). *Planning Guidance for Wiltshire: Bat Special Areas of Conservation (SAC)*. York, UK: Natural England and Towbridge, UK: Wiltshire Council. 24 pp. [Online: Available at: http://www.wiltshire.gov.uk/bat-special-areas-of-conservation-planning-guidance-for-wilthshire.pdf] (Accessed August 9, 2017).

NFPA (2016). *NFPA 520: Standard on Subterranean Spaces, 2016 Edition*. Quincy, MA: National Fire Protection Association. 17 pp. [Online: Available at: http://catalog.nfpa.org/NFPA-520-Standard-on-Subterranean-Spaces-2016-Edition-P1329.aspx?icid=B484] (Accessed January 30, 2017).

Ng, K.C., Roberts, K.J. and Ho, Y.K. (2013). Rock caverns - unlimited space for future development. In: *Proceedings of the HKIE Geotechnical Division 33rd Annual Seminar 2013, Geotechnical Aspects of Housing Supply and Development,*

Hong Kong (eds. R. Lueng, B. Ieong, D. Lai et al.), 19–31. Hong Kong: Hong Kong Institution of Engineers.

NRC (2013). *Underground Engineering for Sustainable Urban Development*. Washington, DC: National Academy Press. 230 pp. [Online: Available at: https://www.nap.edu/catalog/14670/underground-engineering-for-sustainable-urban-development] (Accessed: December 28, 2016).

Pedersen, F.F., Refsgaard, A., Pedersen, Ph.G. et al. (2016). *White paper: Greater Water Security with Groundwater – Groundwater Mapping and Sustainable Groundwater Management*. Copenhagen, Denmark: The Rethink Water Network, Danish Water Forum and State of Green. 18 pp. [Online: Available at https://stateofgreen.com/en/publications/greater-water-security-with-groundwater/] (Accessed June 24, 2020).

Price, S.J., Ford, J.R., Campbell, S.D.G. and Jefferson, I. (2016). Urban futures: the sustainable management of the ground beneath cities. *Engineering Geology Special Publications*, 27: 19–33. https://doi.org/10.1144/EGSP27.2.

RÉSO (2017). Montreal Underground City. [Website: https://montrealundergroundcity.com] (Accessed June 24, 2020).

Rosenbaum, M.S., McMillan, A.A., Powell, J.H. et al. (2003). Classification of artificial (man-made) ground. *Engineering Geology* 69: 399–409. https://doi.org/10.1016/S0013-7952(02)00282-X.

Rosie, G. (2010). Global trail of disputes behind company at centre of tram row. *The Times, December 18, 2010*.

Ross, S., Wallace, M.I., Ho, Y.K. et al. (2014). Recent studies into the feasibility of development rock cavern facilities in a dense urban city. In: *Proceedings of the ACUUS 2014 conference "Underground Space: Planning, Administration and Design Challenges", Seoul, South Korea*, 441–449.

Rowson, J. (2009). Cologne: groundwater extraction method probed. *New Civil Engineer, March 19, 2009*. [Online: Available at: www.nce.co.uk/cologne-groundwater-extraction-method-probed/1995535.article] (Accessed July 16, 2017).

Rudland, D.J., Lancefield, R.M. and Mayell, P.N. (2001). *Contaminated Land Risk Assessment: A Guide to Good Practice*. London, UK: Construction Industry Research and Information Association (CIRIA). 159 pp.

Rudolph, M. (2001). Sinking of a titanic city. *Geotimes, July 2001*. [Online: Available at: http://www.geotimes.org/july01/sinking_titanic_city.html (Accessed February 25, 2017).

Sandberg, H. (2003). Three dimensional partition and registration of subsurface space. *Israel Law Review* 37(1): 119–167. [Online: Available at: https://ssrn.com/abstract=704941] (Accessed June 2020).

Sandersen, P.B.E., Vangkilde-Pedersen, T., Jørgensen, F. et al. (2016). A national 3D geological model of Denmark:

condensing more than 125 years of geological mapping. In: *Three-Dimensional Geological Mapping: Workshop Extended Abstracts* (eds. K.E. MacCormack, L.H. Thorleifson, R.C. Berg and H.A.J. Russell), 71–77. Edmonton, AL: AER/AGS, Special Report 101 [Online: Available at: https://ags.aer.ca/document/SPE/ SPE_101.pdf] (Accessed June 24, 2020).

Siems, M.M. (2007). Legal origins: reconciling law & finance and comparative law. *McGill Law Journal* 52: 55–81. https://dx.doi.org/10.2139/ssrn.920690.

Sterling, R.L. and Godard, J.P. (2000). Geoengineering considerations in the optimum use of underground space. In: *GeoEng2000: An International Conference on Geotechnical & Geological Engineering, Vol.1: Invited Papers*, 708–720. Boca Raton, FL: CRC Press.

Sterling, R., Admiraal, H., Bobylev, N. et al. (2015). Sustainability issues for underground spaces in urban areas. *Proceedings of the Institution of Civil Engineers – Urban Design and Planning* 165 (4): 241–254. https://doi .org/10.1680/udap.10.00020.

Stoter, J., Ploeger, H., Roes, R. et al. (2017). Registration of multi-level property rights in 3D in The Netherlands: two cases and next steps in further implementation. *ISPRS International Journal of Geo-Information* 6 (6): 158. https:// doi.org/10.3390/ijgi6060158..

Taves, M. (2014). In Kansas City, it's the rise of the underground. *Wall Street Journal*, November 26, 2014. [Online: Available at: https://www.wsj.com/articles/in-kansas-city-its-the-rise-of-the-underground-1416950522] (Accessed January 31, 2017).

Turner, R. (2010). Workers stole steel parts before Cologne archive collapse, prosecutors say. *Deutsche Welle, September 2, 2010.* [Online: Available at: http://www.dw.com/en/ workers-stole-steel-parts-before-cologne-archivecollapse-prosecutors-say/a-5232539] (Accessed July 16, 2017).

Vähäaho, I. (2014a). Underground space planning in Helsinki. *Journal of Rock Mechanics and Geotechnical Engineering* 6: 387–398. https://doi.org/10.1016/j.jrmge .2014.05.005.

Vähäaho, I. (2014b). *Urban Underground Space: Sustainable Property Development in Helsinki*. Helsinki, Finland: City of Helsinki, Real Estate Department, Geotechnical Division. 48 pp. [Online: Available at: https://www.hel.fi/static/kv/ Geo/urban underground space print.pdf] (Accessed December 17, 2016).

Van Campenhout, I., de Vette, K., Schokker, J. and Van der Meulen, M.J. (2016). *Rotterdam*. COST TU1206 Sub-Urban Report TU1206-WG1-013. 93 pp. [Online: Available at: http://sub-urban.squarespace.com/s/TU1206-WG1-013-Rotterdam-City-Case-Study_red-size.pdf] (Accessed July 16, 2017).

Van der Meulen, M.J., Doornenbal, J.C., Gunnink, J.L. et al. (2013). 3D geology in a 2D country: perspectives for geological surveying in the Netherlands. *Netherlands Journal of Geosciences* 92 (4): 217–241. https://doi.org/10 .1017/S0016774600000184.

Van der Meulen, M.J., Campbell, S.D.G, Lawrence, D.J., González, R.C.L. and Van Campenhout, I.P.A.M. (2016a). *Out of Sight, Out of Mind? Considering the Subsurface in Urban Planning – State of the Art*. COST TU1206 Sub-Urban Report TU1206-WG1-001. 49 pp. [Online: Available at: http://sub-urban.squarespace.com/s/TU1206-WG1-001-Summary-report-Out-of-sight-out-of-mind.pdf] (Accessed July 16, 2017).

Van der Meulen, M.J., Kiden, P., Maljers, D. et al. (2016b). Systematic geomodeling: what does it actually imply? In: *Three-Dimensional Geological Mapping: Workshop Extended Abstracts* (eds. K.E. MacCormack, L.H. Thorleifson, R.C. Berg and H.A.J. Russell), 83–87. Edmonton, AL: AER/AGS, Special Report 101 [Online: Available at: https://ags.aer.ca/document/SPE/ SPE_101.pdf] (Accessed June 24, 2020).

Wallace, M.I. and Ng, K.C. (2016). Development and application of underground space use in Hong Kong. *Tunnelling and Underground Space Technology* 55: 257–279. https://doi.org/10.1016/j.tust.2015.11.024.

Wallis, S. (2009). Köln—speculation and anger in aftermath. *Tunnel Talk, March 9, 2009.* [Web page: http://www .tunneltalk.com/Cologne-collapse-Mar09-Deadly-collapse-in-Cologne.php] (Accessed 28 December, 2016).

Webster, G.S. (1914). Subterranean street planning. In: *Annals of the American Academy of Political and Social Science: Housing and Town Planning* (eds. E.R. Johnson, C.L. King, R.C. McCrea, et al.), 200–207. Philadelphia, PA: American Academy of Political and Social Sciences.

Wikipedia (2017a). *Cologne Stadtbahn*. San Fransisco, CA: Wikimedia Foundation. [Web page: https://en.wikipedia .org/wiki/Cologne_Stadtbahn] (Accessed July 16, 2017).

Wikipedia (2017b). *Strataca*. San Fransisco, CA: Wikimedia Foundation. [Web page: https://en.wikipedia.org/wiki/ Strataca] (Accessed August 6, 2017).

Wikipedia (2017c). *PATH (Toronto)*. San Fransisco, CA: Wikimedia Foundation. [Web page: https://en.wikipedia .org/wiki/PATH_(Toronto)] (Accessed July 20, 2017).

Wikipedia (2017d). *Underground City, Montreal*. San Fransisco, CA: Wikimedia Foundation. [Web page: https:// en.wikipedia.org/wiki/Underground_City,_Montreal] (Last accessed July 20, 2017).

Williamson, I.P. (2001). Land administration "best practice", providing the infrastructure for land policy implementation. *Land Use Policy* 18 (4): 297–307. https://doi.org/10.1016/S0264-8377(01)00021-7.

World Bank (2006). *Water resources management in Japan: policy, institutional and legal issues*. Washington, DC: World Bank. vi + 24 pp. [Online: Available at: http://documents.worldbank.org/curated/en/558331468051297615/Water-resources-management-in-Japan-policy-institutional-and-legal-issues] (Accessed August 6, 2017).

Zweigert, K. and Kötz, H. (1998). *An Introduction to Comparative Law, 3rd edition*. Oxford, UK: Oxford University Press. 744 pp.

4

The Economic Case for Establishing Subsurface Ground Conditions and the Use of Geological Models

Jennifer Gates

ARUP, London W1T 4BQ, UK

4.1 Introduction

Unexpected geological conditions are among the principal sources of risk in building and construction projects. Because geological materials are often irregularly arranged and highly variable in their properties, geotechnical risk is inversely proportional to the level of detail and accuracy of the geological information. In 1993, the Institution of Civil Engineers in the UK stated that the ground itself is a vital but poorly defined element of all structures; safe and economical design of structures can only be achieved when ground conditions are fully understood (Site Investigation Steering Group 1993). Fifteen years later, Chapman (2008) argued that construction problems related to ground conditions still occurred too frequently, at huge delay and cost to projects, and that often the problems could have been avoided. Recognizing this, the UK Government recently provided guidance to redress the systematic underestimation of cost and duration of work, which appears to be caused by a systematic "optimism bias" that afflicts the appraisal process for major projects (HM Treasury 2018). As far as this relates to ground conditions, the recent uptake of 3-D geological information into Building Information Modeling (BIM) by the construction sector is a positive development. This has progressed to the extent that this book treats it as an emerging field.

A good ground model and the associated understanding of the ground constraints and opportunities at a site provides fundamental information for progressing any construction project. This understanding is usually the first thing a project team will pursue, with an engineering geological desk study and geotechnical investigation often on the critical path during the early stages of a project. However, in this stage of the project, there may be many consultants and contractors from various disciplines competing for this budget at a time when funds may be tight and often borrowed. The requirement for gaining an understanding of the ground at a site should start as early as possible in order to understand how the ground constraints and opportunities at the site will be incorporated into the planning and specification of a project. By the bidding stage, when construction companies compete commercially for a project, it is important to have available a reasonably clear understanding of the likely ground conditions because the appraisal of ground-related risks is vital to offering a competitive bid. Clients typically have limited budgets, and winning the project while managing the commercial risk is the primary objective.

The first geotechnical elements of a project are planned and scoped using an engineering geological desk study. This is followed by a ground investigation (GI) which, by definition, takes place during the outset of a project when optimism is great, but when a project scope and characteristics may still not be defined very well. As the project proceeds, subsequent stages of ground investigations can sometimes be afflicted by deficient transfer of information through the project life-cycle. Limited information transfer also occurs between adjacent projects, and with staff turnover and rotation of engineers from project to project, many companies have a short half-life of knowledge. The lack of continuity from team to team and from project to project potentially results in the repetition of mistakes and missed opportunities. Ideally, data should be collected and circulated within the industry to benefit future projects by improved anticipation of potential hazards, constraints, and opportunities. However, so far, this has only been legislated at a country scale in the Netherlands (see Section 4.5.6) and systemized elsewhere at major infrastructure project scale (see Sections 4.5.2–4.5.4) or urban region scale (see Section 4.5.5). Case histories are an important and valuable part of the geotechnical engineering and soil mechanics literature and several are presented in Section 4.5 of this chapter.

Applied Multidimensional Geological Modeling: Informing Sustainable Human Interactions with the Shallow Subsurface, First Edition.
Edited by Alan Keith Turner, Holger Kessler, and Michiel J. van der Meulen.

Chapman (2008) concluded that the first important step toward resolving these issues is establishing better communication by engineering geologists and geotechnical engineers to their clients using an understandable language. The focus is to persuade those clients of the merits of timely and well-scoped geotechnical investigations and now this can be extended to include the development of a 3-D geological model. Chapman restricted his considerations to the management of financial risks within the overall design and construction process; however, he noted that projects where overall design and construction processes are skimped on may share characteristics with those that suffer the sorts of organizational breaches that lead to accidents and regulatory oversight.

Section 4.2 of this chapter describes the geotechnical investigation process and Section 4.3 explores the economic case for understanding subsurface ground conditions and the role of geological models in reducing geotechnical uncertainty. When the relative cost of geotechnical input versus the potential cost of unforeseen events is assessed, it becomes clear that even an expensively obtained ground model is good value for money. It can be hugely beneficial to have clarity before any significant financial or engineering decisions are made, reducing the unforeseen events which a project team may encounter during a project. Section 4.4 discusses mitigation options and the importance of risk-reduction processes, geotechnical code requirements, and guidelines while performing a complete geotechnical site investigation.

Section 4.5 then discusses all these considerations in the context of several case histories which cite experiences from industry. Several large firms have begun to internally evaluate the cost and efficiency aspects of using 3-D geological models within the workflow of design and construction of infrastructure projects.

4.2 The Nature of Geotechnical Investigations

Site investigation consists of two main activities: (i) a desk study, and (ii) a ground investigation. The desk study gathers information about a site and the ground conditions from documentary sources and includes a summary of all the ground-related hazards that could potentially affect the project throughout its design, construction, and over the design life. The ground investigation includes the physical investigation of a site using geophysics, boreholes, trial pits, and other intrusive and non-intrusive methods, followed by laboratory testing of samples and installation of monitoring systems to establish the ground, groundwater, and other environmental conditions. Especially for larger

and more complex sites, a GIS (Geographic Information System) or geological model become an important component for the planning of these site investigation tasks and for visualizing and interpreting the compiled information.

4.2.1 Geotechnical Investigations for Management of Geotechnical Risk

The planning and specification of geotechnical investigations are ultimately governed by a combination of the financial motives and interests of the project developer, the preferred design-construction process, the codes and good practice guidance defined by the relevant national or industry regulations, and the contractual arrangements between the different parties involved. However, the management of geotechnical risk is the overall requirement of any ground investigation.

Chapman (2008) defined three main types of developer:

- Commercial developers creating developments essentially to make a profit;
- Self-developers creating offices or manufacturing facilities to improve business operations; and
- State developers creating national infrastructure such as roads, railways, schools, and hospitals.

Each of these main types of developer will adopt slightly different approaches to their planning and undertaking of ground investigations because these are influenced by differing combinations of governing factors.

Design and construction contracts are undertaken using a range of traditional contracting arrangements including: one- and two-stage design-build, and various public–private partnerships. These influence the organization, timing, and characteristics of the geotechnical investigation process because they change the assignment and perception of geotechnical risk among the participants. Section 4.4.1 of this chapter discusses these risk allocation issues.

Van Staveren (2006) introduced the GeoQ process, a formal structure for good geotechnical practice and geotechnical risk management which emphasizes the advantages of early and comprehensive data gathering. The GeoQ approach distinguishes six generic project phases (see Figure 4.1). For any project, five of the phases form a logical sequence beginning with the feasibility phase, continuing through the pre-design, design, and construction phases, and terminating with the operation/maintenance phase. The position of the sixth contracting phase depends on the type of construction contract. In case of a traditional contracting arrangement, the design, and the construction are typically undertaken under separate contracts and therefore the design and initial ground investigation

Figure 4.1 The six GeoQ phases. After van Staveren (2006).

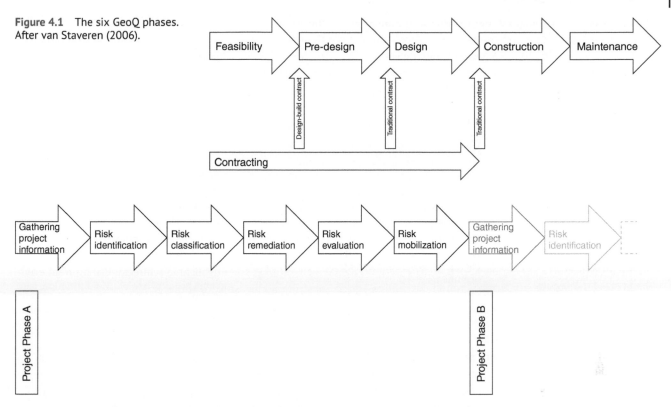

Figure 4.2 The six GeoQ risk management steps are performed sequentially within each project phase, and the cycle is repeated in subsequent phases. After van Staveren (2006); Chapman (2012).

contracting will occur well before construction. However, for design-build contracts the contracting phase will occur before the design phase or even before the pre-design phase. In these later situations, the ground investigations will occur within the design-build contract. Figure 4.1 illustrates these options with three arrows from the contracting phase.

A series of six risk management steps are performed within each GeoQ phase (Figure 4.2). While the six GeoQ phases are flexible, the six GeoQ risk management steps are considered to be quite rigid. In each project phase these six steps are repeated in order to implement a comprehensive ground-related risk management process (van Staveren 2006). Table 4.1 compares the GeoQ risk management steps to normal project design stages and provides some typical examples of the type of geotechnical issues that may be identified and communicated to allow the associated geotechnical risk to be managed appropriately.

4.2.2 How Geological Models Sit Within the Geotechnical Investigation Process

In his seminal paper on the subject, Fookes (1997) described the use of "geological models" on engineering projects. Fookes et al. (2000, 2001) refined this concept as "total geological history" which states site conditions reflect the entire geological history of the area or site. Subsequent discussions on the role of geological models are described by Knill (2003), Harding (2004), Culshaw (2005), Sullivan (2010), Haff (2013), and Parry et al. (2014). Knill (2003) differentiated three model types: *Geological Models, Ground Models*, and *Geotechnical Models*. Baynes et al. (2005) differentiated two types of geological model – *Conceptual*, and *Observational* – and applied the "total engineering" process to railway infrastructure in Western Australia. A web-based interactive tutorial is available to support this method (Davison et al. 2003). The International Association for Engineering Geology and the Environment (IAEG) Commission 25 reported on the use of "engineering geological models", defined a model as *"an approximation of reality created for the purpose of solving a problem"*, and postulated that the model is effectively a hypothesis to be tested (Parry et al. 2014).

Parry et al. (2014) propose the conceptual model approach be used to generate an initial engineering geological ground model at the desk study stage of a project by combining the knowledge of the proposed engineering construction and the geologist's knowledge of the site. Legacy data are used if available. The observational model approach is used to further develop a more complete picture of the site by including engineering parameters

Table 4.1 A simplified example of GeoQ steps during normal project stages.

GeoQ step	Normal project stage	Typical example
1. Gathering project information	*Desk study*	Collecting historical and geological maps and other information
2. Risk identification		Discovery of old maps showing a filled-in watercourse crossing the site
3. Risk classification	Normally done during the *Ground investigation* stage, when the magnitude and extent of the hazard can be quantified	Prove that the channel is up to 10 m in depth and infilled with soft organic alluvium
4. Risk remediation	Normally done during *design*, when the development can protect itself against the main risks	Design of different foundations for that part of the site; inclusion of methane protection measures for basement
5. Risk evaluation	Prior to *tender*, collection of unmitigated risks so they can be transferred to the contractor as part of the contract documents	Assess remaining extent of issues connected with channel
6. Risk mobilization – filing in the risk register		Highlight to contractor that his plant may get bogged down in alluvial formation; care of piling rig stability

Source: Chapman (2012).

and other data that can support engineering decisions. Development of the observational model involves an iterative procedure which allows the model to evolve as more of the underlying ground conditions are determined and serves as a precursor to subsequent analytical (numerical) model calculations or simulations (Figure 4.3).

Burland (1987) introduced the soil mechanics triangle; subsequently the concept was generalized as the geotechnical triangle (Burland 2006), including reference to an "appropriate model". The triangle displays the four interlinked but essential components of site investigation and geotechnical design (Figure 4.4): (i) the *ground profile*, (ii) the *measured behavior* of the ground, (iii) an *appropriate model*, and (iv) *empirical procedures and well-winnowed experience*. The *ground profile, measured behavior,* and *appropriate model* components are represented at the apexes of an equilateral triangle, while the *empirical procedures and well-winnowed experience* is linked to all three at the center of the triangle. The *ground profile* component combines the geological modeling and site investigation activities; it is placed at the apex of the triangle to reflect its crucial importance. Similarly, the *empirical procedures and well-winnowed experience* component is located at the center of the triangle because it is an essential aspect of all geotechnical engineering and relates to the other three activities. Each of the activities in the triangle has its own distinct methodology and rigor, but the aspects are interlinked. For geotechnical engineering to be successful the triangle should be kept in balance and all aspects, including the "appropriate model" should be considered

properly. Burland (2012) illustrates the application of the geotechnical triangle by revisiting the well-known case history of the underground car park at the Palace of Westminster in London.

However, the wider application of geological models to geotechnical investigations has only relatively recently become common and accepted as a component of geotechnical good practice. Major projects now systematically employ geological models and there are many examples which demonstrate the benefits of geological models in the planning, specification, and interpretation of ground investigation and the management of geotechnical risk (Fookes and Shilston 2001; Baynes et al. 2005; Munsterman et al. 2008). Section 4.5 of this chapter contains a selection of short discussions by geotechnical firms that provide recent examples of their use of geological models on major projects. Several of these examples are cross-referenced to more detailed case studies in subsequent chapters.

4.2.3 Potential Impact of Geotechnical Risks

After reviewing data undertaken for many projects, Chapman (2012) concludes that significant delays due to ground conditions probably occur some 17–20% of the time in UK projects (Table 4.2). Although it is certainly the case that some of these delays could have been prevented by obtaining a good ground model and a thorough review of the possible geotechnical risks, others could potentially never have been foreseen. Managing risks for

Figure 4.3 Geological models are iteratively developed from conceptual and observational approaches and serve as precursors to analytical models to support engineering design. After Parry et al. (2014).

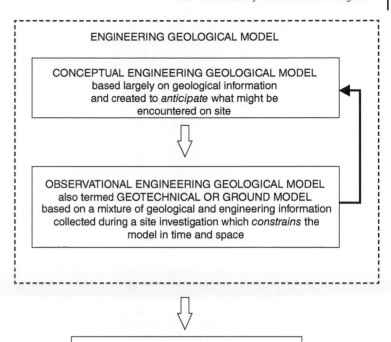

Figure 4.4 Burland's geotechnical triangle. Modified from Burland (2012).

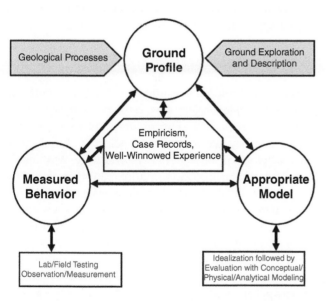

the "unknown unknowns" is difficult; these events are not amenable to empirical or statistical understanding.

Taleb (2010) introduced the concept of a "Black Swan" event for unimagined occurrences. The name is based on the unassailable belief by people in the Old World that all swans were white, confirmed by centuries of empirical evidence, until black swans were encountered in Australia in 1697. Black Swan events are outliers that fall outside the realm of regular expectations because prior experience has not encountered them; they may have extreme impacts; and once they have happened, they become obvious and are readily explained with hindsight. Taleb distinguishes between "mild randomness" and "wild randomness", suggesting Black Swan events happen all too frequently and tend to dominate the success of many endeavors. Chapman (2012) notes that many ground risks are potentially a wildly random type of Black Swan because their consequences can be so severe. Thus, it is important that the possibility of unexpected events should not be discounted in geotechnical assessments. It is possible by good planning, and through use of a 3-D geological model, to minimize the likelihood of a Black Swan event. However, by their very nature, the incidence of Black Swans can never be entirely eliminated. They often arise because engineers only rely on previous experience and don't consider what *might* happen. A sense of humility is needed by those who manage complex geotechnical projects and it is important to appreciate the possibility of the range of catastrophic discoveries that can arise at any apparently innocuous site.

Table 4.2 Summary of impacts of geotechnical problems on construction projects reported by Chapman.

Source	Projects evaluated	Findings
National Economic Development Office (1983)	5000 UK industrial building construction projects	50% of projects were completed at least 1 month late; 38% of these delays were due to ground problems.
National Economic Development Office (1988)	8000 UK commercial development projects	34% of projects were completed up to 1 month late; another 33% were completed more than 1 month late (therefore 67% were delayed). 17% of those delayed by more than 1 month were due to ground problems.
National Audit Office (2001)	Audit of public sector construction projects	70% of projects were delivered late and 73% exceeded tender price
Van Staveren and Chapman (2007)	Review of financial impact of all construction failures in the Netherlands	Estimated cost of failures is 5–13% of annual €70-billion expenditure of Dutch construction industry
Chowdhury and Flentje (2007)	Overview of role of modeling in geotechnical risk management, especially for landslides, slope stability, and foundations, based on multiple published sources	Geotechnical problems encountered during the design phase in 57% of projects, and during the construction phase of 38% of projects

Based on data provided by Chapman (2012).

4.3 Benefits of Using 3-D Models and Establishing Subsurface Ground Conditions

Latham (1994) stated, *"No construction project is risk free. Risk can be managed, minimized, shared, transferred, or accepted. It cannot be ignored"*. Chapman (2012) explains that geotechnical risks can be wildly random, with consequences that are hugely disproportionate to the initial cost of an appropriately comprehensive geotechnical investigation. A well-resourced desk study and ground investigation is often justified because even mildly random geotechnical problems have consequences that far outweigh the very modest cost of a good ground investigation. However, it is important to recognize that the risk can never be eliminated entirely.

The dangers of geotechnical risk are enhanced by overconfidence and excessive reliance on the experience of an individual or even an individual company, or by the insistence of designers or clients that some risks need not be mitigated. Chapman (2012) concludes that management of geotechnical risk by contingency funding is an approach fraught with peril and that management by avoidance and/or mitigation of risks is much more likely to produce the desired outcome.

A 3-D geological model can help to avoid these pitfalls by providing a single source of well-documented information relating to ground risk, or by highlighting where information is scarce and further investigation is required to mitigate this risk, thereby reducing the contingency finding required.

4.3.1 Cost of Geotechnical Investigations

In the past, traditional estimating practices tended to produce a single project cost estimate; but this single number masks the critical uncertainty and implies a sense of precision beyond what can be achieved during planning, scoping, or early design phases. The Washington State Department of Transportation (2014) Project Risk Management Guide shows that estimates have a *base cost* component and a *risk* (or uncertainty) component. Base cost is defined as the likely cost of the planned project if no significant problems occur. Once the base cost is established, a list of uncertainties – called a "risk register" – is created for both opportunities and threats. Risk assessment replaces general and vaguely defined contingency concepts with explicitly defined risk events which define the probability of occurrence and consequences of each potential risk event. Figure 4.5 illustrates how project information develops over time. As a project evolves, increased knowledge provides better information on the complexity, and hence costs, of various project elements and defines those elements which require significant additional resources. As cost estimates are reviewed and validated, the base cost for the project is more precisely determined and the uncertainty is reduced but not eliminated.

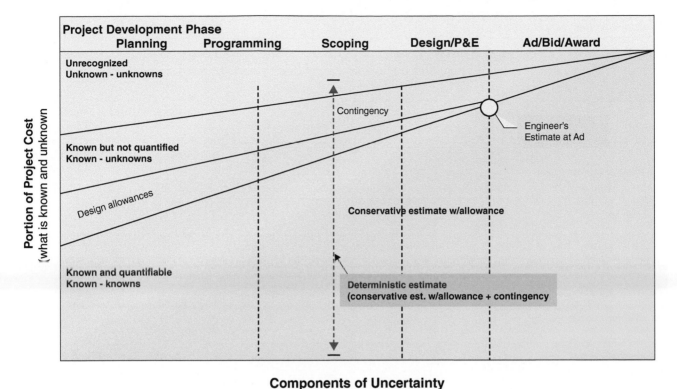

Components of Uncertainty

Figure 4.5 Evolution of project knowledge through project development. Modified from Washington State Department of Transportation (2014).

On a typical building project, the cost of generating a comprehensive ground model through research and geotechnical investigation is a barely significant proportion of the overall lifetime cost to the developer of a building (Figure 4.6). Chapman (2008, 2012) used a typical building project in London as an example to demonstrate how the entire cost of design and construction is only 4% of the total nominal 25-year lifetime cost of a building. Only 7% of this very small proportion is associated with construction of the substructure; thus, only 0.28% of the lifetime building cost is attributed to the substructure, including desk study and ground investigation.

Even the most thorough ground investigation will only provide information on a virtually negligible proportion of the ground volume with which a project will interact. For example, a typical ground investigation for a building site which follows the guidelines of Eurocode 7 Part 2 (European Committee for Standardization 2007) will only recover some 0.03% of the soil beneath the site and only a fraction of that will be examined by a professional geologist or engineer (Chapman 2008, 2012).

Considering these figures, it is inconceivable that any project should be willing to withhold investment on the relatively trivial cost of these activities which occur right at the outset of a project, given that the ground is almost certainly any project's highest source of uncertainty, risk, and potential cost, and when establishing a comprehensive ground model can offer significant savings.

4.3.2 Geotechnical Baseline Report

A geotechnical baseline report (GBR), or "Ground Reference Conditions" effectively sets out in a contractual, report format, the 3-D ground model to be assumed for design. GBRs are required at the outset of a project, and defined as *"integral and formative information on which tenders shall be based and the client shall take responsibility for the information issued"* (ITIG 2012).

It provides the basis for comparison between ground conditions assumed during the tender for pricing by the contractor, and those encountered during construction. The GBR identifies and attempts to quantify the hazards so as to allow for consistent pricing across the tenders and ensure that deviations from expected conditions are easily measurable. The aim is to not only inform the contractor of the available ground model, but to reduce future costs associated with disputes, providing clarity on when extra payment is due.

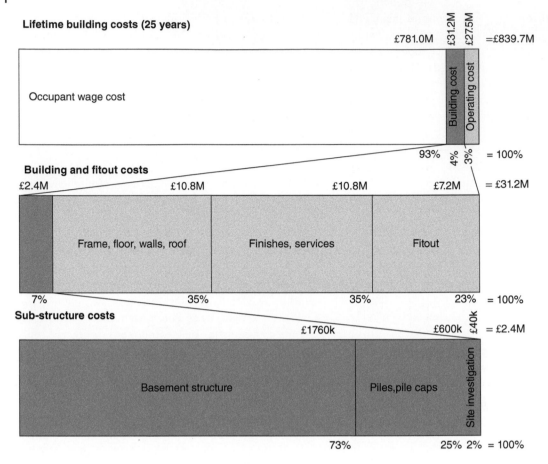

Figure 4.6 Breakdown of costs for a typical building (Chapman 2008).

4.4 Processes, Codes, and Guidelines for Establishing Subsurface Conditions and Managing Risk

The management of geotechnical risk is guided by project management processes, codes, and good practice guides, as defined by government or industry regulations.

4.4.1 Risk Reduction Strategies to Manage Deficient Ground Information

A standard strategy for risk management is to devise a risk register for a project. Risk registers often use a "risk matrix" to qualitatively evaluate individual events. Scores of the adverse consequences are listed in a series of columns and the likelihood (or probability) of occurrence in a series of rows. When four categories are used for the columns and rows, this yields a matrix with 16 cells. Typically, the consequences are labeled *severe, moderate, slight,* and *negligible,* while the likelihood categories are labeled *high, medium, low, and negligible.* The combination of consequence and likelihood allows definition of qualitative risk levels such

as *high, medium,* or *low* (Lee and Jones 2004). Sometimes a numerical scheme is used to define the consequence and likelihood categories. In some cases, the values of the cells are computed by multiplication of the relevant category values. However, while these risk-matrix procedures provide a comparison of different relative risk conditions, they represent a gross simplification of reality. They represent a systematic procedure based on professional intuition and judgment, a process long defended as appropriate in many cases, particularly by the engineering community.

Insurance is a common method of risk management by transferal; many consider insurance to be the cheapest risk management option. However, underground construction, and particularly tunneling, is highly competitive with a trend toward design-build contracts; it involves high-risk construction methods, tight construction schedules, and low financial budgets. Wannick (2006), from Munich Re Insurance Group, reviewed the experience with tunneling projects in the decade 1994–2005 and found 19 major failures resulted in insurance losses of approximately US$600 million. This was obviously an untenable situation for the insurance industry. The high-frequency of major tunnel losses resulted in insufficient premium income to pay for all the losses, with repair costs often exceeding the original total estimate of the construction costs. In response, a *Joint Code of Practice for Risk Management of Tunnel Works* was developed as a professional risk management tool (British Tunnelling Society 2003). It sets minimum standards for risk assessment and ongoing risk management procedures for tunneling projects and defines clear responsibilities to all parties involved in tunnel projects. Application of this Code is now compulsory to obtain insurance cover for all tunnel projects.

4.4.2 Investments to Mitigate Against Deficient Ground Information

A project team may consider that a lower tender price leaves more for contingency and that the funds are safer remaining in their pockets than in the pockets of their contractors or consultants. It is easier to concentrate on the direct capital expenditure (CAPEX) saving and not on the potential consequential losses and operating expenditure (OPEX). This approach may have been validated many times in the past when the typical events have been dealt with as they occurred with contingency funds. However, due to the wildly random nature of some geotechnical and geological risks, the so-called "Black Swan" event, this may not be a justifiable risk management approach. Whereas large projects can accommodate relatively large contingencies, a team with a more modest budget is likely to have a smaller contingency available. Projects of all scales are vulnerable to cost over-runs if the geotechnical risks are not managed. Selection of the lowest-tendered contractor price cannot therefore be considered the most appropriate option for the management of risk. Projects undertaken with the lowest-tendered price rely on certainty to an even greater degree and may not have the budget they need to invest where the risk is greatest and where the opportunity to manage these risks is most valuable.

If a site investigation contractor is appointed on the basis of lowest cost, and that contractor's saving comes through minimizing the scale of the ground investigations, employing inexperienced site supervisors, or by limiting the time the supervisor spends on site, adverse consequences are likely to be ultimately be realized later. The geotechnical team have a responsibility to provide geotechnical advice and notification of these kinds of risks at the right stages of a project. Constraints associated with the ground model and with foundation construction can potentially require major changes to the rest of a project, and geotechnical advisors need to be aware of the timescales over which the rest of the project is being designed. If the ground model

data are found to be insufficient or unreliable, a much more expensive conservative design will be adopted, or the site will require multiple re-visits to confirm further assumptions. Both of these consequences will involve much larger costs than any misconceived cost savings from the planning, specifying, and undertaking of an appropriate quality ground investigation.

4.4.3 Code Requirements

In Europe, the process of planning a ground investigation and generating a ground model at the outset of a project has been codified and formalized. At the turn of the millennium, a set of codes were developed by the European Community with the aim of *"elimination of technical obstacles to trade and the harmonisation of technical specifications"*. These Eurocodes supplemented and replaced the various existing codes for each country and provided common rules for design and construction. National standards for each country are still incorporated by inclusion of a National Annex, and other complementary guides or standards. In these, parameters which are left open in the Eurocode for national choice can be defined along with non-contradictory complementary information. Eurocode 7 Part 1 and 2 (European Committee for Standardization 2004, 2007) address Geotechnical Design and provide guidance on the selection of ground investigation methods in different stages (Table 4.3).

The American system of codes differs from that adopted in European countries. Whereas Eurocodes are the prevalent documents used for all design across Europe, in America, there is no one unifying code, but instead various guidelines are available for different engineering circumstances. Large federal agencies such as the Federal Highway Administration (FWHA) and the Naval Facilities Engineering Command (NAVFAC), as well as State Departments of Transportation, have the budgets to document their requirements and so have created codes for others to use. These are then adopted as guidelines or informal expectations for various situations.

The British Tunnelling Society and the Association of British Insurers jointly prepared *The Joint Code of Practice for Risk Management of Tunnel Works in the UK* (British Tunnelling Society 2003). As described in Section 4.4.1, an international version of this code, which included modifications to the UK code to resolve some provisions that were not appropriate or legal in all nations, regions, or cities, was subsequently developed by the International Tunnelling Association and the International Association of Engineering Insurers (ITIG 2012). These two codes provide a professional risk management tool that sets minimum standards for risk assessment and ongoing risk

Table 4.3 Example of the selection of ground investigation methods at different stages.

		Preliminary Investigations[1]	Design Investigations[2]	Control Investigations[3]
Fine soil	**Pile Foundation**	CPT, SS, DP, SE FVT or SPT OS TP, PS, OS GW	SS, CPT, DP, SR FVT, SPT, PIL PS, OS, CS, PMT GWC	PIL, Pile Driving tests, Stress wave measurements GWC, settlements, Inclinometers
	Shallow Foundation	SS, CPT, DP, SR SPT AS, OS, TP GW	SS or CPT, DP FVT, DMT or PMT BJT PS, OS, CS, TP GWC	Check of the soil type Check of the stiffness (CPT) Settlements, Inclinometers, GWC Volume change Potential due to water content change
Coarse soil	**Pile Foundation**	SR, CPT, MWD, PLT CS, AS, TP GW	CPT, DR, SR SPT, DMT, PIL OS, TP GWO	PIL, Pile driving tests, Stress wave Measurements GWC, Settlements Inclinometers
	Shallow Foundation		CPT, DP, SPT, PMT, BJT, DMT, PLT, OS, TP GWO	Check of the soil type Check of the stiffness (CPT, DP, SPT) Settlements
Rock	**Pile or shallow foundation**		SR, MWD, mapping of discontinuities RDT, PMT, BJT TP, CS GWO	Check inclination and discontinuities in the rock and its surface. Check contact between pile toe/ foundation and rock surface. Verify water conditions of flow and pressure

Abbreviations: Field Testing: BJT, Borehole jack test; DP, Dynamic Probing; SR, Soil Rock Sounding; SS, Static sounding (e.g. weight sounding test, WST); CPT(U), Cone penetration test with pore pressure recording; SPT, Standard Penetration test; PMT, Pressure meter test; DMT, Dilatometer test; FVT, Field vane test; PLT, Plate load test; MWD, Measuring while drilling; SE, Seismic measurements; PIL, Pile load test; RDT, Rock dilatometer test

Sampling: PS, Piston Sampler; CS, Core Sampler; AS, Auger Sampler; OS, Open Sampler; TP, Test Pit Sampling

Ground Water Measurements: GW, Ground water measurements; GWO, Ground water measurements with open system; GWC, Ground water measurements with closed system.

Column Title Notes:

1 **Preliminary Investigations** includes desk study of topographical, historical, geological and hydrogeological data, mineral extraction records, aerial photo-interpretation, and review of archives of previous construction works and investigations; followed by site inspection, preliminary geophysical surveys, and preliminary intrusive investigations

2 **Design Investigations** involve the preliminary choice of foundation method.

3 **Control Investigations** verify the choice of foundation method and design procedure and control ground improvement works and their stability during construction

Additional Notes:

Soils include naturally deposited and anthropogenic deposits

Surveying and logging are not included in this chart

Laboratory tests are not presentation on this table.

management procedures required to obtain insurance coverage for all tunnel projects.

In the UK nuclear power sector, the development and use of geological models has become an accepted and necessary component of good practice in the site appraisal process. In the United Kingdom, the Office for Nuclear Regulation (2018a,b) specifically refers to the preparation of a three-dimensional model of geological faults in the vicinity of the site. A major site appraisal that incorporated state-of-good-practice standards was undertaken by the engineering consultancy Arup for the developer, Horizon Nuclear Power, of a proposed nuclear power facility to be located at Wylfa Newydd on Anglesey in Wales. All geological, geophysical, geotechnical, and seismological information was compiled into a three-dimensional GIS database for interpretation. The conceptual geological model and geological fault model developed for the project are described in Free et al. (2018).

It is understood that code requirements for ground models in the UK are likely to be more formalized in the proposed 2020 update to Eurocode 7. Although ground models are referred to in the current version and in BS 5930: 2015 British Standard Code of Practice for Site Investigations (British Standards Institute 2015), the proposed version is likely to recognize the importance of producing a comprehensive ground model by dedicating a section to the topic.

4.5 Examples of the Use of 3-D Geological Models for Infrastructure Projects

In recent decades, many geotechnical consultants and civil engineering contractors have become more aware of the potential value of using 3-D geological models within the workflow of design and construction for infrastructure projects. The UK Government recently mandated use of Building Information Modeling (BIM) for all public-sector construction/design projects (HM Government 2015) This has led to more systematic use of 3-D geological modeling tools, and an evaluation of their capabilities and how they can be integrated into BIM technologies (Grice and Kessler 2015). This section discusses the experiences through case examples, including:

- The international engineering consulting firm CH2M (now Jacobs) specifically established an internal team to evaluate the benefits and potential limitations of employing 3-D geological modeling tools within the mandated use of BIM technologies (see Section 4.5.1).
- The international engineering consulting firm Arup has also been advocating the use of 3-D geological models as a standard component of geotechnical interpretation reporting on projects for some time, and were early adopters of the total geological history concepts proposed by Fookes et al. (2000, 2001). Early examples of geological models being used to plan, specify, and then interpret ground investigations can be found in many Arup projects worldwide since the 1970s. In 2011, Arup were commissioned to undertake the seismic hazard assessment for a proposed nuclear facility in the UK and a 3-D conceptual geological model of the site area was developed as the first task in the planning of the extensive ground investigations to be undertaken for the project (see Section 4.5.2). Successively detailed iterations of the project's 3-D geological model were used for planning, specifying, and costing ground investigations over a period of eight years on the project (2011–2018) and were used extensively as an essential component of the seismic hazard assessment and especially the capable faulting assessment work (Free et al. 2018).
- Dr. Sauer & Partners, a UK consultancy, used a 3-D model during the design and construction of the Crossrail Farringdon Station in London (see Section 4.5.3). This project was successfully completed ahead of schedule by using the 3-D model to manage direct observations of actual ground conditions, thereby materially reducing geotechnical risk while permitting efficient construction.
- TSP Projects (formally Tata Steel Projects) has recently utilized digital conceptual ground models to support the design and construction of rail infrastructure projects (see Section 4.5.4). The ground models were constructed using the GSI3D geological modeling platform, but were converted to CAD products and used to evaluate infrastructure development within a BIM environment.
- For the Silvertown Tunnel project (see Section 4.5.5), Atkins produced 3-D ground models and managed project design and planning within a CAD environment that directly employed BIM principles.
- In the Netherlands, the Dutch government has adopted a new law on subsurface information which aims to reduce failure costs in construction (see Section 4.5.6).

Several of these case studies describing 3-D geological modeling applications are described in greater depth in Chapters 18–25.

4.5.1 Investigating Three-Dimensional Geological Modeling as a Tool for Consultancy

Oliver J. N. Dabson and Ross J. Fitzgerald

Jacobs UK Ltd, London W6 7EF, UK

The applied earth sciences sector of the UK construction industry faced a new challenge in 2015 when the UK Government mandated the use of Building Information Modeling (BIM) for all public-sector construction/design projects (HM Government 2015). However, in most cases, BIM models have not represented subsurface geology and geotechnical conditions (Grice and Kessler 2015). Traditional geological analyses rely on two-dimensional sections and maps to interpret geological scenarios (Kessler et al. 2009), but use of BIM by the construction industry requires high-resolution 3-D datasets (Figure 4.7). These new methods must retain the historical roots that underpin our very understanding of geology.

A small earth engineering task force within Jacobs (then CH2M) was formed to find and test a suitable 3-D geological modeling platform for consultancy end use. After selection of a candidate platform, discussions between Jacobs and the software developers provided feedback on the use of 3-D geological information by consultancies and defined 3-D geological modeling requirements of software suitable for commercial applications. Specific project objectives included:

- Developing a preliminary ground model to assess and critique modeling workflow and output;
- Using results from a ground investigation (GI) results to refine the 3-D ground model;

- Producing derivatives from the ground model, including vertical and horizontal sections, surface plots, and strata thicknesses;
- Investigating geotechnical applications of the modeling platform;
- Investigating how model outputs might supplement factual/interpretive reports;
- Examining compatibilities and interactions with Auto-CAD and ArcGIS;
- Demonstrating the advantages of 3-D modeling over conventional 2-D industry practices; and
- Incorporating 3-D models into a BIM environment.

Evaluation of 3-D subsurface geological modeling packages developed for mining and petroleum industries demonstrated their shortcomings when dealing with the broad scope of surface and subsurface data typically encountered in consultancy projects. However, the British Geological Survey (BGS) and INSIGHT Geologische Softwaresysteme GmbH had produced a user-friendly, knowledge-driven 3-D modeling platform. Initially named "Geological Surveying and Investigation in 3 Dimensions (GSI3D)" and subsequently as "SubsurfaceViewer (SSV)", this SSV/GSI3D platform employs a paradigm based on the fundaments of geological analysis (Kessler et al. 2009). Chapter 10 describes the procedures for 3-D geological framework modeling with the SSV/GSI3D platform. The Jacobs task force adopted and evaluated this platform in three stages. In Stage 1, an initial evaluation of a basic, low-resolution model was undertaken using sample data provided by the software developer. This revealed:

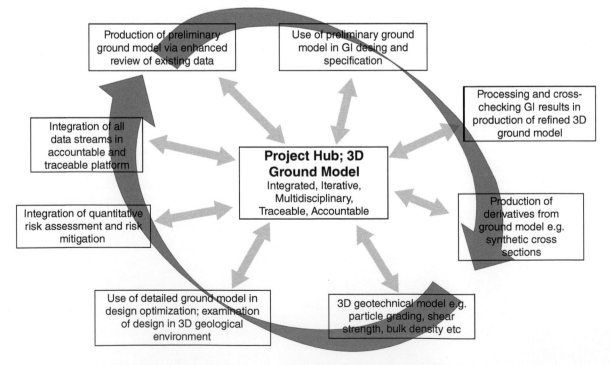

Figure 4.7 3-D model as the nucleus of a typical interactive project workflow as per BIM philosophy.

Box 4.1 Jacobs Evaluation of 3-D Modeling Potential for a Typical Client Project

Jacobs evaluated the potential of 3-D geological models with data from an existing client project in Wembley, North London. This tested how contributions from a 3-D model might supplement factual and interpretative reports produced after completion of an initial ground investigation (GI). The SSV/GSI3D platform produced several preliminary ground models using a variety of subsurface and surface data, including borehole logs, Digital Elevation Models, and geological maps (Figure 4.1.1).

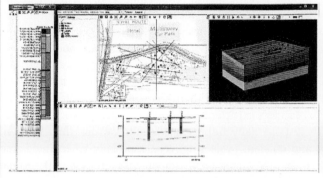

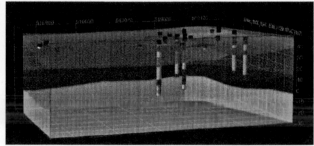

Figure 4.1.1 Screenshot of the modeling interface showing a borehole log, and plan, cross-sectional and 3-D views of the subsurface at the preliminary modeling stage for the Wembley project.

Figure 4.1.2 Screenshot of the refined 3-D model for Wembley, showing boreholes through transparent stratigraphy.

Once a suitable preliminary ground model was created, it was easily further refined with the additional data from the GI (Figure 4.1.2). The operator had full control of geological interpretation and complete traceability of input data sources during this refining process. The model was readily interrogated to produce several subsurface derivative outputs, including user-defined "synthetic" boreholes, "virtual" cross-sections based on modeled strata, and horizontal slices at specified locations and depths.

However, the SSV/GSI3D platform had limitations in producing acceptable displays of these derivative products. It currently does not readily manipulate legends, scale bars, etc. to customize the appearance of these products. This limits their usefulness as illustrations for client reports. This limitation was ameliorated by post-processing the model displays with GIS tools. Figure 4.1.3 shows an example cross-section output produced for the Wembley project after GIS tools were used to add annotations, scale, and legend.

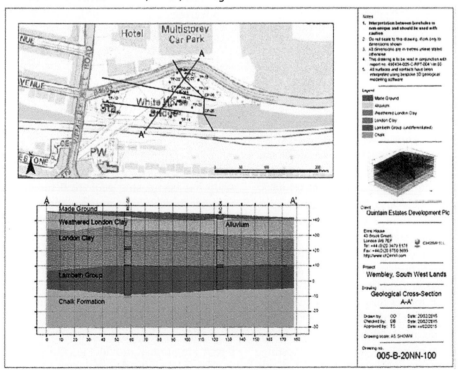

Figure 4.1.3 Example client report illustration: a model cross-section with annotations, legend, and scale added in GIS.

- Integration of surface and subsurface datasets maximized the value of data acquisition;
- The 3-D platform could visualize and interpret subsurface conditions, even with limited data;
- The 3-D model focused ground investigation (GI) toward data-deficient areas, thus reducing project costs by avoiding unnecessary boreholes; and
- The 3-D model aided recognition and characterization of potential geotechnical risks.

In Stage 2, a second 3-D model based on data from an existing client project in Wembley, North London, was used to evaluate how 3-D modeling might supplement initial ground investigation factual and interpretative reports. Examples of this modeling effort are described in Box 4.1.

In Stage 3, the 3-D modeling platform was used on a new project evaluating a coastal landslide complex at Barton-on-Sea, Hampshire in the UK. Details of this modeling effort by C2HM are described as the Barton-on-Sea case study in Chapter 22.

While BIM methodologies largely depend on AutoCAD and ArcGIS software standards, the SSV/GSI3D platform facilitates management of user-interpreted 3-D geological information in a BIM environment. The 3-D ground model thus becomes an ideal hub for integrating and exchanging high-resolution surface and subsurface datasets, as it eliminates the need for bulky hardcopy reports. The 3-D modeling approach provides additional capabilities:

- All legacy and project data and interpretations are stored in a single 3-D workspace, facilitating auditing, and repeat work;
- Iterative modeling of complex geology produces a better conceptual ground model; and
- Integration of geotechnical analysis within the BIM environment supports workflows to evaluate the variability of geotechnical parameters.

A 3-D ground model, compatible with BIM protocols, can store, visualize, and interrogate ground conditions. Because it forms a visual reference for all project data, it adds strategic value to bids and results in competitive advantages. The visual outputs add a "wow factor" to reports providing an additional benefit at no additional cost. They demonstrate proficiency with cutting-edge technology and the ability to provide quality deliverables. Use of 3-D geological models also leads to efficiencies over a project lifespan. When a 3-D geological model serves as a hub for subsurface interpretation, GI efforts can address data deficiencies; the value of GI investments is maximized and unnecessary boreholes are avoided. Aldiss et al. (2012) quantify these savings for the central London area: a detailed geological model may cost somewhere in the region of £10 000 to commission depending on scale and complexity. A single modern borehole, including permissions, mobilization, sampling, logging, and testing, will cost in the order of £1000 per meter. Therefore, if the 3-D model can help target ground investigation so that six planned 25 m boreholes might be reduced to two in order to pinpoint the location of a geological structure, even a model costing an order of magnitude more than the estimated £10 000 cost will pay for itself. The integration and exchange of data, as per the BIM philosophy, supports a more holistic digital representation of the physical and functional characteristics of a construction project. The centralized hub allows multiple teams open access to the database, avoids repeat work downstream and facilitates more effective internal project management. The 3-D modeling software delivers a coherent model to the client, rather than a collection of spatially unconnected sections.

Jacobs's experience demonstrates 3-D geological modeling potentially improves geotechnical data management and efficient project delivery, especially when integrated with BIM technology. Implementation of a 3-D ground model as a project information hub speeds up data interpretation over a greater spatial area, with greater detail, and facilitates efficient data management. New information allows iterative improvement of the ground model. A coherent visualization of ground conditions can be presented to the client. This has clear financial benefits.

Following this feasibility study, the logical next step is development of a cost-benefit analysis for a project using the SSV/GSI3D platform. This requires the consent of not only the client, but also the engineer. Future combinations of 3-D modeling software platforms and data exchange formats may allow improved access to BGS's National Geotechnical Properties Database. This synthesis would permit the generation of superior design products and ultimately a more formal integration of geotechnical engineering disciplines within UK BIM projects.

4.5.2 Three-Dimensional Geological Modeling for a Nuclear Power Facility in Anglesey, Wales, UK, to Enhance Ground Investigation Quality and Optimize Value

Matthew Free, Ben Gilson, Jason Manning, and Richard Hosker

Arup, London W1T 4BQ, UK

Arup has been using geological models as a component of geotechnical interpretation and reporting since the 1970s and 1980s. More recently, Arup (like Jacobs above) established a working group to investigate the range of in-house and proprietary 3-D geological modeling tools that are used across the firm internationally. The aim of this working party was not to recommend one preferred 3-D modeling software or process but to describe the pros and cons of the broad range of different tools available, to provide some guidance on what options might suit different situations, provide guidance on inter-operability

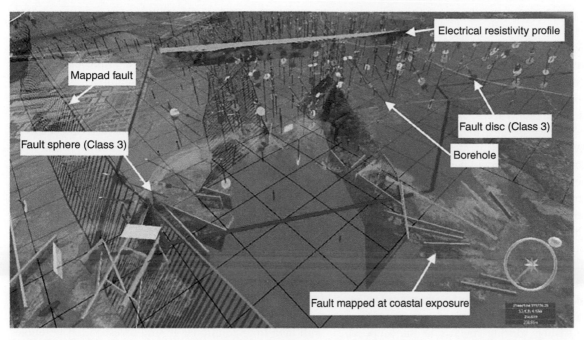

Figure 4.8 A screen capture of the 3-D Geological Fault Model developed for the Wylfa Newydd Nuclear Power Plant project by Arup (Free et al. 2018).

between different tools, as well as provide practical advice on project scales and likely budgets. Interestingly, the question of whether geological modeling would not be considered entirely justifiable on a cost-benefit basis was never discussed given the overwhelming support for geological models as good practice within the firm.

In 2011, Arup was commissioned to undertake the seismic hazard assessment for a proposed nuclear facility in the UK. A 3-D Conceptual Geological Model of the site area was developed as the very first technical task in the planning and specifying of the extensive ground investigations to be undertaken for the project. Ground investigations were undertaken in stages over a period of eight years (2011–2018) and successively more detailed iterations of the project 3-D models were used for planning, specifying, and costing of future stages of ground investigations. They were used extensively as an essential component of the seismic hazard assessment and especially for the capable faulting assessment work (Free et al. 2018). Each stage of the project reporting includes a section describing the 3-D Conceptual Geological Model.

In addition, the project developed a 3-D Geological Fault Model to more accurately characterize geological faulting at the site as required by the capable-faulting component of a seismic hazard assessment. Arup extensively used the 3-D Geological Fault Model throughout the project from initial desk study, to planning, and specification of the ground investigations, through to post investigation interpretations. Geological faults identified by geological

mapping and intrusive and non-intrusive ground investigations were visualized in the 3-D Geological Fault Model and then mapped across the site and in the vicinity of the site to accurately characterize the faults spatially and geometrically. Figure 4.8 illustrates the visualization of the interpreted geological faults.

Age dating of fault gouge samples was undertaken to constrain the last age of movement on the faults as required by the International Atomic Energy Agency (IAEA) guidelines, to demonstrate whether faults are "capable". This systematic process led to increased quality of interpretations and better communication of the interpretations to stakeholders, including the Office of Nuclear Regulation, at each stage of the project. The model was also shared with independent checkers, other consultants working on the project, and with the main contractor and their designers; it provided one source of truth in digital format.

GeoVisionary software produced by Virtalis was used to create the 3-D Geological Fault Model by visualizing various primarily sub-surface datasets; this allowed correlations to be made between interpreted or possible geological fault data points. Attempting to make these correlations in 2-D using cross-sections would not have been feasible. Regional 3-D geological models were also developed to visualize the distribution of regional earthquake hypocenters and all nearby regional geological faults in order to provide a suitable regional context for the geological fault modeling at the site.

The 3-D Geological Fault Model was systematically used in the planning of each ground investigation stage and this process was found to deliver considerable cost savings. A simple but tangible cost saving achieved using the 3-D methodology was to reduce the number of boreholes needed to provide coverage of the site and to determine the geometry of inclined boreholes so they intersected potential geological fault alignments passing through the site. Inclined boreholes were carefully positioned to prevent the occurrence of gaps in the 3-D investigation coverage of the site. It was estimated that this process allowed a reduction in the number of ground investigation boreholes of about 10% over the many stages of ground investigation. In total, more than 35 km of drilling was completed; thus the model potentially produced a very significant saving. In the final stage of investigation, two of the originally proposed 18 additional boreholes were eliminated by incorporating ongoing ground investigation results and updating interpretations into the 3-D Geological Fault Model as the works progressed. This equated to a saving in the order of £130 000. In addition to these direct savings, during reviews to the scope of work, the use of the 3-D Geological Fault Model allowed responses within hours to technical queries. On several occasions, challenging site access or ground conditions were encountered that prevented the completion of planned boreholes to their required depth, or required their relocation. Revised site instructions were quickly provided and agreed to following a review of the 3-D model. This reduced or eliminated substantial amounts of potential standing-time charges by the ground investigation contractor that would have been incurred and ultimately paid for by the client.

4.5.3 Integrating 3-D Models Within Project Workflow to Control Geotechnical Risk

Angelos Gakis[1], Paula Cabrero[1], and David Entwisle[2]

[1] *Dr. Sauer & Partners Ltd, Surbiton, Surrey KT6 6QH. UK*
[2] *British Geological Survey, Keyworth, Nottingham NG12 5GG, UK*

Crossrail, a new underground railway system, will provide a direct East–West connection through the center of London. The Crossrail route includes 21 km (13 mi) of twin-bore tunnels and eight new stations connecting with London national rail stations and the London Tube network. At Farringdon Station, the planned Crossrail tunnels had to be placed around or beneath existing utilities and London Underground transit tunnels. The two platform tunnels are approximately 30 m (100 ft) below the surface and extend between the existing Farringdon and Barbican Underground Stations (Aldiss et al. 2009, 2012; Gakis et al. 2016). This depth placed the tunnels mostly within the Lambeth Group, which underlies the London Clay

Formation, and is known to be a challenging tunneling medium due to its lateral and vertical lithological variation and presence of water-bearing sands. While the running tunnels were constructed using tunnel boring machines (TBMs), the stations were constructed using an open face excavation tunneling method with sprayed concrete linings (SCLs) (Gakis et al. 2015).

Excavation of Farringdon Station was expected to encounter some of the most challenging ground conditions in the whole of Crossrail. In 2009, during the pre-construction phase, the BGS was commissioned to develop a 3-D geological model of the local area (Aldiss et al. 2009, 2012). In 2013, as construction began, this model was handed over to the contractor, a BAM-Ferrovial-Kier Joint Venture (BFK) and their specialized tunneling consultant, Dr. Sauer & Partners (DSP).

As described by a case study in Chapter 23, DSP/BFK integrated 3-D geological modeling into the site supervision workflow. Initially, the BGS model was updated with data from additional ground investigation boreholes and shaft excavations, received between 2009 and 2013. Subsequently, the 3-D model was updated daily with data from the tunnel faces. Thus, the Farringdon Station project provides one of the first examples of the economic benefits of integrating 3-D geological models and their associated constantly evolving geological databases into the workflow of an active ongoing subsurface infrastructure project.

Initial ground investigations were unable to establish a sufficiently detailed, coherent ground model, capable of supporting design and construction phases. The initial 3-D model, developed by the BGS, established a suitable pre-construction ground model. During construction, the model was continuously updated and this increased its reliability and utility (Gakis et al. 2016). Rapid and easy examination of existing records and production of virtual boreholes and sections in areas of interest provided essential support for multiple, additional design changes during construction of the SCL tunnels. In some locations, the detailed knowledge of the geotechnical conditions allowed for significant reductions in lining thickness. As confidence in the expected ground conditions predicted by the 3-D model increased, BFK/DSP significantly reduced the amount of in-tunnel probing by approximately 70%, resulting in large cost savings. When water-bearing sand lenses were encountered, the 3-D model was used to optimize the direction of in-tunnel probes and locations of depressurization wells. Throughout the construction phase, reference to the 3-D geological model materially reduced geotechnical risk, allowed for efficient construction, and resulted in a four-month reduction in the planned construction time. The case study in Chapter 23 details the benefits resulting from integrating a 3-D model into project management.

The Crossrail Farringdon Station project provided a unique, real-world opportunity to test and develop systems and databases capable of dealing with geological interpretations provided by industry. It is vitally important that all data, formats, and nomenclatures, as well as software systems, are compatible and are shared between project partners. At Farringdon, a commissioned BGS geological model provided valuable initial guidance. Subsequently the project provided additional valuable direct observations of actual ground conditions. These were progressively used to produce updated 3-D models. These models and all new observational data must flow back to the national archive for future reuse. The Farringdon data and models have been supplied to the BGS (Gakis et al. 2016).

4.5.4 The Economic Value of Digital Ground Models for Linear Rail Infrastructure Assets in the United Kingdom

Gerard McArdle

TSP Projects, York YO24 1AW, UK

TSP Projects (formally Tata Steel Projects) has recently completed rail infrastructure projects in the UK for which digital subsurface models were used extensively to support the design and construction process. This experience provides a basis for evaluating the potential economic impact of digital ground models for rail infrastructure projects. Introduction of Level 2 Building Information Modeling (BIM) into the rail industry in the UK has encouraged the wider use of digital conceptual ground models. The challenge lies in employing digital conceptual geological models to bring economic benefits to the maintenance and modernization of a rail system that has many engineering challenges.

The success of rail infrastructure construction activities, generally undertaken within tight schedules to minimize disruption to existing rail services, depends on the accurate prediction of ground conditions and geotechnical risks. A digital 3-D model has the potential to be a valuable tool. The linear nature of most rail infrastructure projects, when combined with limited ground investigation opportunities, is a constraint on capturing enough data to produce a digital 3-D model based upon the fieldwork alone. Ground investigations for rail infrastructure projects are often hampered by limited access to locations outside the rail corridor. Even though rail corridors are narrow, significant differences in the natural ground conditions are often found within a corridor, especially at locations previously affected by Quaternary glacial and periglacial processes.

The value of any digital 3-D geological model is largely governed by the skill of the geoscience modeler and availability of information. The original earthworks along UK rail corridors were constructed by non-engineered methods, and many locations have been subjected to various modifications over time. The origin and subsequent modification of these earthworks means a digital geological model (DGM) cannot be constructed by simply extrapolating between exploratory holes. A geologist develops a digital model by interpreting available data and produces a sequence of surfaces defining geological strata; the degree to which these surfaces conform to the original exploratory hole information is critical. If this relationship is not defined, future use of the digital model may be compromised.

TSP Projects has tried two approaches for developing digital 3-D geological models for use in UK rail infrastructure projects:

1. The DGM is provided by a subcontractor, or
2. The DGM is developed in-house utilizing information obtained from the Ground Investigation Report process.

Both approaches have various economic advantages and disadvantages. The economic value of DGMs by either approach depends on the modeler's ability to provide verifiable, assured information. Can the client and others use the ground model in an informed and economically useful manner? This question may be evaluated by assessing the following:

- Is the digital ground model accompanied by an assurance document?
- Is there a BIM execution plan (BEP) that sets out the federated modeling, CAD standard requirements, and model file format?
- How are the ground risks communicated by the digital ground model?
- How has the uncertainty of the ground conditions been communicated?
- Can the digital ground model be located within an appropriate national or project-related topographic grid system?
- What is the accuracy and source of the topographic information?

In summary, the generator of a digital 3-D geological model must appropriately anticipate its application within the rail infrastructure project design and construction process. A digital model accompanied by a suitable assurance document will provide information critical to its immediate application, its long-term utility to future projects, and its use by a contractor to estimate costs and construction

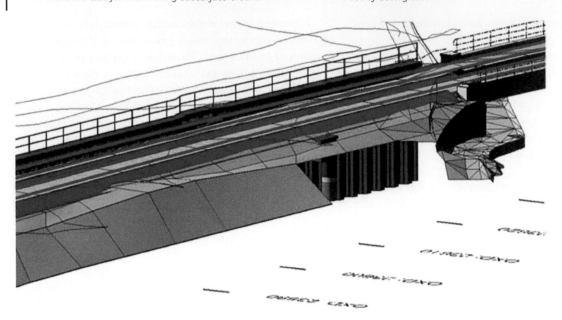

Figure 4.9 Design for a proposed sheet pile wall, including existing ground surface. The image includes a cone penetration test. Colors define subsurface materials: Red: coarse grained made ground; Yellow: alluvium; Orange: river terrace deposits (sand and gravel). Source: TSP Projects.

risks. Use of a DGM to evaluate geotechnical risks requires information provided in a Ground Investigation Report.

In 2015, the BGS developed a digital 3-D geological model for the North Trans-Pennine Electrification Project as a sub-contractor to TSP Projects (Burke et al. 2015). A case study in Chapter 23 provides further details on this project and the development and application of the digital 3-D Geological Model. This approach provided two advantages:

1. The BGS issued a 3-D ground modeling assurance document along with the model (Burke et al. 2015).
2. The BGS demonstrated capability to deliver large-scale digital 3-D geological modeling projects; the requested digital 3-D geological model was delivered after only one month.

However, the high quality of the digital 3-D geological model produced by the BGS posed a potential disadvantage. The 3-D geological model users tended to accept without question the accuracy of the ground conditions predicted by the digital model, potentially resulting in under-estimation of geotechnical risks.

TSP Projects used the alternative in-house approach when a digital 3-D geological model was required for a large station remodeling scheme in Oxfordshire. An embankment, located on the Thames flood plain adjacent to a site of special scientific interest, required widening to allow doubling of an existing single track. Because the project required all design documents to be generated from a 3-D federated BIM model, a digital 3-D geological model was developed using typical CAD methods. Digital fence diagrams (cross-sections) were produced between exploratory holes. These reflected interpreted ground conditions and geotechnical observations reported in the

Ground Investigation Report. This allowed users to assess the information uncertainty without immediate access to any assurance documents. Because the resulting digital 3-D geological model communicated ground engineering risks to the client and contractor, it was possible to select the most appropriate solutions for detailed design purposes, based on cost and program considerations. Figure 4.9 illustrates the design of a sheet pile wall using this digital 3-D geological model.

4.5.5 Employing an Integrated 3-D Geological Model for the Reference Design of the Silvertown Tunnel, East London

Jerome Chamfray[1,2], Simon R. Miles[1], and Gary Morin[3]

[1] *Atkins, Epsom KT18 5AL, UK*
[2] *Currently at Jacobs, London SE1 2QG, UK*
[3] *Keynetix Ltd, Redditch B98 9PA, UK*

In 2012, after many years of consideration, approval was obtained for a new road tunnel crossing the River Thames in east London. The proposed Silvertown Tunnel was justified as it would alleviate severe congestion in the nearby Blackwall Tunnel and support the economic development of east and southeast London. There was also a desire to improve the resilience and reliability of the regional road transport network (Transport for London 2013). In March 2014, WS Atkins plc (commonly known as Atkins) was commissioned to carry out the reference design for the Silvertown Tunnel.

The tunnel alignment faced several design challenges related to the feasibility of tunnel construction and operation due to existing site characteristics. Previously, the area was highly utilized by heavy industry. A now demolished gas works was located near the south portal and underground remnants of redundant and demolished structures remain near both north and south portals. Additionally, several new infrastructure elements, such as Docklands Light Railway Woolwich Branch Viaduct, and the Emirates Air Line cable car crossing, were constructed near the proposed tunnel route. Bored twin tunnels under the river and cut and cover tunnels on both banks of the river near the tunnel portals were specified. The cut and cover sections would require the removal of anticipated abandoned historical foundations and allow for the launching of the TBMs. The increased costs of treating contaminated soils relating to a former gas works on the Greenwich Peninsula made accurate volume calculations a critically important aspect in ensuring cost-effective construction decisions. In addition, the desired tunnel vertical profile (which produced a longitudinal gradient of 4.2% along the north tunnel section, resulting in reduced vehicle emissions and ventilation operating costs) placed the lowest point of the tunnel only slightly below the lowest point of the river channel. This meant that precise knowledge of the geological materials at the site, in 3-D, would help reduce geotechnical risks during construction. Faced with these design challenges, Atkins decided to utilize integrated 3-D modeling for the preliminary design of the Silvertown Tunnel. Chapter 23 contains a case study that describes the approach to this project.

Geological data, from historical records and concurrent ground investigations, were stored within a database that was integrated with a Building Information Modeling (BIM) 3-D CAD environment (AutoCAD Civil 3-D). This allowed for rapid combination, organization, management, and visualization of the geological data in conjunction with existing and proposed above and below-ground structures. When new geotechnical data were entered into the database, those changes were automatically reflected in the 3-D model. The time saved by this synchronization allowed for more experimentation with design alternatives, resulting in refinements and improvements. This integration of geological data with the 3-D BIM model permitted the Atkins design team to: (i) visually understand and evaluate the design alignment; (ii) pinpoint potential construction obstructions; (iii) determine what new site investigations were needed; and (iv) easily generate earthworks quantities for project costing and risk assessment.

For the design of the Silvertown Tunnel, Atkins found increased efficiency, and thus economic benefit, from the use of a fully integrated, multidisciplinary 3-D model that combined BIM and geological data (Atkins 2016). Visualizing ground conditions in a design context resulted in a design that offered reduced project risk and project construction costs. It also provided for the rapid production of 3-D visualizations suitable for public discussions (Figure 4.10).

TRANSPORT FOR LONDON Silvertown Tunnel:
EVERY JOURNEY MATTERS By kind permission of Transport for London

Figure 4.10 Example of 3-D visualization of the Silvertown Tunnel design showing the integration of BIM elements defining existing or proposed infrastructure and geological conditions (shown as layers). The features shown are: (1) Purple Box: cable car station; (2) Gray: concrete, either tunnels or piles; (3) Green: ground surface; (4) Brown: base of made ground; (5) Mustard: base of alluvium; (6) Blue: base of River Terrace; (7) Purple: base of London Clay; (8) Pink: base of Harwich Formation; (9) Dark gray: base of Lambeth Group – laminated beds. Source: Atkins and Transport for London.

Without the use of the 3-D modeling approach, the amount of project risk embedded in the outline design, and thus reflected in the construction estimates (and ultimately the tenders to finance, design, and build it) would be significantly greater than they currently are. The commission from Transport for London was to reduce project risk as far as reasonably practicable prior to tendering for detailed design and construction as a Public Private Partnership (PPP), with a view to reducing as much as possible the risk priced in to the funding estimates, and this was achieved through the use of the 3-D model. Without this, there would certainly be significant risk costs for dealing with (unknown) buried foundations, and possibly even an alternative tunnel alignment taking it deeper under the river at the expense of longer approach ramps (as they are already at their steepest allowable gradient), and consequently increased traffic pollution. These additional perceived risks to the lenders may have pushed the project cost outside of its funding envelope, and it was therefore very important to the client that they be minimized as far as possible.

4.5.6 A New Dutch Law on Subsurface Information to Enable Better Spatial Planning

Martin R.H.E Peersmann[1] and Michiel J. van der Meulen[2]

[1] *Keynetix Ltd, Redditch B98 9PA, UK*
[2] *TNO, Geological Survey of the Netherlands, 3584 CB Utrecht, The Netherlands*

In 2015, the Dutch government adopted a new law on subsurface information (Van der Meulen et al. 2013; Anonymous 2015), which states that:

- Subsurface data acquired with public funds are to be deposited in a single repository, "BRO", which is part of the national systems of "key registers";
- Public bodies have to consult the BRO when they make decisions that affect or pertain to the subsurface; and
- Users of the register who find an error, have to report this in order for the error to be investigated and corrected if necessary.

Basisregistratie Ondergrond (BRO) requires suppliers of data to remain responsible and accountable for data quality; this is an important difference from the operations of previous Dutch geological data and information repositories.

The Dutch "key registers" are designated single government databases for vital public data, including personally identifiable information (PII); identification and ownership of real estate, companies, and vehicles; real-estate value; income, employment relations, and social-security benefits; addresses and buildings; and base topography. The system is designed to avoid the duplication that is inherent to decentralized data management and to reduce the errors this brings.

The BRO is currently being developed from the national geological database known as DINO (TNO 2018) and the Dutch national soil survey database known as BIS (Bodem Informatie Systeem) (WER 2018). This entails a standardization effort and development of the necessary IT systems. The BRO will be operated by the Geological Survey of the Netherlands on behalf of the Ministry of the Interior and Kingdom Relations, which assumed responsibility for all key registers in 2018. The Geological Survey will also be appointed as the legally mandated supplier of three geological models to the BRO. As of 2019, these are *GeoTOP*, *DGM*, and *REGIS-II* (discussed in Chapters 5, 9, 11, and 19). Additional models, such as the *DGM-deep* model, which describes the upper 5 km for energy resource evaluations, may be included in later stages.

The BRO acknowledges the government's reliance on subsurface data and information for a number of important planning and permitting procedures, including: spatial planning for housing, land use, infrastructure, utilities; exploration and production of natural resources; and groundwater management. The subsurface data that are to be collected in the BRO will allow the government to better consider the subsurface in land-use and infra-structure planning, identify opportunities, and risks, take precautionary and mitigation measures when necessary, and manage residual risks.

The decision to start the legislative process was underpinned by a societal cost-benefit analysis (Terpstra et al. 2011; Anonymous 2013), which concluded that there would be a marginally positive business case for the BRO. The analysis was by and large an estimate of general cost-avoidance caused by re-using subsurface data, primarily in construction, expressed as a proportion of total investments included in preparatory ground investigations. However, when the law was eventually adopted, this analysis was deemed not specific enough to actually allocate the investment costs for the BRO, which were budgeted at €50 million.

A more in-depth financial analysis required the assessment of individual construction works. For this purpose, a working group was established within the Ministry of Infrastructure and the Environment, which screened the entire portfolio of national infrastructure projects that were funded from the national infrastructure fund of ~€5 billion/year (Anonymous 2018a), and water safety projects funded from the Delta fund of ~€2 billion/year (Anonymous 2018b). The screening followed a value-engineering and value-management approach that considers cost over project life (Oke and Aigbavboa 2017). For each of these projects, provisions for unforeseen events and conditions were identified for the entire project life cycle from the planning to the maintenance stages, and estimates were made of the share of these conditions that were directly related to, or otherwise attributable to, ground conditions. It was then argued that the BRO could feasibly result in a reduction of 2–5% of subsurface-related failure costs, especially in the early project stages, which would give the BRO a projected net present value of about €80 million in 2019 and rising to €130 million from 2028 onwards. This estimate exceeds by two to three times the estimated total investment costs of about €50 million, and provides an initial annual return on investment of €8 million during the five-year implementation stage. Hence, a funding decision was made, and the BRO project went from the preparatory to the implementation stage.

The benefits are in fact underestimated, because only national-level investments and revenues were considered, and corresponding infrastructure and water works undertaken and co-financed by the provinces and municipalities were not included. The current benefits are obtained by extrapolating current spatial and water management strategies. Much larger benefits are expected from the implementation of the energy transition (thermal energy storage, climate-proofing of cities, emplacing, and replacing energy infrastructures), national housing targets (including the associated infrastructures), and coping with climate change (involving major upgrades of the Dutch flood defense systems). The investments in these plans these add up to tens of billions of euros per year, and the benefits of better subsurface data and information from the BRO are expected to increase proportionally.

Beyond making sure that subsurface data are re-used rather than being shelved in inaccessible project databases or reports, the BRO will be instrumental in building ever-improving models of the subsurface that can be used in early stages of project development, where they are expected to especially improve the quality and reliability of desk studies. Transparency in decision making – the BRO is an open data system – is a key element, allowing for a level playing field for all stakeholders in any project that relies on subsurface data. In this way, BRO will also create a more level playing field for construction companies and their consultants in the bidding stage. They will be able to compete based on engineering ingenuity and efficiency rather than access to information. These developments are expected to boost innovation in 3-D data across domains, for example by paving the way to integrate BIM-models of the built environment with geological models of the substrate, and calling for 4-D models (3-D plus time) that allow modeling of how the subsurface changes during building and construction, or evaluation of groundwater flow or ground subsidence processes.

There are two take-away messages associated with the decision to develop BRO. The first is that there is a strong business case for organizing subsurface information for the construction sector. As there is no reason to assume that the Netherlands operates fundamentally differently than other countries when it comes to national infrastructure planning and development, we argue that the business case would also be valid elsewhere.

The second message is that it took access to confidential, commercially-sensitive, government data to build a convincing business case that provides an appropriate financial overview. This overview could not be achieved by external parties, neither private ones such as construction companies and engineering agencies, nor government agencies without appropriate clearances. This explains the lack of satisfactory business cases in the literature. Rather than convincing the government that thorough ground (pre)investigations are important, the decision process for developing BRO should serve as an incentive to other governments to assess this themselves.

Acknowledgments

Jennifer Gates would first like to thank Tim Chapman for facilitating the opportunity to write on this topic and for his knowledge and guidance that formed the basis of this chapter. Thanks go to the case study contributors, including Oliver Dabson, Ross Fitzgerald, Ben Gilson, Jason Manning, Richard Hosker, Angelos Gakis, Paula Cabrero, David Entwisle, Gerard McArdle, Jerome Chamfray, Simon Miles, Gary Morin, and Martin Peersmann as well as Joe Smith and Dr. Peter Ingram for contributing thoughts in their respective areas of expertise. Thanks also extend to both Matthew Free and Michiel J. van der Meulen for their contributions to the case studies, as well as for their guidance in the content of the chapter. This work was partially funded by the Arup Innovation Fund and so final thanks extend to Arup for supporting this research.

References

Aldiss, D.T., Entwisle, D.C. and R.L. Terrington (2009). *A 3D Geological Model for the Proposed Farringdon Underground Railway Station, central London*. Keyworth, UK: British Geological Survey, Commissioned Report, CR/09/014.

Aldiss, D.T., Black, M.G., Entwisle, D.C. et al. (2012). Benefits of a 3D geological model for major tunnelling works: an example from Farringdon, east-central London, UK. *Quarterly Journal of Engineering Geology and Hydrogeology* 45 (4): 405–414. https://doi.org/10.1144/qjegh2011-066.

Anonymous (2013). *Regels omtrent de basisregistratie ondergrond (Wet basisregistratie ondergrond), Memorie van toelichting Tweede Kamer (in Dutch: Explanatory Notes to the Draft Law Establishing the Key Register for the Subsurface BRO)*. The Hague, Netherlands: Tweede Kamer der Staten-Generaal, vergaderjaar 2013–2014, 33 839, nr. 3. 51 pp. [Online: Available at: https://zoek.officielebekendmakingen.nl/kst-33839-3] (Accessed June 2020).

Anonymous (2015). *Wet Basisregistratie Ondergrond (in Dutch; Law Establishing the Key Register for the Subsurface BRO). Adopted in 2015, Amended in 2018*. The Hague, Netherlands: Ministry of Infrastructure and Water Management. [Online: Available at: http://wetten.overheid.nl/jci1.3:c:BWBR0037095&z=2018-07-28&g=2018-07-28] (Accessed September 2018).

Anonymous (2018a). *Vaststelling van de begrotingsstaat van het Infrastructuurfonds voor het jaar 2018 (in Dutch: Budget Decision for the Infrastructure Fund in 2018)*. Tweede Kamer, vergaderjaar 2017–2018, 34 775 A, nr. 1. 5 pp. [Online: Available at: https://zoek.officielebekendmakingen.nl/kst-34775-A-1.html] (Accessed June, 2020).

Anonymous (2018b). *Vaststelling van de begrotingsstaat van het Deltafonds voor het jaar 2018 (in Dutch: Budget Decision for the Infrastructure Fund in 2018)*. The Hague, Netherlands: Tweede Kamer der Staten-Generaal, vergaderjaar 2017–2018, 34 775 J, nr. 1. 5 pp. [Online: Available at: https://zoek.officielebekendmakingen.nl/kst-34775-A-1] (Accessed June, 2020).

Atkins (2016). Combining geological modelling and BIM for infrastructure. London, UK: Atkins. 2 pp. [Online: Available at: https://www.keynetix.com/wp-content/uploads/2016/11/Atkins-TfL-Silvertown-Tunnel.pdf] (Accessed June 25, 2020).

Baynes, F.J., Fookes, P.G. and Kennedy, J.F. (2005). The total engineering geology approach applied to railways in the Pilbara, Western Australia. *Quarterly Journal of Engineering Geology and Hydrogeology* 64: 67–94. https://doi.org/10.1007/s10064-004-0271-4.

British Standards Institute (2015). *BS 5930: 2015 British Standard Code of Practice for Site Investigations*. London, UK: British Standards Institution. 326 pp.

British Tunnelling Society (2003). *The Joint Code of Practice for Risk Management of Tunnel Works in the UK*. London, UK: British Tunnelling Society. 18p.

Burke, H.F., Hughes, L., Wakefield, O.J.W. et al. (2015). *A 3D geological model for B90745 North Trans Pennine Electrification East between Leeds and York*. Keyworth, UK: British Geological Survey, Commissioned Report, CR/15/04N. 32 pp. [Online: Available at: http://nora.nerc.ac.uk/id/eprint/509777/] (Accessed June 2020).

Burland, J.B. (1987). The teaching of soil mechanics. A personal view. Nash lecture. In: *Proceedings of the IX European Conference on Soil Mechanics and Foundation Engineering, Dublin*, 1427–1447. Rotterdam, Netherlands: A.A. Balkema.

Burland, J.B. (2006). Interaction between structural and geotechnical engineers. *The structural Engineer* 84 (8): 29–37.

Burland, J.B. (2012). Chapter 4: The geotechnical triangle. In: *ICE Manual of Geotechnical Engineering: Volume I* (eds. J. Burland, T. Chapman, H. Skinner and M. Brown), 17–26. London, UK: Institution of Civil Engineers.

Chapman, T. (2008). The relevance of developer costs in geotechnical risk management. In: *Proceedings of the Second BGA International Conference on Foundations* (eds. M.J. Brown, M.F. Bransby, A.J. Brennan and J.A. Knappett), 3–25. Watford, UK: IHS BRE Press.

Chapman, T. (2012). Chapter 7: Geotechnical risks and their context for the whole project. In: *ICE Manual of Geotechnical Engineering: Volume I* (eds. J. Burland, T. Chapman, H. Skinner and M. Brown), 59–73. London, UK: Institution of Civil Engineers.

Chowdhury, R. and Flentje, P. (2007). Perspectives for the future of geotechnical engineering. In: *Invited Plenary Paper, Printed Volume of Plenary Papers and Abstracts, Civil Engineering for the New Millennium, 150 Anniversary Conference*, 59–75. Shibpur, India: Indian Institute of Engineering Science and Technology.

Culshaw, M.G. (2005). From concept towards reality: developing the attributed 3D geological model of the shallow subsurface (The Seventh Glossop Lecture). *Quarterly Journal of Engineering Geology and Hydrogeology* 38: 231–284. https://doi.org/10.1144/1470-9236/04-072.

Davison, L., Fookes, P. and Baynes, F. (2003). Total geological history: a web-based modelling approach to the anticipation, observation and understanding of site conditions. In: *New Paradigms in Subsurface Prediction* (eds. M.S. Rosenbaum and A.K. Turner), 237–252. Berlin: Springer-Verlag, Lecture Notes In Earth Sciences, vol. 99. https://doi.org/10.1007/3-540-48019-6_21.

European Committee for Standardization (2004). *European Standard EN 1997-1. Eurocode 7: Geotechnical Design – Part 1: General Rules.* Brussels: European Committee for Standardization (CEN). 168 pp.

European Committee for Standardization (2007). *European Standard EN 1997-2. Eurocode 7: Geotechnical Design – Part 2: Ground Investigation and Testing.* Brussels: European Committee for Standardization (CEN). 196 pp.

Fookes, P.G. (1997). Geology for engineers: the geological model, prediction and performance. *Quarterly Journal of Engineering Geology and Hydrogeology* 30: 293–424. https://doi.org/10.1144/GSL.QJEG.1997.030.P4.02.

Fookes, P.G. and Shilston, D.T. (2001). Building the geological model: case study of a rock tunnel in SW England. *Engineering Geology Special Publication*, 18: 123–128. https://doi.org/10.1144/GSL.ENG.2001.018.01.18.

Fookes, P.G., Baynes, F.J. and Hutchinson, J.N. (2000). Total geological history: a model approach to the anticipation, observation and understanding of site conditions. In: *Proceedings of the International Conference on Geotechnical and Geological Engineering; GeoEng 2000, Melbourne, Australia, vol 1* (eds. S.H. Chew, G.P. Karunaratne, S.A. Tan et al.), 370–460. Lancaster, PA: Technomic Publishing Co.

Fookes, P., Baynes, F. and Hutchinson, J.N. (2001). Total Geological History: A Model Approach to Understanding Site Conditions. *Ground Engineering, March 2001*, 42–47.

Free, M., Gilson, B., Manning, J. et al. (2018). A 3D geological fault model for characterisation of geological faults at the proposed site for the Wylfa Newydd nuclear power plant, Wales. In: *IAEG/AEG Annual Meeting Proceedings, San Francisco, CA, USA, 2018 – Volume 6: Advances in Engineering Geology, Education, Soil and Rock Properties, Modeling* (eds. A. Shakoor and K. Cato), 245–252. Cham, Switzerland: Springer Nature.

Gakis, A., Salak, P. and St. John, A. (2015). Innovative geotechnical risk management for SCL tunnels. *Proceedings of the Institute of Civil Engineers, Geotechnical Engineering* 168 (5): 385–395. https://doi.org/10.1680/jgeen.14.00070.

Gakis, A., Cabrero, P., Entwisle, D. and Kessler, H. (2016). 3D geological model of the completed Farringdon underground railway station. In: *Crossrail Project: Infrastructure Design and Construction* 3: 431–446. London, UK: Institution of Civil Engineers Publishing.

Grice, C. and Kessler, H. (2015). *Collaborative Geotechnical BIM technologies.* Keyworth, UK: British Geological Survey, Unpublished powerpoint presentation. 19 pp. [Online:

Available at: http://nora.nerc.ac.uk/510823] (Accessed June 25, 2020).

Haff, P.K. (2013). Prediction in geology versus prediction in engineering. *Geological Society of America Special Paper* 502: 127–134. https://doi.org/10.1130/2013.2502(06).

Harding, C. (2004). Site investigation and site conceptual models. The link between geology and engineering. In: *Advances in Geotechnical Engineering: The Skempton Conference, vol. 2* (eds. R.J. Jardine, D.M. Potts and K.G. Higgins), 1304–1315. London, UK: Institution of Civil Engineers Publishing.

HM Government (2015). *Digital Built Britain: Level 3 Building Information Modelling – Strategic Plan.* London, UK: Department for Business, Innovation & Skills. 47 pp. [Online: Available at: https://www.gov.uk/government/uploads/system/uploads/attachment_data/file/410096/bis-15-155-digital-builtbritain-level-3-strategy.pdf] (Accessed June 25, 2020).

HM Treasury (2018). *The Green Book: Central Government Guidance on Appraisal and Evaluation.* London, UK: Treasury Stationery Office. 132 pp. [ONLINE: Available at: https://assets.publishing.service.gov.uk/government/uploads/system/uploads/attachment_data/file/685903/The_Green_Book.pdf] (Accessed August 2018).

ITIG (2012). *A Code of Practice for Risk Management of Tunnel Works.* International Tunnelling Insurance Group. 28p.

Kessler, H., Mathers, S. and Sobisch, H.-G. (2009). The capture and dissemination of integrated 3-D geospatial knowledge at the British Geological Survey using GSI3-D software and methodology. *Computers & Geosciences* 35: 1311–1321. https://doi.org/10.1016/j.cageo.2008.04.005.

Knill, J.L. (2003). Core values: the first Hans Cloos lecture. *Bulletin of Engineering Geology and the Environment* 62: 1–34. https://doi.org/10.1007/s10064-002-0187-9.

Latham, M. (1994). *Constructing the Team: Final Report of the Government/Industry Review of Procurement & Contractual Arrangements in the UK Construction Industry.* London, UK: HM Stationery Office. 130 pp. [Online: Available at: http://constructingexcellence.org.uk/wp-content/uploads/2014/10/Constructing-the-team-The-Latham-Report.pdf] (Accessed August 2018).

Lee, E.M. and Jones, D.K.C. (2004). *Landslide Risk Assessment.* London, UK: Thomas Telford Ltd. 454 pp.

Munsterman, W.P., Ngan-Tillard, D.J.M. and Venmans, A.A.M. (2008). Total Engineering Geological Approach Applied to Motorway Constructions on Soft Soils. In: *Proceedings of the 2nd European Regional Conference of the International Association of Engineering Geology and the Environment (EuroEnGeo 2008), Madrid, Spain*, Paper 42. Madrid: Asociación Española de Geologia Aplicada a la Ingeniería.

National Audit Office (2001). *Modernising Construction*. London, UK: HM Stationery Office. 102 pp. [Online: Available at: https://www.nao.org.uk/report/modernising-construction/] (Accessed June 25, 2020).

National Economic Development Office (1983). *Faster Building for Industry*. London, UK: HM Stationery Office. 111 pp.

National Economic Development Office (1988). *Faster Building for Commerce*. London, UK: HM Stationery Office. 156 pp.

Office for Nuclear Regulation (2018a). *Nuclear Safety Technical Assessment Guide: Expert Panel Report No: GEN-SH-EP-2016-1*. Bootle, UK: Office for Nuclear Regulation. 45 pp. [Online: Available at: http://www.onr.org.uk/operational/tech_asst_guides/ns-tast-gd-013-annex-1-reference-paper.pdf] (Accessed June 25, 2020).

Office for Nuclear Regulation (2018b). *Nuclear Safety Technical Assessment Guide Annex: Seismic Hazards. NS-TAST-GD-013 Revision 1*. Bootle, UK: Office for Nuclear Regulation. 84 pp. [Online: Available at: http://www.onr.org.uk/operational/tech_asst_guides/ns-tast-gd-013.pdf] (Accessed June 25, 2020).

Oke, A.E. and Aigbavboa, C.O. (2017). *Sustainable Value Management for Construction Projects*. Cham, Switzerland: Springer Internatonal Publishing AG. 195 pp. https://doi.org/10.1007/978-3-319-54151-8.

Parry, S., Baynes, F.J., Culshaw, M.G. et al. (2014). Engineering geological models: an introduction: IAEG Commission 25. *Bulletin of Engineering Geology and the Environment* 73: 689–706. https://doi.org/10.1007/s10064-014-0576-x.

Site Investigation Steering Group (1993). *Site Investigation in Construction Part 1: Without Site Investigation Ground is a Hazard*. London, UK: Thomas Telford Ltd. 56 pp.

Van Staveren, M.T. (2006). *Uncertainty and Ground Conditions: A Risk Management Approach*. Oxford: Butterworth Heinemann. 332 pp.

Van Staveren, M.T. and Chapman, T. (2007). Complementing code requirements: managing ground risk in urban environments. In: *Proceedings of the XIV European Conference on Soil Mechanics and Geotechnical Engineering, Madrid*, Spain (eds. V. Cuéllar, E. Dapena, E. Alonso et al.). Rotterdam: Millpress Science Publishers.

Sullivan, T.D. (2010). The geological model. In: *Geologically Active: Proceedings of the 11th Congress of the International Association for Engineering Geology and the Environment, Auckland, New Zealand* (eds. A.L. Williams, G.M. Pinches, C.Y. Chin, et al.), 155–170. London, UK: CRC Press.

Taleb, N.N. (2010). *The Black Swan: The Impact of the Highly Improbable, 2nd edition*. New York, NY: Random House. 444 pp.

Terpstra, S., Van Capelleveen, E., Woltjes, J. et al. (2011). *Maatschappelijke Kosten Baten Analyse van de Basisregistratie Ondergrond (MKBA BRO), fase 1 (In Dutch: Societal Cost-Benefit Analysis for the Key Register of the Subsurface, Phase 1)*. The Hague, Netherlands: Ministry of Infrastructure and the Environment. 105 pp.

TNO (2018). *Data and Information Portal of TNO, Geological Survey of the Netherlands*. [Web site: www.dinoloket.nl] (Accessed November 2018).

Transport for London (2013). *TfL River Crossings - Ground Investigation Desk Study, Preliminary Sources Study Report*. London, UK: Transport for London. 91 pp. [Online: Available at: http://content.tfl.gov.uk/st-river-crossings-ground-investigation-desk-study-pssr.pdf] (Accessed June 25, 2020).

Van der Meulen, M.J., Doornenbal, J.C., Gunnink, J.L. et al. (2013). 3D geology in a 2D country: perspectives for geological surveying in the Netherlands. *Netherlands Journal of Geosciences* 92 (4): 217–241. https://doi.org/10.1017/S0016774600000184.

Wannick, H.P. (2006). *The Code of Practice for Risk Management of Tunnel Works: Future Tunnelling Insurance from the Insurers' Point of View*. PowerPoint Presentation, International Tunneling Association (ITA) Conference Seoul, South Korea. Singapore: Munich Reinsurance Company. 26 slides. [Online: Available at: https://tunnel.ita-aites.org/media/k2/attachments/public/OS_2006_5MunichRe.pdf] (Accessed June 25, 2020).

Washington State Department of Transportation (2014). *Project Risk Management Guide*. Olympia, WA: Washington State Department of Transportation. 122 pp. [Online: Available at: http://www.wsdot.wa.gov/publications/fulltext/cevp/projectriskmanagement.pdf] (Accessed August 2018).

WER (2018). *Data and Information Portal of the National Soil Survey, Wageningen Environmental Research*. [Web site: http://maps.bodemdata.nl] (Accessed November 2018).

Part II

Building and Managing Models

5

Overview and History of 3-D Modeling Approaches

Andrew J. Stumpf[1], Donald A. Keefer[1], and Alan Keith Turner[2,3]

[1] *Illinois State Geological Survey, University of Illinois at Urbana-Champaign, Champaign, IL 61820, USA*
[2] *Colorado School of Mines, Golden, CO 80401, USA*
[3] *British Geological Survey, Keyworth, Nottingham NG12 5GG, UK*

5.1   Introduction

This chapter focuses primarily on layer-based modeling of the subsurface, which aims at a geometrically consistent, three-dimensional representation of various types of stratigraphic surfaces. This is the archetypal geological modeling approach, undertaken to provide a framework or a starting point for a variety of purposes, including groundwater studies and any form of (pre)prospecting. Other, technically more sophisticated, modeling approaches discussed further in this book evolved from layer-based modeling.

Many software platforms for geological modeling were initially developed to support specific market sectors (such as petroleum, mining, or environment applications), and included paradigms, features and tools to address their modeling needs. Over time, the capabilities of the software were expanded to attract users from other markets, but the new focus led to building larger, more complex commercial software modeling platforms. Modeling specialists were needed to manage their data-handling and modeling processes. Because geoscientists should collect and interpret the data, these platforms require coordination between geoscientists and modeling specialists to ensure the models accurately depict geologic systems following the basic laws and principles of science (e.g. law of superposition), and thus support reasonable predictions.

This type of staffing, however, is not always affordable, or desirable, for many public and private geoscience organizations. In addition, licensing costs for many of these products are beyond the budgets of most shallow subsurface projects and many geological surveys. Modeling software selection should consider the needs and perspectives of the geologists who will be using it, as well as its ability to satisfy the desired workflows, project timelines and budgets, and user requirements. In addition, the selected modeling software should provide supplemental products to support the visualization of produced geological models and their distribution to user communities. These issues may be addressed using several methods. For example, some modeling software platforms offering greater flexibility and/or solutions are openly accessible, whereas other options are more restrictive and expensive (Chapters 6, 16, and 17).

For organizations who are embarking on 3-D modeling, the choice of modeling approach depends on the expertise and availability of all staff who will work on the project at hand, the availability and quality of the existing data, the ability to finance and obtain any new data, and specified requirements for data management, interpretation, and model product development (Chapter 6). Understanding the technical backgrounds of project personnel, their preferences for specific mapping methods, and their willingness and ability to integrate new software and workflows is important when selecting appropriate modeling software. For example, geologists primarily experienced in making cross-sections or hand-drawn structure contour maps may have a very difficult time moving to a modeling approach that does not rely on these formats.

As organizations move toward systematic geological modeling, the modeling approaches must be consolidated and formalized, along with the supporting data management, and staff must be chosen and trained accordingly. This is a prerequisite for the required consistency in space and over time when modeling is no longer a project, but an ongoing process.

Geological framework models define the subsurface distribution of assemblages of geological deposits – typically the distribution of units within some type of stratigraphic system. The models also contain structural or tectonic features appropriate to the user requirements and mapping objectives. As recently as 10 years ago, most software

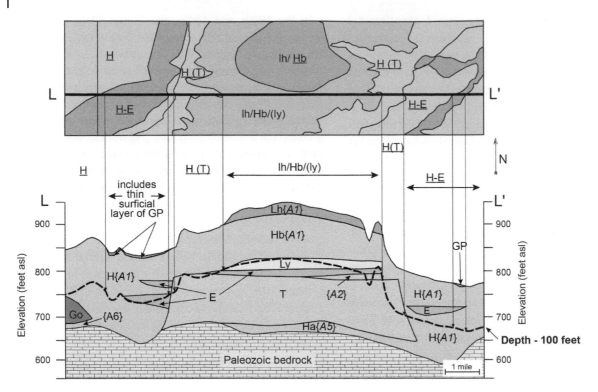

Figure 5.2 Cross-section L–L' of part of the McHenry USGS 7.5-minute quadrangle shown below part of a stack-unit map. The stack units are mapped to a depth of 100 ft, as delineated by the thick black dashed line. For this study, the stack-units were defined and labeled above the cross-section and then transposed onto the map along the lines of section. Vertical exaggeration is 24 times. Modified from Curry et al. 1997. Copyright © 1997 University of Illinois Board of Trustees. Used by permission of the Illinois State Geological Survey.

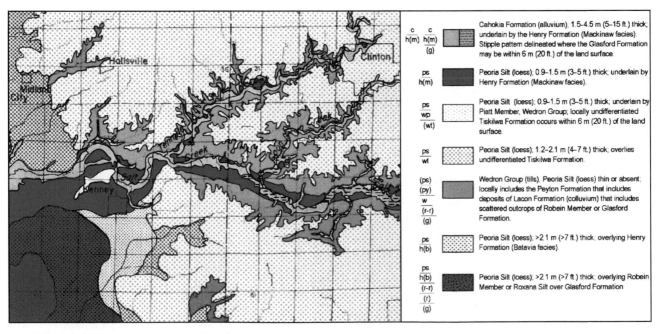

Figure 5.3 Example of a colored stack-unit map. Modified from Hunt and Kempton (1977). © 1977 University of Illinois Board of Trustees. Used by permission of the Illinois State Geological Survey.

5.2.2 Electrical Analog Models

Before high-speed digital computers became available, electrical analog models were used to demonstrate the complexity of groundwater flow systems (Robinove 1962). These models relied on the analogy between Ohm's law of electric current flow and Darcy's law of groundwater movement. Changes in electrical voltage were used to evaluate changes in groundwater heads. The models represented the geography and geology of the studied area with a network of resistors and capacitors with a suitable resolution. The resistors simulate the resistance to groundwater flow, whereas the capacitors represent the storage of water (Figure 5.4). However, because an electrical analog model had to be custom-designed to represent a specific groundwater system, each study required the design and construction of a separate electrical network.

In the mid-1960s, the Illinois State Water Survey (ISWS) created electrical analog models to forecast the consequences of developing nonhomogeneous aquifers in Illinois. These had irregular shapes and boundaries and a wide variety of head and discharge conditions. In contrast to the limited capabilities of early digital computers, analog models at the time offered the versatility needed to analyze complex groundwater flow systems. The models included

(a) Elemental Cube of Porous Media

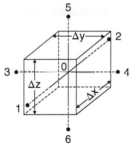

(b) Resistor-Capacitor Analog Node

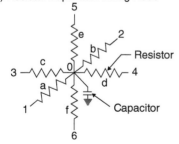

Figure 5.4 Basic concept of electric analog model: (A) elemental cube of porous material forming the aquifer; (B) analog capacitor and resistor network node consisting of six resistors (a–f) and one capacitor. Resistors e and f are proportional to hydraulic conductivity of element in the Δz direction, c and d in Δx direction, and a and b in Δy direction. The capacitor is proportional to the storage coefficient of the element.

relatively simple equipment that were operated at low to moderate costs. Therefore, analog models offered the best option for rapidly and accurately testing the relative merits of alternative schemes for developing groundwater resources (Walton 1964). Two analog models built during this time period are currently on display at the ISWS, located on the University of Illinois at Urbana-Champaign campus. One of these models was run to assess the groundwater flow conditions in the East St. Louis area (Schicht 1965). The other model is one of the first 3-D analog models ever built; it represents part of the Mahomet aquifer near the Cities of Champaign and Urbana, Illinois (Visocky and Schicht 1969). Specifications for the models are provided in Section 5.3.1 of this chapter, as part of the discussion of the early modeling efforts for the Mahomet aquifer.

Another early 3-D electric analog model was developed to study of the water resources of the San Luis Valley in south-central Colorado (Emery 1970, 1982). The model consisted of three layers to simulate the hydrologic boundaries, storage, transmissivity, and interconnections of unconfined and confined multiple aquifers in the valley. The confined aquifer was modeled as two (upper and lower) parts, each approximately 460 m (1510 ft) thick, to represent two major layers separated by a partially impermeable layer, which was represented by interconnections that allowed a low uniform vertical hydraulic conductivity. The Rio Grande and Conejos Rivers were simulated as having partial hydraulic connection with the unconfined aquifer and as constant-head sources. Emery (1970, 1982) provides additional details of this analog model. Initial tests of the model served as a check on the reasonableness of the model; the analog model computed water-level changes in the unconfined aquifer, and potentiometric head changes in the confined aquifer were the same order of magnitude as those observed in the field (Emery 1982).

5.2.3 The Adoption of Digital Mapping Technologies

Geological maps using computer-aided drafting techniques were first produced in London in the early 1970s, resulting in the experimental production of the 1:50 000 scale geologic map of Abingdon (Rhind 1971). Cratchley (1977) used digital methods to produce 3-D engineering geology presentations, principally by maps and sections at 1:15 000 scale, for a 450 km² (175 mi²) area of South Essex. The production of these maps revealed limitations of the existing digital technologies; these were largely resolved by the subsequent evolution of GIS. Broad commercial acceptance of GIS occurred in North America about 1988, with worldwide adoption of GIS technologies for geological mapping rapidly occurring over the next few years.

Applications of subsurface geological modeling have evolved over the past three decades. Initially, models were employed to better represent the 3-D subsurface geological "framework," but more recently the use of additional dimensions representing time, scale, or uncertainty have been proposed (see Chapter 1). This book has the title *Applied Multidimensional Geological Modeling* to include the full range of modeling applications. The evolution of subsurface geological modeling can be readily described in three phases, each lasting about a decade:

- **Phase 1:** In the decade 1985–1995, the primary emphasis was on the question "Can we do it?". In this period, several mostly academic research groups addressed fundamental research topics related to spatial data handling and the representation of geological features. These efforts were initially constrained by existing computer software and hardware limitations; the advent of the first modern computer workstations in the middle of this period encouraged these research efforts.
- **Phase 2:** In the decade 1995–2005, the first commercial 3-D geological modeling software products encouraged further experimentation with geological subsurface modeling by some geological surveys and some mineral and energy groups. They adopted some of the earlier research developments to determine "How do we do it?". Major advances in this period included the testing and establishment of appropriate workflows, interpolation algorithms, and databases that could support available software tools and computer hardware (Smith 2005). These led, in turn, to the rapid evolution of several competing, but increasingly mature and stable, software platforms.
- **Phase 3:** In the decade 2005–2015, and continuing to the present time, the primary emphasis became "Why are we doing it?". The ability to produce sophisticated geological subsurface models was now established as a fundamental capability of many geological surveys. Applications of subsurface geological models were now becoming accepted by various user communities. In some cases, the use of subsurface geological models showed immediate economic benefits.

Subsurface geological models are now being accepted as valuable resource-management and planning tools and, in a few cases, have become mandated as part of official regulation and management environments (Van der Meulen et al. 2013; Section 2.8 of Chapter 2). Thus, as demonstrated throughout this book, the subsurface geological modeling process has become increasingly driven by the demands and expectations of diverse user communities.

5.2.4 Evolution of 3-D Mapping and Modeling Collaborative Forums

In the United States, since the early 1990s, collaboration has been fostered between the Association of American State Geologists (AASG) and the U.S. Geological Survey (USGS) through the federally mandated National Cooperative Geologic Mapping Program (NCGMP). The NCGMP mandates a national emphasis on improved geological mapping through collaborative development of cartographic standards, stratigraphic naming standards, and the compilation of a National Geologic Map Database. Collaboration in evolving the practices for making geologic maps has been through formal discussions and exchanges of technical information and expertise at an annual series of Digital Mapping Techniques (DMT) workshops. The NCGMP, through the DMT forums, led to the creation of national digital cartographic standards for geological maps, in the form of proposed standards. The USGS has extensive web pages providing direct access to the national map catalog, mapping standards, DMT proceedings, and other related resources (http://ngmdb.usgs.gov).

In 1997, the state geological surveys of Illinois, Indiana, Michigan, and Ohio joined with the USGS to create the Central Great Lakes Geologic Mapping Coalition. The coalition was formed to address the need for more detailed geological information in the glaciated midcontinent, where glaciers have deposited a thick cover containing complex sedimentary sequences. From the beginning, the coalition combined their expertise and resources to develop 3-D geological mapping (Berg et al. 2016; Section 2.9 of Chapter 2). In 2008, the original five members expanded to include the state geological surveys of Minnesota, Wisconsin, Pennsylvania, and New York, completing all American constituents of the Great Lakes drainage. The coalition was renamed the Great Lakes Geologic Mapping Coalition (GLGMC). The Ontario Geological Survey (OGS) joined as a non-funded partner in 2012. The GLGMC holds annual two-day meetings to discuss geological applications, new technologies, and advances in subsurface mapping, data dissemination, and data management (http://greatlakesgeology.org).

On a biennial schedule, since 2001, North American and European geologists have attended a series of special workshops addressing 3-D geological mapping and modeling developments held at Geological Society of America annual meetings, and other venues. Berg et al. (2011) has provided a detailed synopsis of accomplishments in the first decade. The ISGS maintains an online resource for all of the presentations at these workshops (https://www.isgs.illinois.edu/three-dimensional-geological-mapping). In 2011, a group of European participants decided to start a European workshop series to help coordinate and exchange information among the geological surveys of Europe. Meetings have subsequently been held at Utrecht in 2013, Edinburgh in 2014, Wiesbaden in 2016, Oréans in 2018 and Bern in 2019. A website makes accessible the presentations, abstracts, and some images from those meetings (http://www.3dgeology.org).

5.3 The Mahomet Aquifer: An Example of Evolving Subsurface Modeling

On multiple occasions, beginning about 1.2 million years ago, glaciers covered much of the Great Lakes region (Larson and Schaetzl 2001). Glacial deposits now blanket a preglacial landscape formed on Mesozoic and Paleozoic sedimentary and older crystalline rocks. The preglacial Teays Bedrock Valley System was first identified in West Virginia and Kentucky (Tight 1903), but was subsequently traced across Ohio and Indiana in the subsurface (Ver Steeg 1946; Norris and Spicer 1958). A similar bedrock valley in the subsurface, named the Mahomet Bedrock Valley (MBV), was identified crossing east-central Illinois (Horberg 1945). The preglacial river occupying the MBV is considered to form part of the Teays river system (Figure 5.5),

the combination being named the Teays-Mahomet river system (Teller and Goldthwait 1991). Recent investigations in east-central Illinois by Stumpf and Dey (2012) uncovered fossil mollusk shells in preglacial alluvial deposits filling a channel in the deepest part of the MBV. They are equivalent to fossil shells of clams and snails found further west in the MBV by Miller et al. (1992) and are a similar species to those in the modern Mississippi and Ohio Rivers.

In east-central Illinois, the MBV is partially filled with deposits of glacial sand and gravel, which form the Mahomet aquifer, an important groundwater resource for 15 counties (Figure 5.6). The tributary Mackinaw Bedrock Valley, which joins the MBV along the Middle Illinois Bedrock Valley, also contains sediments composing a major glacial sand and gravel aquifer, the Sankoty aquifer (Herzog et al. 2003). The Mahomet aquifer in the lower part of the MBV has a width of 6–22 km (4–14 miles) and a thickness of 15–60 m (50–200 ft).

Since the mid-twentieth century, significant research has been undertaken to determine the formation, extent, and characteristics of the MBV and the associated Mahomet aquifer (Horberg 1945; Walton and Prickett 1963; Walton 1964; Visocky and Schicht 1969; Kempton et al. 1982, 1991; Wilson et al. 1994, 1998; Herzog et al. 1995, 2003; Soller et al. 1999; Atkinson 2011; Atkinson et al. 2014;

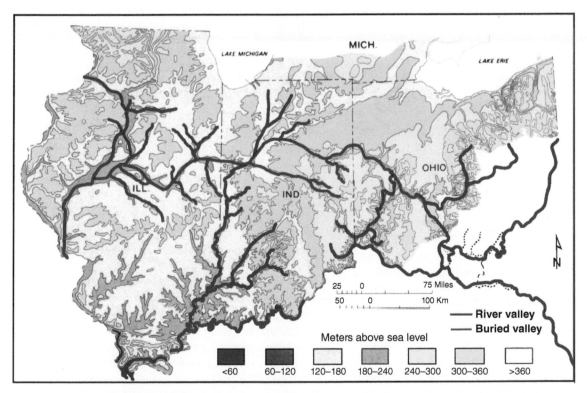

Figure 5.5 Interpreted thalweg of the Teays-Mahomet river system from Brown et al. (2018). © 2018 University of Illinois Board of Trustees; used by permission of the Illinois State Geological Survey. Base map from Melhorn and Kempton (1991). © 1991 Geological Society of America. Used by permission.

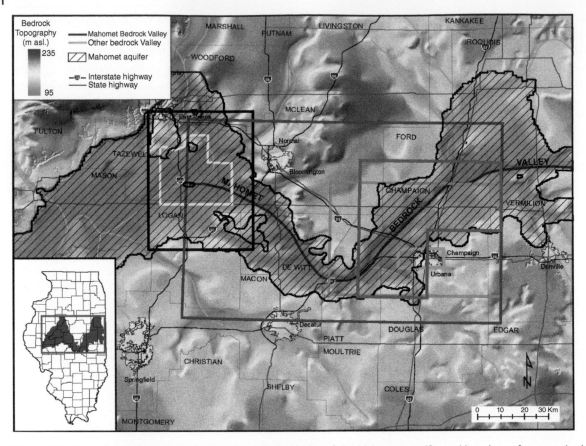

Figure 5.6 Map of east-central Illinois showing the boundaries of the Mahomet aquifer and locations of past geologic and hydrogeologic studies. The two green boxes outline early analog and digital models of the Champaign-Urbana area (Source: Walton and Prickett 1963; Walton 1964; Visocky and Schicht 1969). The red box outlines the area modeled by a joint USGS-ISGS regional geologic study (Soller et al. 1999). The boundaries of the 3-D aquifer and MODFLOW models of Herzog et al. (1995), Wilson et al. (1998), and Herzog et al. (2003) are delineated by the black and yellow boxes, respectively. The blue box outlines the area where the most recent geologic mapping and modeling was undertaken by Stumpf and Dey (2012), and Stumpf and Atkinson (2015). Aquifer boundaries provided courtesy of the Illinois State Water Survey. © 2018 University of Illinois Board of Trustees. Used by permission of the Illinois State Geological Survey.

Stumpf and Dey 2012; Stumpf and Atkinson 2015). These periodic re-evaluations were undertaken to study drought events and applications for increasing groundwater withdrawals from the Mahomet aquifer. Even after all the study, uncertainties remain about the MBV that will necessitate the collection of additional data in the future. The MBV is a large and complex geologic system, and the data presently available indicate large spatial variability in geological and aquifer properties, requiring additional exploration for adequate characterization.

In east-central Illinois, the 150 m (500 ft) bedrock elevation contour approximates the top of MBV and tributary bedrock valleys (Kempton et al. 1991), which have traditionally been used to delineate the extent of the Mahomet aquifer. However, more recent studies (e.g. Stumpf and Dey 2012) suggest the Mahomet aquifer extends beyond these bedrock valleys, bounded by the 195 m (650 ft) contour,

and include deposits of glacial sand and gravel previously assigned to the lower Glasford aquifer.

At its deepest, the MBV is filled with more than 120 m (390 ft) of glacial sediment, assigned to three major stratigraphic units: the Early to Middle Pleistocene (1 200 000–430 000 years BP) Banner Formation, the Middle Pleistocene (200 000–130 000 BP) Pearl and Glasford Formations, and the Late Pleistocene to Holocene (55 000–12 000 BP) Wedron Group (Figure 5.7). Significant hiatuses occurred between these time periods. Considerable sand and gravel layers at the base or top of some lithostratigraphic units were deposited by sediment-laden meltwater draining from advancing or retreating ice margins. Clayey tills and lake sediment form the aquitard units, separating the Mahomet and Sankoty aquifers from overlying aquifers. Stumpf and Dey (2012) and Atkinson et al. (2014) provided details of the geological units within the MBV.

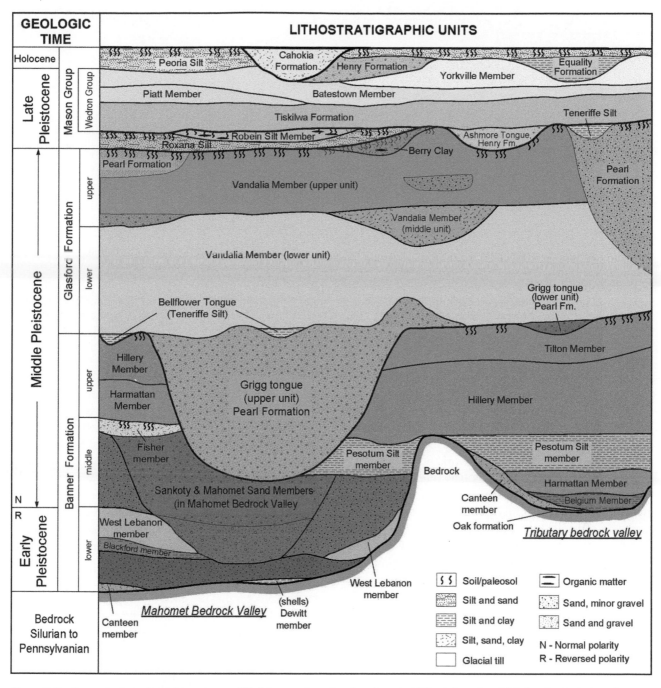

Figure 5.7 Conceptual stratigraphic column of Pleistocene deposits of east-central Illinois based on high-level unit designations from Kempton and Visocky (1992). Column was modified to reflect detailed geological models by Stumpf and Dey (2012), Soller et al. (1999), Herzog et al. (2003), and Atkinson et al. (2014). Source: Stumpf 2018. © 2018 University of Illinois Board of Trustees. Used by permission of the Illinois State Geological Survey.

5.3.1 Early Modeling Efforts

In the early 1960s, ISWS groundwater engineers constructed electric analog models for the Champaign-Urbana area (Figure 5.6) that were used to predict the consequences of further groundwater development, the practical sustained yield of existing wells and pumping centers, and the potential yield of the multiple-aquifer system (Walton and Prickett 1963; Visocky and Schicht 1969). The models were developed after the analog model specifications of Skibitzke (1960). One of these models still exists as an historical display at the ISWS (Figure 5.8). The model was designed to a scale of 1:62 500 (1 in. = 1 mile) and was

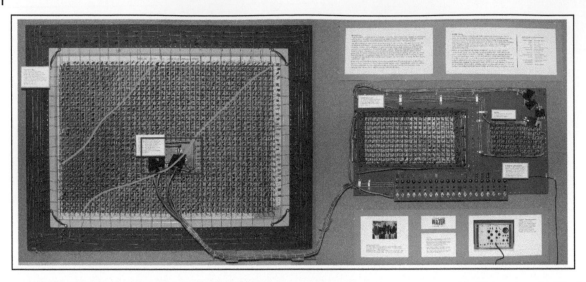

Figure 5.8 Analog model of the Mahomet aquifer in the Champaign-Urbana area as currently displayed at the Illinois State Water Survey. Both the regional model (area bounded by the green box in Figure 5.6) and the later more detailed insert model are displayed with their electrical interconnection. The area where Walton (1964) and Walton and Prickett (1963) undertook a detailed study is outlined by the red boxes. The extent of the Mahomet aquifer is delineated by the light blue lines. © 2018 University of Illinois Board of Trustees. Used by permission of the Illinois State Geological Survey. Photograph by Michael Knapp.

mounted on a board with perforated holes in a 2.5 cm (1 in.) square pattern; thus, the model represented the aquifer with a 1-mile resolution.

The two main aquifers of the multiple-aquifer system were represented with two horizontal arrays of resistors and capacitors. A horizontal network of ground wires represented the water table; this was interconnected by two vertical arrays of resistors to represent leakage through the lower and upper confining beds. Initial analog model results demonstrated that this model grid was too coarse to define conditions accurately in the well fields west of Champaign, so the model was modified and interconnected with a second analog model with a finer grid spacing that represented a smaller area west of, and including, the Champaign-Urbana area. This modified model included electrical connections between the fine mesh of the inset area and the coarse mesh of the surrounding area – an early example of nested models!

5.3.2 Initial 3-D Geological and Hydrogeological Evaluations

In the 1970s and 1980s, several of the stack-unit maps produced by the ISGS, and described in Section 5.2.1, provided 3-D information to Mahomet aquifer investigations (Gross 1970; Hunt and Kempton 1977; Berg and Kempton 1988). A drought in 1988–1989 brought renewed interest in the Mahomet aquifer that led to additional studies by the ISGS, which used GIS technologies to help model development and visualization (Kempton and Visocky 1992;

Berg and Abert 1994; Larson et al. 1997; McLean et al. 1997). These studies provided an extensive information base for a cooperative mapping project involving the ISGS and the USGS (Soller et al. 1999), which evaluated an important section of the MBV (delineated by the red box in Figure 5.6). A stratigraphic database with 177 boreholes (selected for their high quality) provided key stratigraphic control (Soller et al. 1999). These data were supplemented by several thousand additional borehole records, and existing maps and cross-sections to model the top surfaces of five primary stratigraphic units, two minor sand bodies, and the bedrock surface (Figure 5.9).

Each surface was developed iteratively, largely by a manual process that included plotting stratigraphic control data, followed by hand contouring a map of the stratigraphic unit based on an understanding of its regional distribution and geologic origin. This process reflected the limited editing and contouring abilities of the GIS technologies at that time. Once a suitable hand-contoured map of a single surface was approved, it was scanned, and a digital vector map of the contours was developed using ArcGIS. The ArcGIS Topogrid algorithm was used to convert the contour maps into a 100 × 100 m (330 × 330 ft) raster grid. Contouring each surface involved several iterations of editing and revision to ensure they represented correct spatial and stratigraphic relationships (Soller et al. 1999). Once all eight surfaces were constructed, they were converted to a single 3-D model within the EarthVision software (Dynamic Graphics 1995).

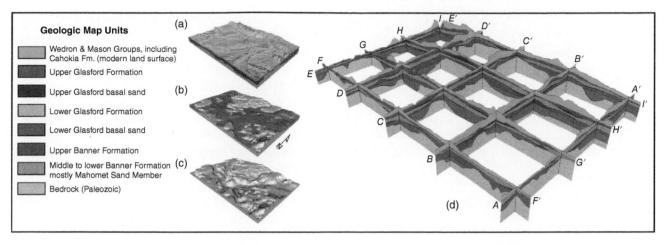

Figure 5.9 The 3-D EarthVision model by Soller et al. (1999): (a) land surface; (b) lower Banner Formation (Mahomet Sand Member); (c) bedrock; (d) isometric view of cross-sections. Modified from Soller et al. 1999. Source of figures: U.S. Geological Survey.

A principal objective of this study was to demonstrate the value of digital 3-D mapping technologies for visualizing in detail the subsurface geology and supporting multiple societal applications, such as groundwater modeling (Soller et al. 1998, 1999). A website that contains animations of the modeled surfaces remains active (http://pubs.usgs.gov/imap/i-2669). The animations permit limited inspection of the 3-D model, interior and static versions are provided of all model products along with additional maps and geological and modeling procedure information.

5.3.3 Recent Geological and Hydrogeological Models

In 1993, following the drought of 1988–1989, the ISGS and ISWS undertook a cooperative study of the sand-and-gravel aquifers in southwest McLean and southeast Tazewell Counties, shown as black and yellow outlines in Figure 5.6. New field investigations expanded the previous studies (Kempton and Visocky 1992; Wilson et al. 1994). The cooperative study involved two major tasks: (i) hydrogeologic characterization of the glacial deposits (Herzog et al. 1995) and (ii) development of a computer-based mathematical model of the groundwater flow (Wilson et al. 1998). Herzog et al. (2003) summarized the results of both tasks. Wilson et al. (1998) provided a detailed description of how the geological subsurface characterization was simplified to produce a hydrogeological characterization and then discretized into regular grid cells to permit groundwater flow modeling (Figure 5.10a,b). The resulting hydrogeological model consisted of eight units; three of these were aquifers. Figure 5.10c is a 3-D representation of this model created by using EarthVision (Dynamic Graphics 1995).

The use of EarthVision in the modeling further automated the process and provided the capability to define non-conformities, where units pinched out. Although this process produced a model with units in the correct stratigraphic sequence, identification and removal of inconsistencies required several iterations – up to 12 revisions of each aquifer layer map and five versions of the 3-D model (Wilson et al. 1998). This 3-D hydrostratigraphic model was converted to a MODFLOW groundwater flow model by using the graphical interface provided in Visual MODFLOW; software developed by the Canadian company Waterloo Hydrogeologic.

The ISWS later constructed a larger regional numerical groundwater flow model of the entire Mahomet aquifer (Roadcap et al. 2011). The geological model of Wilson et al. (1998) was expanded by including data from several additional geologic and hydrogeologic studies (Stumpf and Dey 2012). The study area is delineated by the blue box in Figure 5.6. This MODFLOW model contained 275 rows and 603 columns with a spacing of 402 m (1320 ft) and was developed using several pre- and post-processing software packages. The model contains seven active layers and an inactive layer representing bedrock (Figure 5.11); this is the same structure as used by Wilson et al. (1998). This latest model includes a refined conceptual geological model of east-central Illinois based on new field research (Stumpf and Dey 2012), which determined that erosion by ice or water partially removed the upper Banner tills (lower aquitard) over much of the MBV, creating hydrologic connections between the Mahomet aquifer and overlying lower and upper Glasford aquifers.

Subsequently, Stumpf and Dey (2012) determined that designating the Glasford deglacial unit as an aquifer or aquitard is conceptually misleading, and the concept of a

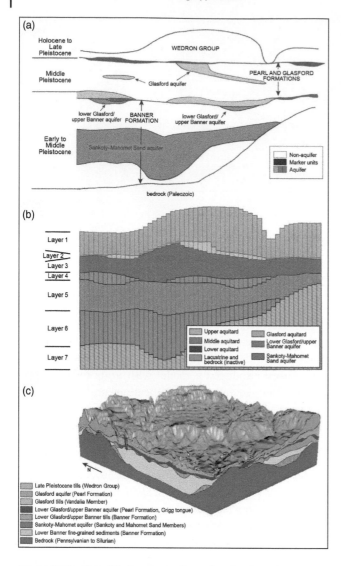

Figure 5.10 Simplified cross-sections showing how the geological and hydrogeologic conditions were simplified to develop the hydrostratigraphy represented in the groundwater flow model (Wilson et al. 1998): (a) derived simplified hydrogeological cross-section. Modified from Wilson et al. 1998; (b) cellular representation within the groundwater flow model. After Wilson et al. 1998; (c) view of 3-D hydrostratigraphic model with a vertical exaggeration of 100 times. Panels A and B modified from Wilson et al. 1998. © 1998 University of Illinois Board of Trustees; used by permission of the Illinois State Water Survey and Illinois State Geological Survey. Panel C modified from Wilson et al. 1998, and Herzog et al. 2003. © 2003 University of Illinois Board of Trustees; used by permission of the Illinois State Water Survey and Illinois State Geological Survey.

"hybrid hydrostratigraphic unit" was introduced to better represent its internal complexity (Atkinson 2011; Atkinson et al. 2014). As a result, this latest model improved the predictive accuracy of groundwater modeling that was performed by Roadcap et al. (2011).

5.4 Digital 3-D Geological Modeling Approaches Discussed in This Book

Chapters 9–12 describe four principal approaches to creating geological framework models: (i) models created from stacked stratigraphic surfaces (Chapter 9); (ii) models created using digital borehole records and interpreted geological cross-sections (Chapter 10); (iii) models created as 3-D cellular voxel arrays (Chapter 11); and (iv) integrated rule-based (implicit) geological models (Chapter 12). Each approach has advantages and limitations, and most of them are supported by multiple software products.

Although these distinctions in modeling approach are important, this is a simplified perspective of the technology. Software applications have evolved to accommodate multiple modeling methods. For instance, several projects have used combinations of the stacked-surface and cross-section approaches, whereas other applications combine stacked surfaces with 3-D voxel modeling. The increasing desire to constrain models with different kinds of geological knowledge has resulted in an increased interest in implicit modeling approaches.

5.4.1 Stacked-Surface Approach to Model Creation

The stacked-surface approach to 3-D geological modeling is probably the most common and is identical to manual methods used to make structure and isopach contour maps for petroleum reservoirs or buried aquifers; Chapter 9 discusses this in some detail. Stacked-surface approaches model the distribution of units within a model domain by defining only the horizon surfaces; these are sometimes described as "flying carpets." In this approach, the top and/or bottom surfaces of individual map units are identified from field or laboratory observations and correlated to units in the geological framework. The modeling software then applies an interpolation algorithm to these correlated map-unit picks to generate the desired surface. This procedure is repeated for all the surfaces in the model. Some software platforms use only unit tops or bases, so the exact procedure is software specific. The space between top and bottom surfaces is not attributed within the computer, so these models result in much smaller files, allowing large models to be managed with fewer computer resources. This can be advantageous, depending on the size of the model domain and the capacity of the computers used for modeling. Third-party software packages, such as MAPublisher, a suite of plug-in tools for Adobe Illustrator (Avenza Systems Inc.), can be used to help make the stacked-surface modeling approach more efficient (Chapter 9).

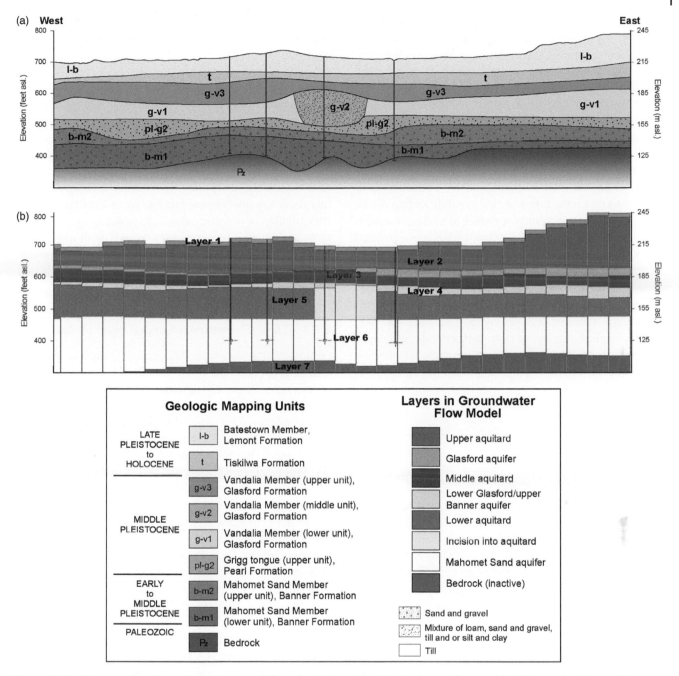

Figure 5.11 Cross-sections from the most recent 3-D geological model and groundwater flow model of Champaign County. The cross-sections were extracted along the west to east transect along County Road 1800N. Modified from Stumpf and Dey (2012). © 2018 University of Illinois Board of Trustees. Used by permission of the Illinois State Geological Survey.

Common interpolation algorithms used by for stacked-surface modeling include inverse distance squared, kriging, minimum tension spline, or radial basis functions (Chapter 13). These algorithms function by estimating the value of the property (i.e. top-surface elevation) at regular points on a 2-D plane across the model domain, resulting in a gridded file structure. This is a common capability of GIS software, which typically manages these grids as raster file types. Each algorithm provides a different function for weighting the importance of neighboring data when estimating the value at each grid node, and most applications allow some user input to determine the size and shape of the neighborhoods. Thus, while the procedure depends on traditional surface interpolation algorithms, these do not have automatically applied constraints to account for geomorphological or sedimentary patterns.

Typically, each surface is independently generated, without any reference to neighboring surfaces. Thus, these unconstrained surfaces are likely to need correction to account for geological constraints, such as allowing specific abrupt stratigraphic changes or ensuring the preservation of morphologic patterns in areas of low data density. Difficulties are most likely to arise when units are locally discontinuous; in areas of low data density, it is impossible to predict map discontinuities, so bounding surfaces can intersect in unintended ways. More sophisticated geological modeling packages have added tools or functions to the basic interpolation capabilities that can be used to provide constraints for stacked-surface methods for accommodating geologic onlap, offlap, and erosional contact geometries, faulting, folding, and discontinuous units. Alternatively, geologists can implement these constraints manually by using common grid or grid-to-grid calculations; these are typically supplied in toolbox options in common GIS and 3-D geological modeling software. The Geological Survey of the Netherlands developed a sophisticated stacked-surface approach to develop their national Digital Geological Model (Gunnink et al. 2013).

5.4.2 Modeling Based on Cross-Sections and Boreholes

A second commonly applied method of creating geological framework models, described in Chapter 10, closely replicates the traditional geological field mapping process. Geologists manually interpret sets of stratigraphic surfaces directly from manually constructed cross-sections. Borehole data are ingested into the modeling software and transects of boreholes are identified and graphically drawn on a 2-D window. Then the modeler typically uses a mouse to digitize the top and/or base surfaces of each stratigraphic unit on the cross-section. The points from these digitized lines, sometimes with borehole-based picks and at other times with additional geophysical data, are then used with an interpolation algorithm to create the stacked surface.

One advantage of this approach is that it allows geologists to draw unit boundaries that reflect the geological characteristics (e.g. specific surface morphologies and geometric intersections) of the inferred geologic systems; thus, the models explicitly reflect the geological knowledge of the geologist and include more realistic geological patterns than models produced by interpolation algorithms. Cross-sections are traditional mapping tools of geologists; this approach is easier for some geologists to use than other methods. In fact, the cross-section-based GSI3D and Groundhog modeling tools developed for/by the British Geological Survey show that the approach is easy to adopt

for senior staff members, who would otherwise probably not have taken up model building. One limitation of this approach is that highly variable geological environments or situations with significant numbers of knowledge-based surface constraints may require a very large number of cross-sections to maintain proper surface morphologies between cross-sections; this can translate into considerable time, effort, and expense. In addition, as it is a deterministic way of modeling, the results depend heavily on the experience and skills of the modeler, which makes the result less straightforwardly reproducible than models that rely more heavily on geostatistics. Some applications use triangulation between cross-section nodes to form the surfaces, rather than uniform grids defined with algorithms that introduce some degree of curvature. Correcting unintended surface intersections and morphologies between sets of triangulated surfaces can be particularly difficult.

5.4.3 Three-Dimensional Gridded Voxel Models

Geological framework models typically map the distribution of major geologic units, which represent assemblages of rock types. These distributions are defined by sets of top and bottom surfaces that bound each unit. Given that most data sets for geological framework modeling are primarily composed of borehole data, these models are not usually well suited to creation by direct 3-D interpolation. However, 3-D interpolation methods are well suited to the creation of geological property models. Most applications of these property models have been applied to address issues arising in the petroleum and mining sectors. These methods also are being used for parameterizing groundwater flow or contaminant transport models (Stafleu et al. 2011; Bianchi et al. 2015) or for building cubes of geophysical properties for petroleum exploration. Chapter 11 describes the development and applications of voxel-based geological models.

In recent years, some advanced software packages have begun employing direct full-volume approaches to modeling rock-type, or facies, distributions. Rock-type models are typically created as a first step in modeling geological property distributions. Rock-type modeling largely relies one of three different 3-D modeling methods, *transition probability* or *Markov chain approach* (Carle and Fogg 1996; Carle et al. 1998), *plurigaussian simulation* (Armstrong et al. 2003) and *multiple-point statistics* (MPS) (Mariethoz and Caers 2014). All three methods have been developed within the geostatistical research community to improve geological property models by incorporating knowledge of the spatial patterns of rock types and have become increasingly popular over the past decade (Phelps and Boucher 2009; Dunkle et al. 2011; Cronkite-Ratcliff et al. 2012). These methods appear to be well-suited to some geological modeling applications; however, they require considerable expertise.

The transition probability method uses transition probabilities, length scale, and proportion information, often gleaned from a map or training image, to define the constraints for spatial relationships between rock types. It then uses indicator semivariograms within a kriging-based interpolation. This method is typically used within a Markov chain/Monte Carlo simulation approach to generate realizations with realistic levels of heterogeneity.

Plurigaussian simulation relies on transitional or adjacency relationships, defined by rock-type rules (captured by geological knowledge that could include training images), and then applies a mixture of Gaussian distributions to create stochastic rock-type simulations, constrained by data, with these relationships. MPS uses 2-D or 3-D training images of the rock types to create large transition probability matrices that are used to guide the simulation of the rock types.

All these methods are typically used to create stochastic simulations (Chapter 13), in which each model (called a realization) is supposed to have an equally probable chance of reflecting the actual geologic system. These methods work for categorical units (e.g. rock type) and effectively assume that each unit can be found vertically and horizontally throughout the model domain – although each rock type may be assigned a very low probability of occurring in some locations. Accordingly, these methods are not well suited to stratigraphic systems, where at any location each unit can only appear once and only within a specific prescribed order.

5.4.4 Integrated Rule-Based (Implicit) Geological Models

Chapter 12 provides additional information on an approach to the creation of geological models that is currently gaining in popularity. Known as "implicit modeling," software products that employ this approach are more advanced and complex to use; they use a combination of radial basis functions for interpolation, with additional geologically-based constraints that are programmed to automatically control the geological plausibility of the resulting models. Because radial basis functions may also be used in stacked-surface modeling approaches, as well as in some more advanced applications, the distinctions between implicit and stacked-surface approaches to model creation are not always clear-cut. It is often difficult to tell, especially when more advanced modeling platforms hide many of the technicalities of the algorithm choices from the user. Although implicit modeling can be useful in any geological setting, it is especially useful for modeling problems and domains that have extreme tectonic deformation, particularly metamorphosed terranes.

References

Armstrong, M., Galli, A.G., Le Loc'h, G. et al. (2003). *Plurigaussian Simulations in Geosciences*. Berlin, GE: Springer, 149 pp. https://doi.org/10.1007/978-3-662-12718-6.

Atkinson, L.A. (2011). *Subsurface Analysis of Late Illinoian Deglacial Sediments in East-Central Illinois, United States, and Its Implications for Hydrostratigraphy*. MSc thesis University of Waterloo, Canada. [Online: Available at: http://www.collectionscanada.gc.ca/obj/thesescanada/vol2/OWTU/TC-OWTU-6458.pdf] (Accessed March 2017).

Atkinson, L.A., Ross, M. and Stumpf, A.J. (2014). Three-dimensional hydrofacies assemblages in ice-contact/proximal sediments forming a heterogeneous 'hybrid' hydrostratigraphic unit in Central Illinois, USA. *Hydrogeology Journal* 22 (7): 1605–1624. https://doi.org/10.1007/s10040-014-1156-7.

Berg, R.C. and Abert, C.C. (1994). Large-scale aquifer sensitivity model. *Environmental Geology* 24 (1): 34–42. https://doi.org/10.1007/BF00768075.

Berg, R.C. and Greenpool, M.R. (1993). *Stack-Unit Geologic Mapping: Color-Coded and Computer-Based Methodology*. Champaign, IL: Illinois State Geological Survey, Circular 552. 11 pp. [Online: Available at: http://hdl.handle.net/2142/42833] (Accessed August 2018).

Berg, R.C. and Kempton, J.P. (1988). *Stack-Unit Mapping of Geologic Materials in Illinois to a Depth of 15 Meters*. Champaign, IL: Illinois State Geological Survey, Circular 542. 23 pp. [Online: Available at: http://hdl.handle.net/2142/96249] (Accessed August 2018).

Berg, R.C., Mathers, S.J., Kessler, H. and Keefer, D.A. [Eds.] (2011). *Synopsis of Current Three-Dimensional Geological Mapping and Modeling in Geological Survey Organizations*. Champaign, IL: Illinois State Geological Survey, Circular 578. 92 pp. [Online: Available at: http://hdl.handle.net/2142/43535] (Accessed August 2018).

Berg, R.C., Brown, S.E., Thomason, J.F. et al. (2016). A multiagency and multijurisdictional approach to mapping the glacial deposits of the Great Lakes region in three dimensions. In: *Geoscience for the Public Good and Global*

Development: Toward a Sustainable Future (eds. G.R. Wessel and J.K. Greenberg), 415–447. Boulder, CO: Geological Society of America, Special Paper 520. https:// doi.org/10.1130/2016.2520(37).

Bianchi, M., Kearsey, T. and Kingdon, A. (2015). Integrating deterministic lithostratigraphic models in stochastic realizations of subsurface heterogeneity: impact on predictions of lithology, hydraulic heads and groundwater fluxes. *Journal of Hydrology* 531: 557–573. https://doi.org/ 10.1016/j.jhydrol.2015.10.072.

Brown, S.E., Thomason, J.F. and Mwakanyamale, K.E. (2018). *The Future of Science of the Mahomet Aquifer*. Champaign, IL: Illinois State Geological Survey, Circular 594. 25 pp. [Online: Available at: http://hdl.handle.net/2142/99079] (Accessed March 2017).

Carle, S.F. and Fogg, G.E. (1996). Transition probability-based indicator geostatistics. *Mathematical Geology* 28 (4): 453–476. https://doi.org/10.1007/BF02083656.

Carle, S.F., LaBolle, E.M., Weissmann, G.S. et al. (1998). Conditional simulation of hydrofacies architecture: a transition probability/Markov approach. In: *Hydrogeologic Models of Sedimentary Aquifers* (eds. G.F. Fraser and J.M. Davis), 147–170. Tulsa, OK: Society for Sedimentary Geology SEPM, Concepts in Hydrogeology and Environmental Geology, vol. 1. https://doi.org/10.2110/ sepmcheg.01.

Cratchley, C.R. (1977). Engineering geology of South Essex. In: *Surface Modelling by Computer, Institution of Civil Engineers Paper 6*, 43–49 (and Discussion pp. 92-99). London: Institution of Civil Engineers.

Cronkite-Ratcliff, C., Phelps, G.A. and Boucher, A. (2012). *A Multiple-Point Geostatistical Method for Characterizing Uncertainty of Subsurface Alluvial Units and Its Effects on Flow and Transport*. Reston, VA: U.S. Geological Survey, Open-File Report 2012-1065. 24 pp. [Online: Available at: http://pubs.usgs.gov/of/2012/1065] (Accessed May 2017).

Culshaw, M.G. (2005). From concept towards reality: developing the attributed 3D geological model of the shallow subsurface (The Seventh Glossop Lecture). *Quarterly Journal of Engineering Geology and Hydrogeology* 38: 231–284. https://doi.org/10.1144/1470-9236/04-072.

Culshaw, M.G. and Price, S.J. (2011). The 2010 Hans Cloos lecture: the contribution of urban geology to the development, regeneration and conservation of cities. *Bulletin of Engineering Geology and the Environment* 70 (3): 333–376. https://doi.org/10.1007/ s10064-011-0377-4.

Curry, B.B., Berg, R.C. and Vaiden, R.C. (1997). *Geologic mapping for environmental planning, McHenry, County, Illinois*. Champaign, IL: Illinois State Geological Survey,

Circular 559. 79 pp. [Online: Available at: http://hdl.handle .net/2142/43551] (Accessed June 2020).

Dearman, W.R. (1991). Czechoslovakian stripe method and parallel developments. In: *Engineering Geological Mapping*, 68–89. Oxford: Butterworth-Heinemann. https://doi.org/10 .1016/b978-0-7506-1010-0.50009-7.

Dunkle, K.M., Anderson, M.P. and Hart, D.J. (2011). Multiple-point geostatistics for creation of 3D hydrostratigraphic models, Outagamie County, WI. In: *Three-Dimensional Geological Mapping: Workshop Extended Abstracts* (eds. H.A.J. Russell, R.C. Berg and L.H. Thorleifson), 23–27. Ottawa, ON: Geological Survey of Canada, Open File Report 6998. https://doi.org/10.4095/ 289609.

Dynamic Graphics (1995). *EarthVision User's Guide*. Alameda, CA: Dynamic Graphics Inc.

Emery, P.A. (1970). *Electric Analog Model Evaluation of a Water-Salvage Plan, San Luis Valley, South-Central Colorado*. Denver, CO: Colorado Water Conservation Board, Ground-Water Circular 14. 11 pp.

Emery, P.A. (1982). A model analysis of ground water in the San Luis Valley, Colorado, USA. In: *Ground-Water Models: Volume I: Concepts, Problems, and Methods of Analysis with Examples of their Application, International Hydrological Programme, Working Group 8.1, Case History 3* (ed. J.D. Bredehoeft), 53–66. Paris, France: UNESCO. [Online: Available at: https://unesdoc.unesco.org/ark:/48223/ pf0000048909] (Accessed June 2020).

Galster, R.W. (1977). A system of engineering geology mapping symbols. *Bulletin of the Association of Engineering Geologists* 14 (1): 39–47. https://doi.org/10.2113/gseegeosci .xiv.1.39.

Gross, D.L. (1970). *Geology for Planning in De Kalb County, Illinois*. Champaign, IL: Illinois State Geological Survey, Environmental Geology Notes 33. 26 pp. [Online: Available at: http://hdl.handle.net/2142/91726] (Accessed April 2017).

Gunnink, J.L., Maljers, D., Van Gessel, S.F. and Menkovic, A. (2013). Digital geological model (DGM): a 3D raster model of the subsurface of the Netherlands. *Netherlands Journal of Geosciences* 92: 33–46. https://doi.org/10.1017/ S0016774600000263.

Herzog, B.L., Wilson, S.D., Larson, D.R. et al. (1995). *Hydrogeology and Groundwater Availability in Southwest McLean and Southeast Tazewell Counties, Part 1: Aquifer Characterization*. Champaign IL: Illinois State Geological Survey and Illinois State Water Survey Cooperative, Groundwater Report 17. 70 pp. [Online: Available at: http://hdl.handle.net/2142/35244] (Accessed March 2017).

Herzog, B.L., Larson, D.R., Abert, C.C. et al. (2003). Hydrostratigraphic modeling of a complex glacial-drift aquifer system for importation into MODFLOW. *Ground*

Water 41 (1): 57–65. https://doi.org/10.1111/j.1745-6584
.2003.tb02568.x.

Horberg, L. (1945). A major buried valley in east-central
Illinois and its regional relationships. *Journal of Geology*
53: 349–359. https://doi.org/10.1086/625294.

Hunt, C.S. and Kempton, J.P. (1977). *Geology for Planning in
De Witt County, Illinois.* Champaign, IL: Illinois State
Geological Survey, Environmental Geology Notes 83. 42 pp.
[Online: Available at: https://www.ideals.illinois.edu/
handle/2142/78872] (Accessed March 2017).

Jessell, M.W. (2015). A History of 3D Geological Modelling.
In: *Proceedings, Saying Goodbye to a 2D Earth,
International Conference, Margaret River,Western Australia*
(2–7 August 2015). 34 slide. [Online: Available at: https://
pdfs.semanticscholar.org/fdfb/
f1887b3926de4236c0cbe756163de46e2323.pdf] (Accessed
December 2018).

Keaton, J.R. (1984). Genesis–lithology–qualifier (GLQ)
system of engineering geology mapping symbols.
Environmental & Engineering Geoscience 21 (3): 355–364.
https://doi.org/10.2113/gseegeosci.xxi.3.355.

Kempton, J.P. and Visocky, A.P. (1992). *Regional Groundwater
Resources in Western McLean and Eastern Tazewell
Counties with Emphasis on the Mahomet Bedrock Valley.*
Champaign, IL: Illinois State Geological Survey and Illinois
State Water Survey Cooperative, Groundwater Report 13.
41 pp. [Online: Available at: http://hdl.handle.net/2142/
35251] (Accessed April 2017).

Kempton, J.P., Morse, W.J. and Visocky, A.P. (1982).
*Hydrogeologic Evaluation of Sand and Gravel Aquifers for
Municipal Groundwater Supplies in East-Central Illinois.*
Champaign, IL: Illinois Department of Energy and Natural
Resources, Cooperative Groundwater Report 8. [Online:
Available at: http://hdl.handle.net/2142/35243] (Accessed
April 2017).

Kempton, J.P., Johnston, W.H., Heigold, P.C. and Cartwright,
K. (1991). Mahomet Bedrock Valley in East-Central
Illinois; topography, glacial drift stratigraphy, and
hydrogeology. In: *Geology and Hydrogeology of the
Teays-Mahomet Bedrock Valley System, Geological Society of
America Special Paper 258* (eds. W.N. Melhorn and J.P.
Kempton), 91–124. Boulder, CO: Geological Society of
America. https://doi.org/10.1130/spe258-p91.

Larson, G. and Schaetzl, R. (2001). Origin and evolution of
the Great Lakes. *Journal of Great Lakes Research* 27 (4):
518–546. https://doi.org/10.1016/s0380-1330(01)70665-x.

Larson, D.R., Kempton, J.P. and Meyer, S. (1997). *Geologic,
Geophysical, and Hydrologic Investigations for a
Supplemental Municipal Groundwater Supply, Danville,
Illinois.* Champaign, IL: Illinois State Geological Survey
and Illinois State Water Survey, Cooperative Groundwater

Report 18. 62 pp. [Online: Available at: http://hdl.handle
.net/2142/35252] (Accessed April 2017).

Mariethoz, G. and Caers, J. (2014). *Multiple-Point
Geostatistics: Stochastic Modeling with Training Images.*
Sussex, UK: Wiley Blackwell, 376 pp. https://doi.org/10
.1002/9781118662953.

McLean, L.R., Kelly, M.D. and Riggs, M.H. (1997). *Thickness
of Quaternary Deposits in McLean County, Illinois.* Illinois
State Geological Survey Open-File Series 1997-1E, map
1:100,000. [Online: Available at: http://hdl.handle.net/
2142/55430] (Accessed April 2018).

Melhorn, W.N. and Kempton, K.P. [Eds.] (1991). *Geology and
Hydrogeology of the Teays-Mahomet Bedrock Valley System.*
Boulder, CO: Geological Society of America, Special Paper
258. 128 pp. https://doi.org/10.1130/spe258.

Miller, B.B., Reed, P., Mirecki, J.E. and McCoy, W.D. (1992).
Fossil molluscs from pre-Illinoian sediments of the buried
Mahomet Valley, central Illinois. *Current Research in the
Pleistocene* 9: 117–118.

Norris, S.E. and Spicer, H.C. (1958). *Geological and
Geophysical Study of the Preglacial Teays Valley in
West-Central Ohio.* Reston, VA: U.S. Geological Survey,
Water-Supply Paper 1460-E. pp. 199–232. [Online:
Available at: https://pubs.usgs.gov/wsp/1460e/report.pdf]
(Accessed March 2017).

Phelps, G. and Boucher, A. (2009). Mapping locally complex
geologic units in three dimensions: the multi-point
geostatistical approach. In: *Three-Dimensional Geologic
Mapping for Groundwater Applications Workshop* (eds. R.C.
Berg, H.A.J. Russel and L.H. Thorleifson), 36–39.
Champaign, IL: Illinois State Geological Survey [Online:
Available at: https://www.isgs.illinois.edu/sites/isgs/files/
files/3Dworkshop/2009/phelps.pdf (Accessed May 2017).

Rhind, D.W. (1971). The production of a multi-colour
geological map by automated means. *Nachrichten aus dem
Karten- und Vermessungswesen* 52: 47–52.

Roadcap, G.S., Knapp, H.V., Wehrmann, H.A. and Larson,
D.R. (2011). *Meeting East-Central Illinois Water Needs to
2050: Potential Impacts on the Mahomet Aquifer and
Surface Reservoirs.* Champaign, IL: Illinois State Water
Survey, Contract Report 2011–08. 188 pp. [Online:
Available at: http://hdl.handle.net/2142/39869] (Accessed
March 2017).

Robinove, C.J. (1962). *Ground-Water Studies and Analog
Models.* Reston, VA: U.S. Geological Survey, Circular 468.
12 pp. [Online: Available at: https://pubs.usgs.gov/circ/
1962/0468/report.pdf] (Accessed March 2017).

Royse, K.R., Reeves, H.J. and Gibson, A.R. (2008). The
modelling and visualization of digital geoscientific data as
a communication aid to land-use planning in the urban
environment: an example from the Thames gateway.

Geological Society Special Publications 305: 89–106. https://doi.org/10.1144/SP305.10.

Schicht, R.J. (1965). *Ground-Water Development in East St. Louis Area, Illinois*. Champaign, IL: Illinois State Water Survey, Report of Investigation 51. 70 pp. [Online: Available at: http://hdl.handle.net/2142/101993] (Accessed January 2019).

Skibitzke, H.E. (1960). Electronic Computers as an Aid to the Analysis of Hydrologic Problems. In: *Proceedings, General Assembly of Helsinki, Commission of Subterranean Waters*, 347–358. Wallingford, UK: International Association of Scientific Hydrology, Publ. no. 52. [Online: Available at: http://hydrologie.org/redbooks/052.htm] (Accessed January 2019).

Smith, I.F. [Ed.] (2005). *Digital Geoscience Spatial Model Project Final Report*. Keyworth, UK: British Geological Survey, Occasional Publication 9. 56 pp. [Online: Available at: www.bgs.ac.uk/downloads/start.cfm?id=535] (Accessed August 2018).

Soller, D.R., Price, S.D., Berg, R.C. and Kempton, J.P. (1998). A method for three-dimensional mapping. In: *Digital Mapping Techniques '98-Workshop Proceedings* (ed. D.R. Soller). Reston, VA: U.S. Geological Survey, Open-File Report, 98-487. [Online: Available at: http://pubs.usgs.gov/of/1998/of98-487] (Accessed June 2020).

Soller, D.R., Price, S.D., Kempton, J.P. and Berg, R.C. (1999). *Three-Dimensional Geologic Maps of Quaternary Sediments in East-Central Illinois*. Reston, VA: U.S. Geological Survey, Geologic Investigations Series Map I-2669, 3 map sheets + text file [Online: Available at: http://pubs.usgs.gov/i-maps/i-2669] (Accessed March 2017).

Stafleu, J., Maljers, D., Gunnink, J.L. and Menkovic, A. (2011). 3D modelling of the shallow subsurface of Zeeland, the Netherlands. *Netherlands Journal of Geosciences* 90: 293–310. https://doi.org/10.1017/S0016774600000597.

Stumpf, A.J. (2018). The Mahomet Bedrock valley: its history and character. *Illinois Geographer* 59 (2): 58–78.

Stumpf, A.J. and Atkinson, L.A. (2015). *Geologic Cross Sections across the Mahomet Bedrock Valley, Champaign, Ford, Mclean, Piatt, and Vermilion Counties, Illinois*. Champaign, IL: Illinois State Geological Survey, Map 19. [Online: Available at: http://hdl.handle.net/2142/89865] (Accessed March 2017).

Stumpf, A.J. and Dey, W.S. [Eds.] (2012). *Understanding the Mahomet Aquifer: Geological, Geophysical, and Hydrogeological Studies in Champaign County and Adjacent Areas*. Champaign, IL: Illinois State Geological Survey, Contract Report 2007-02899. 332 pp. [Online: Available at: http://hdl.handle.net/2142/95787] (Accessed March 2017).

Teller, J.T. and Goldthwait, R.P. (1991). The Old Kentucky River: a major tributary to the Teays River. In: *Geology and Hydrogeology of the Teays-Mahomet Bedrock Valley System*

(eds. W.N. Melhorn and J.P. Kempton), 29–41. Boulder, CO: Geological Society of America, Special Paper 258. https://doi.org/10.1130/spe258-p29.

Tight, W.C. (1903). *Drainage Modifications in Southeastern Ohio and Adjacent Parts of West Virginia and Kentucky*. Reston, VA: U.S. Geological Survey, Professional Paper 13. 111 pp. [Online: Available at: https://pubs.usgs.gov/pp/0013/report.pdf] (Accessed March 2017).

Turner, S. and Dearman, W.R. (1979). Sopwith's geological models. *Bulletin of the International Association of Engineering Geology* 19: 331–345. https://doi.org/10.1007/BF02600498.

Van der Meulen, M.J., Doornenbal, J.C., Gunnink, J.L. et al. (2013). 3D geology in a 2D country: perspectives for geological surveying in the Netherlands. *Netherlands Journal of Geosciences* 92 (4): 217–241. https://doi.org/10.1017/S0016774600000184.

Varnes, D.J. (1974). *The Logic of Geological Maps, with Reference to Their Interpretation and Use for Engineering Purposes*. Reston, VA: U.S. Geological Survey, Professional Paper 837. 48 pp. https://doi.org/10.3133/pp837.

Ver Steeg, K. (1946). The Teays river. *The Ohio Journal of Science* 46 (6): 297–307. [Online: Available at: http://hdl.handle.net/1811/3568] (Accessed June 2019).

Visocky, A.P. and Schicht, R.J. (1969). *Groundwater Resources of the Buried Mahomet Bedrock Valley*. Champaign, IL: Illinois State Water Survey, Report of Investigation 62. 58 pp. [Online: Available at: http://hdl.handle.net/2142/102004] (Accessed January 2019).

Walton, W.C. (1964). Electric analog computers and hydrogeologic system analysis in Illinois. *Groundwater* 2 (4): 38–48. https://doi.org/10.1111/j.1745-6584.1964.tb01785.x.

Walton, W.C. and Prickett, T.A. (1963). Hydrogeologic electric analog computers. *Journal of the Hydraulics Division, American Society of Civil Engineers* 89 (6): 67–91.

Wilson, S.D., Kempton, J.P. and Lott, R.B. (1994). *The Sankoty-Mahomet Aquifer in the Confluence Area of the Mackinaw and Mahomet Bedrock Valleys, Central Illinois: A Reassessment of Aquifer Characteristics*. Champaign, IL: Illinois State Water Survey and Illinois State Geological Survey, Cooperative Ground-Water Report 16. 64 pp. [Online: Available at: http://hdl.handle.net/2142/35245] (Accessed March 2017).

Wilson, S.D., Roadcap, G.S., Herzog, B.L. et al. (1998). *Hydrogeology and Groundwater Availability in Southwest McLean and Southeast Tazewell Counties, Part 2: Aquifer Modeling and Final Report*. Champaign, IL: Illinois State Water Survey and Illinois State Geological Survey, Cooperative Groundwater Report 19. 138 pp. [Online: Available at: http://hdl.handle.net/2142/35254] (Accessed March 2017).

6

Effective and Efficient Workflows

Donald A. Keefer and Jason F. Thomason

Illinois State Geological Survey, University of Illinois at Urbana-Champaign, Champaign, IL 61820, USA

6.1 Introduction

While it is possible to purchase modeling software, organize available project data and just begin modeling, there are significant differences between mapping by hand and mapping via 3-D software. Taking time to define a custom workflow for a digital 3-D modeling project or program will likely produce significant gains in modeling efficiency, product reliability, and the ability to meet project modeling, financial, and timeline goals. This chapter is primarily targeted at organizations or research groups who are in the process of setting up a 3-D modeling project or are reviewing the effectiveness of an existing 3-D modeling program. Throughout this process, it is important to choose 3-D modeling software, design workflows, and develop skills and experience that are compatible with long-term institutional goals.

6.1.1 Understanding the Geologic Modeling Process

A generic geologic modeling workflow (Figure 6.1) clarifies the major computational steps that occur and major iterations that can take place within a single modeling project. On a basic level, geologic modeling can be broken down into five major tasks (Figure 6.1):

1. compilation, collection and processing of existing and new data;
2. description and basic interpretation of materials and features within the data;
3. development of a generalized geologic framework model, including mapping units and relevant structural or other features;
4. correlation of interpreted data with the mapping units from the generalized geologic framework model; and
5. prediction of the distribution of map units and geologic features, and occasionally rock properties, at all unsampled locations.

The first two tasks are typically routine and require experienced attention, while the last two efforts demand more analytical input, requiring the mappers to integrate the data with existing knowledge about sedimentary and structural geology, and constrain this with their personal experiences. The third major task, development of a generalized geologic framework, can be routine if a stratigraphic system and a structural and/or tectonic model for the region are current and available. However, significant creative and analytical input may be required to develop more generalized geologic framework mapping units if the natural geologic complexity cannot be reliably modeled due to either a high degree of spatial variation in the sediments or the presence of overturned or complexly faulted structural regimes. Novel approaches may be required to reliably create a 3-D model if no stratigraphic system or structural and/or tectonic model has been defined for the model domain, if the current stratigraphic system or model needs updating to best meet the current modeling objectives, or if the model domain contains complex geologic structures. The model developer must recognize that a lithostratigraphic unit that is mappable in 2-D is not necessarily mappable in 3-D. In these latter situations, consideration of the time, staffing, and data requirements needed to revise a general geologic framework model needs to be made early in the project timetable.

The modeling process is usually flexible and highly iterative; insights gained in one step can provide feedback to justify new data collection or require re-evaluation of the general framework model or the existing correlations. Iteration typically continues until the geologist is satisfied that

Applied Multidimensional Geological Modeling: Informing Sustainable Human Interactions with the Shallow Subsurface, First Edition.
Edited by Alan Keith Turner, Holger Kessler, and Michiel J. van der Meulen.
© 2021 John Wiley & Sons Ltd. Published 2021 by John Wiley & Sons Ltd.

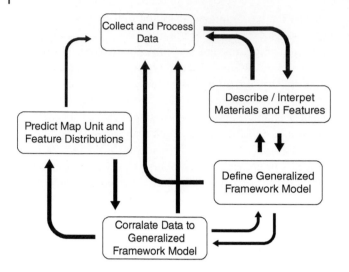

Figure 6.1 Generic 3-D geologic framework modeling workflow. © 2018 University of Illinois Board of Trustees. Used by permission of the Illinois State Geological Survey.

the model agrees with their geologic interpretations and meets the project goals.

Organizations and individuals often develop 3-D geologic modeling workflows heuristically, guided by experience with 2-D mapping projects. While 2-D mapping workflows are typically well established, the shift to 3-D will likely require significant increases in effort and many changes in technology. Taking the time to formally consider the issues involved in 3-D modeling workflows can help avoid unexpected incompatibilities in software, expertise, and timelines, and ensure that project deliverables and software compatibilities have been thought through. The motivation is to avoid time and cost overruns as well as incompatibilities between project resources and geologic setting that can result in missed deadlines and products that do not match project goals.

The ideas presented in this chapter regarding the workflow design process are illustrated using examples from a recent multiple-year project conducted by the Illinois State Geological Survey (ISGS) (Thomason and Keefer 2013). This 3-D geologic modeling project involved the mapping and characterization of the late Quaternary sediments above the bedrock surface across McHenry County, Illinois – a county of 1582 km^2 (611 mi^2) and approximately 308 000 people located in northwest metropolitan Chicago (Figure 6.2). A primary focus of the modeling was to define the distribution and general character of the sand and gravel aquifers throughout the county, which currently provide approximately 60% of municipal water supplies. Reliable characterization and delineation of these aquifers is critical for supporting long-term, sustainable water resource management decisions.

6.1.2 Developing Custom Workflows

The generic modeling process outlined in Figure 6.1 is influenced by decisions related to: (i) operational considerations, (ii) selection of modeling methods and software, (iii) production and distribution of appropriate model products, and (iv) maintenance and upgrading of models. Typical solutions to these issues are illustrated in Sections 6.2–6.6 and through individual case studies presented in Section 6.7.

6.2 Operational Considerations

Six operational considerations can be used to help ensure success of geologic modeling projects. They include: (i) identifying and prioritizing user requirements; (ii) defining mapping objectives; (iii) evaluating the likely geologic complexity within the model domain; (iv) determining

Figure 6.2 McHenry County location map and overview of 3-D geologic framework model. © 2018 University of Illinois Board of Trustees. Used by permission of the Illinois State Geological Survey.

availability of, and management options for, existing data; (v) estimating how much new data will be collected; and (vi) evaluating availability and expertise of personnel.

6.2.1 User Requirements

The contractual requirements of the funding agent and the expected product needs of stakeholders should be among the first items compiled and should be used as a basis for evaluating the remaining operational considerations. These requirements should include the specific problems to be addressed or decisions to be supported by the final model; the spatial modeling boundary (2-D or 3-D); the total budget for the project; a timeline for completion of the project; and a list of deliverables and acceptable product formats. The timeline and budget will control the collection of new data and hiring of new staff members and will directly impact most of the other decisions within any modeling project.

6.2.2 Defining Mapping Objectives

Project mapping objectives reflect funding agent and primary stakeholder requirements. Ensuring that mapping objectives are achieved requires decisions on the following topics.

6.2.2.1 Delineation of Model Domain

The horizontal and vertical extents of the model domain, which are typically determined by funding-agent requirements, must be considered and defined. While the horizontal extent is often based on political boundaries, it is a common practice to extend the mapping area beyond this boundary by a few kilometers, to reduce potential for interpolation errors near the model edges. The lowermost vertical extent of the model is often set to be the top or bottom surface of a specific geologic unit – a boundary that will vary in depth and elevation. Alternatively, the vertical extent can also be set at a constant depth or constant elevation. This choice is typically influenced by the distribution of geologic deposits and structural features that are expected to occur within the model domain, the distribution and quality of data at depth, and the expected end-user requirements.

6.2.2.2 Definition of the General Geologic Framework Model

The general geologic framework model defines the expected succession of geological mapping units, the major structural features, and all other major modelable geologic features. The selected framework should address all previously-identified stakeholder needs. High-resolution models increasingly have mapping units that are lithologies, or rock types, while traditional 3-D geologic models have mapping units that are from either formal or informal stratigraphic systems. Selection of rock-type versus stratigraphic mapping units will be based on the problems that the model is being used to address, the quantity and quality of available data relative to identification of potential mapping units, the natural complexity of the system being modeled, and whether probabilistic modeling methods are expected to be used. This selection of mapping units also will constrain the eventual selection of modeling methods and software.

6.2.2.3 Determination and Representation of the Desired Model Accuracy

In every model, the accuracy of predictions will vary across the model domain. The accuracy of these predictions will depend on several variables, including the distribution, accuracy and types of available data; the degree of generalization, or granularity, that is captured within the definitions of map units and features; the expertise of the project staff; and the real-world complexity, or variability, in map unit or feature distributions that exists below land surface. Variations in the reliability of model predictions, if not clearly documented, can have a significant impact on the risks assumed by decision makers who might rely on the model predictions. Clear documentation and visualization of variations in predictive reliability can give decision makers information to more reliably evaluate their risk exposure and make more informed decisions.

Standard methods have not been developed for characterizing the predictive accuracy, or uncertainty, of the distribution of stratigraphic units within geologic models (Chapter 15), though many researchers continue to explore uncertainty of different attributes within stratigraphic models (e.g. uncertainty in the elevation of the top surface of a stratigraphic unit). While probabilistic modeling methods offer excellent tools for providing structured insight on the reliability or uncertainty of model predictions, the typical sparsity of high-quality data, the lack of validated probabilistic models describing successions of geologic deposits, and the high level of expertise required to use probabilistic models make them difficult to consistently implement. If probabilistic methods are not used, we recommend experienced geologists develop qualitative or semi-quantitative methods for describing the predictive accuracy of geologic framework models, using metrics and methods that fit with the data, geologic framework and final predictions of each model. The results of such exercises can, for example, be visualized as overlay maps of the model area indicating the reliability of predictions – based on the expertise of the modeling team – using categories such as high, medium, and low (Van der Meulen et al. 2007). Modelers should remember that most clients and users will take their 3-D predictions at face value and will not have the expertise needed to independently identify variations in reliability of model predictions. Regardless of format, a clear discussion of variations in the predictive reliability of the model, preferably accompanied by maps or 3-D visualizations, will provide critical insights from the modeling team that can be used to reduce risk exposure by decision makers.

6.2.2.4 Consideration of Formats for Final Deliverables

The project funding agent may require specific products or deliverables to ensure compatibility with anticipated stakeholder model uses. For example, the British Geological Survey has produced 3-D geological models in specified Computer Aided Design (CAD) file formats for some railway infrastructure renewal projects (Burke et al. 2015). In this effort, the geological subsurface information was integrated with infrastructure elements defined by CAD design processes (Chapter 23). Distribution formats for products can include 1-D profiles, 2-D layers, or 3-D volumes of one or more basic mapping units, derivative interpretations from the basic units, or software-specific formats, including CAD-formats, and multimedia formats such as animations, 3-D static images, etc. Any desired model products should be considered during the selection of the modeling software package(s) to ensure that the necessary output files can be generated at acceptable levels of graphic and cartographic quality.

6.2.3 Geologic Setting and Natural Complexity

The inherent complexity that is expected in the distribution of deposits and structural features within the model domain should be evaluated during the preparatory stages of a modeling project. This natural complexity will dictate the amount of data and knowledge needed to achieve accuracy goals for the model predictions. This evaluation should be used to guide decisions on the value of new data, the resolvable model accuracy, and the appropriate modeling methods. The expectations and requirements of the funding agent and stakeholder should be used to influence decisions about the appropriate accuracy of the final model, the quantity, spatial distribution and types of data needed to achieve this predictive accuracy, the project timeline, and the project budget.

Large numbers and a relatively uniform distribution of data are required in areas of complex geology to reliably model important features and variations in map units. This type of data distribution is rarely available, however, so an evaluation of the available sample distribution should be made to help consider how much of the natural complexity can be reliably predicted. Because samples are required to demonstrate the degree of natural complexity that exists in a model domain, every modeling project can benefit significantly from new data that are strategically located to fill gaps in the knowledge about the model domain. Typically, 1-D drillers' log or core descriptions provide observations of rocks encountered. Two-dimensional geophysical profile data can provide insights on the lateral continuity of units and features, and on the nature of horizontal variation of observable rock properties (Jørgensen et al. 2003). Smaller amounts of data are required in less complex geological settings to capture the succession of units and modeled features with comparable accuracy.

6.2.4 Existing Data Availability and Management

Many modeling projects have one or more legacy datasets that are available for modeling subsurface geology. Data that can be evaluated for locational accuracy and reliability of material description are critical for high-quality modeling. It is important to remember that legacy data will have been gathered for reasons other than the current modeling project and may contain biases or measurement issues that are relevant to current modeling goals. If the locational accuracies of many records are suspect, it may be necessary to spend time and money to improve this information. The value of verified locations depends on user requirements, local variations in relief in land-surface elevation, thickness of the model domain, modeling objectives, and, importantly, the range in elevation for the top and bottom surfaces of individual mapping units. For example, in areas where the land surface has significant relief or where geologic units are particularly thin or variable, it can be very important to obtain locations that are within a few tens of meters to the actual location. In these situations, small differences in horizontal placement of an observation can correspond to significant differences in land surface elevation, which can significantly complicate reliable correlation in these situations. Alternatively, if land surface relief is particularly small or geologic units are thick or relatively uniform in thickness, changes in horizontal placement of the data may not have a noticeable impact on the reliability of interpretations in the data or predictions in the model.

A review of the capabilities of available data management software is recommended when evaluating the utility and availability of legacy data sets. While advances in software have reduced the likelihood of incompatibilities between data management and modeling software applications, unexpected incompatibilities can still occur.

Incompatibilities in software can create significant time delays due, for example, to the need for frequent reformatting of data during an iterative modeling workflow. An evaluation must also be made of the capabilities for migrating final geological interpretations and predictions from the finished 3-D model back into project or institutional enterprise databases. Modeling software differ greatly in the functionality they provide for exporting this knowledge. While the migration of final interpretations can often be done after the formal project is completed, the absence of export capabilities that are compatible with investigator or institutional data-management frameworks can make it difficult to capture the knowledge gained from the modeling exercise and integrate it into institutional data or knowledge bases.

6.2.5 Collection of New Data

There are no universally standard practices regarding the collection of new data for modeling projects. For example, many 3-D geologic modeling projects will integrate and rely on the collection and interpretation of new field or laboratory data for their modeling. Others will not collect any new data and will rely exclusively on legacy data. Similarly, many projects use small numbers of high-quality new data collected in the early stages of a modeling project to calibrate less reliable, legacy data. This can be accomplished by locating new data in areas of sparse data, to confirm local or regional trends, or by locating new data in areas of clustered legacy data and complex geology to guide interpretations of the legacy observations. While experienced geologists typically understand the logistical issues of new data collection, it can be helpful to consider this aspect of a 3-D modeling project during this planning stage. For example, planning for the collection, processing, interpretation, and integration of new data must be done early in the project to allow time to integrate the interpretations and resolve any geometric problems. Some types of data have seasonal or weather constraints that limit the times of their collection. If data processing and interpretation are likely to require significant time, this must also be built into project timelines. Before interpretations from new data can be used in modeling, they often must be reformatted for ingestion into the modeling software. Complex data types, such as seismic profiles, may dictate the use of alternate data structures. An evaluation of reformatting needs is relevant because workflow decisions can depend on the data structures, modeling software limitations, and programming/data management skills of project participants. Consideration of these issues during the early planning stages of the project lifecycle can help ensure efficiency during the modeling stage.

6.2.6 Staff Availability and Expertise

Requirements for staffing numbers and expertise in any project will depend on several factors, including: the size of the model domain, complexity of geology, availability of existing data, amount and type of new data to be collected, the selected software capabilities for data management, modeling, and creation of final products, and the type, numbers and formats of the final products that will be created.

In addition to evaluating overall staffing resources, it is critically important to evaluate project team members for their domain expertise and their comfort in learning new software and methods. Younger geologists are often more trainable for use of various interpolation algorithms and modeling software, but they have less overall experience and provide less-informed expertise for a given project. Older geologists, while potentially having significant domain expertise, often have more difficulty in migrating from traditional paper-based mapping methods to software that uses different representations of the data and maps or requires the selection of interpolation parameters that are different from the properties they typically consider. The difficulty older geologists often experience in transitioning to digital modeling software will depend on the architecture and user interface of the modeling software. An example of a successful transition to digital modeling involves the GSI3D/Subsurface Viewer and Groundhog software packages, developed by the German company INSIGHT GmbH and the British Geological Survey. These have simple, user-friendly designs and user interfaces. These packages require the user to select and digitize cross-sections on the screen and have proved to be well accepted by senior staff members who found more sophisticated software packages to be too cumbersome to learn.

Early reflection and discussion about expected project staff, their roles, expertise, and comfort for learning new approaches to modeling, can identify potential problems and help team members develop strategies anticipating problems before they are encountered. Workflows can be adjusted to accommodate staff with significant domain expertise or are less comfortable with using selected modeling software. For example, staff can be encouraged to create hand-drawn contoured surface models that are then scanned and imported into the modeling software.

6.3 Selection of Modeling Methods and Software

Within the context of this book, "modeling method" refers to the parts of the modeling workflow that correlate the data to the formal mapping units and then predict the distribution of these units throughout the model domain. The prediction step usually involves the application of one or more interpolation algorithms. A generalized discussion of the four principal modeling methods and details for selecting the best method are provided in Chapter 5. Chapters 9–12 of this book provide more clarity on how these four approaches can be used to create geologic framework models. Each approach has advantages and limitations, and most are implemented within several different software products. This section provides guidance for selecting a modeling method and modeling software package.

Most modeling software has evolved to accommodate multiple modeling methods and a wide range of geologic scenarios. For instance, combinations of the stacked-surface and cross-section approaches are more common, as are applications that combine stacked surfaces with 3-D voxel modeling. Some applications even allow combination of cross-sections, surfaces, and voxels in the creation of a model. Additionally, the increasing desire to constrain models with different kinds of geologic knowledge has pressured vendors to build in various types of geologic and geometric constraints. Implicit modeling applications, which have become more popular in the past decade, emphasize the role of geologic constraints as integral parts of their software. Other software, employing what are alternatively called explicit methods, have adapted to implement geologic and geometric constraints through function libraries and menu-based toolboxes that provide capabilities such as the subdivision of a model domain based on GIS-based polygon maps, or the adjustment of model-unit distributions using custom grid calculations.

While software capabilities continue to advance, most currently available software applications still provide limited options in modeling methods (see Chapter 5). To ensure successful modeling outcomes, the selection of modeling method and software should be based on a consideration of: (i) the complexity of the modeling units (i.e. stratigraphic units versus rock types) and structural features that need to be explicitly represented in the final model predictions; (ii) staff preferences for specific modeling methods; (iii) software design and ease of use; (iv) staff availability throughout the project lifecycle; (v) expected final products; and (vi) long-term institutional goals for 3-D modeling.

The degree of geologic complexity to be represented in the final model must be considered when evaluating modeling methods and software. Stratigraphic units represent aggregations of deposits and are typically easy to model with any software. The explicit representation of higher-resolution properties, such as within-unit deformation, foliation, or rock-type variations, is avoided when they are embedded within the stratigraphic unit definitions. The choice of method and software is more critical to ensuring model success, if these high-resolution features are to be explicitly modeled. At the coarser scale of geologic features, several modeling packages can accommodate high-angle faulting and mild to moderate folding of formations. Relatively few packages can reliably model heavily faulted settings, overturned folds, or metamorphic terrain. If these complex structural features are anticipated to occur within potential model domains, and if they need to be explicitly represented in the geologic models, they should be considered when evaluating modeling software packages. If software with the desired capabilities cannot be acquired for a project, mapping units can be defined to acknowledge the presence of un-represented high-resolution features and cartographic symbols used to identify the locations of structural features. Additionally, representations of complex structural features can be added to cross-sections or other visualizations, using image editing software, while not being fully represented in the digital 3-D model.

Each software package should be evaluated based on the background, technological aptitude, and mapping-style preferences of the geologists who will be using it. Geologists experienced with mapping from cross-sections or hand-drawn structure contour maps, for example, may have a very difficult time adapting to a modeling approach that does not allow the interactive use of these representations as part of the modeling process. Alternatively, poorly designed software might allow interaction with cross-sections or contour maps, but the implementation might make the interface too hard to manage for most geologists.

Software trial periods can be helpful in identifying incompatibilities between the functionality and ease of use of a software package and the model requirements and staff capabilities. Once a package is selected, formal training can help dramatically shorten the learning curve for users and can help establish valuable relationships between modeling staff and software technical support. The most sophisticated commercial software platforms were developed for petroleum or mining applications and may require a dedicated modeling specialist to manage data-handling and modeling processes, leaving geoscientists to collect and interpret the data and work with the modeler to ensure map-unit predictions fit their interpretations. If this level of staffing commitment is not supportable, these higher end packages might need to be avoided.

The expected range of products from the final geologic model should also be considered during the software evaluation process. Some geologic modeling software offers many options for image capture and map export formats, while others are much more restrictive. In some situations, modeling software cannot provide all the export options that are needed. In these situations, supplemental software might be used to edit images or visualize grid files exported from a modeling application. Section 6.4 of this chapter discusses considerations related to products and product distribution in more detail.

As part of the software evaluation process, it is worthwhile evaluating long-term institutional goals for 3-D geologic modeling. Long-term programs can benefit from workflow automation, version control, integration with enterprise data management applications, and integration with product/model delivery priorities. However, costs for modeling software that provides these features can be significant. For many institutions, it can be difficult to anticipate the growth of 3-D modeling programs and changes in program funding. Custom generated applications, developed in-house or by a contractor, might provide more affordable solutions to these needs. Custom solutions also can help ensure institutional flexibility if changes in staffing, resources, or stakeholder demands require workflow changes that are incompatible with the capabilities of modeling applications. Many 3-D modeling software vendors will develop custom tools under contract that work with their products and meet specific institutional needs. Matching program goals with software capabilities allows identification of incompatibilities or unmet needs before software is purchased, helping to ensure success of the modeling program.

6.4 Products and Distribution

Advances in technology have resulted in a nearly-constant evolution of options in products created from 3-D modeling projects. Traditional products from 3-D models include cross-sections, 2-D maps, 3-D images, borehole logs, and reports. Advances in 3-D visualization and data-mining tools enable the creation of multimedia products that can combine animation, slide-based presentations, video, and audio commentary.

Geologic modeling software vendors increasingly provide options for bundling models, data, interpretations, and pre-selected views of the model into standalone viewing apps. Some vendors charge a fee to the model-builder for bundling the content into these viewers, while others make a freely-distributed, non-editing version of the modeling software available. Some of these software/data bundles can be made easily available to stakeholders and model users via project websites. The user interfaces of these standalone applications vary from simple to complex, so if these tools are priorities for the project, the user-friendliness needs to be evaluated prior to purchase.

Other low-cost or widely accessible interactive distribution formats include 3-D PDF products that open on virtually any digital interface, or KMZ files that can be displayed in Google Earth. An example of the 3-D PDF capability is the interactive model of the northern part of the Assynt Culmination in the Moine Thrust belt in the highlands of Scotland (Figure 6.3). A free 3-D PDF file may

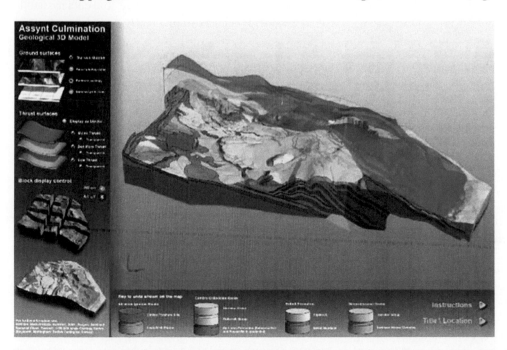

Figure 6.3 View of the 3-D PDF display of the BGS Assynt Culmination model. Orange surface is the Moine Thrust. Some model display tools shown on the left side. A mouse is used to rotate or zoom the model. Source: British Geological Survey.

be downloaded from the British Geological Survey (BGS) webpage http://www.bgs.ac.uk/downloads/start.cfm?id=2693. The BGS-generated 3-D PDF interface allows the user to rotate and zoom the model view, "remove" selected model components, and use a "cross-section tool" to cut into the model to view its internal structure. It is important to note that many of these tools are developed by mapping organizations to support their programs, and their development is guided by institutional priorities. Over the long term, this mapping-institution development model might make these products less reliable for external users because continued development is likely to leave external-user needs unsupported, since internal priorities guide feature development and may diverge from external-user priorities.

Identification and prioritization of deliverables, including product formats, and delivery methods should be evaluated in the planning stages of a project. This allows establishment of realistic milestones and project timelines. The specification of deliverables must consider user requirements, realistic evaluation of availability and expertise of personnel, and required computer resources. Larger programmatic goals and modeling software capabilities should also be considered to ensure technological trends are identified and translated to institutional goals. These considerations and goals are evaluated and prioritized based on budgets and timelines. It is not unusual for desired deliverables or delivery methods to be cost prohibitive. Effective and efficient project workflows benefit from this level of planning.

While 3-D visualization-based products are attractive and help even non-technical users more intuitively understand the modeled geologic systems, 2-D maps and map layers are still superior products for supporting any decision making. Map products in digital formats can be easily included in stakeholder GIS systems, but even hard copy maps are invaluable for assisting in decisions related to any planning or land-use problem. Ideally, stakeholders will have a suite of 2-D map layers, 3-D static, and 3-D interactive visualizations available to help ensure an understanding of the geologic framework and to help guide how that insight might be applied to specific decisions.

6.5 Model Maintenance and Upgrades

Maintenance of 3-D geologic models can include long-term management of the data used in modeling, the interpretations assigned to those data, the predictions made from these interpretations, and any historical knowledge used to explicitly map or define distributions of units and properties. Long-term management decisions should consider the time constraints under which these products might be needed in the future and the likelihood of updates being made to any of these products. Updating can include processes including data-quality improvements (e.g. from inclusion of more accurate locational coordinates), validation of observations or attributes through nearby sample or outcrop analysis, collection of new data from site investigations, re-interpretations of data, modifications to pre-existing map-unit classification systems, or updated predictions in the distribution of specific geologic units.

Increasingly, 3-D geological models are being specifically included in statutes, regulations, and policies, resulting in an increased need for effective long-term model and data management, possibly including the need to revise portions of the original model. For example, a new law has commissioned the Geological Survey of the Netherlands to build and maintain a national database for subsurface data and information, which Dutch government bodies are obliged to use when making policies or decisions that interact with the subsurface. This has required modification and redesign of a substantial portion of the geological survey operations (van der Meulen et al. 2013; Chapter 2 this volume). These long-term needs typically fall outside the project and contractual obligations related to creating individual 3-D models. However, their consideration during project planning may identify critical constraints on decisions regarding software selection, data management, and staffing. Long-term needs for model and data availability and accuracy can be complex; early planning and needs assessments help ensure the adequacy of organizational structures and resources for creating effective and efficient maintenance and updating workflows.

In addition to the maintenance needs necessitated by inclusion of models in statute, larger modeling programs are likely to have additional management and maintenance needs as contiguous modeling projects are completed. For example, the British Geological Survey has created numerous local and regional models designed to address the requirements of various projects and users. It became apparent that a national model would not only be useful in addressing national problems, it would provide a "national framework" that allows the integration of local and regional models. Combining and "nesting" the natural hierarchy of models can promote greater operational efficiency, and add value to existing models and databases (Waters et al. 2016). The Illinois and Ontario geological surveys, by contrast, have long-term programs but limited staffing support, resulting in maintenance strategies that do not currently include centralization or rectification of adjoining interpretations and results.

6.6 Illinois State Geological Survey 3-D Modeling Workflows

For more than 25 years, the Illinois State Geological Survey (ISGS) has been developing 3-D geologic models that address different stakeholder priorities. Recent 3-D geologic modeling projects include simulations of deep bedrock systems, as part of a large CO_2 sequestration research program, and models of unconsolidated Quaternary sediments, that provide support to planners, and groundwater managers. These projects generally target mapping of specific lithostratigraphic or hydrostratigraphic units. Some of the sequestration-related projects also involve modeling of rock property distributions such as porosity and permeability.

The metropolitan Chicagoland area has been the subject of a series of 3-D modeling projects by the ISGS to support groundwater resource protection and management. While the ISGS efforts utilize a wide range of software and workflows for 3-D geologic modeling, the recently completed 3-D model for McHenry County provides an example of the philosophical priorities emphasized at the ISGS. The McHenry County 3-D modeling project demonstrates the application of ideas outlined in this chapter; alternate approaches used in other ISGS projects are noted as appropriate.

McHenry County, located northwest of Chicago, is completely dependent on groundwater for its water supply; nearly 60% of municipal water supplies in the county are derived from glacial sand and gravel aquifers while the remainder use groundwater from bedrock aquifers. The eastern portion of McHenry County is undergoing significant groundwater-resource stress due to suburban expansion, while the western portion of the county needs a better understanding of their aquifer systems to support sustainable groundwater supply decisions. Bedrock aquifer resources, though available, are significantly depleted throughout northeastern Illinois, and planners have prioritized targeting of these shallow sand and gravel resources for future developments.

The surficial and near-surface geology of McHenry County reflects late Pleistocene continental glaciations and subsequent reworking and weathering. The landscape includes some of the most prominent preserved moraines and outwash fans in the Midwestern United States. The shallow subsurface sediments consist of ice-marginal, subglacial, and proglacial deposits consisting of moraines, till, and outwash which are derived from at least two successions of older (late Illinois Episode) deposits and three younger (Wisconsin Episode) deposits, which cumulatively range in thickness between 3 and 150 m (10 and 492 ft). The glacial and post-glacial deposits overlie a regionally present fractured carbonate bedrock unit; this bedrock surface contains occasional bedrock valleys, with as much as 100 m (328 ft) of relief (Thomason and Keefer 2013). Three major sand and gravel aquifers occur in McHenry County: (i) a lowermost sand and gravel aquifer that is often in contact with the underlying fractured carbonate bedrock; (ii) a middle sand and gravel aquifer system associated with the first advance of glaciers during the last Ice Age; and, (iii) an extensive surficial sand and gravel aquifer that includes late-stage glacial outwash deposits overlain locally by modern alluvial deposits. Locally, the lower two aquifers are in unconformable contact and are occasionally separated by a paleosol. While the workflows for these county-scale hydrostratigraphic models vary by project, Figure 6.4 illustrates the generalized workflow used to model the Quaternary sediments in McHenry County.

Figure 6.4 Generalized workflow for 3-D modeling of Quaternary sediments in McHenry County at the Illinois State Geological Survey. Source: © 2018 University of Illinois Board of Trustees. Used by permission of the Illinois State Geological Survey.

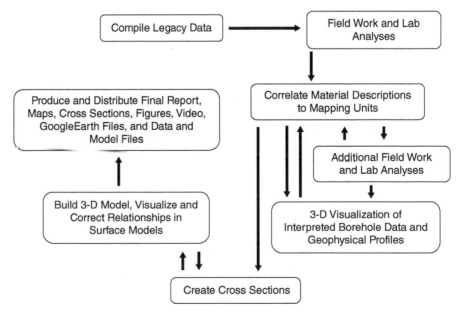

6.6.1 Project Objectives

The initial goals and tasks in the McHenry County Project reflected the funding and time constraints, and the end-user needs that were set by the County. The original tasks included: (i) conversion of a 10-year-old set of paper-based geological-maps into a fully digital 3-D model that described the distribution of sand and gravel aquifers and associated confining units; (ii) production of GIS-compatible digital map layers of aquifer distributions that could be used by county planners; and (iii) the collection of a limited amount of new data to support interpretation of existing water well databases and characterization of the succession of geologic materials. Model results were intended to guide subdivision development based on water supply potential, identify areas of potential over-competition for groundwater resources, and provide a framework and hydrogeologic insights for groundwater flow models that were being developed to assist the County with long-term water supply management.

After the first year of the project, additional funds were obtained that resulted in new tasks, including the collection of additional borehole and geophysical profile data, and development of higher-resolution modeling in three small areas in the County. These tasks improved the understanding some of the more complex geologic relationships across the county and improved the predictions of aquifer distribution and inter-aquifer connectivity in these high-resolution areas.

6.6.2 Project Schedule

The McHenry County project was initially funded jointly by the County and the ISGS with a three-year duration (2008–2010) and a budget that supported staffing, software, a limited number of geophysical profiles and a few new borings to obtain core samples. After the first year, additional funds from the U.S. Geological Survey, State of Illinois, and the ISGS allowed the project goals to expand, as mentioned above. With these additional funds, the project timeline was extended by three years, allowing increased collection of county-wide data, and creation of the three higher-resolution inset models.

6.6.3 Project Staffing Considerations

The range in project needs and funding levels for ISGS 3-D modeling projects has resulted in a variety of staffing choices. Planning, groundwater, and petroleum related projects have typically relied on one lead geologist doing the modeling, with limited support from drillers, geophysicists, lab assistants, GIS/cartography specialists, web programmers, and database managers. The specific project goals, available funds, and timeline tend to dictate the kind and quantity of new data collection. These projects also typically rely on student hourly employees for support in data cleaning and core description. The better-funded CO_2-sequestration projects have had significantly more staff to support geologic modeling and reservoir simulation. These projects are typically led by a petroleum engineer. Logistical assistance is provided by a project manager; geological modelers and/or geostatisticians, geologists, and petroleum engineers support geological and reservoir modeling efforts; while a database manager, editor, and web programmer provide technical support.

For the McHenry County project, a Quaternary geologist was the project Principal Investigator, team leader and primary modeler. This geologist had responsibility for project planning, coordination of field work, leading, and coordinating description of all newly collected core samples, managing project databases, interpreting all data, building the 3-D geologic model, creating final products, and leading and coordinating writing of all reports. Additional staff for this project included: (i) a co-Principal Investigator/hydrogeologist who led the proposal development, and assisted with core descriptions, installation of groundwater observation wells and writing the final report; (ii) a drilling team that drilled all boreholes, and assisted in the collection of core samples and installation of groundwater observation wells; (iii) geophysicists and field assistants who collected, processed and interpreted 2-D resistivity and seismic profiles; (iv) graduate students who assisted with field work and developed some of the inset models; (v) a database manager who assisted with data downloading from the enterprise database; and (vi) a GIS programmer who developed programs for visualizing selected profiles from the completed 3-D geologic model. The team leader and primary modeler for the McHenry County modeling project was comfortable adapting to new digital, software-based geologic modeling workflows. His capability in learning new software and methods helped ensure project success by guiding the physical creation of the model, precluding the need for additional support staff, at what would have been a significant additional cost.

6.6.4 Software Selection

The long history and wide programmatic reach of the ISGS 3-D modeling efforts have resulted in a wide range of geologic modeling software choices. Software platforms for the petroleum and CO_2 sequestration programs have included Stratamodel, Landmark Decision Space, Petra, ZMap, Petrel, and Isatis. Planning/zoning and groundwater driven programs have relied on EarthVision, RockWorks,

Surfer, Isatis, ArcGIS, MapInfo, Illustrator/MaPublisher, SubsurfaceViewer, and GeoScene3D. Enterprise database management at the ISGS has included a large Oracle database of point-located information. Project-level data are typically downloaded from Oracle and managed locally in MS Access or MS Excel. ArcGIS and Illustrator are used throughout the institution as the go-to platforms for creation of cartographic products. Some projects use ArcGIS for 3-D visualization of data and surface models.

The McHenry County model was developed using a suite of three geologic modeling software packages: ArcGIS (ESRI, Redlands, CA, USA), GeoVisionary (Virtalis Ltd., Cheshire, UK), and SubsurfaceViewer (INSIGHT GmbH, Cologne, Germany). The model creation was enhanced by the use of custom tools developed by an ISGS programmer. These custom tools supplemented and integrated the built-in standard ArcGIS toolsets to allow rapid 3-D visualization of attributes associated with the data (Figure 6.5a) and "on-the-fly" geological interpretation and interpolation of geological contacts (Figure 6.5b). Using these applications, more than 11 000 stratigraphic correlations were made for water-well driller's logs. GeoVisionary,

an immersive 3-D visualization package, was used to provide high-resolution 3-D visualization that integrated LiDAR-based DEMs, geologic and soil map layers, and ArcGIS-generated borehole data using color-coded cylinder representation (Figure 6.6). SubsurfaceViewer, a 3-D geological mapping software package that builds models from cross-sections and boreholes (Kessler et al. 2009; Chapter 10 this volume), was the main model-building application and the primary software used by stakeholders to interact with the final 3-D model. SubsurfaceViewer was used to create 32 key cross-sections based on the highest-quality drilling and geophysical data (Figure 6.7). Numerous additional supporting cross-sections were added to delineate and constrain geologic deposits in areas of either complex geometry or low data density.

6.6.5 Data Assessment

Over two field seasons (2008–2009), the ISGS mapping and drilling teams collected 31 continuous lithologic cores to bedrock. Gamma ray and electrical resistivity logs were collected in the boreholes at each location. The ISGS geophysics and mapping teams also collected approximately 51 km (32 mi) of seismic and 39 km (24 mi) of electrical resistivity profiles. These geophysical profiles provided transect-based data that provided important insights into the horizontal variations of the glacial sediments (Figure 6.8).

Water well driller's logs (~22 000 records) are an abundant source of geologic information in McHenry County. While the quality of these water well records is a recognized concern for geologic mappers (Logan et al. 2001), ISGS geologists have found them helpful for identifying major sedimentologic trends – particularly for identifying the upper portions of sand and gravel aquifers. The ISGS workflow incorporated a twofold approach to check and improve the quality of these records, through lithologic standardization and location verification. The non-standard lithologic descriptions in water well driller's logs were standardized into 17 standardized lithologies. This improved the speed and consistency of comparing different water-well driller's descriptions to each other and to the more detailed and reliable ISGS core descriptions. In a separate effort, locations for approximately 80% of the well records in the McHenry County project database were evaluated and, when determined to be incorrect, adjusted to the best location available. These corrections used street addresses and tax parcel information together with web-based mapping (i.e. GoogleMap) (Figure 6.9). Although expensive and time consuming, location correction was identified as very successful in improving

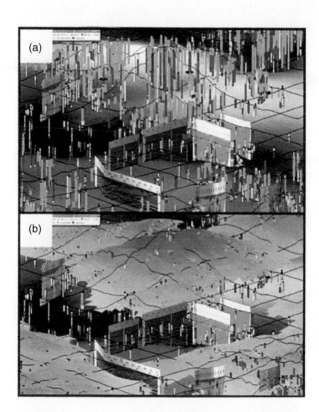

Figure 6.5 Model development using ArcScene software. (a) 3-D view of a portion of the stratigraphic correlation step; (b) interpolated surface of the top of one stratigraphic unit. Source: Modified from Thomason and Keefer 2011. © 2011 University of Illinois Board of Trustees. Used by permission of the Illinois State Geological Survey.

Figure 6.6 Completed McHenry County model showing topography and surficial geology, viewed from the southwest using GeoVisionary software. Green and purple shades identify fine-grained deposits (till or lake sediments); orange and yellow shades identify coarse-grained deposits (outwash and modern alluvium). Source: Modified from Thomason and Keefer 2011. © 2011 University of Illinois Board of Trustees. Used by permission of the Illinois State Geological Survey.

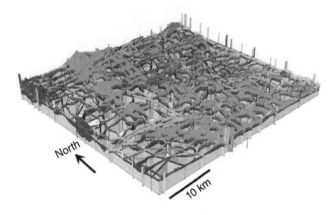

Figure 6.7 Selected cross-sections displayed above the bedrock surface grid. Source: © 2018 University of Illinois Board of Trustees. Used by permission of the Illinois State Geological Survey.

water-well data quality and thereby improving the resulting model – improved locations resulted in improved assignment of land surface elevations, which improved the elevations of each geologic deposit described in the logs.

The initial resolution for the McHenry County model was based on an earlier, more limited mapping effort in the County (Curry et al. 1997), which mapped all deposits on 1:24 000 scale base maps (Thomason and Keefer 2013). At this resolution, if sufficiently sampled, geologic deposits as small as 1.5 km (1 mi) in size could be reliably represented in the model. In the higher-resolution inset models, added to the project in year two, the model resolution allowed identification and representation of features as small as

0.75 km (half-mile) in size. In situations where sufficient data were available (e.g. geophysical data or high-density well records), local geologic deposits or features were resolvable to an even higher resolution (e.g. 0.25 km in size). To reduce interpolation errors near the county borders, the model domain was extended to include a 1.5 km (1 mi) buffer surrounding the county.

6.6.6 Project Deliverables

Project deliverables for the McHenry County project included a final report, standardized 2-D surface grid files for all stratigraphic top and bottom surfaces and thickness maps, graphic images of important cross-sections and seismic and resistivity profiles, and data files of geologic and downhole geophysical logs. In the other ISGS 3-D modeling projects, deliverables are project dependent and can include additional files of 3-D voxel-based models of rock properties, rock types or stratigraphic units, maps of aquifer thickness and aquifer sensitivity to contamination, data files for specific data types (e.g. SEGY files of seismic profiles), or videos of the 3-D models. In recent years, the standard delivery system has been the ISGS website, where links for all downloadable content is provided. Hard-copy versions of reports and maps can often be ordered from the ISGS website (http://isgs.illinois.edu).

6.6.7 Post-Project Model Management

The ISGS does not currently standardize the post-project management, maintenance, or archiving of 3-D geologic

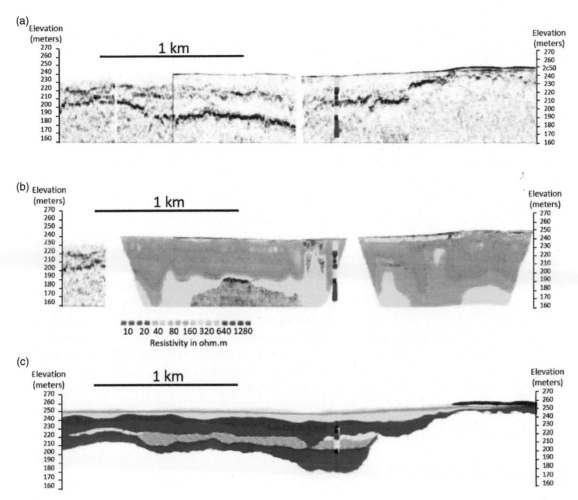

Figure 6.8 Geophysical profiles provide information on the glacial deposits: (a) seismic profile; (b) electrical resistivity profile; and (c) geological interpretation. Source: © 2018 University of Illinois Board of Trustees. Used by permission of the Illinois State Geological Survey.

models and data. Lead geologists are responsible for archiving and managing model and deliverable files on file servers and are encouraged to upload stratigraphic interpretations for well log and core description interpretations into the enterprise database. To date, most 3-D models have not undergone testing and implementation phases that would produce additional data or justify re-evaluation of the interpretations and modeled map-unit distributions.

6.7 Modeling Workflow Solutions by Other Organizations

Chapter 2 discusses the role of geological survey organizations in developing and managing 3-D geologic framework models, with special emphasis on the 3-D modeling experiences of the British Geological Survey, the Geological Survey of the Netherlands, and of the Great Lakes Geologic Mapping Coalition – a consortium of nine American state and Canadian provincial geological surveys and the U.S. Geological Survey. However, there are many additional successful 3-D modeling programs, in public institutions, which have resolved operational challenges. These programs are primarily in geological survey organizations, but they also include university faculty who conduct 3-D modeling as part of their research and teaching. The implementation and degree of success with which private companies have integrated 3-D modeling programs is impossible to evaluate. The growth of 3-D modeling software applications suggests that at least some consultants and corporations view this technology as valuable.

This section briefly summarizes the experiences of four organizations – one university and three geological survey organizations – that provide examples of workflow solutions to issues discussed earlier in this chapter. Additional relevant examples are provided in Chapter 8.

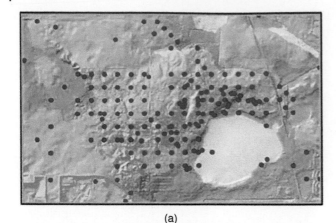

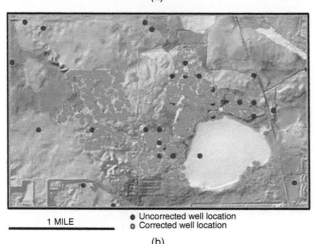

(a)

(b)

1 MILE

● Uncorrected well location
○ Corrected well location

Figure 6.9 Example of the impacts of verifying water-well locations. (a) Initial approximate locations provided by public records; (b) corrected, "best available" locations, as defined by GPS, street address, or tax parcel information. Modified from Thomason and Keefer 2011. © 2011 University of Illinois Board of Trustees. Used by permission of the Illinois State Geological Survey.

6.7.1   University of Waterloo, Department of Earth and Environmental Sciences

At the University of Waterloo, Ontario, Canada, Dr. Martin Ross has developed a geological modeling program to support his applied Quaternary geology research. Models have been applied to various problems such as sustainable groundwater resource development and land use management (Ross et al. 2002, 2005; Atkinson et al. 2014) and geospatial modeling for seismic site response (Nastev et al. 2016a,2016b). His modeling projects are conducted mostly by teams of graduate students with the workflow being driven by the project-specific research goals. Chapter 12 contains a technical discussion of Ross's modeling methods.

Student training is a priority – students are responsible for most of the modeling and creation of deliverables. The

durations of most geomodeling projects are designed to fit within the MSc or PhD research timeframes of two to four years. Ross maintains his own personal backups of models and allows the funding agents to handle their own long-term archival needs. A more robust archival system, while somewhat advantageous, is currently not available.

Ross's workflows begin with data preparation, including collection of existing data, acquisition of new information from field surveys, data quality assurance, and data standardization (Figure 6.10). The initial focus is on a geomorphic analysis and surficial geologic mapping, combined with analysis of existing subsurface data. Sedimentological interpretations and facies models are used to provide insights, and cross-sections are used to help define the spatial distributions of mapping units. Next, gaps in data distributions are identified and questions are developed regarding surficial and subsurface geologic environments. The collection of new data is often limited, but when funding is available new data are collected to resolve the questions at priority locations. Sedimentologic interpretations from surficial mapping are extended into the subsurface through development of upper and/or lower surfaces of all stratigraphic units. The process is iterative; re-evaluation and interpretation continue until the most likely solution is found.

Ross selected GOCAD/SKUA software for his modeling because its use of smooth interpolators and minimum-maximum thickness constraints are well suited for modeling thin and discontinuous Quaternary deposits. In addition, its high-quality visualizations, accessible academic licensing, and free model-viewing software

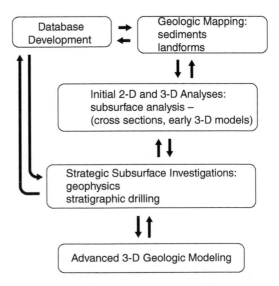

Figure 6.10 Typical Ross-team workflow for modeling Quaternary deposits for a regional hydrogeology project. © 2018 University of Illinois Board of Trustees. Used by permission of the Illinois State Geological Survey.

were advantages. GOCAD has an outstanding integration of property modeling algorithms and tools, which introduce geologic spatial constraints using more geologically-intuitive tools. Recent projects have further benefited from this choice as collaborators have also been GOCAD users.

While the full 3-D model is always the primary product, 2-D maps and cross-sections of various deposits are also important deliverables. Three-dimensional visualizations enhance understanding of subsurface material distributions, but 2-D maps are still needed to support spatially-constrained decision making. The 3-D models support volumetric and spatial calculations that can be displayed on maps and cross-sections and are used to generate images for publication. Models are distributed primarily to research sponsors and are discussed in journal articles, with a few being documented in government and other open-access reports. Ross has produced updates based on additional information on sediment distribution in a previously-modeled area; these updates have been distributed by email to the supplier of the new information.

6.7.2 Delaware Geological Survey

The Delaware Geological Survey (DGS) has conducted two county-scale projects mapping the distribution of surficial and subsurface geological units (Andres and Kleinbeil 2006; McLaughlin and Velez 2006; McLaughlin et al. 2014) to support groundwater resource assessment and management needs of state agencies. These projects have allowed the creation of 3-D geological models for aquifers and confining units in these counties, which are used for groundwater pumping analyses. A detailed case study of DGS geologic modeling in this program is included in Chapter 19.

The DGS projects demonstrate an approach to 3-D modeling workflows at state geological surveys with small project budgets and few staff. The projects typically involve a lead geologist or hydrogeologist and a junior geoscientist or two. Essential new subsurface geological data are generally collected for these projects through drilling a small number of test holes at key locations. However, because of limitations in timelines, budget, and staffing, the DGS subsurface mapping efforts principally use existing data, mostly water-well driller's descriptions. The stratigraphic picks made for these boreholes are managed in a centralized database.

Within the DGS 3-D modeling workflow (Figure 6.11), geologic materials descriptions are compiled for the study area, then correlated to the mapping units. Mapping-unit distributions – as defined by top and bottom surface models – are interpolated into a series of stacked 2-D surfaces using the Radial Basis Function method in ArcGIS Geostatistical Analyst. These are visualized in 3-D, analyzed, and, using grid algebra tools, cross-checked with a land-surface digital elevation model (Delaware Geological Survey 2008) and with grids of adjacent stratigraphic surfaces. Grids are iteratively adjusted as needed to ensure that resulting unit surfaces interact in a geologically-realistic manner and that distributions of the units fit the data and geologic interpretations. The DGS uses ArcGIS as their software platform for all surface modeling, 3-D visualization and product development. Even though ArcGIS does not support a robust 3-D subsurface data model, geologists using ArcGIS were able to construct fully-populated 3-D models with stratigraphic map units that are characteristic of the state's relatively undeformed coastal plain province. ArcGIS was chosen due to cost, ease of use, sophisticated interpolation algorithms, and grid algebra toolbox, which provided a single integrated software environment for the full project workflow.

The state environmental regulatory agency requested DGS assistance in defining the distribution and thickness of aquifers and aquitards. To meet their needs, deliverables focused on 2-D maps of aquifer elevation and thickness, cross-sections showing the relationship of aquifers and aquitards, and maps of existing water well development within the various aquifers. The data files generating the maps and resultant map products were included in DGS reports to the State and are actively used by state personnel for well permitting and groundwater allocation decisions.

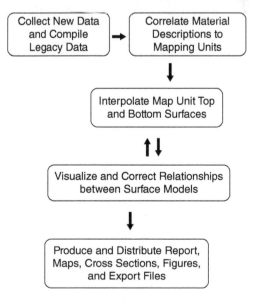

Figure 6.11 Generalized workflow for 3-D modeling of coastal province deposits at the Delaware Geological Survey. © 2018 University of Illinois Board of Trustees. Used by permission of the Illinois State Geological Survey.

The small size of the DGS and the lack of a formal 3-D mapping program has resulted in a less structured approach to maintenance and archiving of 3-D mapping project results. The correlations of borehole log descriptions to mapping units are retained in the central database, and grid files describing map unit surfaces and any graphics files of cross-sections or map products are backed up and managed by the project's lead geologist. This strategy meets the DGS needs while requiring only limited technical specialization.

6.7.3 Ontario Geological Survey

The Ontario Geological Survey (OGS) has a long-term, provincially-funded initiative focused on source water protection through 3-D Quaternary sediment mapping: maps of aquifer and aquitard distribution support land use planning and sustainable water supply management. Funding for each three- to five-year project is significant, with much of it targeted to collection of new data. The guiding priority of the initiative is the creation of a high-quality 3-D interpretation of the subsurface. A wide range of products are designed for various stakeholder groups, including consultants, policymakers, planners, and the public.

Mapping projects – staffed by a fully dedicated lead geologist, and assisted on a part-time basis by a geophysicist, a data manager, a graphic artist, and student assistants – start with a compilation of existing published and unpublished information and a reconnaissance field season (Figure 6.12). The reconnaissance field work is used to develop an understanding of the regional Quaternary history, check existing surficial mapping, and locate sites for geophysical surveys and initial drilling, which is conducted over subsequent field seasons. Particle size and geochemical analyses are conducted using targeted subsamples from the core collected during drilling.

Modeling of regional-scale mapping units involves the iterative creation and evaluation of surfaces and cross-sections. The primary modeling software used is Datamine Studio, using proprietary scripts generated for the OGS under contract with Datamine, that improve the efficiency and effectiveness of the OGS modeling workflow. OGS chose Datamine Studio as their primary modeling software because:

- It linked well to Microsoft Access databases.
- It provided geologic constraints to assist in modeling surfaces.
- It allowed for efficient creation of 2-D surfaces and 3-D voxel models.
- It had tools for volumetric calculations
- It could export easily to various ASCII and ESRI formats.
- It provided high-quality visualizations of models and cross-sections.

Although only two geologists are involved with the mapping, product development and reporting, the OGS staff have established a wide array of products – all of which are available as free downloads from the website OGSEarth, or on CD/DVD by mail for a nominal fee. Deliverables are generated at several steps in the project and are marketed to provide information to stakeholders as the project advances. Summary Field Reports are compiled and published annually to provide consultants with an early understanding of the new data – the reports do not

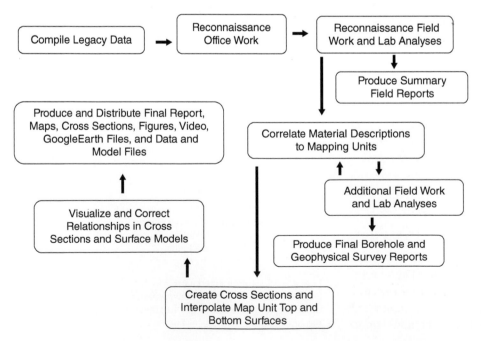

Figure 6.12 Generalized workflow for 3-D modeling of Quaternary sediments at the Ontario Geological Survey. © 2018 University of Illinois Board of Trustees. Used by permission of the Illinois State Geological Survey.

include detailed analytical work or detailed core descriptions. Detailed borehole data are released, with products targeted towards consultants and more technical stakeholders, after all analytical results are reviewed. Release of gridded geophysical maps and data follow interpretation of the geophysical surveys. A robust Final Report is published following model completion, with components targeted to various stakeholder groups. The final report contains a comprehensive discussion of data, methods, interpretations, predicted model results, characterization of each mapping unit, and a discussion of aquifer vulnerability to contamination and recharge potential. Technical data are formatted according to the needs of various stakeholder groups. Products derived specifically from the 3-D model include top-surface grid files for all mapped units, and cartographically-standardized map plates of aquifers and important lithostratigraphic units. Google Earth files are created for the boreholes, pre-selected cross-sections, and all map unit surfaces. Occasionally, OGS geologists will create and publish 3-D videos of the 3-D models.

The OGS groundwater mapping initiative is structured to provide one-time characterizations for each region. There are no current plans to update interpretations or predictions as new data become available or to integrate new data into the regional database. Currently, there is no structured procedure for archiving or maintaining 3-D models generated within this program. Models and data in final format are stored and made available from the website, and copies are maintained on the lead geologist's computer.

6.7.4 Geological Survey of Denmark and Greenland

In 1998, the Danish Parliament passed a plan to dramatically improve the characterization and protection of groundwater resources. Subsequently, the regional water resource authorities were tasked with geologic mapping and hydrologic modeling to ensure protection of groundwater resources. High-value water abstraction areas were targeted for detailed mapping – an area covering almost $17\,800$ km^2, or 41% of the land area of Denmark. A water-use tax of €0.04/m^3 paid for the mapping program – providing approximately €250 000 000 over the life of the program. These funds were used to pay for helicopter-borne, time-domain electromagnetic (HTEM) geophysical surveying, drilling, water sampling and hydrogeologic modeling (Thomsen 2011). The Environmental Agency in Denmark managed this program and oversaw consultants who collected and processed the HTEM data and created the geologic models from these datasets. The Geological Survey of Denmark and Greenland (GEUS) developed HTEM collection standards, developed a small number of 3-D geologic models from select HTEM datasets, provided technical assistance in the development and evaluation of software modeling methods, and managed a national database of completed 3-D geologic models.

At GEUS, a typical 3-D modeling team includes 5–10 staff on various aspects ranging from data cleaning, processing and management, data interpretation, geological modeling and product development. Because there are no significant funds for additional data collection, the workflow on 3-D modeling projects typically follows the pattern of data compilation, interpretation and correlation to stratigraphic framework, and creation of 3-D lithostratigraphic models. Data are compiled from the HTEM surveys, which have been collected along transects spaced approximately 250 m apart. An exploratory borehole with sample collection is located approximately every 4 km^2 (Thomsen 2011). These data are used to calibrate and correct interpretations of rock and sediment types characterized within the HTEM data. While the interpretation and correlation steps in the modeling workflow are iterative and allow for updating and revision, there is no room in the schedule or budget for additional data collection later in the project.

GEUS relies primarily on the GeoScene3D geologic modeling software application to interpret and correlate the HTEM and borehole data and to construct the 3-D geologic models. GeoScene3D was chosen because it is formally integrated within the groundwater mapping program at a national level – interacting directly with the national HTEM (GERDA) and water well (Jupiter) databases. Importantly, GeoScene3D is capable of directly loading the very large geophysical survey data sets that are generated by this program – HTEM data sets for a given 3-D model can include 500 000 resistivity soundings – and has machine learning algorithms integrated into the software to simplify interpretation of these large data sets. GeoScene3D is used to create both surface-based 3-D stratigraphic models and 3-D voxel models based on interpretation of resistivity and other geologic properties.

Products developed in GEUS-led geologic modeling projects include a report describing the mapping units and geological setting of the model area and a full description of the final model, interpretations of the HTEM profiles, selected cross-sections through the model, maps of top surfaces and thicknesses of modeled units, updates to the GERDA and Jupiter databases from data collected for the project, and a digital version in GeoScene3D format of the final 3-D geologic model. The data interpretations, modeled surfaces, and report are made available through the national database of 3-D geologic models, and the HTEM and borehole data are available through the GERDA and Jupiter databases, respectively.

6.8 Creating a Custom Workflow

The overarching goals of geologic framework modeling projects include describing the morphology and distribution of mapping units and structural features. Project objectives and programmatic priorities are the major constraints determining mapping units and how the modeling results will be used. The five case studies demonstrate how variations in funding and staffing can be used to create different strategies for modeling workflows within the larger programmatic goals. These examples also highlight how continuity in modeling programs tends to accompany evolutions in how products are managed, maintained, distributed, and archived.

Technology has supported 3-D geological modeling for over 25 years, and modeling continues to evolve with the introduction of major new modeling technologies, improved software capabilities, imposition of new legal and operational requirements reflecting the digital information revolution, and societal economic and environmental priorities. The ideas and examples presented in this chapter provide a framework that can help launch a new modeling program or review and re-evaluate the direction and resource investment of an established program. Evaluation of specific needs and potentials of a long-term commitment to a 3-D modeling program are critical steps to developing a workflow that fits specific situations.

Acknowledgments

The authors acknowledge the assistance of colleagues active in the development of 3-D geological models in writing Section 6.7. The following individuals have reviewed and corrected the following:

- Section 6.7.1: Dr. Martin Ross, Department of Earth and Environmental Sciences, University of Waterloo, Waterloo, ON, Canada
- Section 6.7.2: Dr. Peter McLaughlin, Delaware Geological Survey, University of Delaware, Newark, Delaware, USA.
- Section 6.7.3: Abigail Burt, Ontario Geological Survey, Sudbury, Ontario, Canada.
- Section 6.7.4: Dr. Flemming Jørgensen, formerly at the Geological Survey of Denmark and Greenland, now at Environment Division, Central Denmark Region, Viborg, Denmark.

References

Andres, A.S. and Klingbeil, A.D. (2006). *Thickness and Transmissivity of the Unconfined Aquifer of eastern Sussex County, Delaware*. Newark, DE: Delaware Geological Survey, Report of Investigations No. 70. 19 pp. 1 plate.

Atkinson, L.A., Ross, M. and Stumpf, A.J. (2014). Three-dimensional hydrofacies assemblages in ice-contact/proximal sediments forming a heterogeneous 'hybrid' hydrostratigraphic unit in East-Central Illinois, USA. *Hydrogeology Journal* 22: 1605–1624. https://doi.org/10.1007/s10040-014-1156-7.

Burke, H.F., Hughes, L., Wakefield, O.J.W. et al. (2015). *A 3D Geological Model for B90745 North Trans Pennine Electrification East between Leeds and York*. Keyworth, UK: British Geological Survey, Commissioned Report CR/15/04N. 32 pp. [Online: Available at: http://nora.nerc.ac.uk/id/eprint/509777/] (Accessed July 2018).

Curry, B., Berg, R.C. and Vaiden, R.C. (1997). *Geologic Mapping for Environmental Planning, McHenry County, Illinois*. Champaign, IL: Illinois State Geological Survey, Circular 559. 79 pp. 4 Plates. [Online: Available at: http://library.isgs.illinois.edu/Pubs/pdfs/circulars/c559.pdf] (Accessed July 2018).

Delaware Geological Survey (2008). *Continuous 3-meter DEM of the State, DMRASTER.DE_DEM3*. Newark, DE: Delaware Geological Survey.

Jørgensen, F., Sandersen, P.B.E. and Auken, E. (2003). Imaging buried quaternary valleys using the transient electromagnetic method. *Journal of Applied Geophysics* 53 (4): 199–213. https://doi.org/10.1016/j.jappgeo.2003.08.016.

Kessler, H., Mathers, S. and Sobisch, H.-G. (2009). The capture and dissemination of integrated 3D geospatial knowledge at the British Geological Survey using GSI3D software and methodology. *Computers and Geosciences* 35 (6): 1311–1321. https://doi.org/10.1016/j.cageo.2008.04.005.

Logan, C., Russell, H.A.J. and Sharpe, D.R. (2001). *Regional Three-Dimensional Stratigraphic Modelling of the Oak Ridges Moraine Area, Southern Ontario*. Ottawa, ON: Geological Survey of Canada, Current Research CR2001-D1. 30 pp. https://doi.org/10.4095/212119.

McLaughlin, P.P. and Velez, C.C. (2006). *Geology and Extent of the Confined Aquifers of Kent County, Delaware*. Newark, DE: Delaware Geological Survey, Report of Investigations No. 72. 40 pp. 1 plate. [Online: Available at: https://www.dgs.udel.edu/publications/ri72-geology-and-extent-confined-aquifers-kent-county-delaware] (Accessed July 2018).

McLaughlin, P.P., Tomlinson, J.L. and Lawson, A.K. (2014). *Aquifers and Groundwater Withdrawals, Kent and Sussex Counties, Delaware*. Newark, DE: Delaware Geological Survey, Contract Report. 116 pp.

Van der Meulen, M.J., Broers, J.W., Hakstege, A.L. et al. (2007). Surface mineral resources. In: *Geology of the Netherlands* (eds. T.E. Wong, D.A.J. Batjes and J. De Jager), 317–333. Amsterdam, NL: Royal Netherlands Academy of Arts and Sciences (KNAW).

Van der Meulen, M.J., Doornenbal, J.C., Gunnink, J.L. et al. (2013). 3D geology in a 2D country: perspectives for Geological Surveying in the Netherlands. *Netherlands Journal of Geosciences (Geologie en Mijnbouw)* 92 (4): 217–241. https://doi.org/10.1017/S0016774600000184.

Nastev, M., Parent, M., Benoit, N. et al. (2016a). Regional VS30 model for the St. Lawrence Lowlands, Eastern Canada. *Georisk* 10 (3): 200–212. https://doi.org/10.1080/17499518.2016.1149869.

Nastev, M., Parent, M., Ross, M. et al. (2016b). Geospatial modelling of shear-wave velocity and fundamental site period of quaternary marine and glacial sediments in the Ottawa and St. Lawrence Valleys, Canada. *Soil Dynamics and Earthquake Engineering* 85: 103–116. https://doi.org/10.1016/j.soildyn.2016.03.0060267-7261/.

Ross, M., Parent, M., Lefebvre, R. and Martel, R. (2002). 3D geologic framework for regional hydrogeology and land-use management; a case study from southwestern Quebec, Canada. In: *Three-Dimensional Geological Mapping for Groundwater Applications, Workshop Extended Abstracts* (eds. R.C. Berg and L.H. Thorleifson), 52–55. Ottawa, ON: Geological Survey of Canada Open File 1449. https://doi.org/10.4095/299506.

Ross, M., Parent, M. and Lefebvre, R. (2005). 3D geologic framework models for regional hydrogeology and land-use management: a case study from a quaternary basin of Southwestern Quebec, Canada. *Hydrogeology Journal* 13 (5–6): 690–707. https://doi.org/10.1007/s10040-004-0365-x.

Thomason, J.F. and Keefer, D.A. (2011). Geologic framework modeling for groundwater applications in Northeast Illinois. In: *Three-Dimensional Geological Mapping; Workshop Extended Abstracts* (eds. H.A.J. Russell, R.C. Berg and L.H. Thorleifson), 79–82. Ottawa, ON: Geological Survey of Canada, Open File Report 6998, https://doi.org/10.4095/289609.

Thomason, J.F. and Keefer, D.A. (2013). *Three-Dimensional Geologic Model of Quaternary and Holocene Sediments, McHenry County, Illinois*. Champain, IL: Illinois State Geological Survey, Contract Report.

Thomsen, R. (2011). 3D groundwater mapping in Denmark based on calibrated high-resolution airborne geophysical data. In: *Three-Dimensional Geological Mapping; Workshop Extended Abstracts* (eds. H.A.J. Russell, R.C. Berg and L.H. Thorleifson), 79–82. Ottawa, ON: Geological Survey of Canada Open File Report 6998. https://doi.org/10.4095/289609.

Waters, C.N., Terrington, R.L., Cooper, M.R. et al. (2016). *The Construction of a Bedrock Geology Model for the UK: UK3D_v2015*. British Geological Survey Report OR/15/069. 22 pp. [Online: Available at: http://nora.nerc.ac.uk/id/eprint/512904/1/UK3D Metadata Report_Submission copy.pdf] (Accessed February 2018).

7

Data Sources for Building Geological Models

Abigail K. Burt[1], Phillip Sirles[2], and Alan Keith Turner[3,4]

[1] Ontario Geological Survey, Sudbury, ON P3E 6B5, Canada
[2] Collier Consulting, Lakewood, CO 80215, USA
[3] Colorado School of Mines, Golden, CO 80401, USA
[4] British Geological Survey, Keyworth, Nottingham NG12 5GG, UK

7.1 Introduction

The efficient development of any subsurface geological model requires the identification and acquisition of the best available data, as well as the maintenance of subsurface databases and resulting models within a suitable data management environment. Chapter 5 introduces four model-building approaches. Chapters 9–12 discuss these in greater detail. Chapter 6 emphasizes the importance of adopting efficient workflows to support the construction of models that meet user requirements.

This chapter discusses the identification of valid data for 3-D model development that is fit-for-purpose and collected in a cost-effective manner. The chapter begins by exploring the difference between data and information, then describes alternative data classifications and the advantages and disadvantages of using legacy data sets. The remaining sections of the chapter describe sources of elevation data, surficial and subsurface geological data, and geophysical data. Chapter 8 continues with discussions on the importance of appropriate management procedures for both data and models. Chapter 15 explores how data quality determines the accuracy and precision of models – an important component of model construction.

7.2 Defining and Classifying Data

Data are raw facts or observations and may be *quantitative* (measurements that are recorded numerically) or *qualitative* (descriptive). The measured sample depths and grain-size data shown in columns A and B of Figure 7.1 are examples of quantitative data. Sketches and descriptions of landforms are qualitative data (Figure 7.2). Quantitative data are typically easily imported into computer modeling software. Qualitative data, such as the sketch of a dipping anticline shown in Figure 7.2, typically assist with model conceptualization during 3-D model development; they help define the rules and relationships which the 3-D model must honor.

7.2.1 Data Versus Information

The terms *data* and *information* are frequently used interchangeably although they are different. When context, or interpretations, are provided, data becomes *information*, which can then be used for decision-making and/or problem-solving (Liew 2007). Geologists use their expert knowledge and geological theories to analyze and interpret raw data, interpolating between data points (Clarke 2004). Geological maps, cross-sections, and geotechnical reports interpreting data are examples of information, which in numerical form can be used in much the same way as quantitative data. Because most geological models are constructed using a combination of data and information, these sources are collectively referred to as "data sources," or "data" in this chapter.

The borehole log in Figure 7.1 demonstrates the close relationship between data and information. The lithology is summarized by a combination of quantitative and qualitative data; columns A and B present quantitative (numerical) definitions of measured sample depths and grain-size percentages, while column C contains non-numerical (qualitative) lithologic descriptions such as "fine sand" or "bedrock" and column D contains numerical (quantitative) percentage logs that summarize laboratory grain size analyses shown in column B along with percentages of four pebble lithology classes. The final column E contains interpretive information defined by the geologist based on a combination of field descriptions

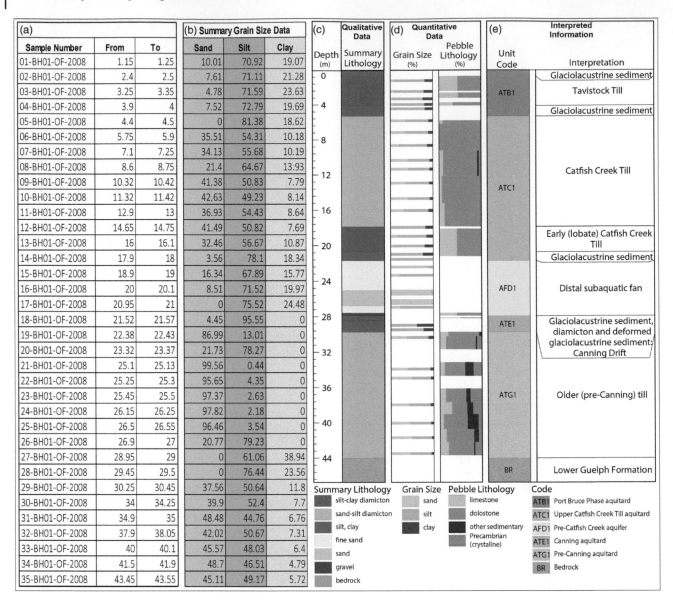

Figure 7.1 Comparison of data and information from a continuously cored borehole in Ontario, Canada. (a) Measured sample depths; (b) Summary quantitative grain size data (percent sand, silt and clay); (c) graphically displayed qualitative data (field descriptions); (d) graphically displayed quantitative data (grain size and pebble lithologies); (e) information (hydrostratigraphic units and interpreted depositional environments and formations). Modified from Burt and Dodge (2016). Copyright © Queen's Printer for Ontario, 2016. Used by permission of the Ontario Geological Survey.

and laboratory analyses in columns A–D; the unit code identifies hydrostratigraphic units used for 3-D modeling while the interpreted depositional setting assigns specific formation names for the various tills.

Interpreting data makes it easier for non-specialists to use; however, interpretations are subjective and may change as our understanding of geological systems are refined or new data become available. For example, a shallow borehole log may report silty clay with a trace of gravel. Without supporting data, such as degree of compaction, stratigraphic unit, or landform classification, this sediment may be interpreted as either alluvium deposited by a river or as a fine-grained till deposited by a glacier.

7.2.2 Classifying Data

Data may be classified in several ways, including:

- Spatial location and extent using points, lines, and polygons;
- Framework versus property data;
- Elevation, surficial and subsurface data.

Figure 7.2 Field sketch of a dipping anticline.

7.2.2.1 Spatial Location and Extent Using Points, Lines, and Polygons

Geographic information systems (GISs) provide the tools required to create and disseminate digital versions of traditional 2-D maps (Aronoff 1989). Many GIS employ topological principles to relate various map elements defined by points, lines, and polygons. Points define sites of individual measurements or observations (Aronoff 1989) such as boreholes, water wells, springs, or sinkholes. Points may also define the locations of groups of observations that are too small to be presented individually. Lines are defined by sequences of points and are used to represent geologic contacts, strike lines, fold axes, fault traces, topographic contours, roads, streams, or the geophysical survey transects discussed below (e.g. seismic, ground-penetrating radar, electro-magnetic). Polygons define areas with common characteristics. For example, they are used to represent physiographic landform classes, geologic formations, lakes, rivers, wetlands, land ownership, and administrative units.

In addition to map elements defined by topologically related points, lines, and polygons, most GIS use regularly spaced grids or triangulated networks (TINs) to represent interpolated surfaces defining terrain, water tables, or subsurface stratigraphy. As outlined in Chapter 10, these interpolated surfaces provide only a 2.5-D representation: as they have a single z-value per x, y-pair, they can neither fully envelope a volume, nor can they represent complex geologic structures that involve over-turned folds or repeated strata due to faulting.

These concepts can be extended to 3-D subsurface modeling workflows (Soller and Berg 2001). Observations from individual boreholes are now represented by 3-D points defined by three coordinates (x, y, and z). While individual cross-sections derived from geological interpretations or seismic reflection profiles can be represented by standard 2-D data sets, if the cross-sections are to be correctly located within a 3-D model, they must be converted to 3-D polylines. Similarly, 2-D polygons will become 3-D surface or volume elements, while 3-D meshes define discretized models (Chapter 13).

7.2.2.2 Framework Versus Property Data

Data can be classified according to how it is used in building a framework model or populating the model domain with properties. A framework model divides the model domain into, for example, lithostratigraphic, hydrostratigraphic or geotechnical units, depending on the intended application. The base and top of the modeled units, as well as faults, are represented by surfaces and mark a change in bulk or average properties, just as lines do on a traditional 2-D geological map. A framework model is primarily geometric, but to some extent it could also be considered a property model because between its bounding surfaces the diagnostic bulk or average properties associated with the modeled unit apply. This means that average property values can be assigned to geological units by using standardized tables for clay, silt, sand, sedimentary and crystalline rocks, or facies interpreted from drillers logs (Royse et al. 2009; Dunkle et al. 2009).

However, many model applications require more detailed evaluations of the internal variability within individual geological units. For example, 3-D groundwater flow or contaminant transport models require hydraulic properties such as conductivity, transmissivity, permeability, porosity, and density. Determining geotechnical responses to proposed engineering works may require the definition of characteristics such as plasticity, shrinkage, penetration, and shear strength. This requires the framework model to be subdivided or discretized into a network of small cells, each of which is assigned a unique value for each property. Chapter 13 discusses the development of regularly spaced 2-D grids or 3-D meshes required to assign high-resolution quantitative properties data to define the internal heterogeneity of individual geological units. It also discusses the application of stochastic modeling to assess the uncertainties in the assignment of estimated properties values in discretized property models. Chapter 11 describes in detail how a framework model is used to populate a voxel model with properties while keeping the interpolations confined to unit boundaries and each unit having its own spatial correlation structure.

7.2.2.3 Elevation, Surficial, and Subsurface Data

Elevation, surficial, and subsurface data form the principal data types for model building. Elevation data describe the shape of the earth's surface and may be represented according to local (project-specific), national, or international datums such as mean sea level. Accurate elevation data is critical as the ground surface, either on land or under water, forms the top surface of many geological models and is used to "hang" subsurface data when creating 3-D models (McLaughlin et al. 2015).

Because the Earth's surface is accessible, surficial data provide detailed descriptions of the near-surface sediments or rocks. In contrast, because subsurface observations are much more difficult to obtain, subsurface datasets typically contain relatively widely spaced and limited descriptions of sediments and bedrock. Yet, subsurface data are vital for all 3-D models. Depending on the focus of the 3-D modeling exercise, surficial data may form an important dataset or may be irrelevant. Ensuring that model layers align with a simplified rendering of surficial geology is important for 3-D sediment models produced for land use planning and groundwater flow modeling applications (Burt and Dodge 2016). In contrast, 3-D bedrock geology models of the same areas may simply strip away the overburden and use the bedrock surface as the model top or lump the overburden into a single, undifferentiated unit (Logan et al. 2020).

7.3 Legacy Data

Geological survey organizations usually develop regional 3-D geological models from existing, or legacy, data which may belong to the survey or have been acquired from other sources; such data may be supplemented with new field data obtained specifically for the project. There are many advantages to using legacy datasets, but there are also pitfalls. Data can be used with confidence for model development if legacy datasets with locations, elevations, geological formations (contacts, material types), and analytical results (grain size, geochemistry) defined in georeferenced and standardized digital formats have been subjected to a quality assurance or vetting process (Barnhardt et al. 2001). Although it is a time consuming, laborious, and therefore expensive, process to prepare legacy data that is not housed in a database (Russell et al. 2015), the use of such legacy data may still produce considerable cost savings (Arnold et al. 2001). This is important because many projects do not have the resources to acquire new data. Projects with workplans that include new data collection will also benefit from incorporating legacy datasets as the collection of new data can be targeted to resolve uncertainties or questions that are discovered during initial model-building efforts. Other advantages of using legacy datasets are the ability to document change (subsidence, compaction, mass wasting, mineral extraction), and to model pre-existing conditions or conditions at locations that are no longer accessible. Use of interpreted legacy datasets has an additional potential benefit of gaining insights from diverse sources.

Figure 7.3 shows the locations of legacy water well, petroleum, and geotechnical records, as well as new geological and geophysical data, acquired for the Ontario Geological Survey (OGS) Niagara Peninsula 3-D sediment model (Burt 2017; OGS 2014; Dietiker et al. 2019). Nearly 750 line-kilometers (470 line-miles) of ground gravity and seismic surveys and 130 measured outcrops and hand auger and probe holes were used to select optimal exploratory drilling locations that would define landform-sediment associations. A total of 99 continuously cored boreholes, producing a total of 3200 m (10 500 ft) of core, were acquired to facilitate the interpretations of legacy datasets. Within the 5000 km² (1950 mi²) project area, the legacy datasets consisted of 28 high-quality archived field observations and boreholes, 13 552 geotechnical records, 7561 petroleum records, and 33 105 water well records.

Legacy datasets are not without limitations and modelers and their clients must use caution when creating and using models that are based on legacy data. Lithology, the most basic of geological properties, is often based on visual estimation – this may be a source of operator-dependent error of uncertain magnitude. Furthermore, lithology is often defined by classification schemes, which can vary over time, and may change depending on location, and often differ among organizations and applications. For example, common terms such as "fine sand" and "coarse sand" are used in most sediment classifications, but

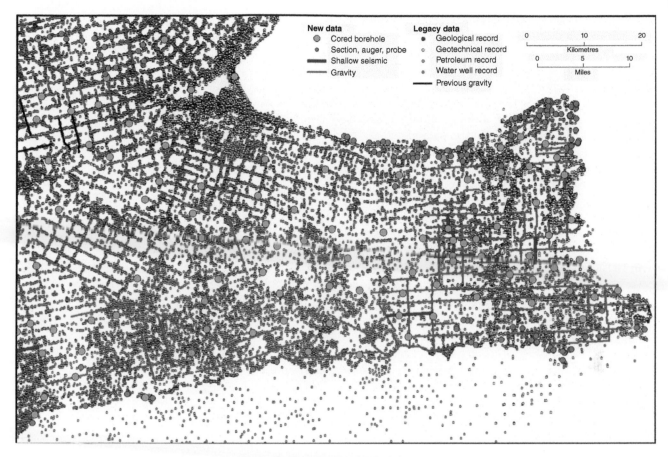

Figure 7.3 Distribution of legacy datasets (geological, geotechnical, petroleum, and water well records) and new geological and geophysical data (gravity and seismic surveys, measured outcrops and hand auger and probe holes and continuously cored boreholes) used to construct a 3-D sediment model within the Niagara Peninsula of southern Ontario, Canada. Data from Burt (2017), Burt et al. (2018), Ontario Geological Survey (2014), Ontario Ministry of Environment, Conservation and Parks (2013), Ontario Ministry of Natural Resources and Forestry (2019), Dietiker et al. (2019). Copyright © Queen's Printer for Ontario, 2019. Used by permission of the Ontario Geological Survey.

different classifications often use considerably different class boundaries. Uncertainty in lithological data is important when creating 3-D models because this uncertainty propagates into any derived properties such as hydraulic conductivity, porosity, or aggregate potential.

Legacy data sets may contain observations that are sporadic or clustered, and were probably obtained for multiple purposes; thus, the observations may not include the level of detail or types of observations required to create an accurate model. Legacy observations may focus on the wrong part of the geological record or may only extend to shallow depths. Assessing the quality of the data and the errors or uncertainty that may exist in defined spatial locations or geological observations may be difficult (see Chapter 15). Poor location control, both of spatial position and elevation, is common, particularly in older datasets where GPS technology was not available. The errors and uncertainty of location are compounded when the wrong projection or coordinate system is used to incorporate the

data into a database (Figure 7.4a). Changes in geological conceptualizations, including revisions to interpretations of regional geology, will produce errors and uncertainty in geological models; additional sources of errors and uncertainty result from the use of unknown and possibly incorrect techniques used to make measurements or geological observations, or if insufficient data and unknown or inappropriate techniques are used during interpolation or extrapolation (Clarke 2004).

A wide variety of sources may provide legacy datasets. Many public agencies, especially geological surveys, make their data holdings available free of charge while others operate under a business model that provides free access to basic data, but charge a fee for enhanced or interpreted data. Data that are in the private sector, such as site assessments or environmental impact reports prepared by consulting companies, can be difficult to access due to business interests or client confidentiality agreements (Russell et al. 2015; Kessler et al. 2011).

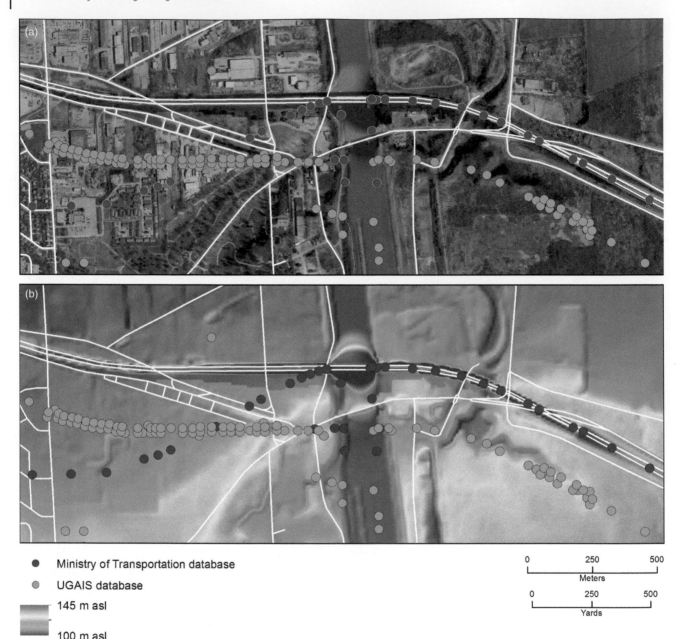

Figure 7.4 Geotechnical borehole records obtained from different sources for a bridge over the Welland Canal in Ontario, Canada. The road network is shown in white. (a) Displaying the boreholes on ortho imagery shows that the Ministry of Transportation (MTO) GEOCRES database portrays the data (blue dots) in its proper location whereas the records from an urban geotechnical database of subsurface information (orange dots) are in a different coordinate system. (b) The boreholes were drilled prior to construction of the bridge and ramps. If elevation data are obtained from a recent digital surface model that includes the ramps, the borehole elevations will not correspond to the ground surface. Data from Ontario Geological Survey (2019). Copyright © Queen's Printer for Ontario, 2019. Used by permission of the Ontario Geological Survey.

7.4   Elevation Data

Surface elevation data, in the form of a digital terrain model (DTM) is a critical dataset for 3-D modeling (Kessler et al. 2011). Early DTMs were created by digitizing contours and spot elevations shown on topographic maps; the accuracy

and precision of the resulting DTM depended on the age and quality of the base map. Mapping of elevations at the national level began in the UK in the 1740s; these early surveys used compasses and chains, sketching in features by eye. A national geodetic leveling survey began in the 1840s (Ordnance Survey 2019) and the pace of geological mapping was greatly accelerated after systematic topographic surveying provided the essential base layer (Chapter 2). In the United States, low resolution field-based topographic maps were developed during the late nineteenth century from hand-sketched contours and control points obtained using tape and compass traverses and aneroid barometers (Usery et al. 2009). Subsequent advancements in surveying techniques, such as the use of plane tables, and later photogrammetry using aerial photography, improved the accuracy of contour mapping (Usery et al. 2009).

Computer-assisted cartography and GIS-based seamless mapping using differential GPS, satellite interferometry and airborne LiDAR data have increased accuracy and resolution of the terrain data available (Jones et al. 2009; Usery et al. 2010). A DTM derived from a LiDAR survey may combine the elevation of the naturally occurring or anthropogenically modified ground surface including vegetation (tree canopy) and built features (buildings) as found in digital surface models (DSMs), or the data may be processed to remove vegetation and built features to create a smooth, bare-earth digital elevation model, or DEM (GISGeography 2018). DEM and bathymetric datasets are publicly available in many countries and may be obtained from agencies such as the United States Geological Survey (USGS) and the National Oceanic and Atmospheric Administration (NOAA) Office for Coastal Management, and the European Space Administration (ESA). Vector and raster elevation datasets are available in formats suitable for use in GIS and modeling environments.

For geological models of coastal locations, or if there are lakes, large rivers, or other water bodies within the model area, the DTM must be adjusted to account for the depth of the water bodies. Bathymetric data, ranging from historic lead line soundings obtained from nautical charts to 50 cm resolution raster datasets derived from multibeam surveys, can be used to extend land-based DTMs into large bodies of water as shown in Figure 7.5 (European Marine Observation and Data Network 2019; Logan et al. 2020; Matile et al. 2002). This is particularly important where borehole or geophysical data are used to extend model boundaries offshore (Logan et al. 2020).

Although terrain data are becoming more accurate, and DTMs are defined by smaller grids, it is not always possible to filter out the human imprint such as urban sprawl and expanding transportation corridors which modify the landscape. Natural processes of erosion and deposition also constantly modify the landscape. Because the DTM often forms the datum for the geological data used in model construction, a changed landscape can result in contacts being placed at the wrong elevation. Figure 7.4b shows the locations of boreholes drilled prior to construction of the

bridge and ramps. If elevation data are obtained from a recent DSM that includes the ramps, the borehole elevations will not correspond to the ground surface. In cases where changes to the land surface are ongoing, a better geological model will result if terrain data match the age of the subsurface data. Where land-surface modification is limited to only parts of the model area, an effective approach is to create a mosaic from two or more datasets, using the best data for each sub-area (Ross et al. 2015).

Depending on the scale of the model, high resolution DTM data may not be required and may unnecessarily increase the size of the model files and adversely affect the model construction process. For this reason, many DTMs must be resampled before being used to create geological models. Figure 7.6 illustrates how model type, spatial resolution, vertical accuracy, and age of the source data can affect a resampled terrain model. Figure 7.6a–d show four DTMs derived from multiple sources created between 2000 and 2018; they show increased spatial resolution from 30 m to 50 cm with corresponding increased vertical accuracy from ~5 m to ~10 cm. A landfill is clearly visible at the middle left of Figure 7.6a,c,d; it has been filtered out of the dataset in Figure 7.6b, forming a DEM. Figure 7.6e–h are resampled versions of the data in Figure 7.6a–d with a 100 m cell size. There is little to distinguish between these resampled terrain datasets; so, depending on the resolution and intended purpose of the 3-D geological model, investing in LiDAR may not be necessary. The older dataset, shown in Figure 7.6b–f, may be the most suitable terrain data option for creating a subsurface geological model because the landfill has been filtered out. Alternatively, if the resolution of a model allows for the representation of anthropogenic deposits, then not only landfills, but also levees, road embankments, foundation layers and similar deposits could be included, and LiDAR may be the most suitable data source.

7.5 Surficial and Subsurface Geological Data

Many sources of surficial and subsurface geological data may be used to construct 3-D geological models; each data source provides information with variable quality, depth of penetration, and resolution (Figure 7.7). The model builder must consider which data sources are the most useful and fit-for-purpose; then direct available resources to acquiring and standardizing the selected data to produce the desired model. For example, petroleum records (Figure 7.7l–m) provide limited useful data when constructing sediment models but are a key dataset for constructing bedrock models.

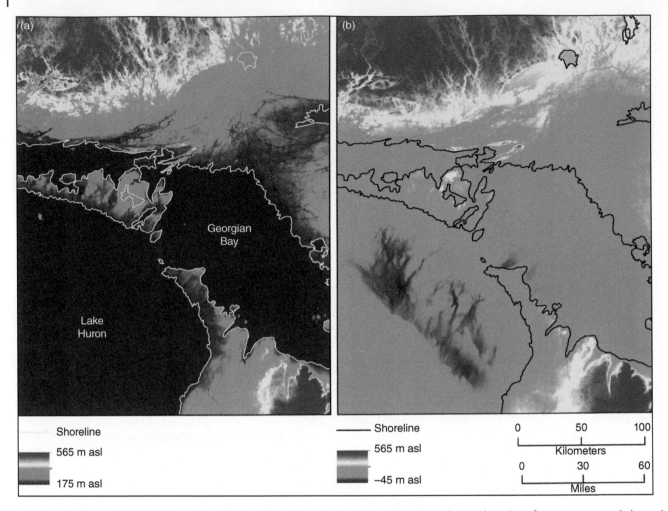

Figure 7.5 DTMs for a portion of Ontario, Canada. On the right, bathymetric data show the continuation of escarpments and channels under Lake Huron and Georgian Bay. Map data from: Ontario Ministry of Natural Resources and Forestry (2005), Ontario Ministry of Natural Resources and Forestry (2010a). Copyright © Queen's Printer for Ontario, 2019. Used by permission of the Ontario Geological Survey.

7.5.1 Geological Survey Data

Geological surveys typically have legacy data holdings in a variety of formats that include maps and associated databases, analytical data, reports and academic contributions, borehole cores and samples, and unpublished material such as field notes and sketches (Burt and Dodge 2016; Royse et al. 2009; Russell et al. 2015). Acquired by trained geologists as part of routine mapping or thematic projects, the data represent decades of investment and experience. Geological surveys may also be repositories for data originally acquired by other agencies, for example old mine plans, and associated cross-sections and borehole data. In recent years, 3-D models at national, regional and local scales have been produced, based on the data contained in legacy holdings. Such models also serve as important data sources when new models are developed.

7.5.1.1 Map Data

For years, geological surveys have published geological maps at a variety of scales and with a range of themes targeting different audiences. Maps provide data on bedrock formations, sediment packages, fault frameworks, geological structures defined by strike and dip measurements, and results of geochronological and geochemical analysis (Jacobsen et al. 2011). Digital geological maps are produced by GIS methods using points and lines to show the locations of mineral deposits, outcrops, and faults and polygons to define the extents of various geological units. Where maps exist only as printed paper products or in scanned PDF formats, they must be digitized, to create a digital map geodatabase that can be processed by GIS methods. In recent years, many geological surveys have begun to produce and release seamless digital map products; these have been reprocessed to resolve edge matching issues along tile borders and differences in legends among

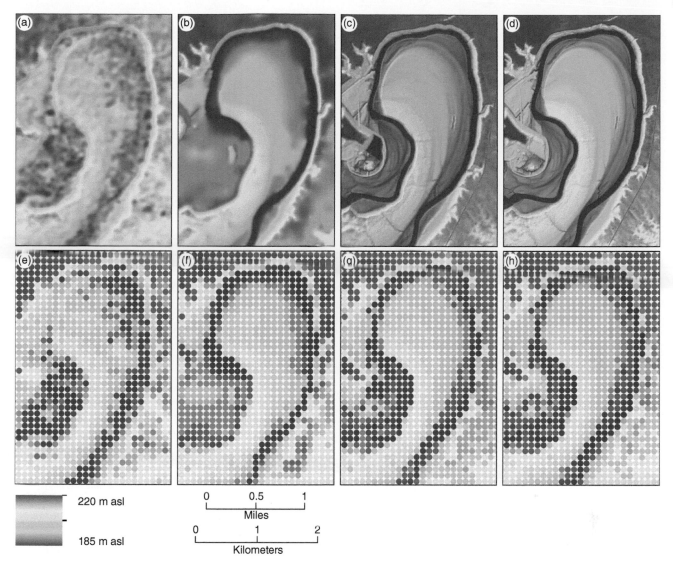

Figure 7.6 Eight versions of terrain data for a section of the Grand River floodplain near Brantford, Ontario that illustrate the effects of horizontal and vertical resolution, date (time) of data acquisition, and resampling. Versions (a)–(d) show terrain data draped over a hillshade image, with a progressive increase in spatial resolution from 30 m to 50 cm and a corresponding increase in vertical accuracy from ±5 m to ±10 cm. The elevation data in version (b) has been reprocessed to remove the landfill shown in the middle left of versions (a), (c), and (d). Versions (e)–(h) are resampled versions of (a)–(d) using a 100 m cell size. Version details: (a) Ontario Radar Digital Surface Model (ORDSM) with a 30 m spatial resolution and 5 m vertical accuracy (Ontario Ministry of Natural Resources and Forestry 2010a). (b) Ontario Provincial Digital Elevation Model – Version 2.0.0 raster dataset with a 10 m resolution interpolated from contour line, DTM, spot height and water datasets (Ontario Ministry of Natural Resources and Forestry 2005). (c) Southwestern Ontario Orthophotography Project (2010) – Digital Surface Model raster dataset with a 2 m spatial resolution generated from off-leaf, 20 cm resolution stereo imagery (Ontario Ministry of Natural Resources and Forestry 2010b). (d) LiDAR Digital Terrain Model (2016–2018) bare-earth 50 cm spatial resolution raster dataset with a 10 cm non-vegetated vertical accuracy (Ontario Ministry of Natural Resources and Forestry 2018). Copyright © Queen's Printer for Ontario, 2019. Used by permission of the Ontario Geological Survey.

the constituent maps. Seamless digital maps make the incorporation of map data into the modeling process much easier.

Many jurisdictions publish complementary map series that define surficial and bedrock geology separately. Figure 7.8 compares a series of digital maps for a small area in the Province of Ontario, Canada (Ontario). Figure 7.8a

is a digital surficial geology map which defines the composition and depositional environments of the surficial unconsolidated sedimentary units while areas of exposed bedrock are mapped as undifferentiated Paleozoic rocks. The accompanying geodatabase contains definitions of composition properties and other characteristics for each polygon, which have been used to recode the polygons

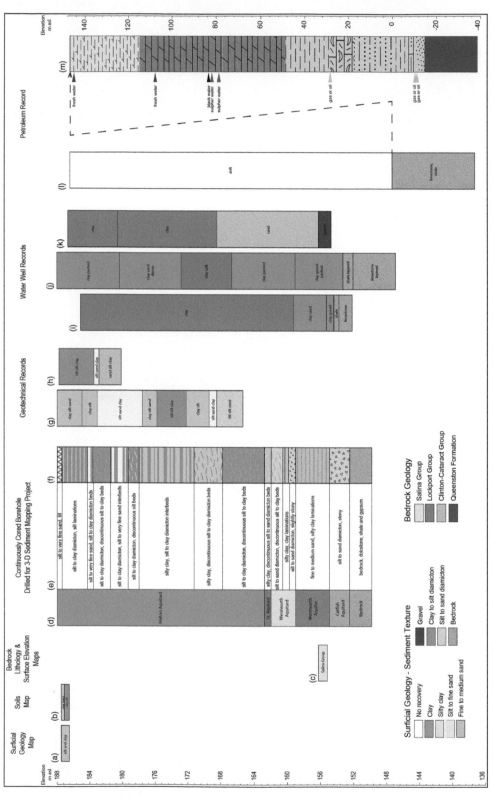

Figure 7.7 Comparison of surficial and subsurface geological data obtained from different sources within a 250 m radius illustrating the typical depth, resolution and area of focus. (a) Surficial geology and (b) soils maps provide high resolution data on the upper 1–2 m. (c) In this area, bedrock lithology and surface elevation maps provide low resolution data on the predicted subcropping formations. (d–f) Continuously cored boreholes provide high quality data on the Quaternary sediment cover and upper few meters of bedrock. (g,h) Geotechnical records provide data on the upper portion of the sediment cover. In this illustration, the primary sediment types have been interpreted from data extracted from the records. Comparison with the continuously cored borehole record shows some of the difficulties with these records. (i–k) Water well records provide low resolution, and often low quality, data for the full sediment package. (l,m) Petroleum record provides little useful data when constructing sediment models but form the main data source for constructing bedrock models. Note the scale change on (M). Data from: (a) Ontario Geological Survey (2010); (b) Ontario Ministry of Agriculture, Food and Rural Affairs (2015); (c) Armstrong and Dodge (2007); (d–f) Burt (2017), Gao et al. (2007), (d–f) Burt (2017), (g,h) Ontario Geological Survey (2019); (i–k) Ontario Ministry of Environment, Conservation and Parks (2013); (l,m) Ontario Ministry of Natural Resources and Forestry (2019). Copyright © Queen's Printer for Ontario, 2019. Used by permission of the Ontario Geological Survey.

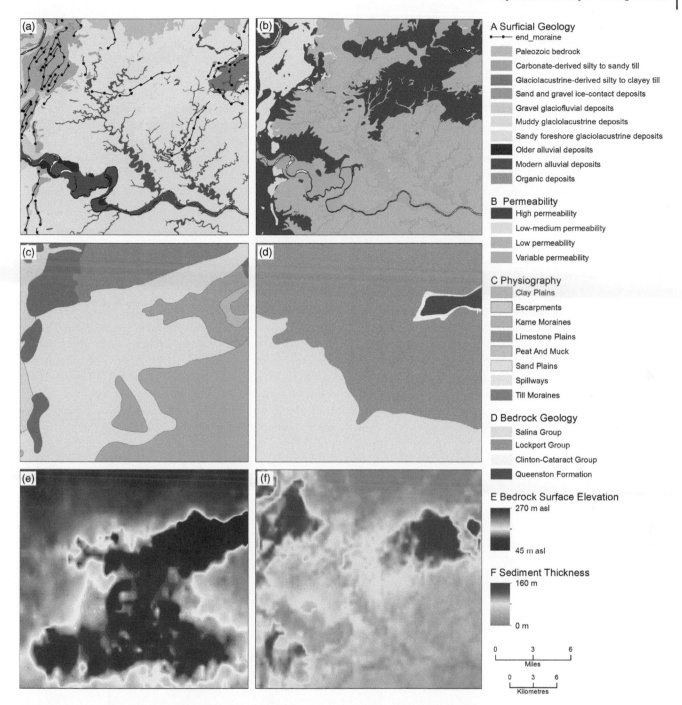

Figure 7.8 Pan-provincial digital maps available from the Ontario Geological Survey. Surficial geology map showing (a) texture and depositional environment, (b) permeability, (c) low resolution physiography, (d) bedrock formations, (e) bedrock surface, and (f) sediment thickness. Map data from: (a,b) Ontario Geological Survey (2010); (c) Chapman and Putnam (2007); (d) Armstrong and Dodge (2007); (e,f) Gao et al. (2007). Copyright © Queen's Printer for Ontario, 2019. Used by permission of the Ontario Geological Survey.

according to their permeability (Figure 7.8b). Figures 7.8a and b illustrate relatively large-scale, high-resolution mapping units; however, less-detailed regional map units often supply useful information, particularly when modeling large areas. Figure 7.8c is an example of a regional map

which shows generalized physiographic units such as moraines, plains, and escarpments, while Figure 7.8d shows the extent of bedrock units. In this area of thick drift, the bedrock contacts are speculative with interpretations based on subsurface data sources rather than direct observation. The geodatabase for this map includes primary and secondary lithologies, geological formations (as shown in Figure 7.8d), and age. However, this map does not indicate whether the rock outcrops (exposed at surface) or subcrops (buried beneath the sediment) and it does not define how deeply the bedrock surface is buried by the surficial deposits. The depth to bedrock can be defined by first developing a map showing the elevation of the bedrock surface (Figure 7.8e) and then creating a sediment thickness map by subtracting bedrock surface elevations from ground elevations (Figure 7.8f). Although shown here as 2-D maps, bedrock surface elevations and sediment thicknesses form simple 3-D models which are basic components of more comprehensive 3-D subsurface models.

Geology maps capture more than just areas of common rock types or sediment textures; they convey the mapper's interpretation of the stratigraphy and geological history of an area. This is demonstrated by comparing legacy and recently updated surficial geology maps for an area in Ontario, Canada (Figure 7.9). The new map (Figure 7.9b) has a higher resolution and more accurately represents exposed bedrock, till, sand, and gravel. It also provides additional details of the depositional environment (beach vs littoral sands) and the stratigraphy of the deposits (the sub-till sands shown in red are older).

The information presented on geological maps must be captured at a resolution appropriate to the model scale and in the format required by the modeling software (Matile et al. 2002; Burt and Dodge 2016). Lithostratigraphic schemes presented on 2-D maps may need to be adjusted for 3-D modeling purposes (Van der Meulen et al. 2013). Digital map polygons may be recoded to define desired model unit attributes and converted to regularly gridded point files for use in modeling software packages (Figure 7.10; Bajc and Newton 2005). Alternatively, the linework on maps may be captured as a series of themes or layers (Mathers et al. 2011).

7.5.1.2 Boreholes

Cored boreholes, when logged and interpreted by a trained geoscientist, are one of the highest quality data sets available for 3-D modeling. Continuous coring is an expensive and sometimes slow process, and many modeling projects do not have the financial resources or time allowances to conduct extensive drilling programs. Cheaper drilling methods, such as mud rotary or auger with split spoon collection of selected intervals, may be used. However, the reduced cost of these methods must be weighed against the loss of information for non-cored intervals. Figure 7.11 shows examples of core obtained using several drilling

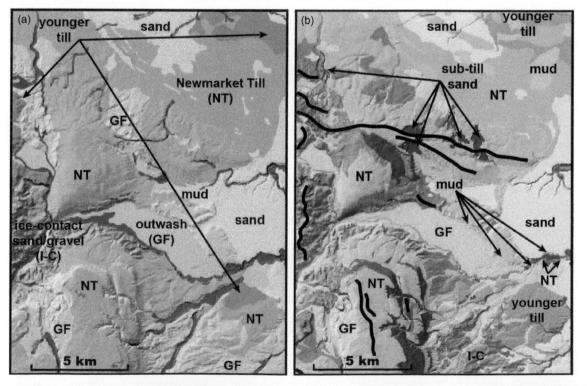

Figure 7.9 Comparison of legacy (a) and new (b) surficial geology maps of the Niagara Escarpment area south of Georgian Bay. Modified from Mulligan et al. (2019); original sources are Burwasser (1974); Mulligan (2017). Copyright © Queen's Printer for Ontario, 2019. Used by permission of the Ontario Geological Survey.

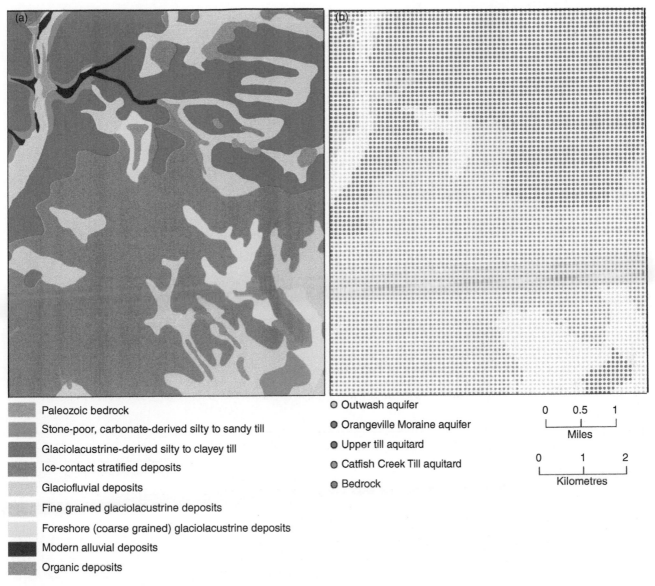

Paleozoic bedrock

Stone-poor, carbonate-derived silty to sandy till

Glaciolacustrine-derived silty to clayey till

Ice-contact stratified deposits

Glaciofluvial deposits

Fine grained glaciolacustrine deposits

Foreshore (coarse grained) glaciolacustrine deposits

Modern alluvial deposits

Organic deposits

○ Outwash aquifer

◐ Orangeville Moraine aquifer

◑ Upper till aquitard

◔ Catfish Creek Till aquitard

◒ Bedrock

0   0.5   1
Miles

0   1   2
Kilometres

Figure 7.10   Examples of two digital formats for a portion of the high-resolution surficial geology map of the Orangeville-Fergus 3-D sediment model area. (a) Map polygons have been classified according to texture and depositional environment; (b) the map polygons have been recoded to represent hydrogeologic units and converted to a regularly spaced 3-D point file. (a) map data from Ontario Geological Survey (2010) – Copyright © Queen's Printer for Ontario, 2019; (b) modified from Burt and Dodge (2016) – Copyright © Queen's Printer for Ontario, 2016. Used by permission of the Ontario Geological Survey.

methods. Figures 7.11a–d are cores from a mud rotary rig equipped with a Christensen core barrel; depending on the materials encountered, this method can provide the highest quality information. Figures 7.11e–h show examples of core recovered from a hollow stem auger equipped with a sampler, rotosonic coring, and split spoons sampling. These methods typically produce lower quality, less easily interpreted, core. Project-specific drilling programs allow drill locations to be selected where they can best establish bedrock-sediment-landform associations, refine

stratigraphic relationships, answer specific questions such as the nature of buried-bedrock valley fills, or assist with the interpretation of geophysical surveys. Careful placement of new boreholes will assist in the development of conceptual geological models (see Box 7.1) and facilitate the interpretation of older and lower quality datasets. This is particularly important in projects where low-quality records, such as water wells, are the primary data source. Legacy geological borehole records may provide accurate data at no cost to the project but may not be situated in optimal locations.

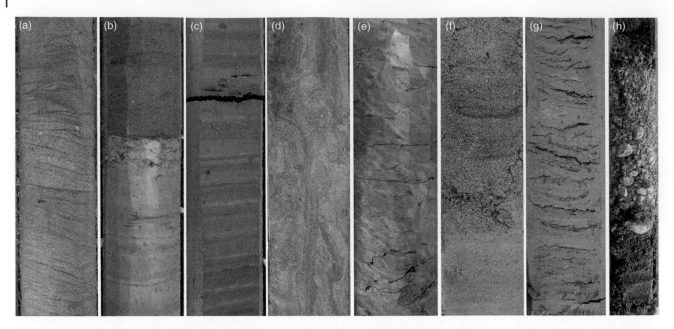

Figure 7.11 Examples of core drilled for OGS 3-D sediment mapping projects in Ontario, Canada. (a–d) High quality, undisturbed core drilled using a mud rotary rig equipped with a Christensen core barrel. The deformation visible in the middle of core B and throughout core D are natural and not drilling related disturbances. The sediment structures allow the depositional setting of sediments to be interpreted. (e) Example of core twisted and sheared during drilling with a hollow stem auger equipped with a sampler. (f, g) Rotosonic core. The core was deformed during drilling; however, some primary structures are still discernable. (h) Split spoon core. This core is highly disturbed, and only major sediment packages are visible. Copyright © Queen's Printer for Ontario, 2019. Used by permission of the Ontario Geological Survey.

Box 7.1 Ontario Geological Survey Drilling Procedures for 3-D Modeling Projects

The Ontario Geological Survey (OGS) conducts a multi-year drilling program in connection with individual regional scale 3-D modeling projects. For example, the OGS drilled 99 project-specific boreholes as part of a 5000 km^2 (1950 mi^2) regional-scale 3-D sediment modeling project on Niagara Peninsula. A total of 3200 m (10 500 ft) of continuous core was recovered from these boreholes (Burt 2017).

The drilling program is designed to assess the unconsolidated Quaternary sediment cover and upper 1.5–5 m (5–15 ft) of bedrock, so mud-rotary drills with 1.5 m (5 ft) samplers retrievable by wireline are used to recover continuous core. The drill is equipped with Christensen core barrels which produce 8.5 cm (3.35 in) diameter core. Upon recovery, the core is logged and photographed at 0.25 m (10 in) increments. Representative samples are collected every 1.5 m (5 ft) or when significant changes in lithology occur. These samples are analyzed to determine grain-size, carbonate and heavy mineral content, radiocarbon dates, and changes in paleo-ecology. In clay-rich areas, a pocket penetrometer is used to perform field penetration tests. The graphic borehole log in Figure 7.1.1 (located on page 147) summarizes these observations. For the Niagara Peninsula project, the boreholes provided both data and information that confirmed interpretations of legacy data from over 54 500 water wells, petroleum exploration logs, and shallow geotechnical investigations.

Selected boreholes were converted to monitoring sites by conservation authority and municipal partners (Campbell and Burt 2015). In these boreholes, installation of a 6.3 cm (2.5 in) diameter threaded, flush-joint, polyvinyl chloride (PVC) pipe with 1.5 or 3 m (5 or 10 ft) long slotted screens permits future monitoring of groundwater levels and chemical composition. All remaining boreholes are sealed with bentonite clay to prevent subsequent contamination of groundwater supplies.

Downhole geophysical surveys were conducted by the Geological Survey of Canada in select monitoring wells. Downhole compression (Vp) and shear (Vs) seismic logs are used to calibrate nearby seismic data and convert time profiles to true depths while apparent conductivity and magnetic susceptibility logs are used to characterize lithological variations within and between sediment packages (Burt and Crow 2019). Regional groundwater temperature variations were determined from high-resolution fluid temperature logs (Crow et al. 2017).

BOX 7.1 Ontario Geological Survey Drilling Procedures for 3-D Modeling Projects (Cont'd)

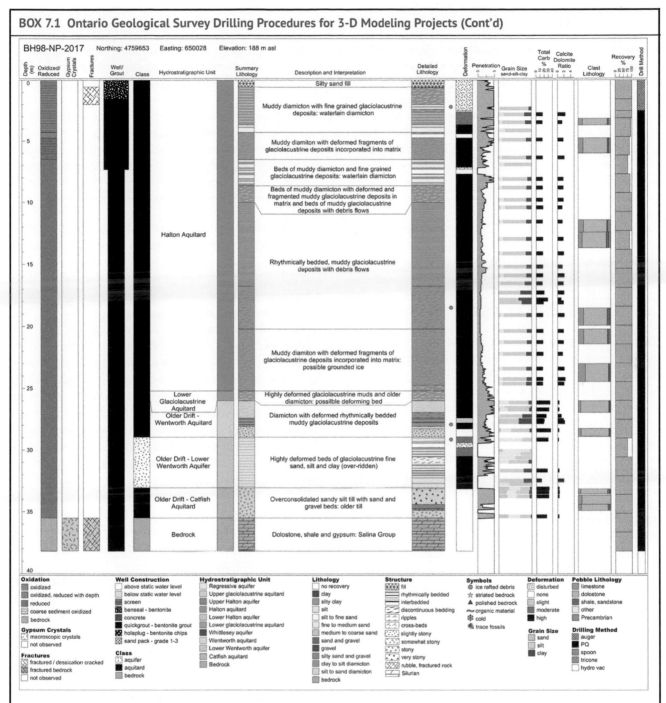

Figure 7.1.1 Detailed graphic log for a new borehole drilled on the Niagara Peninsula, Ontario, Canada. (Burt 2017). Copyright © Queen's Printer for Ontario, 2017. Used by permission of the Ontario Geological Survey.

7.5.1.3 Analytical Databases

Many geological surveys release databases of analytical results from rock, sediment, and water sampling programs. The data are collected as part of sediment or bedrock mapping projects to provide baseline data for mineral exploration, or for groundwater-themed projects (see Box 7.2). Originally published as tables of data forming appendices to written reports, in recent years analytical data are more commonly released in digital formats. These datasets contain the quantitative data required to assign property values to cells in a discretized model, or to assign general values to units in a framework model. Analytical data can also be used to interpret stratigraphic relationships. For example, tills form important regional-scale

---

Box 7.2 Ontario Ambient Groundwater Geochemistry Project

Since 2007, the ambient groundwater geochemistry project in Ontario, Canada, has been sampling selected ground-water wells across the populated portions of the province. Untreated water samples are collected from bedrock and overburden wells located on a 10×10 km grid. The project has been able to correlate the geochemical characteristics of groundwater to the chemical characteristics of specific aquifers.

To date, approximately $120\,000$ km^2 ($46\,330$ mi^2) of southern Ontario, and parts of northern Ontario, have been characterized. The results have been the subject of dozens of graduate theses and reports. Major discoveries include a 1400 km^2 (540 mi^2) contiguous area of karstic breathing wells in the Middle Devonian Lucas Formation (Freckelton 2013) and a 2500 km^2 (965 mi^2) area of elevated iodine levels that are associated with marine clays; some iodine concentrations exceed 10 times the tolerable daily intake for adults (Lemieux et al. 2018).

---

marker units in framework 3-D models developed in glaciated portions of North America (Bajc and Shirota 2007; Burt and Dodge 2016). Till formations may be distinguished by analyzing the texture and geochemical characteristics of matrix sediments, heavy mineral assemblages, and pebble lithologies.

7.5.1.4 Reports and Academic Contributions

Reports, posters, oral presentations, and academic journal papers describe data collection and analytical methods, results, and interpretations as well as providing recommendations for further study. Although best used to develop, or refine, an understanding of the geologic history of an area, these sources may contain cross-sections or tables of analytical data. Cross-sections may be sketched to convey a conceptualization of the regional geology (Figure 7.12a) or may be constructed using field observations (Figure 7.12b).

7.5.1.5 3-D Models

A growing number of geological surveys are producing publically accessible 3-D models. The intended purposes, scales (Figure 7.13), and outputs of these models vary, but all can be used as sources for new modeling efforts (Keller et al. 2009; MacCormack 2018; Sandersen et al. 2015). For example, the Alberta Geological Survey distributes individual layers and complete models as ASCII grids, digital stratigraphic picks, a Petrel project file, and a 3-D visualization tool. The OGS releases regional-scale 3-D sediment models accompanied by an abbreviated version of the subsurface database used for model construction, structural contour and isopach maps of modeled surfaces, raster data sets, comma-delimited ASCII grid files, analytical data files, and detailed reports (Burt and Dodge 2016).

7.5.1.6 Accessibility

Access to data holdings varies considerably among organizations. In some countries, access is affected by data security and confidentiality restrictions. Some organizations allow images of maps to be viewed without charge in their offices or by using internet browsers but charge a fee to download digital maps or obtain print copies. Other organizations provide free access to maps, both as images and as fully attributed geodatabases containing point, line and polygon data, as well as reports and analytical data. A business model adopted by some surveys distinguishes between data and information holdings, allowing free access to data but charging for information.

Discoverability and data retrieval are challenges that many organizations continue to grapple with (Burt et al. 2018). The 20-year review of the Canadian Urban Geology Automated Information System (UGAIS), a 1970s era project by the Geological Survey of Canada to collect and collate geological and geotechnical borehole data for 27 Canadian cities in an effort to improve efficiency, highlights some of the difficulties with accessing legacy datasets (White and Karrow 1997). Some cities continued to use and maintain the database after the project was terminated while others could not even locate their database (White and Karrow 1997). Belanger (1998) suggested multiple factors contributed to this lack of continued use of the UGAIS database, including a reliance on paper formats, the use of early mainframe computers making the database cumbersome to use, duplication and the existence of multiple databases, difficulties integrating different datasets, and a lack of updates. Fortunately, several organizations are now beginning to recognize that data holdings are their biggest asset and have improved data management procedures to develop standardized corporate databases (Royse et al. 2009; Chapter 8).

Web portals that allow geoscience data to be viewed and downloaded have proven an effective means for clients to access geological survey data. Launched in 2007, One-Geology is an example of a portal that allows multiple geological surveys to share and deliver geoscience map information to a worldwide audience while the data remain with the original organization (http://www.onegeology .org/portal/home.html). In addition to geometric harmonization across boundaries where necessary, a distributed map service requires unified standards for data and

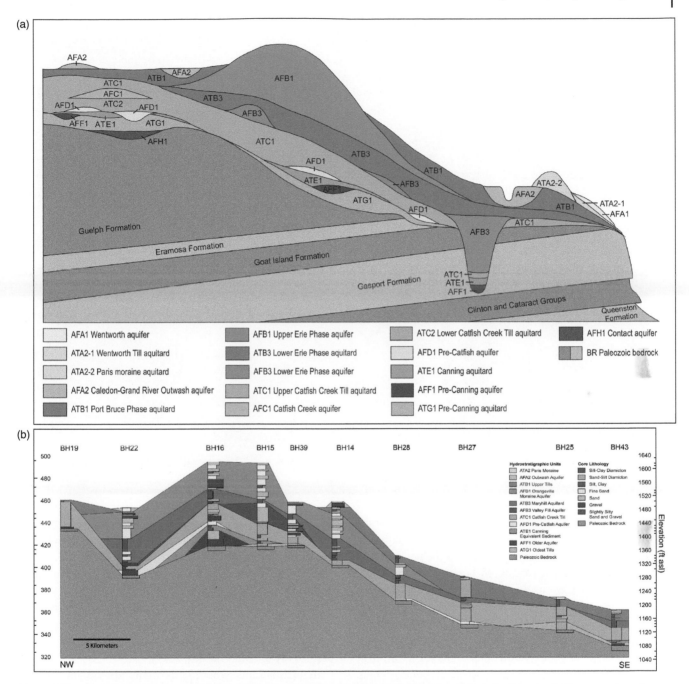

Figure 7.12 Two cross-sections along transects perpendicular to the Orangeville Moraine in Ontario, Canada. (a) Conceptual geology cross-section; (b) Cross-section constructed from cored holes. Sources: (a) Burt and Dodge (2016) – Copyright © Queen's Printer for Ontario, 2016; (b) modified from Burt (2018) – Copyright © Queen's Printer for Ontario, 2018. Used by permission of the Ontario Geological Survey.

metadata structure. OneGeology uses a markup language, GeoSciML, that is based on international standards for internet data exchange.

7.5.2 Soil Data

Pedology, the study of soils, defines five principal soil formation factors: climate, organisms, topography, parent material and time (Joffre 1936; Kellog 1936). Soil surveys examine only relatively shallow depths, usually no more than 2 m (6 ft), and soil maps can be used to identify near-surface geological features. There are many national soil classification systems, which have a common denominator in the World Reference Base (WRB) for Soil Resources (FAO 2015). The WRB classifies soils into

AGS 3D Geological Framework Models

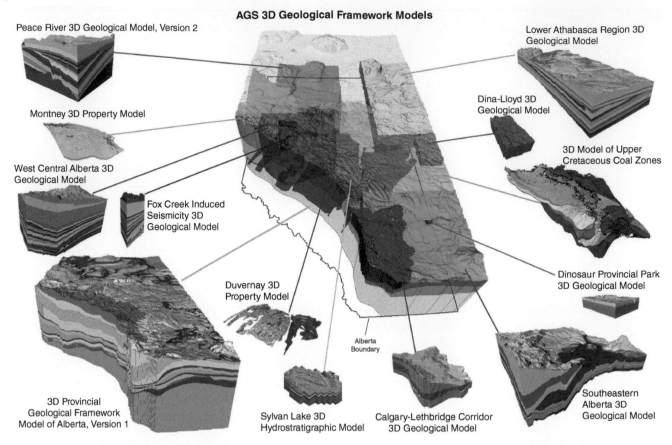

Peace River 3D Geological Model, Version 2

Montney 3D Property Model

West Central Alberta 3D Geological Model

Fox Creek Induced Seismicity 3D Geological Model

Lower Athabasca Region 3D Geological Model

Dina-Lloyd 3D Geological Model

3D Model of Upper Cretaceous Coal Zones

Dinosaur Provincial Park 3D Geological Model

Alberta Boundary

3D Provincial Geological Framework Model of Alberta, Version 1

Duvernay 3D Property Model

Sylvan Lake 3D Hydrostratigraphic Model

Calgary-Lethbridge Corridor 3D Geological Model

Southeastern Alberta 3D Geological Model

Figure 7.13 Variety of provincial and sub-model scale 3-D models that are contained within the Alberta Geological Survey's 3-D Geological Framework. (MacCormack 2018).

32 reference soil groups with supplementary qualifiers according to diagnostic horizons, properties and materials. In any soil description, information about the parent material or the lowest, least weathered horizon (often termed C-horizon) is the most useful for 3-D geologic modeling.

Soils maps are often more detailed than surficial geology maps making them particularly useful for site-specific models (Mathers et al. 2011; Prokopovich 1984). Figure 7.14 compares surficial geology and soils maps for a thin drift area of southern Ontario. The surficial geology map (Figure 7.14a) shows, in pink, areas of exposed or thinly covered bedrock. The map has been draped over a high resolution hillshade and a close examination reveals a series of narrow ridges. It may be expected that thicker, and potentially different, sediments will be found in the intervening lows. Many of these details are revealed on the soils map (Figure 7.14b). In this area, the soils map database and report define soils series characteristics that can be used to provide more detailed sedimentary texture properties than are provided by the surficial geology map and geodatabase.

In many countries, agricultural agencies, not geological surveys, are responsible for soils surveys. Results were distributed as paper maps and reports until the advent of GIS technologies led to the creation of geospatial soils databases (CanSIS 2019; Natural Resources Conservation Service 2019a, 2019b). The availability of digital data sources has expanded the potential applications of soil survey data. The Minnesota Geological Survey has recently issued a state-wide surficial geology map and associated soil texture database which is largely based on digital soil-survey records (Lusardi 2017).

In jurisdictions where soils maps are not already in digital formats or supporting databases do not lend themselves to easy reclassification into geological materials, careful consideration should be given to the scale and focus of the model before allocating resources to the acquisition and standardization of soils maps. Since they will only provide information on surficial materials, soils maps may not be useful for deeper modeling projects (Column b in Figure 7.7). Also, regional-scale models may not benefit from precise descriptions of surficial materials, while near-surface or site-specific modeling projects may find soils data invaluable.

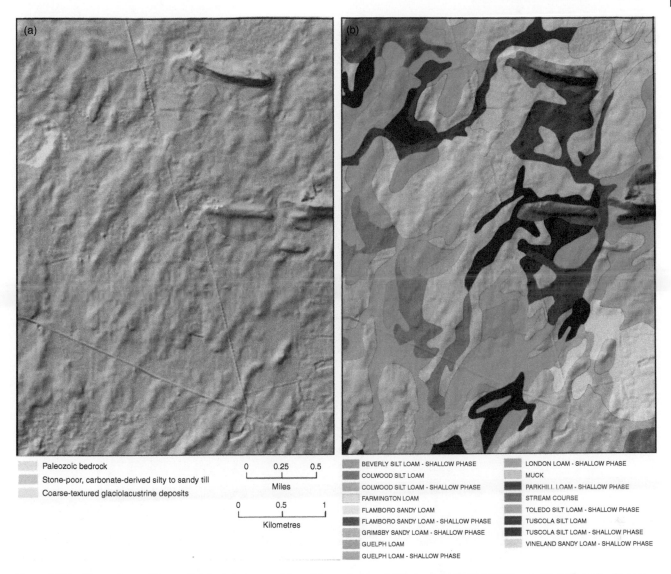

Figure 7.14 Comparison of the surficial geology map (a) and the soils map (b) for an area of thin drift in southern Ontario, Canada. The surficial geology map does not discriminate between bedrock exposed along narrow ridges and the sediments deposited in the intervening lows. The soils map defines soils series that can be recoded to provide more detailed material texture properties using databases and accompany reports. Map data from: (a) Ontario Geological Survey (2010), Ontario Ministry of Agriculture, Food and Rural Affairs (2015). Copyright © Queen's Printer for Ontario, 2019. Used by permission of the Ontario Geological Survey.

7.5.3 Geotechnical Data

Geotechnical reports form important components of many environmental impact assessments for landfills, pits, quarries, new housing or industrial estates, and for construction projects for infrastructure (railways, roads, bridges, tunnels, utility corridors) and building foundations. These geotechnical reports are valuable data sources for geological models, and their use in 3-D modeling aligns well with the fact that the built environment is becoming an increasingly important area of model application (Chapters 4, 23, and 25). Older reports can be particularly useful as they provide insights into the characteristics of sediments underlying built up areas that can no longer be readily accessed.

Geotechnical borehole logs may contain sediment and rock descriptions, penetration test results (used to determine the load bearing strength of shallow geological units), downhole geophysical logs, sampled intervals and monitoring well construction details. Geotechnical test results, static water levels, pumping test results, geochemical analyses and geological interpretations may be found in associated reports and collectively are an important source of property data. Cone penetration test (CPT) results may be more abundant than borehole descriptions as they are cheaper and easier to obtain. The Geological Survey of the

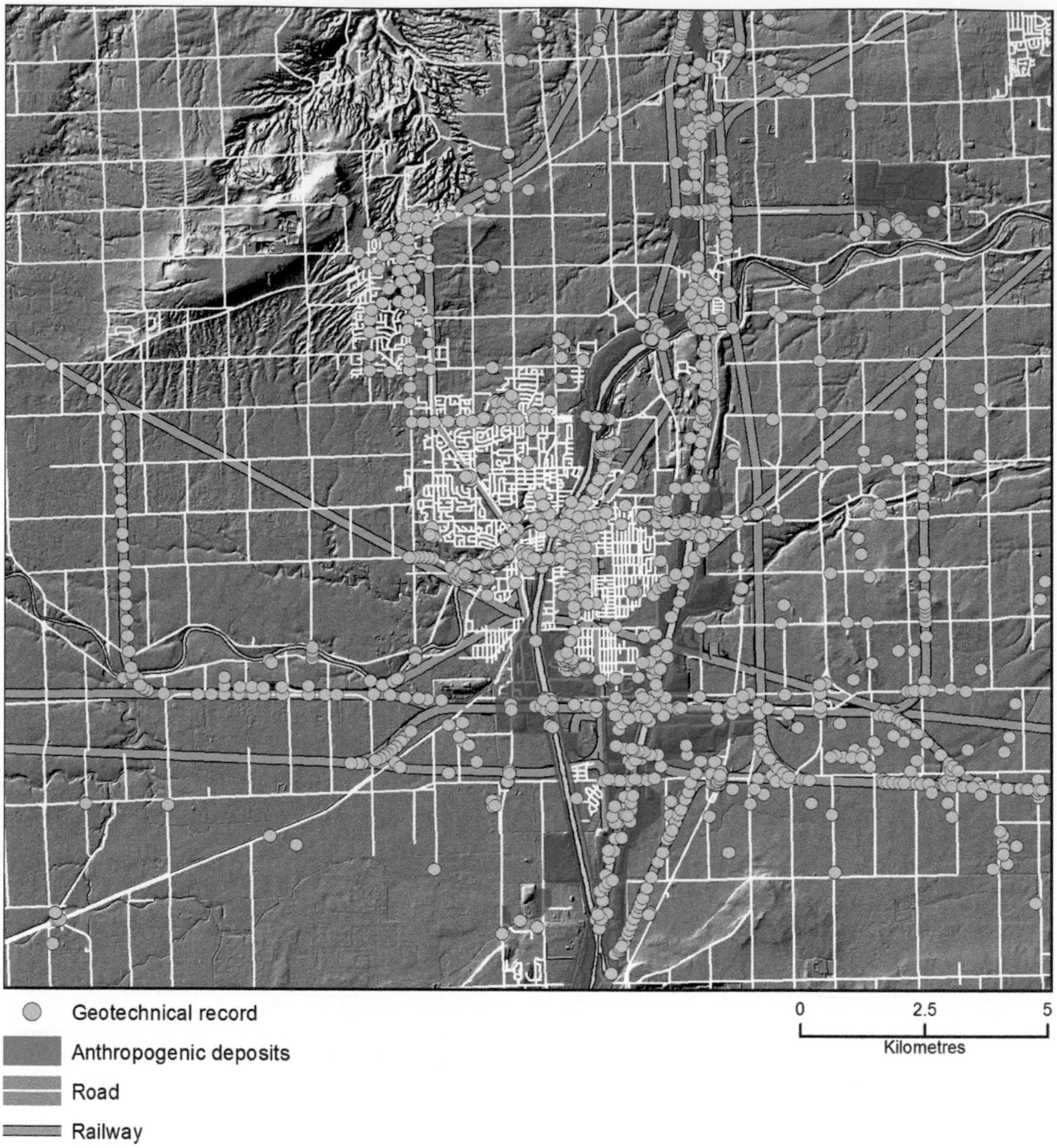
Geotechnical record
Anthropogenic deposits
Road
Railway
0
2.5
5
Kilometres

Figure 7.15 Distribution of geotechnical records in the urban geology automated information system (UGAIS) database around Welland, Ontario. It is evident that data are clustered in urban areas and along the main transportation corridors and the Welland canal (surrounded by anthropogenic deposits). Copyright © Queen's Printer for Ontario, 2019. Used by permission of the Ontario Geological Survey.

Netherlands has successfully incorporated CPT data into shallow 3-D models by defining the relationship between CPT responses and geological characteristics observed in nearby cored boreholes (van Maanen et al. 2017).

There are challenges associated with using geotechnical databases. The data are typically clustered in or near urban areas or along major transportation routes (Figure 7.15). Many projects (especially older projects) use local grids and datums for location and elevation. The boreholes are often shallow, ranging from a few meters to a few tens of meters. Depending on the purpose of the original study, the borehole logs may target the overburden or bedrock, but not both. The logs may be produced by junior staff with limited geological background or experience; this means that the interpretations may reflect a poor understanding of geological processes and history. Even experienced field personnel may be hampered by the drilling methods used. Budget restrictions rarely allow retrieval of undisturbed, continuous core necessary for detailed sedimentological interpretations.

7.5.4 Water Well Data

In many jurisdictions, it is a legal requirement for water well drillers to submit their logs capturing data on location, geological materials encountered, well construction and water quality to a regulatory body. Records dating back to the late 1800s may be available as digital databases, scanned documents, paper records, or a combination of all three.

Water well records form the largest dataset used in many shallow 3-D models (Arnold et al. 2001; Bajc and Dodge 2011; Belanger 2001; Tipping and Meyer 2001). Depending on the intended purpose of the model, obtaining groundwater data from water well databases, well-testing reports or water-use records may be beneficial (Eaton et al. 2001). These datasets may include water withdrawal data, well screen elevation, static water levels, and hydrogeological data such as pump tests, slug tests, and aqueous geochemistry data (McLaughlin et al. 2015, Burt and Dodge 2011). Geological data are typically stored as a series of material descriptions and qualifiers. Examination and interpretation of individual well records may not be practical in regional-scale projects. Instead, automated or semi-automated procedures are used to summarize and translate the hundreds of unique terms into standardized material descriptions (Burt and Bajc 2005; Keefer 2011; Russell et al. 1998).

While water well records have long been recognized as one of the most plentiful, but challenging, sources of subsurface data (Arnold et al. 2001), the records typically have a sporadic distribution. Water wells are plentiful in the more densely populated and farming areas but may be scarce or absent in urban areas with public water supplies, in forested areas or parks where there is no demand for wells, or in areas where groundwater is not suitable for drinking or irrigation. Furthermore, the location information associated with individual records may be of low resolution or incorrect. The large number of Ontario wells shown in Figure 7.16 that are plotted outside the Province, or within lakes, clearly demonstrates inaccurate location information. Low resolution location data result from the practice of using section or lot centroids or from estimating locations from maps (Figure 7.17). Some jurisdictions are engaged in the expensive and time-consuming process of verifying and correcting location data using field-based programs where wells are located using hand-held GPS devices, or by desk-top programs using street addresses, tax parcel information, and web-based mapping (Keefer 2011). Another approach is to accept the low-quality location data and determine general trends within a series of 5 km wide sections that are viewed in 2-D (Figure 7.18; Keller et al. 2005).

Other challenges are related to the validity of geological information contained within the records. Water wells

Figure 7.16 Ontario water wells plotted using location data obtained from the Ministry of Environment, Conservation and Parks water well information system. Inaccurate location data are demonstrated by wells plotting outside of the province or within lakes. Data from: Ontario Ministry of Environment, Conservation and Parks (2013). Copyright © Queen's Printer for Ontario, 2019. Used by permission of the Ontario Geological Survey.

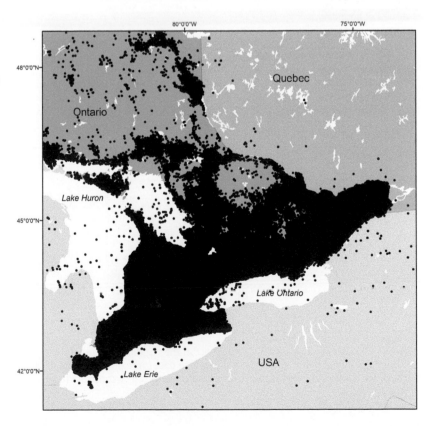

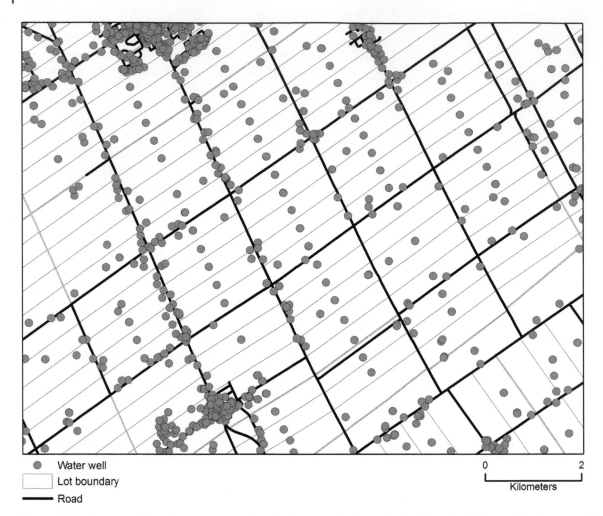

Water well
Lot boundary
Road

0   2
Kilometers

Figure 7.17 Close-up of Ontario water wells plotted using location data obtained from the Ministry of Environment, Conservation and Parks water well information system. Note the large number of wells that are plotted in the middle of properties instead of adjacent to the road network. Water wells are drilled as close as possible to houses or barns, and so this indicates that locations in the database are displaced by as much as 1 km. Data from: Ontario Ministry of Environment, Conservation and Parks (2013). Copyright © Queen's Printer for Ontario, 2019. Used by permission of the Ontario Geological Survey.

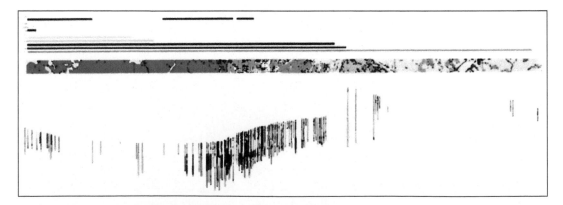

Figure 7.18 Typical 5 km east–west transect in the Winnipeg region, showing rock unit extent across the top, surficial geology and drill hole locations in the middle, and subsurface data (primarily drill hole data) in profile view below. This information was used to hand draw a series of cross-sections which were then used to construct the 3-D model. Source: (Keller et al. 2005).

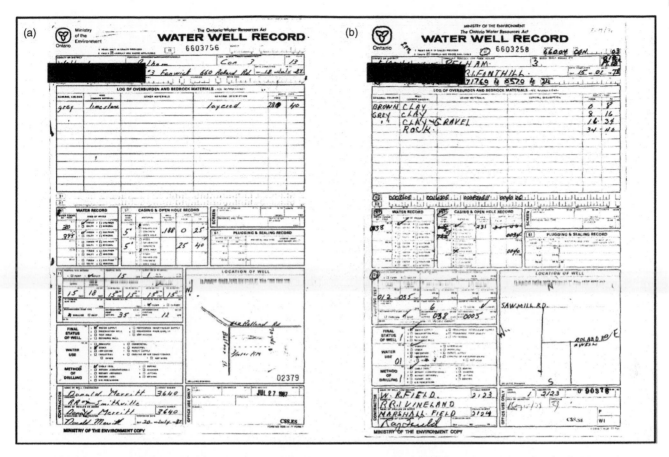

Figure 7.19 Examples of scanned Ontario water well records from two adjacent properties. (a) Overburden material descriptions were omitted from the well record. (b) Material descriptions in this well record can be interpreted as glaciolacustrine mud overlying muddy diamicton. Source: Ontario Ministry of Environment, Conservation and Parks (2013). Copyright © Queen's Printer for Ontario, 2013. Used by permission of the Ontario Geological Survey.

are drilled as quickly as possible, observations are often made by well drillers rather than geologists, observations are made from fluctuations in circulation pump pressures, drilling speeds, or cuttings (not core). The logs may not be completed for days or even weeks following completion of the drilling process. In most cases, water wells are terminated once they encounter a suitable water source so in areas with productive shallow overburden aquifers, this results in wells that only penetrate the uppermost geological units (Logan et al. 2005; Burt and Dodge 2011). The level of detail recorded by well drillers varies considerably; some drillers record all major sediment packages encountered while others may simply record 'overburden' and "bedrock" or "clay" and "sand," viewing their observations as proprietary information that gives them a competitive edge. Figure 7.19 shows well records downloaded from Ontario's water well information system for two adjacent properties. In the first record, the driller simply noted 'bedrock' resulting in little useable data. The second record provides more details and a geologist working in the area may interpret "brown clay" as oxidized muddy glaciolacustrine sediments, "gray clay" as reduced muddy glaciolacustrine sediments, "gray clay and gravel" as diamicton and 'rock' as bedrock.

7.5.5 Petroleum Data

Petroleum records housed by public agencies provide key datasets for the production of 3-D models focused on sedimentary bedrock. The data available range from historic cable-tool driller's reports or formation picks to boreholes described from drill cuttings or core, as well as additional data including geophysical logs, results of geochemical analyses, archived drill cuttings, and thin sections. Water bearing intervals and depths of oil and gas horizons may also be recorded. Scout tickets are short reports about wells that include location, total depth drilled, logs, formation tops, and production status.

Petroleum records are subject to many of the same constraints previously discussed for geotechnical and water well records. The distribution of records is confined to producing oil and gas fields or areas of exploration interest.

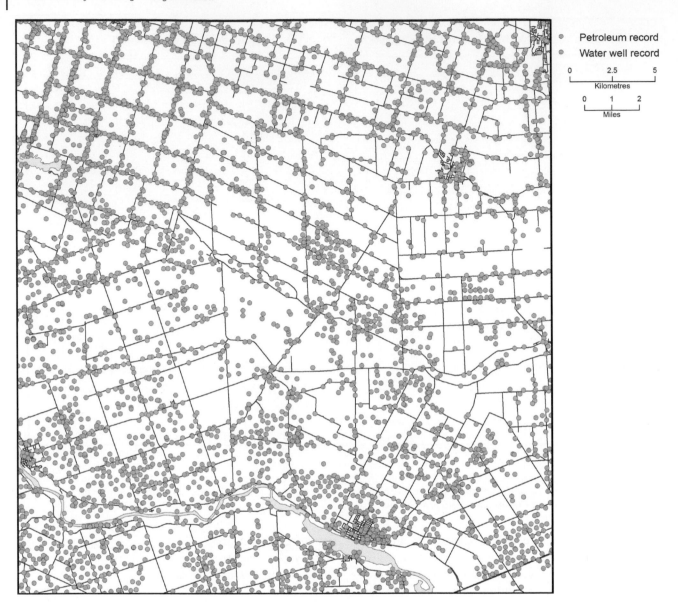

Figure 7.20 Location of petroleum records (gray dots) and water well records (blue dots) in Ontario, Canada. Note that the petroleum records are concentrated in the central and southern portion of the map area and the water well records are concentrated in the north. Data from: Ontario Ministry of Environment, Conservation and Parks (2013); Ontario Ministry of Natural Resources and Forestry (2019). Copyright © Queen's Printer for Ontario, 2019. Used by permission of the Ontario Geological Survey.

This is illustrated in Figure 7.20 which compares locations of petroleum and water well records in an area of southern Ontario, Canada. The petroleum records are concentrated in the central and southern portions of the map where there is a gas/oil pool. Elevations and depths in oil and gas records often only provide approximate location information; wells are not necessarily vertical and so the measured depth is often different from the true vertical depth. Furthermore, a variety of datums are used including mean sea level, ground level, the kelly bushing, and the floor of the drill rig. Older records may also be in feet instead of meters.

Even when the record includes the true vertical depth, a known datum, and units of measure have been recorded, using petroleum records to model the sediment cover can be unreliable; oil and gas records may not identify the top of bedrock until solid rock is reached, ignoring heavily weathered or fractured intervals. Furthermore, the composition and even the thickness of the overburden is of little interest to well drillers. This is illustrated in Figure 7.7 where the bedrock surface was measured at 116.5 ft. below surface (152.3 m asl) in a continuously cored borehole (column f in Figure 7.7) and estimated at 130.5 ft. below surface

(148 m asl) in an adjacent petroleum well. Reinterpreting formation picks within the context of revised stratigraphy presents additional challenges.

7.6 Geophysical Data

Geophysical data are important data sources for 3-D geological models because they provide subsurface information over large distances and below areas that are inaccessible to direct observation or sampling. However, they provide large-scale perspectives and indirect material property measurements. The direct measurements of physical properties of soil or rock described in Section 7.5 may be used to calibrate the indirect measurements from geophysical surveys. When calibrated by physically and analytically derived data sources, geophysically derived data about material properties provide broader insight into the lateral and vertical distribution and continuity of overburden soils, shallow bedrock weathering, and rock variability. Multiple geophysical surveys, for example combining gravity and resistivity data, produce complementary information on material properties required to develop robust 3-D geologic models.

Geophysical measurements can be conducted in one, two, or three dimensions. While they are most routinely used for exploring the deep surface, especially for hydrocarbons, geophysical methods are applied to the evaluation of shallow unconsolidated deposits and rock for engineering and environmental applications as discussed in this book (USACE 1995; Wightman et al. 2003; Burger et al. 2006). Geophysical methods generally require a material contrast to successfully delineate layering and/or structure; this contrast may be physical (density, stiffness, unit weight, etc.) or chemical (water content and type, presence of natural or anthropogenic chemicals, etc.). Geophysical methods must be carefully chosen on the basis of the geological setting and the survey target, otherwise they are likely to produce misleading results (Sirles 2006).

Legacy geophysical data may provide a good starting point for 3-D geologic subsurface models. However, because the vast majority of legacy geophysical data were collected for mineral- and energy-resource exploration, they were designed to disregard the near-surface deposits and provide higher quality information on deeper strata. Thus, while these data often are available from agency or commercial sources, their information on the near-surface is typically of limited value so they are typically used only for deeper 3-D models. All legacy geophysical data sources should be carefully evaluated before being integrated in a 3-D model; primary concerns are positional accuracy, the standard of practice at the time of data collection, and

data processing procedures. If the data are going to be used for purposes other than those of the original survey, reprocessing the raw data may be required.

Near-surface geophysical surveys typically investigate to a maximum depth of approximately 300 m (1000 ft). Successful use of information from geophysical surveys by 3-D geological modeling projects requires careful attention to the selection of the appropriate geophysical methods and to the design and implementation of the survey. Wightman et al. (2003) created a method-selection matrix that allows non-geophysicists to work with a trained geophysicist to determine what method will provide the required data to the required depth by assessing the benefits and limitations of each method for that application.

The most commonly used geophysical survey methods are: Seismic (Section 7.6.1), Resistivity (Section 7.6.2), Electromagnetic (EM) (Section 7.6.3), Gravity (Section 7.6.4), Ground Penetrating Radar (GPR) (Section 7.6.5), and Borehole Geophysical Measurements (Section 7.6.6). For engineering and environmental investigations, Sirles (2006) showed that the most commonly utilized near-surface geophysical methods are seismic and resistivity, followed by GPR. Most geophysical methods can be further refined by deploying more than one technique to acquire the appropriate model data. The following sections describe the methods and their constituent techniques.

7.6.1 Seismic Survey Method

Seismic surveys may be conducted by two contrasting survey approaches: body-wave surveys and surface-wave surveys. Body wave surveys are broken into refraction and reflection techniques. Both techniques have been used for over five decades to investigate shallow subsurface conditions (Redpath 1973; Steeples and Miller 1988) and, for a nearly a century, in energy resource exploration (Waters 1992). Surface wave techniques have been used for about 30 years and can be broken down into spectral analysis of surface waves (SASW) and multi-channel analysis of surface waves (MASW) as described by Ivanov et al. (2009), either of which can be acquired in active or passive survey mode.

7.6.1.1 Seismic Refraction Surveys

Seismic refraction surveys map the contrasts in seismic velocity of either the compressional (P) wave or shear (S) wave; these correlate with a material's bulk compressibility (Young's modulus) or stiffness (Shear Modulus). Refraction surveys are affected by changes in lithology, degree of cementation or induration, water content, and weathering or fracturing in rock. The physics of refraction is governed by Snell's Law (Telford et al. 1990) which requires the body-wave velocity to increase with depth.

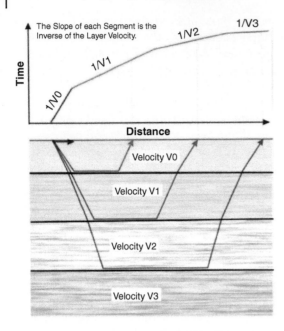

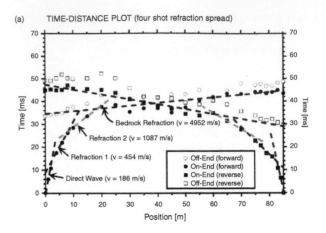

(a)   TIME-DISTANCE PLOT (four shot refraction spread)

(b)   TOMOGRAPHIC INVERSION (five shot refraction spread; arrows)

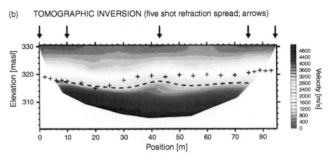

Figure 7.21 Seismic Refraction Method. The ray paths and time distance graph over four-layered ground are shown. The inverse of the slopes of the line segments on the time–distance graph provide the velocities of the refractors (Wightman et al. 2003). Source of figure: (U.S. Department of Transportation, Federal Highway Administration).

Figure 7.22 Results of seismic refraction measurements collected near Guelph. (a) Delay-time interpretation, (b) tomographic inversion model. The + symbols on the tomographic inversion correspond to estimated bedrock depths using delay-time technique; the dotted line represents the interpreted bedrock interface based on a velocity (v) threshold of $3200\,\mathrm{m\,s^{-1}}$. Modified from Steelman et al. (2018).

Refraction surveys require survey lines that extend about four times the desired depth of investigation. Refraction data are analyzed by the travel-time method (Figure 7.21), or by the tomographic inversion method (Sheehan et al. 2005). Steelman et al. (2018) analyzed refraction data by both the travel-time and tomographic inversion methods (Figure 7.22). When analyzed by the travel-time method at a series of stations along a survey line, refraction results provide the depths of layer boundaries beneath the line. Depending on the application, refraction tomography can be utilized along 2-D lines or with 3-D grids (Ditmar et al. 2001). The most useful applications of refraction are to determine depth to rock beneath unconsolidated overburden, as this defines the "top of rock." Since the 1960s, P-wave velocities have been used to determine bedrock rippability (Caterpillar 2000). P-wave velocities are affected by water saturation in unconsolidated sediments because saturation changes the bulk compressibility of the soil skeleton, increasing the P-wave velocity dramatically, as shown in Figure 7.23 (Valle-Molina 2006). In coarse-grained soils, P-wave refraction can often map the water table. S-wave velocities are not affected by the presence of fluids in the pore space, so they can identify soil strata in saturated soil conditions. The widely used American Society for Testing and Materials Standard D-5777-18 (ASTM 2018) defines appropriate standards for conducting seismic refraction surveys.

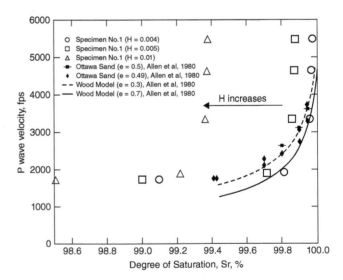

Figure 7.23 Comparison of relationship between degree of saturation (S_r) and P-wave velocity. Valle-Molina (2006).

7.6.1.2 Seismic Reflection Surveys

Seismic reflection surveys map contrasts in seismic impedance; the product of seismic velocity and the soil or

Figure 7.24 The seismic reflection method (Wightman et al. 2003). Source of figure: (U.S. Department of Transportation, Federal Highway Administration).

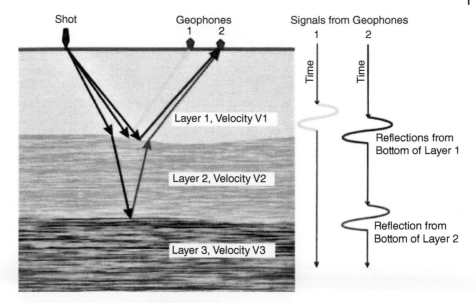

rock density (Waters 1992; Pugin et al. 2009; Figure 7.24). Utilizing reflection survey results is more straightforward than refraction results because the reflectors correspond to layering and structure, not variations in material velocity. However, data acquisition and especially the data processing are considerably more complex. When the geology is favorable, seismic reflection is a very good tool for producing detailed 2-D profile and 3-D volumetric images to define stratigraphy and structure. The seismic reflection technique is commonly used for hydrocarbon exploration. Some countries request exploration data from the operators and make them publicly available. Unfortunately, most software that supports a seismic interpretation and modeling workflow is developed for and used by the hydrocarbon industry and is prohibitively expensively for organizations in other sectors. Very shallow reflection surveys also suffer from difficulties in imaging the upper 10–20 m (30–60 ft).

To reduce the expense of reflection surveys, and to get better near-surface resolution in unconsolidated soil deposits, mobile vibrating energy sources are used in combination with an array of geophones, termed a land-streamer, which can be towed along the ground (Pullan et al. 2013). Such equipment can be configured to acquire P-wave or S-wave data (Pugin et al. 2002, 2004, 2006, 2007; Pullan et al. 2013). Using vibratory energy sources in surveys along roads allows for filtering out vehicle vibrations and allows for the economical collection of high-quality data (Figure 7.25).

7.6.1.3 Surface Wave Surveys

Two types of surface waves, known as Rayleigh and Love waves, propagate with unique particle motions when interacting with the Earth's surface. In general, Rayleigh

Figure 7.25 Microvib and SH-wave geophones in operation along Interstate 70 in Ohio (Wightman et al. 2003). Source of photo: Bay Geophysical, Inc.

waves result from a combination of P-wave and vertical S-wave energy, while Love waves result from horizontal S-wave energy interacting with the free surface (the ground surface). Both Rayleigh and Love waves can propagate below stiff layers located at or near the ground surface. Stokoe et al. (1994) showed that imaging beneath very thick concrete was a distinct advantage of the surface wave technique. Surface waves do not decay with distance as fast as body waves and thus carry the vast majority of the energy felt on the ground surface following earthquakes. Body and surface waves are always present whenever seismic data are collected but the amplitude of the surface waves is always higher, making the surface wave technique well-suited to noisy sites or urban environments.

Although slower than body waves, surface waves propagate at a speed that is empirically related to the S-wave velocity (Stokoe et al. 1994). This relationship is used

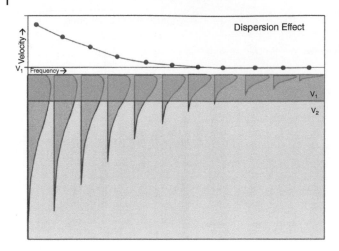

Figure 7.26 Illustration of the surface wave dispersion effect. Image courtesy of Jacob Sheehan.

extensively in geotechnical investigations to measure shear-wave velocity, thus mapping material stiffness. Although shallow subsurface materials do not always have increasing stiffness with depth, and thus increasing seismic velocity, surface waves are capable of mapping less-stiff strata beneath higher stiffness materials. This characteristic of surface waves results from the fact that these waves are dispersive; that is, the phase velocity of waves varies according to their frequency. Because lower frequency

waves have relatively longer wavelengths, they penetrate more deeply into the Earth than the higher frequency waves. Under normal geologic conditions where the seismic velocities increase with depth due to overburden stress, the longer wavelength (low frequency) waves travel faster than the shorter wavelength (high frequency) waves so the different frequencies exhibit classic phase velocity dispersion when processed in the frequency domain, unlike the time domain of body wave data processing (Figure 7.26).

Before 1999, the primary technique to analyze dispersive-wave behavior was SASW; this produces a one-dimensional model of S-wave velocity. However, more recently, MASW has become the routinely used analysis technique for mapping the 1-D and 2-D distribution of S-wave velocities (Song et al. 1989; Miller et al. 1999).

MASW is used to map the lateral and vertical distribution of S-wave velocities in the subsurface by generating 1-D shear-wave soundings (Park et al. 1999; Xia et al. 2000) or, by combining multiple 1-D soundings to produce 2-D profiles (Figure 7.27; Miller et al. 1999). Passive surface wave measurements do not need any sources with triggered mechanisms, but instead measure the ambient seismic-wave energy recorded at a site, which can include micro-earthquakes (microtremors), vehicles, trains, vibrating foundations and induced impacts such as those used in body-wave surveys (but with no time-zero triggers). Louie (2001) reports the use of passive MASW seismic

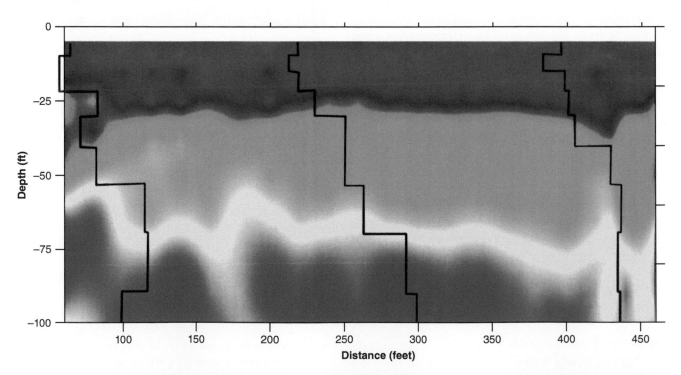

Figure 7.27 Example of a 2-D shear-wave velocity model generated from MASW, showing three of the 15 1-D shear-wave soundings that were used to generate the 2-D gridded/smooth model. Image courtesy of Jacob Sheehan.

techniques that employed deep sources of energy, including microtremors, to image as deep as 2 km. MASW is currently used for a wide range of applications, including groundwater and geotechnical assessments such as seismic hazard classification of sites according to the International Building Code (International Code Council 2017) which uses the average S-wave velocity for the upper 30 m (Vs30) as the primary parameter.

7.6.2 Resistivity Survey Method

Resistivity surveys map the contrast in electrical resistivity of earth materials (or its inverse, conductivity). Measuring the resistivity of earth materials is governed by Ohm's Law (Telford et al. 1990). Resistivity surveys are often referred to as Direct-Current (DC) resistivity, Electrical Resistivity Imaging (ERI), or Electrical Resistivity Tomography (ERT). The method measures the apparent resistivity of the subsurface by applying a known current between two "current" electrodes and observing the potential difference across two "potential" electrodes (Figure 7.28a). When an electrode array is repeatedly expanded, the current penetrates deeper into the ground; the resulting sounding curve shows the resistivity values as a function of electrode spacing (Figure 7.28b). The resistivity of earth materials is most affected by water content, chemistry of the ground water (or contaminants), fines content (in soils), and rock type; for example, shales are typically low resistivity and limestone is usually a highly resistive rock type.

Key elements to a successful resistivity survey are the spacing between the electrodes and choice of array configuration (Figure 7.29). Different arrays provide either better vertical resolution (Schlumberger), or lateral resolution (Wenner), or both (Dipole-Dipole). Resistivity has gained popularity since the advent of Multi-Electrode Resistivity (MER) systems. Traditional four-pin surveys took significant time and labor expense to conduct. Although MER systems activate only four electrodes for each measurement, MER systems provide the ability to select the array configuration and make multiple array configuration measurements; this capability permits the use of tomography to provide lateral and

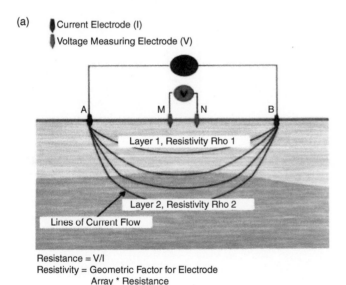

Figure 7.28 Basic DC resistivity four-pin electrode configuration (a) Data recording geometry (Schlumberger illustrated here), (b) resulting sounding curve (Wightman et al. 2003). Source of figure: (U.S. Department of Transportation, Federal Highway Administration).

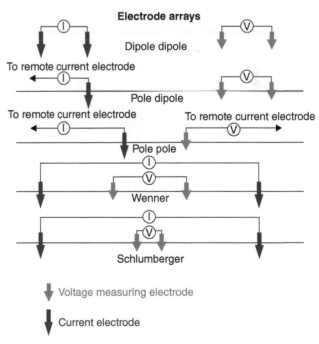

Figure 7.29 Various resistivity electrode array configurations: (top to bottom) Dipole-Dipole, Pole-Dipole, Pole-Pole, Wenner and Schlumberger. Modified after Wightman et al. (2003). Source of figure: (U.S. Department of Transportation, Federal Highway Administration).

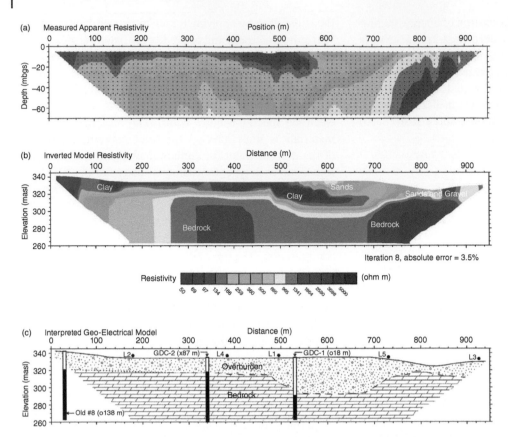

Figure 7.30 Example 2-D profile for ~1 km long ERT survey. (a) Measured apparent resistivity pseudo-section; (b) inverted resistivity earth model (called a geoelectric section); and (c) interpreted geologic model. (Steelman et al. 2018).

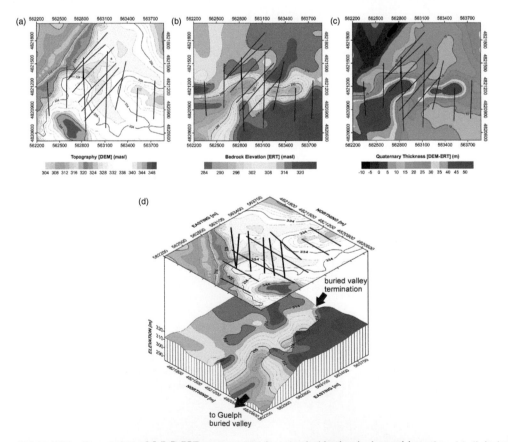

Figure 7.31 Generation of 2.5-D ERT earth model for a buried bedrock channel based on (a) digital terrain model, (b) ERT-derived bedrock elevation model (black lines represent ERT transects), and (c) vertically exaggerated interpretation of bedrock channel. Modified from Steelman et al. (2018).

vertical images of resistivity variations in soil and rock (Loke et al. 2013; Oldenburg and Li 2005; Møller et al. 2006). Each 2-D resistivity survey generates an apparent resistivity pseudo-section, a vertical section representing resistivity as a function of depth and electrode spacing(s) (Figure 7.30a). The pseudo-section is inverted using tomography to generate the most reasonable distribution of resistivity that matches the measured data (Figure 7.30b). Lithologic interpretation requires geologic information to constrain the modeling process; this allows layers to be assigned to geoelectric units (Figure 7.30c). Volumetric (2.5-D) geoelectric models can be generated from multiple inverted 2-D pseudo-sections collected along a network of intersecting cross-sections (Figure 7.31). However, deploying MER systems with large grids of electrodes permits acquisition of true 2-D resistivity data which can be processed with a 3-D tomography code; these 3-D ERT data best represent the subsurface, particularly with large lateral geologic variations (Chambers et al. 2006).

7.6.3 Electromagnetic Survey Method

There are numerous electromagnetic (EM) techniques, but the most commonly used techniques are Time-Domain electromagnetic (TDEM) and Frequency Domain electromagnetic (FDEM). These techniques have been used extensively in the mineral exploration industry (Denith and Mudge 2014) and for assessing groundwater resources (Danielsen et al. 2003). With either technique, the instruments vary the primary magnetic field, either in time (TDEM) or continuously (FDEM). This generates eddy currents in conductive earth materials due to the principle of electromagnetic induction and Faraday's Law (Telford et al. 1990). The receiver system coil measures the strength of the resulting secondary magnetic field (Figure 7.32). Both techniques obtain data about the subsurface conductivity (or inversely the resistivity) of earth materials from static or moving platforms (Sørensen et al. 2006; Nabighian and Macnae 1991). Both ground-based and airborne measurements are possible. Airborne surveys allow efficient surveys of large areas.

7.6.3.1 Time Domain Electromagnetic Surveys (TDEM)

TDEM, also referred to as Transient Electromagnetic Method (TEM), utilizes a transmitter coil to turn on and off a steady DC current and a receiver coil to measure the time decay of voltage (i.e. rate of change) of the secondary magnetic field resulting from the eddy currents (Figure 7.33a). The rate of electromagnetic energy diffusion is indicative of the conductivity of different subsurface materials (Figure 7.33b). The physical size of the transmitter loop and the number of turns and current driven through the

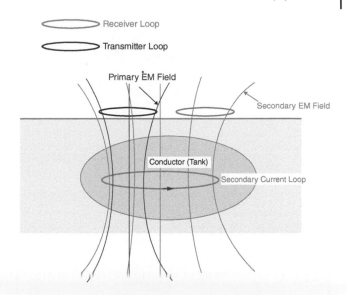

Figure 7.32 Concept diagram for electromagnetic methods. An electrical current in the transmitter loop generates a primary EM field (shown in red) which creates currents in subsurface conductors (green object). The conductors generate secondary EM fields (shown in green) that are detected by the receiver loop which produces a signal proportional to the conductivity of the object (Wightman et al. 2003). Source of figure: (U.S. Department of Transportation, Federal Highway Administration).

loop (i.e. the moment) dictate the depth of TEM investigations, which can vary from 10 to 1000 m (30–3000 ft). Large metallic objects on the surface, or ambient electromagnetic noise from power lines and grounded fences, can negatively impact TDEM survey data quality. Additionally, transmitter and receiver system response must be calibrated during a survey to identify any system drift; this requires careful pre-processing of the data to resolve. TDEM measurements produce one-dimensional soundings acquired at specified spacings; these are inverted to match measured or geologic layer resistivities, then gridded and plotted as 2-D geo-electric sections (Abraham et al. 2018). These resulting EM data are very useful for generating 3-D geologic models.

Jørgensen et al. (2003) used TDEM techniques integrated with exploration boreholes to provide confirmatory information for delineating buried valleys in Denmark. This approach allowed ground-based TDEM surveys to successfully delineate many buried valleys and, despite some resolution limitations, was a cost-effective overview of large areas by providing interpretation of overall geological structures. The successful delineation with the TDEM technique was largely due to the fact that the buried valleys were cut into clay-dominated (conductive) sediments which are filled with sandy (resistive) Quaternary sediments; thus, they produced a strong EM response.

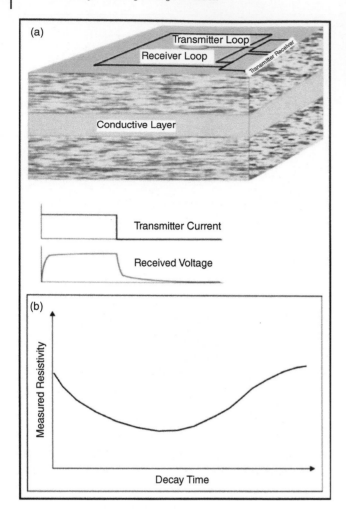

Figure 7.34 FDEM ground survey using a Dualem421 instrument which has dual-oriented coils at spacings of 4, 2, and 1 m. Source: www.dualem.com (Public domain).

Figure 7.33 Time Domain Electromagnetic Sounding. (a) Field TDEM measurement procedure, with examples of the transmitted and received signals; (b) a TDEM Sounding curve. The measured resistivity is plotted against the decay time. Modified after Wightman et al. (2003). Source of figure: (U.S. Department of Transportation, Federal Highway Administration).

instrumentation measures both the phase shift and the corresponding amplitude shift as the quadrature and in-phase components of system response. When measurements of each component are plotted independently as a function of distance along multiple transects, the data result in 2-D plan maps. Recent hardware improvements permit the use of multiple frequencies, multiple coil spacings, and orthogonal coil orientations (Figure 7.34). The inversion of these FDEM data produces 2-D geo-electric sections and 3-D volumetric geo-electric models as shown in Figure 7.35, where clay lenses were mapped below the roadway (Pfeiffer and Hanna 2009). Only the quadrature component is used for geo-electric imaging and is more useful for 3-D geologic model development. The in-phase system response readily detects the presence of buried ferrous or non-ferrous metals due to their extremely high inductive response.

7.6.3.2 Frequency Domain Electromagnetic Surveys

FDEM also utilizes a transmitter and receiver coil to induce and record the primary and secondary magnetic fields from eddy currents generated by conductive earth materials. However, the FDEM transmitter coil induces a sinusoidally varying current at a specific frequency (or specific frequencies). Depth of investigation for FDEM depends on the frequency of the EM signal(s), the spacing between the transmitter and receiver coil(s), and the orientation of the transmitter-receiver coils. With FDEM, the resulting eddy currents produce secondary EM field readings at the receiver coil(s) which are shifted both in phase and amplitude. These shifts are caused by the mutual inductance of the conductive properties of the subsurface. The FDEM

7.6.3.3 Airborne Electromagnetic Surveys

Shallow subsurface 3-D geological models have been built incorporating data from airborne electromagnetic (AEM) surveys (Jørgensen et al. 2013; Korus et al. 2017). AEM surveying was largely developed in Denmark for groundwater resource evaluations (Sørensen and Auken 2004; Siemon 2006; Auken et al. 2009). Airborne systems utilize the same EM principles as the ground based TDEM and FDEM surveys discussed in Sections 7.6.3.1 and 7.6.3.2; they may use either time or frequency varying instruments. Either fixed-wing or helicopter platforms are deployed (Figure 7.36).

For the best resolution of the subsurface, AEM surveys use low-altitude flights, which may require pre-clearance

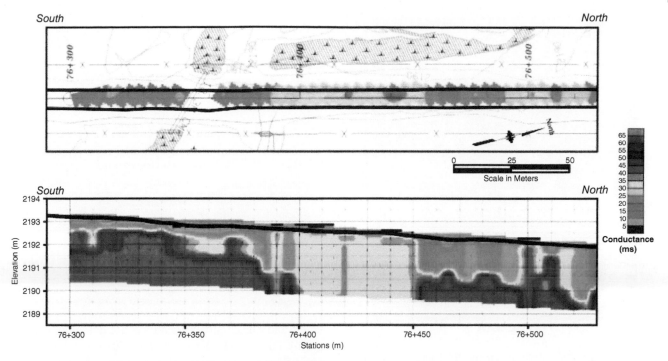

Figure 7.35 FDEM data example of mapping clay beneath roadways. Upper plot is a plan view with 1-m (3.2 ft) stripped away, and the lower plot is a 2-D cross-section. Cool colors represent sandy and silty soils, hot colors represent clayey soil deposits Pfeiffer and Hanna (2009). Source of Figure (U.S. Department of Transportation, Federal Highway Administration).

Figure 7.36 A Eurocopter AS 350 conducts an airborne electromagnetic survey for the U.S. Geologic Survey. Source: U.S. Geological Survey (Public domain), via Wikimedia Commons.

with the local authorities and public announcements. Because cultural noise from power lines, pipelines or railway tracks adversely affect AEM surveys, these surveys are most effective in rural areas with low relief. Jørgensen et al. (2013) describes the use of an AEM survey combined with a high-resolution seismic survey and borehole geophysical logging to provide an integrated 3-D geophysical evaluation of the salinity distribution of groundwater resources. Figure 7.37 presents example results from an AEM survey used to locate sand and gravel deposits in Saskatchewan, Canada. Sapia et al. (2014) discuss the relative merits of ground-based and airborne TEM methods based on surveys conducted in Manitoba, Canada; while Abraham et al. (2018) presents vast amounts of AEM model results over the central United States for characterizing the framework of regional aquifers. The Alberta Geological Survey has undertaken multiple AEM surveys to model the distribution and physical attributes of sediment- and bedrock-aquifer complexes in glacial terrain (Slattery and Andriashek 2012).

7.6.4 Gravity Surveys

Gravity surveys can be conducted on-foot over discrete targets and small sites for geological, environmental, geotechnical, industrial and resource mining applications. Or they can be performed over large areas using airborne techniques to evaluate large-scale geologic features and assess mineral, oil and gas, and ground water resources.

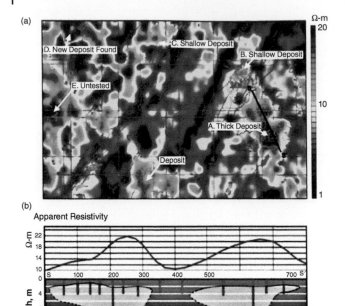

(a)

(b)
Apparent Resistivity

Drill Section

Distance, m

Figure 7.37 Airborne electromagnetic (AEM) survey example. (a) Map showing AEM results for a survey conducted for sand and gravel assessment in Saskatchewan, Canada. (b) Section view comparing exploratory drilling results and resistivity values over sand and gravel deposits. Modified after Wightman et al. (2003). Source of figure: Fugro Airborne Surveys.

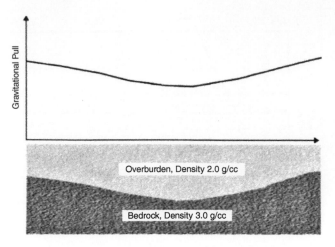

Figure 7.38 Gravitational pull over a bedrock depression (Wightman et al. 2003). Source of figure: (U.S. Department of Transportation, Federal Highway Administration).

Both ground-based and airborne gravity surveys have provided useful data for 3-D geological model development; the size and characteristics of the area being surveyed dictates the choice of acquisition method. As with all potential field geophysical methods, gravity survey results are non-unique and thus must be verified and calibrated with additional geologic information. Usually this supporting information is supplied by rock core recovered from exploratory boreholes or from geophysical borehole logging (Section 7.6.6).

7.6.4.1 Ground-based Gravity Surveys

Gravity surveys measure density variations of subsurface materials by obtaining a series of individual gravity observations at stations located along a series of profiles or a grid. In general, shallower and smaller geologic targets require shorter distances between stations. The intensity of the force of gravity due to changes in near-surface material properties is superimposed on the much larger force of gravity due to the total mass of the earth. Thus, field measurements must be interpreted to define gravity anomalies related to geologic conditions (Figure 7.38). Within a study area, density contrasts between rock types are often sufficient to cause observable gravity responses. Karst conditions, presence of voids in abandoned mine

lands, and buried stream channels are common shallow gravity targets, but the method is also extensively used for resource exploration. For example, the Geoscience Australia Geophysical Archive Data Delivery System website provides access to data sets from several Australian National, State and Territory Government gravity surveys that have been undertaken for mineral resource exploration (Geoscience Australia, 2020).

Gravity is measured in Gals; a unit of measure named after Galileo Galilei. The Gal is defined as an acceleration of one hundredth of a meter per second squared ($0.01 \, m/s^2$). Identification of anomalous areas within a gravity survey project area requires more precise units; the milligal (mGal) and microgal (μGal) are commonly used. Gravity meters, or gravimeters, used in ground-based gravity surveys are capable of accuracies of 5–20 μGal (Hinze et al. 2005). Depending on the scale of the target and its depth of burial, detecting small targets may require the use of microgravity measurements which are subjected to a series of repeat station measurements and significant corrections made during processing (Long and Kaufmann 2013; Hinze et al. 2005).

Because gravity is strongly influenced by the position and height of observation, accurate gravity surveys require precise information on the location and elevation of each individual station. Use of survey-grade Differential Global Positioning System instruments and sub-centimeter precision procedures can provide sufficiently accurate values for location and elevation. Observed gravity values must also be corrected for instrument drift, Earth tides, latitude, free air, Bouguer density, and terrain (topography). These corrections are described in many standard geophysical texts (Parasnis 1972; Telford et al. 1990; West 1992; Burger

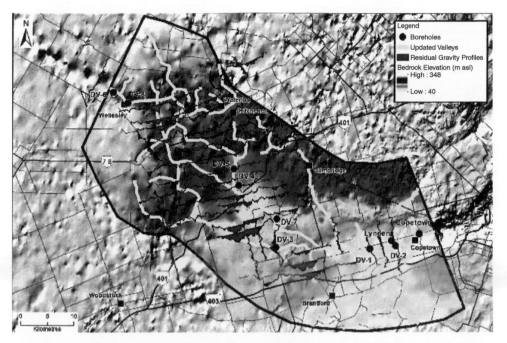

Figure 7.39 Gravity surveys successfully defined buried bedrock valley near Dundas Ontario (Bajc et al. 2018). Copyright © Queen's Printer for Ontario, 2018. Used by permission of the Ontario Geological Survey.

et al. 2006). Gravity signals coming from near-surface geologic features are only a very small fraction of the earth's gravitational field; they are anomalous values superimposed on the responses due to larger or deeper sources. Therefore, it is often necessary to remove a regional, or background, field (Gupta and Ramani 1980; Mickus et al. 1991).

Recent 3-D modeling programs conducted by the OGS in southern Ontario have used gravity surveys to identify changes in buried bedrock elevation (Bajc and Rainsford 2010; Burt and Rainsford 2010). While gravity surveys can be highly effective in identifying locations of potential bedrock valleys buried by glacial deposits (Figure 7.39), they provide limited insight into the characteristics of Quaternary infill and stratigraphy (Greenhouse and Monier-Williams 1986). Without additional information, gravity survey data contain significant uncertainty concerning true bedrock elevations; the OGS has conducted project-specific exploratory drilling programs to provide the required calibration information to constrain the gravity model(s).

7.6.4.2 Airborne Gravity Surveys

Airborne gravity surveys are conducted using aircraft or helicopters along flight lines commonly 150–200 m above the ground and spaced around 2–2.5 km apart. They sense geological changes with a relatively high-frequency response that is calibrated with supplementary ground-based gravity measurements. To resolve accuracy issues due to the height and motion of the platform with this survey technique, airborne gravity surveys routinely employ the gradiometric technique, which relies on the simultaneous measurement of gravity at two or more closely separated sensor positions. This is accomplished with stabilized gravimeters that compensate for nearly all the extraneous forces. Gradiometer surveys produce a spatial definition of the gravity gradient tensor field (gradient vector).

Airborne gravity surveys have been conducted over extensive areas where they provide information on subsurface structure for mineral resource and groundwater evaluations. Murray and Tracey (2001) question the quoted accuracy of airborne gravity measurements proposed by survey providers. While the accuracy of airborne gravity surveys over mountainous terrain has been a concern, Verdun et al. (2003) reported good results for a 1998 survey in the Swiss and French Alps. Howard et al. (2018) provide a detailed description of Australian airborne gravity surveys. Drenth et al. (2013) describe how an airborne gravity survey conducted in 2012 in southwestern Colorado investigated the subsurface structural framework that influences groundwater hydrology and known seismic hazards.

7.6.5 Ground Penetrating Radar

GPR is a high-resolution geophysical method which utilizes the propagation of high frequency (10–2000 MHz)

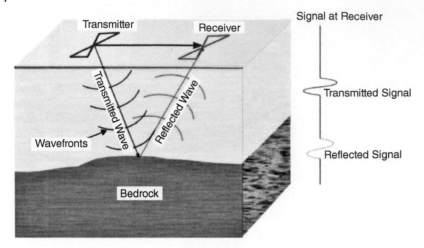

Figure 7.40 Ground Penetrating Radar schematic (Wightman et al. 2003). Source of figure: (U.S. Department of Transportation, Federal Highway Administration).

electromagnetic (EM) waves. GPR systems utilize a transmitter that radiates short pules of electromagnetic energy into the subsurface. Contrasts in the EM properties of earth materials cause some energy to be reflected back to the surface. The amplitude and phase of these reflections are recorded as a time-series by a receiver antenna (Figure 7.40). By moving the GPR system across the ground surface, 2-D sections of the earth materials response are generated (Figure 7.41). If many sets of transmitter–receiver pairs are used simultaneously, tomography procedures create 3-D GPR volumetric images. Jol (2009) provides detailed descriptions of GPR theory and applications. Additional information on GPR methods is provided by Smith and Jol (1995), Jol and Bristow (2003), Machguth et al. (2006), Conyers (2013), and Utsi (2017).

GPR maps contrasts in dielectric permittivity as the diagnostic property of earth materials, with stronger contrasts producing higher amplitude reflections. Variation of dielectric permittivity in sediments is mainly caused by

mineralogy and pore-fluid content (Olhoeft 1998; Martinez and Byrnes 2001; Neal 2004; Daniels 2004, chapter 4). Typical vertical resolutions for GPR surveys are sub-meter; higher frequencies produce higher resolutions but achieve less depth-of-investigation. With modern GPR systems capable of high pulse-transmit rates, lateral resolutions are scalable from tens of meters down to sub-meter depths below ground. Multi-channel GPR systems have been developed with 40–70 transmitter-receiver pairs, as well as step-frequency GPR hardware; these are capable of creating 3-D GPR images of the subsurface using standard tomographic inversion processing (Figure 7.42).

The primary limitation of GPR is signal attenuation. Attenuation of these high-frequency EM waves is mainly a function of antenna frequency because higher-frequency waves have higher rates of attenuation and bulk electrical conductivity of earth materials (Annan 1973; Jol 2009; Bradford 2007). Due to this, depths of investigation for GPR surveys are typically on the order of several meters

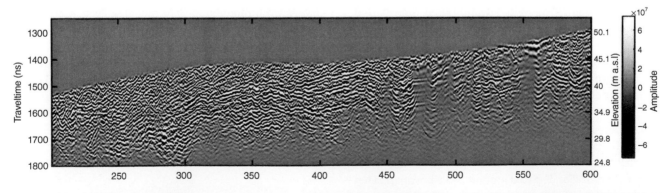

Figure 7.41 GPR reflection profile acquired in a carbonate environment using a 100 MHz system. Note investigation depth reaches over 10 m. Modified from Bowling (2017).

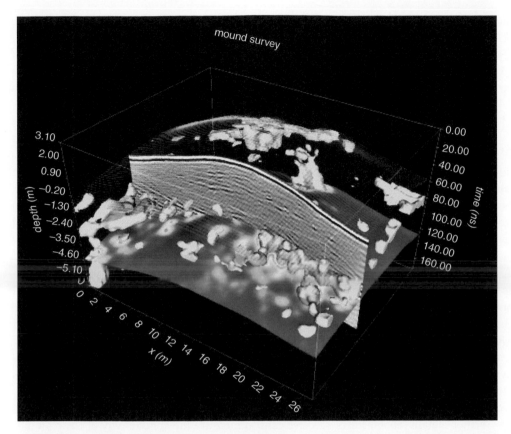

Figure 7.42 3-D GPR model for environmental application. Source of photo: www.gprslice.com (public domain).

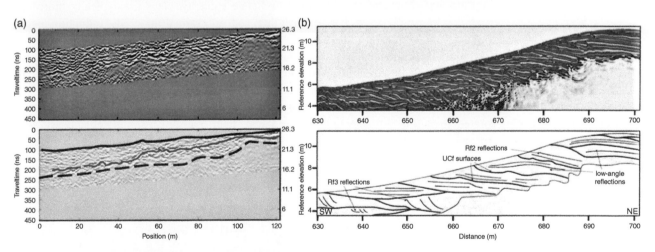

Figure 7.43 Two applications of ground penetrating radar data. (a) Structure mapping in carbonate geologies. Source: Modified from Bowling (2017); (b) Eolian protodune stratigraphy mapping. Modified from Phillips et al. (2019).

into the subsurface. Because higher conductivities cause more attenuation, GPR investigations perform poorly in areas underlain by clay-rich materials, in coastal areas with salty groundwater, or where permafrost is present. Sandy sediments that are relatively devoid of mineralization and are relatively dry or contain low salinity pore-fluids are well suited to GPR investigation.

Despite the attenuation limitations, GPR provides very-high resolution images of small structures, stratigraphy and anomalous conditions (karst) which makes GPR a useful data source for shallow 3-D geologic models. GPR has been applied to a wide range of geological settings including; eolian (Bennett et al. 2009; Rozar 2015; Phillips et al. 2019), carbonate karst (Martinez et al. 1998; Jorry and Biévre

2011; Bowling et al. 2018), and glacial (Møller and Jakobsen 2002; Bakker and Van der Meer ; Møller and Jørgensen 2006; Rutishauser et al. 2016) environments. GPR has been successfully used for understanding glacier thickness, geologic structures, lithology determination, and water content estimation (Figure 7.43).

7.6.6 Borehole Geophysics

Borehole geophysics provides more continuous records than can be obtained from sample descriptions and analysis. Because sampling boreholes is expensive, obtaining detailed continuous borehole geophysical data adds significant value to an exploration program for a minimal cost. Borehole logs are routinely used to calibrate and correlate other nearby surface geophysical information. For decades geophysical borehole techniques have been a very effective data source for lithologic mapping and development of 3-D geological models (Bajc and Hunter 2006; Bajc et al. 2014), and they are playing an expanded role in geological, geotechnical and hydrogeological investigations. There are two broad geophysical categories of borehole measurements – borehole logging (conducted in open holes) and in-hole seismic logging (conducted in cased holes). Labo (1987) provides further details about the logging tools and how the data are evaluated for specific formation properties such as porosity, lithology, or chemistry.

7.6.6.1 Borehole Geophysical Logging

Borehole geophysical logging supplies continuous data related to in-situ properties of soils and rocks, their contained fluids, and well construction (Doveton and Prentsky 1992; Keys 1997; Wonik and Hinsby 2006). It supplements visual logs and may be interpreted to define lithology, thickness, continuity, and properties of aquifers and confining beds as part of groundwater investigations (Kobr et al. 2005). Geophysical logging allows for interpretation and correlation of key stratigraphic horizons by means of pattern recognition.

The most common borehole geophysical logging tools are: natural gamma (gamma ray), induction conductivity, magnetic susceptibility, resistivity, and sonic/acoustic; however, many additional logging technologies exist (Labo 1987). These nonradioactive logging techniques are frequently used for subsurface formation characterization because they do not require specialized licensure allowing for widespread application (Crow et al. 2018). Figure 7.44 provides examples of geophysical borehole logs for several typical near-surface formations in Southern Ontario (Crow et al. 2018).

Natural gamma (or gamma ray) logs measure naturally occurring radioactivity. The most common natural radionuclides in rock and sediment are potassium (K), uranium (U), and thorium (Th). Subtle changes in total count rates are used in a qualitative manner to estimate variation in grain size because clays (and shales) generally exhibit higher count levels than coarse or sandy materials (Crow et al. 2018).

Induction conductivity logging tools use transmit and receive coils to measure the conductivity of the materials encountered during drilling. Owing to their increased cation exchange capacity, clays are typically more conductive than coarser grained materials. Induction logging is very useful for assessing ground water chemistry; in particular, for identifying salt-water intrusion in coastal aquifers, the extent of contamination plumes, and the presence of saline or alkaline groundwater.

Magnetic susceptibility logging records changes in the magnetic field caused by magnetically susceptible minerals; these are characteristic of igneous or

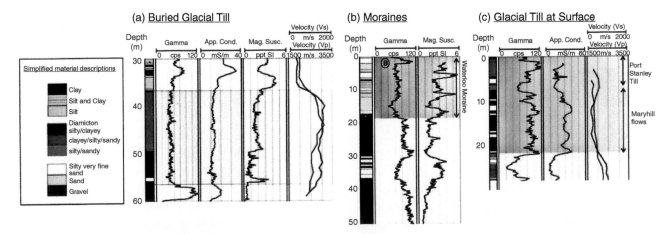

Figure 7.44 Borehole logs from the Waterloo, Ontario region with responses characteristic of (a) buried glacial till, (b) moraine sediments, and (c) glacial till at the surface. Modified from Crow et al. (2018).

metamorphic rocks, or sediments derived from them. Therefore, magnetic susceptibility logs provide an indication of sediment provenance in the glacial deposits in Canada that are derived from the Canadian Shield Precambrian rocks (Crow et al. 2018).

Resistivity logging is primarily used for lithologic and formation fluid evaluation. Resistivity logs, including self-potential (SP) logs, are used to evaluate the effective porosity, relative permeability and the lithology of soil and rock utilizing cross-plot techniques (multiple log data and formation properties assessed together); this has proved extremely valuable to hydrocarbon exploration (Asquith and Gibson 2004).

Acoustic or sonic logs provide information on formation properties related to strength, density and modulus. These logs measure the formation matrix velocity; measurements within approximately 0.3 m (1-ft) of the borehole requires use of borehole compensated sonic devices that account for borehole rugosity and size variation.

7.6.6.2 In-hole Seismic Geophysical Logging
In-hole seismic geophysical logging techniques yield one-dimensional vertical profiles of the P- and S-wave velocities (Figure 7.45). These data support calibration of surface seismic measurements and calculation

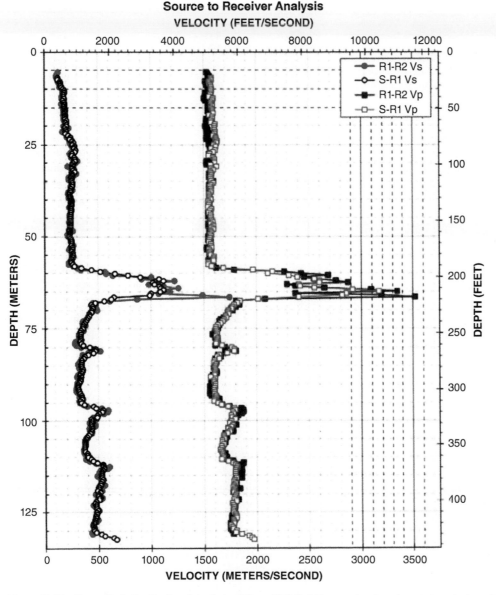

Figure 7.45 Example 1-D velocity plot, derived from OYO P-S Suspension Logging system, but applicable for CS, DS, and Suspension logging application. (Diehl et al. 2006).

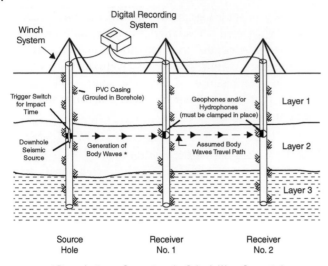

* Dependent upon Source type the Seismic Wave Generated could be P-, SV-, or SH Body Waves

Figure 7.46 Schematic of cross hole geophysical logging (Wightman et al. 2003). Source of figure: (U.S. Department of Transportation, Federal Highway Administration).

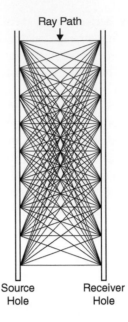

Figure 7.47 Source and receiver locations for a borehole tomographic survey (Wightman et al. 2003). Source of figure: (U.S. Department of Transportation, Federal Highway Administration).

of attenuation coefficients and damping, which can be extremely valuable for fault investigations, probabilistic seismic hazard assessments, and evaluation of vibrating foundation conditions (Sirles 1986). Along with knowledge of the in-situ unit weight of the soil and rock surrounding the borehole(s), this information allows determination of the low-strain elastic moduli (Bulk, Shear, Young's, and Lame's constants and the Poisson's Ratio) which are important in engineering design. Three in-hole techniques dominate the near-surface seismic borehole measurements: Crosshole, Downhole and P-S suspension logging. ASTM standards exist for both Crosshole (D4428) and Downhole (D7400) techniques (ASTM 2014, 2019, respectively).

Crosshole seismic (CS) testing is conducted in a straight ray field set-up with either two or three in-line borings (Figure 7.46). Crosshole seismic tomography (CST) is an adaptation of CS measurements that utilizes a higher-density ray coverage between a series of borings to analyze the travel-times using tomography inversion software (Figure 7.47). CST results obtained from multiple boreholes in a nonlinear arrangement and 3-D tomographic inversion software can be visualized in 2-D (plane) or 3-D (volumetric) images (Figure 7.48). Both CS and CST surveys require a deviation survey be performed in each bore hole so that accurate distances can be calculated between the boreholes.

Downhole seismic (DS) and P-S Suspension logging require only one drill hole and can be conducted to greater depths than either CS or CST. Neither DS nor Suspension logging require borehole deviation surveys because interval travel-time measurements are made in the same hole between geophones at different depths. Suspension logging has the best resolution of thin layers of any of the in-hole seismic systems because of the closely spaced (1 m) upper and lower three-component geophones used in this set-up (Figure 7.49). CS, DS, and P-S logging techniques are accepted as the highest resolution seismic information provided for the near-surface characterization.

Acknowledgments

Sections 7.1–7.5 are published with the permission of the director of the Ontario Geological Survey. The authors gratefully acknowledge the assistance of many colleagues in providing information on the various data sources and data collection options. Kei Yeung (OGS) is thanked for his assistance with assembling and managing data from a wide variety of sources. Comments and suggestions by Andy Bajc (OGS) improved Sections 7.1–7.5.

Figure 7.48 Tomograms showing: (a) low-velocity zones in bedrock near socketed piles and (b) caisson on top of bedrock. (Wightman et al. 2003). Source of figure: (U.S. Department of Transportation, Federal Highway Administration).

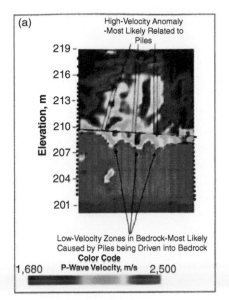

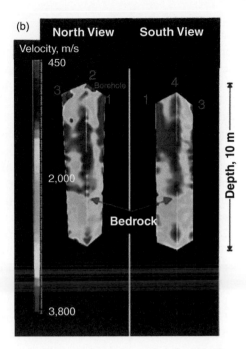

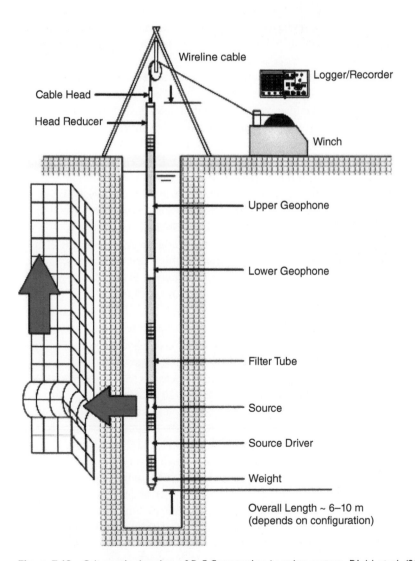

Figure 7.49 Schematic drawing of P-S Suspension Logging system. Diehl et al. (2006).

References

Abraham, J.D., Asch, T.H. and Cannia, J.C. (2018). Ten Years of AEM in Nebraska: Stitching Surveys Together. In: *Abstracts from AEM 2018, Kolding, Denmark, June 16–20, 2018*. 4 pp.

Allen, F.A., Richart, F.E. and Woods, R.D. (1980). Fluid wave propagation in saturated and nearly saturated sands. *Journal of Geotechnical Engineering* 106 (3): 235–254.

Annan, A.P. (1973). Radio interferometry depth sounding: part I - theoretical discussion. *Geophysics* 38 (3): 557–580.

Armstrong, D.K. and Dodge, J.E.P. (2007). *Paleozoic Geology of Southern Ontario*. Sudbury, ON: Ontario Geological Survey, Miscellaneous Release—Data 219. 5 pp. [Online: Available at: http://www.geologyontario.mndm.gov.on.ca/mndmfiles/pub/data/imaging/mrd219//MRD219.pdf] (Accessed July 12, 2020).

Arnold, T.L., Friedel, M.J. and Warner, K.L. (2001). Hydrogeologic inventory of the upper Illinois River basin – creating a large data base from well construction records. In: *Geological Models for Groundwater Flow Modeling: Workshop Extended Abstracts* (eds. R.C. Berg and L.H. Thorleifson). Champaign, IL: Illinois State Geological Survey, Open File Series 2001-1, 1–5. [Online: Available at: http://library.isgs.illinois.edu/Pubs/pdfs/ofs/2001/ofs2001-01.pdf] (Accessed July 15, 2020).

Aronoff, S. (1989). *Geographic Information Systems: A Management Perspective*. Ottawa Canada: WDL Publications. 294 pp.

Asquith, G.B. and Gibson, C.R. (2004). *Basic Well Log Analysis for Geologists, 2e*. Tulsa, OK: The American Association of Petroleum Geologists, Methods in Exploration Series, vol 16. 244 pp. https://doi.org/10.1306/Mth16823.

ASTM (2014). *ASTM D4428/D4428M-14, Standard Test Methods for Crosshole Seismic Testing*. West Conshohocken, PA: ASTM International. 7 pp.

ASTM (2018). *ASTM D5777–18, Standard Guide for Using the Seismic Refraction Method for Subsurface Investigation*. West Conshohocken, PA: ASTM International. 14 pp.

ASTM (2019). *ASTM D7400/D7400M-19, Standard Test Methods for Downhole Seismic Testing*. West Conshohocken, PA: ASTM International. 11 pp.

Auken, E., Christiansen, A.V., Westergaard, J.A. et al. (2009). An integrated processing scheme for high resolution airborne electromagnetic surveys, the SkyTEM system. *Exploration Geophysics* 40: 184–192. https://doi.org/10.1071/EG08128

Bajc, A.F. and Dodge, J.E.P. (2011). *Three-Dimensional Mapping of Surficial Deposits in the Bradford-Woodstock Area, Southwestern Ontario*. Sudbury, ON: Ontario Geological Survey, Groundwater Resources Study 10. [Online: Available at: http://www.geologyontario.mndm.gov.on.ca/mndmfiles/pub/data/imaging/GRS010/GRS010.pdf] (accessed July 15, 2020).

Bajc, A.F. and Hunter, J.A. (2006). *Results of 2003–2004 Overburden Drilling Programs in the Region of Waterloo, southwestern Ontario*. Sudbury, ON: Ontario Geological Survey, Miscellaneous Data Release 205. [Online: Available at: http://www.geologyontario.mndm.gov.on.ca/mndmfiles/pub/data/imaging/MRD205//MRD205.pdf] (Accessed July 15, 2020).

Bajc, A.F. and Newton, M.J. (2005). 3-D modelling of quaternary deposits in waterloo region, Ontario: a case study using Datamine studio® software. In: *Three-Dimensional Geological Mapping for Groundwater Application; Geological Society of America Annual Meeting, Salt Lake City, Utah* (eds. H.A.J. Russell, R.C. Berg and L.H. Thorleifson), 5–10. Ottawa, ON: Geological Survey of Canada, Open File 5048. https://doi.org/10.4095/221818.

Bajc, A.F. and Rainsford, D.R.B. (2010). Three dimensional mapping of quaternary deposits in the southern part of the county of Simcoe, southern Ontario. In: *Summary of Field Work and Other Activities, 2010* (eds. Ayer, J.A., Easton, R.M., Beakhouse, G.P., et al.). sudbury, ON: Ontario Geological Survey, Open File Report 6260: 30/1-30/10. [Online: Available at: http://www.geologyontario.mndm.gov.on.ca/mndmfiles/pub/data/imaging/OFR6260//OFR6260.pdf] (Accessed July 15, 2020).

Bajc, A.F. and Shirota, J. (2007). *Three-Dimensional Mapping of Surficial Deposits in the Regional Municipality of Waterloo, Southwestern Ontario*. Sudbury, ON: Ontario Geological Survey, Groundwater Resources Study 3. 41 pp.

Bajc, A.F., Russell, H.A.J. and Sharpe, D.R. (2014). A three-dimensional Hydrostratigraphic model of the Waterloo moraine area, southern Ontario, Canada. *Canadian Water Resources Journal* 39 (2): 95–119. https://doi.org/10.1080/07011784.2014.914794.

Bajc, A.F., Marich, A.S., Priebe, E.H. and Rainsford, D.R.B. (2018). Evaluating the groundwater resource potential of the Dundas buried Bedrock Valley, Southwestern Ontario: an integrated geological and hydrogeological case study. *Canadian Journal of Earth Sciences* 55 (7): 659–676. https://doi.org/10.1139/cjes-2016-0224.

4 Structure of a Pleistocene Push Moraine Revealed by GPR: The Eastern Veluwe Ridge, the Netherlands. *Geological Society Special Publications* 211: 143–151. https://doi.org/10.1144/GSL.SP.2001.211.01.12

Barnhardt, M.L., Hansel, A.K. and Stumpf, A.J. (2001). Developing the database for 3-D modeling: acquiring, assembling, verifying, assessing, interpreting, and integrating source data. In: *Geological Models for Groundwater Flow Modeling: Workshop Extended Abstracts* (eds. R.C. Berg and L.H. Thorleifson), 6. Champaign, IL: Illinois State Geological Survey, Open File 2001-1. [Online: Available at: http://library.isgs.illinois.edu/Pubs/pdfs/ofs/2001/ofs2001-01.pdf] (Accessed July 25, 2020).

Belanger, J.R. (1998). Urban geology of Canada's National capital area. In: *Urban Geology of Canadian Cities* (eds. P.F. Karrow and O.L. White), 365–384. St. John's, NL: Geological Association of Canada, Special Paper 42.

Belanger, J.R. (2001). Geological model for groundwater flow studies, greater Ottawa, Canada. In: *Geological Models for Groundwater Flow Modeling: Workshop Extended Abstracts* (eds. R.C. Berg and L.H. Thorleifson), 7–9. Champaign, IL: Illinois State Geological Survey, Open File 2001-1. [Online: Available at: http://library.isgs.illinois.edu/Pubs/pdfs/ofs/2001/ofs2001-01.pdf] (Accessed July 25, 2020).

Bennett, M.R., Cassidy, N.J. and Pile, J. (2009). Internal structure of a barrier beach as revealed by ground penetrating radar (GPR): Chesil Beach, UK. *Geomorphology* 104 (3–4): 218–229. https://doi.org/10.1016/j.geomorph.2008.08.015.

Bowling, R. (2017). *Applications of Ground Penetrating Radar to Structural Analysis of Carbonate Terraces on the Island of Bonaire, Caribbean Netherlands.* Master's thesis Texas A&M University. 231 pp. [Online: Available at: http://hdl.handle.net/1969.1/165766] (Accessed August 13, 2020).

Bowling, R.D., Laya, J.C. and Everett, M.E. (2018). Resolving carbonate platform geometries on the island of Bonaire, Caribbean Netherlands through semi-automatic GPR facies classification. *Geophysical Journal International* 214 (1): 687–703. https://doi.org/10.1093/gji/ggy175.

Bradford, J.H. (2007). Frequency-dependent attenuation analysis of ground-penetrating radar data. *Geophysics* 72 (3): J7–J16. https://doi.org/10.1190/1.2710183.

Burger, H.R., Sheehan, A.F. and Jones, C.H. (2006). *Introduction to Applied Geophysics: Exploring the Shallow Subsurface.* New York, NY: W. W. Norton and Company. 554 pp.

Burt, A.K. (2017). Digging deep on the Niagara peninsula: a drilling update. In: *Summary of Field Work and Other Activities 2017*, 24-1–24-16. Sudbury, ON: Ontario Geological Survey, Open File Report 6333. [Online: Available at: http://www.geologyontario.mndm.gov.on.ca/mndmfiles/pub/data/imaging/ofr6333//ofr6333.pdf] (Accessed August 13, 2020).

Burt, A.K. (2018). Three-dimensional hydrostratigraphy of the Orangeville moraine area, Southwestern Ontario, Canada. *Canadian Journal of Earth Sciences* 55: 802–828. https://doi.org/10.1139/cjes-2017-0077.

Burt, A.K. and Bajc, A.F. (2005). Three-dimensional groundwater mapping in complex moraine sediments: current projects in Ontario, Canada. *Geological Society of America, Abstracts with Program* 37 (7): 144.

Burt, A.K. and Crow, H.L. (2019). Insights from combined interpretation of sediment cores and geophysical logs in the Niagara peninsula, southern Ontario. In: *Regional-Scale Groundwater Geoscience in Southern Ontario–An Ontario Geological Survey, Geological Survey of Canada, and Conservation Ontario Open House* (eds. H.A.J. Russell, D. Ford, S. Holysh and E.H. Priebe), 4. Ottawa, ON: Geological Survey of Canada, Open File 8528. https://doi.org/10.4095/313544.

Burt, A.K. and Dodge, J.E.P. (2011). *Three-Dimensional Modelling of Surficial Deposits in the Barrie–Oro Moraine Area of Southern Ontario.* Sudbury, ON: Ontario Geological Survey, Groundwater Resources Study 11. [Online: Available at: http://www.geologyontario.mndm.gov.on.ca/mndmfiles/pub/data/imaging/GRS011/GRS011.pdf] (Accessed August 13, 2020).

Burt, A.K. and J.E.P. Dodge (2016) *Three-Dimensional Modelling of Surficial Deposits in the Orangeville-Fergus Area of Southern Ontario.* Sudbury, ON: Ontario Geological Survey, Groundwater Resources Study 15. [Online: Available at: http://www.geologyontario.mndm.gov.on.ca/mndmaccess/mndm_dir.asp?type=pub&id=GRS015] (Accessed August13, 2020).

Burt, A.K. and Rainsford, D.R.B. (2010). The Orangeville moraine project: buried valley targeted gravity study. In: *Summary of Field Work and Other Activities, 2010* (eds. J.A. Ayer, R.M. Easton, G.P. Beakhouse et al.), 31-1–31-6. Sudbury, ON: Ontario Geological Survey, Open File Report 6260. [Online: Available at: http://www.geologyontario.mndm.gov.on.ca/mndmfiles/pub/data/imaging/OFR6260//OFR6260.pdf] (Accessed August 13, 2020).

Burt, A.K., Biswas, S., Rainsford, D. et al. (2018). Bottoms up: developing a new bedrock surface for the Niagara peninsula. In: *Regional-Scale Groundwater Geoscience in Southern Ontario: An Ontario Geological Survey, Geological Survey of Canada, and Conservation Ontario Geoscientists Open House* (eds. H.A.J. Russell, D. Ford, E.H. Priebe and S. Holysh), 6. Ottawa, ON: Geological Survey of Canada, Open File 8363. https://doi.org/10.4095/306489.

Burwasser, G.J. (1974). *Quaternary Geology of the Collingwood-Nottawasaga Area, Southern Ontario.* Sudbury, ON: Ontario Geological Survey, Preliminary Map 919, scale 1:50 000.

Campbell, J.D. and Burt, A.K. (2015). Filling groundwater data gaps in the Niagara region to assist decision making processes. In: *Summary of Field Work and Other Activities 2015* (eds. R.M. Easton, A.F. Bajc, S.M. Hamilton et al.), 36-1–36-8. Sudbury, ON: Ontario Geological Survey, Open File Report 6313. [Online: Available at: http://www.geologyontario.mndm.gov.on.ca/mndmfiles/pub/data/imaging/ofr6313//OFR6313.pdf] (Accessed August 13, 2020).

CanSIS (2019). *Canadian Soil Information Service (CanSIS).* Website: http://sis.agr.gc.ca/cansis] (Accessed February 22, 2019).

Caterpillar (2000). *Handbook of Ripping, 12*e. Peoria, IL: Caterpillar, Inc. 32 pp.

Chambers, J.E., Kuras, O., Meldrum, P.I. et al. (2006). Electrical resistivity tomography applied to geologic Hydrogeologic, and engineering inbestigations at a former waste disposal site. *Geophysics* 71 (6): B231–B239. https://doi.org/10.1190/1.2360184.

Chapman, L.J. and Putnam, D.F. (2007) *Physiography of Southern Ontario.* Sudbury, ON: Ontario Geological Survey, Miscellaneous Release—Data 228. [Online: Available at: http://www.geologyontario.mndm.gov.on.ca/mndmaccess/mndm_dir.asp?type=pub&id=MRD228] (Accessed August 13, 2020).

Clarke, S. (2004). *Confidence in Geological Interpretation: A Methodology for Evaluating Uncertainty in Common Two and Three-Dimensional Representations of Subsurface Geology.* Keyworth, UK: British Geological Survey, Internal Report IR/04/164. 41 pp. [Online: Available at: http://nora.nerc.ac.uk/id/eprint/509482/] (Accessed August 13, 2020).

Conyers, L.B. (2013). *Ground-Penetrating Radar for Archaeology, 3*e. Lanham MD: AltaMira Press 241 pp.

Crow, H.L., Olson, L.C., Burt, A.K. and Brewer, K.D. (2017). *Downhole Geophysical Logs in Quaternary Sediments of the Niagara Peninsula, Southern Ontario.* Ottawa, ON: Geological Survey of Canada, Open File 8252, 32 pp. https://doi.org/10.4095/304203.

Crow, H.A., Hunter, J.A., Olson, L.C. et al. (2018). Borehole geophysical log signatures and stratigraphic assessment in a Glacial Basin, southern Ontario. *Canadian Journal of Earth Sciences* 55 (7): 829–845. https://doi.org/10.1139/cjes-2017-0016.

Daniels, D.J. [Ed.] (2004). *Ground Penetrating Radar, 2*e. London, UK: The Institution of Electrical Engineers, IEE Radar Series 15. 734 pp.

Danielsen, J.E., Auken, E., Jørgensen, F. et al. (2003). The application of the transient electromagnetic method in Hydrogeophysical surveys. *Journal of Applied Geophysics* 53: 181–198. https://doi.org/10.1016/j.jappgeo.2003.08.004.

Denith, M. and Mudge, S.T. (2014). *Geophysics for the Mineral Exploration Geoscientist.* Cambridge, UK: Cambridge University Press. 439 pp. https://doi.org/10.1017/CBO9781139024358.

Diehl, J.G., Martin, A.J. and Steller, R.A. (2006). Twenty Year Retrospective on the OYO P-S Suspension Logger. In: *Proceedings, 8th U.S. National Conference on Earthquake Engineering, April 18–22, San Francisco, California,* Paper No. 319.

Dietiker, B., Pugin, A.J.-M., Burt, A. et al. (2019). *High-resolution Shallow Seismic Reflection Profiles for Groundwater Studies in the Niagara Peninsula Region, Ontario.* Ottawa, ON: Geological Survey of Canada, Open File 8561. 49 pp. https://doi.org/10.1016/j.jappgeo.2003.08.004.

Ditmar, P., Penopp, J., Kasig, R. and Makris, J. (2001). Interpretation of shallow refraction seismic data by reflection/refraction tomography. *Geophysical Prospecting* 47 (6): 871–901. https://doi.org/10.1046/j.1365-2478.1999.00168.x.

Doveton, J.H. and Prentsky, S.E. (1992). Geological applications of wireline logs: a synopsis of developments and trends. *The Log Analyst* 33 (3): 286–303.

Drenth, B.J., Abraham, J.D., Grauch, V.J.S. and Hodges, G. (2013). *Digital Data from the Great Sand Dunes Airborne Gravity Gradient Survey, south-central Colorado.* Reston, VA: U.S. Geological Survey, Open-File Report 2013–1011. 5 pp. https://doi.org/10.3133/ofr20131011.

Dunkle, K.M., Mickelson, D.M., Anderson, M.P. and Fienen, M.N. (2009). Troy Valley glacial aquifer: 3D Hydrostratigraphic model aiding water management in southeastern Wisconsin, USA. In: *Three-Dimensional Geological Mapping: Workshop Extended Abstracts; Geological Society of America Annual Meeting, Portland, Oregon* (eds. R.C. Berg, H.A.J. Russell and L.H. Thorleifson), 1–4. Champaign, IL: Illinois State Geological Survey, Open File Series 2009-4. [Online: Available at: https://isgs.illinois.edu/publications/ofs2009-4] (Accessed August 13, 2020).

Eaton, T.T., Feinstein, D.T., Bradbury, K.R. and Krohelski, J.T. (2001). Estimating parameters for a complex regional 3D groundwater flow model in southeastern Wisconsin. In: *Geological Models for Groundwater Flow Modeling: Workshop Extended Abstracts* (eds. R.C. Berg and L.H. Thorleifson), 15–17. Champaign, IL: Illinois State Geological Survey, Open File Series 2001-1. [Online: Available at: https://isgs.illinois.edu/publications/ofs2001-1] (Accessed August 13, 2020).

European Marine Observation and Data Network (2019). *What is EMODnet?* [Website: http://www.emodnet.eu/what-emodnet] (Accessed February, 2019).

FAO (2015). *World Reference Base for Soil Resources 2014, International Soil Classification System for Naming Soils and Creating Legends for Soil Maps – Update 2015.* Rome, Italy: UN Food and Agriculture Organization, World Soil Resources Reports 106. 192 p.

Freckelton, C.N. (2013). *A Physical and Geochemical Characterization of Southwestern Ontario's Breathing Well Region.* Master's thesis University of Western Ontario. 242 pp. [Online: Available at: https://ir.lib.uwo.ca/etd/1105] (Accessed February, 2019).

Gao, C., Shirota, J., Kelly, R.I. et al. (2007). *Bedrock Topography and Overburden Thickness Mapping, Southern*

Ontario. Sudbury, ON: Ontario Geological Survey, Miscellaneous Release—Data 207. [Online: Available at: http://www.geologyontario.mndm.gov.on.ca/mndmaccess/mndm_dir.asp?type=pub&id=MRD207] (Accessed August 13, 2020).

Geoscience Australia (2020). *Geophysical Archive Data Delivery System*. Symonston, Australia: Geoscience Australia. [Website: http://www.geoscience.gov.au/cgi-bin/mapserv?map=/nas/web/ops/prod/apps/mapserver/gadds/wms_map/gadds.map&mode=browse] (Accessed July 15, 2020).

GISGeography (2018). *DEM, DSM & DTM Differences – A Look at Elevation Models in GIS*. [Online: Available at: https://gisgeography.com/dem-dsm-dtm-differences] (Accessed February, 2019)

Greenhouse, J.P. and Monier-Williams, M. (1986). A gravity survey of the Dundas buried valley west of Copetown, Ontario. *Canadian Journal of Earth Sciences* 23: 110–114. https://doi.org/10.1139/e86-012.

Gupta, V.K. and Ramani, N. (1980). Some aspects of regional-residual separation of gravity anomalies in a precambrian terrain. *Geophysics* 45 (9): 1412–1426. https://doi.org/10.1190/1.1441130.

Hinze, W.J., Aiken, C., Brozena, J. et al. (2005). New standards for reducing gravity data: the north American gravity database. *Geophysics* 70 (4): J25–J32. https://doi.org/10.1190/1.1988183.

Howard, S.H.D., Brett, J., Lane, R. et al. (2018). Airborne gravimetry takes off in the Western Australia 'Generation 2' reconnaissance gravity mapping project. *ASEG Extended Abstracts* 1: 1–8. https://doi.org/10.1071/ASEG2018abM3_2E.

International Code Council (2017). *2018 International Building Code (ICC IBC-2018)*. Washington, DC: International Code Council. xxvi + 726 pp.

Ivanov, J., Tsoflias, G., Miller, R.D. and Xia, J. (2009). Practical aspects of MASW inversion using varying density. *Symposium on the Application of Geophysics to Engineering and Environmental Problems* 22: 171–177. https://doi.org/10.4133/1.3176692.

Jacobsen, L.J., Glynn, P.D., Phelps, G.A. et al. (2011). U.S. geological survey: a synopsis of three-dimensional modeling. In: *Synopsis of Current Three-Dimensional Geological Mapping and Modeling in Geological Survey Organizations* (eds. R.C. Berg, S.J. Mathers, H. Kessler and D.A. Keefer), 69–79. Champaign, IL: Illinois State Geological Survey, Circular 578. [Online: Available at: https://isgs.illinois.edu/publications/c578] (Accessed August 13, 2020).

Joffre, J.S. (1936). *Pedology*. New Brunswick, NY: Rutgers University Press, 575 pp.

Jol, H.M. (Ed.) (2009). *Ground Penetrating Radar Theory and Applications*, 1e. Amsterdam, NL: Elsevier. 544 pp. https://doi.org/10.1016/B978-0-444-53348-7.00019-3.

Jol, H.M. and Bristow, C.S. (2003). GPR in sediments: advice on data collection, basic processing and interpretation, a Good practice guide. *Geological Society, London Special Publications* 211: 9–27.

Jones, R.R., McCaffrey, K.J.W., Clegg, P. et al. (2009). Integration of regional to outcrop digital data: 3D visualisation of multi-scale geological models. *Computers & Geosciences* 35: 4–18. https://doi.org/10.1016/j.cageo.2007.09.007.

Jørgensen, F., Andersen, H.L., Sandersen, P. et al. (2003). Geophysical investigations of buried valleys in Denmark: an integrated application of transient electromagnetic and reflection seismic surveys, and exploratory well data. *Journal of Applied Geophysics* 53: 215–228. https://doi.org/10.1016/j.jappgeo.2003.08.017.

Jørgensen, F., Møller, R.R., Nebel, L. et al. (2013). A method for cognitive 3D geological voxel modelling of AEM data. *Bulletin of Engineering Geology and the Environment* 72: 421–432. https://doi.org/10.1007/s10064-013-0487-2.

Jorry, S.J. and Biévre, G. (2011). Integration of sedimentology and ground-penetrating radar for high-resolution imaging of a carbonate platform. *Sedimentology* 58 (6): 1370–1390. https://doi.org/10.1111/j.1365-3091.2010.01213.x.

Keefer, D.A. (2011). Illinois state geological survey: a modular approach for 3-D mapping that addresses economic development issues. In: *Synopsis of Current Three-Dimensional Geological Mapping and Modeling in Geological Survey Organizations* (eds. R.C. Berg, S.J. Mathers, H. Kessler and D.A. Keefer), 53–59. Illinois State Geological Survey Circular 578. [Online: Available at: https://isgs.illinois.edu/publications/c578] (Accessed August 13, 2020).

Keller, G., Matile, G., Thorleifson, H. and Malolepszy, Z. (2005). 3D geological model of the Red River Valley, central North America. In: *Three-Dimensional Geological Mapping for Groundwater Application; Geological Society of America Annual Meeting, Salt Lake City, Utah* (eds. H.A.J. Russell, R.C. Berg and L.H. Thorleifson), 35–38. Ottawa, ON: Geological Survey of Canada, Open File 5048. https://doi.org/10.4095/221885.

Keller, G.R., Matile, G.L.D. and Thorleifson, L.H. (2009) Progress in Three-Dimensional Geological Mapping in Manitoba and the Eastern Prairies. In: *Report of Activities 2009*, 207–213. Winnipeg, MB: Manitoba Innovation, Energy and Mines, Manitoba Geological Survey. [Online: Available at: https://www.manitoba.ca/iem/geo/field/roa09pdfs/GS-21.pdf] (Accessed August 13, 2020).

Kellog, C.E. (1936). *Development and Significance of the Great Soils Groups of the United States*. Washington, DC: U.S. Dept. of Agiculture, Misc. Publication 229. 40 pp.

Kessler, H., Mathers, S.J. and Keefer, D.A. (2011). Logistical considerations prior to migrating to 3-D geological modeling and mapping. In: *Synopsis of Current Three-Dimensional Geological Mapping and Modeling in Geological Survey Organizations* (eds. R.C. Berg, S.J. Mathers, H. Kessler and D.A. Keefer), 39–44. Champaign, IL: Illinois State Geological Survey, Circular 578. [Online: Available at: https://isgs.illinois.edu/publications/c578] (Accessed August 13, 2020).

Keys, W.S. (1997). *A Practical Guide to Borehole Geophysics in Environmental Investigations*. New York, NY: Routledge. 192 pp. https://doi.org/10.1201/9781315136417.

Kobr, M., Mares, S. and Paillet, F. (2005). Geophysical well logging – borehole geophysics for hydrogeological studies: principles and applications. In: *Hydrogeophysics* (eds. Y. Rubin and S.S. Hubbard), 291–331. Amsterdam, Netherlands: Springer.

Korus, J.T., Joeckel, R.M., Divine, D.P. and Abraham, J.D. (2017). Three-dimensional architecture and hydrostratigraphy of cross-cutting buried valleys using airborne electromagnetics, glaciated central lowlands, Nebraska, USA. *Sedimentology* 64 (2): 553–581. https://doi.org/10.1111/sed.12314.

Labo, J. (1987). *A Practical Introduction to Borehole Geophysics - An Overview of Wireline Well Logging Principles for Geophysicists*. Tulsa, OK: Society of Exploration Geophysicists, Geophysical Reference Series. 340 pp.

Lemieux, A.J., Hamilton, S.M. and Clark, I.D. (2018). Allochthonous sources of iodine and organic carbon in an eastern Ontario aquifer. *Canadian Journal of Earth Sciences* 56 (3): 209–222. https://doi.org/10.1139/cjes-2018-0082.

Liew, A. (2007). Understanding data, information, knowledge and their inter-relationships. *Journal of Knowledge Management Practice* 8 (2). [Online: Available at: http://www.tlainc.com/articl134.htm (Accessed 24 July 2017).

Logan, C., Russell, H.A.J. and Sharpe, D.R. (2005). *Regional 3-D Structural Model of the Oak Ridges Moraine and Greater Toronto Area, Southern Ontario (Version 2.1)*. Ottawa, ON: Geological Survey of Canada, Open File 5062. 27 pp. https://doi.org/10.4095/221490.

Logan, C., Russell, H.A.J., Mulligan, R. et al. (2020). A three-dimensional surficial geology model of Southern Ontario. In: *Southern Ontario Groundwater Project 2014-2019 Summary Report* (eds. H.A.J. Russell and B.A. Kjarsgaard), 49–64. Ottawa, ON: Geological Survey of Canada, Open File 8536. https://doi.org/10.4095/321084.

Loke, M.H., Chambers, J.E., Rucker, D.F. et al. (2013). Recent developments in the direct-current geoelectrical imaging method. *Journal of Applied Geophysics* 95: 135–156. https://doi.org/10.1016/j.jappgeo.2013.02.017.

Long, L.T. and Kaufmann, R.T. (2013). *Acquisition and Analysis of Terrestrial Gravity Data*, 1e. Cambridge University Press. 192 pp.

Louie, J.N. (2001). Faster, better: shear-wave velocity to 100 meters depth from refraction microtremor arrays. *Bulletin of the Seismological Society of America* 91 (2): 347–364. https://doi.org/10.1785/0120000098.

Lusardi, B.A. (2017). *1:100,000 Surficial Geologic Texture Database*. St. Paul, MN Minnesota Geological Survey, Open File Report 17–01. [Online: Available at: http://hdl.handle.net/11299/191889] (Accessed November 2018).

Van Maanen, P-P., Schokker, J., Harting, R. and De Bruijn, R. (2017). Nationwide Lithological Interpretation of Cone Penetration Tests using Neural Networks. *Geophysical Research Abstracts* 19: EGU2017-8473. [Online: Available at: https://meetingorganizer.copernicus.org/EGU2017/EGU2017-8473.pdf] (Accessed November 2018).

MacCormack, K.E. (2018). Developing a 3D geological framework program at the Alberta geological survey; optimizing the integration of geologists, Geomodellers, and Geostatiticians to build multi-disciplinary, multi-scalar, Geostatistical 3D geological models of Alberta. In: *Three-Dimensional Geological Mapping: Workshop Extended Abstracts, Geological Society of America Annual Meeting, Vancouver, Canada, June 16–17, 2018* (eds. R.C. Berg, K.E. MacCormack, H.A.J. Russell and L.H. Thorleifson), 64–67. Champaign, IL: Illinois State Geological Survey, Open File Series 2018-1. [Online: Available at: https://isgs.illinois.edu/publications/ofs2018-1] (Accessed August 13, 2020).

Machguth, H., Eisen, O., Paul, F. and Hoelzle, M. (2006). Strong spatial variability of snow accumulation observed with helicopter-borne GPR on two adjacent alpine glaciers. *Geophysical Research Letters* 33 (13): L13503. https://doi.org/10.1029/2006GL026576.

Martinez, A. and Byrnes, A.P. (2001). Modeling dielectric-constant values of geologic materials: an aid to ground-penetrating radar data collection and interpretation. *Current Research in Earth Sciences, Kansas Geological Survey Bulletin* 247 (1): 1–16. [Online: Available at: https://journals.ku.edu/mg/article/view/11831] (Accessed August 13, 2020).

Martinez, A., Kruger, J.M. and Franseen, E.K. (1998). Utility of ground-penetrating radar in near-surface, high-resolution imaging of Lansing-Kansas City (Pennsylvanian) limestone reservoir Analogs. *Current Research in Earth Sciences, Kansas Geological Survey Bulletin* 241 (3): 43–59. [Online: Available at: https://journals.ku.edu/mg/article/view/11830] (Accessed August 13, 2020).

Mathers, S.J., Kessler, H. and Napier, B. (2011). British geological survey: a Nationwide commitment to 3-D geological modeling. In: *Synopsis of Current Three-Dimensional Geological Mapping and Modeling in Geological Survey Organizations* (eds. R.C. Berg, S.J. Mathers, H. Kessler and D.A. Keefer), 25–30. Champaign, IL: Illinois State Geological Survey, Circular 578. [Online: Available at: https://isgs.illinois.edu/publications/c578] (Accessed August 13, 2020).

Van der Meulen, M.J., Doornenbal, J.C., Gunnink, J.L. et al. (2013). 3D Geology in a 2D Country: perspectives for Geological Surveying in the Netherlands. *Netherlands Journal of Geosciences* 92 (4): 217–241. https://doi.org/10.1017/S0016774600000184.

Matile, G.L.D., Keller, G.R., Pyne, D.M. and Thorleifson, L.H. (2002). Development of methods for 3-D geological mapping of southern Manitoba Phanerozoic Terrane. In: *Report of Activities 2002, Manitoba Industry, Trade and Mines*, 274–282. Winnipeg, MB: Manitoba Geological Survey.

McLaughlin, P.P., Tomlinson, J.L. and Lawson, A.K. (2015). Attributing groundwater withdrawals to aquifers using 3-D geological maps in Delaware, USA. In: *Three-Dimensional Geological Mapping: Workshop Extended Abstracts; Geological Society of America Annual Meeting, Baltimore, Maryland* (eds. K.E. MacCormack, L.H. Thorleifson, R.C. Berg and H.A.J. Russell), 57–62. Edmonton, AL: AER/AGS, Special Report 101. [Online: Available at: https://www.ags.aer.ca/document/SPE/SPE_101.pdf] (Accessed August 13, 2020).

Mickus, K.L., Aiken, C.L.V. and Kennedy, W.D. (1991). Regional-residual gravity anomaly separation using the minimum-curvature technique. *Geophysics* 56 (2): 279–283. https://doi.org/10.1190/1.1443041.

Miller, R.D., Xia, J., Park, C.B. and Ivanov, J.M. (1999). Multichannel analysis of surface waves to map bedrock. *The Leading Edge* 18: 1392–1396. https://doi.org/10.1190/1.1438226.

Møller, I. and Jakobsen, P.R. (2002). Sandy till characterized by ground penetrating radar. In: *Ninth International Conference on Ground Penetrating Radar. Proceedings of SPIE 4758* (eds. S.K. Koppenjan and H. Lee), 308–312. Bellingham, WA: The International Society for Optics and Photonics (SPIE).

Møller, I. and Jørgensen, F. (2006). Combined GPR and DC-resistivity imaging in hydrogeological mapping. In: *Proceedings of 11th International Conference on Ground Penetrating Radar, June 19–22, 2006, Columbus Ohio, USA* (ed. J.J. Daniels), 5. Columbus, OH: The Ohio State University.

Møller, I., Sørensen, K.I. and Auken, E. (2006). Geoelectrical methods. In: *Groundwater Resources in Buried Valleys – a Challenge for Geosciences* (eds. R. Kirsch, H.M. Rumpel, W. Scheer and H. Wiederhold), 77–87. Hannover, Germany: Leibnitz Institute for Applied Geosciences. [Online: Available at: https://www.liag-hannover.de/fileadmin/user_upload/dokumente/grundwassersysteme/burval/buch/i-x.pdf] (Accessed August 13, 2020).

Mulligan, R.P.M. (2017). *Quaternary geology of the Collingwood Area*. Sudbury, ON: Ontario Geological Survey, Preliminary Map P.3815, scale 1:50 000.

Mulligan, R.P.M., Bajc, A.F., Burt, A.K. et al. (2019). *Landform and Sediment Mapping for Modelling, Land Use, and Recharge Assessment*. Poster: GSC/OGS/CO-Partnership Workshop 2019, Guelph, Ontario.

Murray, A.S. and Tracey, R.M. (2001). *Best Practice in Gravity Surveying*. Canberra, Australia: Geoscience Australia. 43 pp. [Online: Available at: http://pid.geoscience.gov.au/dataset/ga/37202] (Accessed November 2018).

Nabighian, M.N. and Macnae, J.C. (1991). Time domain electromagnetic prospecting methods. In: *Electromagnetic Methods in Applied Geophysics, vol. 2* (ed. M.N. Nabighian), 427–520. Tulsa, OK: Society of Exploration Geophysicists.

Natural Resources Conservation Service (2019a). *Soil Data Access: Query Services for Custom Access to Soil Data*. Washington, DC: U.S. Department of Agriculture. [Website: https://sdmdataaccess.nrcs.usda.gov] (Accessed February 22, 2019).

Natural Resources Conservation Service (2019b). *Web Soil Survey*. Washington, DC: U.S. Department of Agriculture. [Website: https://websoilsurvey.sc.egov.usda.gov/App/HomePage.htm] (Accessed February 22, 2019).

Neal, A. (2004). Ground-penetrating radar and its use in sedimentology: principles, problems and progress. *Earth-Science Reviews* 66: 261–330. https://doi.org/10.1016/j.earscirev.2004.01.004.

Oldenburg, D.W. and Li, Y. (2005). Inversion for applied geophysics: a tutorial. In: *Near Surface Geophysics* (ed. D.K. Butler), 89–150. Tulsa, OK: Society of Exploration Geophysicists.

Olhoeft, G.R. (1998). Electrical, magnetic, and geometric properties that determine ground penetrating radar performance. In: *Proceedings of the Seventh International Conference on Ground Penetrating Radar*, 177–182. Lawrence, KS: Radar Systems & Remote Sensing Laboratory, University of Kansas.

Ontario Geological Survey (2010). *Surficial Geology of Southern Ontario*. Sudbury, ON: Ontario Geological Survey, Miscellaneous Release—Data 128–Revised. [Online: Available at: http://www.geologyontario.mndm.gov.on.ca/mndmaccess/mndm_dir.asp?type=pub&id=MRD128-REV] (Accessed August 13, 2020).

Ontario Geological Survey (2014). *Ontario Geophysical Surveys, Ground Gravity Data, Grid and Point Data (ASCII and Geosoft* ® *Formats) and Vector Data, Niagara Area.* Sudbury, ON: Ontario Geological Survey, Geophysical Data Set 1073. [Online: Available at: http://www.geologyontario .mndm.gov.on.ca/mndmaccess/mndm_dir.asp? type=pub&id=GDS1073] (Accessed August 13, 2020).

Ontario Geological Survey (2019). *Geotechnical Boreholes.* [Website: https://www.mndm.gov.on.ca/en/mines-and-minerals/applications/ogsearth/geotechnical-boreholes] (Accessed April, 2019).

Ontario Ministry of Agriculture, Food and Rural Affairs (2015). *Soil Survey Complex.* [Website: https://geohub.lio .gov.on.ca/datasets/ontarioca11::soil-survey-complex] (Accessed August 13, 2020).

Ontario Ministry of Environment, Conservation and Parks (2013). *Water Well Information System (Well Location and Summary).* Peterborough, ON: Government of Ontario. [Website: https://www.ontario.ca/dataset/well-records] (Accessed February, 2019).

Ontario Ministry of Natural Resources and Forestry (2005). *Provincial Digital Elevation Model - Version 2.0.0.* Peterborough, ON: Government of Ontario.

Ontario Ministry of Natural Resources and Forestry (2010a) *Ontario Radar Digital Surface Model (FL).* Peterborough, ON: Government of Ontario. [Online: Available at: https:// data.ontario.ca/dataset/ontario-radar-digital-surface-model-fl (Accessed February, 2019)

Ontario Ministry of Natural Resources and Forestry (2010b) *Southwestern Ontario Orthophotography Project (2010) - Digital Surface Model.* Peterborough, ON: Government of Ontario. [Online Available at: https://geohub.lio.gov.on.ca/ datasets/southwestern-ontario-orthophotography-swoop-2010 (Accessed February, 2019).

Ontario Ministry of Natural Resources and Forestry (2018) *User guide, LiDAR – Digital Terrain Model (2016–18) LIO Dataset.* Peterborough, ON: Government of Ontario. 49 pp. [Online: Available at: https://geohub.lio.gov.on.ca/ datasets/776819a7a0de42f3b75e40527cc36a0a] (Accessed August 13, 2020).

Ontario Ministry of Natural Resources and Forestry (2019). *Ontario Petroleum Database System.* Peterborough, ON: Government of Ontario. [Online: Available at https://data .ontario.ca/dataset/petroleum-wells] (Accessed January 2019).

Ordnance Survey (2019) [Online: Available at: https://www .ordnancesurvey.co.uk/about/overview/history.html]. Ordnance Survey History. Southampton, UK: Ordnance Survey. [Web page: https://www.ordnancesurvey.co.uk/ about/overview/history.html] (Accessed February, 2019)

Parasnis, D.S. (1972). *Principles of Applied Geophysics, 2e.* London, UK: Chapman & Hall. 214 pp.

Park, C.B., Miller, R.D. and Xia, J. (1999). Multichannel analysis of surface waves. *Geophysics* 64: 800–808. https:// doi.org/10.1190/1.1444590.

Pfeiffer, J. and Hanna, K. (2009). *Clay Seam Mapping with Electromagnetic Induction.* Washington, DC: Federal Highway Administration, Technical Report FHWA-CFL/TD 05–003. 94 pp.

Phillips, J.D., Ewing, R.C., Bowling, R. et al. (2019). Low-angle Eolian deposits formed by Protodune migration, and insights into Slipface development at White Sands dune field, New Mexico. *Aeolian Research* 36: 9–26. https://doi.org/10.1016/j.aeolia.2018.10.004.

Prokopovich, N.P. (1984). Use of agricultural soil survey maps for engineering geologic mapping. *Bulletin of the Association of Engineering Geologists* 21 (4): 437–447.

Pugin, A.J.M., Larson, T. and Phillips, A. (2002). Shallow high-resolution shear-wave seismic reflection acquisition using a land-streamer in the Mississippi River floodplain: potential for engineering and hydrogeologic applications. In: *Proceedings, SAGEEP'02 – Symposium on the Application of Geophysics to Engineering and Environmental Problems, February 10–14 Las Vegas, Nevada.* Denver, CO: Environmental and Engineering Geophysical Society. 7 pp.

Pugin, A.J.M., Larson, T.H., Sargent, S.L. et al. (2004). Near-surface mapping using SH-wave and P-wave seismic land-streamer data Acquisition in Illinois, U.S. *The Leading Edge* 23 (7): 677–682. https://doi.org/10.1190/1.1776740.

Pugin, A.J.M., Sargent, S.L. and Hunt, L. (2006). SH- and P-wave Seismic Reflection using Landstreamers to Map Shallow Features and Porosity Characteristics in Illinois. In: *Proceedings, SAGEEP 2006 - Symposium on the Application of Geophysics to Engineering and Environmental Problems,* 1094–1109. Denver, CO: Environmental and Engineering Geophysical Society. https://doi.org/10.3997/2214-4609-pdb.181.113.

Pugin, A.J.M., Hunter, J.A., Pullan, S.E. et al. (2007). Buried-channel imaging using P- and SH-wave shallow seismic reflection techniques, examples from Manitoba, Canada. In: *Proceedings, SAGEEP 2007 - Symposium on the Application of Geophysics to Engineering and Environmental Problems,* 1330–1338. Denver, CO: Environmental and Engineering Geophysical Society. https://doi.org/10.4133/1.2924641.

Pugin, A.J.M., Pullan, S.E. and Hunter, J.A. (2009). Multicomponent high-resolution seismic reflection profiling. *The Leading Edge* 28: 1248–1261. https://doi.org/ 10.1190/1.3249782.

Pullan, S.E., Pugin, A.J-M., Hinton, M.J., Burns, R.A. et al. (2013). *Delineating Buried Valleys in Southwest Manitoba using Seismic Reflection Methods: Cross-Sections over the Medora-Waskada, Pierson and Killarney Valleys (2006–07).* Ottawa, ON: Geological Survey of Canada, Open File 7337. 54 + iii pp. https://doi.org/10.4095/292392.

Redpath, B.B. (1973). *Seismic Refraction Exploration for Engineering Site Investigations*. Vickburg, MS: U.S. Army Corps of Engineers, Waterways Experiment Station, Technical Report E-73-4. 63 pp.

Ross, M., Parent, M. and Taylor, A. (2015). A modelling strategy to develop a regional quaternary geological model across rural and urban areas and administrative Borders using existing geological information. In: *Three-Dimensional Geological Mapping: Workshop Extended Abstracts; Geological Society of America Annual Meeting, Baltimore, Maryland* (eds. K.E. MacCormack, L.H. Thorleifson, R.C. Berg and H.A.J. Russell), 51–55. Edmonton, AL: AER/AGS, Special Report 101. [Online: Available at: https://www.ags.aer.ca/document/SPE/SPE_101.pdf] (Accessed August 13, 2020).

Royse, K.R., Rutter, H.K. and Entwisle, D.C. (2009). Property attribution of 3D geological models in the Thames gateway, London: new ways of visualising Geoscientific information. *Bulletin of Engineering Geology and the Environment* 68 (1): 1–16. https://doi.org/10.1007/s10064-008-0171-0.

Rozar, E.J. (2015). *Defining Antecedent Topography at Coral Pink Sand Dunes, Kane Couty, Utah: The Influence of Structural Controls on Dune-Field Boundary Conditions and Holocene Landscape Evolution*. Master's thesis, Boise State University. 96 pp. [Online: Available at: https://scholarworks.boisestate.edu/td/987/] (Accessed August 13, 2020).

Russell, H.A.J., Brennand, T.A., Logan, C. and Sharpe, D.R. (1998). Standardization and Assessment of Geological Descriptions from Water Well Records, Greater Toronto and Oak Ridges Moraine Areas, Southern Ontario. *Geological Survey of Canada Current Research* 1998–E: 89–102. https://doi.org/10.4095/209947.

Russell, H.A.J., Brodaric, B., Keller, G.R. et al. (2015). A perspective on a three-dimensional framework for Canadian geology. In: *Three-Dimensional Geological Mapping: Workshop Extended Abstracts; Geological Society of America Annual Meeting, Baltimore, Maryland* (eds. K.E. MacCormack, L.H. Thorleifson, R.C. Berg and H.A.J. Russell), 21–31. Edmonton, ON: AER/AGS, Special Report 101. [Online: Available at: https://ags.aer.ca/document/SPE/SPE_101.pdf] (Accessed August 13, 2020).

Rutishauser, A., Maurer, H. and Bauder, A. (2016). Helicopter-borne ground-penetrating radar investigations on temperate alpine glaciers: a comparison of different systems and their abilities for bedrock mapping. *Geophysics* 81 (1): WA119–WA129. https://doi.org/10.1190/geo2015-0144.1.

Sandersen, P.B.E., Vangkilde-Pedersen, T., Jorgensen, F. et al. (2015). A national 3D geological model of Denmark: condensing more than 125 years of geological mapping. In: *Three-Dimensional Geological Mapping: Workshop Extended Abstracts; Geological Society of America Annual Meeting, Baltimore, Maryland* (eds. K.E. MacCormack, L.H. Thorleifson, R.C. Berg and H.A.J. Russell), 71–77. Edmonton, AL: AER/AGS, Special Report 101. [Online: Available at: https://ags.aer.ca/document/SPE/SPE_101.pdf] (Accessed August 13, 2020).

Sapia, V., Viezzoli, A., Jørgensen, F. et al. (2014). The impact on geological and hydrogeological mapping results of moving from ground to airborne TEM. *Journal of Environmental and Engineering Geophysics* 19 (1): 53–66. https://doi.org/10.2113/JEEG19.1.53.

Sheehan, J.R., Doll, W.E. and Mandell, W.A. (2005). An evaluation of methods and available software for seismic refraction tomography analysis. *Journal of Environmental and Engineering Geophysics* 10: 21–34. https://doi.org/10.2113/JEEG10.1.21.

Siemon, B. (2006). Frequency-domain helicopter-borne electromagnetics. In: *Groundwater Resources in Buried Valleys – a Challenge for Geosciences* (eds. R. Kirsch, H.M. Rumpel, W. Scheer and H. Wiederhold), 89–98. Hannover, Germany: Leibnitz Institute for Applied Geosciences. [Online: Available at: https://www.liag-hannover.de/fileadmin/user_upload/dokumente/grundwassersysteme/burval/buch/i-x.pdf] (Accessed August 13, 2020).

Sirles, P.C. (1986). *Shear-wave velocity and attenuation analysis of liquefiable soils in the south Truckee Meadows, Washoe County, Nevada*. Master's thesis University of Nevada, Reno. 157 pp. [Online: Available at: http://hdl.handle.net/11714/1784] (Accessed August 13, 2020).

Sirles, P.C. (2006). *NCHRP Synthesis 357 – Use of Geophysics for Transportation Projects*. Washington, DC: Transportation Research Board. 108 pp. [Online: Available at: http://onlinepubs.trb.org/onlinepubs/nchrp/nchrp_syn_357.pdf] (Accesses August 13, 2020).

Slattery, S.R. and Andriashek, L.D. (2012). *Overview of Airborne-Electromagnetic and Magnetic Geophysical Data Collection using the GEOTEM Survey north of Calgary, Alberta*. Edmonton, AL: ERCB/AGS, Open File Report 2012–09. 168 pp. [Online: Available at: https://ags.aer.ca/document/OFR/OFR_2012_09.pdf] (Accessed August 13, 2020).

Smith, D.G. and Jol, H.M. (1995). Ground penetrating radar: antenna frequencies and maximum probable depths of penetration in quaternary sediments. *Journal of Applied Geophysics* 33: 93–100. https://doi.org/10.1016/0926-9851(95)90032-2.

Soller, D.R. and Berg, R.C. (2001). A method for three-dimensional mapping, merging geologic interpretation, and GIS computation. In: *Geological Models for Groundwater Flow Modeling: Workshop Extended Abstracts* (eds. R.C. Berg and L.H. Thorleifson), 47–50. Illinois State Geological Survey, Open File 2001-1. [Online:

Champaign, IL: Available at: https://isgs.illinois.edu/publications/ofs2001-1] (Accessed August 13, 2020).

Song, Y.Y., Castagna, J.P., Black, R.A. and Knapp, R.W. (1989). Sensitivity of Near-surface Shear-Wave Velocity Determination from Rayleigh and Love Waves. *SEG Technical Program Expanded Abstracts* 1989: 509–512. https://doi.org/10.1190/1.1889669.

Sørensen, K.I. and Auken, E. (2004). SkyTEM - a new high-resolution helicopter transient electromagnetic system. *Exploration Geophysics* 35: 191–199. https://doi.org/10.1071/EG04194.

Sørensen, K.I., Christiansen, A.V. and Auken, E. (2006). The transient electromagnetic method (TEM). In: *Groundwater Resources in Buried Valleys – a Challenge for Geosciences* (eds. R. Kirsch, H.M. Rumpel, W. Scheer and H. Wiederhold), 65–75. Hannover, Germany: Leibnitz Institute for Applied Geosciences. [Online: Available at: https://www.liag-hannover.de/fileadmin/user_upload/dokumente/grundwassersysteme/burval/buch/i-x.pdf] (Accessed August 13, 2020).

Steelman, C.M., Arnaud, E., Pehme, P. and Parker, B.L. (2018). Geophysical, geological, and hydrogeological characterization of a tributary buried Bedrock Valley in southern Ontario. *Canadian Journal of Earth Sciences* 55 (7): 641–658. https://doi.org/10.1139/cjes-2016-0120.

Steeples, D.W. and Miller, R.D. (1988). Seismic reflection methods applied to engineering, environmental, and groundwater problems. In: *Proceedings, 1st EEGS Symposium on the Application of Geophysics to Engineering and Environmental Problems*, 409–461. Houten, Netherlands: European Association of Geoscientists & Engineers. https://doi.org/10.3997/2214-4609-pdb.214.1988_005.

Stokoe, K.H. (II), Wright, S.G., Bay, J.A. and Roesset, J.M. (1994). Characterization of geotechnical sites by SASW method. In: *Geotechnical Characterization of Sites* (ed. R.D. Wood), 15–26. New Delhi, India: Oxford and IBH Publishing Co.

Telford, W., Geldart, L. and Sheriff, R. (1990). *Applied Geophysics*, 2e. Cambridge, UK: Cambridge University Press. 744 pp.

Tipping, R.G. and Meyer, G.N. (2001). Mapping buried glacial deposits in Washington County, Minnesota – applications to hydrogeologic characterization of glacial terrains. In: *Geological Models for Groundwater Flow Modeling: Workshop Extended Abstracts* (eds. R.C. Berg and L.H. Thorleifson), 55–57. Champaign, IL: Illinois State Geological Survey, Open File 2001-1. [Online: Available at: https://isgs.illinois.edu/publications/ofs2001-1] (Accessed August 13, 2020).

USACE (1995). *Geophysical Exploration for Engineering and Environmental Investigations*. Washington, DC: U.S. Army Corps of Engineers, Engineering and Design Manual EM 1110-1-1802. 204 pp.

Usery, E. L., Varanka, D. and Finn, M.P. (2009). *125 Years of Topographic Mapping. USGS history, Part 1: 1884-1980.* Arcnews, Fall 2009. [Online: Available at: https://www.esri.com/news/arcnews/fall09articles/125-years.html] (Accessed February, 2019).

Usery, E. L., Varanka, D. and Finn, M.P. (2010). *125 Years of Topographic Mapping. USGS history, Part 2: From the Dawn of Digital to The National Map.* Arcnews, Winter 2009/2010. [Online: Available at: https://www.esri.com/news/arcnews/fall09articles/125-years.html] (Accessed February, 2019).

Utsi, E.C. (2017). *Ground Penetrating Radar: Theory and Practice*, 1e. Oxford, UK: Butterworth-Heinemann. 224 pp.

Valle-Molina, C. (2006). *Measurements of Vp and Vs in Dry, Unsaturated and Saturated Sand Specimens with Piezoelectric Transducers*. PhD dissertation, University of Texas at Austin. 386 pp. [Online: Available at: http://hdl.handle.net/2152/2616] (Accessed August 13, 2020).

Verdun, J., Klingele, E.E., Bayer, R. et al. (2003). The alpine Swiss-French airborne gravity survey. *Geophysical Journal International* 152: 8–19. https://doi.org/10.1046/j.1365-246X.2003.01748.x.

Waters, K.H. (1992). *Reflection Seismology: A Tool for Energy Resource Exploration*, 3e. Malabar, FL: Krieger Publishing Company. 554 pp.

West, R.E. (1992). The land gravity exploration method. In: *Practical Geophysics II* (ed. R. Van Blaircom), 194–195. Spokane Valley, WA: Northwest Mining Association.

White, O.L. and Karrow, P.E. (1997). Urban geology: a Canadian perspective. In: *Engineering Geology and the Environment, Volume 4. Proceedings: International Symposium of Engineering Geology and the Environment* (eds. P.G. Marinos, G.C. Koukis, G.C. Tsiambaos and G.C. Stournaras), 3439–3450. Rotterdam, Netherlands: A.A. Balkema Publishers.

Wightman, W.E., Jalinoos, F. Sirles, P. and Hanna, K. (2003). *Application of Geophysical Methods to Highway Related Problems*. Washington, DC: Federal Highway Administration, Technical Manual FHWA-IF-04-021. 742 pp.

Wonik, T. and Hinsby, K. (2006). Borehole Logging in Hydrogeology. In: *Groundwater Resources in Buried Valleys – a Challenge for Geosciences* (eds. R. Kirsch, H.M. Rumpel, W. Scheer and H. Wiederhold), 107–122. Hannover, Germany: Leibnitz Institute for Applied Geosciences. [Online: Available at: https://www.liag-hannover.de/fileadmin/user_upload/dokumente/grundwassersysteme/burval/buch/i-x.pdf] (Accessed August 13, 2020).

Xia, J., Miller, R.D., Park, C.B. et al. (2000). Comparing shear-wave velocity profiles from MASW with borehole measurements in unconsolidated sediments, Fraser River Delta, B.C., Canada. *Journal of Environmental and Engineering Geophysics* 5 (3): 1–13. https://doi.org/10.4133/JEEG5.3.1.

8

Data Management Considerations

Martin L. Nayembil

British Geological Survey, Keyworth, Nottingham NG12 5GG, UK

8.1 Introduction

Economic and technological trends of the past two decades have caused several geological survey organizations (GSOs) to transition from the production of traditional geological maps to a digital environment that produces 3-D geological models (see Chapter 2). The GSO user communities increasingly demanded or expected to receive information in digital formats. Concurrently, the evolving computer hardware and software encouraged the adoption of digital procedures; as these became more affordable and sophisticated, increasingly detailed geological models could be developed.

Many GSOs converted existing legacy data sources to digital formats. Although many were only basic digital scans, wherever possible data were converted to retrievable formats that allowed the data to be re-analyzed and used to create new digital products, including geological models. Managing these new digital data sources was critically important if GSOs were to efficiently undertake their missions as repositories of geological information and sources of useful geological advice.

The efficient production of geological models was recognized as an important component for achieving corporate goals. Geological modeling is not merely a technological process. The successful implementation of geological modeling within any GSO depends upon:

- Development of effective and efficient workflows, as discussed in Chapter 6;
- Availability of adequate data sources from a wide variety of legacy and recently-acquired sources, as discussed in Chapter 7; and
- A robust data management environment that supports the identification, use, and re-use of data sources, models, and model products.

This chapter addresses data management in support of geological modeling. All GSOs that are engaged in 3-D modeling have had to organize their data and store their outputs. Even though each such organization will have honored the basic principles of data management, there is a large variation in the way data management is undertaken. The operational setting determines which data types are relevant, while the available resources define the appropriate level of effort for data management. Rather than attempting to distill generalities from a comprehensive analysis of multiple GSOs, we draw mainly from data management experiences at the British Geological Survey (BGS). The BGS is an all-round, well-funded organization that can serve as a case example for a coherent description of all relevant aspects of data management. However, the BGS has participated in many collaborations with other GSOs and shared development strategies with them. Shared experiences have led to the mutual exchange of data management successes (and difficulties!); the discussions in this chapter reflect those broadly accepted "truths" about data management in support of geological modeling activities. Section 8.7 discusses data management at the Geological Survey of Denmark and Greenland (GEUS). This adds to the British case because this particular survey relies, to a greater extent than other GSOs, on geophysical data for shallow 3-D modeling, and has designed its data management accordingly. Section 8.8 discusses transboundary modeling projects, highlighting the interoperability problems that arise when GSOs, each with its own standards and databases, join forces in a single modeling effort.

At first, the BGS relied on individual topical digital data sources that contained specific data classes and were not too voluminous. The first geological modeling projects were developed individually by modeling teams who located and acquired legacy data, supplemented these sources with new observational data, created a project-based database that could be accessed by the chosen modeling software, created the geological model, and transmitted the completed model to the sponsoring user in mutually agreed-to formats. At this stage, there was little or no standardization among the different topical databases or among the models.

However, as the data volumes grew, and the number of completed models increased, it became apparent that it was becoming increasingly difficult to efficiently access the stored topical data, or to recover and re-use the project databases from earlier projects, nor were the existing models always readily available for further development. Geological modeling was becoming a core function of the BGS; these inefficiencies were hampering the achievement of corporate goals. Corporate standards were needed to manage the enterprise-wide varied data sets, to manage the modeling process, and to catalog and maintain access to completed models so that they could supply information needed by subsequent projects.

Data management includes the process of acquiring, validating, storing, organizing, and retrieving data. The definition of data provenance and the provision of metadata describing data characteristics improve the quality and reliability of derived information and knowledge products (Figure 8.1). Data management is a large complex topic that has been extensively discussed by several authors, including West and Fowler (1996), Date (2004), Whitten et al. (2004), Simison and Witt (2005), and West (2011). Section 8.2 of this chapter summarizes data management methods.

The application of data management to geological modeling is addressed under four topics. Section 8.3 discusses the management of databases that contain geological source data required to develop geological framework models. Section 8.4 discusses the management of the resulting geological framework models. Section 8.5 discusses the management of databases defining geological properties

that are required to convert geological framework models into process models, which are described in Chapter 14. Section 8.6 discusses the management of these process models. The chapter concludes with several illustrative examples demonstrating the role of data management in providing efficiency to geological modeling applications.

8.2 Data Management Methods

As computer technology evolved, storage of vast volumes of data in random access storage became feasible, along with the ability to rapidly analyze these data in real time. *Integrated data management* offers the necessary foundation, strategy, and structure needed to ensure that data is managed as an asset within an organization, allowing data to be transformed into meaningful information. Integrated data management principles should be applied to all data assets, both digital and non-digital. Data being digital doesn't mean it's automatically well-managed and of high quality, data governance and security principles need to be applied to ascertain its quality. The ability to relate and manage diverse geoscience data sets provides robust evidence for decisions made.

8.2.1 Standards and Best Practice

The development of any data and information platform should follow the definitions and best practice implementation standards for integrated data management. Adoption of best practices for data management allows for the easier linkage, maintenance, and development of multiple sections of the information architecture and data assets. The International Standards Organization (ISO), Open Geospatial Consortium (OGC), INSPIRE (an EU initiative to establish an infrastructure for spatial information in Europe), and the official British Standards produced by the BSI Group for the United Kingdom promote best

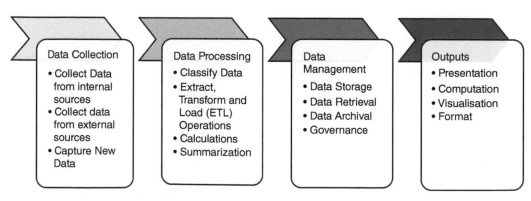

Figure 8.1 Data life cycle.

practices that support general interoperability between differing technologies, systems, and datasets.

8.2.2 The Database System

Information may be stored and accessed in centralized or distributed corporate databases. A centralized database system has advantages related to the management, security, and cost of the system and to the ease of access to the data within the system. Because components are co-located, a centralized database system is easier to manage with a higher degree of control, is more cost-effective to support and maintain, and has better performance and scalability with auditing and traceability. By reducing duplication of data across multiple distributed sources, a centralized database enhances data integrity, helps coordinate data management procedures, and provides a structured data environment that promotes consistency in data querying. The centralized environment offers a single source of data access to everyone. It also significantly reduces organizational risk from a legislative and legal compliance perspective by offering immediate and long-term access and preservation of data assets.

A distributed database system is usually stored on multiple computers; these may be located in the same physical location or dispersed over a network. Replication and duplication of data across the various components help ensure such a database system stays up-to-date. The distribution of data and applications within a distributed database system may provide some potential advantages over a traditional centralized database system. However, possible variations in adopted standards within portions of a distributed network increase the complexity, and therefore the cost, of security and integrity control, although the increasing standardization of communication and data access protocols may mitigate these concerns.

8.2.3 Data Modeling

Data modeling is a key component of integrated data management for geological modeling. It involves professional data modelers working closely with geoscientists and other potential users of the information systems to define and analyze the data requirements needed to support the efficient construction, retention, and delivery of geological models while utilizing the corporate geological survey information systems. Data modeling thus defines structure and management of new and legacy data required to develop geological models, the management of the resulting geological framework models, and the integration of properties data with geological models to produce process models requested by internal and external users.

Sections 8.3–8.6 provide details of these data modeling activities.

Consistent system-wide use of a data model ensures data compatibility, allowing multiple applications to seamlessly share data. Data modeling relies on two closely related concepts – *data model* and *database schema*. While a data model describes the required data classes and their relationships, a database schema is the code or the database structure expressed in a formal language, such as Structured Query Language (SQL), which is supported by the database system that implements data storage on a specific database platform.

Brodie and Schmidt (1982) and Simison and Witt (2005) describe three kinds of data-model (or schema) that constitute the official *American National Standards Institute (ANSI) data modeling architecture*. The *conceptual model* describes the model scope by specifying the kinds of facts or propositions that can be expressed with the model. The *logical model* describes the information structure using tables, columns, object-oriented classes, and XML tags. The *physical model* defines the physical means used to store data. The different perspectives of these three models allow them to be relatively independent; changes in storage technology will not affect either the logical or the conceptual schema and the table/column structure of the database can change without (necessarily) affecting the conceptual schema. However, the structures must remain consistent across all schemas of the same data model.

Figure 8.2 illustrates this data modeling process along with the constraints placed on it that allow the data and data models to be shared and redeveloped to form new and improved products. Some typical external inputs to the process are shown on the left of the figure. The blue arrows show how standards, defined on the right side, influence the various stages of the data modeling process. Once a standard is incorporated into any stage of the process, that standard influences all subsequent stages. The actual data modeling process involves *activities*, which involve the creation or updating of the conceptual, logical, and physical models, and production of *deliverables*, which are the conceptual, logical, and physical schemas and ultimately the development of one or several relational databases. The red arrows show that, while the data modeling process progresses through the sequence of conceptual, logical, and physical development stages, it typically involves some iterations between the models and the schemas.

8.2.4 Relational Databases

Relational databases are the most widely used type of digital database; they are based on the *relational model of data*, first proposed in 1970 (Codd 1970). Relational

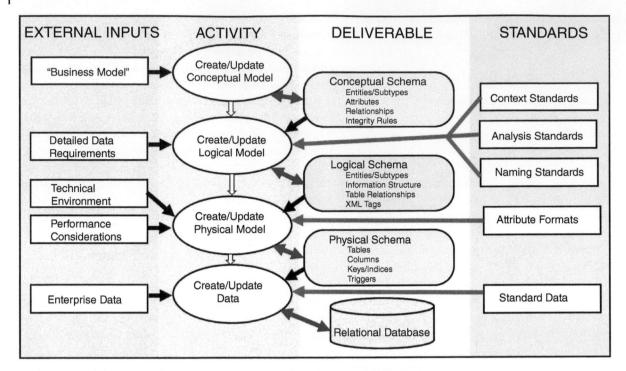

Figure 8.2 Flowchart of the data modeling process that incorporates external constraints and standards to ensure the databases developed by the process are robust and capable of supporting multiple geological modeling applications.

databases are developed and maintained using relational database management systems (RDBMS); virtually all relational database systems use SQL for querying and maintaining them (Brodie and Schmidt 1982; Oracle Corp. 2018). Table 8.1 summarizes the terminology used by SQL and RDBMS to define the basic elements of the relational data model.

The relational data model organizes data into one or more tables (or "relations") containing columns and rows, with a unique key identifying each row (Figure 8.3). Relationships are logical connections between different tables. Relational databases define rows as "tuples" or "records" and columns as "attributes" or "fields." Generally, each table/relation represents one "entity type" such as boreholes or laboratory samples. The rows represent instances of that type of entity; for example, each row of a borehole table represents a unique borehole. The columns represent values attributed to that instance; for example,

three borehole location coordinates (x, y, and z), drilling date, and driller name. Each row in a table has its own "unique key"; for example, each borehole is assigned an ID number. Rows in a table can be linked to rows in other tables by adding a column to hold these unique keys within each row of the linked row – these columns are known as "foreign keys." Codd (1970) showed that data relationships of arbitrary complexity can be represented by this simple set of concepts.

The relational model specifies that the tuples (finite ordered lists) of a relation have no specific order and that the tuples, in turn, impose no order on the attributes. Applications access data by specifying queries, which use operations such as *select* to identify tuples, *project* to identify attributes, and *join* to combine relations. Relations can be modified using the *insert*, *delete*, and *update* operators. All observational data within a relational database are stored and accessed as a series of "base relations" (or

Table 8.1 Important SQL and relational database terms.

SQL term	Relational database term	Description
Row	Tuple (or record)	A data set representing a single item
Column	Attribute (or field)	A labeled element of a Tuple (e.g. "Geologic Formation")
Table	Relation	A set of Tuples sharing the same Attributes – a set of columns and rows
View (or result set)	Derived relation	Any set of Tuples – a data report from a RDBMS in response to a query

Figure 8.3 Relational database terminology.

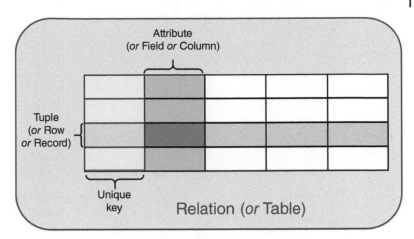

"base tables"). New relations (or tables) may be developed by applying relational operations to existing relations (or tables); these are "derived relations," "views," or "queries."

8.2.5 Entity-Relationship Diagrams

An entity-relationship diagram (ERD), or entity relationship model, is a graphical representation of the relationships among the objects, locations, concepts, or events stored within an information system (Chen 1976). ERDs support the design of normalized relational databases; they are not capable of accurately representing semi-structured or unstructured data. ERDs are generally used to depict conceptual, logical, or physical data models and schemas. Figure 8.4 illustrates a relatively simple,

high-level ERD for the BGS borehole data model (Watson et al. 2014). The rectangular boxes represent entities (Borehole Index, Borehole Log, Log Properties, and Log Source). The lines connecting the boxes represent the relationship between entities, while text defines the entity characteristics. A *cardinality notation* defines the type (one-to-one, one-to-many, or many-to-many) and status of each relationship (is it optional or mandatory?). The BGS borehole data model has only one-to-many relationships.

8.2.6 Normalization Process

Database normalization is the process of restructuring a relational database in accordance with a series of so-called *normal forms* in order to reduce data redundancy and

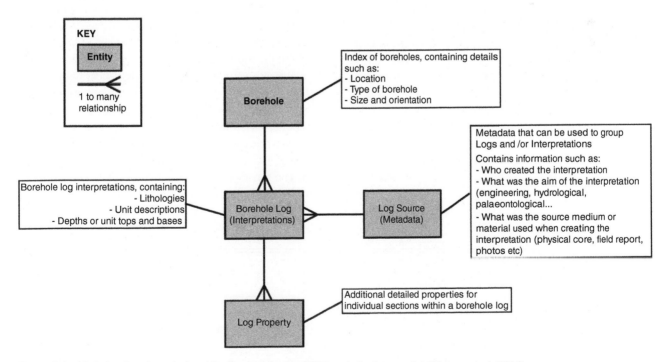

Figure 8.4 High-level entity-relationship diagram for the BGS borehole data model (Watson et al. 2014).

improve data integrity (Kent 1983). The concept of normalization and what is now known as the first normal form (1NF) was first defined by E.F. Codd in 1970; this was subsequently expanded with definitions of the second normal form (2NF) and third normal form (3NF) (Codd 1970, 1971). Although additional more sophisticated normal forms have been defined, a relational database is often described as "normalized" if it meets 3NF criteria which ensure that database relations (or tables) are free of insertion, update, and deletion anomalies. The 3NF design reduces the duplication of data, thus minimizing storage volumes while ensuring "referential integrity" of the database. However, the 3NF increases the number of relations (or tables) forming a database. Extraction of information thus often requires many "joins" among the stored relations (or tables) and may become slow, and if many relations are to be joined, it may be prohibitively slow (Shin and Sanders 2006). Table 8.2 summarizes the functionalities of the first three normalized forms and compares them to an "un-normalized" relational database.

8.2.7 Denormalization Process

Connolly and Begg (2014) define denormalization as a process where normalized tables are refined and consolidated into secondary simplified data structures with controlled redundancy to maximize performance. Denormalization should not be confused with the unnormalized form. Denormalization should only take place after a satisfactory level of normalization has taken place and constraints and/or rules have been created to deal with the inherent anomalies in the design. For example, all the relations should be in 3NF. If the denormalized structure is large and contains many data values from several normalized data sources, coordinated technical procedures should be invoked to keep the denormalized versions of the data synchronized with the source databases. These procedures should be performed on a regular basis, such as on a weekly or monthly schedule, or they can be invoked on demand.

Figure 8.5 illustrates a typical implementation of denormalization to provide geological modeling teams with rapid and coherent access to multiple geological data sources held in multiple normalized relational databases. The BGS has employed denormalization strategies when developing its PropBase database; this allows geological property information from various databases, each with its own relational structure, to be provided efficiently by a single consistent point of access. A single comprehensive information source permits greater efficiency in geological model development (Kingdon et al. 2013, 2016).

8.2.8 Extract, Transform, Load (ETL) Processes

Extract, transform, load (ETL) processes extract data from the source databases, enforce data quality and consistency standards, conform data so that information from separate sources can be used together, and deliver data in a presentation-ready format so that application developers can build applications and end users can make decisions (Kimball and Caserta 2004). ETL systems are an important component of the denormalization process. Data extraction involves extracting data from homogeneous or heterogeneous sources, while data transformation processes convert data entities into a proper storage format/structure for the purposes of querying and analysis. Data loading involves the insertion of data into a target database, such as a data warehouse.

8.2.9 Data Warehousing

A data warehouse is a repository of an organization's electronically stored data that is organized to facilitate easy access to the data by users. Data warehouses usually contain multiple databases, each designed to house standardized, structured, consistent, integrated, correct, "cleaned," and timely data, extracted from various operational systems within an organization. The integrated data warehouse environment provides an enterprise-wide perspective while the data structures support critical reporting and analytic requirements. Metadata and data management and retrieval procedures are an essential component

Table 8.2 Supported functionality of unnormalized and normalized relational databases up to third normal form.

Supported functionality	UNF (1970)	1NF (1970)	2NF (1971)	3NF (1971)
Primary key	BASIC	YES	YES	YES
No repeating groups	BASIC	YES	YES	YES
Atomic columns	NO	YES	YES	YES
No partial dependencies	NO	NO	YES	YES
No transitive dependencies	NO	NO	NO	YES

Figure 8.5 A denormalized database that combines selected data from multiple normalized databases provides geological modeling teams with a rapidly accessible comprehensive data source.

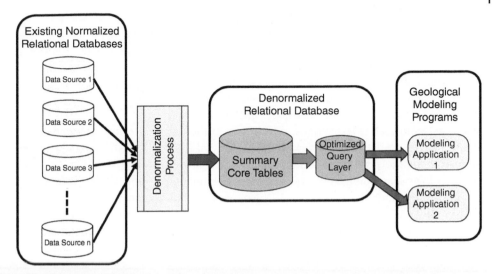

of a data warehouse system. Kimball et al. (2008) describes metadata as "the DNA of the data warehouse."

8.2.10 The Important Role of Metadata

Metadata affects multiple aspects of information management. Thus, many organizations have developed formal definitions of metadata; for example, the *Metadata Primer* issued by the British National Information Standards Organization (Riley 2017) and the report by the Association for Library Collections & Technical Services (ALCTS) Committee on Cataloging Task Force on Description and Access (ALCTS 2000). The ALCTS Task Force reviewed 46 potential metadata definitions before developing its official formal definition of metadata. It also addressed two closely related topics – metadata schemas and interoperability – and developed definitions for each.

Kimball et al. (2008) refers to three main categories of metadata: *technical metadata, business metadata,* and *process metadata. Technical metadata* defines the data structures such as tables, fields, data types, indexes, and partitions in the relational engine, as well as databases, dimensions, measures, and data mining models. It also defines the data model and the reports, schedules, distribution lists, and user security rights. *Business metadata,* also referred to as *discovery metadata,* describes the data in more user-friendly terms; it defines the data holdings, their source, their content, and their relationship to other data. *Process metadata* is used to describe the results of various data management operations. For example, it provides key data about ETL processing, such as: start time, end time, CPU seconds used, disk reads, disk writes, and rows processed. This metadata becomes invaluable when troubleshooting the ETL or query process.

8.3 Managing Source Data for Modeling

The principal aspects of data management are:

1. Managing the multiple data sources,
2. Managing the connectivity among the data sources, and
3. Managing the model outputs.

This section discusses the first two aspects; Sections 8.4 and 8.5 discuss the management of geological models. All three aspects are critical to ensuring effective geological modeling programs; the difficulty in managing data sources has been summarized as follows:

> *The most time-consuming part of the modeling process is the data interoperability and access. BGS has large holdings of data but it is never certain where all the data is stored and in many cases the data needs to be manually transformed into something the software can use. If we can get the data in more efficiently and include all the data, the models will improve dramatically. (Kessler et al. 2016; Sect 6.4, Page 59).*

8.3.1 Data from Multiple Data Sources

Geological surveys exist to provide information describing the subsurface to governments, industry, and the public. Geological surveys have established and manage relatively large and diverse databases that store not only data resulting from their own exploration activities, but also large amounts of data collected by others for many commercial and regulatory purposes. These archival databases are a critical resource when undertaking comprehensive geological modeling programs. However, considerable investment is required to support the ongoing maintenance and management of these databases. For example, the BGS has invested well over tens of millions of pounds sterling since the late 1970s designing and building geoscience databases.

The BGS currently supports about 100 separate corporate relational databases covering key geoscientific subject areas. These contain a variety of sources, ranging from highly variable legacy data to recently acquired geological data, including: boreholes, borehole logs, borehole cores, geophysics, geochemistry, paleontology, mineral occurrence, geotechnics, hydrogeology, geohazards, and offshore geological investigations, as well as numerous controlled vocabularies (e.g. rock classification scheme, lithostratigraphic lexicon). Specific additional databases hold geological mapping data, scanned legacy documents, images, sensor measurements, and 3-D geological objects (Nayembil and Baker 2015). Other geological surveys hold similar data collections. Section 8.7 describes the integrated data management approach of the Geological Survey of Denmark and Greenland, which extensively employs a variety of geophysical data sources to build 3-D geological framework models (Møller et al. 2009a,b).

The BGS Geoscience Data Hub (GDH) has developed and evolved over many years to become an integrated multi-tiered data architecture containing all its geoscientific data. Figure 8.6 provides a high-level overview of the GDH. The large number of external data sources that the BGS is mandated to manage are summarized on the left of Figure 8.6. Data accessioning processes convert these sources and combine them when required with internal BGS data sources to create a series of mostly normalized relational databases within the GDH. The GDH also contains additional file stores and a data archive. Geological models are stored in a separate geological object store

(GOS), which is described in Section 8.4. Data warehouses and application program interfaces provide convenient access to the GDH for a variety of users and applications.

8.3.2 Managing the Connectivity among Data Sources

Geological framework modeling typically requires data to be extracted from diverse datasets; a typical model requires data defining the land surface, locations of boreholes and borehole logs existing digital geological maps, and cross-sections. Often these data must be collated or combined with supplemental hydrogeological or geophysical information and descriptions of engineering properties.

Uniform vocabularies are required if different scientists, databases, and clients are to communicate, compare, integrate, and re-use multiple data sources during the modeling process. However, vocabularies are generally difficult to establish; scientists by their nature like bespoke classifications for their own purposes. This results in the development of many similar classifications; even minor differences in these classifications create difficulties in merging data from multiple sources. Section 8.8 describes several recent European collaborations to resolve data consistency issues encountered during the development of transnational (transboundary) geological models.

The BGS Geoscience Data Hub User was developed to facilitate data access. Related information stored in multiple normalized databases is combined and transformed by ETL processes to produce denormalized data

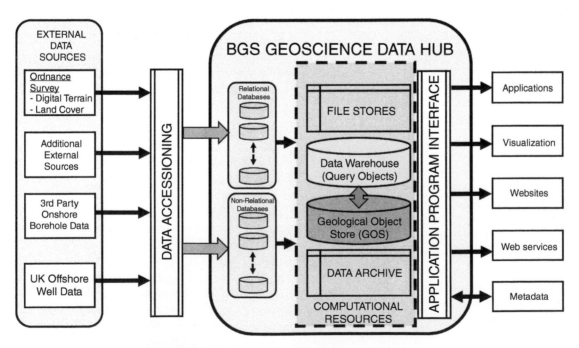

Figure 8.6 A high-level overview of the BGS Geoscience Data Hub.

warehouses. These make the accessing and extraction of data by geological modeling teams much more efficient. The process is especially important in handling archived geological modeling data within the GOS, which is a component of the Geoscience Data Hub. Section 8.4 discusses this in more detail.

The availability of metadata also enhances user access to data. Entity-relationship (ER) diagrams are available for BGS topical databases. Section 8.2 used the borehole ER diagram (Figure 8.4) as an example of the basic structure of a key geological modeling data source. The geochemistry ER diagram (Figure 8.7) shows the slightly more complex structure typical of databases holding multiple properties measurements. A similar ER diagram may represent

geotechnical properties data; however, much geotechnical data are received or transmitted in the AGS4 digital data transfer format (Association of Geotechnical and Geoenvironmental Specialists 2017). This has made the acquisition and quality assurance of these data sources much faster.

8.3.3 Facilitating Sharing of Database Designs

A three-year (2011–2014) knowledge exchange project, *OpenGeoscience Data Models*, undertaken by the BGS with support from the Natural Environment Research Council, set out to encourage an open sharing of geoscience data models amongst geological surveys, industry, and academics (Watson et al. 2014). The data model was recognized to be the key to successful information management,

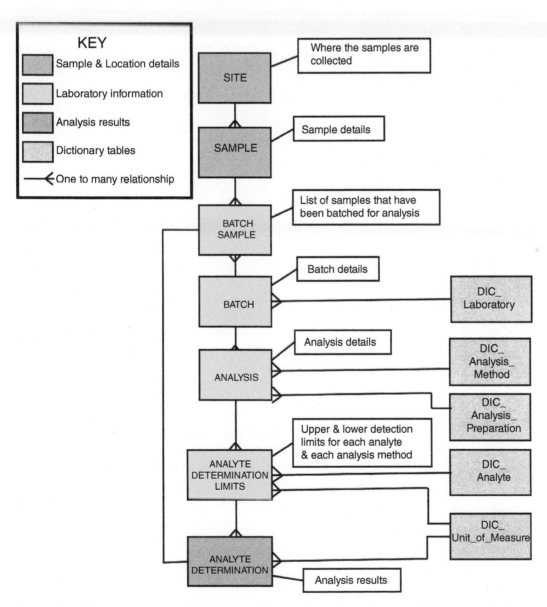

Figure 8.7   High-level entity-relationship diagram for a geochemistry data model (Watson et al. 2014).

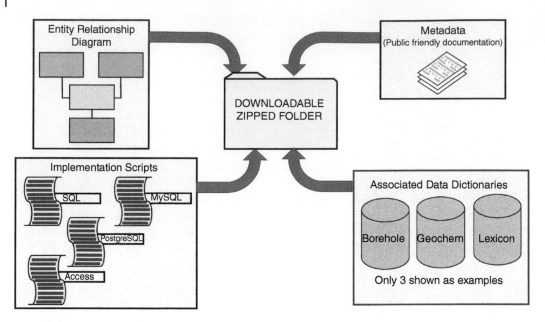

Figure 8.8 The components included in a downloadable database design package. (After Watson et al. 2014)

yet prior interactions between the BGS and numerous geological surveys revealed a lack of data management experience (Watson et al. 2014). Thus, the project created an online repository for data model designs to provide access to open, ready-to-use, freely available, database designs.

The project produced a number of downloadable database design packages. These provided relational data models in the form of: (i) entity relations (ER) diagrams; (ii) metadata (plain English documents available in MS Word and PDF to maximize access); (iii) implementation scripts; and (iv) example data and links to associated dictionary data sources (Figure 8.8).

Each database design package includes an explanation of how users can create their own versions of the databases using either a supplied pre-populated Microsoft Access file (a *.mdb file) or by requesting SQL implementation examples for MySQL, Oracle, PostgreSQL, and SQL Server (British Geological Survey 2013).

To date, the BGS has completed four downloadable models: *borehole index*, *borehole interpretations*, *lithostratigraphical lexicon*, and *geochemistry*. These may be downloaded from the BGS Open Geoscience Data Models Website. Additional models have been proposed for *collections management*, *hydrogeology*, *landslides*, and *rock classification*. Their development requires financial support from users. Some additional designs developed by project partners and associated organizations have been hosted on the websites of these providers with links provided on the EarthDataModels.org model library (British Geological Survey 2017).

8.4 Managing Geological Framework Models

During the initial, project-based, stage of model development at the BGS, completed 3-D geological models were simply stored as data files on a local network, in native software formats, without any attached metadata. This practical low-cost solution required no specialized tools or databases. However, this straightforward approach to model storage has several limitations: long-term model recovery and re-use is difficult, file structures are potentially vulnerable to accidental modification, and the incorrect identification of the latest model versions may result in future model development being based on inaccurate data and concepts.

The BGS undertook the first steps toward improving the situation by establishing requirements for metadata documents and spreadsheets to be developed and stored alongside the model data, and continues to use such a system for work-in-progress. While this is a practical day-to-day solution for project teams, it does incur manual overhead to maintain and organize the databases. As the number of completed geological models increased, the shortcomings of the file-based archiving system became apparent and the BGS began to develop an enterprise-wide data management approach to make the models more widely accessible.

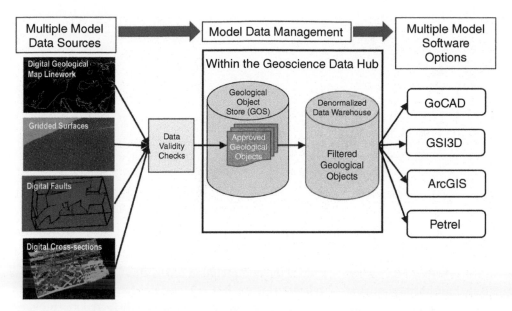

Figure 8.9 The Geological Object Store (GOS) is an agnostic Oracle Spatial database of geological modeled objects. It is part of the BGS Geoscience Data Hub. Geological objects include geological map line-work, gridded surfaces, faults, and cross-sections. All objects are versioned within the database.

8.4.1 BGS Model Database Design Principles

The BGS has developed a prototype corporate 3-D model storage database called the *Geological Object Store* (GOS). The GOS design employs the concept of *geological objects*. As illustrated by Figure 8.9, the GOS is a part of the Geoscience Data Hub (Section 8.3), and the geological objects produced by the various modeling workflows undergo validity checks before being accepted into the GOS.

Because the majority of geological modeling within the BGS has been undertaken using the map and cross-section approach (see Chapter 10), the GOS was developed to manage the interpreted linework resulting from these models. Although they formed the most abundant existing model data, managing these data posed a complex technical challenge.

Related sets of linework form collections representing classes of geological object, rather than being disaggregated into their individual attributed vertices. There are four main geological object classes: (i) geological map linework, (ii) gridded surfaces, (iii) fault planes, and (iv) cross-sections (Figure 8.9). All of these geological objects store 2-D spatial information. Cross-section correlation linework is defined in the plane of the cross-section itself, forming a 2-D "vertical map." Map linework, whether at the surface, or in the sub-surface, is also only held in 2-D; it is expected that the individual modeling packages will resolve the third dimension at calculation time. Each individual line within a geological linework object has only minimal attribution – normally a coded rock layer value

based on the BGS lexicon of named rock units and/or rock classification scheme (British Geological Survey 2018a,b). Each geological object has basic metadata attached, including a name and audit information. The audit information includes version numbers, timestamps, and user IDs so that the iteration history and provenance of all geological objects can be tracked over time.

8.4.2 Versioning Existing Models

Geological models developed following the map and cross-section approach are easy to iterate and refine by undertaking additional geological interpretation work. This makes it critically important for the GOS to support complex sequences of model versions and accompanying metadata.

The GOS database versioning system is based on that of software source-code versioning. The database maintains a constantly incrementing global version number that can be stamped onto objects being saved into the database by the geologist. New objects are automatically assigned a unique geological object ID number. Revised objects are stored as "difference-only" objects; only those parts of the geometries that have changed are stored. This allows the efficient storage and management of small changes to complex objects within the database. It also allows extraction of a detailed audit trail of an object's history. The versioning system uses a check-out/check-in motif; geologists wishing to edit objects in the database perform

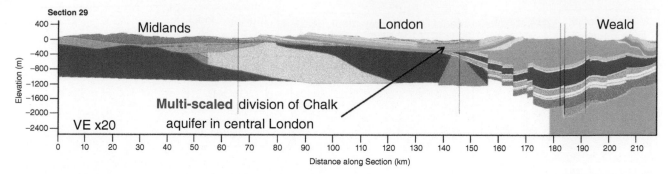

Figure 8.10 Detail of BGS national geological model cross-section 29. Extra detail within the Chalk Group aquifer (shades of green) beneath London is derived from the more detailed GSI3D London model, while the increased section depth in the Weald Basin is derived from BGS published maps and reports. (Mathers 2011).

a simple 'check-out' operation which locks the object to their user ID. Other workers can still download the locked objects, but they cannot save edits to those objects until the lock is relinquished. When a geologist completes an edit, a simple 'check-in' records the new object and includes a comment. In addition to this versioning system, an approval mechanism has been implanted that allows individual objects (or groups of objects) to be moved through a multi-step approval chain. This is a critical part of the GOS database because it allows work-in-progress to exist alongside approved work.

Project geologists may decide to "check out" non-approved objects to support their day-to-day work. Product managers will typically download read-only copies of fully approved objects. As noted in Figure 8.9, these will normally be subjected to a *denormalization* process and stored in a *data warehouse* database reformatted to suit one of several alternative modeling software products. These data can then be sent to clients directly, they can be used to create new 3-D geological models, or can be used to develop or constrain derivative process models, as described in Section 8.6.

8.4.3 Creating New Models Based on Existing Models – "Model Interoperability"

Beginning in 2009, the BGS released a series of 3-D national geological models (Waters et al. 2015). These models consist of a series of interconnected cross-sections that were largely developed by interrogating legacy source data and geological framework models stored within the GOS; these were primarily regional resolution GOCAD models based on the BGS subsurface memoir series of sedimentary basins, while higher resolution GSI3D models provided information for the London Basin and southern East Anglia-Essex. The GSI3D models are relatively shallow, so they provide information on only the uppermost bedrock units. Where multiple resolution models

were available, the highest resolution model usually guided the interpretation unless this was known to be less reliable (Figure 8.10). Metadata documentation for the individual models and published BGS reports provided quality-assurance information.

More detailed sections and data can be keyed to stratigraphic units of the national geological model using a nested or child-parent approach (Mathers et al. 2011). As shown in Figure 8.11, a national model can impose coherency among regional models. In this case, the major stratigraphic boundaries defined in the national model are incorporated into the higher resolution models. Alternatively, detailed models can be generalized and incorporated into lower resolution regional models. Both these procedures offer significant efficiencies to the creation of new models. The imposition of data management protocols provides the standardization that allows for such model interoperability.

8.5 Managing Geological Properties Data and Property Models

Many GSOs hold large volumes of subsurface property data derived from numerous diverse sources. Knowledge of subsurface heterogeneities of porosity and rock strength provide fundamental information required when assessing the opportunities and risks to subsurface development. These assessments are of great, and increasing, societal and economic importance.

However, the subsurface property data are typically stored in multiple relational databases to ensure their long-term continuity. These have, by necessity, complex structures. Obtaining the required property data for a 3-D model may require interrogation of 10–20 major databases, each with multiple and diverse datatypes and containing millions of records. Each dataset involves a different search process and the extracted information has to be laboriously

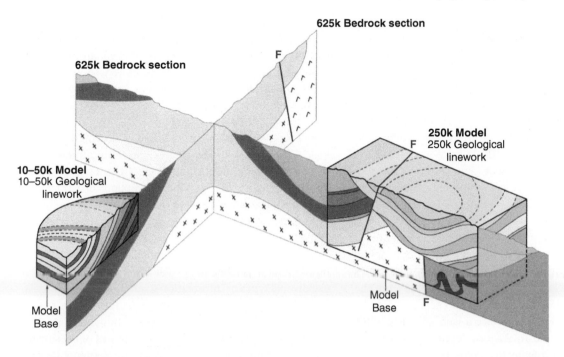

Figure 8.11 The imposition of data management protocols allows for model interoperability and ensures coherence among models with differing levels of detail and resolution. (Mathers et al. 2014).

reformatted to provide consistency. Thus, several days of data acquisition effort are required before the development of any initial 3-D property model can begin. In response to this problem, the BGS has developed *PropBase,* a generic denormalized data warehouse that stores property data and facilitates rapid standardized data discovery and access (Kingdon et al. 2013, 2016; Nayembil et al. 2014).

8.5.1 Characteristics of Property Data Sources and Models

Subsurface physical properties databases typically are derived from three main sources:

- Laboratory analyses from cores and geological samples collected by GSOs and from geotechnical, petroleum, and water resource industry sources.
- Geophysical logging data, which provide derived proxies of physical properties.
- Geotechnical testing.

Each dataset will have been collected at different times, for different purposes, under differing environmental conditions, using distinct collection methods, and archived in specified dataset structures. Building physical property models from such diverse data sources therefore requires multiple data extraction operations and leveling these varied extracted results. However, regardless of source, all property information has a minimum set of common properties:

- A location measured in three dimensions (+/– time-stamps).
- Measured values, units, and margins of error.
- Metadata describing the acquisition, analysis, storage, and processing.

Provided data can be expressed in this format, multiple datasets can be integrated (Shaw 2006).

Geological framework models represent the 3-D subsurface geological units as shells or volumes. Many model applications require a definition of subsurface physical properties along with information defining their spatial heterogeneity and the uncertainty associated with these estimates. This is achieved by capturing physical, hydrological, and other property information from diverse sources and then creating gridded (discretized or voxelated) versions of the framework models to store the subsurface properties data and define the subsurface heterogeneity. These new models are termed *property models* and are in turn used to constrain the development of various types of numerical *process models* such as groundwater models (Chapter 14).

8.5.2 Applications within the British Geological Survey

Over the last 25 years, the BGS has held property data in multiple normalized structured databases. The large number of topical databases, each with distinct data structures and vocabularies, made the extraction and collation of the required properties data to develop any properties model a complex and inefficient process. As a solution, the BGS developed PropBase to provide a single integrated access point for 2-D and 3-D physical and chemical property data, along with associated metadata (Kingdon et al. 2016).

PropBase is a data warehouse system that uses ETL processes to integrate the multiple property data sources into a simplified data model so that the data are compatible and accessible from a single interface. The system consists of data tables that form the core of a simplified data structure, coding routines that use ETL processes at regular intervals to ensure that PropBase information is synchronized with the sources, and a second-tier partitioned denormalized data access QueryLayer that serves as the data access point for all applications. The PropBase core data tables are only used to collate and harmonize the data from the heterogeneous data sources and are never accessed directly by client applications. There is also a web service to allow for machine-to-machine interaction, enabling other software systems to interrogate the datasets directly and to visualize and manipulate them. Kingdon et al. (2016) and Nayembil et al. (2014) provide detailed information about the PropBase data model.

PropBase includes a highly optimized data access *Query-Layer* to facilitate access by multiple applications using PropBase Web services. The QueryLayer contains several millions of spatially enabled property datapoints from 10 separate primary databases containing 50 different data types. The vocabulary set supports 557 different property types enabling semantic interoperability. The PropBase Query Layer's implementation has enabled rapid data discovery, visualization, and interpretation of geological data, massively simplifying the process of populating 3-D framework models with physical properties and facilitating the study of intra-unit heterogeneity. By enabling property data searches across multiple databases PropBase has facilitated new scientific research, previously considered impractical. PropBase is easily extended to incorporate 4-D (time series) data and is providing a baseline for new "big data" monitoring projects.

8.6 Managing Process Models

Geological framework models frequently serve as the information source for further application-specific analyses that utilize *process models*, as described in Chapter 14. Process models extend the dimensionality of the 3-D models beyond their spatial coordinates to analyze time-varying phenomena; they involve 4-D modeling based on three spatial dimensions plus time. Applications of process models include modeling and predicting the movement of fluids or the variation of pressure and temperature in the subsurface, or evaluating stresses and strains around mines, tunnels, and underground structures. These applications use finite-difference, finite-element, or discrete-element techniques to numerically analyze time-varying flow and transport conditions or conduct kinematic analyses.

Process models thus require *discretization* of the 3-D geological framework model to create an appropriate grid or mesh for these numerical analyses. Chapter 13 discusses discretization procedures. Gridded versions of 3-D geological framework models enable the use of stochastic simulations to assess the uncertainty of property distributions in the subsurface (Chapter 15). Key issues common to these modeling applications are how much complexity should be represented in the derived gridded model, and how best to use diverse types of information to enhance these models.

The data management requirements for process models thus involve the management of the individual grids or meshes. While these objects require the use of the attribution, versioning, and approval system implemented for the geological framework models, as described in Section 8.4, their geometries must be handled differently. The meshes or grids are truly static objects; unlike a cross-section in a framework model, they will never be iterated directly because each model revision will result in a logically new object. Therefore, these objects may be treated as distinct geological objects and stored individually. It is desirable to store the mesh or grid geometry in both native and agnostic form as binary objects, and to employ relational database structures for their versioning and to define their spatial extents to facilitate their discovery. The spatial extent may be defined by deriving some form of coverage polygon or shapefile for the object on-the-fly, with the result stored in the relational database.

8.7 Integrated Data Management in the Danish National Groundwater Mapping Program

In Denmark, regional groundwater mapping was initiated in the 1990s when demand for groundwater resources increased due to urban development. At the same time, concerns were raised about the dangers of pollution from industrial and agricultural sources. In 1999, the Danish Government initiated the National Groundwater Mapping Program with the objective to significantly intensify hydrogeological mapping so as to improve the protection of the Danish groundwater resources (Thomsen et al. 2004).

Danish groundwater supplies are produced from aquifers within heterogeneous glacial deposits. The National

Groundwater Mapping Program has employed a variety of legacy (archival) and new data sources to develop regional hydrogeological assessments. Ground-based and airborne geophysical measurements have contributed significantly to the mapping of aquifers in Denmark. Three-dimensional regional hydrogeological models were constructed by combining these geophysical data sources with additional borehole information, such as lithological descriptions, geophysical logs, and data on water chemistry and hydraulic parameters.

The integrated data management system is set up to handle a variety of geophysical data sources as well as other data sources, such as borehole logs, and descriptions of material properties (Figure 8.12). The data management system contains software that readily processes and interprets new geophysical data or re-processes and re-interprets geophysical data acquired over long periods by different companies with different instruments. This is of great value for creating 3-D regional hydrogeological

models, as well as for future mapping and administrative purposes (Møller et al. 2009a,b).

The development of a national GEophysical Relational DAtabase (GERDA), hosted by Geological Survey of Denmark and Greenland (GEUS), began in the early 2000s. GERDA contains a wide variety of geophysical data obtained from both ground-based and airborne transient electromagnetic (TEM) geophysical surveys (Møller et al. 2009a,b). Various kinds of 1-D and 2-D models resulting from inversion of electrical and electromagnetic data and all information about data acquisition, data processing, and inversion are also stored. This facilitates the re-processing of data and makes transparent the inversion and interpretation of the data.

GEUS also hosts another database (Jupiter) for borehole data. Jupiter contains information about geological and lithological descriptions, groundwater levels, and water-quality observations. Both the Jupiter and GERDA databases have web-based graphical user interfaces

Figure 8.12 The GEUS integrated system of databases and program packages handling geophysical data and geological modeling. The arrows show the flow of data between the geophysical database GERDA, the borehole database Jupiter, the Aarhus Workbench program package, the Geoscene3D visualization and modeling tool, and the geological model database Modeldb. (Møller et al. 2009a).

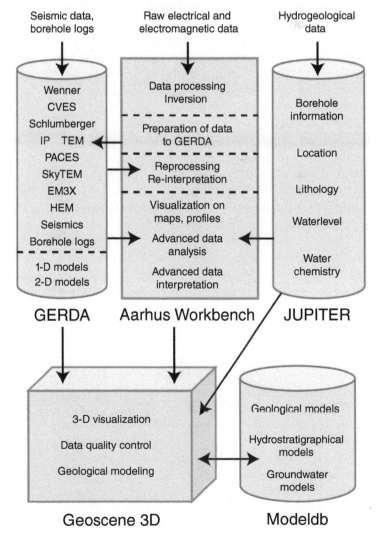

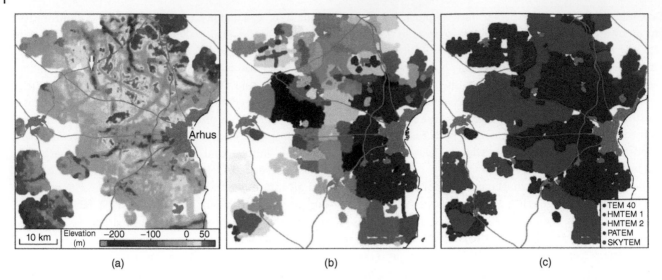

Figure 8.13 Data coverage and results from an area in eastern Jylland. (a) Map showing the position of the top surface of the deepest low-resistivity layer in the area relative to sea level. The data are from 94 individual mapping projects (indicated with different colors in panel b), that used five different TEM methods (c). (Møller et al. 2009a).

(http://jupiter.geus.dk and http://gerda.geus.dk) where users can search for and download data free-of-charge (Møller et al. 2009a).

The integrated data management system (Figure 8.12) contains two software packages and one additional geological model database (Modeldb). The Aarhus Workbench developed by Aarhus University (http://hgg.au.dk/software/aarhus-workbench) and provided by Aarhus GeoSoftware (https://www.aarhusgeosoftware.dk) permits the processing of the geophysical data in GERDA and its combination with other hydrogeological data in the Jupiter database. The Aarhus Workbench combines modules for handling, processing, inverting, interpreting, and visualizing electrical and electromagnetic data in a common GIS platform with a common database. These modules permit the processing of the geophysical data in the GERDA database without having to know the complicated data model of GERDA or construct explicit database queries. By using the GIS platform within the Aarhus Workbench, various types of maps are easily compiled from the geophysical data.

Geoscene3D, developed by the Danish firm I-GIS (https://geoscene3d.com/), is a new 3-D geological modeling and visualization tool designed to process and combine the geophysical data stored in GERDA and the borehole information stored in Jupiter. The resulting 3-D models are stored in the geological model database (Modeldb), which may contain several different 3-D models: geological models, hydrogeological models, and groundwater models (Møller et al. 2009a).

The integrated data management system in relation with a TEM test site providing absolute calibration of all TEM (Foged et al. 2013) allows electromagnetic and electrical data to be displayed without showing any discrepancies at borders between adjacent projects, even when the data have been acquired by different companies, with different instruments and methods, and at different times. A $50 \times 60 \, \text{km}^2$ area in eastern Jylland provides an example of the strength of the integrated data management system (Figure 8.13). The most prominent features found in the area are a large number of buried valleys incised into the Palaeogene clay deposits. The 3-D surface of the deepest low-resistivity model layer represents these Palaeogene clay deposits except in the north-eastern corner, where it represents salty pore-water in Danian limestone (Figure 8.13a). The buried valleys show no direct correlation to the overall topography. Large parts of this area are covered by ground-based single-site TEM soundings (about 83 000 soundings in total) and the surface of the deepest low-resistivity model layer in Figure 8.13a is based on the interpretation of these TEM soundings. Yet these TEM soundings were collected during more than 90 mapping campaigns (Figure 8.13b) and with five different TEM methods (Figure 8.13c) over a time span of more than 10 years.

8.8 Transboundary Modeling

The mandates of geological surveys generally restrict their geological investigations, and thus their geological models, to within their national boundaries. However, geological structures often cross international boundaries and the evaluation of geological resources sometimes requires the use of data for locations in more than one country. As many geological surveys in Europe are now developing and applying geological models, several examples of transboundary collaborative modeling have already occurred. Such modeling requires the use of data and definitions based on long-standing national protocols, so the modeling naturally encounters interoperability issues. Resolving these depends on robust data management practices. There are several recent European examples of international collaboration and sharing of transboundary information.

Section 8.8.1 discusses the H3O program, a series of projects with Dutch, Flemish and German partners, that resolved the inconsistencies in existing subsurface information sources that prevented efficient groundwater resource management in this border region. The projects tested two approaches for transboundary model harmonization: an integrated data management protocol with the national geological surveys working in parallel to common standards and a process where specific modeling tasks are assigned to various team-leaders from individual organizations.

Section 8.8.2 features the Polish-German TransGeoTerm Project which evaluated the potential development of shallow geothermal resources in the Neisse valley area by coordinating the national data sources and creating an integrated geological model (Görne and Krentz 2015), resolving transboundary inconsistencies by creating new collaborative databases that supported and integrated 3-D modeling effort.

Section 8.8.3 discusses the EU-sponsored GeoMol Project, a multinational information coordination effort among six countries to establish a networked geoscience data store for the Alpine Foreland basins located to the north and south of the Alps. GeoMol retained the existing national data sources but established a networked

data store that supported an interoperable, harmonized, multi-dimensional geoscientific information process across multiple national domains. It also developed web-based search, retrieval, and visualization tools to provide widespread access to this information by many classes of user, while operating within existing data privacy and security protocols.

8.8.1 The H3O Program: Toward Consistency of 3-D Hydrogeological Models Across the Dutch-Belgian and Dutch-German Borders

The Roer Valley Graben contains important drinking water sources for the Netherlands, Flanders, and Germany. Inconsistencies between subsurface information prevent efficient groundwater resource management in this border region. A series of projects with the prefix "H3O" for "Hydrogeologische 3D modellering Ondergrond" (English translation: "Hydrogeological 3-D modeling of the Subsurface") began in 2012–2014 with "H3O – Roer Valley Graben" by Dutch and Flemish partners. It created 3-D geological and hydrogeological models of the unconsolidated Cenozoic sediments in the southern Dutch and Flemish part of the Roer Valley Graben (Deckers et al. 2014). A second Dutch-Flemish project, called "H3O – De Kempen," started in 2015 (Vernes et al. 2018) followed by a Dutch-German project, called "H3O – ROSE" (ROer valley graben South-East) in 2016. A fourth project, called "H3O – Roervalley Graben North-West" began in 2017 to model the remaining Dutch part of the Roer Valley Graben. Ultimately, the combined and integrated results of all four H3O projects will provide consistent geological and hydrogeological models of the entire Roer Valley Graben. While these projects mainly focus on the Flemish–Dutch and Dutch–German border areas of the Roer Valley Graben

Table 8.3 Characteristics and advantages of the "Sewing" and the "Knitting"-approaches to harmonize hydrogeological models at the border (Van der Meulen 2016).

	"Sewing" approach	"Knitting" approach
Characteristics	• Working in parallel • Division by area • Common standards • Joined result	• Working together • Division of tasks • Shared standards • Joint result
Advantages	• Pragmatic, robust • Suits GSO[a] mandates • Repeatable, updatable	• Efficient use of resources • Learning opportunities • Integrated result
Optimized for	• Long-term viability	• Result
Applications in the H3O Projects	• H3O – ROSE	• H3O – Roer Valley Graben • H3O – De Kempen

a) GSO stands for Geological Survey Organization.

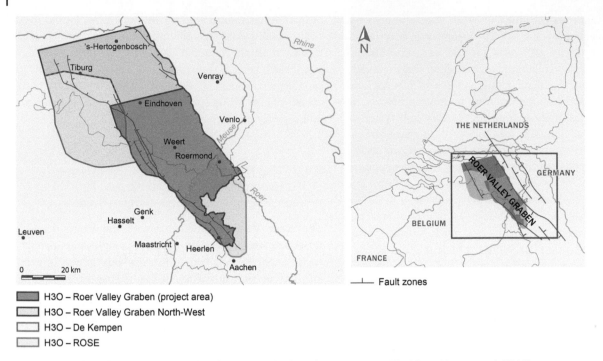

H3O – Roer Valley Graben (project area)
H3O – Roer Valley Graben North-West
H3O – De Kempen
H3O – ROSE

Figure 8.14 Location of the Roer Valley Graben and H3O project areas. (Modified from Vernes et al. 2016).

(Figure 8.14), they provide examples of international collaboration and sharing of transboundary information.

Van der Meulen (2016) distinguishes two approaches for transboundary model harmonization, the "Sewing Approach" and "Knitting Approach." Each has specific characteristics (Table 8.3). The "sewing" approach means the national geological surveys work in parallel to common standards, while the "knitting" approach represents a short-term joint project using shared standards, with specific tasks assigned to various team-leaders from individual organizations. The two Dutch-Flemish projects used the "Knitting" approach. The Geological Survey of the Netherlands (TNO) did the modeling work of the first project. In the second project this task was split between the partners; the Geological Survey of the Netherlands (TNO) created shallow geological and hydrogeological models and took responsibility for the time-depth conversion of seismic records for the deep models, while the Flemish organization VITO created the deep geological models. The shallow and deep models were subsequently combined to form the final model. The H3O – ROSE project chose the "sewing" approach, with both partners basing the models on a combination of Dutch and German interpreted borehole and seismic data.

The Roer Valley Graben transboundary area is on the borders of several existing 3-D geological and hydrogeological models: these include four Dutch models, two Flemish models, and one German model (Table 8.4). The

(A) Geological model trans-boundary mismatch

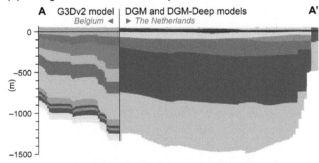

(B) Hydrogeological model trans-boundary mismatch

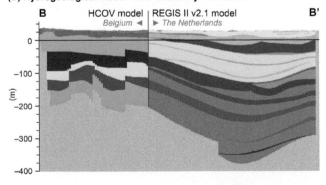

Figure 8.15 Transboundary mismatches between Flemish and the Dutch models. (a) Geological models G3Dv2 (Belgium) and DGM and DGM-Deep combined (the Netherlands); (b) hydrogeological models HCOV (Belgium) and REGIS II (the Netherlands). Figure 8.16 shows locations of these cross-sections. Only the upper 400 m of the hydrogeological models are shown to illustrate the difference in detail. (Deckers et al. 2014).

Table 8.4 Characteristics of the most recent Dutch, Flemish, and German 3-D models of the Roer Valley Graben.

	Model name	Description	Depth range	Coverage
Dutch models	GEOTOP	High-resolution voxel model (100 m × 100 m × 0.5 m)	30–50 m	Regional, completion of Roer Valley Graben in 2018
	NL3D	Low-resolution voxel model (250 m × 250 m × 1 m)	Corresponds to DGM units	National
	REGIS II (regional geohydrological information system)	Layered hydrogeological model with 125 hydrogeological units	To 500 m; 1200 m in Roer Valley Graben	National
	DGM (digital geological model)	Layered geological model with 34 geological units	To 500 m; 1200 m in Roer Valley Graben	National
	DGM-deep (digital geological model deep)	Layered geological model with 11 geological horizons	To 5 km; based on seismic data	National; onshore and offshore
Flemish models	G3Dv2	Layered geological model with 39 geological units	To 5 km	Flanders and the Brussels capital region; onshore
	HCOV	Layered hydrogeological model	To 1.9 km	Flanders and parts of the neighboring countries/regions
German model	3D-Modell Rurscholle-Nord	Layered model with 15 geological and hydrogeological horizons	Up to 900 m	German part of the Roer Valley Graben

(hydro)stratigraphic classification differences used to create these national models resulted in obvious mismatches at the national border (Figures 8.15 and 8.16).

The H3O projects used a five-stage process (Figure 8.17) to create transboundary, up-to-date, 3-D geological and hydrogeological models of the Roer Valley Graben. Iterations of some stages were required when interpretations of the available information revealed refinement of the model structure was required.

The Roer Valley Graben, is filled with up to 1800 m of Cenozoic sediments. Modeling these deeply buried sediments relied on borehole data to define the shallower deposits and seismic profiles to define the deeper units (Figure 8.18). Data sources were gathered for an area extending several kilometers outside the model area to ensure the model is accurate near its edges. Sparsely available deep boreholes were used from an even larger area. Boreholes selection was based on depth, location, quality of lithological descriptions, and availability of geophysical borehole logs. Seismic profiles were selected on the quality of their records for the Cenozoic deposits and their age. Older data sources were digitized when necessary.

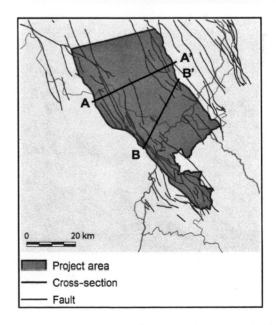

Figure 8.16 Locations of cross-sections shown in Figures 8.15 and 8.22 (Deckers et al. 2014).

The data were made available in the different national coordinate systems used by the partners.

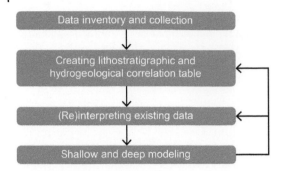

Figure 8.17 The four-stage model development process used by the H3O Project. The arrows show possible iterations of the middle stages to refine the model.

As existing Dutch and Belgian stratigraphic systems have considerable differences, the transborder projects required a common consistent stratigraphic classification scheme. First, the Dutch and Belgian (hydro)stratigraphic units were correlated. Next, a decision was made on the (hydro)stratigraphic classification scheme that would be applied in the project. The Netherlands has a detailed stratigraphic system for the Quaternary, based on extensive field investigation. A highly detailed Belgian stratigraphic classification defines the Miocene and Paleogene units which are located at depth in the Roer Valley Graben but are found at the surface in Belgium. Thus, the Dutch–Flemish H3O projects used the Dutch classification for the Quaternary deposits and the Belgian classification for the Miocene and Paleogene units to produce a coherent stratigraphic correlation chart of the geological and derived hydrogeological units. Several shallow and deep cross-sections were developed by experts from both countries to test these new correlation standards. At trans-border locations, re-interpretation was required to eliminate inconsistencies. Dutch, Flemish, and German parties were responsible for their respective data; extensive communication links between the participants ensured consistency of the re-interpretations.

Two modeling workflows were used by TNO Geological Surveys of the Netherlands, one for the modeling of shallower units and one for deeper units (Figure 8.19). These models relied on different primary data sources,

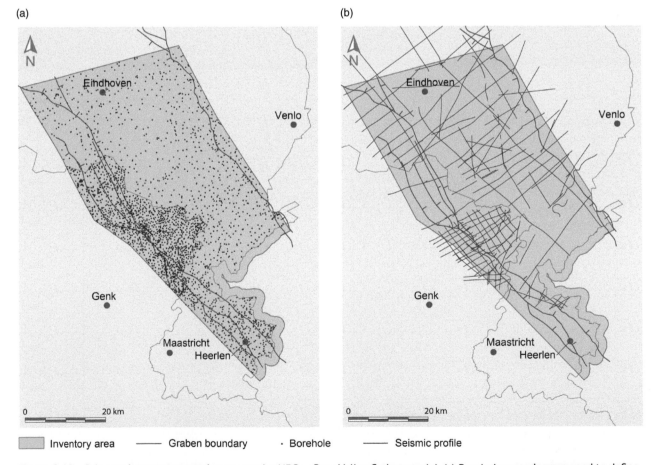

Figure 8.18 Primary data sources used to create the H3O – Roer Valley Graben model. (a) Borehole records were used to define shallower horizons; (b) seismic profiles provided information on deeper horizons. (Modified from Deckers et al. 2014).

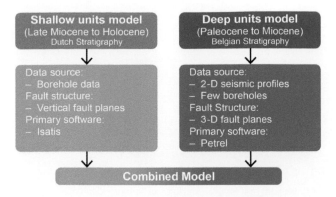

Figure 8.19 H3O modeling workflow develops a combined model by combining models of shallow units and deep units.

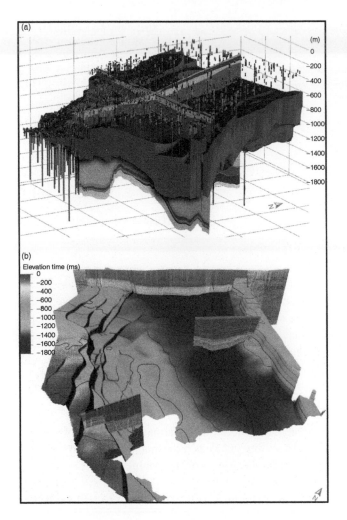

Figure 8.20 Examples of models of (a) shallow units – Stramproy Formation and (b) deep units – base of the Opglabbeek Formation. (Deckers et al. 2014).

incorporated different fault plane definitions, and used different primary software tools. Figure 8.20 provides an example of a shallow model unit and a deep model unit. Interpretations of seismic data, borehole data,

cross-sections, and existing fault maps were used to develop a 3-D fault plane model that included all faults considered to be relevant at the scale of the geological and hydrogeological models.

The models were subsequently combined to form a single model (Figure 8.21). The boundary between the shallow and deep models was chosen at the first horizon that could be interpreted in most seismic data. A geological framework model was created first; a hydrogeological model was derived from the geological model by providing a further subdivision into permeable and less-permeable hydrogeological units. The H3O-projects significantly improved the consistency of (hydro)geological interpretations in this transboundary area (Figure 8.22).

Transboundary modeling serves as a framework for various national groundwater management studies and for more detailed models. The H3O Roer Valley Graben model has been reviewed to ensure its overall consistency and validity for groundwater management purposes by an advisory board, representing regional authorities, research institutes, universities, water supply companies, and water boards. The project results are available on the TNO Netherlands Geological Survey web portal DINOLOKET (http://www.dinoloket.nl) and the Flanders database web portal DOV (https://dov.vlaanderen.be).

Consistency of transboundary models requires easy and free access to primary data sources, especially borehole and seismic data. Differences among existing national lithologic and stratigraphic classification systems, and derivative hydrogeologic classification systems, hamper transboundary analyses. Establishment of consistent classification standards, including standards for describing soil samples and encoding borehole records, would make transboundary modeling much easier. Better data interchanges and coordinate transformations between modeling software products are desirable technical improvements that would benefit interactions between national geological surveys.

8.8.2 The Polish–German TransGeoTherm Project

The TransGeoTherm Project, a co-operative, cross-border project of the Polish Geological Institute (PGI) and the Saxon Geological Survey (LfULG) co-financed by the European Union (EU), evaluated the potential development of shallow geothermal resources along the valley of the Neisse River, which forms the border between the German state of Saxony and Poland. A case study in Chapter 20 provides additional information on this modeling project.

Prior to the TransGeoTherm project, no coherent transboundary information system of geothermal databases existed, so the project collected and standardized specified

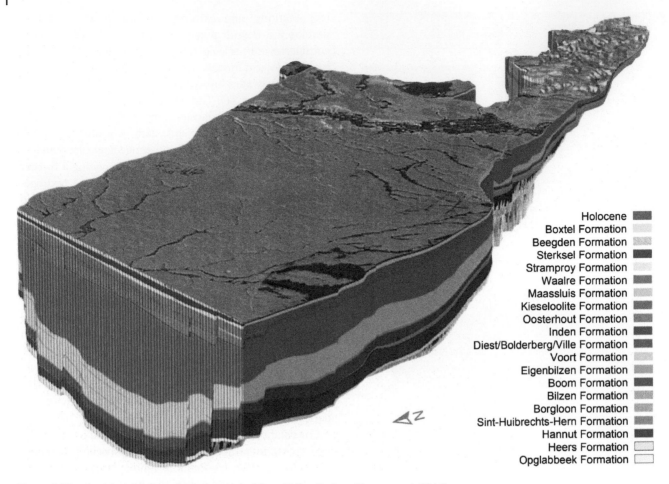

Holocene	■
Boxtel Formation	■
Beegden Formation	▨
Sterksel Formation	■
Stramproy Formation	□
Waalre Formation	▨
Maassluis Formation	▨
Kieseloolite Formation	▨
Oosterhout Formation	■
Inden Formation	■
Diest/Bolderberg/Ville Formation	■
Voort Formation	▨
Eigenbilzen Formation	▨
Boom Formation	■
Bilzen Formation	▨
Borgloon Formation	▨
Sint-Huibrechts-Hern Formation	▨
Hannut Formation	■
Heers Formation	□
Opglabbeek Formation	□

Figure 8.21 Completed 3-D geological model of Roer Valley Graben (Vernes et al. 2016).

geological, hydrogeological, and geothermal data from Polish and German sources. Problems arose with some coordinate transformations and because of the different scales, ages, and contents of these sources (Görne and Krentz 2015). Seven geological cross-sections were developed to support the 3-D modeling; they were oriented northeast–southwest, approximately perpendicular to bedrock geological structures and the extensional tectonic elements of the region. The 3-D geological subsurface model was created using an implicit modeling approach (Chapter 12) as implemented by the SKUA-GOCAD software by Paradigm. Subsequently, numerical modeling methods were applied to the 3-D model to produce a series of maps, illustrating the estimated geothermal conditions at specified depths.

The TransGeoTherm project provides another example of the importance of establishing formal data management practices when geological projects cross administrative boundaries. A lack of interoperability among the input data and data structures caused some modeling challenges and inhomogeneity in the intermediate results. With help

of a 3-D database approach, the final data were stored consistently and made available for further projects. The derived information (3-D model and geothermal maps) is available through a public web portal which the Trans-GeoTherm project has committed to maintaining until at least December 2019.

8.8.3 The GeoMol Project

The GeoMol project ran for almost three years (September 2012 – June 2015) as a partnership of 14 institutions from six member-states of the Alpine Space Cooperation Area. Diepolder and GeoMol Team (2015) provides a detailed description of the GeoMol Project. It evaluated the subsurface geology of the Alpine Foreland Basins located to the north and south the Alpine mountain range (Figure 8.23). The northern Foreland Basin, the Molasse Basin, extends over 1000 km in a southwest to northeast arc that encompasses French, Swiss, German, Austrian, and Czech territories. The southern Foreland Basin, the Po Basin, extends for about 500 km in northern Italy, from Turin to almost Trieste. The actual borders of the GeoMol

(a) Hydrogeological model transboundary mismatch

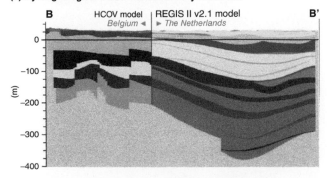

(b) H3O new transboundary hydrogeological model

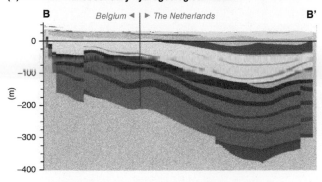

(c) H3O interfaces overlaid on old hydrogeological model

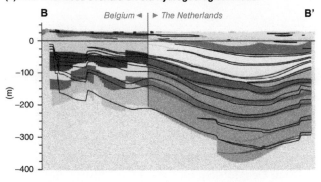

Figure 8.22 Hydrogeological cross-section B-B′ of the Roer Valley Graben located on Figure 8.16. (a) Original national hydrogeological models HCOV (Flanders) and REGIS II (the Netherlands); (b) the completed new combined 3-D hydrogeological model; (c) interfaces of the new hydrogeological model overlaid on the existing national HCOV and REGIS II models. (Deckers et al. 2014).

Project area were determined by both the geological situation and political administrative constraints; the GeoMol Project does not fully cover the Alpine Foreland Basins.

Within the Alpine Foreland Basins, the subsurface, a finite spatial resource, faces increasing competition from many different users. The principal challenge for spatial planning and licensing in this region is the sustainable management of the subsurface, including consideration of increased renewable energy production and storage and

the continuing need for water supplies, raw materials, and waste disposal. Because sustainable management of the subsurface requires 3-D definition of subsurface conditions, the primary objective of the GeoMol Project was the creation of suitable 3-D geological models based on seismic data, scattered, and clustered borehole evidence, and conceptual models of basin evolution.

GeoMol developed a 3-D geological framework model to provide a synoptic reference model for all existing or future detailed models. It covers 55 000 km² of the Foreland Molasse of the Rhône-Alpes region of France, the entire Swiss Molasse Basin, and the Molasse share of Baden-Württemberg and Bavaria in Germany and Upper and Lower Austria. In addition, five areas were selected for more detailed 3-D modeling to assess geopotentials focused on current issues in the respective region. They served as use cases for the application and validation of the GeoMol approach in different geological settings. One of these areas, the Brescia-Mantova-Mirandola area, was located in the southern Foreland Basin; it covers about 5700 km² of the Po Basin. It is part of the most densely populated area in Italy and is threatened by various natural hazards (Diepolder and GeoMol Team 2013).

Because many of the existing national datasets required to produce basin-wide 3-D models contain confidential data, access restrictions require that all model building and further product preparation may only be implemented by the legally mandated regional or national GSO. For a transnational project, geared towards providing harmonized information, this imposes particular requirements on the database content and design. The dissemination of information involving confidential data is subject to different statutory provisions in the partner states, but generally has to balance the conflict between the data protection and freedom of information policies. The GeoMol Project faced an additional challenge. Historically different nomenclatures and subdivisions describing the stratigraphic subdivision of the Alpine foreland basins had evolved within national GSOs. Thus, basin-wide transboundary studies required a semantic harmonization and realignment of stratigraphic definitions before correlations using a uniform lithostratigraphic column could be achieved.

In order to comply with both national data policies and the EU's requirement for seamless transnational information, the GeoMol project implemented a sophisticated approach to 3-D data interoperability and web visualization; the selected software was GST (Geo Sciences in Space and Time) originally developed at Bergakademie Frieberg (Gabriel et al. 2015). Diepolder (2011) and Gabriel et al. (2015) describe the major technical characteristics and principal features of GST; Le et al. (2013) provide

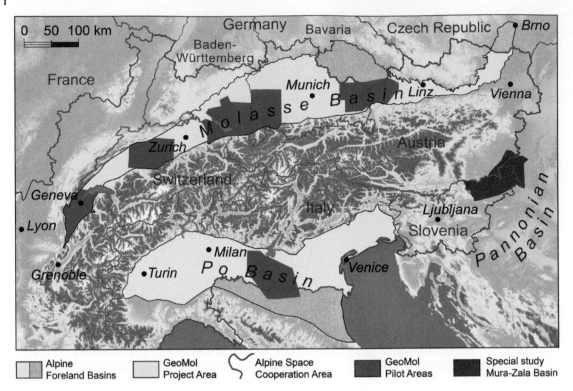

Figure 8.23 GeoMol Project area, including five pilot areas and one Special Study Area in Slovenia. (Diepolder and GeoMol Team 2015).

additional details of the fundamental object-relational data model. GST accesses, visualizes, and manipulates geo-objects using open standards. Common access points at each GSO provide access to locally-stored data sets. This distributed-organized system allows individual GSOs to keep all data locally and to share only the cleared portions of the data, thus adhering to national geo-data access regulations. The models contain all baseline data and are kept confidential at the respective legally mandated GSOs which are responsible for adding any relevant new scientific findings or interpretations, improving existing data holdings, and maintaining the models. Access and data retrieval is only granted on request to those institutions entitled to such data, subject to confidentiality agreements.

As the GeoMol Project approached its conclusion in 2015, a public domain portal was developed to raise public awareness of the subsurface conditions. A GeoMol browser-analyst called *3D-Explorer* was provided to support web-based visualization of 3-D models. Two different role-based access modes were provided; approved organizations were allowed password-protected access to the more detailed 3-D models of the GeoMol pilot areas, while the public-domain portal could only access the 3-D GeoMol basin-wide geological framework model. Both access modes are subject to the users' acceptance of *terms of use* stating the limitations of the data, especially

with respect to its maximum resolution. Chapter 10 of Diepolder and GeoMol Team (2015) provides a comprehensive description of these methods for disseminating the GeoMol project results through an interconnected collaborative environment.

By late 2018, about three and-a-half years after the GeoMol project was concluded, advances in software systems made the Oracle database that supported the original GeoMol Explorer and model distribution system outdated. However, since the GST software (Diepolder 2011; Gabriel et al. 2015) was already a quasi-standard for model storage and exchange among the GeoMol partners, the 3-D browser-analyst based on GSTWeb technology at the Bavarian State GSO was adopted for exploring the GeoMol open-source models (Diepolder et al. 2019). GST software also allows the frictionless exchange with other commonly used model building software packages such as MOVE or Petrel. The very powerful GST browser may be accessed at https://www.3dportal.lfu.bayern.de/webgui/gui2.php. The browser allows the user to rotate the model and customize the display with exploded views, selected surfaces or faults, slices through the model, generation of arbitrary cross-sections and virtual drill holes, and draping geospatial information obtained from other web sources (Figure 8.24). With the next release of GST, the browser will also be capable of depicting gridded volumes.

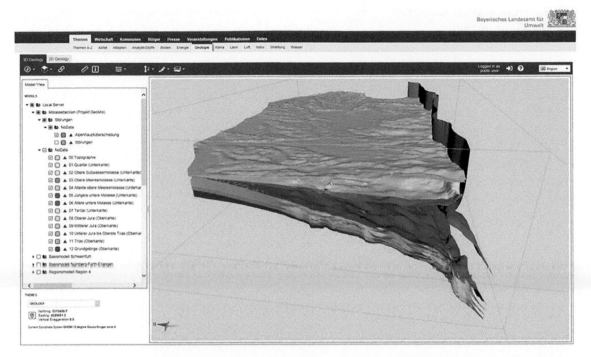

Figure 8.24 A model display produced by the new 3-D browser portraying the eastern slice of the Bavarian Molasse Basin of the GeoMol framework model. Fault planes (except for the Alpine orogenic front wedge in dark green) and the topographic overlay are disabled for clarity. (Diepolder et al. 2019).

Further developments are expected to influence the future distribution of GeoMol models. In January 2017, 45 national and regional European GSOs established a new collaborative research program called GeoERA (http://geoera.eu) to further promote the optimal use and management of the subsurface (GeoERA 2017). Development of a common geoscience information platform capable of integrating up-to-date data, interpretations, and models from multiple distributed sources is one of the GeoERA research tasks (http://geoera.eu/themes/information-platform). In the future, once it is fully operational, this GeoERA Information Platform may provide additional support for the distribution and display of GeoMol models.

Acknowledgments

Section 8.8 discusses data management challenges and solutions undertaken by European geological survey organizations. This information has been provided with the kind assistance of several individuals. Ronald W. Vernes with TNO – Netherlands Geological Survey, Jef Deckers with VITO – Flemish Institute for Technological Research, Bernd Linder with the Geological Survey of North Rhine-Westphalia, and Vanessa M.A. Heyvaert with Royal Belgian Institute of Natural Sciences and the Geological Survey of Belgium provided the text and illustrations for the H3O Project (Section 8.1). Discussion of the Polish-German TransGeoTherm Project (Section 8.2) is based on contributions from Sascha Görne and Ottomar Krentz with the Department of Geology, Saxon State Office for Environment, Agriculture, and Geology, in Freiberg, Germany. Discussion of the GeoMol Project (Section 8.3) is based on contributions from Gerold W. Diepolder with the Bavarian Environment Agency – Geological Survey, Augsburg, Germany.

References

ALCTS (2000). *Committee on Cataloging: Description and Access, Task Force on Metadata, Final Report (CC:DA/TF/Meta data/5)*. Chicago, IL: Association for Library Collections & Technical Services. [Online: Available at: https://www.libraries.psu.edu/tas/jca/ccda/tf-meta6.htm] (Accessed October 2018).

Association of Geotechnical and Geoenvironmental Specialists (2017) *AGS4. Electronic Transfer of Geotechnical and Geoenvironmental Data. Edition 4.0.4 – February* 2017. Bromley, UK: Association of Geotechnical and Geoenvironmental Specialists. 150 pp. [Online: Available at: http://www.agsdataformat.com/datatransferv4/download.php] (Accessed August 2020).

British Geological Survey (2013). *Open Geoscience data models*. Keyworth, UK: British Geological Survey. [Online: Available at: https://www.bgs.ac.uk/services/dataModels/] (Accessed August 2020).

British Geological Survey (2013). *Open Earth Data Models Library*. Keyworth, UK: British Geological Survey. [Online: Available at: http://www.earthdatamodels.org/viewDesigns.html] (Accessed August 17, 2020).

British Geological Survey (2018a). *BGS Lexicon of Named Rock Units Webpage*. Keyworth, UK: British Geological Survey. [Online: Available at: www.bgs.ac.uk/Lexicon] (Accessed October 2018).

British Geological Survey (2018b). *BGS Rock Classification Scheme Webpage*. Keyworth, UK: British Geological Survey. [Online: Available at: www.bgs.ac.uk/bgsrcs] (Accessed October 2018).

Brodie, M.L. and Schmidt, J.W. (1982). Final report of the ANSI/X3/SPARC DBS-SG relational database task group. *Bulletin ACM SIGMOD* 12 (4): 1–62.

Chen, P. (1976). The entity-relationship model – toward a unified view of data. *ACM Transactions on Database Systems* 1 (1): 312–339. [Online: Available at: http://extras.springer.com/2002/978-3-642-59413-7/3/rom/pdf/Chen_hist.pdf] (Accessed October 2018).

Codd, E.F. (1970). A relational model of data for large shared data banks. *Communications of the ACM* 13 (6): 377–387. [Online: Available at: https://www.seas.upenn.edu/~zives/03f/cis550/codd.pdf] (Accessed October 2018).

Codd, E.F. (1971). Further Normalization of the Data Base Relational Model. In: *Data Base Systems. Courant Computer Science Symposium 6* (ed. R. Rustin), 33–64. Englewood Cliffs, N.J.: Prentice Hall.

Connolly, T.M. and Begg, C.E. (2014). *Database Systems: A Practical Approach to Design, Implementation and Management, 6e*. London, UK: Pearson. 1440 pp.

Date, C.J. (2004). *An Introduction to Database Systems, 8e*. London UK: Pearson. 1040 pp.

Deckers, J., Vernes, R.W., Dabekaussen, W. et al. (2014). *Geologisch en hydrogeologisch 3D model van het Cenozoïcum van de Roerdalslenk in Zuidoost-Nederland en Vlaanderen (H3O – Roerdalslenk)* [3D Geological and hydrogeological model of the Cenozoic Roer Valley Graben in southeastern Netherlands and Flanders (H3O – Roer Valley Graben)]. Utrecht, Netherlands: TNO, Geological Survey of the Netherlands / Mol, Belgium: VITO. 108 pp. [Online: Available at: https://www.dinoloket.nl/downloads-project-h3o-roerdalslenk] (Accessed August 17, 2020).

Diepolder, G.W. (2011). 3D modelling at the Bavarian State Geological Survey – examples for cooperation towards 3D standards. In: *Three-Dimensional Geological Mapping Workshop Extended Abstracts* (eds. H.A.J. Russell, R.C. Berg and L.H. Thorleifson), 17–21. Ottawa, ON: Geological Survey of Canada, Open File 6998. https://doi.org/10.4095/289610.

Diepolder, G.W. and GeoMol Team (2013). The project GeoMol: transnational 3D modelling and 3D geopotential assessment in the Alpine Foreland Basins. In: *Three-Dimensional Geological Mapping Workshop Extended Abstracts, Denver, Colorado, October 26, 2013, Minnesota Geological Survey Open File Report OFR-13-2* (eds. L.H. Thorleifson, R.C. Berg and H.A.J. Russell), 29–33. St Paul, MN: Minnesota Geological Survey. [Online: Available at: https://www.isgs.illinois.edu/sites/isgs/files/files/abstracts/Diepolder.pdf] (Accessed October 2018).

Diepolder, G.W. and GeoMol Team (2015). *GeoMol – Assessing Subsurface Potentials of the Alpine Foreland Basins for Sustainable Planning and Use of Natural Resources. Project Report*. Bavarian Environment Agency, Augsburg. 188 pp. [Online: Available at: http://www.geomol.eu/geomol/report/GeoMol_Report_web_reduced.pdf] (Accessed October 2018).

Diepolder, G.W., Pamer, R. and Großmann, J. (2019). Advancements in 3D geological modelling and geo-data integration at the Bavarian State Geological Survey. In: *2019 Synopsis of Current Three-Dimensional Geological Mapping and Modeling in Geological Survey Organizations* (eds. K.l. MacCormack, R.C. Berg, H. Kessler, et al.), 48–61. Edmonton, AL: AEG/AGS, Special Report 112. [Online: Available at: https://ags.aer.ca/publications/SPE_112.html] (Accessed August 17, 2020).

Foged, N., Auken, E., Christiansen, A. and Sørensen, K. (2013). Test-site calibration and validation of airborne and ground-based TEM systems. *Geophysics* 78: E95–E106. https://doi.org/10.1190/geo2012-0244.1.

Gabriel, P., Gietzel, J., Le, H.H. and Schaeben, H. (2015). A new storage approach for 3D modeling. In: *Proceedings of the 17th Annual Conference of the International Association for Mathematical Geosciences, Freiberg, September 5–13, 2015* (eds. H. Schaeben, R.T. Delgado, K.G. van den Boogaart and R. van den Boogaart), 3. Houston, TX: International Association for Mathematical Geosciences.

GeoERA (2017). *Establishing the European Geological Surveys Research Area to deliver a Geological Service for Europe (GeoERA)*. Utrecht, Netherlands: TNO Geological Survey of the Netherlands. 2 pp. [Online: Available at: http://geoera.eu/wp-content/uploads/2017/02/GeoERA_Factsheet_FINAL_20170116.pdf] (Accessed March 6, 2019).

Görne, S. and Krentz, O. (2015). Cross border 3D modelling – challenges and results of a transnational geothermal project. In: *Proceedings of the 17th Annual Conference of the International Association for*

Mathematical Geosciences, Freiberg, September 5–13, 2015 (eds. H. Schaeben, R.T. Delgado, K.G. van den Boogaart and R. van den Boogaart), 883–885. Houston, TX: International Association for Mathematical Geosciences.

Kent, W. (1983). A simple guide to five normal forms in relational database theory. *Communications of the ACM* 26: 120–125. https://doi.org/10.1145/358024.358054.

Kessler, H. Lark, R.M., Monaghan, A., Kingdon, A. et al. (2016). *Benchmarking of 3D Geological Modelling Systems in BGS*. Keyworth, UK: British Geological Survey, Internal Report IR/16/033. 75 pp.

Kimball, R. and Caserta, J. (2004). *The Data Warehouse ETL Toolkit: Practical Techniques for Extracting, Cleaning, Conforming, and Delivering Data*. Indianapolis, IN: Wiley. 491 pp.

Kimball, R., Ross, M., Thornthwaite, W. et al. (2008). *The Data Warehouse Lifecycle Toolkit, 2e*. New York, NY: Wiley. 672 pp.

Kingdon, A., Nayembil, M.L., Richardson, A.E. and Smith, A.G. (2013). *PropBase QueryLayer: A Single Portal to UK Physical Property Databases*. European Geosciences Union General Assembly 2013, Vienna, Austria, 7–12 April 2013. 3p. [Online: Available at: http://nora.nerc.ac.uk/id/eprint/500933/] (Accessed August 17, 2020).

Kingdon, A., Nayembil, M.L., Richardson, A.E. and Smith, A.G. (2016). A geodata warehouse: using denormalisation techniques as a tool for delivering spatially enabled integrated geological information to geologists. *Computers & Geosciences* 96: 87–97. https://doi.org/10.1016/j.cageo.2016.07.016.

Le, H.H., Gabriel, P., Gietzel, J. and Schaeben, H. (2013). An object-relational spatio-temporal geoscience data model. *Computers & Geosciences* 57: 104–115. https://doi.org/10.1016/j.cageo.2013.04.014.

Mathers, S.J. (2011). 3D Geological Mapping (Modeling) in Geological Survey Organisations and the new BGS initiative to build a National Geological Model of the UK. In: *Three-dimensional Geological Mapping Workshop Extended Abstracts* (eds. H. Russell, R. Berg and H.L. Thorleifsen), 45–48. Ottawa, ON: Geological Survey of Canada, Open File Report 6998. https://doi.org/10.4095/289610.

Mathers, S.J., Kessler, H. and Napier, B. (2011). British Geological Survey: A Nationwide Commitment to 3-D Geological Modelling. In: *Synopsis of Current Three-dimensional Geological Mapping and Modeling in Geological Survey Organisations* (eds. R.C. Berg, S.J. Mathers, H. Kessler and D.A. Keefer), 25–30. Champaign, IL: Illinois State Geological Survey.

Mathers, S.J., Terrington, R.L., Waters, C.N. and Leslie, A.G. (2014). GB3D – a framework for the bedrock geology of Great Britain. *Geoscience Data Journal* 1: 30–42. https://doi.org/10.1002/gdj3.9.

Møller, I., Søndergaard, V.H. and Jørgensen, F. (2009a). Geophysical methods and data administration in Danish groundwater mapping. *Geological Survey of Denmark and Greenland Bulletin* 17: 4–44. https://doi.org/10.34194/geusb.v17.5010.

Møller, I., Søndergaard, V.H., Jørgensen, F. et al. (2009b). Integrated management and utilization of hydrogeophysical data on a national scale. *Near Surface Geophysics* 7: 647–659. https://doi.org/10.3997/1873-0604.2009031.

Nayembil, M.L. and Baker, G. (2015). The role of data modelling in a modern geological survey. In: *Proceedings of the 17th Annual Conference of the International Association for Mathematical Geosciences, Freiberg, September 5-13, 2015* (eds. H. Schaeben, R.T. Delgado, K.G. van den Boogaart and R. van den Boogaart), 8. Houston, TX: International Association for Mathematical Geosciences.

Nayembil, M.L., Richardson, A.E., Smith, A.G. and Burden, S. (2014). *PropBase "Warehouse" Architecture* (Unpublished discussion memo). Keyworth, UK: British Geological Survey. 7 pp. [Online: Available at: http://nora.nerc.ac.uk/id/eprint/509988/] (Accessed August 17, 2020).

Oracle Corp. (2018). *A Relational Database Overview*. Santa Clara, CA: Oracle Corporation. [Online: Available at: https://docs.oracle.com/javase/tutorial/jdbc/overview/database.html] (Accessed October 2018).

Riley, J. (2017). *Understanding Metadata: What is Metadata, and What is it For?* Baltimore, MD: National Information Standards Organization (NISO). 49 pp. [Online: Available at: https://groups.niso.org/apps/group_public/download.php/17446/Understanding%20Metadata.pdf] (Accessed October 2018).

Shaw, R.P. (2006). *PropBase Scoping Study*. Keyworth, UK: British Geological Survey, Internal Report IR/06/088. 35 pp. [Online: Available at: http://nora.nerc.ac.uk/id/eprint/7308/] (Accessed August 17, 2020).

Shin, S.K. and Sanders, G.L. (2006). Denormalization strategies for data retrieval from data warehouses. *Decision Support Systems* 42 (1): 267–282. https://doi.org/10.1016/j.dss.2004.12.004.

Simison, G.C. and Witt, G.C. (2005). *Data Modeling Essentials, 3e*. Burlington, MA: Morgan Kaufmann Publishers. 560 pp.

Thomsen, R., Søndergaard, V.H. and Sørensen, K.I. (2004). Hydrogeological mapping as a basis for establishing site-specific groundwater protection zones in Denmark. *Hydrogeology Journal* 12: 550–562. https://doi.org/10.1007/s10040-004-0345-1.

Van der Meulen, M.J. (2016). Opportunities for International/EU cooperation. *3rd European Meeting on 3D Geological Modelling,Wiesbaden, Germany, 16–17 June,*

2016. European 3D Geological Modelling Community. 15 pp (presentation slides). [Online: Available at: http://3dgeology.org/wiesbaden.html] (Accessed August 17, 2020).

Vernes, R.W., Deckers, J. et al. (2018). *Geologisch en hydrogeologisch 3D model van het Cenozoïcum van de Belgisch-Nederlandse grensstreek van Midden-Brabant / De Kempen (H3O – De Kempen) [3D Geological and Hydrogeological Model of the Cenozoic of the Dutch-Belgian border region of Midden-Brabant /the Campine area (H3O – The Campine)]*. Utrecht, Netherlands: TNO, Geological Survey of the Netherlands / Brussels, Belgium: Geological Survey of Belgium / Mol, Belgium: VITO. 285 pp. [Online: Available at: https://www.dinoloket.nl/sites/default/files/2018-07/R11261_H3O_De_Kempen_Final_sec.pdf] (Accessed August 17, 2020).

Vernes, R.W., Deckers, J., Verhaert, G. and Heskes, E. (2016). *The H3O-Project: Closing the Gap Between Our Nationwide (Hydro)Geological Models. 3rd European Meeting on 3D Geological Modelling,Wiesbaden, Germany, 16–17 June, 2016*. European 3D Geological Modelling Community. 17 pp (presentation slides). [Online: Available at: http://3dgeology.org/wiesbaden.html] (Accessed August 17, 2020).

Waters, C.N., Terrington, R.L., Cooper, M.R., Raine, R.J. and Thorpe, S. (2015). *The Construction of a Bedrock Geology Model for the UK: UK3D_v2015*. Keyworth, UK: British Geological Survey, Report OR/15/069. 22 pp. [Online: Available at: http://nora.nerc.ac.uk/id/eprint/512904/1/UK3D Metadata Report_Submission copy.pdf] (Accessed February 2018).

Watson, C., Baker, G. and Nayembil, M.L. (2014). *Open Geoscience Data Models: End of Project Report*. Keyworth, UK: British Geological Survey, Report OR/14/061. 30 pp. [Online: Available at: http://nora.nerc.ac.uk/id/eprint/508791] (Accessed October 2018).

West, M. (2011). *Developing High Quality Data Models*. Burlington, MA: Morgan Kaufmann Publishers. 408 pp. https://doi.org/10.1016/C2009-0-30508-5.

West M. and Fowler, J. (1996). *Developing High Quality Data Models*. London, UK: The European Process Industries STEP Technical Liaison Executive (EPISTLE). 56 pp. [Online: Available at: https://sites.google.com/site/drmatthewwest/publications/princ03.pdf?attredirects=0] (Accessed October 2018).

Whitten, J.L., Bentley, L.D. and Dittman, K.C. (2004). *Systems Analysis and Design Methods, 6e*. Boston: McGraw-Hill. 724 pp.

9

Model Creation Using Stacked Surfaces

Jason F. Thomason and Donald A. Keefer

Illinois State Geological Survey, University of Illinois at Urbana-Champaign, Champaign, IL 61820, USA

9.1 Introduction

Reliable three-dimensional (3-D) conceptualizations of subsurface geology are often constructed by interpreting two-dimensional (2-D) surfaces that represent geologic boundaries or features. These boundaries reflect subsurface stratigraphic relationships between rock and/or sediment geologic units associated with lithologic changes, sedimentological changes or unconformities. They may also represent multi-scale structural or tectonic features such as faults or folds, or they may reflect more transient boundaries such as a water table elevation or a contaminant plume boundary.

The application of stacked-surface modeling is generally efficient, effective, and versatile. However, it has limitations, even when the most comprehensive software and interpolation capabilities are utilized. These limitations are related to the modeling scale, geologic complexity, and density of available data. For example, stacked-surface modeling may not be able to efficiently model the bounding surfaces of thin, discontinuous geologic units without access to large numbers of optimally distributed observations. Furthermore, while some software packages include algorithms that can be manipulated to simulate faults, standard interpolation methods cannot typically accommodate faulted surfaces or complex geologic structures such as intersecting faults or overturned fold systems.

Despite these limitations, the stacked-surface approach has become widely integrated into 3-D geological modeling strategies, especially for modeling and evaluating relatively undeformed, moderately complex, sedimentary geologic environments. Many areas with high population densities are underlain by such geology and so are amenable to stacked-surface geological modeling in support of economically important analyses for groundwater resources, geothermal potential, waste disposal, and subsurface development. The stacked-surface approach will likely become even more effective with continued computing advancements.

This chapter highlights the concepts, methods, and tools used to create 3-D stacked-surface geological framework models and demonstrates effective modeling strategies and applications using stacked surfaces.

9.2 Rationale for Using Stacked Surfaces

The stacked-surface approach has been widely used for many years to conceptualize geology. Ever since geologists began to conceptualize relationships between subsurface rocks and structures, they have defined bounding surfaces for these features by interpolating and contouring between known data locations. These relationships often can be represented effectively and efficiently as sets of 2-D surfaces. For example, the individual 2-D surfaces can be visualized as structural contour maps of stratigraphic surfaces, isopach maps of deposit thicknesses, contoured maps of unconformable surfaces such as top of bedrock, or contours of the water-table or potentiometric surface.

Sedimentary rock strata have boundaries and lateral extents that are functions of their sedimentological, structural, and erosional history. While stratigraphic boundaries in many sedimentary environments are continuous and mappable over broad regional areas, deposits of glacial or fluvial origin have much more variable and discontinuous stratigraphic boundaries. Mapping spatial relationships between sedimentary strata often reveals the presence of folds or faults. Faults, which disrupt sedimentary boundaries, are typically conceptualized as planar features. Despite difficulties in areas of geometric complexity, representing these types of geologic features as a set of 2-D surfaces is often efficient and effective. The process aligns with basic stratigraphic principles of geology; these

Applied Multidimensional Geological Modeling: Informing Sustainable Human Interactions with the Shallow Subsurface, First Edition.
Edited by Alan Keith Turner, Holger Kessler, and Michiel J. van der Meulen.

methods and maps have proved to be part of a robust approach to conceptualizing and representing subsurface geology.

Geographic information systems (GIS) or computer-aided design (CAD) systems incorporate procedures that can evaluate observed geological contacts to construct 2-D surfaces. Widely available GIS software products, such as ArcGIS or Surfer, use a regular gridded matrix to represent a 2-D surface, while CAD software products typically prefer a Triangulated Irregular Network (TIN) to represent a surface. Section 9.3 reviews commonly employed software packages and interpolation methods. Chapter 13 discusses the details of various grid and TIN implementations in 2-D and 3-D.

Geological survey organizations (GSOs) have utilized GIS and CAD software since the late 1980s when computer technology first allowed researchers to rapidly generate digital surfaces that were geomorphically realistic, spatially referenced, and easily accessible (Hamilton and Jones 1992). Thus, many GSOs, when considering how they could most easily adopt 3-D geological modeling, were attracted to the stacked-surface modeling approach. They had access to appropriate software and experienced staff familiar with its operation. They also had considerable legacy digital data sources defining important subsurface geological bounding surfaces.

In addition to developing 3-D geometric depictions of geological frameworks, stacked-surface modeling has widespread natural resource and engineering applications. For example, when stratigraphic relationships in aquifer systems are modeled as a stacked series of regularly-gridded surfaces, they have an inherent similarity to numerical modeling methods. These models are readily incorporated into groundwater-flow analyses that rely on grid-based hydrogeologic boundaries. Similarly, geotechnical engineers use grid-based geological bounding surfaces to constrain stability assessments of rock and earth materials surrounding excavations, to establish the stability of natural and man-made slopes, and for geohazard investigations (Chapter 14).

9.3 Software Functionality to Support Stacked-Surface Modeling

The stacked-surface approach to 3-D modeling typically utilizes regular 2-D grids to represent individual surfaces. This regular and standard file structure allows widely-used GIS software products to support the stacked-surface approach to geologic model creation. For a given workflow or project, selection of a software application should be based on a review of the software features and costs, together with staffing resources, project goals, and other considerations (Chapter 6).

Table 9.1 lists software products that have been found useful for the stacked-surface approach to geologic modeling. Key software functions that help enable the application of a stacked-surface modeling approach include: (i) multiple interpolation options, (ii) grid math tools, (iii) grid correction/modification options, (iv) 3-D visualization, (v) error analysis, (vi) integration of diverse interpretation sources, and (vii) use of specialist third-party applications. While some of the products listed in Table 9.1 offer only limited functionality, they offer cost-effective support for selected tasks in the modeling workflow, and thus satisfy the needs of modelers.

9.3.1 Selection of an Interpolation Algorithm

While the development of suitable 2-D surfaces depends upon reliable and accurate observational data, the selection of an interpolation algorithm and its parameters greatly influences the geometry of the resulting surfaces. Table 9.2 summarizes the characteristics of algorithms commonly used to interpolate elevation or thickness values between observed data locations to create the stacked surfaces. Many GIS software products support a broad range of interpolation algorithms with variable parameter-control options and allow the user to specify combinations of these basic interpolation methods. Chapter 13 discusses the details of alternative interpolation approaches.

Most interpolation algorithms evaluate only scattered observation points, but a few of them allow interpolation in the presence of faults, ridge lines, or sharp valleys; features collectively known as break lines. The ANUDEM (Australian National University) algorithm is one of the few algorithms that can also accommodate contour lines as an input data source (Hutchinson 1989, 2011). This algorithm has many options for controlling surface smoothness and the degree to which the surfaces honor data values. Furthermore, an option in ANUDEM allows for the use of a drainage enforcement algorithm to modify the surface by filling many isolated depressions and creating drainable surface models around inferred or observed stream and lake locations. ESRI has licensed a version of ANUDEM and implemented it in the ArcGIS Topo-to-Raster interpolation algorithm (ESRI 2018).

9.3.2 Grid Math Tools

The stratigraphic consistency of stacked-surface models is maintained and constrained by the hierarchical relationships between stacked surfaces. The identification and

Table 9.1 Software products that have been found useful for creating geologic models using a stacked-surface approach.

Software	Developer	URL
3D Borehole Tools	Illinois State Geological Survey	https://pubs.usgs.gov/of/2014/1167/pdf/ofr2014-1167_carrell-tools-and-techniques.pdf
3D GeoModeller	Intrepid-BRGM	https://www.geomodeller.com/geo/
ArcGIS	ESRI	https://www.esri.com/en-us/arcgis/about-arcgis/overview
ArcHydro	AquaVeo	http://www.aquaveo.com/archydro-groundwater
Argus ONE	Argus Holdings Ltd	http://www.argusone.com
COMSOL	COMSOL	https://www.comsol.com
EarthVision	Dynamic Graphics Inc.	https://www.dgi.com/earthvision/evmain.html
Erdas Imagine	Hexagon	https://www.hexagongeospatial.com/products/power-portfolio/erdas-imagine
Fledermaus	IVS 3D	https://qps.nl/fledermaus/
GeoScene3D	IGIS	https://www.geoscene3d.com
IDL/ENVI	ITT Visual Information Solutions	https://www.ittvis.com
Isatis	Geovariances	https://www.geovariances.com/en/software/isatis-geostatistics-software
Model Viewer	USGS	https://www.usgs.gov/software/model-viewer-a-program-three-dimensional-visualization-ground-water-model-results
MODFLOW, GWT, SUTRA, PHAST, MODELMUSE,USGS	USGS	http://water.usgs.gov/software/lists/groundwater / http://en.wikipedia.org/wiki/MODFLOW
Move	Petroleum Experts Ltd	https://www.petex.com/products/move-suite/move/
Oasis montaj	Seequent	https://www.seequent.com/products-solutions/geosoft-oasis-montaj/
PolyWorks	InnovMetric Software Inc	https://www.innovmetric.com
Quick Terrain Modeler	Applied Imagery	https://www.appliedimagery.com
Rockworks	RockWare	https://www.rockware.com/product/rockworks/
SGeMS	Stanford University	http://sgems.sourceforge.net/?q=node/20
Voxler, Surfer	Golden Software	https://www.goldensoftware.com/products

assessment of intersecting surfaces is an important step that ensures individual surfaces appropriately truncate against other surfaces or model-domain boundaries. Resolving undesired morphologies in sets of surfaces requires either a set of grid math tools or software with built-in options for enforcing specific geologic surface-property constraints.

Grid math tools include a wide range of functions to enforce or create mathematical relationships, either within individual interpolated grids or between two different grids. Functions critical to stacked-surface modeling are: (i) add or subtract specific values from a single grid; (ii) add or subtract one grid from another; (iii) implement a high-pass filter, where two grids are compared and the

highest value is output to a new grid; or (iv) implement a low-pass filter, where the lowest value between two grids is preserved.

High-pass or low-pass filters are commonly used to clip and remove portions of intersecting stacked surfaces. Depending on the intended relationship, portions of a selected surface can be removed if it is either greater than (low-pass) or less than (high-pass) a second, intersecting, surface (Figure 9.1a). High-pass filters preserve the highest of either surface and are used to simulate onlap or intrusive relationships between surfaces. In Figure 9.1b, a high-pass filter is applied to the top of Unit 2, with the top of Unit 1 being the intersecting surface; this truncates the top of

Table 9.2 Common interpolation methods.

Interpolation method	Description and characteristics
Splines	Splines typically provide the smoothest surfaces, with interpolation across the entire surface being fit to one or more polynomial functions.
	Spline-interpolated surfaces do not always fit exactly to the data points; instead they provide a smooth surface that is optimally fit to all the data. The closeness of fit is typically controlled by a parameter often referred to as the "tension" of the surface. A large tension value produces a smooth surface but does not pass though most data points; a small tension value more closely honors the data points, but surface shapes may fluctuate considerably.
Inverse distance weighted	Inverse-distance weighting algorithms typically honor the data values, but where the data are noisy or where surface features are not well sampled, these surfaces can oscillate unrealistically.
	The importance (weight) of neighboring values drop with the inverse of the distance, raised to some power; the most commonly used weighting exponent is 2, creating an inverse distance squared weighting function.
Natural neighbor	Natural neighbor algorithms use Thiessen polygons around data values and grid nodes to create weights based on the specific distance to the points closest to each grid node.
Kriging	Kriging is typically used in a fashion similar to inverse distance algorithms, but the distance weighting function is different. Kriging, inverse distance, and Natural Neighbor algorithms tend to fit data values exactly when they occur at grid node locations.
Triangulation (triangulated irregular networks, or TINs)	Triangulated Irregular Networks (TINs) are vector-based surfaces that are constructed by triangulating between point data to build a network of edge-connected triangles. The surface shape along any edge is assumed to vary linearly between data points. TINs always fit data values exactly.
	TINs are very sparse data structures. Some software applications assign triangular surface values to regular grids to simplify comparison and use with other gridded surfaces.

(a) Grid surfaces defining tops of Units 1 and 2 intersect

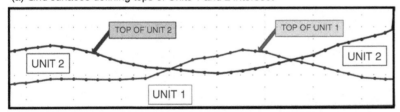

Figure 9.1 Cross-sections illustrating the use of Grid Math Tools to resolve surface intersections. (a) Two intersecting surfaces; (b) high-pass filter allows the top of Unit 2 to truncate into, or drape on top of, the top of Unit 1; (c) low-pass filter allows the top of Unit 2 to truncate the top of Unit 1.

(b) Unit 2 pinches out around Unit 1 (Unit 1 top surface has priority)

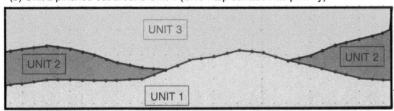

(c) Unit 3 base defined by Unit 2 top (Unit 2 top surface has priority)

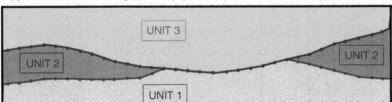

Unit 2 as if Units 2 and 3 were conformably deposited on top of Unit 1. Low-pass filters preserve the lowest value between two surfaces and are used to simulate erosional contacts. In Figure 9.1c, a low-pass filter is applied to the top of Unit 2 with the top of Unit 1 being intersected; this truncates the top of Unit 1 as if Unit 3 truncated it by erosion.

9.3.3 Grid Correction or Modification

Grid correction or modification is another critical process that must be supported by modeling software. The stacked-surface modeling process is iterative and involves the development of a set of independent 2-D surfaces. This process may result in some errors. Individual 2-D surfaces may improperly intersect other surfaces, contain anomalous high or low points, or violate geologic rules. These errors are often due to non-ideal distributions of data, a grid resolution that is too coarse, or mathematical limitations of the interpolation algorithm. To prevent propagation of errors through multiple iterations, the model surfaces should be corrected after every iteration to reflect the surface morphologies and spatial relationships envisioned by the geologist.

Local grid refinement is one form of grid modification. It is used to increase the accuracy of the surface definition by including "hard" and 'soft' data sources. Interpolated surfaces fit exactly to selected "control points" which are hard-data values such as borehole intersections, but the surfaces only approximate soft-data values such as seismic data picks. Kriging the hard data with a locally-varying mean value defined by soft data is commonly used to modify grids and minimize anomalous peaks and valleys in a surface. Alternatively, some interpolation algorithms, such as *Discrete Smooth Interpolation* (Mallet 1992), allow geologists to assign different weights to data of differing quality or accuracy. In other situations, review of individual data values, followed by their deletion or correction, is sometimes required to identify and correct anomalies in resulting grids.

9.3.4 Three-Dimensional Visualization

Three-dimensional visualization, while not required for stacked-surface modeling, is very useful for ensuring model consistency and accuracy. It is also very important for communication: in fact, to many users the visualization is the prime deliverable, rather than the entire model which they might not be able to visualize themselves. Depending on how the visualization software is implemented, any data representing either raw or interpreted values can be visualized in 3-D (Figure 9.2). This visualization allows for a much better understanding and interpretation of the system being modeled. Although improvements in hardware and software libraries have resulted in more

consistent results, different software products often have very different visualization and data integration capabilities. Thus, prior to software selection, the capabilities of various software products to visualize a variety of data types should be evaluated.

9.3.5 Error Analysis

As noted in Table 9.2, several interpolation methods produce surfaces that do not exactly pass through, or match, the observed values at some locations. Each interpolation algorithm, and associated parameter configurations, applies different solutions to generate a surface; this results in variations in how the surface relates to the observed values. The difference between the predicted value of the surface and the observed value at each location may be defined as an error. Because this degree-of-fit is an important consideration for a geologist evaluating a surface model, assessing error associated with an interpolated grid is an important part of model evaluation. Some software products provide menu options clearly labeled for assessing interpolation error, while others offer only a more generalized grid toolbox application that may require multi-step grid analyses to quantify and address error. Chapter 15 addresses model error estimation more fully.

9.3.6 Integration of Diverse Data Sources

All the widely-used GIS software products defined in Table 9.1 allow for interpolation of observed point-locations to create individual surfaces. These observations include outcrops, driller's logs, or core descriptions. Some of the applications allow a wider range of data to be used in the interpolation process, such as information from cross-sections or geophysical profiles, and they may support additional information sources such as contour lines, fault lines, break lines that define surface-feature boundaries, and map-unit contacts on the land or bedrock surface. If a software application does not allow use of all these types of data, tools or applications must be used to convert

Figure 9.2 Three-dimensional visualization of various subsurface data used in stacked-surface modeling (well records, test borings, 1-D and 2-D geophysical data). Visualization in Geovisionary. © 2018 University of Illinois Board of Trustees. Used by permission of the Illinois State Geological Survey.

data from one format to another to produce the desired integrated dataset. For example, digitized line-strings may be converted to individual points that represent geologic contacts at the land surface. Similarly, 2-D polylines can be converted to 3-D polylines to represent, for example, a geologic interpretation of a geophysical profile. Software capabilities should be evaluated and compared with the anticipated diversity of data sources prior to selecting software for stacked-surface modeling.

9.3.7 Specialist Third-Party Applications

Third-party applications can be used to supplement existing software features and capabilities. A third-party extension to Adobe Illustrator, called MAPublisher (Avenza), allows importing and rapid editing of contour maps generated from 2-D surface grids from any surface-modeling software application, such as ArcGIS or Surfer. As part of a 3-D modeling workflow, ESRI's ArcScene can be integrated into a 3-D workflow to visualize stacked surface grids, identify grid shape problems, create contour maps in ArcGIS, import the contours to Illustrator/MAPublisher, rapidly edit the contours, import the revised contours back to ArcGIS, and generate corrected grids from these contour lines (Brown 2013). While seemingly circuitous, this process is significantly more efficient than trying to use other methods to correct surface morphology problems on individual grids.

In addition to existing tools, most organizations will eventually create customized specialist software applications to meet specific needs. Several of such solutions are described in Section 9.6 of this chapter. However, these are vulnerable over longer timeframes to decisions by software vendors to substantially change the host software environment. For example, ESRI will be eliminating support for Visual Basic for Applications (VBA) applications in future ArcGIS software releases. To preserve the utility of these Illinois State Geological Survey (ISGS) applications within the established modeling workflow, they will have to be rewritten in a version of Python if they are to be used with new editions of ArcGIS software.

9.4 Defining the Stacked-Surface Model Framework

Stacked-surface geologic modeling is an inherently iterative process of data management, visualization, analysis, and interpretation. As with other scientific processes, the discovery of knowledge at any step may require revisiting former analyses and interpretations. The iterative process may be initiated by the inclusion of new data, new visualizations, or new interpolations of stacked surfaces, and may be limited or optimized by the versatility of available software. Ultimately, the forward progress of stacked-surface modeling is controlled by effectively gaining new knowledge through repeated testing of hypotheses represented by a conceptual geologic model. As each component of the stacked-surface model is constructed and revised, the model becomes increasingly robust.

The ability to control and check the lateral extent and vertical variability of the sequence of 2-D surfaces in 3-D space is fundamental to modeling stacked-surfaces. It ensures the model represents the subsurface with geologic consistency and accuracy. Each 2-D surface represents the top or bottom of a geologic unit, and these are constrained conceptually by boundaries such as land surface, bedrock topography, or other stratigraphic and/or structural relationships. Thus, a geologically coherent stacked-surface model must collectively integrate, as necessary, model-domain boundaries, geologic map units and associated contacts, relevant data points in 3-D space, cross-section interpretations, and previously-mapped surfaces. The model must also be consistent with stratigraphic relationships in 3-D space. The complexity of the iterative modeling process increases with a broadening spectrum of data sources and types. Fortunately, various strategies can be implemented to assure that these relationships and their geologic conceptualizations are effectively managed in stacked-surface models.

9.4.1 Establishing Critical Model Boundaries

Creation of a stacked-surface geological model begins by defining the top, base, and lateral extent of the model domain. The land surface forms the top of most models; it is defined by existing digital topographic information which usually has a higher resolution than any subsurface horizons. The model base is often constrained at a constant subsurface elevation, which is defined by a gridded surface of appropriate uniform values. However, the model base can be defined by any surface with an appropriate extent, such as the top of a rock sequence not included in the model domain. The lateral model extents are established by assigning appropriate and uniform grid boundaries to all the stacked surfaces.

Most models also contain several lithostratigraphic boundaries that are critical in establishing an accurate representation of surficial and subsurface conditions. These boundaries typically include mapped geologic contacts at the land surface, the boundary between surficial unconsolidated sediments and bedrock, or an erosional marker such as a regionally extensive paleosol. The density of available observations often determines whether a

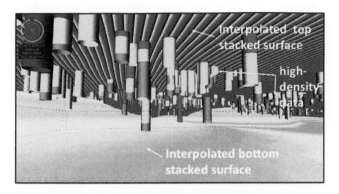

Figure 9.3 Three-dimensional view of subsurface data illustrating a case where the top stacked surface of a geologic unit is tightly constrained by dense data, and the bottom of the unit is poorly defined. Visualization in Geovisionary. © 2018 University of Illinois Board of Trustees. Used by permission of the Illinois State Geological Survey.

lithostratigraphic boundary is defined sufficiently to be useful as a stacked-surface datum – a surface that can be used to control other surfaces. For example, when a regionally-extensive geologic unit is used as a potential water supply, the available geologic information to model the unit, primarily from water-well records, abundantly indicates the top of the unit clearly and consistently. However, the bottom of the unit is often less constrained because most water wells do not penetrate the full thickness of the water-bearing unit. Thus, the top of the modeled stacked surface is more tightly constrained and may be used as a regional datum from which other stacked surfaces may be constrained (Figure 9.3). The 3-D model of the Oak Ridges Moraine (Section 9.6.2) located north of Toronto, Canada is an example of varied data density associated with defining aquifer boundary datums (Logan et al. 2005, 2006; Sharpe et al. 2007).

9.4.1.1 Defining the Land Surface

The land surface is often the most data-rich surface associated with a stacked-surface model. While older digital elevation models (DEMs) were developed from interpolations of topographic maps, the advent of LiDAR has resulted in high density (1–5 m resolution) DEM data for many 3-D modeling applications. Depending on the modeling scale and software capabilities, the full resolution of LiDAR-based DEMs may be greater than required, or even desired, to meet the objectives of a stacked-surface model. Thus, model development may require a coarser resampling of the LiDAR data to be effective for certain modeling applications. This resampling may reduce the fidelity of the topographic detail and must be accommodated to optimize coherency and accuracy of stacked-surface models.

9.4.1.2 Defining Geologic Contacts

Similar to land surface topographic data, geologic data at land surface is the highest density geologic information associated with many stacked-surface models. Conceptually, geologic contacts mapped at the land surface define, in 3-D space, the intersections of stacked-surfaces with the land surface. These contacts link high-density surficial data with the lower density subsurface data. In stacked-surface modeling, these geologic contacts are incorporated into the interpolation workflow by assigning 3-D spatial attributes (X, Y, and elevation) to nodes distributed along the contact lines. Furthermore, to maintain model integrity, the nodes must conform to the land-surface DEM resolution. This allows the mapped geologic contacts to be transformed into 3-D data and incorporated into interpolations of stacked surfaces that intersect land surface. Ultimately, this process incorporates the highest-density geologic data located near the land surface and optimizes the accuracy of the stacked-surface model.

9.4.1.3 Defining the Bedrock Surface

The transition between unconsolidated sediments and lithified rock (which is termed "rockhead" in some countries and "bedrock topography" in others) is another commonly modeled surface that is relatively well constrained in many stacked-surface models (Figure 9.4); this is often a very important surface for hydrological and geotechnical applications. This transition is commonly clearly identified in lithologic drilling logs, downhole geophysical logs, or 2-D geophysical profiles. Thus, the interpolated stacked surface defining bedrock topography often provides a reliable and important datum in 3-D geologic modeling. In some cases, stacked-surface models are developed by combining independently-developed models of the surficial deposits and the bedrock units.

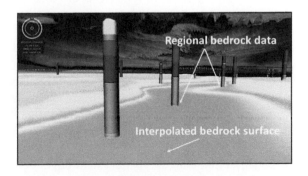

Figure 9.4 Three-dimensional view of a stacked surface representing bedrock topography. Visualization in Geovisionary. © 2018 University of Illinois Board of Trustees. Used by permission of the Illinois State Geological Survey.

9.4.1.4 Defining Subcrop Extents

Outcrops occur where geologic units are exposed at the present-day land surface. Outcrops provide important observational data used to create 3-D geological models. However, development of 3-D geological models also requires the accurate identification of the extent of all contacts between older geologic units and overlying younger geologic units. Termed the "subcrop" of a geologic unit, they occur at unconformities; an especially important unconformity is the buried bedrock surface below younger unconsolidated deposits. Where permeable sand and gravel deposits forming shallow aquifers are in contact with older water-bearing bedrock units, these subcrops provide hydraulic interconnections which may enhance the sustainable yields of the shallow aquifers, but also may expose the bedrock aquifers to contaminants from the surface. In 3-D modeling, subcrops must be mapped appropriately and accommodated in stratigraphic frameworks.

In stacked-surface modeling, subcrops are generally calculated iteratively by comparing the interpolated base of a geologic unit with surfaces defining underlying units. At locations where the interpolated base of a younger geologic unit has a lower elevation than the interpolated surface defining the top of an older underlying unit but is greater than the base of the older unit, the underlying unit must be partially eroded. This situation defines a subcrop area of the older underlying unit. At locations where the base of a younger geologic unit has an elevation that is less than the base of an underlying unit, the underlying unit has been completely eroded. Erosion of a subsurface unit may cause "holes" or gaps in the subcrop extent where the unit is absent. Other types of geologic settings (i.e. glacial deposits) may require more complicated interpretations of processes that result in holes and gaps in the subcrop extent. Nonetheless, these comparisons utilize the concepts of high-pass and low-pass filters discussed in Section 9.3.2 and illustrated in Figure 9.1.

Sometimes, modeling software limitations or modeling objectives may require all geologic units be represented across the entire model domain. Thus, eroded areas actually remain in the model as grid nodes with assigned elevations that are identical to the top of the adjacent underlying unit and the base of the adjacent overlying unit. This allows the unit to exist within eroded areas but with "zero thickness." Some modeling software products allow the eroded areas to be defined as "null" or "no-data" zones. The modeler may define subcrop extents by using polylines that limit the spatial extent of gridding algorithms. In any case, when fully-continuous stacked surfaces are required, the eroded portions of any subsurface geologic unit are surrounded by subcrop areas of the unit.

9.4.1.5 Defining Faults

Because faults compartmentalize geologic strata, 3-D stacked-surface models often must account for fault-networks. The correct interpolation of stratigraphic surfaces across faults depends on understanding the age relationships of the faults and geologic units. Faults disrupt existing strata. They may also create topographic features (fault scarps) that may affect the distribution and characteristics of subsequent geologic units. Strata that are younger than a fault may have different thicknesses and compositions on each side of a pre-existing fault. Thiele et al. (2016) employ spatial and temporal topology of stratigraphic surfaces and faults to evaluate these relationships. In stacked-surface modeling, fault networks should be defined, along with their intersections and terminations, before constructing stratigraphic surfaces (Caumon et al. 2009; Wu et al. 2015).

Modeling faults and fault systems requires careful manipulation of stacked surfaces. Direct observations of fault locations and orientations in the subsurface are rare and usually inadequate to accurately define 3-D fault networks. Thus, direct observations must be supplemented by additional data sources; geophysical surveys are a common supplemental information source (Wu et al. 2015). With sufficient data, faults can be modeled using stacked surfaces by integrating line or polygon boundaries that abruptly limit the extent of interpolated surfaces at fault locations. Alternatively, interpolated surfaces may be truncated abruptly with grid-math clipping boundaries that represent fault locations. If spatially-continuous grids are required, fault geometries may be represented by adding point or contour elevation data to define abrupt, linear changes in the continuous elevation of a stacked surface. If the latter method is used, multiple stacked surfaces may be modeled with similar elevation offsets along a given fault line. A more detailed discussion of modeling faults is available in Section 9.5.3 of this chapter.

9.4.2 Importance of Synthetic Data

Stacked-surface modeling relies on geologic knowledge and intuition. Direct observations provide accurate representations of geologic structures and relationships, but they are commonly clustered and spatially variable. Interpolated surfaces derived solely from direct observations often include a spectrum of anomalies, generalizations, or boundary extents that may be geologically unreasonable. The addition of *synthetic data*, defined as *"any data applicable to a given situation that are not obtained by direct measurement"* (McGraw-Hill 2003),

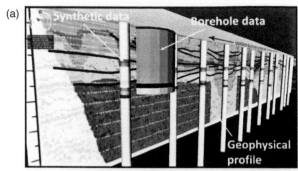

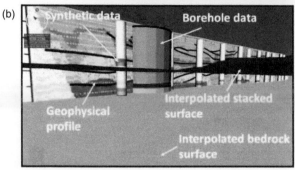

Figure 9.5 (a) Three-dimensional view of synthetic borehole data used to supplement 2-D geophysical-profile interpretations. (b) Stacked surfaces interpolated in part from synthetic data. Visualization in Geovisionary. © 2018 University of Illinois Board of Trustees. Used by permission of the Illinois State Geological Survey.

allows for better control of interpolation boundaries and elevation variability. The use of supplemental synthetic data permits accurate representations of geologically reasonable structures and stratigraphic relationships (Gunnink et al. 2013). Geophysical surveys, synthetic boreholes, and supplemental elevation data are principal sources of synthetic data used in stacked-surface modeling (Figure 9.5).

An accurate representation of the subcrop geometry of all geologic units or features is required to develop a coherent and consistent stacked-surface model. Synthetic data are often used to guide the definition of subcrop areas. Synthetic elevation points or contour data, when added strategically, at varying resolutions and locations, serve to control the extent and elevations of particular stacked-surface interpolations in 3-D space. Synthetic elevation data may be added in specific locations to improve the estimation of regional elevations of stacked surfaces in areas of sparse data, or they may be added to refine detailed geometric relationships between stacked surfaces. Section 9.6.1 discusses an application of synthetic elevation contours in stacked-surface modeling.

9.5 Building Stacked-Surface Geologic Framework Models

Modeling a sequence of gridded surfaces that represent the boundaries of individual geological units requires knowledge and appreciation of the geological evolution of the modeled volume. The chronostratigraphic order provides a fundamental control over the geologic-unit relationships of a stacked-surface model. These relationships follow the *law of superposition*, which states that in any undisturbed sequence of rocks deposited in layers, the youngest layer is on top and the oldest on bottom, each layer being younger than the one beneath it and older than the one above it. Most geological sequences contain gaps in the rock record where either deposition of additional material has not occurred or where erosion has removed portions of the rock record. In addition, structural deformations result in folds and faults or intrusions of younger materials into existing geological units. Conceptualizing these relationships associated with the modeled volume is just as important as the mechanics of mapping individual stacked surfaces.

9.5.1 Establishing the Appropriate Stacking Sequence

Establishing geologically-correct relationships between multiple intersecting stacked surfaces requires establishing an appropriate model-construction order. This order is defined by not only geologic knowledge but is typically the consequence of limitations imposed by software functionality. The model-construction order often differs from the true chronostratigraphic order. Faunt et al. (2010) illustrates the importance of employing model-construction order while accommodating chronostratigraphic relationships when building a stacked-surface 3-D model of Southern Nevada (Figure 9.6). This region included relatively young igneous intrusive bodies (and feeder bodies to volcanoes) and some thrust faults. Their geometries affected the order in which stacked surfaces were imported into the model; they also required the model to be developed as a series of sub-models. For example, the young intrusive bodies (Unit 6 in Figure 9.6a) were constructed in the model first, which contradicts the correct chronostratigraphic sequence (Unit 1 in Figure 9.6b). Thus, the youngest intrusion was assigned to the lowermost ("oldest") geologic unit in the model. In the thrust fault areas, similar units are found above and below thrust horizons (for example, Units 5a and 5b in Figure 9.6b); these were constructed within separate sub-models bounded

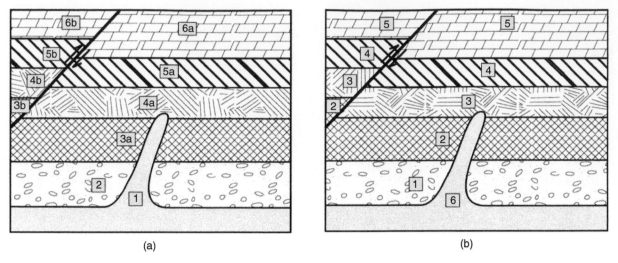

Figure 9.6 Diagrams showing: (a) model-construction order of geologic events and (b) chronostratigraphic order. Yellow numbered tags identify the appropriate unit event order. After Faunt et al. (2010). Source of figure: U.S. Geological Survey.

by thrust-fault surfaces. To complete the model, the sub-models were subsequently combined in the correct chronostratigraphic order.

9.5.2 Stack Adjustment to Represent Unconformities

Surfaces representing *unconformities* (e.g. erosional surfaces) also require adjustment to accommodate areas of non-deposition or erosion. In these cases, the modeler often uses high-pass and low-pass filters (defined in Section 9.3.2) to manipulate grid intersections to preferentially truncate areas of a given geologic unit. For example, a high-pass filter is commonly used to clip stacked surfaces of a modeled unit (e.g. Unit 3 in Figure 9.7) against an underlying datum unit (e.g. Unit 2 in Figure 9.7). Surface-truncation using high-pass filters often results in splitting a geological unit into several isolated components. Alternatively, when localized valleys or channels have been eroded within existing units, the stack-surface modeling process requires use of a low-pass filter (Figure 9.8). Correctly modeling the subsequent deposition of younger deposits within and beyond the channels may require the iterative use of both high-pass and low-pass filters to define the interrelationships among the top and bottom surfaces of these units and isolate components of geologic units as needed.

9.5.3 Stack Adjustment to Represent Faults

In the late 1980s and early 1990s, the developers of 2-D contouring software were aware of the importance of depicting faults to satisfy the requirements of petroleum exploration (Jones et al. 1986; Hamilton and Jones 1992). However, when modeling surfaces in 2-D, only vertical,

normal, or strike-slip faults could be evaluated. Simulating reverse faults or thrust faults required multiple elevation values for a surface at the same location, which was impossible to accommodate in 2-D. Thus, faults were modeled as either single-line traces for vertical faults or by fault polygons when inclined (dipping) faults were encountered. The fault polygons defined areas where the standard surface interpolation and contouring had to be iteratively evaluated. Neither approach adequately represented 3-D fault geometry. Faults were assumed to be *opaque barriers*, where elevation data points on one side of the fault were not used to support surface interpolation on the other side of the fault. This resulted in stratigraphic surfaces near faults being highly inaccurate, or geologically implausible, unless very dense observational data were available. These inaccuracies were compounded when faults intersected multiple stratigraphic horizons (Zoraster 1992).

Fortunately, with the advent of much more powerful graphics workstations around 1989, more sophisticated 3-D modeling approaches became feasible, and software developers introduced several new approaches to modeling faults (Hooper et al. 1992). Zoraster (1992) described three of these approaches: (i) rectangular-gridded fault surfaces, (ii) single-horizon methods, and (iii) multiple-surface triangulated models.

Clarke (1992) used rectangular-gridded fault surfaces to model the Wilmington oil field in Southern California which had relatively dense well control (1200 wells) and 30 significant faults. The alternative single-horizon approach, as defined by Zoraster (1992), initially interpolated individual stratigraphic horizons sequentially without considering faults. A biharmonic filter was used to interpolate the stratigraphic surfaces because it produced

(a) Exploded view of stacked units

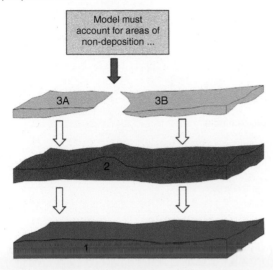

(b) Completed stacked-unit model

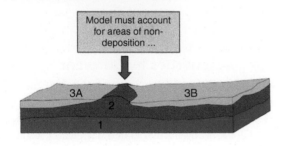

STRATIGRAPHY

3 Youngest

2

1 Oldest

Figure 9.7 Stacked-surface models have to account for areas of non-deposition.

smoothly varying surfaces. Faults were subsequently defined by center-line fault traces, and the stratigraphic surface elevations on each side of the fault were adjusted to account for the assumed fault offset. Inclined fault surfaces were further refined within fault polygons during a subsequent gridding step.

The multiple-surface triangulated models approach integrated a four-step procedure that eliminated most of the re-evaluation stages required by the other approaches. Fault surfaces and their displacements were modeled first. In a second step, elevation data points within each fault block were adjusted so the surfaces assumed pre-faulted positions. In the third step, these un-faulted stratigraphic surfaces were interpolated. In the fourth step, the surfaces within each fault block were adjusted to place them in their faulted positions.

Modern approaches to modeling faults in 3-D stacked-surface models continue to follow these basic approaches. However, more powerful and sophisticated algorithms and techniques have enhanced the efficiency and capability of fault-modeling procedures. More advanced interpolation methods applicable to modeling faulted surfaces defined by sparse data are discussed by de Kemp and Sprague (2003) and Sprague and de Kemp (2005). Frank et al. (2007) used implicit modeling and tetrahedral meshes to analyze complex 3-D geological interfaces defined by irregularly distributed and low-quality data.

Wu et al. (2015) provide a detailed review of recent approaches to fault-network modeling. Because fault locations and orientations are often poorly defined because of sparse direct observations, development of accurate 3-D fault network models requires the integration of multiple data sources to supplement the direct observations (Wu et al. 2015). Cherpeau et al. (2010) propose stochastic simulations of 3-D fault networks to produce alternative realistic fault models and address the uncertainties of fault connectivity, which is applicable at multiple scales. Laurent et al. (2013, 2015) evaluate the displacements and geometries of 3-D fault networks by modeling fault surfaces as curvilinear planes. These planes are parameterized to define displacement amplitudes and the attenuation of deformations in nearby surrounding rock units.

9.5.4 Quantitative Comparisons

Visualization of stacked-surface models helps to ensure modeled surfaces have the correct spatial and stratigraphic relationships, but quantifying those relationships is even more valuable. Variations in visualization quality among software packages make quantitative comparisons between stacked surfaces very beneficial for assessing any systematic and localized problems with accuracy of the stacked-surfaces. Basic algebraic transforms of and between stacked-surfaces are often incorporated into stacked-surface modeling workflows; the results can provide critically important evaluations of model accuracy and of the ability of the model to meet specified user requirements.

When configured as grids with a consistent origin and spacing, the spatial relationships between stacked surface models can be evaluated using basic grid math (Section 9.3.2). Grid-math tools can be used to correct intersecting portions of two or more surfaces, to calculate and check geologic unit thicknesses, and to define boundaries limiting the interpolated extent of geologic units. For example, applying systematic addition or subtraction operations to the elevations of related stacked surfaces can generate thickness maps of individual geologic units or

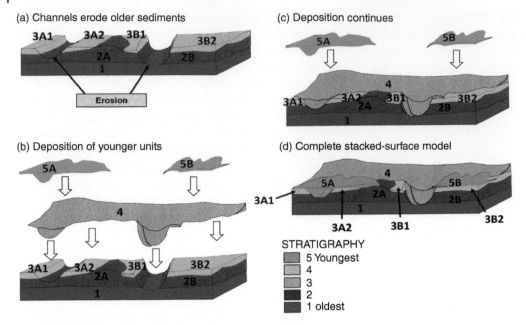

Figure 9.8 Stacked-surface models must represent erosion and deposition processes.

bulk thickness maps of multiple geologic units. Such operations also allow the comparison between two realizations or versions of the same surface to establish whether or not the last iteration is an improvement. Other mathematical operations can produce slope maps for specific surfaces or calculate geological properties, such as transmissivity, that are related to stacked-surface geometries.

Spatial anomalies within stacked surfaces can be addressed and corrected by using grid math procedures to modify surface grids. Interpolated stacked surfaces often contain anomalies; the distribution or thickness of a particular geologic unit may be consistently inaccurate in some areas due to sparse and variable data distributions. Even when synthetic data are used, geologic units may be interpolated beyond geologically-reasonable (depositional) unit boundaries. This can be corrected by grid math procedures removing all occurrences of a unit beyond a specified boundary.

Grid math procedures can be used to evaluate elevations of stacked surfaces to ensure a geologic unit maintains either a minimum or maximum thickness. Minimum unit thicknesses may be important when mapping geologic units for site-scale environmental or engineering implications. Grid math tools are commonly used to check quantitative relationships between stacked-surfaces at different scales. These types of secondary analyses and corrections are important parts of the iterative stacked-surface modeling process and are important tasks that ensure geologically consistent 3-D models.

9.6 Examples of 3-D Framework Modeling Approaches by Different Organizations

Stacked-surface modeling has been applied widely to address geologic problems associated with water resources. The following sections describe selected examples where stacked-surface modeling was effective at building geologic frameworks directly associated with aquifer delineation. These examples also indicate the versatility of stacked-surface models to address derivative relationships between aquifer geometries, hydraulic parameters, and their spatial variability.

9.6.1 Lake County, Illinois

In northeast Illinois, stacked-surface modeling of glacial sediments and sequences has been implemented widely to investigate the geologic framework that influences important groundwater resources. Population growth in several suburban communities in Lake County, located adjacent to Lake Michigan in the northern Chicago-Metropolitan area, have encouraged long-term planning for water resources. Questions arose concerning the viability of increasing use of groundwater supplies compared with the cost of providing additional infrastructure required by increased withdrawal, treatment, and distribution of water from Lake Michigan.

To aid decision-makers and planners, the ISGS built a robust stacked-surface geologic model of Lake County

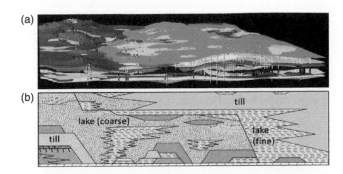

Figure 9.9 (a) Perspective view of preliminary stacked surfaces in the area of Lake County, Illinois (visualization in rockworks) with (b) associated schematic cross-section of the modeled geologic framework.

designed to evaluate the shallow sand and gravel aquifer resources (Brown 2013). These shallow aquifers have been a steadfast water-supply resource for homeowners and municipalities for decades, but geologic details of their distribution and character were poorly known. Stacked-surface modeling was used to map and characterize the complex 3-D spatial relationships of glacial deposits resulting from multiple glacial events that sculpted the modern landscape (Stumpf and Luman 2007). These deposits included a complex assemblage of proglacial lake sediments, fine-grained outwash sediments and glacial diamictons (Figure 9.9). Understanding the detailed distribution of these deposits has helped Lake County decision-makers plan and implement long-term management strategies that address the sustainability and economics of their water resources (Brown 2013).

Numerous surficial geologic mapping studies provided information to describe a robust geologic framework that is consistent with the regional geologic history of the Great Lakes region. As part of the extended Chicago metropolitan area, the geologic history of Lake County has also been studied for decades (Larsen 1973; Riggs et al. 1993; Leetaru 2003). Lake County is underlain by a number of glacigenic sequences, up to 100 m thick. They record the fluctuations of the ice margin of the Lake Michigan glacial lobe during the last glacial period, which occurred between 18 300 and 16 700 years ago (15 000 and 13 500 carbon-14 dating years BP). Along with surficial mapping, hundreds of geologic test borings have been conducted in the area, and there are tens of thousands of water-well records. These sources formed the basis for many detailed interpretations of the variations in regional bedrock geology and topography as well as the stratigraphic relationships between sediments representing multiple glacial episodes.

Despite these abundant data and surficial studies, the 3-D geology of glacial sediments in Lake County had not been developed. Thus, the ISGS was charged with resolving the

3-D geologic framework through stacked-surface modeling. The 3-D stacked-surface model in Lake County was completed by following an iterative process that included: (i) fieldwork and information discovery; (ii) data standardization, visualization, and interpretation; and (iii) implementation of software-modeling strategies to construct and edit stacked surfaces (Stumpf and Luman 2007; Brown 2013). The model included 17 stacked-surfaces that represented both lithostratigraphic and allostratigraphic geologic units. This modeling strategy was necessitated by the existence of numerous proglacial-depositional sequences that were mappable by lithology, location, and morphology.

9.6.1.1 Field Data

Between 2000 and 2013, the ISGS drilled over 200 boreholes by wire-line coring and direct-push methods to obtain additional baseline subsurface data required by the stacked-surface mapping. Downhole geophysical data (natural gamma-ray logs) were obtained from each wire-line borehole, and over 400 additional natural gamma-ray logs were collected at water-well sites in collaboration with water-well drillers and homeowners. More than 21 miles of 2-D geophysical data (ground-penetrating radar, seismic, and electrical resistivity) were collected across the county to supplement the boreholes (Figure 9.10).

9.6.1.2 Standardization, Visualization, and Interpretation

The lithologic logs from more than 20 000 water-well records and boreholes were standardized so that descriptive data could be systematically queried and efficiently visualized in the ESRI ArcScene 3-D viewing environment. In total, 180 000 unique descriptive terms were transformed into a standard lithologic terminology which included about 40 primary lithologies along with associated modifier fields defining color, texture, and consistency.

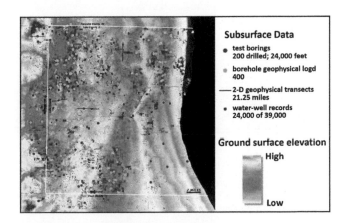

Figure 9.10 Distribution of subsurface data that was used for 3-D stacked-surface mapping in Lake County, Illinois.

After lithologic standardization was complete, water-well, borehole, and downhole geophysical data were visualized in ArcScene by utilizing the custom ISGS 3-D Borehole Tool developed by Carrell (2014) and described below. This supported the iterative selection and population of subsurface data attributes within the 3-D viewing environment. Furthermore, stacked surfaces were iteratively constructed using the 3-D Borehole Tool and the ANUDEM interpolation algorithm (see Section 9.3.1) implemented in the ArcGIS Topo-to-Raster application (ESRI 2018). This workflow efficiently performed surface development in discrete areas by using all available data contiguously in the 3-D environment. Iterative interpolation of stacked surfaces at many scales and subsequent testing of hypotheses based on those surfaces, assisted by effective visualization and rigorous mapping, defined a regionally extensive aquifer deposit in the western portion of Lake County.

9.6.1.3 Stacked-Surface Modeling and Editing

Regional stacked surfaces were constructed and finalized using a combination of software applications and grid math tools. For each geologic unit, subsurface point data were combined with polylines of unit boundaries and elevation contours and interpolated using Topo-to-Raster applications in ArcGIS. Hand-drawn structure contour lines were drawn using Adobe Illustrator with the MAPublisher plug-in, which applies world coordinates to the Adobe artboard, and ultimately creates georeferenced and attributed map-views (Figure 9.11). Thus, the hand-drawn contours were easily exported and integrated into ArcGIS interpolations. They were used primarily to constrain geologic features such as the shape of ice-contact facies,

the lateral boundaries of stream deposits, or the orientation and shape of depositional slopes. Model integrity was assured by controlling the locations and shapes of intersections between stacked surfaces. For example, synthetic-data contours were drawn to extend grid boundaries to strategically intersect a datum grid. Then grid math procedures removed the excess areas of stacked surfaces along deliberate, continuous intersections; the result was a smooth, internally-consistent, set of stacked-surface boundaries.

Ultimately, stacked-surface modeling produced a high-resolution, county-scale 3-D geologic model for Lake County (Figure 9.12). The integration of multiple software packages allowed effective visualization, interpretation, and iteration of a complex glacial geologic framework. By using several innovative stacked-surface modeling techniques, new aquifer geometries were delineated in detail, new interpretations of geologic history were developed and a new understanding of natural resource sustainability in Lake County was attained.

The ISGS staff also developed custom tools to supplement ArcGIS stacked-surface modeling workflows, which often required complicated and inefficient multistep processes. The ISGS developed a set of custom scripts to improve the efficiency of developing and editing cross-sections and 3-D boreholes in the 3-D viewing environment. Carrell (2014) describes two specialist applications: (i) the *Xacto Section Tool* for developing, editing, and managing cross-sections, and (ii) a set of *3D Borehole Tool*s for managing and analyzing lithologic and geophysical borehole logs (Table 9.1). Both of these tools were developed by employing the institutional expertise in the capabilities of ArcGIS.

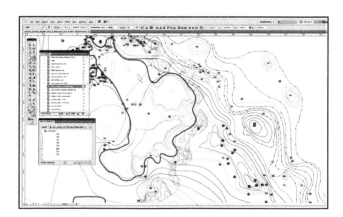

Figure 9.11 Adobe Illustrator art board showing attributed contour lines (multicolored thin lines), outcrop boundaries (thick dark blue line), and other point data used for editing and controlling the interpolation of stacked surfaces. MAPublisher windows are also shown. Modified from Brown (2013). Copyright © 2013 by the Regents of the University of Minnesota. Used by permission of the Minnesota Geological Survey.

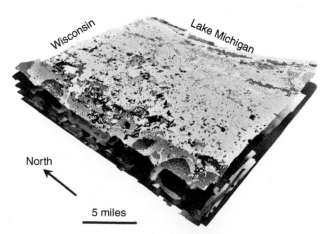

Figure 9.12 Perspective view of the final 3-D stacked-surface model of Lake County, Illinois. Visualization produced in ArcScene. © 2018 University of Illinois Board of Trustees. Used by permission of the Illinois State Geological Survey.

9.6.1.4 Xacto Section Tool

The ISGS used VBA to develop *Xacto Section* as a plug-in to ArcMap (Carrell 2014). The *Xacto section toolbar* allows a user to define a cross-section location within an ArcMap document (Figure 9.13). Then Xacto uses appropriate sets of polyline and point shapefiles to define 2-D cross-section profiles, which are digitally editable in ArcMap or exportable to Adobe Illustrator (with MAPublisher) for cartographic refinement. Completed cross-sections are also exportable as 3-D vector features for viewing and editing in ArcScene (Figure 9.14). Because these cross-sections are spatially referenced, the cross-section measurements remain correct when the ArcMap document is assigned a new desired map scale. Furthermore, when used in combination with the *MAPublisher plug-in for Illustrator*, *Xacto Section* supports the conversion of legacy cross-section vector graphic files into 3-D georeferenced shapefiles (Figure 9.15). This allows the restoration of legacy data sets into valuable quantitative geologic data.

9.6.1.5 3D Borehole Tools

The ISGS also developed *3D Borehole Tools,* which is a collection of 14 applications that enhance the modeling capabilities of ArcScene. The *3D Borehole Tools* recognized that ArcScene provides an attractive platform to support stacked-surface model creation in the 3-D environment. Its interactive 3-D environment readily supports visualization of the geologic subsurface; it has relatively intuitive 3-D navigation tools; it employs the existing ArcGIS data storage formats and workflows; and it includes multiple

(a) Xacto program 2-D output

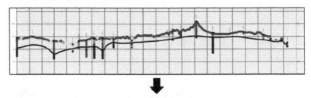

(b) 2-D Cross-section digitally edited in ArcMAP

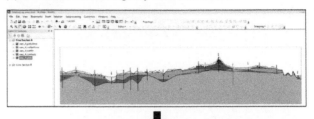

(c) Converted to ArcScene 3-D polygons and lines

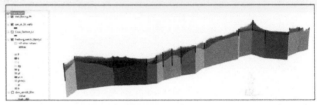

Figure 9.14 A 2-D cross-section profile could be edited in ArcMap and converted into a 3-D shapefile for display by ArcScene. (Carrell 2014).

options for customizing and automating tasks. In addition, an active-user community regularly provides documentation, assistance, and information about customization techniques.

Figure 9.13 The red arrow identifies the Xacto Section toolbar within an example ArcMap document. The blue line represents the location of a cross-section developed with Xacto. (Carrell 2014).

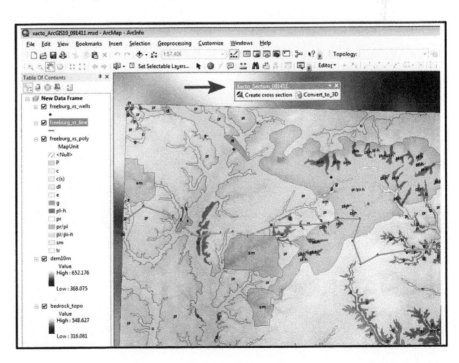

Legacy cross-section graphics   3-D cross-section in ArcMap

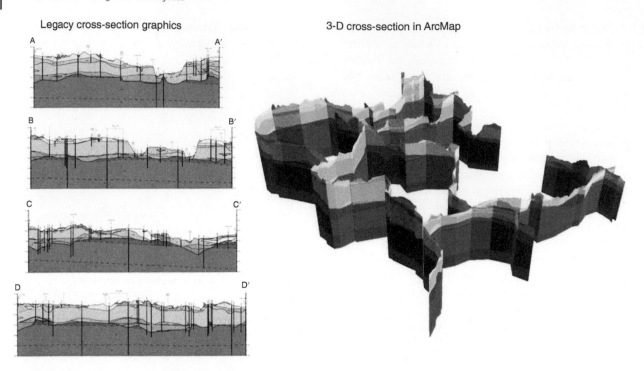

Figure 9.15 Legacy cross-section graphics could be georeferenced with by Xacto in combination with the MaPublisher plug-in for Illustrator. The results were convertible into 3-D cross-sections in ArcMap. (Carrell 2014).

With *3D Borehole Tools*, modelers can visualize lithologic descriptions of water-well records or lithologic borings as 3-D cylinders, display tabular geophysical log data as 3-D lines, add new map attributes, and rapidly create surfaces from queried borehole intervals (Figure 9.16). Additional capabilities support the selection and editing of attributes for multiple segments of boreholes or cores. Based on those selections, *3D Borehole Tools* uses ESRI algorithms to interpolate stacked-surfaces within the 3-D viewing environment and allows the modeler to build and visualize 3-D geologic data, record interpretations, and generate geologic stacked surfaces "on-the-fly," all within a single 3-D environment.

9.6.2 Oak Ridges Moraine, Southern Ontario

Aquifer resource delineation has been an important driver of 3-D geologic modeling efforts in the Greater Toronto Region (Sharpe et al. 1996). The Geological Survey of Canada (GSC) implemented stacked-surface mapping to address the geologic framework of aquifer resources in the Oak Ridges Moraine area in southern Ontario, Canada. As with the Lake County study by the ISGS, 3-D modeling of the glacial geologic framework of the Oak Ridges Moraine area required a rigorous field data acquisition program that included drilling new test holes and collecting 2-D geophysical data (seismic reflection, ground-penetrating radar, and gravity surveys). The study also required

database development and upgrading with supplemental archive data associated with regional lithostratigraphic formations (Logan et al. 2005)

The 3-D geological framework model of the Oak Ridges Moraine area consisted of four geologic units to define the Oak Ridges Moraine and other glacial drift deposits and two stratigraphic surfaces; one surface defined the surface topography, the second defined the buried bedrock topography (Sharpe et al. 1996, 2007). The modeling efforts were

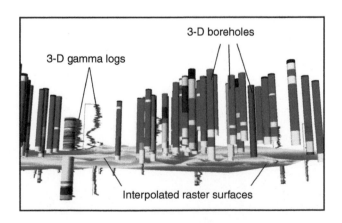

Figure 9.16 A 3-D scene developed by ArcScene shows borehole lines symbolized as tubes, geophysical log graphs as 3-D lines, and raster surfaces interpolated from user-selected borehole segments. (Carrell 2014).

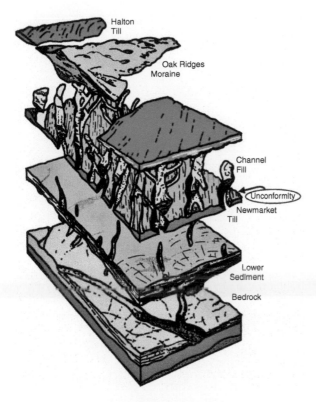

Figure 9.17 Three-dimensional geologic model of the Oak Ridges Moraine area. (Sharpe et al. 2007).

focused on delineating 3-D geometries of the following shallow aquifer settings:

- Surficial aquifer deposits and their hydrogeologic significance.
- Local channelized deposits of sand/gravel and their extent of downcutting.
- Distribution of subsurface sands and their potential for groundwater supply.
- Bedrock valley channels and associated infilling sediments.

The stacked-surface geologic models (Sharpe et al. 1996) very effectively delineated the hydrostratigraphic framework in the context of geologic history and processes (Figure 9.17). New insights were gained about the likely locations of preferred pathways of groundwater flow through the channelized glacial landscape and the implications for long-term groundwater sustainability and resource management. The 3-D stacked-surface model also helped constrain hypotheses regarding landscape genesis and sedimentary architecture of the Oak Ridges Moraine area (Sharpe et al. 2007). The model was used to target geophysical investigations, conceptualize, and visualize regional stratigraphic relationships, and ultimately to relate the glacial history to observed modern glacial processes (Logan et al. 2006).

9.6.3 Regional Aquifer Systems Evaluations by the U.S. Geological Survey

The U.S. Geological Survey (USGS) has used stacked-surface 3-D modeling procedures as part of several Regional Groundwater Availability Studies throughout the United States (USGS 2018). The choice of the stacked surface approach reflected the widespread use of ArcGIS within the USGS and the resultant availability of considerable legacy digital sources defining key geological bounding surfaces.

Many of these assessments are in sedimentary basins, which can be readily modeled using the stacked-surface approach. However, several studies introduced novel adjustments to the basic stacked-surface modeling methods; these defined the critical geological environments affecting the groundwater resources more efficiently. Three examples have been selected: modeling the basalt flows of the Columbia Plateau in the northwest United States, modeling the extensive Williston-Powder River basins in the north-central United States, and the Floridian Aquifer in the southeast United States.

9.6.3.1 Columbia Plateau in the Northwest United States

Burns et al. (2011) created a 3-D framework model for the Columbia Plateau Regional Aquifer System (CPRAS). The geologic model covers approximately 53 000 mi^2 of the Columbia Plateau in portions of Idaho, Oregon, and Washington. The model is designed to predict the distribution of major stratigraphic units that define the regional geometry of the aquifer system (Burns et al. 2011). The Columbia Plateau is composed primarily of a succession of basalt flows, modeled as the Columbia River Basalt Group (CRBG). The CRBG are locally over 15 000 ft thick. Locally discontinuous accumulations of sedimentary valley fill deposits occur between successive flow units. The CRBG is underlain by older bedrock units. Fault displacements compartmentalize the plateau into several regions and cause variations in thickness of some flows. Much of the area is overlain by younger sedimentary units, which were modeled as a single unit, called "Overburden."

An integrated 3-D framework model was developed by modeling the top surfaces of five geologic mapping units: Overburden, Saddle Mountains Basalt, Wanapum Basalt, Grande Ronde Basalt, and Older Bedrock (Figure 9.18). These surfaces were defined by a limited number of well data, existing geologic maps, and land surface elevation models. A regression-based algorithm (LOESS within S-Plus software) evaluated these data to create a preliminary surface grid. Interpolation errors between the observed data points and the surface grid (i.e. residuals)

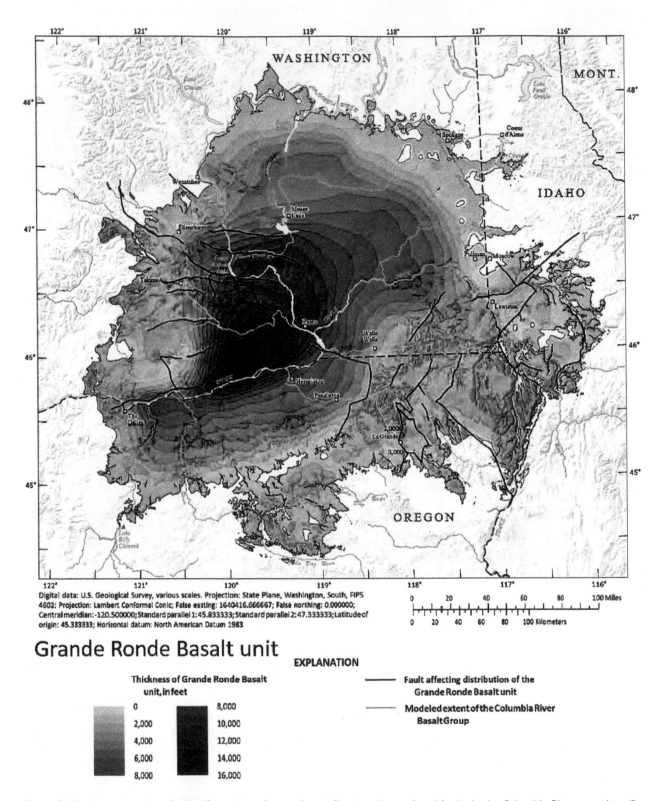

Digital data: U.S. Geological Survey, various scales. Projection: State Plane, Washington, South, FIPS 4602: Projection: Lambert Conformal Conic; False easting: 1640416.666667; False northing: 0.000000; Central meridian: -120.500000; Standard parallel 1: 45.833333; Standard parallel 2: 47.333333; Latitude of origin: 45.333333; Horizontal datum: North American Datum 1983

Grande Ronde Basalt unit

EXPLANATION

Thickness of Grande Ronde Basalt unit, in feet

0	8,000
2,000	10,000
4,000	12,000
6,000	14,000
8,000	16,000

——— Fault affecting distribution of the Grande Ronde Basalt unit

——— Modeled extent of the Columbia River Basalt Group

Figure 9.18 Interpolated stacked surface of the Grande Ronde Basalt and associated faults in the Columbia Plateau region. (Burns et al. 2011).

were evaluated to identify locations of local faulting or other bias. Fault maps, constraining (or guide) points, and local spline interpolations were all used in an iterative process to minimize residuals and incorporate relevant regional geologic knowledge.

Because a smoothing-based interpolation process could not reliably define all of the mapping units, particularly the sedimentary valley-fill deposits of the Overburden unit, a workflow was developed that used the land surface elevation model to identify areas where CRBG rocks were likely to outcrop and areas where the CTBG rocks were likely to be covered by overburden. Land surface elevations defined the correct uppermost CRBG surface in outcrop areas, while the smoothed surface defined the surface of the uppermost CRBG in locations where Overburden was expected. Burns et al. (2011) discusses the entire modeling process and products in detail. Digital versions of the model surfaces are available from a USGS website (see "Data" at http://pubs.usgs.gov/sir/2010/5246), and an interactive map exploration service is available at the same URL under "Interactive Webtool".

9.6.3.2 Williston and Powder River Structural Basins in the North-Central United States

Thamke et al. (2014) used stacked-surface modeling to generate a 3-D hydrogeologic framework and estimate hydraulic parameters of aquifer systems in the Williston and Powder River basins in the north-central United States (Figure 9.19). In this area, groundwater resources are critical for the local energy production industry, so the USGS was charged with modeling aquifer geometries and quantifying groundwater availability from three aquifer systems within the structural basins. These aquifers included (i) the overlying glacial deposits, (ii) the lower Tertiary deposits, and (iii) the Upper Cretaceous deposits. The aquifer systems were discretized into seven hydrogeologic units, which included five aquifers and two confining units.

The workflow involved analyzing 300 downhole electrical-resistivity geophysical logs, many lithologic logs, and other archive data throughout the study area. Interpretations of these data were used as a framework for interpolating stacked surfaces representing the elevations of the tops and bottoms of geologic units. A land-surface terrain model was compiled from existing published data and developed using triangulation methods. The geometry of the surficial glacial deposits was derived by combining published glacial-deposit thickness data with the land-surface terrain model. The stacked surfaces of the underlying hydrogeologic units were developed from combinations of digitized surface contours or anchor points from archive publications and elevation interpolations using the Topo-to-Raster (ANUDEM) Interpolation or

Kernel Interpolation with Barriers applications within ArcGIS. The computed top and base elevations were used to derive the thickness of each hydrogeologic unit. Hydraulic conductivity values for each hydrogeologic unit were estimated from downhole geophysical data and in-situ hydraulic tests. In the Williston Basin, estimates of aquifer transmissivities were calculated by multiplying unit thicknesses from the model by the distributed hydraulic conductivity values; in the Powder River Basin, existing published sources provided equivalent transmissivity data. The Williston-Powder River Basins model (Thamke et al. 2014) is an example of a rigorous workflow that incorporated a wide variety of datasets and multiple interpolation algorithms to produce practical, derivative hydrogeologic products.

9.6.3.3 Floridian Aquifer System in the Southeast United States

The Floridian Aquifer System in the southeast United States underlies an area of about $100\,000$ mi^2 including the entirety of Florida and parts of Georgia, Alabama, and South Carolina (Williams and Kuniansky 2016). Tertiary age carbonate rocks form multiple hydraulically-connected hydrogeologic units that supply water to much of the study area. An overlying surficial aquifer system is the primary water resource in the extreme southern portion of Florida. Defining the elevation of the $10\,000$ mg l^{-1} total dissolved solids freshwater-saltwater boundary is important for evaluating long-term salinity issues (Figure 9.20).

The USGS produced a series of 2-D surfaces to represent the bounding surfaces for a variety of lithologic units, geologic datums, and hydrogeological horizons within the Floridian Aquifer System by interpolating over 500 geophysical and lithologic logs and data from other published reports. The surficial aquifer system was delineated with approximately 6600 data points. Surfaces were interpolated and subjected to a statistical accuracy analysis by comparing observed and interpolated surface elevation values using ArcGIS Topo-to-Raster tools (Williams and Dixon 2015). Additional surfaces represent the top and base of the model/aquifer domain and define a regionally extensive glauconite marker bed that was indicated in geophysical logs. Electrical resistivity logs defined various permeable zones within aquifer units. A surface derived from geophysical logs and water samples defines the elevation of the freshwater–saltwater boundary (Williams and Dixon 2015).

Development of a new numerical groundwater model is continuing at the USGS Caribbean-Florida Water Science Center. The numerical model is constrained by a stacked-surface geological model based on the 2-D surfaces developed by Williams and Dixon (2015). It uses a 1 mi^2 grid resolution to produce better estimates of recharge and

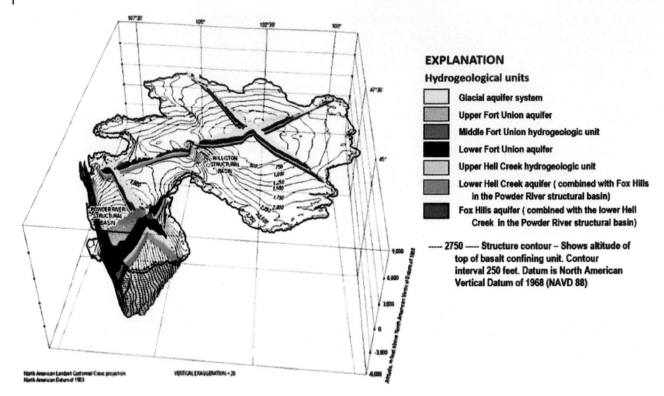

Figure 9.19 Three-dimensional visualization of cross-sections and structure-contour stacked surface in the Williston and Powder River basins. (Thamke et al. 2014).

discharge than previous models and analyzes the relationship between the freshwater-saltwater boundary and other hydrogeologic surfaces to better simulate the spatial and temporal variability of salinity gradients.

9.7 Conclusions

Stacked-surface modeling relies on assembling reliable 2-D interpolations of geologic structures and stratigraphic relationships. The interpolation of 2-D surfaces is inherent in resolving geologic problems, so the stacked-surface approach has become widely integrated into 3-D geological modeling strategies, especially for modeling and evaluating relatively undeformed, moderately complex, sedimentary geologic environments.

Readily available GIS software products incorporate robust interpolation algorithms that are capable of generating 2-D surfaces. Many geological surveys have experience with GIS applications and have extensive legacy data sources defining geological bounding surfaces of major lithostratigraphic and hydrostratigraphic units which are compatible with the stacked-surface approach. In addition, stacked-surface geological models provide constraints that are readily incorporated into groundwater-flow analyses and some numerical modeling applications in geotechnical engineering. Thus, the stacked-surface approach to creating 3-D geological framework models is an attractive option.

Custom-built add-on software provides stacked-surface model developers with capabilities that complement existing GIS software products. Existing custom applications will continue to evolve and progressively optimize the efficiency stacked-surface-model creation. Some custom applications are likely to eventually become integrated in commercial software packages. However, specific needs will continue to require custom solutions, which could usefully be shared amongst geomodeling organizations.

Stacked -surface models are a proven, effective approach to defining 3-D geologic frameworks. Stacked-surface modeling will continue to be applied broadly as an independent modeling strategy, and as a component of more complex and varied 3-D modeling software applications and workflows.

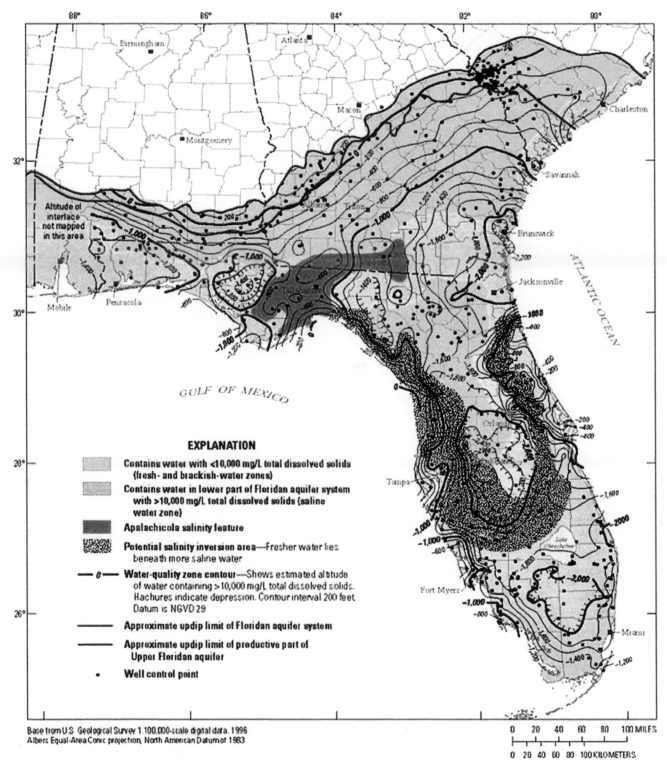

Figure 9.20 Overview of the Floridian Aquifer System in the southeastern United States. Contours show the elevation the 10 000 mg l⁻¹ total dissolved solids freshwater–saltwater boundary within groundwater. (Williams and Kuniansky 2016).

References

Brown, S.E. (2013). Three-Dimensional Geologic Mapping of Lake County, Illinois: No Small Task. In: *Three-Dimensional Geological Mapping Workshop Extended Abstracts, 2013 Annual Meeting, Geological Society of America, Denver, Colorado* (eds. H. Thorleifson, R. Berg and H. Russell), 23–28. St Paul, MN: Minnesota Geological Survey, Open File Report OFR-13-2. [Online: available at: http://hdl.handle.net/11299/159772] (Accessed October, 2018).

Burns, E.R., Morgan, D.S., Peavler, R.S. and Kahle, S.C. (2011). *Three-dimensional Model of the Geologic Framework for the Columbia Plateau Regional Aquifer System, Idaho, Oregon, and Washington.* Reston, VA: U.S. Geological Survey, Scientific Investigations Report 2010–5246. 44 pp. https://doi.org/10.3133/sir20105246.

Carrell, J. (2014). Tools and Techniques for 3D Geologic Mapping in ArcScene: Boreholes, Cross Sections, and Block Diagrams. In: *Digital Mapping Techniques '11–12 Workshop Proceedings* (ed. D.R. Soller), 19–29. Reston, VA: U.S. Geological Survey, Open-File Report 2014–1167. https://doi.org/10.3133/ofr20141167.

Caumon, G., Collon-Drouaillet, P., Le Carlier de Veslud, C. et al. (2009). Surface-based 3D modeling of geological structures. *Mathematical Geosciences* 41 (8): 927–945. https://doi.org/10.1007/s11004-009-9244-2.

Cherpeau, N., Caumon, G., and Lévy, B. (2010). Stochastic simulations of fault networks in 3D structural modeling. *Comptes Rendus Geoscience* 342: 687–694. https://doi.org/10.1016/j.crte.2010.04.008.

Clarke, D. (1992). The gridded fault surface. In: *Computer Modeling of Geological Surface and Volumes* (eds. D.E. Hamilton and T.A. Jones), 141–160. Tulsa, OK: American Association of Petroleum Geologists.

De Kemp, E.A. and Sprague, K.B. (2003). Interpretive tools for 3D structural geological modeling part I: Bézier based curves, ribbons and grip frames. *GeoInformatica* 7 (1): 55–71. https://doi.org/10.1023/A:1022822227691.

ESRI (2018). *How Topo to Raster Works.* Redlands, CA: ESRI. 7 pp. [Online: Available at: http://pro.arcgis.com/en/pro-app/tool-reference/3d-analyst/how-topo-to-raster-works.htm] (Accessed October 8, 2018).

Faunt, C.C., Sweetkind, D.S. and Belcher, W.R. (2010). Chapter E: Three-dimensional hydrogeologic framework model. In: *Death Valley Regional Groundwater Flow System, Nevada and California—Hydrogeologic Framework and Transient Groundwater Flow Model* (eds. W.R. Belcher and D.S. Sweetkind), 161–250. Reston, VA: U.S. Geological Survey, Professional Paper 1711. [Online: Available at: https://pubs.usgs.gov/pp/1711/] (Accessed August 18, 2020).

Frank, T., Tertois, A.-L. and Mallet, J.-L. (2007). 3D-reconstruction of complex geological interfaces from irregularly distributed and noisy point data. *Computers & Geoscience* 33 (7): 932–943. https://doi.org/10.1016/j.cageo.2006.11.014.

Gunnink, J.L., Maljers, D., van Gessel, S.F. et al. (2013). Digital geological model (DGM): a 3D raster model of the subsurface of the Netherlands. *Netherlands Journal of Geosciences* 92: 33–46. https://doi.org/10.1017/S0016774600000263.

Hamilton, D.E. and Jones, T.A. [Eds.] (1992). *Computer modeling of geological surface and volumes.* Tulsa, OK: American Association of Petroleum Geologists. 297 pp. https://doi.org/10.1306/CA1564.

Hooper, N.J., Raven, J.G.M. and Kilpatrick, M.J. (1992). Computer modeling of multiple surfaces with faults: the Ivanhoe filed, outer Moray Firth Basin, U.K. North Sea. In: *Computer Modeling of Geological Surface and Volumes* (eds. D.E. Hamilton and T.A. Jones), 161–174. Tulsa, OK: American Association of Petroleum Geologists. https://doi.org/10.1306/CA1564C12.

Hutchinson, M.F. (1989). A new procedure for gridding elevation and stream line data with automatic removal of spurious pits. *Journal of Hydrology* 106: 211–232. https://doi.org/10.1016/0022-1694(89)90073-5.

Hutchinson, M.F. (2011). *ANUDEM Version 5.3 User Guide.* Canberra, Australia: The Australian National University, Fenner School of Environment and Society. 30 pp. [Online: Available at: http://fennerschool.anu.edu.au/files/usedem53_pdf_16552.pdf] (Accessed October 27, 2018).

Jones, T.A., Hamilton, D.E. and Johnson, C.R. (1986). *Contouring Geological Surfaces with a Computer.* New York: Van Nostrand Reinhold. 314 pp.

Larsen, J.I. (1973). *Geology for planning in Lake County, Illinois.* Illinois State Geological Survey Circular 481, 43 pp.

Laurent, G., Caumon, G., Bouziat, A. and Jessell, M. (2013). A parametric method to model 3D displacements around faults with volumetric vector fields. *Tectonophysics* 590: 83–93. https://doi.org/10.1016/j.tecto.2013.01.015.

Laurent, G., Caumon, G. and Jessell, M. (2015). Interactive editing of 3D geological structures and tectonic history sketching via a rigid element method. *Computers & Geosciences* 74: 71–86. https://doi.org/10.1016/j.cageo.2014.10.011.

Leetaru, H.E. (2003). *3-D Visualization of Bedrock Resources in Lake County, Illinois (1:62,500 Lake County Map Atlas).* Champaign, IL: Illinois Geological Survey, Open File Series 2003–12 Lake-3D. 2 sheets. [Online: Available at: https://www.isgs.illinois.edu/maps/county-maps/three-dimensional-model/lake] (Accessed March 27, 2016).

Logan, C., Russell, H.A.J. and Sharpe, D.R. (2005). *Regional 3-D Structural Model of the Oak Ridges Moraine and Greater Toronto Area, Southern Ontario (Version 2.1)*. Ottawa, ON: Geological Survey of Canada, Open File 5062. https://doi.org/10.4095/220853.

Logan, C., Russell, H.A.J., Sharpe, D.R. and Kenny, F.M. (2006). The role of expert knowledge, GIS and geospatial data management in a Basin analysis, Oak Ridges Moraine, southern Ontario. In: *GIS Applications in the Earth Sciences* (ed. J. Harris), 519–541. Saint John's, NL: Geological Association of Canada.

Mallet, J.L. (1992). Discrete smooth interpolation in geometric modelling. *Computer-Aided Design* 24 (4): 178–191.

McGraw-Hill (2003). *The McGraw-Hill Dictionary of Scientific and Technical Terms, 6e*. New York: McGraw-Hill. 2380 pp.

Riggs, M., Abert, C.A., McLean, M.M., Krumm, R.J. and McKay, E.D. (1993). *Thickness of Quaternary Deposits, North-Central Lake County*. Illinois State Geological Survey Open File Series OFS 1993-10e. 1 map.

Sharpe, D.R., Dyke, L.D., Hinton, M.J. et al. (1996). Groundwater prospects in the Oak Ridges Moraine area, southern Ontario: application of regional geological models. *Geological Survey of Canada Current Research* 1996-E: 181–190. https://doi.org/10.4095/207886.

Sharpe, D.R., Russell, H.A.J. and Logan, C. (2007). A 3-dimensional geological model of the Oak Ridges Moraine area, Ontario, Canada. *Journal of Maps* 3 (1): 239–253. https://doi.org/10.1080/jom.2007.9710842.

Sprague, K.B. and De Kemp, E.A. (2005). Interpretive tools for 3-D structural geological modelling part II: surface design from sparse spatial data. *GeoInformatica* 9 (1): 5–32. https://doi.org/10.1007/s10707-004-5620-8.

Stumpf, A.J. and Luman, D.E. (2007). An interactive 3-D geologic map for Lake County, Illinois, United States of America. *Journal of Maps* 3 (1): 254–261. https://doi.org/10.1080/jom.2007.9710843.

Thamke, J.N., LeCain, G.D., Ryter, D.W., Sando, R. and Long, A.J. (2014). *Hydrogeologic Framework of the Uppermost Principal Aquifer Systems in the Williston and Powder River Structural Basins, United States and Canada (Ver. 1.1 December 2014)*. Reston, VA: U.S. Geological Survey, Scientific Investigations Report 2014–5047. 38 pp. http://doi.org/10.3133/sir20145047.

Thiele, S.T., Jessell, M.W., Lindsay, M. et al. (2016). The topology of geology 1: topological analysis. *Journal of Structural Geology* 91: 27–38. https://doi.org/10.1016/j.jsg.2016.08.009.

USGS (2018). *Regional Groundwater Availability Studies*. Reston, VA: U.S. Geological Survey. [Online: Available at: https://water.usgs.gov/watercensus/gw-avail.html] (Accessed October 27, 2018).

Williams, L.J. and Dixon, J.F. (2015). *Digital Surfaces and Thicknesses of Selected Hydrogeologic Units of the Floridan Aquifer System in Florida and parts of Georgia, Alabama, and South Carolina*. Reston, VA: U.S. Geological Survey, Data Series 926. 24 pp. http://dx.doi.org/10.3133/ds926.

Williams, L.J. and Kuniansky, E.L. (2016). *Revised Hydrogeologic Framework of the Floridan Aquifer System in Florida and parts of Georgia, Alabama, and South Carolina (Ver 1.1 March 2016)*. Reston, VA: U.S. Geological Survey, Professional Paper 1807. 140 pp. http://dx.doi.org/10.3133/pp1807.

Wu, Q., Xu, H., Zou, X. and Lei, H. (2015). A 3D modeling approach to complex faults with multi-source data. *Computers & Geosciences* 77: 126–137. https://doi.org/10.1016/j.cageo.2014.10.008.

Zoraster, S. (1992). Fault representation in automated modeling of geologic structures and geologic units. In: *Computer Modeling of Geological Surface and Volumes* (eds. D.E. Hamilton and T.A. Jones), 123–140. Tulsa, OK: American Association of Petroleum Geologists. https://doi.org/10.1306/CA1564C10.

10

Model Creation Based on Digital Borehole Records and Interpreted Geological Cross-Sections

Benjamin Wood and Holger Kessler

British Geological Survey, Keyworth, Nottingham NG12 5GG, UK

10.1 Introduction

In the stacked-layer modeling workflows discussed in Chapter 9, stratigraphic surfaces are obtained by interpolating between data points derived from stratigraphically interpreted boreholes. In the borehole and cross-section method discussed in this chapter, stratigraphic surfaces are constructed between stratigraphic correlation lines from a network of manually constructed cross-sections. The British Geological Survey (BGS) has adopted and developed this approach as a result of its 2000–2005 strategic Digital Geoscience Spatial Model (DGSM) R&D program (Smith 2005) as one of several methodologies for constructing 3-D geological models. The borehole and cross-section-based method is based on tools and concepts familiar to every geologist – namely, conceptualization, borehole interpretation, mapping, and cross-section drawing (Jones and Wright 1993). Its wide deployment and uptake at the BGS were influenced by a high demand for models in highly variable and complex glaciated terrains and urban areas with highly varied data density and quality, where input data are too sparse or inadequate for methods that rely more heavily on geostatistics. However, the high degree of geologist input makes for an inherently subjective methodology, described by Kessler and Mathers (2004) as "capturing the geologists' vision." This chapter presents the method as deployed at the BGS, in terms of its workflow, applications, advantages, and limitations.

10.1.1 The BGS Cyberinfrastructure

At the BGS, the modeling tools themselves are part of a much wider set of digital systems and work processes, which are described in Chapter 8. Smith (2005) provides a detailed description of the technologies and methodologies that form the basis of this modeling workflow. This process begins with capturing or ingesting data, continues with the model construction process (further detailed below), feeds into the design of derived products and concludes with the storage and delivery of geological data, information, and knowledge. Figure 10.1 illustrates this workflow sequence. Since 2004, the modeling workflow at the BGS has relied partially on the "Geological Surveying and Investigation in 3 Dimensions (GSI3D)" software, which provides a convenient implementation of the borehole and cross-section method (Sobisch 2000; Kessler et al. 2009). The general workflow will be described here with some specific reference to that software.

10.1.2 Geological Surveying and Investigations in 3 Dimensions (GSI3D)

Maps and cross-sections have been used since the origins of geology to show geological relationships, and their combination allows a trained geoscientist to build a 3-D geological model in his or her mind. The GSI3D software tool and methodology takes the classic geological mapping one step further by creating an actual 3-D geological model of an area from a network of cross-sections and the mapped extent of geological units. It was initiated in the 1990s in response to requirements specified in a report by Binot and Röhling (1994) of the Soil and Geological Survey of Lower Saxony (Niedersächsisches Landesamt für Bodenforschung: NLfB), and conceived primarily as a tool for modeling shallow superficial-Quaternary sequences (Hinze et al. 1999; Sobisch 2000).

At the BGS, the starting point for GSI3D modeling is always the existing digital geological map at an appropriate scale, such as 1:50 000 or 1:10 000, effectively carrying an invaluable legacy data set to next-generation geological survey output. Other than that, GSI3D relies upon a pre-defined stratigraphic sequence, a digital elevation model (typically in the form of a triangle mesh), and

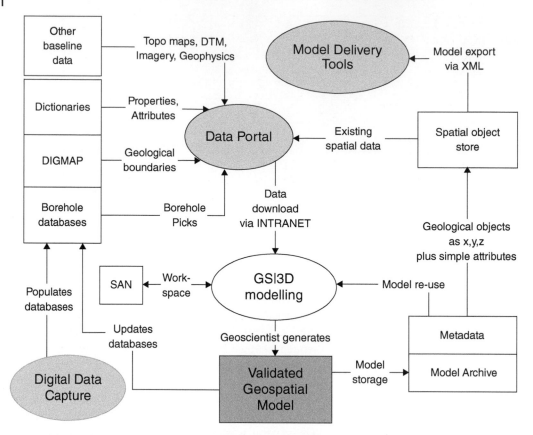

Figure 10.1 GSI3D is only one component of the BGS cyberinfrastructure. (Kessler et al. 2009).

codified borehole data. From this, a user-defined network of interlocking cross-sections and a corresponding set of stratum extent maps are constructed by a geologist. The software then applies a triangulation algorithm to generate a stack of surfaces that can be converted into either a set of closed shells, or a volumetric grid. This model creation process is guided by a graphical interface consisting of four windows (Figure 10.2). The GSI3D workflow is discussed in more detail in Section 10.2.

Culshaw (2005) discusses several early GSI3D modeling studies for urban, coastal, and engineering projects. GSI3D is now routinely deployed in building 3-D models, including a national fence diagram of the entire United Kingdom (Waters et al. 2015), and many regional models including London (Burke et al. 2014, Mathers et al. 2014). GSI3D also supports commercial contracts for clients such as the Environment Agency of England and Wales (Farrant et al. 2016), infrastructure development (Aldiss et al. 2009; Burke et al. 2015), and Local Government (Campbell et al. 2010). While the GSI3D software remains in use at the BGS under license, it is no longer available commercially. However, its successor product Subsurface-Viewer MX, described in Section 10.5.1, is commercially available.

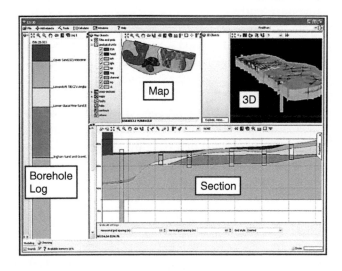

Figure 10.2 GSI3D user interface has four windows – (1) Map, (2) 3-D, (3) Section, and (4) Borehole. From Mathers et al. (2011).

10.1.3 GSI3D at the British Geological Survey

The application of GSI3D at the BGS began shortly after a European Science Foundation conference on 3-D modeling (Rosenbaum and Turner 2003). The implementation of GSI3D within BGS was quite rapid. It was aided by the availability within the BGS of digital geological maps for almost the whole country, a result of a large historic digitization campaign in the 1990s (Jackson and Green 2003). This meant that most of BGS's legacy data holdings were already available digitally as vector or raster data at scales suitable for creating effective geological models for various applications (Jackson and Green 2003). Virtually all BGS's paper records had been scanned and most legacy map data had been geo-registered.

Land-surface topographic information was available from a licensed, nationwide high-resolution Digital Terrain Model (DTM). The DGSM project (Smith 2005; Hatton et al. 2005) established comprehensive databases of borehole locations and borehole logs supported by corporate dictionaries for lithological and stratigraphic terminology. The retrieval and subsequent use of all these data were aided by well-organized data indices and associated metadata.

GSI3D has successfully utilized all these data combined with the wealth of geological knowledge from experienced field geologists and continuing field surveys to produce 3-D geological models. New field surveys use digital field notebooks configured with the BGS SIGMA system (BGS 2015), this facilitates the immediate transfer of the captured data from the field notebooks to the corporate databases. Data for each project are served to the modelers from these corporate databases using a range of internal systems some of which are discussed in Chapter 8. Available borehole records, which have not been digitally codified into the BGS lithostratigraphic scheme (BGS 2019), are re-interpreted and coded up using a Microsoft Access database entry form. Increasingly, borehole data is ingested to these databases from industry donations in the AGS4 digital data transfer standard (Association of Geotechnical and Geoenvironmental Specialists 2017), reducing the future need for digitization by BGS geologists. The data are then visualized and co-validated in GSI3D, whilst the geoscientist may update the corporate borehole or map databases with new interpretations.

10.2 The GSI3D Model Construction Sequence

The GSI3D model construction workflow consists of three stages: (i) model building, (ii) model computation, and (iii) model analysis (Figure 10.3). The GSI3D model building stage depends on a Generalized Vertical Sequence (GVS) file which controls the order in which each geological unit can occur and rejects any relationships drawn in sections that do not correspond to this pre-determined

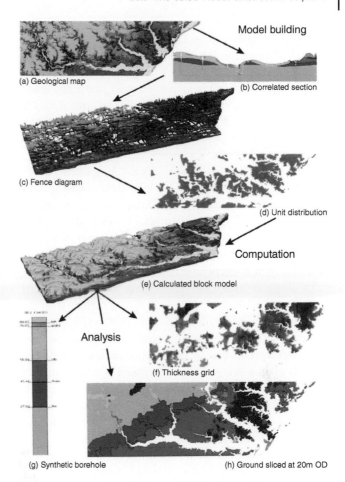

(a) Geological map

Model building

(b) Correlated section

(c) Fence diagram

(d) Unit distribution

Computation

(e) Calculated block model

Analysis

(f) Thickness grid

(g) Synthetic borehole

(h) Ground sliced at 20m OD

Figure 10.3 GSI3D model creation workflow occurs in three stages: model building (a–d), model construction (d–e) and model analysis (f–h). This model comprises some 1200 km² of the Sudbury–Ipswich–Felixstowe area of southern East Anglia, UK. (Mathers 2013).

order. The GVS file is a tabulator-separated ASCII text file produced by the modeler. It evolves as the model develops and ultimately lists all units in their correct and unique stratigraphic order. This order defines the "stack" that is calculated to make the 3-D geological model. A special extension to the format also permits the inclusion of intra-unit "lenses" (outlined below) which sit outside the main GVS framework.

The model building stage utilizes a DTM, geological surface linework and downhole borehole data to enable the geologist to construct a series of intersecting cross-sections by correlating boreholes and the outcrops-subcrops of units to produce a geological fence diagram of the area (Figure 10.3a–c). The alignment and spacing of the cross-sections are entirely user-controlled; they can be drawn anywhere in the map interface, including through the borehole locations. Subsequently, the limits of the units (outcrop plus subcrop) are defined (Figure 10.3d) and these digital maps are referred to as unit envelopes within the GSI3D methodology. Usually, the starting point for these unit envelopes is the existing digital geological map, but the modeler may edit this linework to add subcrop unit extent information that is not held in the existing map. The drawn cross-sections are keyed into the digital geological map by marker points and a linework snapping mechanism within the cross-section view, effectively forming a 3D wireframe of geological contacts.

It is important to design a good alignment of the cross-sections within the model. More cross-sections permit greater control over the shape of the model surfaces by the geologist. In principle, cross-sections should be aligned with due consideration to the overall tectonic structure; specifically, the key cross-sections ought to be more or less perpendicular to key structural features such as dips, folds, and faults. This allows for a much easier interpretation of the geological structure with less need to account for apparent dips. In the simplest case, a series of transects with uniform spacing may be used, but cross-sections can be drawn around the boundary of the model area to ensure the wireframe is neat around the model edges. As further outlined in Section 10.3, it may be necessary to include shorter "helper" cross-sections within the cross-section network in order to resolve detailed features such as channels and lenses (Figure 10.4).

Calculation of a geometric surface or 3D solid model from the wireframe network of interpreted digital cross-sections and the stratum coverage map (unit envelopes) is the final stage in the modeling process. Within GSI3D, this is achieved using a Delaunay triangulation routine which results in triangle meshes or triangulated irregular networks (TINs; see Chapter 13 for a full outline

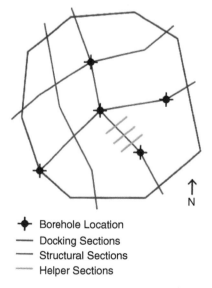

▸ Borehole Location
— Docking Sections
— Structural Sections
— Helper Sections

Figure 10.4 Examples of cross-section design for GSI3D model creation.

of meshing and discretization options). Typically, an initial calculation yields a stack of elevation surfaces representing the top and/or base of the modeled horizons. In the case of GSI3D, the base of a geological unit is calculated starting with the youngest unit in the stack, the tops are derived and composed from overlying unit bases. More sophisticated systems, such as GOCAD, can apply smoothing algorithms that densify or "beautify" the TINs by inserting more triangle vertices according to a proprietary scheme and then interpolating smoothed values to the new vertices.

Volume shells and other full 3-D representations, such as block models, are later derived from this stack as required by shape extrusion or 3-D cube sampling processes. The horizon calculation step can be achieved by various means according to the available software. Whichever technical route is taken, the purpose is fundamentally one of fitting a surface to the cross-section wireframe via some meshing, gridding or interpolation process. Usually this is performed on a per-horizon basis, and may be done across the entire domain, or piecewise (per-cell) if the cross-section alignments are "netty" (Ming et al. 2010). At the simplest level, this can be done using a Geographic Information System (GIS) package or a dedicated surface building package such as Surfer, after either the 3-D linework or a 3-D point-cloud (X, Y, Z) has been imported from the cross-section dataset. In more dedicated geological software, such as GSI3D, the software used to construct the network of cross-sections includes the calculation routine to generate the surfaces.

The detailed mechanics of how a wireframe model is converted into a solid model has been described in various ways and there are various schemes such as set operation-based methods (Jones and Wright, 1993) and horizons-based methods (Lemon and Jones 2003; Kessler et al. 2009). The data structure employed for the model horizons is typically an unstructured TIN because this is well suited to representing the shapes of natural, geological bodies; however, raster grids may also be used, as-in the stacked surfaces methods described in Chapter 9. A TIN may be defined by interpolations of the data point distribution or it can be required to pass through all available data points (borehole picks and cross-section correlation line vertices).

A TIN that is defined by interpolation of the data point distribution can be used as a template, i.e., a common TIN employed by all the model horizons, regardless of the distribution of the observed data points available to each surface (Figure 10.5a). A common TIN has the distinct advantage that, because all horizons share the same mesh topology, extrusions of the surfaces to form a solid model, set operations, and volumetric calculations are very straightforward. However, a major limitation is that a model constructed using a common TIN cannot honor

Figure 10.5 Examples of (a) a common (or template) TIN and (b) surface defined by individual TINs.

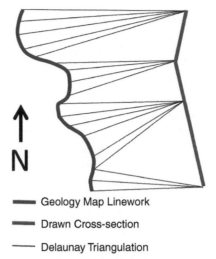

━━━ Geology Map Linework

━━━ Drawn Cross-section

──── Delaunay Triangulation

Figure 10.6 Potential problems result when a TIN is formed by the Delaunay method and point spacings on cross-sections differ greatly from those on map linework defining unit boundaries.

all control points exactly, except in the case where all the boreholes extend through all layers and these borehole data points are the sole source of model geometry, without any cross-section information. Such a model can be assessed as a series of triangular pillars (Bouza-Rodríguez et al. 2014) and these pillars can be subdivided to form a series of tetrahedra (Chapter 13). Thus, this modeling approach can be used to produce one form of unstructured tetrahedral mesh model which may be suitable for some groundwater flow modeling. Raster-based data grids are also very simple to extrude into solid models or to discretize into cellular voxel structures; however, these models contain stepped unit boundaries unless a suitably high resolution is applied, which usually requires additional computation time. Thus, their visualizations appear "pixelated" and are often not desired for visualization purposes.

Special cases set apart, borehole data density decreases with depth, and TINs that honor available data points per horizon have topologies that vary accordingly. The individual TINs more accurately represent the geometry of each horizon, but solid model building, set operations, and volumetric calculations are considerably more complex. The lack of coherence between the triangles defining one horizon and those for vertically adjacent horizons involves considerable computational effort to extrude and clip the different topologies. The use of the Delaunay triangulation method to define the triangles exacerbates these problems because a large number of very thin triangles results along the edges of the computed layers whenever the spacing of the vertices along the cross-sections differs significantly from the boundary polygons defining the geological unit map envelopes (Figure 10.6). This is very likely to occur when boundary polygons are taken from existing digital geological mapping. These thin triangles may produce a good visual result, but they are undesirable if the model is to provide an unstructured mesh for finite element numerical analysis. Resampling the boundary outlines to provide point spacings more similar to those of the cross-sections may help resolve the issue. More acceptable unstructured

meshes can be developed by a series of smoothing operations (Chapter 13). In summary, model users need to understand the model creation process; then they can modify the TINs as required for their applications.

After a model has been constructed, it must be analyzed to check the validity and integrity of the model. As discussed in Chapter 9, the integrity of models created by using GIS surface interpolation and gridding capabilities to generate stacked surfaces can be checked and improved by means of standard GIS raster operations and clipping routines. However, dedicated geological modeling software makes this much simpler. Sophisticated calculation routines in GSI3D cross-reference the geological sequence defined in the GVS-file to ensure consistency and minimum thickness of the units. Horizon distribution maps provide information, usually as a set of polygons, that restricts the extents of units which are not laterally continuous throughout the model due to subaerial erosion or subsurface unconformities. Fitting algorithms evaluate the entire set of available control data and generate a sequence of surfaces through all data points.

Once the model validity and integrity are assured, the model can be used to produce derivative products. Four commonly produced derivative products are: 2-D grids of unit thickness (isopach maps) (Figure 10.3f); synthetic boreholes (Figure 10.3g); horizontal slices through the model at specified depths (Figure 10.3h); and synthetic cross-sections (not illustrated). These products are based on the 3-D model, but are 2-D representations, so can be widely used as illustrations or for further analysis by GIS techniques.

10.3 Model Calculation Considerations

Successful calculation of 3-D solid models from horizons based on a network of interlocking user-defined cross-sections relies on a broad assumption, described by Tipper (1993), that the objects being modeled are horizontally layered and laterally continuous. Depending on the complexity of the geology being modeled and the sophistication of the algorithms and software being applied, a range of issues may arise as this assumption is stretched. In general, problems are likely to occur, regardless of the software or algorithm used to compute the horizons, because the horizons are interpolated independently. This lack of any defined topological relationships among the model layers often results in the surfaces intersecting with one another in non-desirable ways (Turner 2006). Such problems may be resolved with some combination of manual or automated cleaning and editing.

The cross-section-based approach is unable to deal with structurally complex terrains and is mostly used for shallow sedimentary deposits. However, such deposits in the UK can be complex in their own right due to Quaternary glaciations. The following sections discuss four geological scenarios that are difficult to resolve with a cross-section-based approach to modeling.

10.3.1 Laterally Non-continuous Deposits

The basic assumption that each deposit within the model is, or once was, a laterally continuous layer rapidly breaks down when modeling Quaternary geology. Two typical examples are: (i) deposits that have a patchy, non-continuous distribution across the model area, such as solifluction deposits and glacial till, and (ii) channel-style deposits such as alluvium which form ribbon-like layers across the model area. Within the GSI3D methodology, in order to define the deposit, each instance of such deposits

requires a corresponding unit extent envelope (polygon) and at least one drawn cross-section passing through each of those envelope polygons (Figure 10.7a). In the case of channel deposits, multiple helper sections (defined in Figure 10.4) are required to adequately control the channel geometry along its length. Resolving such deposits to a reasonable level of detail can therefore become a laborious process.

Modelers may also want to distinguish lenses of contrasting lithologies within model units, for example a patch of clay within fluvial sands can be hydrogeologically significant. Depending on the distribution of available borehole data and the overall understanding of the geological situation, it may not be possible to accurately depict the location and extent of such lenses. If the general nature of the lenses is well understood, their form may be estimated across the model (Figure 10.7c). Otherwise, they become more of a conceptual element in the modeling, which may be of limited practical use. Accurate definition of such features depends on additional sources of information, such as geophysical surveys. Lenses are difficult to account for during the calculation of the 3-D shapes by the geological software. When working from drawn cross-sections and distribution maps, many cross-sections may be necessary to identify all of the lens structures; this may become laborious because each lens must have at least one cross-section alignment running through it and a boundary polygon to constrain its extent.

10.3.2 Thin Units

Even in models with otherwise low complexity, horizons with a relatively small vertical thickness can pose a geometric problem, especially if they are represented by a coarse set of triangles in a TIN. The intersection of the relatively large triangles forming one surface with those of an adjacent layer leads to a layer model which has poor topological integrity. Where triangles coincide or intersect,

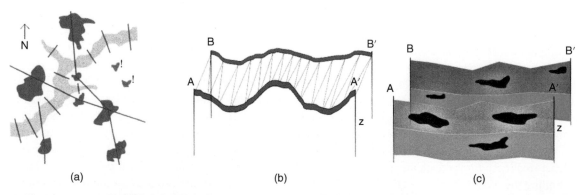

| (a) | (b) | (c) |

Figure 10.7 Three geological situations that are difficult to model using only boreholes and cross-sections. (a) Laterally non-continuous deposits, (b) thin or draping deposits, (c) intra-unit lenses.

the layer effectively has zero or negative thickness. To resolve this, it is generally necessary to increase the density of the cross-sections and also to ensure the drawn linework within each cross-section is more densely populated with vertices that can be used as input to the triangulation routine, thus producing smaller triangles (Figure 10.7b). This can be accomplished by either inserting additional vertices manually, which is very laborious and makes the linework difficult to modify in the future, or by applying a line densification routine within the software. Alternatively, some software products allow the user to specify a minimum thickness to each unit; this will tend to increase the lateral continuity of thin units.

10.3.3 Faults

Faults can be easily represented using cross-sections, but careful consideration ought to be given to the alignment of those cross-sections which pass through faults. Specifically, cross-sections should be aligned as close to perpendicular to the fault plane as is possible to minimize the effects of apparent dip on the fault plane. Cross-sections that pass obliquely through the azimuth of a fault plane, or which cross back and forth across the plane, will lead to complex geometry being required in the digitizing (Figure 10.8). In general, borehole and cross-section information can fairly easily represent vertical (or nearly vertical) fault planes and shallow over-thrust surfaces can be represented as unconformities. Difficulties arise in accurately representing inclined and intersecting faults.

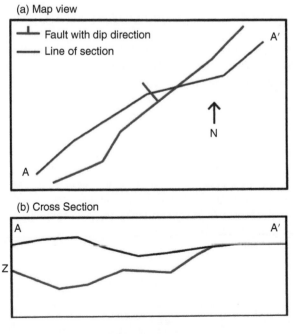

Figure 10.8   Faults oriented obliquely to a model cross-section may have complex cross-section geometry.

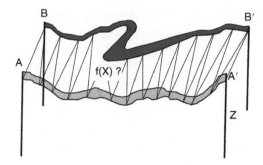

Figure 10.9   Transition to an over-turned fold may be impossible to model with many surface interpolation methods.

10.3.4 Folds

Gentle folds can be easily depicted within individual cross-sections and represented in the model by TINs. However, they pose a serious challenge to the surface calculation in the modeling software. In general, it may be difficult for the interpolation algorithm to accurately blend the shape of the folds between two nearby cross-sections. In the case of overturned folds, the problem becomes acute and additional problems arise from the limitations of the interpolation methods, which may be incapable of handling geometries that have multiple elevation values at a given location (Figure 10.9). This is because most of the algorithms rely on the triangulation or gridding of a set of scattered points on a surface defined by a 2.5-D data structure which only allows a single elevation for the surface at each geographic location.

10.4   Additional Considerations on Using This Methodology

Just as with traditional cross-sections, GSI3D and similar packages allow for the modeling of features that are geologically plausible or even likely, but not (entirely) supported by direct observations. However, manual correlation is inherently subjective, and the geologists must take great care to ensure their interpretations are consistent with the data, other nearby interpretations, and are geologically sensible.

The use of an interlocking network of cross-sections can help produce geologically plausible interpretations because they provide some level of constraint from one cross-section to the next and thus may assist in the spatial correlations. However, it is important that the interpretation and correlation is carried out by a geologist who is very familiar with the regional geology, and all available background data are utilized, including structural measurements and geophysical maps and profiles. In many cases, relatively sparse borehole data may provide the only available information source. Where accurate, high-resolution ground models are required; for example, for engineering purposes at the site scale, a high-quality model will be necessary in order to mitigate risk. High-quality models include a wide range of supporting data and information, such as targeted sampling schemes and geophysical surveys. These are used to inform the conceptual and the detailed geometric understanding of the strata prior to, or in tandem with, the drawing of the cross-sections.

As discussed in Chapter 7, additional down-hole measurements on boreholes and in trial pits, such as the property values determined by geotechnical testing, can be displayed alongside the lithological interpretation and may help to resolve smaller features such as lenses, or to help define key layer boundaries. Other types of down-hole data such as gamma logs and core imagery may also be used to help define model boundaries. Geophysical profiles can be interpreted, digitized, and spatially registered in profile views – they are especially useful for understanding overall structure. These data sources provide insights that are especially valuable for models intended for engineering purposes rather than as regional conceptual models. Spatial heterogeneity is unlikely to be adequately defined by point data sources such as boreholes and trial pits. Structural information in the form of descriptions and field measurements may be projected into the profile view as apparent dip values to guide cross-section development.

Historical borehole data are the primary source used by many geological survey organizations (GSOs) when building their regional geological models and may be the only substantial data source in many areas. New boreholes will provide additional insight and can be used to test hypotheses regarding the nature of the site. They also make the correlation of the drawn cross-sections more valuable and yield a more realistic model. Careful consideration should also be given to the acquisition of new borehole data; their locations should be selected following evaluation of supporting data such as geophysics. In addition, background information such as memoirs, reports, and map marginalia may either inform the conceptual model, or be incorporated directly.

10.5 Other Software Options

Although the basic methodology has been described using the GSI3D software as a convenient example, several options are available for developing models using boreholes and interpreted cross-sections. Commonly available software tools offer solutions to many basic data preparation and management requirements. For example, almost all geologists have access to basic spreadsheet or database software on their desktops. These are potentially powerful data preparation tools as they allow tabular display, filtering and re-formatting of data from new or legacy sources. Sorting functions readily detect outliers.

Many geologists have access to a standard GIS system such as ArcGIS (ESRI) or QGIS (OSGeo 2019), or perhaps access to Computer Aided Design (CAD) systems if they are working in an engineering office. These provide a good environment for data preparation as well as visualization and interpretation. A GIS can be used to construct digitized cross-sections, either as 2-D or 3-D objects, and many third-party applications have been developed to supplement existing GIS software features or capabilities. Carrell (2014) describes how the functionality of ESRI's ArcScene 10 was enhanced by several specialist tools developed at the Illinois State Geological Survey to assist the development and editing of 3-D boreholes and cross-sections. They are described in Sections 9.6.1.4 and 9.6.1.5. Obviously, any such cross-section can be ingested in a GSI3D modeling exercise.

Layer or surface building is possible within GIS or CAD environments, but these are not geological tools and must be augmented by integrity checking routines to detect overlaps and other non-geologically viable geometries. In terms of dedicated modeling environments, for several decades the only available integrated software products for 3-D geological modeling were those developed for the petroleum exploration and mining sectors. They were relatively complex, required considerable experience to be used effectively, and were expensive to purchase and maintain. These systems were designed to operate with the typical data sources and to produce models oriented toward the specific needs of their users. With these characteristics, these software products were not particularly suited to resolving the majority of the modeling tasks facing geological surveys (Rosenbaum and Turner 2003).

However, beginning around 2000, some affordable, fully integrated, and user-friendly software solutions became available that were attractive to geological surveys. These systems allow the modeler to efficiently develop 3-D geological framework models based on information provided by digital cross-sections tied to borehole data (Hinze et al. 1999; Kessler et al. 2009). The main software products using this approach are SubsurfaceViewer MX, GeoScene3D, and Groundhog Desktop.

10.5.1 SubsurfaceViewer MX

SubsurfaceViewer MX (INSIGHT 2016) is a self-contained commercial modeling environment that provides borehole, map and cross-section interpretation capabilities, a model calculation engine, and a 3-D visualization component. SubsurfaceViewer MX is provided by the creator of GSI3D;

it has the same user interface and follows the same basic workflow. With interactive model interrogation functions, SubsurfaceViewer MX allows the geologist to inspect the calculated model both vertically and horizontally. A freely available viewer application allows users to interrogate or explore existing SubsurfaceViewer MX models, but not to edit or revise them (Terrington et al. 2009).

10.5.2 GeoScene3D

GeoScene3D is a Danish geological modeling platform developed by I●GIS A/S, a software developer, in a public-private consortium that includes the Geological Survey of Denmark and Greenland (GEUS), several additional Danish organizations, and major Danish engineering firms. Originating as part of the major Danish groundwater program, GeoScene3D uses a typical user interface showing cross-sections, map views, and 3-D scenes and has expanded from its original focus on groundwater resource assessment to become the standard platform for geoscience data in Denmark.

The GeoScene3D user interface supports a manual, cognitive approach to model creation where several different data types are imported, combined, compared, and interpreted by geologists. GeoScene3D facilitates visualization of a broad range of geoscience data, including a variety of geophysical data and soil and water chemistry data. Due to the extensive Danish experience with airborne and ground-based geophysical surveys, GeoScene3D supports automated modeling based on machine learning and data clustering to assist geologists interpreting resistivity surveys and borehole data. Geological models can be developed using multiple point statistics techniques to evaluate the geological heterogeneity observed within rock units. These more advanced techniques enable the rapid assessment of large-scale geophysical surveys. Specialized tools for voxel modeling allow the building of highly detailed voxel models, based on layered models that can be combined with information on subsurface urban structures such as pipelines or trenches (Figure 10.10).

10.5.3 Groundhog Desktop

Groundhog Desktop, developed by the BGS (Wood et al. 2017), is a free-to-use general-purpose tool that serves as an interface between a standard GIS and a specialized geological modeling application. It offers comprehensive data preparation and filtering tools as well as interactive digitizing functions in both map and cross-section, plus a geology log viewer/editor (Figure 10.11). It supports a range of common data exchange formats, including certain GSI3D, GOCAD, ESRI and industry formats such as LAS, a file format designed for the interchange and archiving of

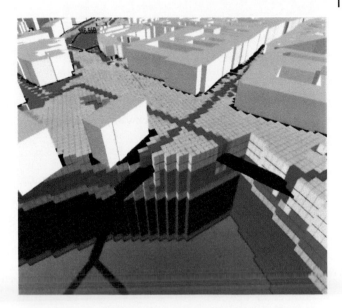

Figure 10.10 GeoScene3D lithologic voxel model showing locations of pipeline trenches and BIM representation of buildings. Image courtesy of I●GIS A/S.

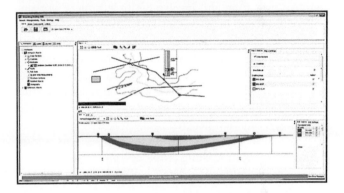

Figure 10.11 Typical view of Groundhog Desktop in use with map linework, cross-sections and boreholes loaded. (Wood et al. 2017).

Lidar point cloud data. Its principal function is to assist the codification of borehole logs and construction interlocking networks of interpreted cross-sections. Because it does not offer 3-D capability, data generated with Groundhog Desktop must be exported to appropriate geological modeling systems for calculations, or to a geological model viewer for visualization.

10.6 Discussion and Conclusions

Whilst cross-section-based approaches use interpretation techniques that are still most often taught to undergraduate geology students using paper-based exercises, the availability of digitizing software now allows the geologist to undertake this subsurface interpretation work more or less digitally in a range of 2-D and 3-D environments. The resulting wireframe can be computed into a 2.5-D stacked surface model via an interpolation algorithm and extruded or sampled to generate a solid 3-D model. Model calculation may be carried out within the geological modeling software, or if suitable modeling software is not available, by exporting the wireframe geometry to a general-purpose surface or solid building package (Surfer, Rockware, CAD, GIS, etc.).

10.6.1 Using the Method

The cross-section approach can be used for the entire model construction, or it can be used in combination with other techniques as part of a hybrid workflow, and many commercially available geological modeling packages permit this sort of working. This may be useful where partial expert-driven control on the model geometry is needed – for example, 3-D control points extracted from a number of digitized cross-sections might be used as one of a number of inputs used to constrain an implicit (numerical) or CAD-based model.

The development of a cross-section model can be highly iterative over time and involve multiple project workers, both within one organization and across organizational boundaries. So long as each worker honors the stratigraphic scheme and adheres to the interlocking aspect of cross-section construction, extension and refinement can take place relatively easily, albeit with considerable manual input. Extensions may increase the coverage of the model by taking it into new geographical areas. Refinements can take the form of additional in-fill cross-sections and/or refinement within existing cross-sections, for example by adding more stratigraphic detail over time.

10.6.2 Advantages of the Method

The cross-section approach has a low barrier to entry and may be particularly attractive to organizations or projects that do not have ready access to high-end modeling packages and/or dedicated modelers and CAD technicians or are starting their journey of digital transformation. The method can also be particularly useful in areas where hard data are sparse or highly heterogeneous (a scenario that limits the application of implicit modeling approaches) because it permits direct input from the geologist in building up the conceptual model. Models built up from user-defined drawn cross-sections are also simple to extend and refine, lending themselves to iterative development over time.

10.6.3 Limitations of the Method

Manual correlation is inherently subjective, and the geologist must take great care to ensure his or her interpretation

is consistent with the data, other nearby interpretations based on the latest published research on the geological evolution of the study area, and is geologically sensible. There is an inherent bias in any cross-section based (explicit) geological model. It is vital that models are accompanied by sound technical metadata if they are to be used effectively by designers and decision makers (Farrant et al. 2016; Kessler et al. 2016; Chapter 8) and the staff who create, modify, review, and utilize the model must be fully trained. It is also highly desirable for model users to have access to the geologist who created the model during any subsequent applications, as this allows the users to draw on the tacit knowledge of the geologists. Early experiences from the Crossrail project at Farringdon Station (Gakis et al. 2016) and continued work on railway improvement projects in the UK (Kessler et al. 2017) indicate that a full integration of a cross-section based approach into a design and engineering project is possible and yields positive outcomes in terms of cost and risk reduction. Detailed descriptions of this project can be found in Chapter 23.

Drawing cross-sections allows a high degree of geologist control on the model but is relatively labor-intensive work. In fact, the effort is proportional to both the volume of the input data and the desired density of the cross-section network. In larger projects, this can be mitigated to some degree by having the project geologist work up the initial version of the model and then hand over the files to a dedicated technical team for tidying up and running consistency checks. Additionally, Chapter 11 offers solutions for the automation of certain steps in borehole interpretation that could possibly also be deployed in cross-section-based approaches.

10.6.4 Anticipated Developments

As discussed throughout the book, modeling technologies and associated tools are diverse and are constantly evolving as new methods are developed, both in terms of algorithms and workflows. Data availability is changing with the promotion of open data and the more widespread application of a variety of data types such as site-scale geophysics and sensor arrays. Cross-section-based methodologies are likely to become increasingly integrated into more holistic modeling workflows as one class of input. Cross-sections remain a very useful representation of a geological conceptual model as discussed by Parry et al. (2014) and will continue to have merit in this regard, whether they are incorporated into 2.5-D or 3-D models, or left as more conceptual representations at the site-scale on smaller projects, for example. Their ease of construction, combined with their pervasiveness, ensures that they will remain an integral part of most ground modeling exercises for the foreseeable future, but there is increasing awareness that cross-sections alone may not be a sustainable modeling technique. Their inherent subjectivity, coupled with the growing availability of other input data and implicit modeling methods, means they are increasingly likely to be used to add a degree of geological interpretation to future semi-automated workflows. They should be integrated into wider business processes and workflows, becoming part of a toolbox with clearly defined interfaces for onwards modeling, design and visualization. A hybrid approach of explicit and implicit geological modeling using boreholes, cross-sections and geophysical data is described by Jørgensen et al. (2013). Furthermore, it will be interesting to see how cross-section methods might be applied to non-stratigraphic modeling, for example as one class of input to 3-D or even 4-D grid-based facies modeling.

References

Aldiss, D.T., Entwisle, D.C. and Terrington, R.L. (2009). *A 3D Geological Model for the Proposed Farringdon Underground Railway Station, central London.* Keyworth, UK: British Geological Survey, Commissioned Report CR/09/014.

Association of Geotechnical and Geoenvironmental Specialists (2017). *AGS4. Electronic Transfer of Geotechnical and Geoenvironmental Data. Edition 4.0.4 – February* 2017. Bromley, UK: Association of Geotechnical and Geoenvironmental Specialists. 150 pp. [Online Available at: http://www.agsdataformat.com/datatransferv4/download.php] (Accessed June, 2019).

BGS (2015). *BGS SIGMA 2015: External User Guide for BGS SIGMA Mobile and Desktop Toolbars.* Keyworth, UK: British Geological Survey, Report OR/15/029. 63 pp. [Online: Available at: http://nora.nerc.ac.uk/id/eprint/514226/] (Accessed August 19, 2020).

BGS (2019). *BGS Lexicon of Named Rock Units.* Keyworth, UK: British Geological Survey. [Online: Available at: www.bgs.ac.uk/lexicon/home.html] (Accessed June, 2019).

Binot, F. and Röhling, H.-G. (1994). Surface 2.5 modeling in the area of Vörden-Hunteburg with ISP/IES/ISM (SATTLEGGER, GEOQUEST, DYNAMIC GRAPHICS) (in German). In: *Studie über die Möglichkeiten Räumlicher Geologischer Modellierung im Niedersaechsisches Landesamt für Bodenforschung,* 73–86. Hannover: Niedersaechsisches Landesamt für Bodenforschung (NLfB) [Lower Saxony State Office for Soil Research].

Bouza-Rodríguez, B., Comesaña-Camposa, A., Menéndez-Díaz, A. and Garcia-Cortes, S. (2014). A novel geometric approach for 3-D geological modelling. *Bulletin of Engineering Geology and the Environment* 73 (2): 551–567. https://doi.org/10.1007/s10064-013-0545-9.

Burke, H., Mathers, S., Williamson, J.P. et al. (2014). *The London Basin Superficial and Bedrock LithoFrame 50 Model.* Keyworth, UK: British Geological Survey, Open Report OR/14/029. 31 pp. [Online: Available at: http://nora.nerc.ac.uk/id/eprint/507607/] (Accessed August 19, 2020).

Burke, H.F., Hughes, L., Wakefield, O.J.W. et al. (2015). *A 3D Geological Model for B90745 North Trans Pennine Electrification East between Leeds and York.* British Geological Survey Commissioned Report CR/15/04N. 32p. [Online: Available at: http://nora.nerc.ac.uk/id/eprint/509777/] (Accessed August 19, 2020).

Campbell, S.D.G., Merritt, J.E., Dochartaigh, B.E.O. et al. (2010). 3D geological models and their hydrogeological applications: supporting urban development – a case study in Glasgow-Clyde, UK. *Zeitschrift der Deutschen Gesellschaft für Geowissenschaften* 161: 251–262. https://doi.org/10.1127/1860-1804/2010/0161-0251.

Carrell, J. (2014). Tools and techniques for 3D geologic mapping in ArcScene: boreholes, cross sections, and block diagrams. In: *Proceedings of the Digital Mapping Techniques 11–12 Workshop* (ed. D.R. Soller), 19–29. Reston VA: U.S. Geological Survey, Open-File Report 2014–1167. https://doi.org/10.3133/ofr20141167.

Culshaw, M.G. (2005). From concept towards reality: developing the attributed 3D geological model of the shallow subsurface (The Seventh Glossop Lecture). *Quarterly Journal of Engineering Geology and Hydrogeology* 38: 231–284. https://doi.org/10.1144/1470-9236/04-072.

Farrant, A.F., Cripps, C., Woods, M. et al. (2016). *Scientific and Technical Report to Accompany the Chess Catchment GSI3D Geological Model.* Keyworth, UK: British Geological Survey, Commissioned Report CR/16/165. 31 pp.

Gakis, A., Cabrero, P., Entwisle, D. and Kessler, H. (2016). 3D geological model of the completed Farringdon underground railway station. In: *Crossrail Project, Infrastructure, Design and Construction, vol. 3* (ed. M. Black), 431–446. London, UK: Thomas Telford Limited and Crossrail.

Hatton, W., Henley, S. and Napier, B. (2005). Multi-dimensional modeling for BGS: the DGSM @ 2005 and beyond. In: *GIS and Spatial Analysis: Proceedings of IAMG '05: The Annual Conference of the International Association for Mathematical Geology* (eds. Q. Cheng and G. Bonham-Carter), 255–260. Houston, TX: International Association for Mathematical Geology.

Hinze, C., Sobisch, H.-G. and Voss, H.H. (1999). Spatial modelling in geology and its practical use. *Mathematische Geologie* 4: 51–60.

INSIGHT (2016). *INSIGHT Geologische Softwaresysteme GmbH Web-Page*. [Online: Available at: http://www.subsurfaceviewer.com] (Accessed February 7, 2016).

Jackson, I. and Green, C. (2003). DigMapGB – the digital geological map of Great Britain. *Geoscientist* 13 (2): 4–7.

Jones, N.L. and Wright, S.G. (1993). Subsurface characterization with solid models. *Journal of Geotechnical Engineering* 119 (11): 1823–1839. https://doi.org/10.1061/(ASCE)0733-9410(1993)119:11(1823).

Jørgensen, F., Rønde Møller, R., Nebel, L. et al. (2013). A method for cognitive 3D geological voxel modelling of AEM data. *Bulletin of Engineering Geology and the Environment* 72 (3–4): 421–432. https://doi.org/10.1007/s10064-013-0487-2.

Kessler, H. and Mathers, S.J. (2004). Maps to models. *Geoscientist* 14 (10): 4–6.

Kessler, H., Mathers, S. and Sobisch, H.G. (2009). The capture and dissemination of integrated 3D geospatial knowledge at the British Geological Survey using GSI3D software and methodology. *Computers & Geosciences* 35: 1311–1321. https://doi.org/10.1016/j.cageo.2008.04.005.

Kessler, H., Terrington, R., Wood, B., Ford, J. et al. (2016). *Specification of In- and Output Data Formats and Deliverables for Commissioned 3D Geological Models*. Keyworth, UK: British Geological Survey, Internal Report OR/16/052. 6 pp. [Online: Available at: http://nora.nerc.ac.uk/id/eprint/515457/] (Accessed August 19, 2020).

Kessler, H., McArdle, G., Burke, H. and Entwisle, D. (2017). Applications of Digital Ground Models to support the Maintenance and Upgrading of Rail Infrastructure. [Speech] At: *Ground Related Risk to Transportation Infrastructure, London, UK, 26-27 Oct 2017*.

Lemon, A.M. and Jones, N.L. (2003). Building solid models from boreholes and user-defined cross-sections. *Computers & Geosciences* 29 (5): 547–555. https://doi.org/10.1016/S0098-3004(03)00051-7.

Mathers, S.J. (2013). *Model Metadata Summary Report for the Ipswich-Sudbury LithoFrame 10-50 Model*. Keyworth, UK: British Geological Survey, Internal Report OR/12/080. 16 pp. [Online: Available at: http://nora.nerc.ac.uk/id/eprint/503113/] (Accessed August 19, 2020).

Mathers, S.J., Wood, B. and Kessler, H. (2011). *GSI3D 2011 Software Manual and Methodology*. Keyworth, UK: British Geological Survey, Internal Report OR/11/020. 152 pp. [Online: Available at: http://nora.nerc.ac.uk/id/eprint/13841/] (Accessed August 19, 2020).

Mathers, S.J., Burke, H.F., Terrington, R.L. et al. (2014). A geological model of London and the Thames Valley, southeast England. *Proceedings of the Geologists' Association* 125 (4): 373–382. https://doi.org/10.1016/j.pgeola.2014.09.001.

Ming, J., Pan, M., Qu, H. and Ge, Z. (2010). GSIS: a 3D geological multi-body modeling system from netty cross-sections with topology. *Computers & Geosciences* 36: 756–767. https://doi.org/10.1016/j.cageo.2009.11.003.

OSGeo (2019). *Discover QGIS*. Chicago, IL: Open Source Geospatial Foundation. [Website: https://www.qgis.org] (Accessed July, 2019).

Parry, S., Baynes, F.J., Culshaw, M.G. et al. (2014). Engineering geological models: an introduction: IAEG commission 25. *Bulletin of Engineering Geology and the Environment* 73: 689–706. https://doi.org/10.1007/s10064-014-0576-x.

Rosenbaum M.S. and Turner, A.K. [Eds.] (2003). *New Paradigms in Subsurface Prediction - Characterization of the Shallow Subsurface Implications for Urban Infrastructure and Environmental Assessment*. Berlin, Germany: Springer, Lecture Notes in Earth Sciences, vol 99. 397 pp. https://doi.org/10.1007/3-540-48019-6_1

Smith, I.F. [Ed.] (2005). *Digital Geoscience Spatial Model Project Final Report*. Keyworth, UK: British Geological Survey, Occasional Publication 9. 56 pp.

Sobisch H.-G. (2000). *Ein digitales räumliches Modell des Quartärs der GK25 Blatt 3508 Nordhorn auf der Basis vernetzter Profilschnitte* [A digital spatial model of the Quaternary at 1:25 000 scale of Sheet 3508 Nordhorn based on intersecting cross-sections]. Aachen, Germany: Shaker Verlag. 113 pp.

Terrington, R.L., Mathers, S.J., Kessler, H. et al. (2009). *Subsurface Viewer 2009: User Manual V1_0.2009*. Keyworth, UK: British Geological Survey, Open Report OR/09/027. 23 pp. [Online: Available at: http://nora.nerc.ac.uk/id/eprint/7195/] (Accessed August 19, 2020).

Tipper, J.C. (1993). Reconstructing three-dimensional geological systems: 1. The topology and geometry of horizons in the subsurface. *Geoinformatics* 4 (3): 199–207. https://doi.org/10.6010/geoinformatics1990.4.3_199.

Turner, A.K. (2006). Challenges and trends for geological modelling and visualisation. *Bulletin of Engineering Geology and the Environment* 65 (2): 109–127. https://doi.org/10.1007/s10064-005-0015-0.

Waters, C.N., Terrington, R., Cooper, M.R. et al. (2015). *The Construction of a Bedrock Geology Model for the UK: UK3D Version 2015*. Keyworth, UK: British Geological Survey, Report OR/15/069. 22 pp. [Online: Available at: http://nora.nerc.ac.uk/id/eprint/512904/] (Accessed August 19, 2020).

Wood, B., Richmond, T., Richardson, J. and Howcroft, J. (2017). *BGS Groundhog ® Desktop Geoscientific Information System v1.8.0 External User Manual*. Keyworth, UK: British Geological Survey, Internal Report OR/15/046. 174 pp. [Online: Available at: http://nora.nerc.ac.uk/id/eprint/511792/] (Accessed July 2018).

11

Models Created as 3-D Cellular Voxel Arrays

Jan Stafleu[1], Denise Maljers[1], Freek S. Busschers[1], Jeroen Schokker[1], Jan L. Gunnink[1], and Roula M. Dambrink[1,2]

[1] *TNO, Geological Survey of the Netherlands, 3584 CB Utrecht, The Netherlands*
[2] *Current address: Wageningen Economic Research, 2595 BM The Hague, The Netherlands*

11.1 Introduction

A geological voxel model schematizes the subsurface in a regular grid of rectangular blocks (termed *voxels, tiles,* or *3-D cells*) each measuring, for instance, 100 by 100 by 1 m (x, y, z) in a Cartesian coordinate system. Each voxel in the model contains multiple attributes describing the stratigraphy (layering) of geological units and the spatial variation of lithology and other parameters within these units. Voxel models can model in great detail the internal heterogeneity and property variability of geological units in the subsurface. Voxel models created using stochastic techniques allow the construction of multiple, equally-probable, 3-D realizations permitting evaluation of model uncertainty. The simple data structure of a voxel model $(x, y, z,$ attribute1, attribute2, …) allows for easy querying and analysis. For example, volume calculations can be performed easily by selecting and counting the voxels that meet certain criteria, such as *"all voxels 1 5 m of the land surface with a greater than 60% probability of containing sand"*. Moreover, the creation of derived products is straightforward. The ability to rapidly construct cross-sections at any chosen location is a great advantage compared with the amount of time needed to construct these cross-sections by hand (Stafleu et al. 2011). Customized 2-D raster maps using vertical voxel-stack analysis are another very useful product (Dambrink et al. 2015; Kruiver et al. 2017a,b).

Voxel modeling was first applied in the earth sciences in the 1990s (Marschallinger 1996), but its full-scale application to large subsurface databases did not start until about 10 years ago (Van Lancker et al. 2017). Applications were mostly for land data and were based on closely spaced borehole data (Van der Meulen et al. 2005; Stafleu et al. 2011; Maljers et al. 2015; Kearsey et al. 2015; Van Haren et al. 2016), dense Standard Penetration Test (SPT) datasets

in urban environments (Ishihara et al. 2013; Tanabe et al. 2015), or other sources of subsurface information such as airborne electromagnetic surveys (Jørgensen et al. 2013; Høyer et al. 2015).

Voxel models must be underpinned by a sound layer-based framework model, which is in many cases derived from the same subsurface dataset, in order to keep properties within geologic (typically lithostratigraphic) units. Examples of such framework models are given in Chapters 9 and 10. When modeling oil and gas reservoirs, the framework model is usually based on the interpretation of seismic data. Van Lancker et al. (2017) used shallow seismic data to produce a layer model that forms the basis of their voxel model of the sand reserves of the Belgian Continental Shelf (BCS).

This chapter describes the different aspects of voxel modeling based on the GeoTOP program developed and run by TNO – Geological Survey of the Netherlands (GDN) (Van der Meulen et al. 2013). The GeoTOP program produces a national voxel model that describes the architecture and properties of the shallow subsurface down to a maximum depth of 50 m below ordnance datum (Stafleu et al. 2011, 2012; Maljers et al. 2015; Stafleu and Dubelaar 2016). In its systematic approach, similar to that of customary 2-D geologic mapping, the GeoTOP program was the first and remains the only one of its kind. It has been running since 2006 and has developed mature workflows that are outlined below, along with applications and use cases.

11.2 Construction of Voxel Models

The GeoTOP modeling procedure described in Section 11.2 is followed by discussions on how to evaluate model uncertainty (Section 11.3). Section 11.4 describes the way

Applied Multidimensional Geological Modeling: Informing Sustainable Human Interactions with the Shallow Subsurface, First Edition.
Edited by Alan Keith Turner, Holger Kessler, and Michiel J. van der Meulen.
© 2021 John Wiley & Sons Ltd. Published 2021 by John Wiley & Sons Ltd.

in which voxel models can be populated with properties such as hydraulic conductivity, geotechnical parameters, organic matter or even absolute geological age. These properties can be used to create derived products aimed at specific applications (Section 11.5); several examples are described in Section 11.6. The chapter concludes with two examples of voxel models developed outside the Netherlands (Section 11.7).

11.2.1   The GeoTOP Model

GeoTOP schematizes the subsurface in a regular grid of rectangular blocks (*voxels*, *tiles* or *3-D grid cells*) each measuring $100 \times 100 \times 0.5$ m (x, y, z). The GeoTOP model is constructed in model areas that roughly correspond to the Dutch provinces (Figure 11.1). To date, the GeoTOP model covers an area of some $23\,300$ km^2, corresponding to approximately 57% of the onshore part of the Netherlands. The GeoTOP model (Version 1.3) is publicly

available at TNO's web portal https://www.dinoloket.nl/ en/subsurface-models (Stafleu and Dubelaar 2016).

GeoTOP contains some $230\,000$ digital borehole descriptions from DINO, the national Dutch subsurface database operated by the GDN, complemented with some $125\,000$ borehole logs from Utrecht University in the central Rhine-Meuse river area. The majority of these boreholes are manually drilled auger holes collected by the GDN during the 1:50 000-scale geological mapping campaigns. Most of the remaining borehole data are from external parties such as groundwater companies and municipalities. Because of the large share of manually drilled boreholes, borehole density decreases rapidly with depth. This implies that, in general, model uncertainty increases with depth.

The upper boundary of the model is derived from the 5 by 5 m cell-size national airborne laser altimetry survey dataset (*AHN2* – Actueel Hoogtebestand Nederland; www .ahn.nl). Information on water depths of rivers, canals, and

Surface geology

Holocene

Coastal deposits

Marine deposits

Fluvial deposits

Peat

Holocene / Pleistocene

Fluvial deposits

Eolian deposits

Pleistocene

Ice-pushed ridges

Glacial deposits

Other deposits

Older deposits

Older deposits

—— Main faults

Model areas

Figure 11.1 Schematic geological map of the Netherlands showing the completed GeoTOP model areas (status 2019).

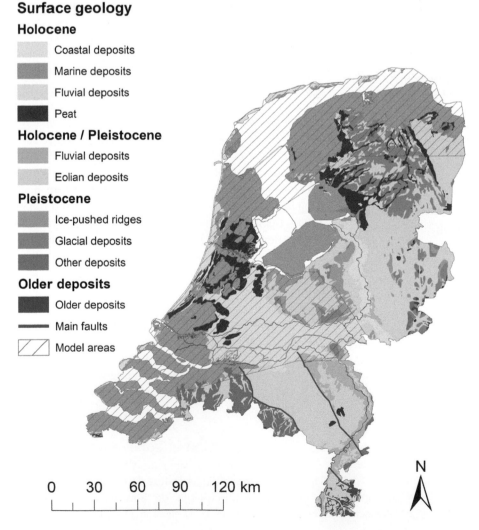

0   30   60   90   120 km

N

large water bodies, such as the Wadden Sea, was obtained from bathymetric survey data.

Information at or close to the land surface comes from the 1:50 000-scale geological maps published by the Geological Survey, the 1:50 000-scale soil map (Steur and Heijink 1991; de Vries et al. 2003), the 1:50 000-scale geomorphological map (Koomen and Maas 2004), and the national land use map developed and maintained by the Dutch national soil survey, hosted by Wageningen University & Research (*LGN5* – Landelijk Grondgebruik Nederland; Hazeu 2005, 2006). These maps were used to define the maximum lateral extent of lithostratigraphic units that occur at or near the surface. The location of the fluvial channel belts of the rivers Rhine and Meuse were derived from maps published by the Geological Survey and Utrecht University (Berendsen and Stouthamer 2001). Existing geological modeling results (raster layers) from Version 2.2 of the layer-based DGM model (Gunnink et al. 2013a) were used to define the maximum lateral extent of selected lithostratigraphic units.

Each voxel in the model contains the most likely estimates of the lithostratigraphic unit the voxel belongs to and the lithologic class (including a grain-size class for sand) that is representative for each voxel (Figure 11.2). These estimates are calculated using stochastic techniques that allow the construction of multiple, equally probable, 3-D realizations of the model as well as an evaluation of model uncertainty (Stafleu et al. 2011).

11.2.2 Modeling Procedure

The GeoTOP modeling procedure involves three steps:

1. Interpretation of borehole descriptions to produce standardized geological units with uniform sediment characteristics, using lithostratigraphic and lithologic criteria (Figure 11.3a).
2. Interpolation of borehole descriptions and construction of 2-D bounding surfaces representing the top and base of the lithostratigraphic units (Figure 11.3b). These

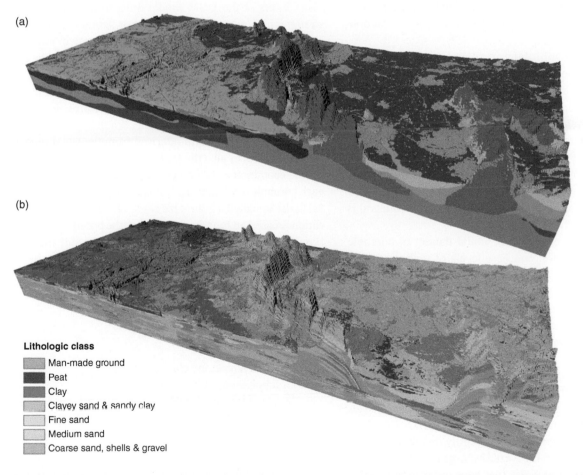

(a)

(b)

Lithologic class

- Man-made ground
- Peat
- Clay
- Clayey sand & sandy clay
- Fine sand
- Medium sand
- Coarse sand, shells & gravel

Figure 11.2 GeoTOP 3-D views of the Gelderse Vallei area in the central part of the Netherlands. (a) Lithostratigraphic units of Holocene and upper Pleistocene formations, members, and beds; (b) lithologic classes. The displayed block measures 62 × 24 km; depth of the base is 50 m below mean sea level; vertical exaggeration is 75×. Modified from Stafleu and Dubelaar (2016).

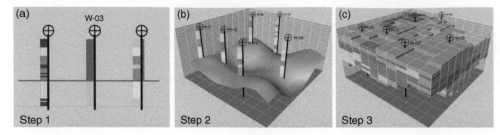

Figure 11.3 Main steps to construct the GeoTOP model. (a) Step 1: Interpretation of the borehole logs. (b) Step 2: Modeling the base of the lithostratigraphic units (raster layers). (c) Step 3: Modeling lithologic classes as voxels. Modified from Stafleu et al. (2011).

surfaces are subsequently used to place each GeoTOP voxel within the correct lithostratigraphic unit.

3. 3-D stochastic interpolation of standard lithologic classes in the borehole descriptions (clay, sand, peat) and, if applicable, sand grain-size class within each lithostratigraphic unit (Figure 11.3c). After this step, a 3-D geological model is obtained.

11.2.2.1 Step 1: Interpretation of Borehole Descriptions

The interpretation of borehole descriptions has two stages: (i) the lithostratigraphic interpretation of borehole descriptions, and (ii) the lithologic classification of borehole descriptions.

The lithostratigraphic interpretation is based on the geological knowledge of the area combined with expert judgment of the feasibility of successfully modeling a unit. For each of the units, a map delineating the maximum lateral extent of occurrence is created. Next, automated procedures consisting of computer programs written in the programming language Python apply a pre-defined set of diagnostic criteria to the borehole descriptions located within the maximum extent of each lithostratigraphic unit. If an interval in the borehole description meets the criteria, this interval is assigned that particular unit.

This procedure produces an initial dataset of borehole locations defining top and base levels of each unit. This dataset is subsequently plotted on a map along with the maximum lateral extent of the unit. An iterative process of successive modifications to the maximum extent and/or the set of criteria revises and improves the resulting interpretations. This iterative process consumes a significant part (~50%) of the total modeling effort. This procedure is followed for many of the Holocene and upper Pleistocene formations, members, and beds. The result is a revised dataset of borehole locations defining top and base levels for each of these lithostratigraphic units.

This procedure works best in roughly the upper 10 m of the subsurface, which is covered by large numbers of good-quality (augered) boreholes that have enough parameters to work with. Because the lithologic contrasts between various units are sharp, especially in Holocene terrains,

they are easily identifiable. Below the upper 10 m, the available borehole data are generally of lesser quality, and the lithological contrasts are more subtle between the generally fluviatile Pleistocene units. Thus, the automated interpretation does not work as well and a separate automated procedure is employed; it assigns the borehole descriptions to alternative lithostratigraphic interpretations based on the national geological stacked-layer DGM model (Gunnink et al. 2013a).

Each borehole is intersected with the top and basal raster layers of the DGM model units to provide a set of stratigraphic interpretations that are geometrically consistent with the DGM model. The interpreted top and base levels defined by the DGM interpretations and by the revised dataset are then combined to form a single dataset containing the stratigraphic interpretation of the borehole description. During this combination procedure, the stratigraphic interpretation based on the borehole descriptions overrules the interpretation derived from the DGM model. GeoTOP also uses also the faults of the DGM model.

The second stage of Step 1 of the modeling process assigns each borehole interval to a lithologic class; each sandy interval is also assigned a grain-size class. The lithologic classification is based on four attributes from the borehole description: (i) main lithology; (ii) admixtures of sand, silt, and clay; (iii) sand median grain-size; and (iv) clay percentage. No lithologic classification is applied to the anthropogenic deposits. The derived lithologic classes are:

- Anthropogenic deposits.
- Organic deposits (peat and gyttja).
- Clay.
- Clayey sand and sandy clay.
- Sand, subdivided into:
  - ○ Fine sand (median grain size 63–150 μm).
  - ○ Medium sand (median grain size 150–300 μm).
  - ○ Coarse sand, gravel, and shells (median grain size >300 μm).

This lithologic classification scheme is also used in the hydrogeological subsurface model REGIS II (Vernes and Van Doorn 2005; Chapter 19) because these lithologic classes are suitable for groundwater flow modeling. The DINO database holds legacy borehole data, which describe lithology according to multiple sediment classifications schemes. Assigning GeoTOP lithologic classes is basically a conversion exercise that needs to deal with data heterogeneity.

11.2.2.2 Step 2: Two-Dimensional Interpolation of Lithostratigraphic Surfaces

The base levels of the lithostratigraphic units found in the boreholes are interpolated to regular grids with a cell size of 100 by 100 m using Sequential Gaussian Simulation (SGS) (Goovaerts 1997; Chilès and Delfiner 2012) as implemented in the Isatis modeling software package from Geovariances. SGS estimates the base level of a unit at a given location based on the values of the data points in a circular search neighborhood and a variogram model describing the spatial correlation. The variogram model ensures that the data most closely correlated with the target cells are given the greatest weight in the interpolation. In addition to the "hard data" from the boreholes, the algorithm also selects values from grid cells within the neighborhood that already have assigned values. By estimating the grid cells in a random order, different grid cells have assigned values at each stage of the simulation. In this way, 100 simulations result in 100 different, but statistically equiprobable, realizations of the base level of each lithostratigraphic unit. From these 100 realizations, mean base surfaces are calculated and subsequently used to construct a single, "most likely" integrated layer model that considers the stratigraphic relationships between the units. In the integrated layer model, top surfaces of each unit are defined by the basal surface of the overlying unit. For many units, the regional trend is removed from the data by first creating a trend surface which represents the base level of the unit on a regional scale. The SGS algorithm was then applied to evaluate the residuals, which define the difference, for each borehole, between the depth of the intersection of the regional surface and the depth of the corresponding stratigraphic transition identified in that borehole.

Several critical decisions are required during construction of the integrated layer model to ensure that geostatistical procedures do not violate geological rules or knowledge. For example, in the GeoTOP workflow, it is almost impossible to preserve integrity of a thin unit when interpolating upper and lower surfaces in close vertical proximity to each other. This problem arises when modeling the Basal Peat Bed at the base of the Holocene series, which is thin (generally <1 m), but (hydro)stratigraphically of key importance. Thus, except in those cases where it is clear that actual erosion has occurred, a minimum thickness of

0.3 m is assigned to the Basal Peat Bed below the base of the overlying units.

After completion of the integrated layer model, the basal, and top surfaces are used to assign the correct lithostratigraphic unit to each voxel within the 3-D model space. All voxels with centroids between the top and base of a lithostratigraphic unit are assigned to that particular unit.

11.2.2.3 Step 3: Three-Dimensional Interpolation of Lithologic Class

The lithologic classes in the boreholes are used as input for a 3-D stochastic simulation procedure within each lithostratigraphic unit. The Sequential Indicator Simulation (SIS) technique (Goovaerts 1997; Chilès and Delfiner 2012) is applied using the Isatis modeling software package. SIS is used because it is a well-established method for simulating lithofacies or lithologic class distributions, it requires modest computation time and is straightforward to implement (Stafleu et al. 2011).

SIS estimates lithologic classes for each voxel within a particular lithostratigraphic unit based on the lithologic class of the surrounding borehole intervals of the same lithostratigraphic unit. The algorithm uses a 3-D ellipsoidal search neighborhood around the target voxel to select the data points and a variogram model which ensures that the data most closely correlated with the target voxel are given the greatest weight. In addition to the "hard data" from the boreholes, the algorithm also selects values from voxels within the neighborhood that were previously assigned values. By estimating the voxels in a random order, 100 simulations result in 100 different, but statistically equiprobable, realizations of 3-D lithologic class distribution.

From these 100 realizations, probabilities of occurrence for each lithologic class are calculated, these provide an indication of model uncertainty (Figure 11.4). In addition, the probabilities are used to compute the most likely lithologic class for each voxel, using the averaging method for indicator datasets described by Soares (1992). The individual realizations remain available for use in applications where uncertainty in the subsurface properties plays an important role – for instance, in groundwater flow modeling.

Especially in the deeper parts of the model, the neighborhood search at a target voxel may not identify any hard data values from boreholes or already simulated voxels. The assignment of lithologic class is then drawn from global proportions of each lithologic class observed in the boreholes. These proportions are assumed to be constant throughout each lithostratigraphic unit. For specific heterogenetic units, a proportion function was used to honor spatial variations in the expected proportions. In the Holocene Naaldwijk Formation for instance, shallow

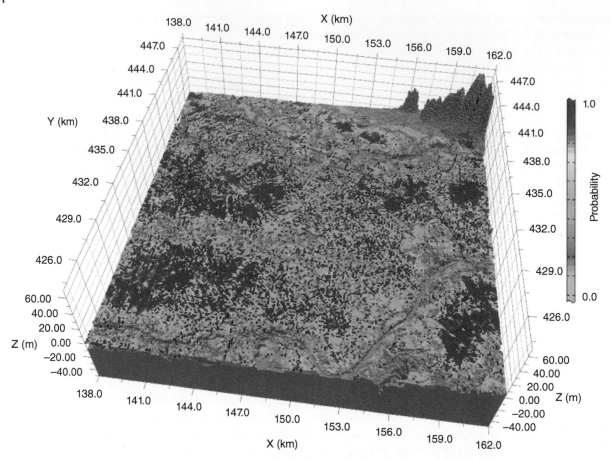

Figure 11.4 Probability of the occurrence of clay in the flood basin area of the river Rhine. Modified from Stafleu and Dubelaar (2016).

tidal flat clays overlie deeper tidal channel sands. This heterogeneity was modeled by applying a *Vertical Proportion Curve* (VPC) describing the expected proportion of sand and clay as a function of depth (Stafleu et al. 2011). Another example is the occurrence of impermeable clay layers within the glacial Peelo Formation, which was modeled using raster layers from Version 2.1 of the hydrogeological model REGIS II (Vernes and Van Doorn 2005). Voxels located within these raster layers were assigned the expected high proportion of clay (Kruiver et al. 2017a,b).

Some borehole data included intervals that have a described lithology 'sand' but lack grain-size data. For this reason, sand grain-sizes were assigned with a multi-step procedure. First, 10 simulations of the distribution of sand versus non-sand sediment were calculated using all available borehole data. Second, for each of the 10 simulations, the voxels assigned a sand lithology were selected. Third, a further 10 simulations were performed on these selected sand voxels, using only the borehole data with known grain-size estimates; this produced a total of 100 simulations of sand grain-sizes. Fourth, a similar two-stage simulation procedure was applied to the non-sand voxels, producing another 100 simulations. Fifth,

the 100 simulations of sand voxels were combined with the 100 simulations of non-sand voxels; this produced 100, statistically equiprobable, simulations of lithologic class distributions.

11.3 Model Uncertainty

The current version of the GeoTOP model covers an area of $23\,300\,\text{km}^2$. Within this area, some 355 000 borehole descriptions are available, which means that only about 7% of the voxels at the land surface actually contain a borehole. Moreover, this number rapidly diminishes with depth because many boreholes are quite shallow. Therefore, the lithostratigraphic unit and the lithologic class of a vast majority of voxels has to be *estimated* on the basis of nearby borehole logs. As described in Section 11.2.2, stochastic interpolation techniques were used to define these estimated properties. These techniques provide multiple, but statistically equally-probable, estimates. For example, if the model is run 100 times, a specific voxel may be estimated with clay 80 times, with clayey sand 10 times, and with fine sand 10 times. The different results indicate a degree of uncertainty of the model estimates of the value of each voxel. In the best case, all model runs result in the same value; in the worst case all possible values occur with the same frequency. Thus, the GeoTOP model contains inherent uncertainty in its assignment of attribute values to voxels.

Chapter 15 discusses the sources of, and alternative methods for, evaluating model uncertainty. In general, the uncertainty contained within any geological subsurface model is the product of five sources: (i) the quality of geological data sources; (ii) the experience of the modeling geologist; (iii) the complexity of the geology being modeled and the scale at which it is conveyed; (iv) the geological modeling methodology; and (v) the way in which the model data are applied.

Since the GeoTOP model is built using a restricted set of pre-defined data sources, has to deal with only a limited number of geological settings, and has been developed by a small group of experienced geologists, several of these sources of uncertainty are kept constant. The GeoTOP model was designed and developed to provide an appropriate and consistent level of reliability, or uncertainty, based on requirements for understanding the shallow Dutch subsurface by anticipated applications (see Section 11.6). However, the stochastic assignment of parameter attributes to each voxel results in some variation in the uncertainty of individual voxels. The extent to which the model successfully estimates the values for each voxel strongly depends on the following factors:

- *The complexity of the geology.* A simple, homogeneous subsurface can be modeled more easily and with fewer borehole descriptions than a complex, heterogeneous subsurface. Differences in complexity may also occur within the lithostratigraphic units.
- *The number of available borehole logs.* Borehole logs are not equally distributed in the Netherlands; the urbanized coastal lowlands and large parts of the Rhine-Meuse area have a high density of boreholes. Other parts of the country, such as the nature reserves on the Veluwe and in the Wadden Sea, have a low density of boreholes. In addition, more data are available for the shallow parts of the subsurface. In general, the limited data deeper than 30 m below the surface strongly reduces the quality of the lithologic class estimates in the GeoTOP model.
- *The quality of the borehole logs.* The borehole logs in the national subsurface database DINO were not collected for use in GeoTOP. Their quality strongly varies, depending on the aim and the method used for the drilling. The best subset is formed by the survey boreholes that were augered for the former 1:50 000 geological mapping program of GDN (discontinued in 1997). These maps were profile type maps (discussed in Chapter 5), which

are essentially shallow 3-D models. Both the data and the maps themselves are an invaluable resource for the successor GeoTOP program.

- *The quality of the constraints used in the modeling.* In some instances, older data sources were used to delineate the extent of the geological units. Information on the lateral extent of deeper units is always less precise than for near-surface units.
- *The modeling algorithm used.* The selected modeling algorithm, as well as its associated parameters, such as the spatial correlation functions (variograms) inferred from the borehole data, will influence the model results.
- *The age of the source data.* Older borehole descriptions and geological and soil maps may not accurately represent the current subsurface conditions. For instance, peat mentioned in a borehole description may have been oxidized over time. Excavations, such as for harbors or dredged channels, or the shifting of active tidal channels in the Wadden Sea, may also result in a considerably changed subsurface.

11.3.1 Probabilities

A probability of occurrence of any lithologic class within a voxel may be calculated from the multiple stochastic model realizations by simply dividing the number of times a particular lithologic class is assigned to a voxel by the total number of realizations. The frequencies can be displayed as a bar graph (Figure 11.5). In a similar way, the probability that a voxel belongs to a certain geological unit may be calculated. For instance, if a voxel is assigned to the Naaldwijk Formation in 75 model simulations out of a total of 100 simulations, then the corresponding probability of the presence of the Naaldwijk Formation is 0.75. Vertical voxel stacks, which can represent *virtual boreholes*, allow visualization of the stratigraphic succession as well as multiple bar graphs of the probabilities (Figure 11.6). These probabilities provide us with a measure of model uncertainty.

11.3.2 Information Entropy

Two-dimensional visualizations of vertical cross-sections cannot show all the probabilities for each voxel in a single view; the user will always be presented with one of the probabilities at a time. To solve this problem, Wellmann and Regenauer-Lieb (2012) proposed to use *information entropy* as a measure of uncertainty in 3-D models. The information entropy of a voxel is a single value ranging from 0 to 1 that can easily be calculated from each of the probabilities of lithologic class. An entropy value of 0 represents no uncertainty, whereas a value of 1 occurs when all lithologic classes are equally likely. Values between 0 and 1

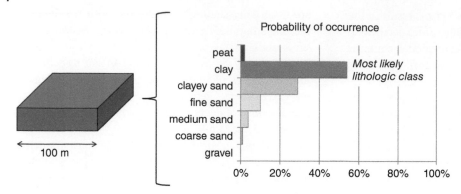

Figure 11.5 A bar graph shows the probabilities of occurrence of each lithology in an individual voxel. In this example, there is a reasonable probability of clay, over 50%, and also a considerable probability of clayey sand. On the other hand, the probability of no sand or peat in this voxel is over 80%.

account for both the number of lithologic classes with a probability higher than 0 (the more classes, the higher the entropy) and the differences amongst the probabilities (the greater the differences, the lower the entropy). Chapter 15 discusses the *information entropy* concept in more detail, including examples from applications by Bianchi et al. (2015) analyzing subsurface deposits in Glasgow. When

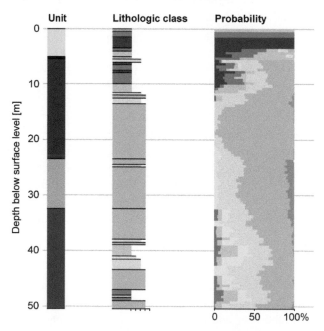

Figure 11.6 Vertical voxel stack taken from the GDN's interactive web portal showing the vertical succession of voxels at a particular location. From left to right: the lithostratigraphic units, the "most likely lithologic class" and a plot of the probabilities for the lithologic classes. In the latter plot we can infer zones with a low uncertainty (coarse sand at 20 m; peat layers near land surface) and zones with a high uncertainty (at depths greater than 32 m). For legend of the lithologic classes see Figure 11.2.

users create a cross-section on GDN's information portal Dinoloket (www.dinoloket.nl), they can toggle between model visualizations and information entropy to get an impression of model reliability.

11.3.3 Borehole Density

Borehole density is another approach to visualizing uncertainty. Many users assume a one-to-one relationship between borehole density and reliability of the model. Although this relationship does not always exist (because a heterogeneous unit may not be fully characterized by a single borehole), it adheres to common sense and is therefore easily understood (Figure 11.7). It provides a useful qualitative assessment of uncertainty – a "rule-of-thumb". Borehole density measures of uncertainty have been used by many organizations. Interesting examples include the dot maps used in Ontario (Bajc and Dodge 2011; Burt and Dodge 2011, 2016) and semi-quantitative evaluations of borehole density, quality, and geological complexity in the United Kingdom (Lelliot et al. 2009). Chapter 15 includes these and other examples.

In the GeoTOP model, borehole density is calculated for horizontal slices through the model, each at a certain height with respect to mean sea level. Standard 2-D GIS procedures counted the number of boreholes available in cells of 5 by 5 km at the depth of each of these horizontal slices. The result is then displayed as a voxel model (Figure 11.7).

11.4 The Value of Adding Property Attributes

Property attributes convert the lithostratigraphic-unit voxel models into powerful instruments for a wide range of applications including groundwater management, ground risk assessment, the planning of infrastructure works, and aggregate assessments.

The spatial variation of many subsurface properties, such as hydraulic conductivity, strongly depends on the

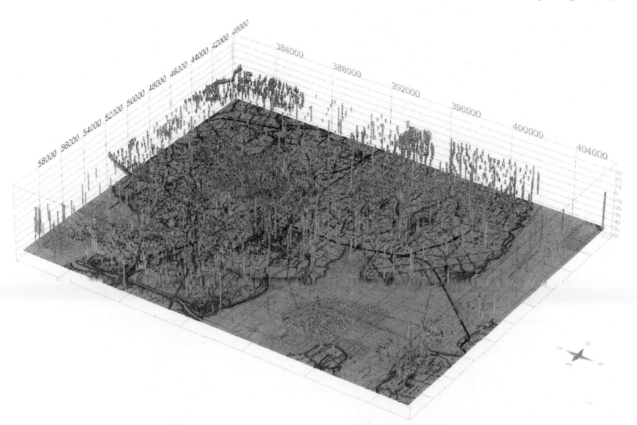

Figure 11.7 Typical variation in borehole density that GeoTOP modeling has to deal with in a random 20 × 25 km size block. There are areas with many boreholes, and areas with few boreholes. In general, borehole density decreases with depth.

two main attributes modeled in GeoTOP and other voxel models: the lithostratigraphic unit and the lithologic class. Hydraulic conductivity can be assigned to each voxel by direct correlation to the combination of lithostratigraphic unit and lithologic class. However, in many cases, depth, an attribute that is obviously available for each voxel, is an additional factor controlling the appropriate property values. One example of this situation is the assignment of appropriate values of permeability and compressibility to organic deposits; both are affected by combinations of lithology and depth. Under these conditions, assignment of an appropriate parameter value to an individual voxel requires assessment of the vertical voxel-stack to which the voxel belongs.

Procedures to add property values may be illustrated with three examples: (i) the determination of *hydraulic conductivity*, which makes the voxel model applicable for groundwater flow studies; (ii) the determination of *shear-wave velocity*, which enables the model to be used in a risk-assessment of human-induced earthquakes; and (iii) defining the *organic matter* content in peat to improve the applicability of the model in land subsidence predictions. The following sections provide details of these procedures.

11.4.1 Hydraulic Conductivity

Hydraulic conductivity is a key parameter in geohydrological modeling. Direct measurements of hydraulic conductivity at the model-scale are often not available. Instead, hydraulic conductivity measurements are derived from pump-tests which are costly and often produce multiple interpretations from empirically derived relationships with grain-size distributions, or by measuring hydraulic conductivity of small volumes of sediments that are extracted from boreholes.

GDN runs a national borehole campaign designed to repeatedly sample each combination of lithostratigraphic unit and lithologic class at different locations and depth ranges. Hydraulic conductivity was measured from the samples using the constant head method (for sandy samples) or the falling head method (for clayey samples), resulting in a unique dataset of geologically classified hydraulic conductivities. However, the volume of the samples ($0.0004\,\text{m}^3$ for sandy samples and even smaller for clayey samples) is several orders of magnitude smaller than the volume of a GeoTOP voxel ($5000\,\text{m}^3$); thus, some form of property upscaling is necessary (Tran 1996; Figure 11.8a). Several upscaling methods exist, but none of these take into account the internal heterogeneity at the model-scale. It is especially important to apply an appropriate upscaling method to units that are thinly layered or have other forms of variation at the sub-unit or sub-voxel scale, for instance tidal deposits which have rapid alternations of sand and clay (Figure 11.8b).

Geological expert-knowledge about the spatial scale at which the alternations of sand and clay within a voxel occur was used to determine the 3-D spatial distribution of lithology within a voxel (Gunnink et al. 2013b). The initial GeoTOP voxel (100 by 100 by 0.5 m) was subdivided into tiny voxels of $0.5 \times 0.5 \times 0.05$ m. Using SIS, 50 spatially correlated models of sand–clay occurrence were produced (Figure 11.9a). For each of these models, a SGS process produced 50 3-D models of hydraulic conductivity for each small voxel according to its specific lithology (sand or clay). This process used values contained within the dataset of geologically classified hydraulic conductivities (Figure 11.9b). An upscaled hydraulic conductivity was calculated for each 100 by 100 by 0.5 m GeoTOP voxel by applying the groundwater flow model iMOD (Vermeulen et al. 2017) with a vertical head of 1 m and "no-flow"-conditions at the lateral boundaries of the voxel.

Figure 11.10 presents some results from the upscaling procedure. The upper row shows the vertical hydraulic conductivity, as derived from the measurements on the small samples, while the bottom graph shows the upscaled vertical hydraulic conductivity for a voxel containing 30% clay and 70% sand. This distribution of sand and clay

(a)

(b)

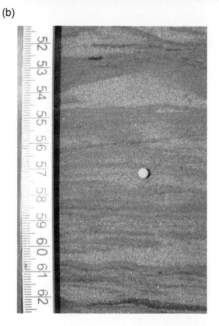

Figure 11.8 Upscaling issues when determining GeoTOP model hydraulic conductivity values: (a) scale difference between sample measurement and dimensions of a GeoTOP voxel; (b) short 0.1 m length of core showing alternating sand and clay, typical of tidal environments, which influence vertical hydraulic conductivity. Modified from Gunnink et al. (2013b).

(a)

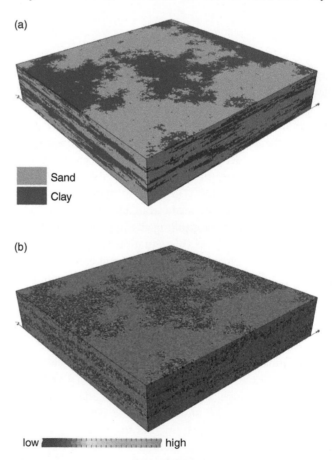

(b)

Figure 11.9 Upscaling hydraulic conductivity within a GeoTOP 100 m × 100 m × 0.5 m voxel. (a) Spatial distribution of sand and clay; (b) hydraulic conductivity conditioned by the sand-clay model. Modified from Gunnink et al. (2013b).

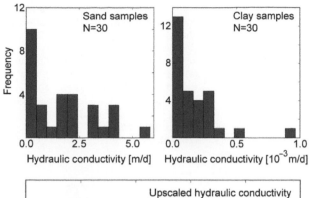

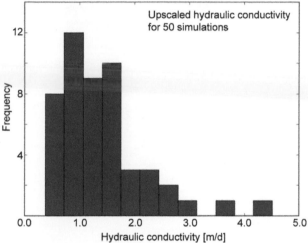

Figure 11.10 Vertical hydraulic conductivity (as measured from the samples) for sand and clay (top row) and upscaled vertical hydraulic conductivity for a voxel containing 30% clay and 70% sand. Modified from Gunnink et al. (2013b).

within the voxel results in an upscaled conductivity that is considerable larger than the conductivity of the clay samples, but not as large as the sandy samples.

Upscaled hydraulic conductivities were assigned to each combination of lithostratigraphy and lithologic class. The hydraulic resistance to vertical flow for each voxel was then determined by multiplying the upscaled hydraulic conductivity by the thickness of each voxel (0.5 m). This procedure resulted in a raster map of the hydraulic resistance of the Holocene deposits (Figure 11.11), which is an important input in regional groundwater flow modeling.

11.4.2 Shear-Wave Velocity

The Groningen gas field in the Netherlands is one of the largest gas fields of Europe and has been in production since the 1960s. Due to the progressive depletion of the reservoir, induced seismic activity has increased in recent years. In 2012, an earthquake of magnitude 3.6 initiated further research into the prediction and management of risks related to human-induced earthquakes.

In risk-assessments of earthquake damage, the shear wave velocity (V_s) for the upper 30 m of the subsurface column (V_{s30}) plays an important role. Kruiver et al. (2017a,b) combined the GeoTOP model of the Groningen area and Seismic Cone Penetration Tests (SCPT's) into a V_s model of the area covering the gas field. Statistical distributions (with mean and standard deviation) of V_s for each combination of lithostratigraphic unit and

Figure 11.11 Hydraulic resistance for vertical flow for Holocene deposits in the SW part of the Netherlands, based on upscaled vertical hydraulic conductivity. Modified from Gunnink et al. (2013b).

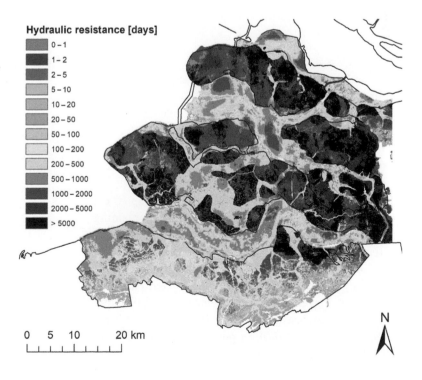

(a)

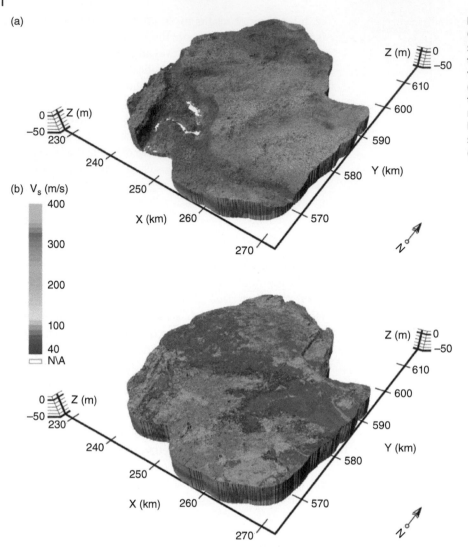

Figure 11.12 GeoTOP model of the Groningen area in 3-D, attributed with shear wave velocity (V_s) estimated from the mean of 100 randomly drawn values from the statistical distribution. (a) The model with all strata younger than the glacial Peelo Formation removed. (b) The full model from 50 m below mean sea level up to land surface. Modified from Kruiver et al. (2017b).

lithologic class derived from 60 SCPTs were used to randomly assign a specific V_s to each voxel in the model (Figure 11.12).

The V_{s30} for each voxel stack was then calculated using the harmonic mean of the V_s of the 60 voxels that cover the upper 30 m and plotted as a raster map. The uncertainty in V_{s30} was determined by repeating this procedure 100 times.

The resulting 3-D V_s model and 2-D V_{s30} map reveal zones with distinct V_{s30} characteristics: areas containing predominantly soft Holocene deposits with low V_{s30} are differentiated from areas with predominantly stiff Pleistocene deposits with high V_{s30}. Previously, only a single V_{s30} value was used for the entire Groningen gas field. Both the new V_{s30} map and vertical voxel stacks attributed with V_s and other sediment properties have been used as input for site amplification predictions (Kruiver et al. 2017a,b).

11.4.3 Organic Matter

Holocene peat is abundantly present in the shallow subsurface of the Netherlands. Due to human-induced lowering of groundwater levels, the peat is experiencing severe volumetric loss by oxidation and compression. This results in land subsidence and CO_2 emissions. To assess the potential of these peat layers to oxidation and compression during future groundwater level lowering, GeoTOP voxels classified as Holocene peat were parameterized with physical properties (Koster et al. 2018a).

Peat is composed of partly decayed vegetation and is therefore rich in organic matter. However, it also contains mineral components, and has many voids that are occupied by water and gas. These produce lateral and vertical variations in peat properties, owing to differences in environmental conditions during peat formation and burial history. The volumes of organic matter in peat are important to predict future oxidation and CO_2 emissions, whereas the void space determines the subsidence potential by induced compression. A three-step process assigned physical properties to peat voxels:

- Peat was classified into 10 classes using empirical relationships between stratigraphic position and vertical trends in dry mass proportions of organic matter and sediments; these relationships were established from 637 peat samples obtained from cored boreholes.
- Each GeoTOP peat voxel was categorized according to these 10 classes. To obtain internal voxel trends, the amount of organic matter and sediments was determined at five vertical positions within each peat voxel.
- Standard values for mass density were subsequently used to convert dry mass proportions of organic matter and sediment into dry volume proportions of organic matter and sediment (Erkens 2009). For each attributed voxel, the ratio between solid and non-solid components was calculated using a compression function that requires definition of the vertical effective stress exerted on the peat (a depth-dependent parameter) and dry mass proportion of organic matter (Den Haan 1992; Koster et al. 2018a). The obtained volumes of solid components were subsequently divided into volumes of organic matter and sediment according to their stratigraphy and proportions of organic matter and sediment.

This parameterization yielded a 3-D overview of the trade-off between volumes of organic matter, mineral components, and void space in Holocene peat voxels within the GeoTOP model (Figure 11.13).

11.5 Derived Products for Applications

Visualizing, exploring, and querying 3-D property models requires sophisticated software packages such as Isatis, GoCAD, and Petrel. These software packages are expensive, often difficult to operate, and mostly not available to the user community. A solution to this problem is to provide users with products derived from the 3-D model. These derived products include: (i) 2-D map products; (ii) vertical "voxel stack" profiles; and (iii) 3-D model products and visualizations.

11.5.1 2-D Map Products

Two-dimensional raster layers of the tops and bases of the lithostratigraphic units can be viewed and analyzed in standard GIS software packages. In addition, 2-D raster maps may be produced that depict a series of horizontal slices through the 3-D voxel model. These maps can display selected voxel attributes such as the lithostratigraphic unit, the lithologic class and probabilities of occurrence of a particular lithologic class at a certain depth below mean sea level or below land surface. Figure 11.14 shows an example of slices through the lithologic class model at depths of 0, −2, −4, −6, and −8 m below mean sea level. In these slices, the position of the tidal channels (sands), adjacent zones of tidal flats (clays), and zones where peat occurs are clearly visible. Other useful 2-D map products are "subcrop maps" where shallower geological units are removed in order to reveal deeper units (Figure 11.15a).

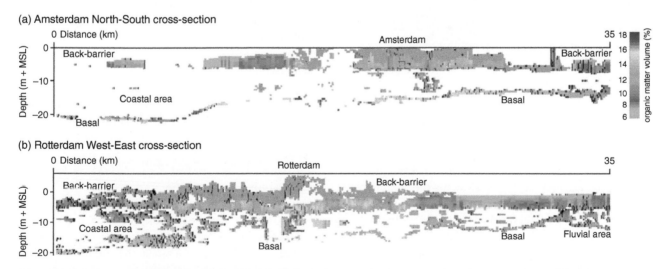

Figure 11.13 Two cross-sections showing Holocene peat voxels attributed with their organic matter volume. Gray voxels indicate man-made ground in the urban areas of Amsterdam (top panel) and Rotterdam (bottom panel), respectively. Peat layers outside urban areas and close to land surface have low volumes (<8%) whereas the deeply buried basal peat layers have the highest (up to 18%). Modified from Koster et al. (2018a).

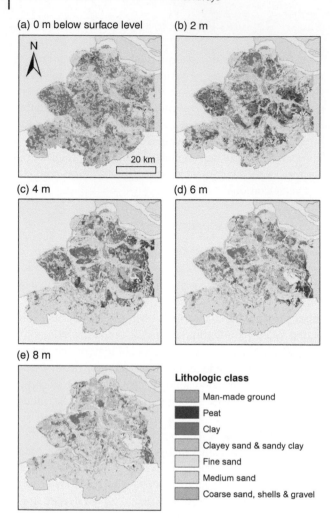

(a) 0 m below surface level

(b) 2 m

N

20 km

(c) 4 m

(d) 6 m

(e) 8 m

Lithologic class

Man-made ground

Peat

Clay

Clayey sand & sandy clay

Fine sand

Medium sand

Coarse sand, shells & gravel

Figure 11.14 Horizontal slices through the voxel model attributed with lithologic class at selected depths below surface level revealing the position of tidal channels within the Holocene lithostratigraphic units present in the SW part of the Netherlands. Modified from Stafleu et al. (2011).

11.5.2 2-D Products from Vertical Voxel Stack Analysis

Other 2-D map products are created by selectively querying the 3-D voxel model using a vertical voxel stack analysis, producing a single summary value that can be plotted as a 2-D raster map (Dambrink et al. 2015). Examples of such 2-D products include: a map showing the cumulative thickness of peat (Figure 11.15b), a generalized lithological map of the upper 8 m (Figure 11.15c), and the hydraulic resistance for vertical flow map, based on the upscaled hydraulic conductivities as described in Section 11.4.1 (Figure 11.11). These examples show how smart selections and combinations of lithostratigraphy and lithologic classes up to a chosen depth result in maps that are tailor-made to user requirements. When needed, additional maps can be

created relatively easily, bringing geological information one step closer to the user's own area of expertise.

11.5.3 3-D Geological Map Products

Detailed stratigraphic-lithologic models such as GeoTOP may be regarded as 3-D geological maps, replacing their 2-D predecessors (especially the aforementioned 1:50 000 geological map series). Schokker et al. (2015) demonstrated how DGM and GeoTOP provide an easily accessible way to obtain a better understanding of the shallow subsurface of the city of Amsterdam. An important step toward a fully interactive and queryable 3-D geological map that is available to the user community is the development of the SubsurfaceViewer software developed by INSIGHT GmbH (Sobisch 2011). The software is delivered free of charge via TNO's web portal in conjunction with the GeoTOP voxel model (Stafleu et al. 2013). The SubsurfaceViewer includes both a classical map view and an interactive 3-D view. The voxels visualized in the 3-D view may be queried by making simple selections on lithostratigraphic units, lithologic class, and/or ranges of probabilities; for example, all Holocene voxels with a probability greater than 65% of containing peat may be identified and displayed. De Mol (2018) developed a more sophisticated web-based voxel-query tool that allows for complex analyses. This web-based approach is attractive to a wider range of users than the desktop-based SubsurfaceViewer.

11.6 Examples of Applications

The GeoTOP model has supported several important applications including: (i) evaluating subsurface geotechnical conditions during the preliminary planning of infrastructure improvements; (ii) assessing land subsidence due to compaction of clay and peat; (iii) aggregate resource assessments; (iv) defining the lithologic characteristics of Holocene channel belt systems; and (v) potential environmental impacts of dredging operations.

11.6.1 Geotechnical Applications

In the planning stage of large infrastructure works, such as the construction of tunnels and highways, GeoTOP provides insight into the expected load-bearing capacity of the subsurface. Especially in the west of the country, the depth of the Pleistocene sands, the composition of the Holocene sediments and the location and thickness of the fluvial channel belt sands are important parameters in the calculation of costs and in the planning of additional subsurface research. The expected composition of the

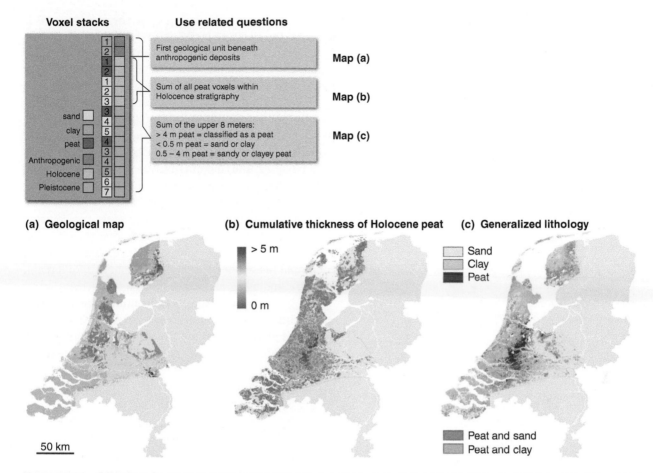

Figure 11.15 Examples of a secondary products created by a vertical voxel stack analysis: (a) geological map, showing geological units at land surface; (b) cumulative thickness of Holocene peat; and (c) generalized lithology of the upper 8 m. Modified from Dambrink et al. (2015).

subsurface may also be a factor in the choice of the road trajectory.

Figure 11.16 shows an example of a light-rail tunnel in the city of Rotterdam with a vertical succession typical for the coastal lowlands: Pleistocene sands of varying grain sizes, followed by Holocene, clayey tidal deposits intercalated with peat layers and capped with a thick sequence of man-made ground. To save construction costs, the larger part of the tunnel is drilled in the Pleistocene sands. Within the Pleistocene sands, areas containing coarse-grained sand present more favorable drilling conditions than areas with fine-grained sand. Geotechnical engineers are therefore particularly interested in probability displays (Figure 11.16b), which gives them the opportunity to more efficiently plan the borehole campaign for gathering additional data. However, each infrastructure project will eventually need additional site-specific investigations of the subsurface (Figure 11.16c). The scale of the GeoTOP model is not appropriate in the construction phase of an infrastructure project.

11.6.2 Land Subsidence

Many urbanized peat-rich coastal-deltaic plains are subsiding due to compression and oxidation of near-surface Holocene peat layers. Examples include the province of Friesland in the Netherlands (Nieuwenhuis and Schokking 1997), the Venice Lagoon (Gambolati et al. 2006) and the Pontina plain (Serva and Brunamonte 2007) in Italy, and the Sacramento–San Joaquin Delta in California (Drexler et al. 2009; Deverel et al. 2016). Land subsidence results in relative sea-level rise, increasing flood risk, salt water intrusion, and damaged infrastructures (Higgins 2016). This form of subsidence is often human-induced, a consequence of phreatic groundwater level lowering and increased surface loading (Figure 11.17). Land subsidence caused by compaction of clay and peat and by oxidation of peat strongly depends on the lithologic composition of the shallow subsurface and on the elevation of the groundwater table.

In the Netherlands, GeoTOP has become an important tool in predicting future land subsidence on a regional

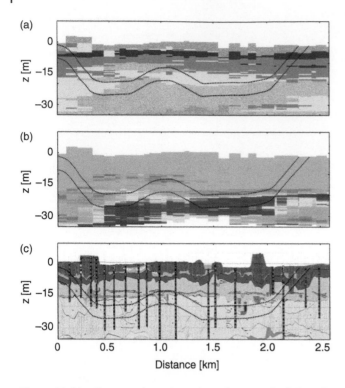

Figure 11.16 Cross-sections along the trajectory of a light-rail tunnel in the city of Rotterdam overlain with the tunnel trajectory. (a) Lithologic class distribution according to the GeoTOP model. The tunnel is constructed well into the Pleistocene sands, except for the station in the center where the tunnel is located closer to land surface. For legend see Figure 11.14. (b) The probability of coarse sand from the GeoTOP model. (c) Geotechnical profile drawn by hand using additional cone penetration tests. Source: Courtesy Municipality of Rotterdam.

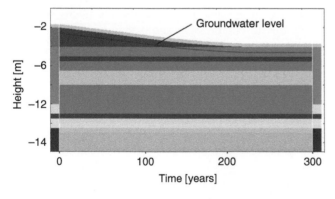

Figure 11.17 Prediction of land subsidence due to oxidation of near-surface peat in a single vertical voxel stack over a period of 300 years. To support dairy farming the phreatic groundwater level is artificially kept at a constant depth below land surface, exposing near-surface peat to air. Consequently, peat oxidizes, and land subsides until all near-surface peat has disappeared. For legend see Figure 11.14.

scale. The land subsidence map of Utrecht province is based on the GeoTOP model (Van der Schans 2012). The Flevoland subsidence map (De Lange et al. 2012) is an example of a prediction based on a voxel model similar to GeoTOP. An online interactive map of land subsidence in the Netherlands based on GeoTOP was published by Climate Adaptation Services (2017). Employing the parameterization of Holocene peat voxels in GeoTOP with organic matter content described in Section 11.4.3, Koster et al. (2018b) quantitatively evaluated the effects of lowering the present-day phreatic groundwater level using predictive functions for peat compression and oxidation. In this assessment, phreatic groundwater levels were lowered by 0.5 m, and the resulting peat compression and oxidation over a period of 30 years were determined. The model area comprised the major cities of Amsterdam and Rotterdam, and surrounding agricultural lands (Figure 11.18).

This predictive modeling has identified the increased susceptibility of agricultural areas to subsidence relative to urbanized areas. Peat in the Dutch low-lying agricultural areas often occurs near the surface, and consequently is

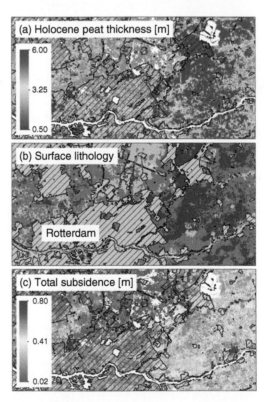

Figure 11.18 Maps showing: (a) accumulated Holocene peat thickness; (b) surface lithology (for legend see Figure 11.14); and (c) predicted land subsidence over a 30-year period in response to a 0.5 m groundwater lowering in the city of Rotterdam with surrounding agricultural lands. Modified from Koster et al. (2018b).

very prone to oxidation, whereas peat layers in urbanized areas are protected from aeration by overlying anthropogenic deposits. Furthermore, this artificial fill causes peat in cities to become compacted with less void space, and therefore to have a lower compressibility than peat in agricultural areas.

11.6.3 Aggregate Resource Assessment

Building and construction require large volumes of aggregate (granular mineral material) produced from natural sand or gravel or solid rock components. Aggregate resource assessments rely on (i) the extent to which the occurrence and grain size of sand and gravel are resolved; and (ii) the proper representation of clay and peat layers. The former estimates the volume and quality of the resource, the latter predicts overburden and intercalations affecting exploitability (Van der Meulen et al. 2005; Section 24.2).

Since 2005, aggregate resource assessments of sand and gravel in the Netherlands have been based on voxel models. Two of these voxel models with national coverage were specifically aimed at resource assessments and resulted in an online mineral resources information system (www .delfstoffenonline.nl, Van der Meulen et al. 2005, 2007). A third model is derived from the voxel model GeoTOP (Stafleu et al. 2011, 2012). All three models are primarily based on the same borehole data. Maljers et al. (2015) systematically compared the three models (Figure 11.19). They clearly showed how progress was made in the extent to which additional soft data (e.g. lithostratigraphic interpretations, geological map data and lithofacies concepts) were considered. In addition, ever-increasing computing power enabled a higher resolution (i.e. smaller voxels)

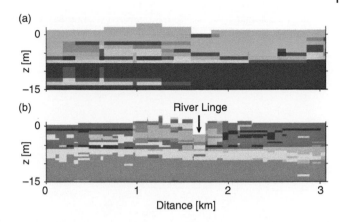

Figure 11.19 Cross-section through two generations of voxel models used in aggregate resource assessments. (a) The share of aggregate for each voxel (blue = low shares; red = high shares) obtained from the Delfstoffen online model (www .delfstoffenonline.nl; Van der Meulen et al. 2005). (b) Discrete lithologic classes derived from GeoTOP model. GeoTOP model shows a much better resolution of the sandy channel belt of the River Linge, a tributary of the Rhine. For legend see Figure 11.14. (Maljers et al. 2015).

in each new model. Both developments were significant, and each successive model better supported resource assessments.

11.6.4 Defining Holocene Channel Belt Systems

GeoTOP data support geoscientific research. An example is the identification of grain-size trends in the Rhine-Meuse delta, which are characterized by a complex of Holocene fluvial channel belt systems (Stafleu et al. 2009; Stafleu and Busschers 2017) (Figure 11.20).

A preliminary 3-D analysis of a single channel belt in the western part of the Rhine-Meuse delta (Figure 11.21) showed a clear downstream increase in percentages of

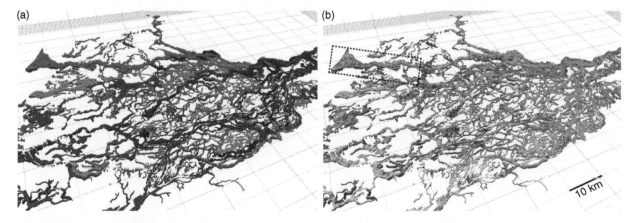

Figure 11.20 Channel belts in the Holocene Rhine-Meuse delta (the Netherlands) visualized in 3-D. (a) Channel belt units. Colors represent different generations (relative ages). (b) Channel belt lithology and grain-size variation. Colors represent lithologic classes (for legend see Figure 11.14). Inset shows position of Figure 11.21. Modified from Stafleu and Busschers (2017).

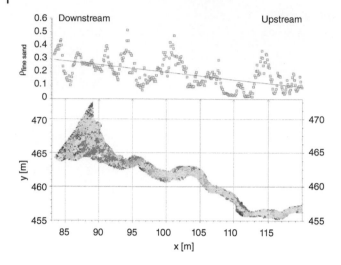

Figure 11.21 Channel belt lithology and grain-size variation (here depicted as the probability of occurrence of fine-sand) along the Roman-age Oude Rijn channel belt (western Rhine-Meuse delta). (Van der Meulen et al. 2013).

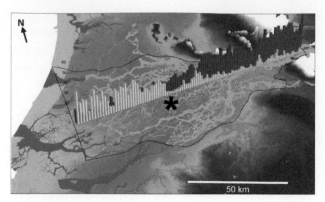

Figure 11.22 Downstream changes in percentages of fine-sand (yellow bars) and coarse-sand (red bars) in the channel belts of the Holocene Rhine-Meuse delta. Fluvial (yellow) channel belts and estuarine-marine (blue) channel belts are depicted within the context of flood basin and tidal flat sediments (green), coastal barrier sediments (yellow) and (elevated) Pleistocene surface. To minimize edge effects, the analysis was restricted to the area indicated by the dotted line. (Stafleu and Busschers 2017).

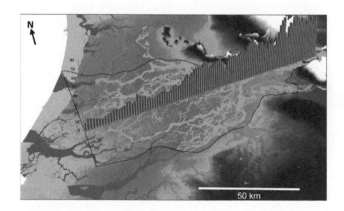

Figure 11.23 Downstream change in Channel Deposit Proportion (CDP) in the Holocene Rhine-Meuse delta. (Stafleu and Busschers 2017).

fine-grained sand and a decrease in percentages of the coarser fractions. This observation, illustrating the potential of the GeoTOP model to identify grain-size trends in sandy lowland rivers, was the trigger for performing a full-scale 3-D analysis of the entire delta.

The "most likely" lithologic class from the GeoTOP model was used to perform a 3-D spatial trend analysis of both lithology and sand grain-size in the Rhine-Meuse delta. The model was analyzed in 1 km wide north–south oriented slices. Several types of analyses were carried out; two examples are presented here. An analysis of the sand grain-size in the combined fluvial and marine-estuarine channel belts shows a clear nearly linear decrease in the percentage of coarse-sand (shown as red bars) from 35% to 45% in the east to about 10% in the west (~0.3% decrease per km) (Figure 11.22). The amount of fine-sand (shown as yellow bars) remains constant in the eastern part of the delta (~5%) but rapidly increases to values of 15–30% west of x-coordinate 120 000 (indicated by a star in Figure 11.22). Values for medium-sand (not plotted) remain constant.

The changes in sand grain-size reflect a combination of a westward decreasing gradient, reworking of older Pleistocene sediments, and an increasing influence of tidal processes. Analysis of the *Channel Deposit Proportion* (CDP) revealed an almost linear downstream decrease in CDP from 0.6–0.7 in the east to 0.1–0.2 in the west (~0.005 decrease per km) (Figure 11.23). These CDP values correspond well with published transect-based estimates of CDP in the eastern part of the delta (Gouw 2008). These results may serve as constraints for hydrocarbon reservoir modeling where this type of information is often sparsely available.

11.6.5 Dredging Activities

Dredging activities in rivers may accidentally lead to a "hydraulic short circuit", i.e., the creation of an unwanted connection between the river and a deeper, (semi-)confined aquifer. Geometries of the Holocene fluvial channel belt modeled in GeoTOP (Figure 11.20) were used to assess such risk along the River Vecht in conjunction with a planned dredging operation (Figure 11.24).

The River Vecht is a minor distributary of the Rhine-Meuse Delta. Under the recent, managed flow regime, fine-grained sediments have been deposited on the riverbed. These overlie the river's older sandy Holocene channel belt deposits, which may or may not cut into an underlying, highly permeable Upper Pleistocene fluvial deposit. The latter deposit was not to be reached during

Figure 11.24 Risk map of the river Vecht near Utrecht, the Netherlands, showing portions of the river course prone to a "hydraulic short circuit" allowing seepage of water from the river into the adjacent flood basin.

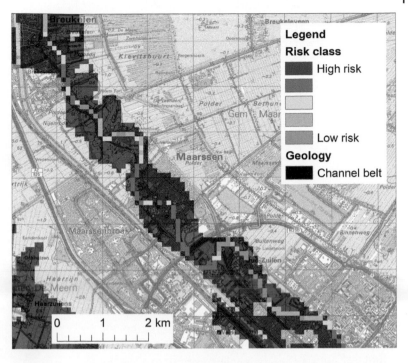

dredging, so the operator intended to stay well within the confinement of the Holocene channel belt. A risk map was constructed by adding the planned dredging depth to the depth of the riverbed and comparing the resulting new bathymetry with the base level of the Holocene channel belt in GeoTOP. Sections of the river where the new bathymetry was close to the base of the channel belt were marked as high-risk. This allowed the operator to concentrate additional geotechnical investigations in areas that mattered, which is much more cost efficient than sampling the riverbed in its entirety.

11.7 Voxel Models Outside the Netherlands

The GeoTOP modeling workflow has been applied in several areas outside the Netherlands. Two examples are: (i) a detailed voxel model of the shallow subsurface of the Tokyo Lowland area, Japan, developed in collaboration with the Geological Survey of Japan (Stafleu et al. 2016); and (ii) a voxel model of the Quaternary sands of the Belgian part of the North Sea (Van Lancker et al. 2017; Hademenos et al. 2019), developed in response to the scope of the Belgian Science Policy project TILES (*Transnational and Integrated Long-term Marine Exploitation Strategies*). The following sections provide details.

11.7.1 Tokyo Lowland Area, Japan

Stafleu et al. (2016) presented a detailed 3-D geological voxel model of the Holocene sediments of the Tokyo Lowland area, central Japan. The model was developed for two reasons: (i) to investigate to what extent TNO's voxel modeling approach, developed for the lowlands of the Netherlands, is also applicable in other geological settings; and (ii) to study the relation between the accumulated thickness of soft Holocene mud and the amount of damage caused by earthquakes.

The Tokyo Lowland is situated in a Neogene sedimentary basin near the triple junction of the North American, Pacific, and Philippine tectonic plates. The basin is filled with Neogene and Quaternary sediments up to a thickness of 3 km. In the upper 70 m of the basin, thick sequences of soft Holocene sediments occur which are assumed to have played a key role in the spatial variation of damage intensity during the 1923 Kanto earthquake (Magnitude 7.9–8.3). Historical records show this earthquake destroyed large parts of the Tokyo urban area, which at that time was largely made up of wooden houses. Although the epicenter was 70 km to the southwest of Tokyo, severe damage occurred north of the city center, presumably due to ground motion amplification in the soft Holocene sediments in the shallow subsurface. In order to assess the presumed relation between the damage pattern of the 1923 earthquake and the occurrence of soft Holocene sediments in the shallow subsurface, a 3-D geological voxel model of the central part of the Tokyo Lowland was constructed.

Earlier geological modeling work of the Tokyo Lowland area was carried out by Tanabe et al. (2015). They created a 3-D voxel model by using the N-values and lithologic class in the borehole descriptions of Tanabe et al. (2008a,b) directly in the 3-D interpolation method of Ishihara et al.

(2013). Stafleu et al. (2016) incorporated more geological knowledge into this model by introducing two intermediate modeling steps derived from the GeoTOP workflow.

First, some 10 000 borehole descriptions (gathered for geomechanical purposes), were subdivided into geological units that have uniform sediment characteristics, using both lithologic and geomechanical criteria. These borings contained Standard Penetration Test (SPT) data, which are obtained by recording the number of standard hammer blows required to drive a sample tube 150 mm (6 in.) into the ground. The sum of the number of blows required for the second and third 150 mm (6 in.) of penetration is termed the *standard penetration resistance* or the *N-value*. The N-value provides an indication of the density of the ground; it is used in many empirical geotechnical engineering formulae.

Second, 2-D bounding surfaces were constructed, representing tops, and bases of the geological units. These surfaces were used to place each voxel (100 by 100 by 1 m) within the correct geological unit. The N-values and lithologic units in the borehole descriptions were subsequently used to perform a 3-D stochastic interpolation of N-value and lithologic class within each geological unit (Figure 11.25).

Using a vertical voxel stack analysis, a map was created showing the accumulated thickness of soft muds in the Holocene succession. A comparison of this map with a published map of the damage-ratio of wooden houses that were destroyed during the Kanto earthquake in 1923 (Kaizuka

and Matsuda 1982), shows a good correlation between zones of maximum destruction and the occurrence of the so-called "zero" muds, the latter representing the sediments most sensitive for ground motion amplification (Figure 11.26).

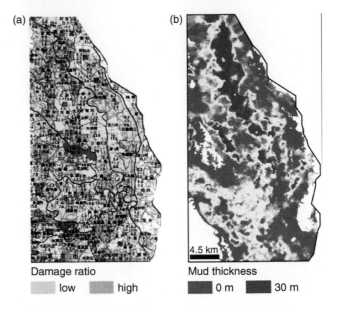

Figure 11.26 (a) Map of the damage-ratio of wooden houses (Modified after Kaizuka and Matsuda 1982). (b) Map showing the accumulated thickness of soft muds in the 3-D model of Figure 11.25. Note the remarkable relation between zones of maximum damage and the occurrence of thick mud sequences which are most sensitive for ground motion amplification. Modified from Stafleu et al. (2016).

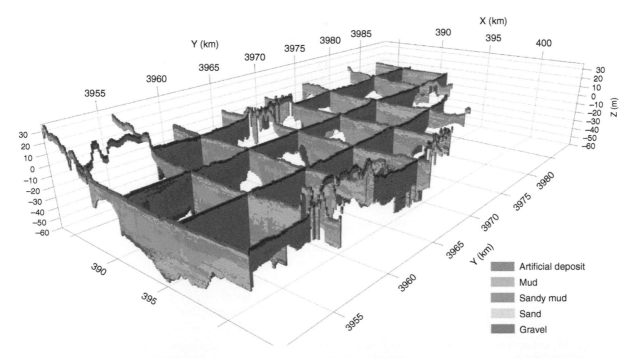

Figure 11.25 Voxel model of the Tokyo Lowland area showing the 3-D distribution of soft muds within the Holocene succession. Modified from Stafleu et al. (2016).

(a)

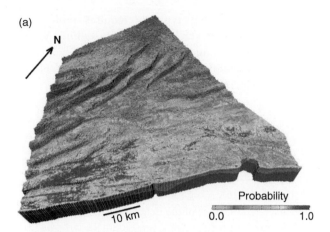

(b)

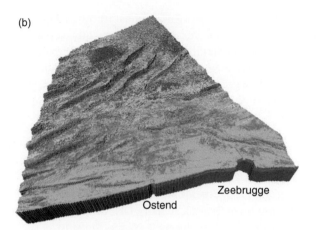

Figure 11.27 Probabilities of occurrence in the voxel model of the Belgian Continental Shelf based on data from Hademenos et al. (2019. (a) Fine-grained sand. (b) Medium-grained sand. Medium-grained sands are mostly located further offshore whereas fine grained sands predominate nearshore.

The results show that the 3-D geological voxel modeling approach presented here is able to make a spatial analysis of earthquake damage sensitivity in the Tokyo Lowland. This makes the GeoTOP workflow a promising tool for seismic hazard assessments in other areas in Japan or elsewhere where detailed insights into earthquake damage from historical records are absent. The results were obtained during a staff exchange project, which allowed for the exchange of 3-D modeling expertise within GDN and the seismic hazard expertise of the Japanese survey. This type of cooperation can be recommended as an excellent way to share knowledge and expertise.

11.7.2 The Belgian Part of the North Sea

The Quaternary marine sands of the Belgian Continental Shelf (BCS), which covers 3454 km² of the North Sea, are a valuable resource for the construction industry and for the reinforcement of the 67 km long Belgian coast. A voxel model of this area, developed in response to Belgian Science Policy TILES project (Van Lancker et al. 2017; Hademenos et al. 2019), was subsequently complemented with an online query-tool (De Mol 2018; De Tré et al. 2018). This tool enables users to perform volume calculations which are crucial for the management of raw materials in the marine environment. In addition to estimating resources, the voxel model supports a numerical modeling suite that simulates hydrodynamics, sediment transport, and seabed morphology over time (Terseleer et al. 2017).

Offshore borehole data are limited, and their density decreases rapidly away from the coast. In contrast, high-resolution shallow seismic data are widespread, and therefore constitute the ideal source of information for the construction of a stratigraphic-layer based model (Van Heteren et al. 2014). Within the BCS, extensive seismic data permitted interpretation of horizons from approximately

6650 km of seismic lines. These were used to generate the bounding surfaces of five different Quaternary and Palaeogene stratigraphic units in the voxel model (De Clercq et al. 2016; Hademenos et al. 2019). Approximately 1770 borehole descriptions were interpreted manually to define stratigraphic and lithologic classes (Hademenos et al. 2019). The lithologic classification uses the Wentworth (1922) grain-size classes, making it suitable for resource estimations. A voxel model with a resolution of $200 \times 200 \times 1$ m was created using the GeoTOP workflow as described in Section 11.2.2.3.

The results show the marine sands of the BCS as a scarce resource occurring in a thin cover of Quaternary sands on top of a Palaeogene substratum, which is mainly composed of clay. The layer-based model reveals detailed structural elements such as a sandy infill of the nearshore Ostend paleo-valley as well as Holocene tidal sandbanks with a Pleistocene core. The voxel model emphasizes the scarcity of the heavily sought-after medium to coarse sands. These are mostly located further offshore whereas fine grained sands and clay predominate nearshore (Figure 11.27). Furthermore, some of the largest volumes of medium sand are located in areas that are currently reserved for offshore wind farms.

11.8 Conclusions

Voxel modeling is a powerful tool for representing subsurface properties in 3D models, capturing substratigraphic variations and features that are hard or impossible to represent in stacked-surface models (Chapter 9). Nonetheless, voxel models should not be considered as a replacement of stacked surface models, but rather a further refinement, where voxel modeling utilizes stacked surface modeling to confine properties to (litho)stratigraphic units.

A basic geological voxel typically has lithostratigraphic and lithologic attributes. Voxels can subsequently be populated with any parameter that correlates with combinations of these basic attributes, and in some cases depth. Featured examples include geotechnical and geohydrological studies and minerals surveys.

Creating voxel models typically requires more data and information than stacked-surface modeling. When modeling large areas, automation is important in various stages of the modeling workflow in order to be able to handle the model inputs.

References

Bajc, A.F. and Dodge, J.E.P. (2011). *Three-dimensional mapping of surficial deposits in the Brantford–Woodstock area, Southwestern Ontario*. Sudbury, ON: Ontario Geological Survey, Groundwater Resources Study 10. [Online: Available at: http://www.geologyontario.mndm.gov.on.ca/mndmaccess/mndm_dir.asp?type=pub&id=GRS010] (Accessed August 17, 2020).

Berendsen, H.J.A. and Stouthamer, E. (2001). *Palaeogeographic Development of the Rhine-Meuse Delta, the Netherlands*. Assen, The Netherlands: Van Gorcum. 268 pp.

Bianchi, M., Kearsey, T. and Kingdon, A. (2015). Integrating deterministic Lithostratigraphic models in stochastic realizations of subsurface heterogeneity: impact on predictions of lithology, hydraulic heads and groundwater fluxes. *Journal of Hydrology* 531: 557–573. https://doi.org/10.1016/j.jhydrol.2015.10.072.

Burt, A.K. and J.E.P. Dodge (2011). *Three-dimensional modelling of surficial deposits in the Barrie–Oro moraine area of southern Ontario*. Sudbury, ON: Ontario Geological Survey, Groundwater Resources Study 11. [Online: Available at: http://www.geologyontario.mndmf.gov.on.ca/mndmaccess/mndm_dir.asp?type=pub&id=GRS011] (Accessed August 17, 2020).

Burt, A.K. and Dodge, J.E.P. (2016). *Three-dimensional modelling of surficial deposits in the Orangeville–Fergus area of southern Ontario*. Sudbury, ON: Ontario Geological Survey, Groundwater Resources Study 15. [Online: Available at: http://www.geologyontario.mndm.gov.on.ca/mndmaccess/mndm_dir.asp?type=pub&id=GRS015] (Accessed August 17, 2020).

Chilès, J.-P. and Delfiner, P. (2012). *Geostatistics – Modeling Spatial Uncertainty*. Hoboken, NJ: Wiley. 699 pp.

Climate Adaptation Services (2017). *Climate Impact Atlas*. Bussum, Netherlands: Foundation Climate Adaptation Services. [Website: https://www.klimaateffectatlas.nl] (Accessed August 17, 2020).

Dambrink, R.M., Maljers, D., Schokker, J. and Stafleu, J. (2015). Tailor-made maps generated from 3D geological voxel models (Abstract). *Geological Society of America Abstracts with Programs* 47 (7): 406. [Online: Available at: https://gsa.confex.com/gsa/2015AM/webprogram/Paper267283.html] (Accessed August 17, 2020).

De Clercq, M., Chademenos, V., Van Lancker, V. and Missiaen, T. (2016). A high-resolution DEM for the Top-Palaeogene surface of the Belgian Continental Shelf. *Journal of Maps* 12: 1047–1054. https://doi.org/10.1080/17445647.2015.1117992.

De Lange, G., Gunnink, J.L., Houthuessen, Y. and Muntjewerff, R. (2012). *Bodemdalingskaart Flevoland* (in Dutch). De Bilt, Netherlands: Grontmij, Report GM-0042778. 58 pp.

De Mol, R. (2018). Sand decision support involves flexible visualization and querying tools (Abstract). *TILES Final Conference on Marine Sands as a Precious Resource, June 1, 2018, Brussels, Belgium*.

De Tré, G., De Mol, R., Van Heteren, S. et al. (2018). Data quality assessment in volunteered geographic decision support. In: *Mobile Information Systems Leveraging Volunteered Geographic Information for Earth Observation. Earth Systems Data and Models, vol. 4* (eds. G. Bordogna and P. Carrara), 173–192. Cham, Switzerland: Springer.

De Vries, F., de Groot, W.J.M., Hoogland, T. and Denneboom, R. (2003). *De Bodemkaart van Nederland digitaal; toelichting bij inhoud, actualiteit en methodiek en korte beschrijving van additionele informatie* (in Dutch). Wageningen, Netherlands: Alterra, Rapport 811. 45 pp. [Online: Available at: https://edepot.wur.nl/21850] (Accessed August 17, 2020).

Den Haan, E.J. (1992). The formulation of virgin compression of soils. *Géotechnique* 42: 465–483. https://doi.org/10.1680/geot.1992.42.3.465.

Deverel, S.J., Ingrum, T. and Leighton, D. (2016). Present-day oxidative subsidence of organic soils and mitigation in the Sacramento-San Joaquin Delta, California, USA. *Hydrogeology Journal* 24: 569–586. https://doi.org/10.1007/2Fs10040-016-1391-1.

Drexler, J.Z., De Fontain, C.S. and Deverel, S.J. (2009). The legacy of wetland drainage on the remaining peat in the Sacramento-San Joaquin Delta, California, USA. *Wetlands* 29: 372–386. https://doi.org/10.1672/08-97.1.

Erkens, G. (2009). *Sediment Dynamics in the Rhine Catchment: Quantification of Fluvial Response to Climate Change and Human Impact*. PhD thesis Utrecht University.

278 pp. [Online: Available at: https://dspace.library.uu.nl/handle/1874/36680] (Accessed August 17, 2020).

Gambolati, G., Putti, M., Teatini, P. and Stori, G.G. (2006). Subsidence due to peat oxidation and impact on drainage infrastructures in a farmland catchment south of the Venice Lagoon. *Environmental Geology* 46: 814–820. https://doi.org/10.1007/s00254-006-0176-6.

Goovaerts, P. (1997). *Geostatistics for Natural Resources Evaluation*. New York: Oxford University Press. 483 pp.

Gouw, M.J.P. (2008). Alluvial architecture of the Holocene Rhine-Meuse delta (the Netherlands). *Sedimentology* 55: 1487–1516. https://doi.org/10.1111/j.1365-3091.2008.00954.x.

Gunnink, J.L., Maljers, D., Van Gessel, S.F. et al. (2013a). Digital geological model (DGM): a 3D raster model of the subsurface of the Netherlands. *Netherlands Journal of Geosciences* 92: 33–46. https://doi.org/10.1017/S0016774600000263.

Gunnink, J.L., Stafleu, J., Maljers, D. and Hummelman, H.J. (2013b). Hydraulic parameterization of 3D subsurface models: from measurement-scale to model-scale. In: *Workshop Extended Abstracts of the 2013 Annual Meeting, Geological Society of America, October 26 2013, Portland, Oregon, USA* (eds. H.A.J. Russell, R.C. Berg and L.H. Thorleifson), 35–39. Edmonton, AL: AER/AGS, Special Report 101. [Online: Available at: https://ags.aer.ca/document/SPE/SPE_101.pdf] (Accessed August 17, 2020).

Hademenos, V., Stafleu, J., Missiaen, T. et al. (2019). 3D subsurface characterisation of the Belgian continental shelf: a new voxel modelling approach. *Netherlands Journal of Geosciences* 98: E1. https://doi.org/10.1017/njg.2018.18.

Hazeu, G.W. (2005). *Landelijk Grondgebruiksbestand Nederland (LGN5). Vervaardiging, nauwkeurigheid en gebruik* (in Dutch). Wageningen, Netherlands: Alterra, Rapport 1213. 92 p. [Online: Available at: https://edepot.wur.nl/17654] (Accessed August 17, 2020).

Hazeu, G.W. (2006). Land use mapping and monitoring in the Netherlands (LGN5). In: *Conference Proceedings, 2nd EARSeL Workshop on Land Use and Land Cover, 28–30 September 2006, Bonn, Germany* (ed. M. Braun), 323–329. Bonn, Germany: Center for Remote Sensing of Land Surfaces (ZFL).

Higgins, S.A. (2016). Review: advances in Delta-subsidence research using satellite methods. *Hydrogeology Journal* 24: 587–600. https://doi.org/10.1007/s10040-015-1330-6.

Høyer, A.S., Jørgensen, F., Sandersen, P.B.E. et al. (2015). 3D geological modelling of a complex Buried-Valley network delineated from borehole and AEM data. *Journal of Applied Geophysics* 122: 94–102. https://doi.org/10.1016/j.jappgeo.2015.09.004.

Ishihara, Y., Miyazaki, Y., Eto, C. et al. (2013). Shallow subsurface three-dimensional geological model using borehole logs in Tokyo Bay Area, Central Japan (In Japanese with English abstract). *Journal of the Geological Society of Japan* 119: 554–566. https://doi.org/10.5575/geosoc.2013.0019.

Jørgensen, F., Rønde Møller, R., Nebel, L. et al. (2013). A method for cognitive 3D geological voxel modelling of AEM data. *Bulletin of Engineering Geology and the Environment* 72: 421–432. https://doi.org/10.1007/s10064-013-0487-2.

Kaizuka, S. and Matsuda, I. [Eds.] (1982). *Active Tectonics, Geomorphic Division of the Tokyo Metropolitan area, Damage Ratio due to the Kanto Earthquake of 1923* (in Japanese). Tokyo: Naigai Chizu 48 pp, 1 map sheet (1 : 200,000).

Kearsey, T., Williams, J., Finlayson, A. et al. (2015). Testing the application and limitation of stochastic simulations to predict the lithology of glacial and fluvial deposits in Central Glasgow, UK. *Engineering Geology* 187: 98–112. https://doi.org/10.1016/j.enggeo.2014.12.017.

Koomen, A.J.M. and Maas, G.J. (2004). *Geomorfologische Kaart Nederland (GKN); Achtergronddocument bij het landsdekkende digitale bestand* (in Dutch). Wageningen, Netherlands: Alterra, Rapport 1039. 38 p. [Online: Available at: https://edepot.wur.nl/40241] (Accessed August 17, 2020).

Koster, K., Stafleu, J., Cohen, K.M. et al. (2018a). Three-dimensional distribution of organic matter in coastal-deltaic peat: implications for subsidence and carbon dioxide emissions by human-induced peat oxidation. *Anthropocene* 22: 1–9. https://doi.org/10.1016/j.ancene.2018.03.001.

Koster, K., Stafleu, J. and Stouthamer, E. (2018b). Differential subsidence in the urbanized coastal-deltaic plain of the Netherlands. *Netherlands Journal of Geosciences* 97 (4): 215–227. https://doi.org/10.1017/njg.2018.17.

Kruiver, P.P., Van Dedem, E., Romijn, R. et al. (2017a). An integrated shear-wave velocity model for the Groningen gas field, the Netherlands. *Bulletin Earthquake Engineering* 15: 3555–3580. https://doi.org/10.1007/s10518-017-0105-y.

Kruiver, P.P., Wiersma, A., Kloosterman, F.H. et al. (2017b). Characterisation of the Groningen subsurface for seismic hazard and risk modelling. *Netherlands Journal of Geosciences* 96: 215–233. https://doi.org/10.1017/njg.2017.11.

Lelliot, M.R., Cave, M.R. and Wealthall, G.P. (2009). A structured approach to the measurement of uncertainty in 3D geological models. *Quarterly Journal of Engineering Geology and Hydrogeology* 42: 95–105. http://dx.doi.org/10.1144/1470-9236/07-081.

Maljers, D., Stafleu, J., Van der Meulen, M.J. and Dambrink, R.M. (2015). Advances in constructing regional geological voxel models, illustrated by their application in aggregate resource assessments. *Netherlands Journal of Geosciences* 94: 257–270. https://doi.org/10.1017/njg.2014.46.

Marschallinger, R. (1996). A voxel visualization and analysis system based on AutoCAD. *Computers & Geosciences* 22: 379–386. https://doi.org/10.1016/0098-3004(95)00093-3.

Nieuwenhuis, H.S. and Schokking, F. (1997). Land subsidence in drained peat areas of the Province of Friesland. *Quarterly Journal of Engineering Geology and Hydrogeology* 30: 37–48. https://doi.org/10.1144/GSL .QJEGH.1997.030.P1.04.

Schokker, J., Bakker, M.A.J., Dubelaar, C.W. et al. (2015). 3D subsurface modelling reveals the shallow geology of Amsterdam. *Netherlands Journal of Geosciences* 94: 399–417. https://doi.org/10.1017/njg.2015.22.

Serva, L. and Brunamonte, F. (2007). Subsidence in the Pontina Plain, Italy. *Bulletin of Engineering Geology and the Environment* 66: 125–134. https://doi.org/10.1007/s10064-006-0057-y.

Soares, A. (1992). Geostatistical estimation of multi-phase structure. *Mathematical Geology* 24: 149–160. https://doi.org/10.1007/BF00897028.

Sobisch, H.-G. (2011). One viewer to publish 3D subsurface models of any modelling origin. *Geological Society of America Abstracts with Programs* 43 (5): 525. [Online: Available at: https://gsa.confex.com/gsa/2011AM/ webprogram/Paper195886.html] (Accessed August 17, 2020).

Stafleu, J. and Busschers, F.S. (2017). Analysing lithological and grain-size trends using a 3D voxel model: a case study from the Holocene Rhine-Meuse Delta. In: *Conference Proceedings, 79th EAGE Conference & Exhibition, June 2017 – Workshop Programme, Paris, France*, 12–15. Houten, Netherlands: European Association of Geoscientists & Engineers (EAGE).

Stafleu, J. and C. W. Dubelaar (2016). *Product specification subsurface model GeoTOP*. Utrecht, Netherlands: TNO, Report 2016 R10133 v1.3, 53 pp. [Online: Available at: https://www.dinoloket.nl/sites/default/files/file/ dinoloket_toelichtingmodellen_20160606_tno_2016_ r10133_geotop_v1r3_english.pdf] (Accessed August 17, 2020).

Stafleu, J., Busschers, F.S., Maljers, D. and Gunnink, J.L. (2009). Three-dimensional property modeling of a complex Fluvio-deltaic environment: Rhine-Meuse Delta, the Netherlands. In: *Workshop Extended Abstracts of the 2009 Annual Meeting, Geological Society of America, October 17 2009, Portland, Oregon, USA* (eds. R.C. Berg, H.A.J. Russell and L.H. Thorleifson), 47–50. Champaign, IL: Illinois State Geological Survey, Open Files Series 2009-4 [Online: Available at: https://isgs.illinois.edu/publications/ofs2009-4] (Accessed August 17, 2020).

Stafleu, J., Maljers, D., Gunnink, J.L. et al. (2011). 3D modelling of the shallow subsurface of Zeeland, the Netherlands. *Netherlands Journal of Geosciences* 90 (4): 293–310. https://doi.org/10.1017/S0016774600000597.

Stafleu, J., Maljers, D., Busschers, F.S., Gunnink, J.L. et al. (2012). *GeoTOP modellering* (in Dutch). Utrecht, Netherlands: TNO, Rapport TNO-2012-R10991. 216 pp. [Online: Available at: https://www.dinoloket.nl/sites/ default/files/file/TNO_2012_R10991_GeoTOP_ modellering_v1.0.pdf] (Accessed August 17, 2020).

Stafleu, J., Sobisch, H.-G., Maljers, D., Hummelman, J. et al. (2013). Visualization and dissemination of 3D geological property models of the Netherlands. *Geophysical Research Abstracts* 15: EGU2013-8770. [Online: Available at: https:// meetingorganizer.copernicus.org/EGU2013/EGU2013-8770.pdf] (Accessed August 17, 2020).

Stafleu, J., Busschers, F.S., and Tanabe, S. (2016). Shallow subsurface control on earthquake damage patterns: first results from a 3D geological voxel model study (Tokyo Lowland, Japan). *Geophysical Research Abstracts* 18: EGU2016-8023. [Online: Available at: https:// meetingorganizer.copernicus.org/EGU2016/EGU2016-8023.pdf] (Accessed August 17, 2020).

Steur, G.G.L. and Heijink, W. (1991). *Bodemkaart van Nederland, Schaal 1:50 000. Algemene begrippen en indelingen, 4e uitgebreide uitgave* (in Dutch). Wageningen, Netherlands: DLO Staring Centrum. 72 pp. [Online: Available at: https://edepot.wur.nl/117853] (Accessed August 17, 2020).

Tanabe, S., Nakanishi, T., Kimura, K. et al. (2008a). Basal topography of the alluvium under the northern area of the Tokyo lowland and Nakagawa lowland, Central Japan (In Japanese with English abstract). *Bulletin of the Geological Survey of Japan* 59: 497–508. https://doi.org/10.9795/ bullgsj.59.497.

Tanabe, S., Ishihara, Y. and Nakashima, R. (2008b). Sequence stratigraphy and paleogeography of the alluvium under the northern area of the Tokyo lowland, Central Japan (In Japanese). *Bulletin of the Geological Survey of Japan* 59: 509–547. https://doi.org/10.9795/bullgsj.59.509.

Tanabe, S., Nakanishi, T., Ishihara, Y. and Nakashima, R. (2015). Millennial-scale stratigraphy of a tide-dominated Incised Valley during the last 14 kyr: spatial and quantitative reconstruction in the Tokyo lowland, Central Japan. *Sedimentology* 62: 1837–1872. https://doi.org/10 .1111/sed.12204.

Terseleer, N., Hademenos, V. Missiaen, T., Stafleu, J. et al. (2017). A continuum of knowledge from measurements to modelling as management support for the exploitation of marine aggregates, Belgian part of the North Sea. In: *Hydro17 Conference Handbook, Rotterdam 14–16 November 2017*, 66. Plymouth, UK: International Federation of

Hydrographic Societies. [Online: Available at: https://hydro17.com/doc/Hydro17_Conference_Handbook.pdf] (Accessed August 18, 2020).

Tran, T. (1996). The 'missing scale' and direct simulation of block effective properties. *Journal of Hydrology* 183: 37–56. https://doi.org/10.1016/S0022-1694(96)80033-3.

Van der Meulen, M.J., Maljers, D., Van Gessel, S.F. and Veldkamp, J. (2005). Aggregate resources in the Netherlands. *Netherlands Journal of Geosciences* 84: 379–387. https://doi.org/10.1017/S0016774600021193.

Van der Meulen, M.J., Maljers, D., Van Gessel, S.F. and Gruijters, S.H.L.L. (2007). Clay resources in the Netherlands. *Netherlands Journal of Geosciences* 86: 117–130. https://doi.org/10.1017/S001677460002312X.

Van der Meulen, Doornenbal, J.C., Gunnink, J.L. et al. (2013). 3D geology in a 2D country: perspectives for geological surveying in the Netherlands. *Netherlands Journal of Geosciences* 92: 217–241.

Van der Schans, M.L. (2012). *Phoenix 1.0 – Deelrapport 3: Vervaardiging en evaluatie regionale bodemdalingsapplicatie westelijk deel Provincie Utrecht / HDSR* (in Dutch). De Bilt, Netherlands: Grontmij, Rapport MvdS315624-s. 38 pp.

Van Haren, T., Dirix, K., De Koninck, R. et al. (2016). An interactive voxel model for mineral resources: loess deposits in Flanders (Belgium). *Zeitschrift der Deutschen Gesellschaft für Geowissenschaften (ZDGG)* 167: 363–376. https://doi.org/10.1127/zdgg/2016/0096.

Van Heteren, S., Meekes, J.A.C., Bakker, M.A.J. et al. (2014). Reconstructing North Sea palaeolandscapes from 3D and high-density 2D seismic data: an overview. *Netherlands Journal of Geosciences* 93: 31–42. https://doi.org/10.1017/njg.2014.4.

Van Lancker, V., Francken, F., Kint, L. et al. (2017). Building a 4D voxel-based decision support system for a sustainable Management of Marine Geological Resources. In: *Oceanographic and Marine Cross-Domain Data Management for Sustainable Development* (eds. P. Diviacco, A. Leadbetter and H. Glaves). Hershey, PA: IGI Global.

Vermeulen, P.T.M., Burgering, L.M.T., Roelofsen, F.J. et al. (2017). *iMOD User Manual. Version 4.0*. Delft, the Netherlands: Deltares. 758 pp. [Online: Available at: https://content.oss.deltares.nl/imod/imod40/iMOD_User_Manual_V4_0.pdf] (Accessed August 18, 2020).

Vernes, R.W. and Van Doorn, Th.H.M. (2005). *Van Gidslaag naar Hydrogeologische Eenheid – Toelichting op de totstandkoming van de dataset REGIS II* (in Dutch). Utrecht, Netherlands: Netherlands Institute of Applied Geosciences TNO, Report 05-038-B. 105 pp. [Online: Available at: https://www.dinoloket.nl/sites/default/files/file/dinoloket_toelichtingmodellen_20131210_01_rapport_nitg_05_038_b0115_netversie.pdf] (Accessed August 18, 2020).

Wellmann, J.F. and Regenauer-Lieb, K. (2012). Uncertainties have a meaning: information entropy as a quality measure for 3-D geological models. *Tectonophysics* 526–529: 207–216. https://doi.org/10.1016/j.tecto.2011.05.001.

Wentworth, C.K. (1922). A scale of grade and class terms for clastic sediments. *The Journal of Geology* 30: 377–392. [Online: Available at: https://www.jstor.org/stable/30063207] (Accessed August 18, 2020).

12

Integrated Rule-Based Geomodeling – Explicit and Implicit Approaches

Martin Ross[1], Amanda Taylor[1], Samuel Kelley[1], Simon Lopez[2], Cécile Allanic[2], Gabriel Courrioux[1], Bernard Bourgine[2], Philippe Calcagno[2], Séverine Carit[2], and Sunseare Gabalda[2]

[1]*Department of Earth and Environmental Sciences, University of Waterloo, Waterloo, ON N2L 3G1, Canada*
[2]*Bureau de Recherches Géologiques et Minières (BRGM), 45060 Orléans Cedex 02, France*

12.1 Introduction

Three-dimensional geological modeling is the representation of subsurface stratigraphic, structural, and physical property attributes. Knowledge of the subsurface is, however, limited. Direct observations are restricted to a few exposures (outcrops), excavations, and boreholes. These may be supplemented by information provided by indirect, mainly geophysical, methods. Consequently, once all possible data have been collected and integrated into an appropriate database, the geologist has to perform the important intellectual task of synthesizing the entire set of observations and knowledge in the form of a geological model. Over the last three decades, many methods and software products have been developed to assist the geologist in building geological models and carrying out spatial analyses of geological properties (Houlding 1994; Mallet 2002; Calcagno et al. 2008).

12.1.1 Explicit Geomodeling with Geological Constraints

A common strategy is to represent subsurface geology using an *explicit modeling* approach, as discussed in Chapters 9 and 10. This approach consists of generating a Geological Framework Model (GFM) which represents geological interfaces (e.g. stratigraphic contacts) by a series of bounding non-intersecting surfaces generated directly or indirectly from a given set of data points with *x*, *y*, *z* coordinates. The data-point locations are typically based on geological information and expert interpretation; some are from direct observations (e.g. outcrops, borehole markers), while others may be from lines or contours from a series of interpreted geological cross-sections and geological maps for example, as discussed by Kessler et al. (2009). The

surfaces are either *parametric* – created using polynomials (Fisher and Wales 1992) – or *polygonal,* represented by a mesh or network of segments connecting vertices (Baker 1987). Even though these surfaces describe the outer shell of 3-D geological features, their data structure remains two-dimensional. The surfaces are typically generated independently from each other, which can lead to geological inconsistencies such as intersecting geological interfaces. Solutions exist to mitigate these issues. Those discussed in this chapter involve incorporating additional geological information to further constrain the interpolation. The goal is to constrain the interpolation of these surfaces by making interpretations that follow geological principles, or rules, as much as possible. Once the surfaces are free of geological inconsistencies, they isolate 3-D regions which can then be visualized using boundary representation methods (Caumon et al. 2004) or by generating volumetric grids that conform to the bounding surfaces (Mallet 2002; Görz et al. 2017). This approach is best suited to the construction of deterministic models of stratified basins that have not been heavily deformed, or the modeling of the shallow stratigraphic architecture of discontinuous units constrained by geological maps and sparse borehole data (Ross et al. 2005). However, the approach has limitations. First, the freedom that this knowledge-driven approach gives to the geomodeler may require multiple surface-editing operations, especially in the most complex settings. Second, the approach is partly subjective and several interpretative biases may be introduced; this limits the understanding of uncertainties (Laurent 2016). Chapter 15 also discusses these issues.

12.1.2 Implicit Geomodeling

Generating reproducible and auditable geological models automatically from a variety of observations and analyzing

Applied Multidimensional Geological Modeling: Informing Sustainable Human Interactions with the Shallow Subsurface, First Edition.
Edited by Alan Keith Turner, Holger Kessler, and Michiel J. van der Meulen.
© 2021 John Wiley & Sons Ltd. Published 2021 by John Wiley & Sons Ltd.

its uncertainty may seem impossible, but recent advances in modeling and artificial intelligence offer promising avenues for further development. The *implicit* modeling approach is one family of methods that bring geomodeling closer to this goal. Implicit geological modeling consists of extracting bounding surfaces as equipotentials of a 3-D scalar field computed from scattered data points (Calcagno et al. 2008; Collon and Caumon 2017). The function is first estimated at the data locations, and then it is interpolated on a set of grid points across the model domain (Martin and Boisvert 2017) using techniques such as Radial Basis Functions (RBFs) (Carr et al. 2001; Hillier et al. 2014) or co-kriging (Lajaunie et al. 1997). Surfaces representing points of a constant value (iso-surfaces) can then be extracted automatically using techniques such as "marching cubes" (Lorensen and Cline 1987) and "marching tetrahedra" (Doi and Koide 1991; Bagley et al. 2016). With this approach, the insight of a geologist remains critically important as certain data attributes may still be subject to interpretation and several geological constraints, such as orientation data, should be considered (Calcagno et al. 2008; Gonçalves et al. 2017; Laurent 2016).

Implicit geomodeling is especially useful for representing complexly deformed geological units, such as salt diapirs, intrusive bodies, some structures found in igneous and metamorphic terranes, and faulted stratigraphic sequences. Implicit models will satisfy important geological rules as well as orientation data (Calcagno et al. 2008). For example, the process can ensure that lithological units fill the subsurface space with no gaps and spaces and without bounding surface cross-overs. However, as with all geomodeling approaches the technique comes with some limitations. As pointed out by Collon and Caumon (2017), addressing and computing discontinuities such as faults and unconformities remains challenging. Discontinuous units of highly variable thickness may require several steps to be modeled properly and some may still need to be mapped explicitly (Martin and Boisvert 2017). The two broad approaches (explicit vs. implicit) are thus not mutually exclusive (Collon and Caumon 2017) but are increasingly deployed together as described by Jørgensen et al. (2013).

12.1.3 Scope of Chapter

The strategies described above to build geological models using explicit or implicit approaches, or some combination of the two, have been implemented in a number of commercial geomodeling software suites, which offer numerous functionalities and a variety of algorithms and workflows. It is beyond the scope of this chapter to present these algorithms in detail; more information can be found in the cited specialized publications in the list of references.

Section 12.2 presents an overview of the main interpolation methods generally available in geomodeling software. Section 12.3 presents an overview of the SKUA GOCAD system, while Section 12.4 demonstrates how GOCAD tools and procedures may be applied to modeling shallow discontinuous Quaternary stratigraphic units using explicit methods with added stratigraphic rules and constraints. Section 12.5 provides an overview of the Geological Data Management (GDM) Software Suite and GeoModeller, both developed by the *Bureau de Recherches Géologiques et Minières* (BRGM – French Geological Survey).

12.2 Interpolation Methods

Different interpolation methods are employed by a wide variety of modeling software products. Some algorithms were specifically developed for geomodeling applications and one of them, Discrete Smooth Interpolation (DSI), is briefly described in Section 12.2.1. Other popular techniques are also briefly described including Inverse Distance Weighting (IDW) and Radial Basis Functions (RBF) in Section 12.2.2 and kriging in Section 12.2.3. Interpolation is based on Tobler's first law of geography: "*features that are close to each other are more related than those that are further apart*". Interpolation is fundamental to volume discretization and thus is further discussed in Chapter 13. Geomodeling utilizes these interpolation techniques to produce continuous surfaces from diverse scattered and sparse observations (outcrop contacts, borehole horizons, geological structural measurements, interpreted horizons on cross-sections, geophysical profiles, etc.) and to spatially model physical and chemical properties.

12.2.1 Discrete Smooth Interpolation (DSI)

During the 1970s, computer-assisted design (CAD) established numerous boundary representation (B-Rep) techniques to efficiently represent 3-D objects by defining their boundaries with parameterized surfaces (called "primitives"), then assembling them to build volumetric objects. A piecewise/composite creation of surfaces may be accomplished with Bézier surfaces, B-Spline functions, or Non-Uniform Rational B-splines. These methods provide for convenient and efficient generation of curves and smooth surfaces. Since these features are generated parametrically, only a small amount of data are required to define complex curves and surfaces (Figure 12.1). Complex 3D objects are constructed using Constructive Solid

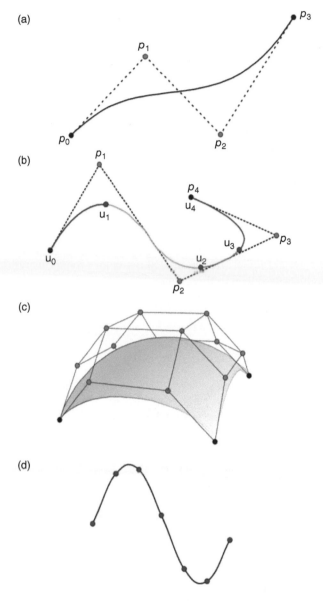

Figure 12.1 Four methods generating curved smooth surfaces: (a) 2-D Bezier curve; (b) 2-D B-spline curve; (c) 3-D surface modeling using Bezier and B-spline procedures; (d) 2-D Radial Basis Function (RBF). Modified from various online sources, such as https://bsplines.org.

Geometry methods which use Boolean operators (union, intersection, complementary, etc.) to join components.

These methods efficiently manage and display engineered objects, such as machine components, which are described by many precise dimensions and have well-defined specific boundaries. In contrast, the geometry of subsurface geological features is typically not known a priori and the goal of the modeling exercise is to predict and describe the shape from limited scattered information of variable precision. Thus, it is not surprising that problems arose when these CAD techniques were applied

to geological modeling (Fisher and Wales 1992; Kelk and Challen 1992).

Mallet (1991, 1992, 2002) recognized the new methods to represent geological objects and proposed the Discrete Smooth Interpolation (DSI) algorithm as an integral part of the original GOCAD software. The DSI algorithm uses a minimization of a roughness criterion to generate smooth curves and surfaces while strictly honoring "hard" constraints (e.g. x, y, z position of a stratigraphic marker), and respecting "soft" constraints (e.g. a stratigraphic thickness min-max range based on regional knowledge) "as much as possible" (Mallet 2002). DSI accomplishes this by splitting a global complex problem into a series of localized linear problems, which then become DSI constraints. The localized problems are then linked as smoothly as possible into a unique solution (Mallet 2002).

12.2.2 Inverse Distance Weighting and Radial Basis Functions

IDW and RBFs are interpolation techniques used in geomodeling to create surfaces that pass exactly through every data point or measured value (Hillier et al. 2014). IDW uses a search neighborhood to consider several data points of known values surrounding the location where a prediction will be calculated. With IDW, the influence of the measured points on the prediction is a function of their distance to the prediction location, which is at the center of the search neighborhood circle or ellipsoid. Specifically, weights are inversely proportional to the linear distance separating data points to the prediction location raised to a power value. Values predicted by IDW are restricted to within the range of values of the dataset. In other words, maximum and minimum values are always at sampled locations. This is an important limitation of the technique as it is unlikely that maximum and minimum values are directly observed or sampled.

In a nutshell, an RBF is a function of the distance from a point of origin. RBFs are calculated across the spatial domain and then linearly combined to approximate functions or data which are known only at sampled locations. In geomodeling, RBFs are used to produce smooth curves and surfaces that pass exactly through sampled data points. One advantage of RBFs over IDW is that predicted values between data points can be above the maximum or below the minimum measured values.

12.2.3 Kriging

The initial concept of kriging was developed by Danie Krige (hence the name "kriging") for the South African mining industry (Krige 1951). Georges Matheron subsequently formalized the concept, making kriging a core component of a broader topic – geostatistics – which was defined as "*the application of probabilistic methods to regionalized variables*" (Matheron 1963). Geostatistics profoundly influenced the practice of geological modeling. The geostatistical approach defined a process that constrained the geometry of geological bodies based only on observations. Regionalized variables account for the intuitive idea that observations that are near each other are likely to have similar values. Kriging applies a consistent probabilistic model – the variogram – to characterize this spatial structure (Chilès and Delfiner 2012).

Kriging is a form of generalized linear regression that formulates an optimal spatial estimator by accounting for the observation volume and stochastic dependence among data. Results are only reliable within the volume defined by the limits of the observed data.

Basic kriging theory assumes the data have "second order stationarity"; this means that: (i) the mean is constant and is independent of location within the region, and (ii) the covariance (or variogram) only depends on the distance between any two values and changes with location. There are several instances of attributes (e.g. water depths, geothermal temperature gradients, sedimentological textures) that have a clear, systematic variation. Hence, models that presume second-order stationarity are inadequate for characterizing these "nonstationary" attributes. As a consequence of these considerations, kriging has developed to include a family of spatial data interpolation estimators: ordinary kriging, universal kriging, indicator kriging, co-kriging, and others.

Ordinary kriging was a first formulation of improved simple kriging that was applicable to sample populations with a constant, but unknown, mean and no regional trend. Universal kriging, also known as kriging with trend, is a full generalization that removes the requirements that the attribute must have a constant mean and that the user must know the value of the mean. Indicator kriging allows for interpolation of categorized numerical data using binary values defined by a threshold (value = 1 if greater than threshold, value = 0 if less than threshold). If several thresholds are defined, the method becomes Multiple-Indicator kriging. This can then be applied to non-numeric data, such as lithological parameters, or can be used to categorize numerical data into multiple categories without having to pre-define the property distribution. This has made it one of the most widely-applied grade estimation techniques in the minerals industry. Co-kriging is the multivariate version of ordinary kriging. Co-kriging deals simultaneously with two or more attributes defined over the same domain to calculate estimates of, or predictions for, poorly sampled variable(s)

by evaluating a well-sampled and related variable (the co-variable). The variables should be highly correlated (positive or negative). Not all attributes must be measured at all locations, but an adequate number of collocated measurements for each pair of attributes is necessary for accurate analysis.

Further discussions on kriging or geostatistics are available in many sources, including Chilès and Delfiner (2012), Isaaks and Srivastava (1989), Journel (1983), Journel and Deutsch (1998) and Matheron (1963). In the late 1970s, BRGM and Ecole des Mines de Paris developed BLUEPACK, a geostatistical software package (Galli and Renard 1986; Renard 1990) which has evolved into a successor called ISATIS (Geovariances 2016). Geostatistical packages are used in the GeoTOP modeling process (see Chapter 11) and are important components of the software products described in this chapter.

12.3 SKUA-GOCAD Geomodeling System

SKUA-GOCAD is a geomodeling software suite that deploys a series of specialized modules and workflows that allow a user to visualize spatial data in 3D "camera" space, generate various geometrical objects and grids, and model geological and geophysical properties, while honoring a set of rules and constraints such as those related to sequence stratigraphy.

Professor Jean-Laurent Mallet and his research group initiated the GOCAD software in 1989 (Mallet 1991). GOCAD was developed to model, manage, and visualize complexly deformed (overturned and highly faulted) geological surfaces. GOCAD elected to employ a modified CAD environment running on graphics workstations, which had just been developed. As discussed in Section 12.2.1, its core is the DSI algorithm that uses smooth curves and surfaces to define the bounding limits of geological objects and interpolate internal properties (Mallet 1989, 1992, 1997).

The DSI algorithm applies constraints reflecting geological knowledge and information to minimize the dispersion of observations about a generated smooth surface, while strictly honoring hard constraints defined by accurate observations and data and respecting soft constraints, which may represent useful geological information with an inherent degree of uncertainty, subjectivity, or imprecision. GOCAD uses several specific terms to describe the components used in this process (Table 12.1). The basic geometric object is a *node*, which is any point defined by x, y, z coordinates. *Control Nodes* represent hard constraints; these nodes will not move during the DSI operation on an object. *Control Points* are the soft constraints; the constraints are

Table 12.1 Defined GOCAD terms and objects.[a]

Gocad class of terms and objects	Gocad name	Definition
General Terms	Node	A node is a single point within an object.
	Control point	A node set as a linear "soft" constraint. Used to pull an object during the DSI[b] operation.
	Control node	A node on an object that is set as a "hard" constraint; a control node is "locked" and does not move during the DSI operation.
	Border	A boundary or a portion of a boundary of a Surface or Solid Object.
	Region	A group of selected nodes on an object.
	Part	An object can consist of separate disconnected parts.
	Shooting Direction	The linear direction set for any linear constraints
2-D Objects	PointsSet	A set of disconnected nodes with X, Y, Z coordinates.
	Curve	A set of nodes connected by segments. Curves can be open (such as in a line) or closed (such as in a polygon).
	Well	A line or well-path containing information such as borehole logs and stratigraphic markers (contacts).
	Surface	A surface consists of a network of triangular faces; each face is defined by three segments (edges) connecting three nodes. Coordinates (x, y, z) and user-defined properties are stored at the nodes.
	2D Grid	A regular grid defined by a point of origin (x0, y0, z0) and two corner points or step-vectors and number of cells along two axes or directions.
3-D Volumetric Objects	Voxet	A regular 3-D grid made-up of identical volume elements called cells. Properties are stored at the centroid called grid points. It is typically used to represent geophysical data. It is defined by an origin (x0, y0, z0) and three corner points or step vectors and number of cells in three directions
	SGrid	A type of 3-D irregular hexahedral mesh or grid. An SGrid can have an overall curvilinear shape that better conforms to the underlying structural model than regular grids such as voxets.
	Solid	A 3-D object used for modeling closed volumes with tetrahedra.

a) Several other objects are available in Gocad for various purposes. Most of them are associated to specialized add-on modules. The reader is referred to the specific GOCAD-SKUA documentation for details.
b) DSI: Discrete Smooth Interpolation (see text for details).

defined by points and *shooting direction* which together are used to pull the nearby nodes of an object in a specified direction during the DSI operation. This systematic approach to representing subsurface data and geological objects consistently with points, lines, surfaces, and volumetric grids has been a distinctive feature of GOCAD.

During the 1990s, the GOCAD software was updated and transferred to a new software company, which became responsible for further development and implementation, commercialization, maintenance, and support of the software platform; this company was acquired by Paradigm in 2006, which later became part of Emerson (www.pdgm.com). Academic GOCAD research continues, however, with the RING-GOCAD Consortium based at the School of Geology at Université de Lorraine in France.

In 2008, Paradigm introduced SKUA, the first commercial implementation of the GeoChron concept, originally developed by Mallet (2004). This concept was developed to accomplish the difficult task of interpolating sedimentary properties (e.g. facies) related to the depositional environment in a basin that was heavily deformed post-deposition. Based on this concept, a model of the stratigraphic architecture and petrophysical properties is developed before considering fault deformations and generating a faulted stratigraphic volumetric mesh.

The 3-D GeoChron model coordinates correspond to the paleo-geographic coordinates of any point at its time of deposition, before any tectonic deformation. The coordinates are thus not aligned to fault networks; instead, mesh layers are orthogonal to stratigraphic horizons and conform to flexural deformations of folds. This is achieved by numerical computation of a chronostratigraphic transform (Caumon and Mallet 2006; Tertois and Mallet 2007). The transformations are developed by mapping any subsurface point onto its image in a flattened depositional space.

Moyen et al. (2004) demonstrated how, by using the GeoChron model, it is possible to compute a high-resolution regularly structured grid model that honors the stratigraphy. The regular structure of such a transformed model can then be used for geostatistical computations. This is important because:

- it provides the correct spatial structure to model properties related to the depositional environment across what is now a complex structural model;
- these property distributions can then be transformed to unstructured grids to represent the deformed basins and used for various analyses such as for fluid-flow simulations.

The GeoChron concept relies on implicit surfaces (also known as level sets) on a tetrahedral mesh (Moyen et al. 2004). SKUA also provided new reservoir gridding algorithms (Jayr et al. 2008; Gringarten et al. 2008). More recent developments include the SKUA Structure Uncertainty product (Tertois et al. 2010), which is based on prior research by the RING-GOCAD Consortium to enable the stochastic perturbation of implicit structural models (Caumon and Mallet 2006; Caumon et al. 2007; Tertois and Mallet 2007).

While the original GOCAD utilized explicit surface modeling solutions to create wireframe models using the DSI algorithm, the incorporation of the SKUA modules allow for implicit, volumetric modeling that can build an entire 3-D model in a single pass. Three-dimensional models can be validated for geological consistency and quickly updated to honor new datasets, allowing for models that are flexible to changing data densities.

12.4 Modeling Shallow Discontinuous Quaternary Deposits with GOCAD

The development of SKUA-GOCAD has been mostly focused on solving specific problems related to modeling deformed sedimentary basins and their physical properties. However, it has maintained the modules of the original GOCAD software suite, which have also been used to deal with other types of problems, such as modeling thin and highly discontinuous stratigraphic layers of the shallow subsurface. These problems are often encountered when modeling near-surface continental deposits, which are important for applications involving environmental geosciences (e.g. hydrogeology), mineral exploration

under sedimentary cover, and geohazards (e.g. earthquake ground motion studies).

Modeling Quaternary continental deposits generally involves describing thin and discontinuous layers representing short-period (high frequency) sedimentary cycles. For example, in North America, Northern Europe, and Siberia, multiple glacial and interglacial episodes have produced a complex sedimentary record of inter-layered and inter-fingered glacial, fluvial, lacustrine (locally marine), and eolian deposits. The subsurface can include buried valleys which often contain important groundwater aquifers (Bajc et al. 2018; Korus et al. 2017); many Quaternary units exhibit considerable lateral variability due to facies changes. Construction of a 3-D model requires consideration of the changing Quaternary geological environments; a successful model results from the development of a conceptual model derived from interpretation of all available information and a regional geological analysis.

Availability of data and the scale of the model define the selection of geological units expressed in a model. For example, when modeling a large area, smaller units may be grouped together to simplify the model content. Any such simplification of the subsurface must consider the anticipated use of the 3-D model so that critical information is retained.

12.4.1 Modeling Approach

Model development begins with data preparation. This includes collection of existing data, acquisition of new information from field surveys, and quality assurance and data standardization (Figure 12.2, Tables 12.2 and 12.3). Typically, all data are georeferenced in the same coordinate system and imported into GOCAD as unassigned objects (e.g. points, wells, curves, surfaces). A stratigraphic column is built where the user defines the stratigraphy of the model domain and specifies the nature of the main stratigraphic contacts (i.e. erosional, conformable). The various objects are then assigned to their respective features, such as a geological horizon.

The model is initiated by developing some initial approximate lithostratigraphic surfaces defining the top of each horizon. Since Quaternary models on land typically start at the ground surface, topography is often developed first. Digital Elevation Model (DEM) data form a regularly spaced *PointsSet* (Table 12.1) which is readily converted into a triangulated surface with the original DEM points forming triangle vertices. In many cases, however, the model will be at a lower resolution than the original DEM. In these cases, a flat surface with the desired cell-size is created first, and then modified using the DSI algorithm (Section 12.2.1) with the DEM data as *Control Points*, a vertical *Shooting*

Figure 12.2 Typical data sources and data preparation steps before modeling Quaternary deposits for a regional hydrogeology project. Modified from Ross et al. (2007).

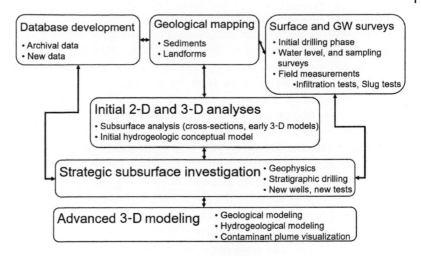

Table 12.2 Commonly used datasets, their dimensions, and their sources.

Data set	Dimensions	Common sources
Geological maps	2	Geological Surveys
Digital elevation models	3	Geological Surveys
Boreholes	1	Geological Surveys, Industry
Water-Wells	1	Geological Surveys
Geophysical datasets	2 or 3	Geological Surveys, Industry
Cross-sections	2	Geological Surveys, Industry
Aerial imagery	2	Geological Surveys, Industry

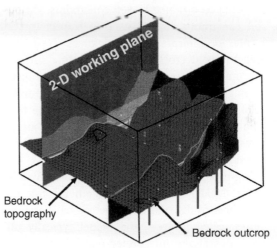

Figure 12.3 Interpretive cross-sections, map polygons and borehole markers provide a distributed set of constraints to the DSI algorithm. Modified from Ross et al. (2005).

Table 12.3 Validation criteria for existing boreholes.

Data type	Reliability factor	Definition	Criterion
Stratigraphic, geotechnical, and water well logs described following accepted scientific methodology	5	Highly reliable	Original logs and reports are available for checking procedures. Continuous and discontinuous samples are available.
	4	Reliable	Original logs and reports are available. Some wells are still accessible in the field and their location have been verified.
Geotechnical boreholes Water wells	3	Less reliable	Original logs are not available, but there is no apparent geological inconsistency with nearby reliable data.
Water wells	2	Poorly reliable	Typical problems: Source elevation does not match DEM elevation; Conflicting stratigraphy reported.
Water wells	1	Unreliable	Multiple problems with descriptions of elevation, lithology, etc.

Modified from Ross et al. (2005).

Direction, and surface border constraints. Polygons representing geological units at the surface are imported from available geological maps and are projected onto the DEM surface to define outcrop areas for each geological unit.

An initial 3-D model can be built by developing a series of triangulated discrete surfaces representing the upper boundary of each horizon in the model. This initial model supports decision-making in the early stages of an investigation (Ross et al. 2005). This step helps identify problematic areas where the data could be erroneous or insufficient. As data validation and new datasets become available, the surfaces are refined using the DSI algorithm and improved to produce a final model (Ross et al. 2005). In GOCAD, a user can set *Control Nodes* (Table 12.1) where a surface is defined and positioned accurately with all other nodes being constrained by soft constraints. Various possible soft constraints can be used to respect certain geological rules or interpretation; thickness constraints may be used to avoid cross-overs of bounding surfaces or to ensure the thickness of a horizon remains realistic based on geological information and knowledge. Usually, for models of Quaternary deposits, the base of the final model is the bedrock surface on which many younger discontinuous surfaces pinch-out, while the top is defined by the surficial topography (Ross et al. 2005).

By following this procedure, a model is represented by a series of interlocked surfaces that together define the bounding contacts of discontinuous geological horizons. In large multifaceted projects, the model typically utilizes various datasets and interpreted features, including borehole top marker beds, contacts interpreted from geophysical profiles, geological maps and cross-sections. Because many Quaternary deposits have irregular spatial distributions and often are missing from some boreholes, geological maps together with a series of interpretative cross-sections provide a very useful source of geological information, which assists in making the 3-D model adhere to an overall geological conceptual model. As shown by Figure 12.3, these cross-sections can provide a distributed set of constraints to the DSI algorithm.

One important challenge with the approach described above is the definition of subsurface pinch-out borders (Ross et al. 2005). This is typically done using a deterministic approach whereby subsurface pinch-outs are mapped by the user explicitly (i.e. by drawing curves onto surfaces), but probabilistic approaches (Desbarats et al. 2001; Benoit et al. 2018) can also be used to help guide the mapping by identifying the areas of the subsurface where the unit is likely to have zero thickness based on available data.

Representing the discontinuous character of Quaternary units is useful, but it creates issues for end-users because many applications that justify the development of

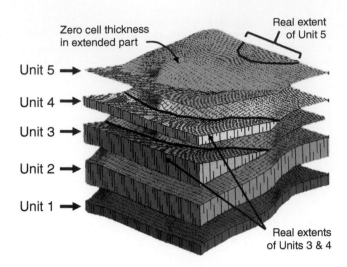

Figure 12.4 Example of five-layer model with discontinuous geological units; the surfaces representing the units are represented by zero thickness layers where the geological units are absent. Modified from After Ross et al. (2004a).

a 3-D geological model, including property modeling and groundwater flow modeling, require continuous surfaces. One common solution to this problem is to create continuous surfaces from an initial reference surface (typically topography or bedrock surface). The copied reference surface with its new set of constraints is then partly modified with the DSI to fit the top of the new horizon. This method produces continuous surfaces to model discontinuous geological horizons; where the top surface is at the same elevation as an underlying surface everywhere the horizon is considered absent at that location; it will thus have zero thickness in the model (Figure 12.4).

The final stage in the model building process is to convert the sequence of surfaces into a volumetric Geologic Framework Model (GFM). The surfaces ideally form closed volumes representing geologic bodies; the surfaces acting as "dividing walls" that isolate 3-D regions (Mallet 2002). Figure 12.5 shows three volumetric representation options for a typical GFM: (i) boundary representation; (ii) regular grid with cubic cells or voxels; and (iii) curvilinear stratigraphic grid with hexahedral cells. The boundary representation (often called "B-rep") is a method for visualizing geological bodies and obtaining their volume. The GFM must be topologically perfect; the surfaces must be "sealed" to form closed volumes and the line defining the intersection of two surfaces must be unique and free of gaps; it is critical that the topology defined by the surfaces be unambiguous. Once the surface topology has been validated, horizons become empty volumes or "shells" represented in 3-D space.

However, producing a consistent B-rep from a GFM with numerous and highly detailed pinching surface borders

Figure 12.5 Three volumetric representation options for a typical GFM. Modified from Ross et al. (2007).

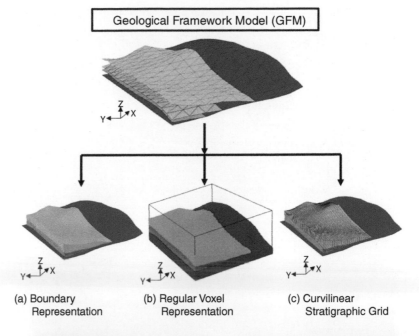

(a) Boundary Representation

(b) Regular Voxel Representation

(c) Curvilinear Stratigraphic Grid

Figure 12.6 Example of semi-regular grid with prismatic cells. Ross et al. (2007).

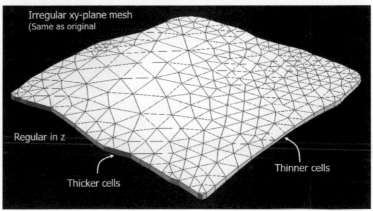

may require laborious operations. Experience has shown that B-rep tools quite often fail at creating a sealed model when the GFM is defined by a series of thin discontinuous (pinching) surfaces extending over several hundred square kilometers. In many of these cases, which are common in continental Quaternary basins, the best way to build consistent volume representations in SKUA-GOCAD may still be through the generation of volumetric grids from the continuous surface approach described above. Grids are also required for property (e.g. facies) modeling.

SKUA-GOCAD software has extensive capabilities for creating and managing regular voxel grids *(Voxets)* and curvilinear stratigraphic grids *(Sgrids)* (Table 12.1). Both are commonly used for 3-D visualization and various property analyses using geostatistical methods and for fluid flow simulation applications (Ross et al. 2005). However, a third type of grid, a semi-regular grid with prismatic cells (Figure 12.6), is more often used by finite-element

groundwater flow models; these require the use of specialty modules for interoperability with GOCAD (Ross et al. 2007).

12.4.2 Hydrostratigraphic Modeling in Eastern Canada

A regional hydrogeologic survey in the St. Lawrence Lowlands undertaken between 1999 and 2002 produced, or collected, information from multiple sources, including: two Quaternary geology maps (Bolduc and Ross 2001a,b; Savard 2013), detailed geological cross-sections, over 5000 borehole logs, and geophysical interpretations using ground penetrating radar (GPR) and shallow seismic reflection surveys (Ross et al. 2005; Ross et al. 2006). Based on these information sources, a volumetric GFM was constructed to provide a consistent stratigraphic reconstruction of the Quaternary sediments which control recharge rates to the

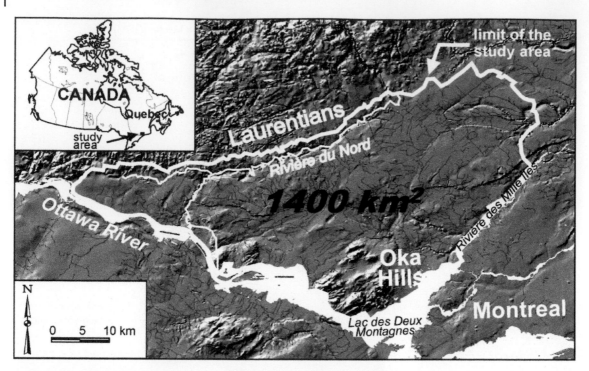

Figure 12.7 Location map of the eastern Canada study area. (Ross et al. 2002).

underlying regional fractured-bedrock aquifer (Ross et al. 2002, 2004a, 2005). The GFM provided new insights into the regional Quaternary geology including buried granular aquifers, the buried bedrock topography, and the regional hydrogeology (Ross et al. 2005). The GFM was then used to assess the vulnerability of the regional fractured bedrock aquifer to contamination (Ross et al. 2004b).

The study area is a Quaternary glaciated basin covering about 1400 km^2 (540 mi^2) located between the Laurentian Highlands, the Ottawa River, and other St. Lawrence River tributaries in southwestern Quebec, Canada (Figure 12.7). This basin consists of a low-relief clay plain incised by paleochannels of the proto-Ottawa River as well as modern rivers and streams, and a drumlinized till plain. Most of the study area is within the Paleozoic St. Lawrence sedimentary platform consisting of gently folded and faulted clastic and carbonate rock successions (Savard 2013). The Laurentian Highlands, the margin of the Canadian Shield, form the northern basin boundary; the study area also includes the Oka Hills, an isolated Shield section with Cretaceous intrusions. In rural and semi-rural parts of the area, the population depends largely on fractured sedimentary bedrock aquifers for water supply; large portions of the region have undergone rapid suburban growth near the cities of Montreal and Ottawa (Savard 2013).

Bedrock is overlain by up to 150 m (500 ft) of Quaternary sediments that mainly include Holocene postglacial sediments and Late Wisconsinan proglacial and subglacial units of variable thickness; older Quaternary units only occur locally. Glacial diamictons (till) and marine fines (silt/clay) are the most widespread sediments (Figure 12.8). A volumetric GFM was constructed following the procedures described in Section 12.4.1. The completed GFM has nine units; eight Quaternary units and an undifferentiated bedrock unit. This represents a simplification of the relatively complex Quaternary stratigraphy; the model emphasizes the distribution of major units, as is appropriate for regional hydrogeological studies.

The geological units were further evaluated and five hydrostratigraphic units were defined (Figure 12.9). The various sandy near-surface Holocene units form a relatively discontinuous localized "upper aquifer." Below an erosional unconformity, Quaternary marine clays form an extensive aquiclude that covers about 60% of the model area. Older buried glaciofluvial channels form some locally significant aquifers (Figure 12.9c). These channels are mostly eroded into older tills which form a regional aquitard that overlies fractured sandstone and carbonate sedimentary rocks. These constitute the most widely used regional water source (Savard 2013).

12.5 BRGM Geomodeling Software

The *Bureau de Recherches Géologiques et Minières* (BRGM – French Geological Survey) has developed two geomodeling

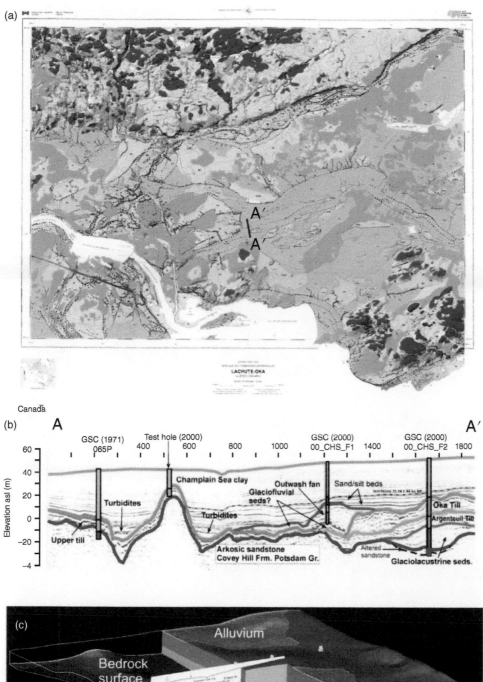

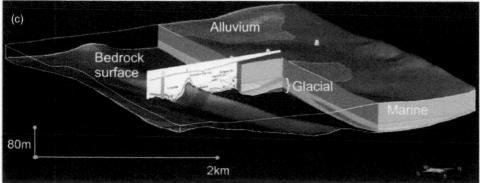

Figure 12.8 (a) Surficial geology between the Laurentian Highlands, the Ottawa River and other St. Lawrence River tributaries in south-western Quebec, Canada (after Bolduc and Ross (2001a)); (b) interpreted seismic profile (see A-A' on map in (a)) showing variations in bedrock topography, till thickness, and a confined glaciofluvial aquifer; (c) a 3-D sub-model of part of the buried valley. Pink and red are bedrock units, green is till, blue units are marine, orange is glaciofluvial, and yellow units are alluvium. Modified after Ross et al. (2006).

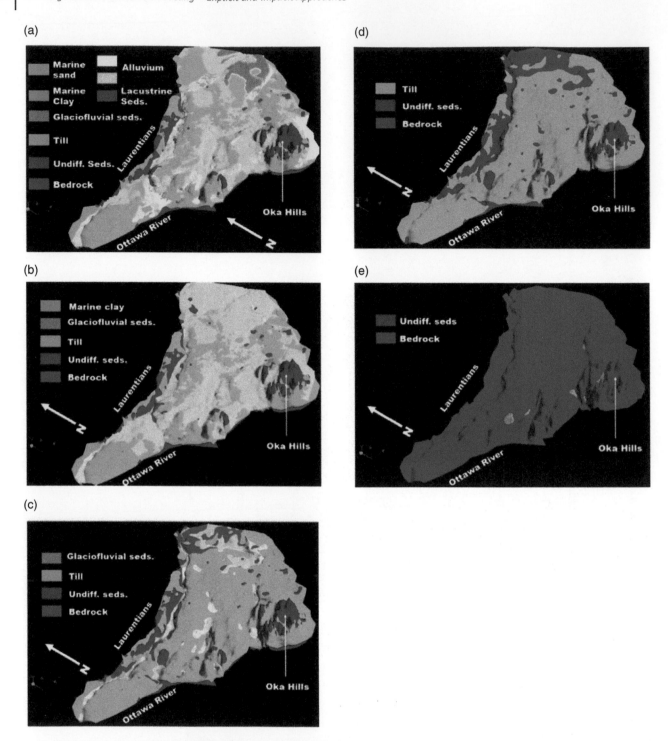

Figure 12.9 Five views of the Geological Framework Model: (a) Completed geological framework model; (b) the regional aquiclude composed of till and marine clay below the shallow surface aquifers; (c) the buried granular aquifers; (d) the lower regional aquitard; (e) top of bedrock which forms the fractured bedrock aquifer. After Ross et al. (2004a).

software systems. The GDM Software Suite provides an explicit modeling approach while GeoModeller supports implicit modeling.

12.5.1 GDM Software Suite

BRGM developed the GDM Software Suite by expanding the BLUEPACK geostatistical toolkit, co-developed with the Mines ParisTech (École Nationale Supérieure des Mines de Paris). The GDM Software Suite provides a sophisticated explicit geological modeling approach with access to geostatistical tools to exploit and make full use of geoscientific data and geological models. GDM is designed to interpolate maps defining strata thicknesses or depths in areas of relatively simple structure. It includes several automatic data coherence tests that are indispensable for efficiently managing large volumes of information describing drill holes and geological maps. With a defined stratigraphic column (Figure 12.10a), GDM can apply geological rules representing erosion or deposition to create 3-D geological models from sequences of interpolated surfaces, while taking faults and boundaries into account. GDM produces a selection of dissemination products, including multilayer maps and vertical sections, drill hole strip logs, and 3-D views (Figure 12.10b,c). GDM contains five modules (Table 12.4); they are now available as a set of online services making it possible, for example, to interface with geographic information systems.

12.5.2 GeoModeller

In the 1990s, BRGM developed the *"Editeur Géologique"* (Geology Editor) to support its geological mapping/modeling activities. This evolved to become GeoModeller, which was subsequently commercialized by Intrepid

Table 12.4 GDM Software Suite modules and their applications.

Module	Application
GDM Standard	Data representation and modeling
GDM ArcGIS	Visualization of data and geological models, and use of the geostatistical methods in ArcGIS
GDM MultiLayer	Three-dimensional geological modeling
GDM Viewer	Visualization of data and geological models
GDM Web Services	Representation of geological data and models on the Web

Geophysics, Melbourne, Australia (Gibson et al. 2013). GeoModeller is designed as a geological mapping aid to model complex geological structures. Implicit 3-D geological models are developed from integrated data sets and a variety of products can be produced (Figure 12.11).

The implicit geological modeling approach allows construction of 3-D geological surfaces and volumes which honor geological observations and/or cartographic data. In the 1990s, joint research at the BRGM and Mines ParisTech recognized that orientation observations were more frequently recorded than geological formation contacts; thus, ignoring them represented an unacceptable loss of information on geological structures. An algorithm (Lajaunie et al. 1997; Calcagno et al. 2008) combines these two main data types (Figure 12.12). The algorithm implicitly defines geological unit boundaries as iso-value points within a 3-D scalar field constrained by the orientation data introduced as normalized potential gradients. Universal co-kriging (multivariable interpolation) of data

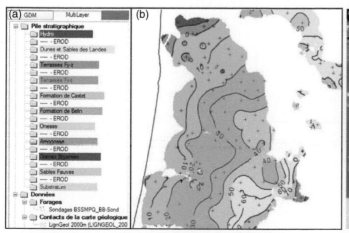

Figure 12.10 GDM Software Suite examples from the Aquitanian Basin, south-west France. (a) Standard stratigraphic column; (b) 2-D Isopach (thickness) map; (c) 3-D model view. Used with permission of BRGM.

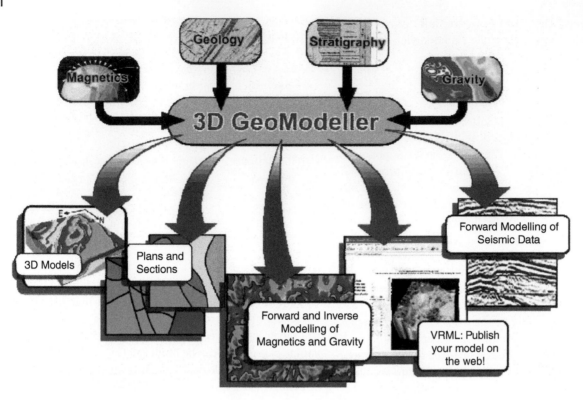

Figure 12.11 Input data and classical output results used for geological modeling. (Lacquement et al. 2011).

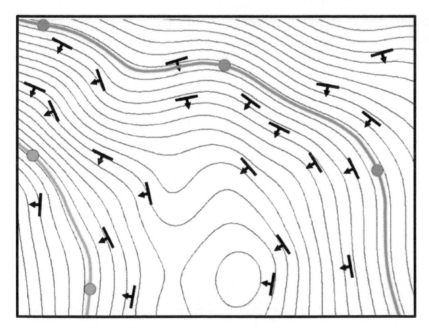

Figure 12.12 Map of known geologic contacts (green dots) for formations belonging to a single series, and structural orientation observations used to compute the potential field (contoured). The potential field is always orthogonal to the (structural) orientation vector. Two iso-potentials of the potential field passing through the two sets of geologic contact points define formation boundaries. This geologic model honors the contact data and the orientation data. After McInerney et al. (2005).

representing both the geological contacts and the structural orientations produces an implicit function that defines a potential field; the method has become known as *"potential-field geological modeling"* (Lajaunie et al. 1997; Calcagno et al. 2008; McInerney et al. 2005, 2007).

A GeoModeller concept, termed the *geological pile*, allows the aggregation of these boundaries using a rule-based approach that honors stratigraphic relationships, such

as erosion or intrusion, and the chronology of the fault network. The *geological pile* uses the topology of the series of interfaces to define the spatial architecture and relationships of geological interfaces without taking into account their chronological order. Because the interface orientations can be viewed as a time arrow, the *geological pile* is often viewed as a stratigraphic column. However, it is definitely not the same; for example, with an overthrust

Figure 12.13 Geological modeling by potential field interpolation; (a) geological contacts (green and purple dots) and strike/dip orientations, (b) formation contacts correspond to potential field iso-value surfaces, (c) fault displacements defined by discontinuities in the drift of the universal kriging calculations. Used with permission of BRGM.

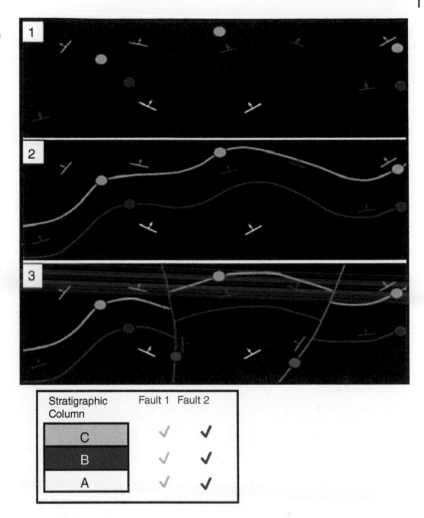

structure, for modeling purposes, the same formation can be duplicated in the *geological pile*, with one version of the formation being "above" the other in the pile. Intruded series are placed at top of the geological pile, regardless of their age.

Since all formations with a common geological deformation history can be expected to exhibit similar geometries, GeoModeller uses the term "*series*" to define sequences of formations with common histories. The *geological pile* defines the formations belonging to each series, and independent potential field functions are computed for each series. This allows the satisfactory reconstruction of structures which have experienced the same deformation sequence, based on a relatively few surface observations. The original interpolation method did not allow discontinuities to be considered. However, once the geometry of faults is also defined with specific potential fields, fault displacements are deduced automatically by introducing discontinuities into the drift of the universal kriging calculations (Figure 12.13).

A GeoModeller 3-D model is completed by combining the individual models for each series. The *geological pile* establishes spatial relationships of the different series. The series boundary characteristics define geological erosional or depositional processes; these influence the overall 3-D model architecture (Figure 12.14). Thus, the *geological pile* controls construction of the 3-D geological model and synthesis of the associated knowledge. It allows automatic combination of independently constructed series.

12.5.2.1 3-D Model Creation and Validation

The GeoModeller process of 3-D model creation and validation involves several steps (Figure 12.15), beginning with the collection of multiple data sources. Field geological surveys provide lithological observations, including locations of observed contacts between geological units, and structural measurements; these include orientations of planar features (stratifications of sedimentary rocks, metamorphic schistosity, or igneous foliations) and linear features (lineations due to deformation). These observations are supplemented by digital topographic data (DEM),

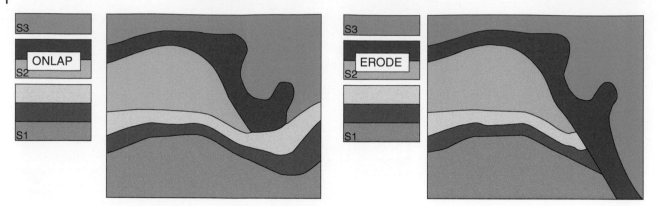

Figure 12.14 A 2-D view of a geologic section with three series of geologic formations. Independent interpolations of each series will produce a unique complete geologic model only after reference to the *geological pile*, which encodes the spatial relationship of formations and series. On the left, series S2 "onlaps," and truncates against the older S1 series. On the right, the "erode" option causes series S2 to cut across older formations. (McInerney et al. 2005).

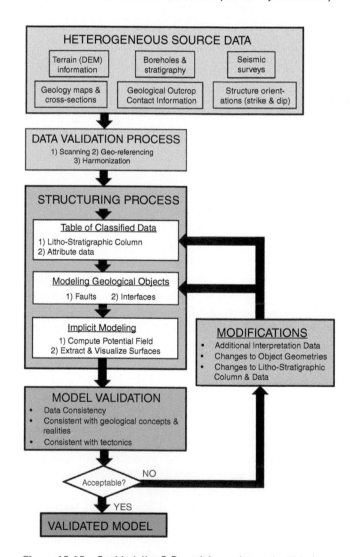

Figure 12.15 GeoModeller 3-D model creation and validation process. Used with permission of BRGM.

and information from existing boreholes, geological maps, and cross-sections. All data sources undergo a data validation process to produce an integrated information base with consistent georeferencing.

The model creation is not complete until the model is validated. This involves evaluating the data to ensure the data are internally consistent, accurately represent geological concepts and observations, and are consistent with the tectonic history. These evaluations involve not only a visual assessment of the completed model (Lacquement et al. 2011), but also geostatistical tests of uncertainty (Aug 2004; Chilès et al. 2004). The availability of new subsurface data usually requires further model validation and adjustment. The model may be further refined by comparing it to gravimetric or magnetic geophysical measurements (Gibson et al. 2013; Guillen et al. 2008; McInerney et al. 2005, 2007). When these validation tests indicate inconsistencies, model modifications are undertaken employing new geological hypotheses, additional data, modifications to the fault network and other object geometries, and adjustments to the *geological pile*, as shown in Figure 12.15. Once the validation tests produce acceptable results, a validated model will be produced and is ready for applications.

12.5.2.2 Student Training in 3-D Mapping using GeoModeller

For a decade, geology students have created 3-D geological models employing the GeoModeller approach during two-week field courses in the coal basin of Alès, in southern France. The study area covers $220 \, km^2$ ($25 \, mi^2$) in the southern part of the French Central Massif located between Alès and Bessèges. Courrioux et al. (2015) discuss the geology of this area, which has relatively complex structures. Three rock sequences – Cambro-Ordovician metamorphic rocks, Carboniferous sedimentary rocks, and Mesozoic cover – are separated by major unconformities and have been subjected to several tectonic events (Figure 12.16).

The purpose of this training is to bring together the traditional practice of map making and the techniques of 3-D geological modeling. Working in groups assigned to different geographical areas, the students are faced with the conventional exercises of geology: quality of observations, reconnaissance of lithological facies, analysis of structures and micro-structures, and acquisition of structural data. This is complemented by the daily entry of the field data into the GeoModeller 3-D geological modeling system. The

process allows the staged formulation of hypotheses and the strategic planning of the areas requiring additional field observation. In this way, the 3-D models and, by extension, the maps of each area are built up iteratively by incorporating new data as it is acquired.

The initial model (Figure 12.17) is based on relatively few surface observations of locations of geologic contacts and some structural orientation data. The 2-D map is, by definition, consistent with the 3-D model and topography. This first step designs the overall architecture of the model. Starting from this model, progressively more detailed data can be incorporated.

Figure 12.16 Example of a completed 3-D model, map, and sections for the entire Alès study area based on field data acquisition by students. Carboniferous units are yellow and brown; Mesozoic units are purple (Triassic) and blue (Jurassic). (Courrioux et al. 2015).

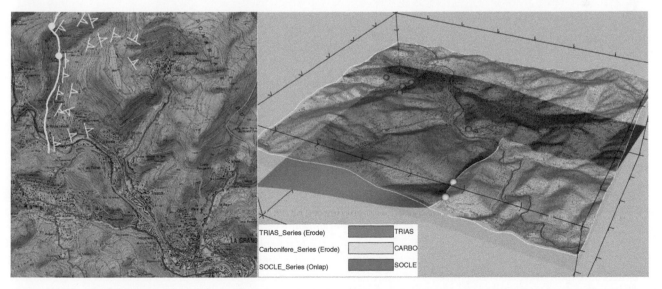

Figure 12.17 The initial model based on surface observations of geological contacts and bedding orientations for the base of Triassic (purple) and base of Carboniferous (yellow). Base of Triassic is an erosional surface. 3-D contours on the map are derived from the 3-D model. Used with permission of BRGM.

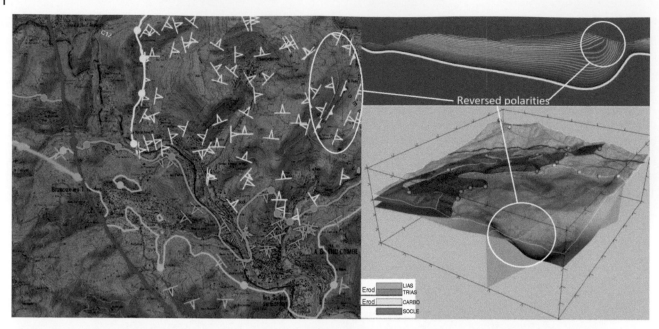

Figure 12.18 A more refined model following completion of the field survey includes two faults (red and green). The Triassic and Liassic units are placed in a single series (purple and blue). Reversed polarities of bedding planes define a deepening of the Carboniferous basin. Because the potential field is defined over the entire space, bedding attitudes within the basin can be represented in addition to the base of the Carboniferous (yellow). The cross-section extracted from the model (upper right) clearly shows the associated trend lines derived from the potential field interpolation. Used with permission of BRGM.

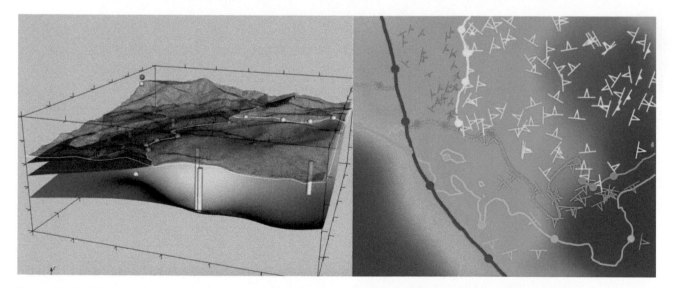

Figure 12.19 Final model after incorporation of drill holes and comparison to a gravity map. The strong negative gravity anomaly (blue) supports the interpreted deepening of the basin in the previous model. Used with permission of BRGM.

Figure 12.18 shows a more refined model based on additional field observations. Noticeably, reversed polarities of bedding planes indicate that the Carboniferous layers have suffered some amount of shortening and folding deformation, which forces the eastern part of the basin to be deepened. Two faults have also been introduced. Additional constraints are required for the Mesozoic units; the Triassic and Liassic rocks are placed in a single series, allowing them to be interpolated simultaneously and forcing them

to behave sub-parallel during interpolation. Subsequently, available drill hole observations were incorporated to produce a final model (Figure 12.19). These constitute the only reliable direct constraints at depth. They confirm, and even amplify, the predicted deepening of Carboniferous layers developed in the previous model (Figure 12.18). At this stage it is also possible to compare the geological model with geophysical survey images. For example, gravimetric survey data show a strong negative anomaly in the eastern part of the Carboniferous basin. This agrees with the model representation of a gentle and regular deepening from the west border, followed by a more abrupt change in the east. Performing a geology/gravity inversion would allow for further validation or adjustment of the model.

12.5.2.3 Capabilities of GeoModeller

GeoModeller develops 3-D subsurface models by replicating the traditional geological mapping process. The geologist/modeler builds a representation of the geology by progressively assessing additional observations. As each new observation is evaluated and understood, the staged approach supports refinement of the conceptual geological model and ensures validation of geological interpretation. Use of GeoModeller enables direct construction of three-dimensional geological objects which, de facto, when intersected with the topography, produce consistent cross-sections and maps. This consistency (or coherence) is critically important; maps produced with this method are constrained to honor all the data without approximation. In common with all implicit approaches, model computation is very rapid; this encourages model updating when new data become available. However, rendering of 2-D or 3-D model views is slower than with explicit models, with the speed depending on the desired resolution.

From its inception, GeoModeller has allowed association of geological data with gravimetric or magnetic geophysical data. The direct calculation of gravity or magnetic fields based on the geological model, followed by a comparison of measured and modeled responses, permits a refined interpretation and validation of the geological model (Gibson et al. 2013; Guillen et al. 2008; McInerney et al. 2005, 2007). Monte Carlo methods for inversion of geophysical measurements allow model adjustments or development of alternative geological models, and may suggest new geological hypotheses (Guillen et al. 2008).

The geostatistical formalism underlying the use of co-kriging for interpolating the potential field permits quantification of the model uncertainty (Aug 2004). With the aid of co-kriging variances, it is possible to map the probability of interface locations, identifying areas of high uncertainty. Although involving considerable computation times, stochastic simulation techniques, including combined Markov Chain/Monte Carlo methods, allow further evaluation of model constraints; these aspects are described in Chapter 15.

12.6 Conclusions

Initial approaches to 3-D geological modeling focused on techniques that built surfaces and grids representing geological boundaries and modeled the distribution of properties within geological units. Many techniques developed over the last three decades are still being used in one way or another to build geological surfaces and 3-D models. Recent developments have increasingly focused on approaches, such as those described in this chapter, which account for geological principles at each step during model creation. Surfaces defining several geological boundaries are no longer constructed independently from each other and subsequently edited to correct multiple geometrical violations of stratigraphic succession or other important geological principles. Methods and software have thus evolved to streamline the geological model creation process directly from maps, cross-sections, and data points, while incorporating geological rule-based tools to reduce manual editing.

These advances, along with improvements to software workflows, help make geological models more reproducible and auditable. This is a critical step toward achieving another important goal; allowing geologists to rapidly build several different plausible models and to evaluate their uncertainty. Explicit and implicit approaches have been part of this evolution and can be used separately or in combination in order to build consistent geological models. The next generation of modeling approaches will probably take advantage of the developments in artificial intelligence. Specifically, "machine learning" algorithms that improve the geological modeling solution and the overall performance with experience will soon be implemented to assist the modeler in building complex models from an ever-increasing quantity and variety of data.

References

Aug C. (2004). *Modélisation Géologique 3D et Caractérisation des Incertitudes par la Méthode du Champ de Potentiel.* PhD Thesis, École Nationale Supérieure des Mines de Paris. 198 pp. [Online: Available at: https://pastel.archives-ouvertes.fr/pastel-00001077/] (Accessed August 19, 2020).

Bagley, B., Sastry, S.P. and Whitaker, R.T. (2016). A marching-tetrahedra algorithm for feature-preserving meshing of piecewise-smooth implicit surfaces. *Procedia Engineering* 163: 162–174. https://doi.org/10.1016/j.proeng.2016.11.042.

Bajc, A.F., Marich, A.S., Priebe, E.H. and Rainsford, D.R.B. (2018). Evaluating the groundwater resource potential of the Dundas buried bedrock valley, southwestern Ontario: an integrated geological and hydrogeological case study.

Canadian Journal of Earth Sciences 55 (7): 659–676. https://doi.org/10.1139/cjes-2016-0224.

Baker, T.J. (1987). Three-dimensional Mesh Generation by Triangulation of Arbitrary Point Sets. In: *Proceedings, AIAA 8th Computational Fluid Dynamics Conference*, 255–271. Reston, VA: American Institute of Aeronautics and Astronautics. https://doi.org/10.2514/6.1987-1124.

Benoit, N., Marcotte, D., Boucher, A. et al. (2018). Directional hydrostratigraphic units simulation using MCP algorithm. *Stochastic Environmental Research and Risk Assessment* 32 (5): 1435–1455. https://doi.org/10.1007/s00477-017-1506-9.

Bolduc, A.M. and Ross, M. (2001a). *Surficial Geology, Lachute-Oka, Quebec*. Ottawa, ON: Geological Survey of Canada, Open File 3520. 1 map sheet (1 : 50,000). https://doi.org/10.4095/212599.

Bolduc, A.M. and Ross, M. (2001b). *Surficial Geology, Laval, Quebec*. Ottawa, ON: Geological Survey of Canada Open File 3878. 1 map sheet (1 : 50,000). https://doi.org/10.4095/212709.

Calcagno, P., Chilès, J.P., Courrioux, G. and Guillen, A. (2008). Geological modelling from field data and geological knowledge. Part I. modelling method coupling 3D potential-field interpolation and geological rules. *Physics of the Earth and Planetary Interiors* 171: 147–157. https://doi.org/10.1016/j.pepi.2008.06.013.

Carr, J.C., Beatsdon, R.K., Cherrie, J.B. et al. (2001). Reconstruction of 3D objects with radial basis functions. In: *SIGGRAPH '01: Proceedings of the 28th annual conference on Computer graphics and interactive techniques* (ed. L. Pocock), 67–76. New York: Association for Computing Machinery. https://doi.org/10.1145/383259.383266.

Caumon, G. and Mallet, J.L. (2006). 3D Stratigraphic models: representation and stochastic modelling. In: *Proceedings, IAMG2006: XIth International Congress, Paper S14-08*. Houston, TX: International Association for Mathematical Geology. 4 pp. [Online: Available at: https://www.ring-team.org/index.php/research/publications?view=pub&id=2054] (Accessed August 19, 2020).

Caumon, G., Lepage, F., Sword, C.H. and Mallet, J.-L. (2004). Building and editing a sealed geological model. *Mathematical Geology* 36 (4): 405–424. https://doi.org/10.1023/B:MATG.0000029297.18098.8a.

Caumon, G, Tertois, A.L. and Zhang, L. (2007). Elements for stochastic structural perturbation of stratigraphic models. In: *Proceedings EAGE Conference on Petroleum Geostatistics, Cascais, Portugal, cp-32-00002*. Houten, Netherlands: European Association of Geoscientists & Engineers. https://doi.org/10.3997/2214-4609.201403041.

Chilès, J.-P. and Delfiner, P. (2012). *Geostatistics: Modeling Spatial Uncertainty, 2e*. Hoboken, NJ: Wiley. 734 pp.

Chilès, J.-P., Aug, C., Guillen, A. and Lees, T. (2004). Modelling the geometry of geological units and its uncertainty in 3D from structural data: the potential-field method. In: *Proceedings of Orebody Modelling and Strategic Mine Planning, Perth, WA, 22-24 November 2004* (ed. R. Dimitrakopoulos), 313–320. Carlton, Australia: Australian Institute of Mining and Metallurgy. [Online: Available at: http://citeseerx.ist.psu.edu/viewdoc/download?doi=10.1.1.583.213&rep=rep1&type=pdf] (Accessed August 19, 2020).

Collon, P. and Caumon, G. (2017). 3D geomodelling in structurally complex areas – implicit vs. explicit representations. In: *Proceedings, 79th EAGE Conference and Exhibition 2017*. Houten, Netherlands: European Association of Geoscientists & Engineers. 5 pp. https://doi.org/10.3997/2214-4609.201701144.

Courrioux, G., Bourgine, B., Guillen, A. et al. (2015). Comparisons from multiple realizations of a geological model: implication for uncertainty factors identification. In: *Proceedings, IAMG2015: 17th Annual Conference of the International Association for Mathematical Geosciences* (eds. H. Schaeben, R. Tolosana Delgado, K.G. van den Boogaart and R. van den Boogaart), 59–66. Houston, TX: International Association for Mathematical Geosciences.

Desbarats, A.J., Hinton, M.J., Logan, C.E. and Sharpe, D.R. (2001). Geostatistical mapping of Leakance in a regional Aquitard, oak ridges moraine area, Ontario, Canada. *Hydrogeology Journal* 9 (1): 79–96. https://doi.org/10.1007/s100400000110.

Doi, A. and Koide, A. (1991). An efficient method of triangulating Equi-valued surfaces by using tetrahedral cells. *IEICE Transactions on Information and Systems* E74-D (1): 214–224.

Fisher, T.R. and Wales, R.Q. (1992). Rational splines and multi-dimensional geologic modeling. In: *Three-Dimensional Computer Graphics in Modeling Geologic Structures and Simulating Processes* (eds. R. Pflug and J.W. Harbaugh), 17–28. Heidelberg: Springer-Verlag, Lecture Notes in Earth Sciences vol 41. https://doi.org/10.1007/BFb0117780.

Galli, A. and Renard, D. (1986). *BLUEPACK Geostatistical Background*. Fontainebleu, France: ENSMP Centre de Géostatistique, Internal Report.

Geovariances (2016). *ISATIS Software Beginners Guide*. Avon, France: Geovariances. 234 pp.

Gibson, H., Sumpton, J., Fitzgerald, D. and Seikel, R. (2013). 3D modelling of geology and gravity data: summary workflows for minerals exploration. *Australia Institute of Geoscientists Bulletin* 57: 24–26.

Gonçalves, Í.G., Kumaira, S. and Guadagnin, F. (2017). A machine learning approach to the potential-field method for implicit modeling of geological structures. *Computers and Geosciences* 103: 173–182. https://doi.org/10.1016/j.cageo.2017.03.015.

Görz, I., Herbst, M., Börner, J.H. and Zehner, B. (2017). Workflow for the integration of a realistic 3D geomodel in process simulations using different cell types and advanced scientific visualization: variations on a synthetic salt Diapir. *Tectonophysics* 699: 42–60. https://doi.org/10.1016/j.tecto.2017.01.011.

Gringarten, E., Arpat, B., Jayr, S. and Mallet, J.-L. (2008). New geologic grids for robust geostatistical modeling of hydrocarbon reservoir. In: *Proceedings Eighth Geostatistics Congress, vol. 2* (eds. J.M. Ortiz and X. Emery), 647–656. Santiago, Chile: Gecamin. [Online: Available at: https://www.pdgm.com/resource-library/articles-and-papers/archive/new-geologic-grids-for-robust-geostatistical-model/] (Accessed August 19, 2020).

Guillen, A., Calcagno, P., Courrioux, G. et al. (2008). Geological modelling from field data and geological knowledge. Part II. Modelling validation using gravity and magnetic data inversion. *Physics of the Earth and Planetary Interiors* 171: 158–169. https://doi.org/10.1016/j.pepi.2008.06.014.

Hillier, M.J., Schetselaar, E.M., de Kemp, E.A. and Perron, G. (2014). Three-dimensional modelling of geological surfaces using generalized interpolation with radial basis functions. *Mathematical Geosciences* 46: 931–953. https://doi.org/10.1007/s11004-014-9540-3.

Houlding, S.W. (1994). *3D Geoscience Modeling: Computer Techniques for Geological Characterization*. New York: Springer. 308 pp.

Isaaks, E.H. and Srivastava, R.M. (1989). *An Introduction to Applied Geostatistics*. New York: Oxford University Press. 561 pp.

Jayr, S., Gringarten, E., Tertois, A.L. et al. (2008). The need for a correct geological modelling support: the advent of the UVT-transform. *First Break* 26: 73–79.

Jørgensen, F., Rønde Møller, R., Nebel, L. et al. (2013). A method for cognitive 3D geological voxel modelling of AEM data. *Bulletin of Engineering Geology and the Environment* 72: 421–432. https://doi.org/10.1007/s10064-013-0487-2.

Journel, A.G. (1983). Nonparametric estimation of spatial distributions. *Mathematical Geology* 15 (3): 445–468. https://doi.org/10.1007/BF00894777.

Journel, A.G. and Deutsch, C.V. (1998). *GSLIB Geostatistical Software Library and Users Guide, 2e*. New York: Oxford University Press. 369 pp.

Kelk, B. and Challen, K. (1992). Experiments with a CAD system for spatial modelling of geoscientific data. *Geologisches Jahrbuch*, A122: 145–153.

Kessler, H., Mathers, S. and Sobisch, H.G. (2009). The capture and dissemination of integrated 3D geospatial knowledge at the British Geological Survey using GSI3D software and methodology. *Computers & Geosciences* 35: 1311–1321. https://doi.org/10.1016/j.cageo.2008.04.005.

Korus, J.T., Joeckel, R.M., Divine, D.P. and Abraham, J.D. (2017). Three-dimensional architecture and hydrostratigraphy of cross-cutting buried valleys using airborne electromagnetics, glaciated central lowlands, Nebraska, USA. *Sedimentology* 64 (2): 553–581. https://doi.org/10.1111/sed.12314.

Krige, D.G. (1951). A statistical approach to some basic mine valuation problems on the Witwatersrand. *Journal of the Chemical, Metallurgical and Mining Society of South Africa* 52 (6): 119–139. [Online: Available at: https://hdl.handle.net/10520/AJA0038223X_4792] (Accessed August 19, 2020).

Lacquement, F., Courrioux, G., Ortega, C. and Thuon, Y. (2011). *Réalisation du Modéle Géologique 3D de la Pointe de Givet (08) – Étape 1: Caractérisation des Potentialités d'Exploitation des Eaux Souterraines*. Orléans, France: Bureau de Recherches Géologiques et Miniéres, Report BRGM/RP-60384-FR. 52 pp. [Online: Available at: https://www.brgm.fr/projet/modelisation-3d-geologie-pointe-givet] (Accessed August 19, 2020).

Lajaunie, C., Courrioux, G. and Manuel, L. (1997). Foliation fields and 3D cartography in geology: principles of a method based on potential interpolation. *Mathematical Geology* 29 (4): 571–584. https://doi.org/10.1007/BF02775087.

Laurent, G. (2016). Iterative thickness regularization of stratigraphic layers in discrete implicit modeling. *Mathematical Geosciences* 48 (7): 811–833. https://doi.org/10.1007/s11004-016-9637-y.

Lorensen, W.E. and Cline, H.E. (1987). Marching cubes: a high resolution 3D surface construction algorithm. *Computers and Graphics* 21 (4): 163–169. https://doi.org/10.1145/37402.37422.

Mallet, J.-L. (1989). Discrete smooth interpolation. *ACM Transactions on Graphics* 8 (2): 121–144.

Mallet, J.-L. (1991). GOCAD: a computer aided design program for geological applications. In: *Three-Dimensional Modeling with Geoscientific Information Systems* (ed. A.K. Turner), 123–142. Dordrecht, the Netherlands: Kluwer Academic Publishers, NATO ASIC vol. 354. https://doi.org/10.1007/978-94-011-2556-7.

Mallet, J.-L. (1992). Discrete smooth interpolation in geometric modelling. *Computer-Aided Design* 24 (4): 178–191. https://doi.org/10.1016/0010-4485(92)90054-E.

Mallet, J.-L. (1997). Discrete modeling for natural objects. *Mathematical Geology* 29 (2): 199–219. https://doi.org/10.1007/BF02769628.

Mallet, J.-L. (2002). *Geomodeling*. Oxford, UK: Oxford University Press. Applied Geostatistics Series. 624 pp.

Mallet, J.-L. (2004). Space-time mathematical framework for sedimentary geology. *Mathematical Geology* 36: 1–32. https://doi.org/10.1023/B:MATG.0000016228.75495.7c.

Martin, R. and Boisvert, J.B. (2017). Iterative refinement of implicit boundary models for improved geological feature reproduction. *Computers and Geosciences* 109: 1–15. https://doi.org/10.1016/j.cageo.2017.07.003.

Matheron, G. (1963). Principles of geostatistics. *Economic Geology* 58: 1246–1266.

McInerney, P., Guillen, A., Courrioux, G. et al. (2005). Building 3D geological models directly from data? A new approach applied to Broken Hill, Australia. In: *Proceedings, Digital Mapping Techniques '05, Baton Rouge, Louisiana* (ed. D.R. Soller), 119–130. Reston, VA: U.S. Geological Survey. [Online: Available at: https://pubs.usgs.gov/of/2005/1428/mcinerney/] (Accessed August 19, 2020).

McInerney, P., Goldberg, A., Calcagno, P. et al. (2007). Improved 3D geology modelling using an implicit function interpolator and forward modelling of potential field data. In: *Proceedings of Exploration 07: Fifth Decennial International Conference on Mineral Exploration* (ed. B. Milkereit), 919–922. Toronto, ON: Decennial Mineral Exploration Conferences. [Online: Available at: http://www.dmec.ca/ex07-dvd/E07/pdfs/69.pdf] (Accessed August 19, 2020).

Moyen, R., Mallet, J-L., Frank, T., Leflon, B. and Royer, J.J. (2004). 3D-parameterization of the 3D geological space – the Geochron model. In: *Proceedings ECMOR IX: 9th European Conference on the Mathematics of Oil Recovery*. Houten, Netherlands: European Association of Geoscientists and Engineers. 8 pp. https://doi.org/10.3997/2214-4609-pdb.9.A004.

Renard, R.D. (1990). Bluepack 3-D and its use in the petroleum industry. *Proceedings, Petroleum Computer Conference, 25–28 June 1990, Denver CO*, 197–204. Richardson, TX: Society of Petroleum Engineers. https://doi.org/10.2118/20352-MS.

Ross, M., Parent, M., Lefebvre, R. and Martel, R. (2002). 3D geologic framework for regional hydrogeology and land-use management; a case study from southwestern Quebec, Canada. In: *Three-Dimensional Geological Mapping for Groundwater Applications: Workshop extended abstracts* (eds. L.H. Thorleifson and R.C. erg), 52–55. Ottawa, ON: Geological Survey of Canada, Open File 1449. https://doi.org/10.4095/299506.

Ross, M., Parent, M., Martel, R. and Lefebvre, R. (2004a). Towards seamless interactions between geologic models and hydrogeologic applications. *Workshop Abstracts, 49th Annual Meeting, Geological Association of Canada, Mineralogical Association of Canada, St. Catherines, Ontario, Canada*. Champaign, IL: Illinois State Geological Survey. 5 pp. [Online: Available at: https://isgs.illinois.edu/2004-ontario-canada] (Accessed August 19, 2020).

Ross, M., Martel, R., Parent, M. et al. (2004b). Assessing bedrock aquifer vulnerability in the St. Lawrence lowlands (Canada) using downward Advective times from a 3D model of surficial geology. *Geofisica Internacional* 43 (4): 591–602.

Ross, M., Parent, M. and Lefebvre, R. (2005). 3D geologic framework models for regional hydrogeology and land-use management: a case study from a quaternary basin of southwestern Quebec, Canada. *Hydrogeology Journal* 13: 690–707. https://doi.org/10.1007/s10040-004-0365-x.

Ross, M., Parent, M., Benjumea, B. and Hunter, J. (2006). The late Quaternary stratigraphic record northwest of Montréal: regional ice sheet dynamics, ice stream activity and early deglacial events. *Canadian Journal of Earth Sciences* 43: 461–485. https://doi.org/10.1139/e05-118.

Ross, M., Martel, R., Parent, G. and Smirnoff, A. (2007). Geomodels as a key component of environmental impact assessments of military training ranges in Canada. In: *Three-dimensional geologic mapping for groundwater applications: Workshop extended abstracts* (eds. L.H. Thorleifson, R.C. Berg and H.A.J. Russell), 59–62. St Paul, MN: Minnesota Geological Survey, Open File 7-4. [Online: Available at: http://hdl.handle.net/11299/109020] (Accessed August 19, 2020).

Savard, M.M. [Ed.] (2013). *Canadian inventory of groundwater resources: integrated regional hydrogeological characterization of the fractured aquifer system of southwestern Quebec*. Ottawa, ON: Geological Survey of Canada, Bulletin 587. 114 pp. [Online: Available at: http://publications.gc.ca/pub?id=9.576477&sl=0] (Accessed August 19, 2020).

Tertois, A.L. and Mallet, J.-L. (2007). Editing faults within tetrahedral volume models in real time. *Geological Society of London, Special Publication* 292 (1): 89–101. https://doi.org/10.1144/SP292.5.

Tertois, A.L., Mallet, J-L., Gringarten, E. and Haouesse, A. (2010). Assessing geometric uncertainties in solid earth models. In: *Proceedings, 72nd EAGE Conference and Exhibition incorporating SPE EUROPEC 2010*. Houten, Netherlands: European Association of Geoscientists & Engineers. 5 pp. https://doi.org/10.3997/2214-4609.201400975.

13

Discretization and Stochastic Modeling

Alan Keith Turner

Colorado School of Mines, Golden, CO 80401, USA
British Geological Survey, Keyworth, Nottingham NG12 5GG, UK

13.1 Introduction

The International Association for Engineering Geology and the Environment (IAEG) Commission C25 began its review of the use of geological models within engineering geology by defining a model as *"an approximation of reality created for the purpose of solving a problem"* (Parry et al. 2014). Geological framework models integrate available observations to provide a realistic 3-D definition of the distribution and geometry of subsurface geological features. Framework models may represent geological formations or minor and localized variations such as individual geological strata or facies. The complexity of the geological environment being modeled and the amount of data available dictate the detail represented within any framework model. Framework models also reflect their anticipated applications and the time and budget available for model construction.

Chapters 9–12 describe the four most common approaches used to build geological framework models. Chapters 9 and 10 discuss two common "explicit modeling" approaches; these depend on the experience of the geologist or modeler to interpret the available geodata and define a series of lithostratigraphic surfaces, and usually involve a series of manual interventions to create the model. Chapter 11 discusses the development of the GeoTOP discretized property model of the near-surface geological units in the Netherlands (Stafleu et al. 2011). Based on layered framework models, GeoTOP is an extensive example of the use of the discretization and property modeling procedures discussed in Section 13.7 of this chapter. Chapter 12 discusses the "implicit modeling" approach which at least partially automates the model creation process by employing various sophisticated interpolation algorithms to ensure fast and consistent volume generation (Chilès et al. 2004).

Collon and Caumon (2017) compare the explicit and implicit modeling approaches.

Most explicitly defined geological framework models describe the boundaries of geo-objects using 2-D regular gridded or triangulated surfaces; a few explicit models and all implicit models use functions, such as splines, to define the surfaces (Mallet 2002; Collon et al. 2015). A coherent model representation is achieved when the volumes of the geo-objects are completely confined and partitioned by these surfaces without holes and overlaps (Caumon et al. 2004).

Geological framework models support the conceptualization of subsurface properties and processes being investigated by process models. For some simulations, the geological units can be assigned average, or typical, property measurements (Royse et al. 2008). However, this simple attribution of "bulk" property values ignores any internal heterogeneity. If knowledge of this internal heterogeneity is required by the simulation, the process model must have the framework model volumes subdivided, or *discretized*, into multiple small unit volumes.

Simulations that assess groundwater resources (Wycisk et al. 2009; Burt and Dodge 2016), or sustainable management of a geothermal reservoir (Kaiser et al. 2013; Alcaraz et al. 2015), or stress and strain analysis for design of tunnel linings (Cherry et al. 1996; Spyridis et al. 2013) require additional information about the distribution of internal heterogeneity of the fluid flow, heat flux, or stress field. The gridded or triangulated surfaces in framework models developed with the stacked-surface approach (Chapter 9) allow assignment of lateral variations of physical properties, but not vertical internal variations.

Thus, there is a need to easily convert framework models to property models that can describe the 3-D heterogeneity of physical properties within geological units. These

Applied Multidimensional Geological Modeling: Informing Sustainable Human Interactions with the Shallow Subsurface, First Edition.
Edited by Alan Keith Turner, Holger Kessler, and Michiel J. van der Meulen.
© 2021 John Wiley & Sons Ltd. Published 2021 by John Wiley & Sons Ltd.

property models must be structured in formats amenable to simulation methods used to model processes. Although technical solutions exist to help these conversions, smooth interactions between geological framework models, property models, and process model applications are often hampered by differing priorities and scales of the models and the applications (Watson et al. 2015).

Most 3-D geological framework models define lithostratigraphy and structures at the formation level. These models often contain complex geometries which undergo considerable simplification during creation of a groundwater flow model due to practical considerations of computational speed and efficiency. The lack of a generic grid exported from typical 3-D geological framework models that can be evaluated by commonly used numerical groundwater modeling packages precludes the automatic propagation of changes made to the geological model into the groundwater model. The reliance on a series of manual data conversion procedures that are prone to inadvertent user-induced errors has resulted in many groundwater models being created with little or no reference to a related 3-D geological framework model (Watson et al. 2015).

Although a generic grid export from typical 3-D geological framework models amenable to commonly used numerical groundwater modeling packages is not yet available, some existing commercial groundwater investigation software systems partially address the problem by including methods to create hydraulic property grids that are representative of and derived from modeled geological units. Examples include Visual MODFLOW 3D Builder, an add-on module for the Visual MODFLOW software (Scientific Software Group 2019), conversion of property attributes in GOCAD geological models to grid formats used by the Groundwater Modeling System (Ross et al. 2005), and GOFEFLOW, which generates FEFLOW compliant data from GOCAD modeled surfaces (Smirnoff et al. 2011). However, none of these solutions fully addresses the conceptual issue of transferring geologically controlled properties to process models; they are designed to work with specific software packages and are focused on the needs of defined applications (Watson et al. 2015). Mustapha (2011) discusses how to discretize 2-D and 3-D complex fractured geological models in formats suitable for flow and transport simulations while maintaining the geometric integrity of the geological framework.

13.2 Grids and Meshes

The terms *Grid* and *Mesh*, and the *Gridding* and *Meshing* procedures for creating them, have been used interchangeably in discussions of discretization processes. For clarity, in this chapter, the terms *Grid* and *Gridding* will refer to the subdivision of 2-D surfaces, and the terms *Mesh* and *Meshing* will refer to the subdivision of 3-D volumes.

Numerous published sources address the methods of grid and mesh generation and their use in process models. Two comprehensive book sources are Thompson et al. (1998) and Frey and George (2008). Thompson et al. (1998) comprehensively addresses the generation of structured and unstructured grids (and meshes) and their use for numerically solving partial differential equations by finite element, finite volume, finite difference, and boundary elements methods. Frey and George (2008) provide a comprehensive survey of the different algorithms and methods used to generate unstructured tetrahedral meshes. Figure 13.1 illustrates the basic terminology used to describe features of grids or meshes, namely:

- *Cell*: basic element within a grid or mesh.
- *Cell Center*: center of a cell.
- *Edge*: boundary of a face.
- *Face*: boundary of a cell.
- *Node*: corner of cell in grid or mesh.

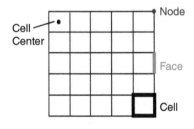

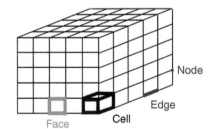

Figure 13.1 Terminology defining basic features of grids and meshes.

Two additional commonly used terms are not shown in Figure 13.1: *Zone*, which refers to a group of nodes, faces, and cells, and *Domain*, which refers to a group of zones.

The use of grids and meshes in 3-D geological property modeling involves subdividing, or "discretizing", the modeled volume within each geological unit into multiple individual cells; different cell types are required depending

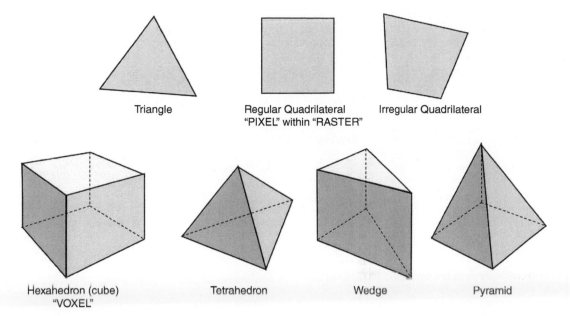

| Triangle | Regular Quadrilateral
"PIXEL" within "RASTER" | Irregular Quadrilateral |

| Hexahedron (cube)
"VOXEL" | Tetrahedron | Wedge | Pyramid |

Figure 13.2 Basic elements forming 2-D grids or 3-D meshes.

on the choice of numerical simulation procedures (Zehner et al. 2015). During discretization, interpolation algorithms determine the volume where each grid or mesh node is located, and then assign an identification number associated with that geologic unit to each node. Because different 3-D framework model software products use various strategies to represent these volumes, discretization software must be able to utilize multiple framework model data formats (Miller et al. 2007). Table 13.1 summarizes the characteristics of the two major classes of grids or meshes: structured and unstructured. Both classes of grids or meshes are composed of the same basic elements illustrated in Figure 13.2.

13.3 Structured Grids and Meshes

In a structured grid or mesh, all interior nodes have the same number of adjacent elements. The simplest form of a structured grid is a regular pattern of points forming the nodes of a regular quadrilateral grid (Figure 13.3a top). However, structured grids are not necessarily orthogonal. Irregular quadrilateral grids (Figure 13.3a bottom) are structured as long as none of the connections are broken and none of the grid elements are inverted. The uneven side dimensions allow the irregular quadrilateral grids to conform to more complex shapes, but they have limitations if used to represent more complex geometries. Numerical schemes for solving equations on these grids often have problems with accuracy, as the stretched grid becomes non-orthogonal.

Unstructured grids or meshes allow any number of elements to meet at a single node (Figure 13.3b). This allows for much greater flexibility in representing complex geometries, but this flexibility comes with a computational cost. Because there is no implied or logical relationship of one element and its neighbors, connectivity lists must be developed to relate grid/mesh elements to nodes.

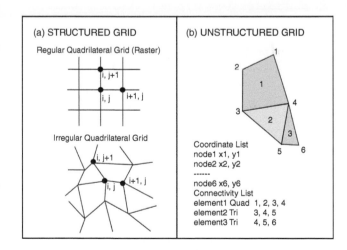

Figure 13.3 Terminology defining basic features of grids and meshes.

13.3.1 Structured Grids

The most common structured grid format consists of uniformly sized square cells (quadrilaterals) that are often stored in a raster format with cell locations defined by cell centroids (Figure 13.4b). This allows for relatively compact

Table 13.1 Classification of grids and meshes.

STRUCTURED		UNSTRUCTURED	
Type	**Description**	**Type**	**Description**
2-D Grids			
Regular quadrilateral (RASTER)	Composed entirely of equally sized quadrilaterals – usually squares but can be rectangles – forming an orthogonal grid (see Figure 13.3).	Triangulated Irregular Network (TIN)	Network of variably-shaped and oriented triangles – often defined using the Delauney criterion (see Figure 13.3)
Regular triangular network	Composed of regularly shaped and identically-sized triangles – often created by adding diagonals to subdivide regular quadrilateral elements	Hybrid grid	Combination of triangles and quadrilaterals
Irregular quadrilateral	Composed entirely of flexibly sized and oriented quadrilaterals to form a non-orthogonal grid (see Figure 13.3).		
Quadtree grid	Individual orthogonal grid elements are recursively subdivided by halving their dimensions to split each square into four smaller squares until the desired resolution is reached		
Hexagonal grid	The use of uniform hexagons, rather the squares, can improve the quality of visualization as shapes become smoother		
3-D Meshes			
Hexahedral mesh (VOXEL)	Composed entirely of cubical elements which have six sides – thus hexahedral. Most common form consists of uniformly-sized cubes (voxels) creating an orthogonal mesh. However, cubes may deform, especially in vertical direction to create a mesh that conforms to the geometry of geological strata	Tetrahedral mesh	The tetrahedron is the 3-D equivalent of a triangle – so a tetrahedral mesh is the 3-D version of a TIN. Basic tetrahedrons can form wedges or pyramids (see Figure 13.2); or may be derived from wedges by decomposition.
Octree mesh	Individual cubic elements are recursively subdivided by halving their dimensions to split each cube into eight smaller cubes until the desired resolution is reached	Hybrid mesh	Contains one of the following: • Any combination of tetrahedra, prisms, and pyramids. • Boundary layer mesh: prisms at walls and tetrahedra everywhere else. • Hexcore: hexahedra in center and other cell types at walls.

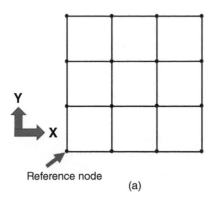

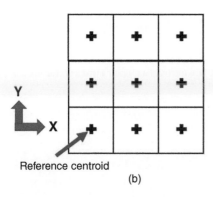

Figure 13.4 Grid location definitions; (a) by corner nodes, (b) by centroids.

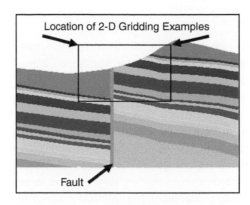

Figure 13.5 A typical 2-D geological cross section – the rectangular region outlined is used to illustrate various grid formats in subsequent illustrations. After Turner (2016); Carl Gable, Los Alamos National Laboratory.

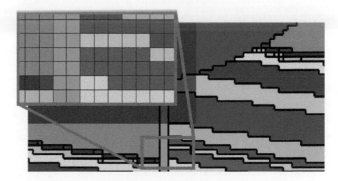

Figure 13.6 Basic regular quadrilateral (Raster) grid. After Turner (2016); Carl Gable, Los Alamos National Laboratory.

Figure 13.7 Quadtree grid. After Turner (2016); Carl Gable, Los Alamos National Laboratory.

data storage because a simple "header" information block can define the number of rows and columns forming the raster grid, the scan-order of the cell values, the cell dimensions, and the x and y coordinates of the base cell centroid. With this information the location of all cells in the raster can be computed and their coordinates need not be stored explicitly. However, if an irregular quadrilateral structure as defined by Figure 13.3a is required, then the location of each cell must be explicitly defined by its nodes, as shown in Figure 13.4a, and larger file sizes result. Some applications convert the quadrilateral cells to a series of regular triangles by adding diagonals to subdivide the quadrilateral elements. This also requires the explicit definition of the coordinates of the nodes.

Figure 13.5 shows a portion of a typical geological cross-section with several strata and a fault. Examples of various grid structures will be illustrated within the rectangular window located on the cross-section. Quadrilateral structured grids approximate geological discontinuities by staircase-like boundaries (Figure 13.6). The quadtree

format results when individual orthogonal grid elements are recursively subdivided by halving their dimensions to split each square into four smaller squares until the desired resolution is reached (Figure 13.7).

13.3.2 Structured Meshes

The majority of commercially available modeling products divide the volume into discrete regular hexahedral cubes that define volume elements, or "voxels." Voxels offer practical and commercial advantages because they provide a simple data structure upon which more specialized applications can be built fairly rapidly. However, unless the cell dimensions are very small, the discretization process may destroy important geometric details and even relatively low-resolution voxel models may produce large data files; a 3-D voxel model involving only $100 \times 100 \times 100$ cells results in 1-million cells. Common resolutions are in the order of tens to hundreds of meters horizontally and decimeters to meters vertically. Common modeling efforts, which cover hundreds to thousands of square kilometers, will have millions to hundreds of millions of cells. Consequently, various methods of variable-sized voxels have been proposed.

Octree meshes are the 3-D equivalent of 2-D quadtree grids; individual cubic elements are recursively subdivided by halving their dimensions to split each cube into eight smaller cubes until the desired resolution is reached. To ensure numerical accuracy in finite element simulations, the difference in element sizes between adjacent octree elements is usually limited to one level; this produces a cascading effect and a "balanced octree." As with the 2-D quadtrees, the octree representation provides for rapid indexing of the individual cells within the database. Although octrees provide some data compression, they must be redeveloped whenever the overall framework geometry is changed; this may introduce some significant development-time penalties during iterative model construction. Voxel or octree models will display rough blocky surfaces reflecting their cellular structure, and this may be a distraction in some visualizations (Figure 13.8).

Many geological elements have much greater variability in the vertical direction compared with their lateral extents. This is especially true in sedimentary environments, where sub-horizontal laterally continuous strata may show much greater consistency of properties compared with variations between strata in the vertical direction. Thus, some commercial 3-D modeling products (for example, Stratamodel) offer partly deformable "geocellular" voxels (Denver and Phillips 1990). Figure 13.9 provides an example of such a model.

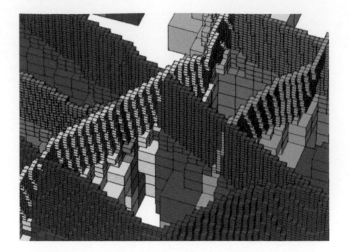

Figure 13.8 Blocky visualization typical of voxel or octree models. After Turner (2016); Carl Gable, Los Alamos National Laboratory.

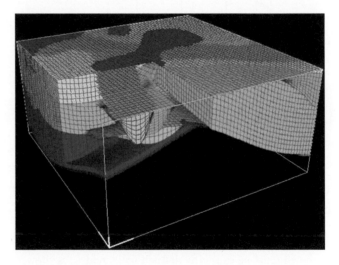

Figure 13.9 A geocellular model with multiple stratigraphic geobjects. After Turner (2016).

13.4 Unstructured Grids and Meshes

Owen (1998) surveyed alternative methods for generating and modifying unstructured grids and meshes. He evaluated methods for generating triangular and quadrilateral 2-D grids and tetrahedral and hexahedral 3-D meshes.

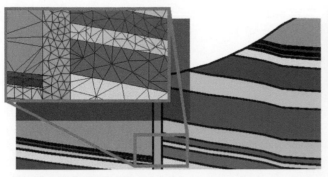 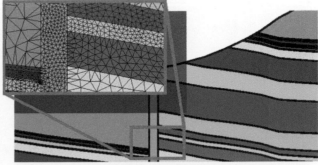

Figure 13.10 Coarse and fine unstructured grids – both conform to the geological boundaries, but the finer grid allows a more detailed assessment of internal heterogeneity. After Turner (2016); Carl Gable, Los Alamos National Laboratory.

13.4.1 Unstructured Grids

Unstructured grids consist of an irregular pattern of grid points with neither a predefined topology nor fixed cell geometry (Zehner et al. 2015). Unstructured cells must be described by an explicit definition of the locations of the corner nodes of each cell and by a topological model that describes the neighborhood relations of the cells. The simplest type of unstructured grid is an irregular triangulated network, or TIN. It can adapt to complicated geometries, since the triangles can represent sharp forms of geo-bodies. Each triangle belongs to a single geological unit, and connected facets of the TIN conform to the geological unit boundaries. In fact, as shown by Figure 13.10, a typical two-dimensional unstructured mesh provides an accurate geometrical definition of geological units over a range of grid densities. Finer resolution of the interior of geological units permits a more detailed evaluation of inherent property heterogeneity but does not materially change the visualization of the unit boundaries. Thus, the TIN format is extensively used in CAD modeling systems.

13.4.2 Unstructured Meshes

More recently, some groundwater modeling systems have utilized unstructured meshes that are not constrained by having to have a constant node and face structure and can link with finite element models (Gable 2006). This provides added flexibility during the development phase for a model that reflects subsurface conditions, but this flexibility comes at a price – added computational demands and very slow model construction – unless sophisticated "mesh builder" software is employed.

Unstructured 3-D meshes may be based on a variety of fundamental elements, including tetrahedra, hexahedra, and dodecahedra. However, the vast majority of

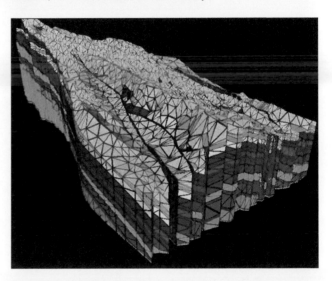

Figure 13.11 Tetrahedral unstructured mesh model of Yucca Mountain unsaturated zone. After Turner (2016); Carl Gable, Los Alamos National Laboratory.

unstructured meshes utilize tetrahedra (Figure 13.11). Unstructured meshes are particularly useful in modeling fracture discontinuities, where finite element models of hydrology or stress/strain relationships are the primary importance (Figure 13.12). For these and other reasons, a number of research teams have invested considerable effort in developing unstructured 3-D mesh systems, and "mesh builders" (Gable et al. 1996a,b). Recently, the Geological Survey of the Netherlands experimented with unstructured hexahedral meshes, in which voxel size vary with model uncertainty (Van Lancker et al. 2017).

13.5 Considerations that Influence Grid and Mesh Design

The design of a grid or mesh typically involves a tradeoff between mesh size (number of nodes and elements) and

Figure 13.12 A 3-D unstructured mesh model of a faulted layered sequence. After Turner (2016); Carl Gable, Los Alamos National Laboratory.

the level of detail represented by the mesh (Gable et al. 1995, 1996a,b). A grid or mesh with more elements will capture more detail but will require more computation time and computer memory. A grid or mesh with fewer elements uses less computation time, but geometric details may be lost. Geologic units with large volumes or regions that are far from the area of interest typically can be represented with less detail, and so are captured with larger mesh elements. Thin geological units or fracture zones typically are delineated with finer mesh elements. Other considerations in representing the geology with a computational mesh include geometries that lack symmetry, have a wide range of length scales, or may contain very thin layers that must be preserved as continuous layers.

The physical basis employed in the simulation may impose numerical constraints that further restrict mesh geometry. Some simulations can solve problems only on orthogonal, uniformly spaced meshes and thus require structured grids or meshes. Other simulations use grids or meshes with special adaptive refinement strategies that can resolve complex geometries; these usually require unstructured grids or meshes. Both low-resolution and high-resolution unstructured grids or meshes may represent geologic interfaces equally well, however, a high-resolution mesh may be required to resolve concentration gradients when solving contaminant transport problems because smaller elements can more accurately represent steep gradients (Figure 13.10) and permit more accurate particle tracking assessments (Bower et al. 2005; Gable 2006).

Zehner et al. (2015) defined three requirements for unstructured tetrahedral meshes to define a complex geological model:

- *Sufficient mesh quality for running a process simulation.* The tetrahedra should not be too acute-angled, because

numerical instabilities can occur. Therefore, the shape of each tetrahedron has to be checked and improved if necessary.
- *Incorporation of geometry for defining boundary conditions and constraints.* The formulation of boundary conditions for the numerical simulation often requires certain points or lines to be part of the mesh. The workflow should be able to add these objects as constraints to the tessellation.
- *Local adaption or refinement.* Local refinement of the mesh should be possible in the vicinity of physical sources and sinks to avoid numerical errors during the simulation.

13.6 Grid and Mesh Generation and Refinement

Grid and mesh generation is a tedious, time consuming, and error-prone process if the considerations just discussed are to be resolved. This is especially the case if the grids are required to both accurately define the geological framework and support numerical simulations using finite difference or finite element procedures for models with complex structures such as faults and truncations of lithostratigraphic layers. Automated grid and mesh generation tools can help resolve the basic development of a suitable grid or mesh. Additional post-generation grid or mesh refinement procedures help to ensure the grid or mesh is optimally configured to support numerical simulations.

13.6.1 Grid and Mesh Generation Tools

Numerous grid and mesh generating packages are described in the literature, including both "freeware" codes and commercial packages. The Cubit meshing package (Sandia National Laboratory 2019) contains many of the latest approaches including capabilities of generating quadrilateral meshes of triangular regions in 2-D or 3-D (Figure 13.13) and subdividing a 3-D tetrahedral mesh into hexahedral elements (Figure 13.14).

Gable et al. (1996a,b) describe the use of grid and mesh generating tools developed at Los Alamos National Laboratory that produce finite element grids which maintain the geometric integrity of input geological volumes, surfaces, and geologic properties. The core functions of GEOMESH utilize capabilities developed in the X3D and LAGriT grid generation packages (Trease et al. 1996; Los Alamos National Laboratory 2019). The GEOMESH tool utilizes an optimal Delaunay algorithm to produce triangular grids for 2-D cross-sections and tetrahedral meshes for regional models that include faults and fractures; these can be converted to quadrilateral grids, or hexahedral meshes if required. GEOMESH also permits adaptive grid refinement

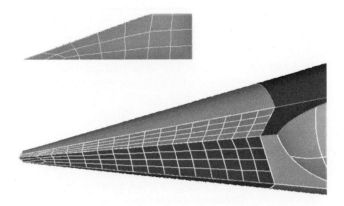

Figure 13.13 Cubit filled "3-sided" regions with quadrilateral elements in 2-D and 3-D.

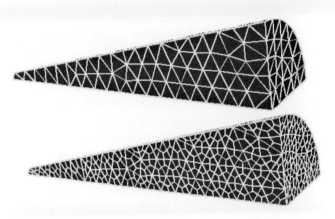

Figure 13.14 Example of Cubit conversion of 3-D tetrahedral mesh to hexahedral elements.

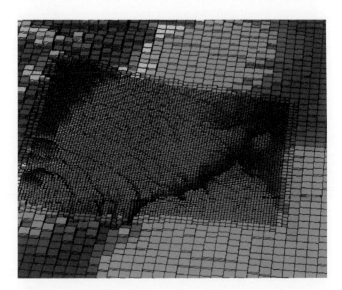

Figure 13.15 Example of finer mesh merged with a coarser mesh.

in three dimensions and contains tools to glue, merge, and insert finer grids into regional models (Figure 13.15).

13.6.2 Post-processing of Grids or Meshes

It is rare that any mesh generation algorithm will be able to define a grid or mesh that is optimal without some form of post-processing to improve the overall quality of the elements (Owen 1998). The main categories of mesh improvement are: (i) smoothing, (ii) clean-up, and (iii) refinement. Smoothing includes any method that adjusts node locations while maintaining the element connectivity; clean-up processes adjust the element connectivity; and refinement reduces element sizes in areas of primary interest.

13.6.2.1 Smoothing

Most smoothing procedures involve some form of iterative process that repositions individual nodes to improve the local quality of the elements. A wide variety of smoothing techniques have been proposed. Most smoothing procedures will iterate through all the internal nodes in the mesh several times until no individual node has moved more than a specified tolerance. The simplest and most straightforward technique is Laplacian smoothing (Field 1988). This method places an internal node in the mesh at the average location of any node connected to it by an edge; it can be applied to any element shape. Canann et al. (1998) provides an overview of other smoothing techniques that iteratively reposition nodes based on a weighted average of the geometric properties of the surrounding nodes and elements.

Some codes use optimization-based smoothing techniques that evaluate the quality of the surrounding elements and move the node to achieve an optimal location. Canann et al. (1998) and Freitag et al. (1995) present optimization-based smoothing algorithms. Because optimization-based smoothing techniques often require increased computational time, Canann et al. (1998) recommend a combined Laplacian/optimization-based approach. Owen (1998) provides details on additional smoothing techniques.

13.6.2.2 Clean-up Processes

There are a wide variety of clean-up processes that improve the quality of the mesh by adjusting element connectivity. Criteria to improve either the shape or the topology of the grid or mesh elements are generally used in conjunction with smoothing (Freitag and Ollivier-Gooch 1997). For triangular grids, simple diagonal swaps are often performed to achieve shape improvement. Each interior edge in the triangulation is evaluated to determine the position of the edge that improves the overall or minimum shape metric of its two adjacent triangles. The Delaunay criteria can also be used to determine if "flipping an edge" improves the shape of two triangle elements (Box 13.1). Joe (1995)

Box 13.1 Definition of Delaunay Triangle Flipping

Within a triangular grid, adjacent triangles will have a common edge. For example, in Figure 13.1.1-A, triangles ABD and BCD have the common edge BD. Since the sum of the angles α and γ is more than 180°, this triangulation does *not* meet the Delaunay condition. In addition, because each of the circumscribing circles contains more than three vertices (Figure 13.1.1-B), this triangulation does *not* meet the Delaunay condition.

This problem can be resolved by using a flipping technique. If two triangles do not meet the Delaunay condition, switching the common edge BD for the common edge AC produces two triangles that do meet the Delaunay condition (Figure 13.1.1-C). This operation is called a flip and can be generalized to three and higher dimensions.

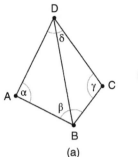  

(a)   (b)   (c)

Figure 13.1.1 A visual definition of Delaunay Triangle Flipping. (a) Sum of angles $\alpha + \gamma > 180°$ *not* Delaunay condition. Wikipedia (2020a). (b) Circumcircles contain > 3 points *not* Delaunay condition. Source: Wikipedia (2020b). (C) Flipped triangles meet Delaunay condition. Source: Wikipedia (2020c).

presents a series of local transformations designed to improve the element quality of tetrahedral meshes. These include swapping two adjacent interior tetrahedra that share a face to create three tetrahedra and reducing three tetrahedra to two tetrahedra.

13.6.2.3 Refinement

Refinement involves any operation that effectively reduces the local element size. A reduced element size may be required to capture a local physical phenomenon, or it may be done simply to improve the local element quality. Starting with a coarse mesh, refinement continues until the desired nodal density is achieved. Refinement is often part of an adaptive solution process, where the results from a previous solution provide criteria for mesh refinement (Joe 1991).

13.7 Stochastic Property Modeling

The term *stochastic* is derived from the Ancient Greek word *stochastikos* which means "skillful in aiming" and related to the words *stochazesthai* (to aim at, guess at) and *stochos* (target, aim, guess) (Merriam-Webster 2019). Because stochastic models are based on random trials, they provide variable answers while deterministic models always produce the same output for a given starting condition (Origlio 2019).

Whitten (1977) introduced the role of stochastic models for evaluating randomness in geological materials and

processes by reviewing applications of stochastic models in Russian and French research. The use of stochastic simulation to produce realistic heterogeneous geological models and for studying the uncertainty inherent in these models was extensively reported in the following decades (Haldorsen and Damsleth 1990; Deutsch and Journel 1992; Rautman and Flint 1992; Guardiano and Srivastava 1993). Theoretical developments in geostatistics and stochastic modeling that emerged throughout the past decade have produced important methods for petroleum reservoir modeling (Deutsch 2002; Hardy and Hatløy 2005; Coburn et al. 2006; Pyrcz and Deutsch 2014), for groundwater and engineering applications (Bianchi et al. 2015; Kearsey et al. 2015; Williams et al. 2019), and for assessing potential geothermal energy resources (Dirner and Steiner 2015).

Geostatistical models of subsurface heterogeneity involve both deterministic and stochastic approaches (Rautman and Flint 1992; Koltermann and Gorelick 1996; de Marsily et al. 2005). Modeling the distribution of geological facies, or the distribution of physical properties, is based on "hard" data defining lithology or other measurable properties obtained by direct observations or from boreholes (Figure 13.16a). The number and location of these hard data sources is usually inadequate to fully characterize the geological system of interest, so additional "soft" data consisting of indirect observations of geological properties, as well as qualitative and interpretative information from geophysical surveys or conceptualizations of the depositional system, are used to provide supplementary information.

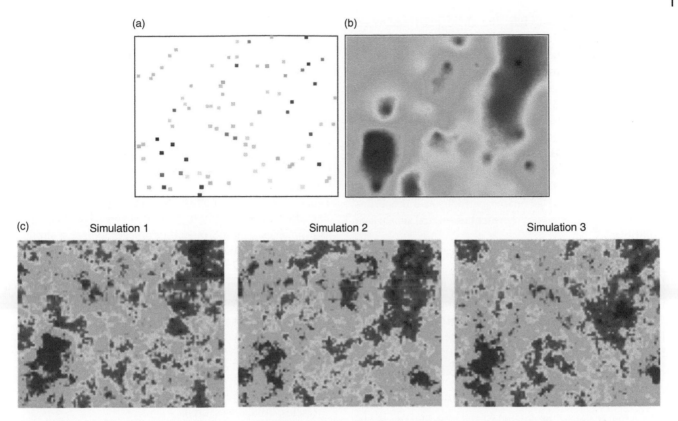

Figure 13.16 (a) Conditioning point data example; (b) Result produced by kriging; (c) Three results produced by stochastic simulation. Modified after Tahmasebi (2018).

The most commonly applied deterministic geostatistical approach, kriging, relies on the analysis of spatial covariance (defined by the variogram). The resulting simulations are not able to reproduce complex patterns (Deutsch and Journel 1992) and the process often produces excessively smooth results (Figure 13.16b). Stochastic approaches develop multiple, equally probable, realizations of the subsurface to describe smaller-scale phenomena that cannot be accurately observed or modeled by deterministic methods (Figure 13.16c). An additional advantage of stochastic modeling is that it provides information about the uncertainty of the model representation of physical-properties and geological structures; both are important considerations for many model applications (Caers 2011; Dirner and Steiner 2015). Because stochastic modeling involves the production of multiple realizations and each realization requires some computing time and resources, determining the appropriate number of realizations to achieve an acceptably reliable estimate of the heterogeneity is of practical interest. Jakab (2017) describes some statistical methods that can be used to determine when a sufficient number of model realizations have been produced. Several useful comparisons of the various stochastic modeling techniques have been published (Yao 2002; Lee et al. 2007; Dell'Arciprete et al. 2012; Koch et al. 2014).

Over the past 10–15 years, the growing popularity of stochastic modeling has included two broad categories of modeling techniques – pixel-based simulation methods and object-based simulation methods. Each category contains several techniques and each category has advantages and limitations that tend to make it more attractive for particular applications.

13.7.1 Pixel and Voxel Based Stochastic Simulation Methods

For the last 20 years, the established approach to modeling rock properties has been to utilize pixel and voxel based simulation methods in a two-stage process (Stafleu et al. 2011; Chapter 11). The first stage is to deterministically develop a model of the geological units and define a relatively small set of mutually exclusive, but all-inclusive, lithofacies categories. The second stage involves the stochastic evaluation of all pixels or voxels in a stacked 2-D or 3-D model, respectively, to populate each lithofacies with rock properties such as porosity or permeability. Pixels or voxels corresponding to the locations of geological observations (hard data sources) are first assigned firm rock properties. All remaining pixels and voxels are assessed in a random sequence. Variograms that represent two-point statistics are used to reproduce volume fractions and trends, but these pixel and voxel based simulation methods cannot reproduce complex and realistic geological structures.

The two-stage approach is required because the geostatistical kriging procedure requires "second-order stationarity" which specifies a uniform mean and covariance so that the autocorrelation function depends solely on the degree of separation of the observations in time or space. This requires the analysis to be restricted to homogeneous geological environments such as represented by individual lithofacies. If distinctly different lithofacies are analyzed together, the generated petrophysical properties will have values that do not represent the actual subsurface conditions.

Two commonly utilized pixel and voxel based stochastic simulation methods are Sequential Gaussian Simulation, or "SGS" (Dimitrakopoulos and Luo 2004), and Sequential Indicator Simulation, or "SIS" (Deutsch and Journel 1992; Yao 2002; Deutsch 2006). Because these methods do not produce realistic descriptions of facies connections, they are often further subjected to a Simulated Annealing ("SA") procedure (Deutsch and Cockerham 1994; Deutsch and Wen 1998).

The transition probability approach (TPROGS) models the spatial structure of geological data by Markov chain models that relate the transition probabilities among facies to distance in three dimensions (Carle et al. 1998; Weissmann et al. 1999; Ritzi 2000; Lee et al. 2007; He et al. 2014). Geological knowledge is considered during the determination of the coefficients of these functions which are related to interpretable geological properties such as proportions of each facies, and their mean lengths, connectivity, and juxtaposition tendencies.

13.7.1.1 Sequential Gaussian Simulation (SGS)

The SGS method is widely used to evaluate quantitative (numerical) data (Dimitrakopoulos and Luo 2004; Chen et al. 2013). SGS involves an eight-step process:

1. Transform the observed hard data to a normal form with zero mean and unit variance;
2. Assign transformed hard data to appropriate pixels or voxels, this forms the initial conditioning data;
3. Randomly select a pixel or voxel for assessment; if this has been assigned a value, select another;
4. Use kriging to estimate a mean and standard deviation at the selected pixel or voxel based on surrounding data and the variogram to generate a local conditional probability distribution *(lcpd)*;
5. Select a random a value from the *lcpd* and set the value of the selected pixel or voxel to that value;
6. Include the newly simulated value as part of the conditioning data;

7. Repeat Steps 4–6 until all pixels or voxels have values; and
8. Back-transform the realization into the original data form.

It is important that the selection of pixels or voxels in Step 3 be governed by a random process; this is controlled by a randomly selected seed value. Multiple equally probable realizations will result from a series of randomly generated seed values.

13.7.1.2 Sequential Indicator Simulation (SIS)

The SIS method is used to evaluate categorical data such as lithofacies. The SIS modeling process follows a similar multi-step simulation sequence as SGS, but each data category is defined as "1" when present and "0" when absent, and Indicator Kriging is used to generate the variograms (Deutsch and Journel 1992; Yao 2002; Deutsch 2006). Figure 13.17 illustrates the SIS procedure.

The GeoTOP model (Chapter 11) was developed using SIS. Stafleu et al. (2011) summarized the procedure as follows:

1. Define n mutually exclusive and all-inclusive categories (facies, lithoclasses, etc.);
2. Recode each category to presence (1) or absence (0) and re-locate the category to the closest voxel;
3. At an unvisited location, perform kriging interpolation for all individual categories, resulting in a probability for each category;
4. Build a cumulative density function for all categories combined;
5. Draw a random number between 0 and 1 and determine the category that belongs to this value; assign this category to the voxel;
6. Add this category to the dataset and start again at Step 3.

Figure 13.18 illustrates the process in Steps 3–6. These steps are repeated at a randomly selected sequence of cells until all locations are visited, producing categorical values for all locations. The SIS method ensures the hard data are honored and the spatial correlation structure is maintained.

Multiple, equally probable, but different, realizations result from changing seed values at the start of the SIS procedure. Post-processing a series of these simulations allows the calculation of an estimate of the probabilities of occurrence of each lithofacies category at each location. The main advantage of this technique lies in the availability of these probabilities of occurrence of individual lithofacies to be considered in assigning petrophysical properties to be used in simulations of groundwater flow or subsidence (Stafleu et al. 2011; Chapter 11).

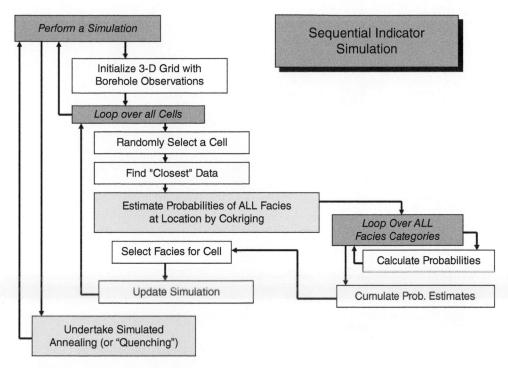

Figure 13.17 Flowchart of Sequential Indicator Simulation process.

Figure 13.18 Example of Sequential Indicator Simulation in 2-D. Modified after Carle (1997).

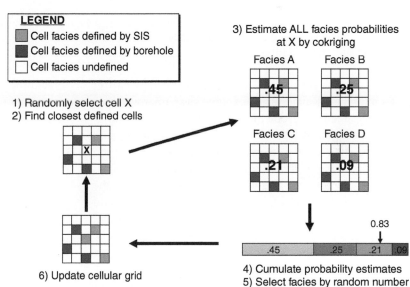

13.7.1.3 Simulated Annealing (SA)

The SGS and SIS methods provide a good starting point for establishing probabilistic lithofacies characteristics; each facies has appropriate volumetric proportions but its geometric dimensions in *x*, *y*, and *z* directions have not been considered. The prediction of flow paths and travel times of subsurface fluids, an important consideration in reservoir analysis of groundwater flow and transport modeling, is adversely affected when appropriate estimates of continuity and connectivity of individual lithofacies are ignored during the modeling process. Simulated annealing (SA) often finds a very good solution, even in the presence of noisy data, to the problem of adding some geometric constraints to SGS or SIS realizations (Carle, 1997).

The SA method is thus named because it mimics the process undergone by misplaced atoms in a metal when it is heated and then slowly cooled. It utilizes an iterative, trial-and-error process to modify an initial model image by swapping the locations of randomly selected pairs of cells. The swap is accepted if it improves the model's

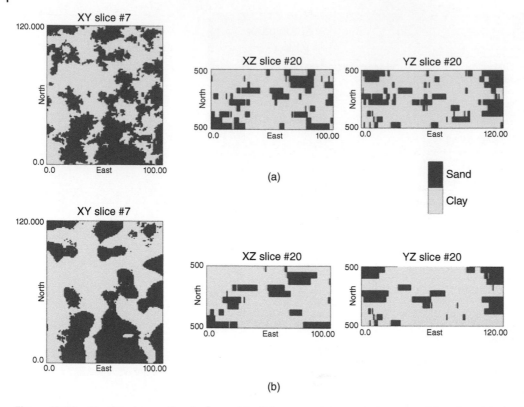

Figure 13.19 Simulated annealing (or "quenching") improves the geometric connectivity of models produced by sequential indicator simulation. (a) Slices through facies model produced by sequential indicator simulation. (b) Slices through facies model after simulated annealing. (Srinivasan 2017).

conformity to a set of threshold criteria and rejected if it does not. The SA method allows some swap improvements to be rejected to prevent the occurrence of a local minimum, a well-known problem with annealing techniques (Deutsch 1992). After a sequence of swaps have been concluded, the threshold criteria are adjusted (equivalent to the temperature cooling in metal annealing) so that the exchanges occur only when they meet stricter criteria and the process continues until a final image is produced that closely conforms to the all the model characteristic criteria. Figure 13.19 is an example of the improved geometric connectivity of lithofacies resulting from a SA process.

During the annealing process, the lithofacies of pixels or voxels at borehole locations remain unchanged while those of nearby cells are constrained and will show only slight changes. Cells further from hard data points are less constrained; their assigned facies are more likely to be adjusted. This process reflects the greater uncertainty of the correct lithofacies assignment at locations far removed from actual direct observations.

13.7.1.4 Transition Probability-based Stochastic Modeling

Carle and Fogg (1996, 1997), Carle et al. (1998), and Carle (1999) introduced the concept of employing a three-dimensional Markov chain model to incorporate conceptual geological information (soft data) with transition rates measured from hard data to form a geologically realistic model of spatial variability. Carle (1999) formally documented the supporting Transition Probability Geostatistical Software (T-PROGS).

The T-PROGS models the spatial structure of geological data by Markov chain models to relate transition probabilities to distance. Although this is a two-point geostatistical method, this approach has several advantages over traditional indicator kriging methods. First, it considers asymmetric juxtaposition tendencies, such as fining-upwards sequences. Second, it does not rely exclusively on empirical curve fitting to develop the indicator variogram model; it includes a conceptual framework for incorporating geologic interpretations into the development of cross-correlated spatial variability. This is advantageous because geologic data are typically only adequate to develop a model of spatial variability in the vertical direction. The conceptual framework links fundamental observable attributes, such as mean lengths, material proportions, anisotropy, and juxtapositioning, with Markov chain model parameters (Carle et al. 1998; Weissmann et al. 1999; Ritzi 2000; Lee et al. 2007; He et al. 2014).

Lithofacies

- Sand with fines
- Clay
- Clean sand
- Silt

Figure 13.20 Example of a T-PROGS generated 3-D grid with four lithofacies. Source: U.S. Army Corps of Engineers, Coastal and Hydraulics Laboratory.

The T-PROGS software produces a model containing a set of lithofacies distributed on a 3-D grid (Figure 13.20). Each lithofacies is conditioned to the borehole data and the materials proportions and transitions between the boreholes follow the trends observed in the borehole data. Recent studies by Lee et al. (2007) and Dell'Arciprete et al. (2012) report that T-PROGS produces results that are comparable with other stochastic modeling methods such as MPS, SGS, and SIS. T-PROGS has been successfully applied to groundwater resource evaluations in several sedimentary environments (Carle et al. 1998; Weissmann and Fogg 1999; Koch et al. 2014; He et al. 2014; Bianchi et al. 2015).

13.7.2 Object-based Stochastic Simulation Methods

Object-based stochastic simulation methods offer an alternative procedure for constructing categorical models. Rather than creating a model on a cell-by-cell basis, they utilize geological objects that have a genetic significance. There are two categories of object-based stochastic simulations – the older established Boolean simulation methods (Haldorsen and Damsleth 1990; Deutsch and Wang 1996; Holden et al. 1998) and the more recently developed multiple point statistics (MPS) methods (Strebelle 2006; Kessler

2012; Pyrcz and Deutsch 2014; Kim et al. 2018; Tahmasebi 2018).

13.7.2.1 Boolean Simulation Methods

Boolean simulation methods require the modeler to specify the proportions and basic shape that describe the geometry of each depositional facies and then to choose parameters that describe the shape. For example, fluvial channels may be defined by a sinuous pattern in map view and as half-elliptical shapes in cross-section. Some algorithms allow specification of the relative positions and relationships of geologic bodies. For example, braided stream channels may cross each other, crevasse splays are attached to channel forms, some geological bodies tend to cluster while others are dispersed and, in some cases, there may be a characteristic minimum separation between individual bodies. If observed hard data points are to be honored, these locations serve as control points to "anchor" some geological objects and these are simulated before the intervening regions. Once the shape distribution parameters and position rules are chosen, the simulation is performed in five steps:

1. Fill the reservoir model with a background lithofacies, such as shale;
2. Randomly select a starting point in the model;

3. Randomly select one lithofacies shape and draw it according to the location, shape, size, and orientation rules;

4. Check to see whether the shape conflicts with any control data or with previously simulated shapes, and if it does, reject it and go back to Step 3;

5. Check to see whether the global lithofacies proportions are correct, and if they are not, return to Step 2.

Once the object-based Boolean simulation has been completed, pixel-based methods may be applied to simulate petrophysical properties within the geologic bodies. The Boolean simulation approach is currently used for petroleum reservoir modeling. Geologists find it particularly satisfactory because the shapes look geologically realistic and are based on geometries and facies relationships that have been measured. However, Boolean simulation requires a large number of input parameters and prior knowledge to select the parameter values. While the initial Boolean-type algorithms could not always honor all the control data due to computational and shape definition limitations, recent developments have greatly alleviated these issues (Deutsch and Tran 2002).

Geel and Donselaar (2007) employed object-based Boolean simulations to generate multiple, equiprobable realizations of the spatial distribution of sediment properties in a tidal estuarine succession within the Holocene Holland Tidal Basin in the Netherlands. This is a well-studied geological environment. They selected a 6 km × 5 km × 25 m deep volume where there were many cone penetration tests and five boreholes. The model contained four lithofacies associated with tidal channels which formed an intricate network of juxtaposed and vertically stacked sand bodies embedded in mud dominated sub-tidal sediments. Figure 13.21 illustrates the base case realization which specified channels 400 m wide and 8 m

deep. The sensitivity of the model to the input parameters was analyzed by varying the tidal channel width between 100 and 800 m and the channel thickness between 2 and 8 m. The sensitivity analysis highlighted the importance of sand-dominated tidal flats in providing lateral connectivity. The model and the sensitivity study utilized Roxar's industry-standard STORM and RMS modeling software (Hardy and Hatløy 2005).

13.7.2.2 The Multiple Point Statistics (MPS) Approach

MPS was developed by Guardiano and Srivastava (1993). In recent years, issues related to CPU time and improved graphical representation of the models have been resolved and have led to rapidly growing applications of MPS for reservoir modeling. MPS has several advantages, including providing realistic facies distributions, honoring well data, and easy conditioning with secondary data (Hashemi et al. 2014; Pyrcz and Deutsch 2014; de Carbalho et al. 2016). MPS offers an attractive alternative to the challenge of building variogram models for each facies, direction, and reservoir properties when only a few direct observations are available (Strebelle 2002; Caers and Zhang 2004). Coburn et al. (2006) explain and demonstrate the advantages and disadvantages associated with alternative MPS methods.

The MPS methods do not develop spatial statistics by using variograms; rather the spatial structure is defined by a conceptual tool – a training image, which may be two- or three-dimensional (Pyrcz et al. 2008). Providing a representative training image, or a set of training images, is the biggest challenge when applying MPS. Training images may be constructed from outcrop data, from Boolean simulations developed as described in Section 13.7.2.1, or from process-based methods that mimic the geological process that formed the geological units (Tahmasebi 2018). Outcrops provide a unique and direct representation of geological features and clearly illustrate the geometry and

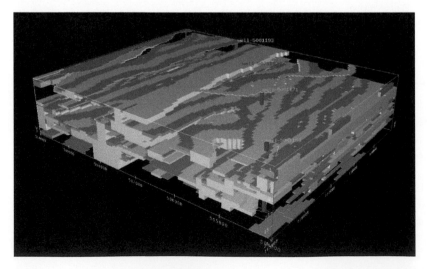

Figure 13.21 3-D view of the base case stochastic simulation model of a 6 km × 5 km × 25 m deep volume of the Holocene Holland Tidal Basin in the Netherlands showing tidal channels in yellow and fringing tidal flats in orange. The mud-dominated inter-channel lithofacies (the background facies) has been filtered out. The five wells on which the model is conditioned are shown. (Geel and Donselaar 2007).

Figure 13.22 The Multiple Point Statistics procedure uses a training image to develop several domain templates; these define the specific correlation patterns which define appropriate MPS templates that control geological unit geometries. (Srinivasan 2017).

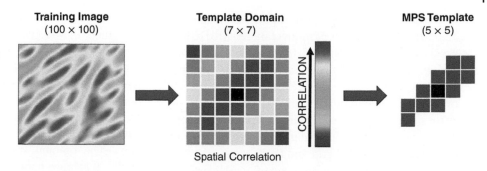

Training Image
(100 × 100)

Template Domain
(7 × 7)

CORRELATION

Spatial Correlation

MPS Template
(5 × 5)

spatial continuity of lithofacies in 2-D sections. Boolean simulation methods are one of the best and most accessible sources for training images. The use of process-based training images is computationally expensive, requiring considerable calibration. Regardless of its source, the training image embodies the geological concepts for shape and distribution of lithofacies (Strebelle 2002). As shown by Figure 13.22, the training image is scanned through a search window to develop one or more domain templates. These define the specific correlation patterns utilized to build a cumulative distribution function to simulate each cell (Strebelle 2002). Because the training image controls the result of facies modeling, it is essential that the training image is validated to ensure reliable property modeling. Strebelle (2006) describes the development of reservoir models that incorporate complex geological and structural information based on MPS methods that combine features of both pixel-based algorithms and object-based techniques. This simulation method utilized the Multiple-Point Statistics Simulation Program, SNESIM, which offers both speed and efficiency (Strebelle and Cavelius 2013).

13.7.3 Assessing Stochastic Model Uncertainty

Because stochastic models consider the inherent randomness in the data by generating multiple, equally probable, realizations of the geological properties of the subsurface, statistical analysis of a large number of these realizations can quantify the uncertainty of the model predictions. Stafleu et al. (2011) discuss how the GeoTOP model uses the probability of the presence of various lithofacies within stratigraphic units to provide an uncertainty estimate and influence the determination of petrophysical properties. Chapter 11 provides additional details of the GeoTOP model.

Chapter 15 provides an extended discussion of all aspects of geological model uncertainty. It describes how Lark et al. (2014) analyzed the results of an experiment to assess the influence of modeler experience on the uncertainty of cross-section interpretations by using stochastic methods to generate several thousand independent realizations of the lithostratigraphic surfaces. Comparisons of estimated

and observed values defined interpretation errors which were used to compute a 95% confidence interval for one of the interpretations.

Dubrule (1994) defined three types of uncertainties resulting from stochastic modeling: (i) algorithm related, (ii) soft data related, and (iii) stochastic related. Algorithm-related uncertainties result from the mathematical parameterization, assumptions, and simplifications of the selected algorithm. Fortunately, there is a considerable variety of stochastic algorithms and approaches to choose from (Deutsch 2002); so the decision to adopt a particular modeling strategy should consider the inherent uncertainty associated with that process. Uncertainties related to the soft data are derived from a poor knowledge of the conditioning parameter values used by the modeling algorithm. They should be assessed during the selection of outcrop analogs and the establishment of the geological conceptual model (Dalrymple 2001; Falivene et al. 2006). Stochastic-related uncertainties are related to the inability of the multiple equiprobable realizations to totally reflect the inherent heterogeneity of the lithofacies and the petrophysical properties. Collecting and using more hard data to build facies models can reduce stochastic-related uncertainties (Falivene et al. 2006).

13.8 Conclusions

In the context of geological modeling, discretization involves the subdivision of the model volume into multiple small-unit volumes. These are most often used to represent substratigraphic property variations within the units of a framework model. Discretization also allows conversion of geological framework models into a cellular form that is required by many model applications. However, there is a considerable variety of 2-D grids or 3-D meshes that can be specified during the discretization process; different types of grids or meshes are required depending on the choice of numerical simulation procedures. In fact, the design of a grid or mesh typically involves a tradeoff between mesh size (number of nodes and elements) and the level of detail represented by the mesh. Additional design constraints exist when the grids or meshes are required to both accurately define the geological framework and support numerical simulations using finite difference or finite element procedures.

Grid and mesh generation is a tedious, time-consuming, and error-prone process – especially for models with complex structures such as faults, and truncations of lithostratigraphic layers. Automated grid and mesh generation tools can help resolve the basic development of a suitable grid or mesh. Additional post-generation grid or mesh refinement procedures help to ensure the grid or mesh is optimally configured to support numerical simulations.

Discretization is required to implement stochastic modeling approaches. Over the past 10–15 years, the growing popularity of stochastic modeling has included two broad categories of modeling techniques – pixel and voxel based simulation methods and object-based simulation methods. Each category contains several techniques and each category has advantages and limitations that tend to make it more attractive for particular applications. The optimal stochastic modeling strategy is, of course, a balance between the amount of realism desired and practical constraints such as the time available, computation capabilities, simplicity of the method, and costs.

References

Alcaraz, S.A., Chambefort, I., Pearson, R. and Cantwell, A. (2015). An Integrated Approach to 3-D Modelling to Better Understand Geothermal Reservoirs. In: *Proceedings World Geothermal Congress 2015*. Bonn, Germany: International Geothermal Association. 7 pp. [Online: Available at: https://www.geothermal-energy.org/cpdb/record_detail .php?id=23105] (Accessed August 21, 2020).

Bianchi, M., Kearsey, T. and Kingdon, A. (2015). Integrating deterministic lithostratigraphic models in stochastic realizations of subsurface heterogeneity. Impact on predictions of lithology, hydraulic heads and groundwater fluxes. *Journal of Hydrology* 531 (3): 557–573. https://doi.org/10.1016/j.jhydrol.2015.10.072.

Bower, K.M., Gable, C.W. and Zyvoloski, G.A. (2005). Grid resolution study of ground water flow and transport. *Groundwater* 43 (1): 122–132. https://doi.org/10.1111/ j.1745-6584.2005.tb02291.x.

Burt, A.K. and Dodge, J.E.P. (2016). *Three-dimensional modelling of surficial deposits in the Orangeville–Fergus area of southern Ontario*. Sudbury, ON: Ontario Geological Survey, Groundwater Resources Study 15. 155p. [Online: Available at: http://www.geologyontario.mndm.gov.on.ca /mndmaccess/mndm_dir.asp?type=pub&id=GRS015] (Accessed August 21, 2020).

Caers, J. (2011). *Modeling Uncertainty in the Earth Sciences*. New York, NY: Wiley. 246 pp. https://doi.org/10.1002/ 9781119995920.

Caers, J. and Zhang, T. (2004). Multiple-point geostatistics: a quantitative vehicle for integrating geology analogs into multiple reservoir models. *AAPG Memoir* 80: 383–394. https://doi.org/10.1306/M80924C18.

Canann, S.A., Tristano, J.R. and Staten, M.L. (1998). An approach to combined Laplacian and optimization-based smoothing for triangular, quadrilateral, and quad-dominant meshes. In: *Proceedings, 7th International Meshing Roundtable, Dearborn, Michigan* (ed. L.A. Freitag), 479–494. [Online: Available at: http://citeseerx.ist.psu. edu/viewdoc/summary?doi=10.1.1.88.9616] (Accessed August 21, 2020).

Carle, S.F. (1997). Implementation schemes for avoiding artifact discontinuities in simulated annealing. *Mathematical Geology* 29: 231–244. https://doi.org/10.1007/BF02769630.

Carle, S.F. (1999). *T-PROGS: Transition Probability Geostatistical Software, Version 2.1*. Davis, CA: University of California. 78 pp. [Online: Available at: http://gmsdocs.aquaveo.com/t-progs.pdf] (Accessed August 21, 2020).

Carle, S.F. and Fogg, G.E. (1996). Transition probability-based indicator geostatistics. *Mathematical Geology* 28 (4): 453–476. https://doi.org/10.1007/BF02083656.

Carle, S.F. and Fogg, G.E. (1997). Modeling spatial variability with one and multidimensional continuous-lag Markov chains. *Mathematical Geology* 29 (7): 891–918. https://doi.org/10.1023/A:1022303706942.

Carle, S.F., LaBolle, E.M., Weissmann, G.S. et al. (1998). Conditional simulation of hydrostratigraphic architecture: a transition probability/Markov approach. In: *Hydrogeologic Models of Sedimentary Aquifers* (eds. S.G. Fraser and J.M. Davis), 147–170. Broken Arrow, OK: SEPM Society for Sedimentary Geology. https://doi.org/10.2110/ sepmcheg.01.147.

Caumon, G., Lepage, F., Sword, C.H. and Mallet, J.-L. (2004). Building and editing a sealed geological model. *Mathematical Geology* 36 (4): 405–424. https://doi.org/10.1023/B:MATG.0000029297.18098.8a.

Chen, F., Chen, S. and Peng, G. (2013). Using sequential Gaussian simulation to assess geochemical anomaly areas of lead element. In: *Computer and Computing Technologies in Agriculture VI – CCTA 2012, IFIP Advances in*

Information and Communication Technology, vol. 393 (eds. D. Li and Y. Chen), 69–76. Berlin, Germany: Springer. https://doi.org/10.1007/978-3-642-36137-1_9.

Cherry, T.A., Gable, C.W. and Trease, H.E. (1996). 3-dimensional wells and tunnels for finite element grids. In: *Proceedings of the 5th International Conference on Numerical Grid Generation in Computational Fluid Dynamics and Related Fields* (eds. B.K. Soni, J.F. Thompson, H. Hauser and P.R. Eiseman). Mississippi State University Press. 12 pp. [Online: Available at: https://lagrit.lanl.gov/pdfs/NGG96.pdf] (Accessed August 21, 2020).

Chilès, J.-P., Aug, C., Guillen, A. and Lees, T. (2004). Modelling the geometry of geological units and its uncertainty in 3D from structural data: the potential-field method. In: *Proceedings of Orebody Modelling and Strategic Mine Planning* (ed. R. Dimitrakopoulos), 313–320. [Online: Available at: http://citeseerx.ist.psu.edu/viewdoc/summary?doi=10.1.1.583.213] (Accessed August 21, 2020).

Coburn, R.C., Yarus, J.M. and Chambers, R.L. [Eds.] (2006). *Stochastic Modeling and Geostatistics: Principles, Methods, and Case Studies, Volume II*. Tulsa, OK: The American Association of Petroleum Geologists. 409 pp. https://doi.org/10.1306/CA51063.

Collon, P. and Caumon, G. (2017). 3D geomodelling in structurally complex areas: implicit vs. explicit representations. In: *Proceedings, 79th EAGE Conference and Exhibition 2017*. Houten, Netherlands: European Association of Geoscientists & Engineers. 5 pp. https://doi.org/10.3997/2214-4609.201701144.

Collon, P., Steckiewicz-Laurent, W., Pellerin, J. et al. (2015). 3D Geomodelling combining implicit surfaces and Voronoi-based re-meshing: a case study in the Lorraine Coal Basin (France). *Computers and Geosciences* 77: 29–43. https://doi.org/10.1016/j.cageo.2015.01.009.

Dalrymple, M. (2001). Fluvial reservoir architecture in the Stratfjord formation (northern North Sea) augmented by outcrop analog statistics. *Petroleum Geoscience* 7: 115–122. https://doi.org/10.1144/petgeo.7.2.115.

De Carbalho, P.R.M., Da Costa, J.F.L., Rasera, L.G., and Varella, L.E.S. (2016). Geostatistical facies simulation with geometric patterns of a petroleum reservoir. *Stochastic Environmental Research and Risk Assessment* 31: 1805–1822. https://doi.org/10.1007/s00477-016-1243-5.

Dell'Arciprete, D., Bersezio, R., Felletti, F. et al. (2012). Comparison of three geostatistical methods for hydrofacies simulation: a test on alluvial sediments. *Hydrogeology Journal* 20 (2): 299–311. https://doi.org/10.1007/s10040-011-0808-0.

De Marsily, G., Delay, F., Goncalves, J. et al. (2005). Dealing with spatial heterogeneity. *Hydrogeology Journal* 13 (1): 161–183. https://doi.org/10.1007/s10040-004-0432-3.

Denver, L.F. and Phillips, D.C. (1990). Stratigraphic geocellular modeling. *Geobyte* 5 (1): 45–47.

Deutsch, C.V. (1992). *Annealing Techniques Applied to Reservoir Modeling and the Integration of Geological and Engineering (Well Test) Data*. PhD Dissertation, Stanford University. 306 pp. [Online: Available at: http://citeseerx.ist.psu.edu/viewdoc/download?doi=10.1.1.330.7685&rep=rep1&type=pdf] (Accessed August 21, 2020).

Deutsch, C.V. (2002). *Geostatistical Reservoir Modeling*. New York, NY: Oxford University Press. 376 pp.

Deutsch, C.V. (2006). A sequential indicator simulation program for categorical variables with point and block data: block SIS. *Computers & Geosciences* 32 (10): 1669–1681. https://doi.org/10.1016/j.cageo.2006.03.005.

Deutsch, C.V. and Cockerham, P.W. (1994). Practical considerations in the application of simulated annealing of stochastic simulation. *Mathematical Geology* 26 (1): 67–82. https://doi.org/10.1007/BF02065876.

Deutsch, C.V. and Journel, A.G. (1992). *GSLIB: Geostatistical Software Library and User's Guide*. New York, NY: Oxford University Press. 340 pp.

Deutsch, C.V. and Tran, T.T. (2002). FLUVSIM: a program for object-based stochastic modeling of fluvial depositional systems. *Computers & Geosciences* 28 (4): 525–535. https://doi.org/10.1016/S0098-3004(01)00075-9.

Deutsch, C.V. and Wang, L. (1996). Hierarchical object-based stochastic modeling of fluvial reservoirs. *Mathematical Geology* 28 (7): 857–880. https://doi.org/10.1007/BF02066005.

Deutsch, C.V. and Wen, X.H. (1998). An improved perturbation mechanism for simulated annealing simulation. *Mathematical Geology* 30 (7): 801–816. https://doi.org/10.1023/A:1021722508504.

Dimitrakopoulos, R. and Luo, X. (2004). Generalized sequential Gaussian simulation on group size and screen-effect approximations for large field simulations. *Mathematical Geology* 36: 567–591. https://doi.org/10.1023/B:MATG.0000037737.11615.df.

Dirner, S. and Steiner, U. (2015). Assessing reservoir uncertainty with stochastic facies modeling of a hydrothermal medium enthalpy reservoir (upper Jurassic carbonates of the southern German Molasse Basin). In: *Proceedings World Geothermal Congress 2015*. Bonn, Germany: International Geothermal Association. 11 pp. [Online: Available at: https://www.geothermal-energy.org/cpdb/record_detail.php?id=23349] (Accessed August 21, 2020).

Dubrule, O. (1994). Estimating or choosing a geostatistical model? In: *Geostatistics for the Next Century* (ed. R. Dimitrakopoulos), 3–14. Dordrecht, Netherlands: Kluwer Academic Publishers. https://doi.org/10.1007/978-94-011-0824-9_1.

Falivene, O., Arbues, P., Gardiner, A. et al. (2006). Best practice stochastic facies modeling from a channel-fill Turbidite sandstone Analog (the quarry outcrop, Eocene Ainsa Basin, Northeast Spain). *AAPG Bulletin* 90 (7): 1003–1029. https://doi.org/10.1306/02070605112.

Field, D.A. (1988). Laplacian smoothing and Delaunay triangulations. *Communications in Applied Numerical Methods* 4: 709–712. https://doi.org/10.1002/cnm.1630040603.

Freitag, L.A. and Ollivier-Gooch, C. (1997). Tetrahedral mesh improvement using swapping and smoothing. *International Journal for Numerical Methods in Engineering* 40: 3979–4002. https://doi.org/10.1002/(SICI)1097-0207(19971115)40:21%3C3979::AID-NME251%3E3.0.CO;2-9.

Freitag, L., Jones, M. and Plassmann, P. (1995). An efficient parallel algorithm for mesh smoothing. In: *Proceedings, 4th International Meshing Roundtable, Albuquerque, New Mexico* (ed. T. Tautges), 47–58. [Online: Available at: http://citeseerx.ist.psu.edu/viewdoc/summary?doi=10.1.1.46.8970]

Frey, P.J. and George, P.-L. (2008). *Mesh Generation: Application to Finite Elements*, 2e. Oxford, UK: Hermes Science Publishing. 814 pp.

Gable, C.W. (2006). *Mesh Generation for Geological Applications: Why it is Different than Planes, Trains and Automobiles*. 2006 Workshop on Challenges and Opportunities at the Interfaces of Scientific Computing and Computational Geodynamics. Davis, CA: Computational Infrastructure for Geodynamics (CIG). 35 slides. [Online: Available at: https://geodynamics.org/cig/files/6514/1158/9117/Gable_Carl.pdf] (Accessed 22 August, 2020).

Gable, C.W., Cherry, T.A., Trease, H.E. and Zyvoloski, G.A. (1995). *GEOMESH Grid Generation*. Los Alamos, NM: Los Alamos National Laboratory, Report LA-UR-95-4143. 15 pp.

Gable, C.W., Trease, H.E. and Cherry, T.A. (1996a). Geological applications of automatic grid generation tools for finite elements applied to porous flow Modeling. In: *Proceedings of the 5th International Conference on Numerical Grid Generation in Computational Fluid Dynamics and Related Fields* (eds. B.K. Soni, J.F. Thompson, H. Hauser and P.R. Eiseman). Mississippi State University Press. 9 pp. [Online: Available at: https://meshing.lanl.gov/lagrit/publications/num_grid.pdf] (Accessed August 22, 2020).

Gable, C.W., Trease, H.E. and Cherry, T.A. (1996b). Automated grid generation from models of complex geologic structure and stratigraphy. In: *Proceedings Volume, Third International Conference Integrating GIS and Environmental Modeling, Santa Fe, NM* (eds. M. Goodchild, L. Steyaert, B. Parks, et al.), 475–487. Fort Collins, CO: GIS World Books.

Geel, C.R. and Donselaar, M.E. (2007). Reservoir modelling of heterolithic tidal deposits: sensitivity analysis of an object-based stochastic model. *Netherlands Journal of Geosciences* 86 (4): 403–411. https://doi.org/10.1017/S0016774600023611.

Guardiano, F. and Srivastava, M. (1993). Multivariate Geostatistics: beyond bivariate moments. In: *Geostatistics Tróia '92. Quantitative Geology and Geostatistics, vol 5* (ed. A Soares), 133–144. Dordrecht, Netherlands: Kluwer Academic Publishers. https://doi.org/10.1007/978-94-011-1739-5_12

Haldorsen, H.H. and Damsleth, E. (1990). Stochastic Modeling. *Journal of Petroleum Technology* 42: 404–412.

Hardy, D. and Hatløy, A. (2005). The changing face of reservoir modelling. *First Break* 23: 63–66.

Hashemi, S., Javaherian, A., Ataee-pour, M. et al. (2014). Channel characterization using multiple-point Geostatistics, neural network, and modern analogy: a case study from a carbonate reservoir, Southwest Iran. *Journal of Applied Geophysics* 111: 47–58. https://doi.org/10.1016/j.jappgeo.2014.09.015.

He, X., Koch, J., Sonnenborg, T.O. et al. (2014). Transition probability-based stochastic geological Modeling using airborne geophysical data and borehole data. *Water Resources Research* 50: 3147–3169. https://doi.org/10.1002/2013WR014593.

Holden, L., Hauge, R., Skare, Ø. and Skorstad, A. (1998). Modeling of fluvial reservoirs with object models. *Mathematical Geology* 30: 473–496. https://doi.org/10.1023/A:1021769526425.

Jakab, N. (2017). Stochastic Modeling in geology: determining the sufficient number of models. *Central European Geology* 60 (2): 135–151.

Joe, B. (1991). GEOMPACK – a software package for the generation of meshes using geometric algorithms. *Advances in Engineering Software* 56 (13): 325–331. https://doi.org/10.1016/0961-3552(91)90036-4.

Joe, B. (1995). Construction of three-dimensional improved-quality triangulations using local transformations. *SIAM Journal of Scientific Computing* 16: 1292–1307. https://doi.org/10.1137/0916075.

Kaiser, B.O., Cacace, M. and Scheck-Wenderoth, M. (2013). 3D coupled fluid and heat transport simulations of the northeast German Basin and their sensitivity to the spatial discretization: different sensitivities for different mechanisms of heat transport. *Environmental Earth Sciences* 70: 3643–3659. https://doi.org/10.1007/s12665-013-2249-7.

Kearsey, T., Williams, J., Finlayson, A. et al. (2015). Testing the application and limitation of stochastic simulations to predict the lithology of glacial and fluvial deposits in Central Glasgow, UK. *Engineering Geology* 187: 98–112.

Kessler, T., (2012). *Hydrogeological characterization of low-permeability clayey tills - the role of sand lenses.* PhD thesis Technical University of Denmark. 62 pp. [Online: Available at: https://backend.orbit.dtu.dk/ws/files/7999138/Timo_C_Kessler_PhD_thesis_WWW_version.pdf] (Accessed August 22, 2020).

Kim, K.H., Lee, K., Lee, H.S. et al. (2018). Lithofacies Modeling by multipoint statistics and economic evaluation by NPV volume for the early cretaceous Wabiskaw member in Athabasca Oilsands area, Canada. *Geoscience Frontiers* 9 (2): 441–451. https://doi.org/10.1016/j.gsf.2017.04.005.

Koch, J., He, X., Jensen, K.H. and Refsgaard, J.C. (2014). Challenges in conditioning a stochastic geological model of a heterogeneous glacial aquifer to a comprehensive soft data set. *Hydrology and Earth System Sciences* 18: 2907–2923. https://doi.org/10.5194/hess-18-2907-2014.

Koltermann, C.E. and Gorelick, S.M. (1996). Heterogeneity in sedimentary deposits: a review of structure-imitating, process-imitating, and descriptive approaches. *Water Resources Research* 32 (9): 2617–2658. https://doi.org/10.1029/96WR00025.

Lark, R.M., Thorpe, S., Kessler, H. and Mathers, S.J. (2014). Interpretative modelling of a geological cross section from boreholes: sources of uncertainty and their quantification. *Solid Earth* 5: 1189–1203. https://doi.org/10.5194/se-5-1189-2014.

Lee, S.Y., Carle, S.F. and Fogg, G.E. (2007). Geologic heterogeneity and a comparison of two Geostatistical models: sequential Gaussian and transition probability-based Geostatistical simulation. *Advances in Water Resources* 30 (9): 1914–1932. https://doi.org/10.1016/j.advwatres.2007.03.005.

Los Alamos National Laboratory (2019). *GEOMESH: X3D Grid Generation for Geological Applications (Handout).* Los Alamos, NM: Los Alamos National Laboratory. 4 pp. [Online: Available at: https://meshing.lanl.gov/text/geomesh.handout.pdf] (Accessed June 2019).

Mallet, J.-L. (2002). *Geomodeling.* Oxford, UK: Oxford University Press, Applied Geostatistics Series. 624 pp.

Merriam-Webster (2019). *Definition of Stochastic.* Springfield, MA: Merriam-Webster. [Online: Available at: https://www.merriam-webster.com/dictionary/stochastic] (Accessed June 2019).

Miller, T.A., Vesselinov, V.V., Stauffer, P.H. et al. (2007). Integration of geologic frameworks in meshing and setup of computational Hydrogeologic models, Pajarito Plateau, New Mexico. In: *New Mexico Geological Society Fall Field Conference Guidebook - 58 Geology of the Jemez Region II* (eds. B.S. Kues, S.A. Kelley and V.W. Lueth), 121-128. Socorro, NM: New Mexico Geological Society. [Online: Available at: https://nmgs.nmt.edu/publications/guidebooks/downloads/58/58_p0492_p0500.pdf] (Accessed August 22, 2020).

Mustapha, H. (2011). G23FM: a tool for meshing complex geological media. *Computers & Geosciences* 15: 385–397. https://doi.org/10.1007/s10596-010-9210-6.

Origlio, V. (2019). Stochastic. In: *MathWorld* (ed. E.W. Weisstein). [Online: Available at: http://mathworld.wolfram.com/Stochastic.html] (Accessed June 2019).

Owen, S.J. (1998). A survey of unstructured mesh generation technology. In: *Proceedings, 7th International Meshing Roundtable, Dearborn, Michigan* (ed. L.A. Freitag), 239–267. [Online: Available at: http://citeseerx.ist.psu.edu/viewdoc/summary?doi=10.1.1.34.5079] (Accessed August 22, 2020).

Parry, S., Baynes, F.J., Culshaw, M.G. et al. (2014). Engineering geological models: an introduction: IAEG commission 25. *Bulletin of Engineering Geology and the Environment* 73: 689–706. https://doi.org/10.1007/s10064-014-0576-x.

Pyrcz, M.J. and Deutsch, C.V. (2014). *Geostatistical Reservoir Modeling, 2e.* New York, NY: Oxford University Press. 430 pp.

Pyrcz, M.J., Boisvert, J.B. and Deutsch, C.V. (2008). A library of training images for fluvial and Deepwater reservoirs and associated code. *Computers & Geosciences* 34 (5): 542–560. https://doi.org/10.1016/j.cageo.2007.05.015.

Rautman, C.A. and Flint, A.L. (1992). Deterministic Processes and Stochastic Modeling. In: *Proceesings, Second Annual International High-Level Radioactive Waste Management Conference*, 1617–1624. La Grange Park, IL: American Nuclear Society.

Ritzi, R.W. (2000). Behavior of indicator Variograms and transition probabilities in relation to the variance in lengths of Hydrofacies. *Water Resources Research* 36 (11): 3375–3381. https://doi.org/10.1029/2000WR900139.

Ross, M., Parent, M. and Lefebvre, R. (2005). 3D geologic framework models for regional hydrogeology and land-use management: a case study from a quaternary basin of southwestern Quebec, Canada. *Hydrogeology Journal* 13: 690–707. https://doi.org/10.1007/s10040-004-0365-x.

Royse, K.R., Reeves, H.J. and Gibson, A.R. (2008). The modelling and visualization of digital Geoscientific data as a communication aid to land-use planning in the urban environment: an example from the Thames gateway. *Geological Society Special Publications* 305: 89–106. https://doi.org/10.1144/SP305.10.

Sandia National Laboratory (2019). *CUBIT Mesh Generation Toolkit.* [Online: Available at: http://endo.sandia.gov/SEACAS/CUBIT/Cubit.html] (Accessed April 2019).

Scientific Software Group (2019). *Visual MODFLOW 3D Builder - Detailed Description.* Salt Lake City, UT: Scientific Software Group. [Online: Available at: http://www

.scientificsoftwaregroup.com/pages/detailed_description .php?products_id=222] (Accessed April 2019).

Smirnoff, A., Blouin, M., Paradis, S.J. and Ross, M. (2011). Transferring geological properties from 3D geomodels to groundwater models with GOFEFLOW. In: *Proceedings, Geohydro 2011*. Canadian Quaternary Association / Richmond, BC: IAH Canadian National Chapter. 6 pp.

Spyridis, P., Nasekhian, A. and Skalla, G. (2013). Design of SCL structures in London. *Geomechanics and Tunnelling* 6 (1): 66–79. https://doi.org/10.1002/geot.201300005.

Srinivasan, S. (2017). Modeling of complex reservoirs: An excursion into Multiple Point Geostatistics and new data integration paradigms (I). In: *Computational Issues in Oil Field Applications Tutorials MARCH 21-24, 2017*. Los Angeles, CA: UCLA, Institute for Pure & Applied Mathematics. 67 slides. [ONLINE: Available at: http://www.ipam.ucla.edu/abstract/?tid=14369& pcode=OILTUT] (Accessed March 25, 2020).

Stafleu, J., Maljers, D., Gunnink, J.L. et al. (2011). 3D modelling of the shallow subsurface of Zeeland, the Netherlands. *Netherlands Journal of Geosciences* 90 (4): 293–310. https://doi.org/10.1017/S0016774600000597.

Strebelle, S.B. (2002). Conditional simulation of complex geological structures using multiple-point statistics. *Mathematical Geology* 34 (1): 1–21. https://doi.org/10.1023/A:1014009426274.

Strebelle, S.B. (2006). Sequential simulation for Modeling geological structures from training images. In: *Stochastic Modeling and Geostatistics: Principles, Methods, and Case Studies, Volume II* (eds. R.C. Coburn, J.M. Yarus and R.L. Chambers), 139–150. Tulsa, OK: The American Association of Petroleum Geologists. https://doi.org/10.1306/1063812CA53231.

Strebelle, S. and Cavelius, C. (2013). Solving speed and memory issues in multiple-point statistics simulation program SNESIM. *Mathematical Geosciences* 46: 171–186. https://doi.org/10.1007/s11004-013-9489-7.

Tahmasebi, P. (2018). Multiple point statistics: a review. In: *Handbook of Mathematical Geosciences* (eds. B.S. Daya Sagar, Q. Cheng and F. Agterberg), 613–643. Cham, Switzerland: Springer. https://doi.org/10.1007/978-3-319-78999-6_30.

Thompson, J.F., Soni, B.K., and Weatherill, N.P. [Eds.] (1998). *Handbook of Grid Generation*. Boca Raton, FL: CRC Press. 1136 pp.

Trease, H.E., George, D., Gable, C.W. et al. (1996). The X3D grid generation system. In: *Proceedings of the 5th International Conference on Numerical Grid Generation in Computational Fluid Dynamics and Related Fields* (eds. B.K. Soni, J.F. Thompson, H. Hauser and P.R. Eiseman) 237–244. Jackson, MS: University Press of Mississippi.

Turner, A.K. (2006). Challenges and trends for geological modelling and visualisation. *Bulletin of Engineering Geology and the Environment* 65 (2): 109–127. https://doi.org/10.1007/s10064-005-0015-0.

Van Lancker, V.R.M., Francken, F., Kint, L. et al. (2016). Building a 4D Voxel-Based Decision Support System for a Sustainable Management of Marine Geological Resources. In: *Oceanographic and Marine Cross-Domain Data Management for Sustainable Development* (eds. P. Diviacco, A. Leadbetter and H. Glaves), 224–252. Hershey, PA: IGI Global. https://doi.org/10.4018/978-1-5225-0700-0.ch010.

Watson, C., Richardson, J., Wood, B. et al. (2015). Improving geological and process model integration through TIN to 3D grid conversion. *Computers & Geosciences* 82: 45–54. https://doi.org/10.1016/j.cageo.2015.05.010.

Weissmann, G.S. and Fogg, G.E. (1999). Multi-scale alluvial fan heterogeneity Modeled with transition probability Geostatistics in a sequence stratigraphic framework. *Journal of Hydrology* 226 (1–2): 48–65. https://doi.org/10.1016/S0022-1694(99)00160-2.

Weissmann, G.S., Carle, S.F. and Fogg, G.E. (1999). Three-dimensional Hydrofacies Modeling based on soil surveys and transition probability Geostatistics. *Water Resources Research* 35 (6): 1761–1770. https://doi.org/10.1029/1999WR900048.

Whitten, E.H.T. (1977). Stochastic models in geology. *The Journal of Geology* 85 (3): 321–330.

Wikipedia (2020a). Scheme for Delauny geometry by Nü es, licensed under CC BY-SA 3.0. In: *Delaunay triangulation*. San Fransico, CA: Wikimedia Foundation. [Online: Available at: https://en.wikipedia.org/wiki/Delaunay_triangulation#/media/File:Delaunay_geometry.png] (Accessed August 20, 2020).

Wikipedia (2020b). Point inside circle - Delaunay condition broken by Jespa, licensed under CC BY-SA 4.0. In: *Delaunay triangulation*. San Fransico, CA: Wikimedia Foundation. [Online: Available at: https://en.wikipedia.org/wiki/Delaunay_triangulation#/media/File:Point_inside_circle_-_Delaunay_condition_broken.svg] (Accessed August 20, 2020).

Wikipedia (2020c). Edge Flip - Delaunay condition ok by Jespa, licensed under CC BY-SA 4.0. In: In: *Delaunay triangulation*. San Fransico, CA: Wikimedia Foundation. [Online: Available at: https://en.wikipedia.org/wiki/Delaunay_triangulation#/media/File:Edge_Flip_-_Delaunay_condition_ok.svg] (Accessed August 20, 2020).

Williams, J.D.O., Dobbs, M.R., Kingdon, A. et al. (2019). Stochastic modelling of hydraulic conductivity derived from geotechnical data; an example applied to Central

Glasgow. *Earth and Environmental Science Transactions of the Royal Society of Edinburgh* 108 (2–3): 141–154. https://doi.org/10.1017/S1755691018000312.

Wycisk, P., Hubert, T., Gossel, W. and Neumann, C. (2009). High-resolution 3D spatial modelling of complex geological structures for an environmental risk assessment of abundant mining and industrial Megasites. *Computers & Geosciences* 35: 165–182. https://doi.org/10.1016/j.cageo.2007.09.001.

Yao, T. (2002). Integrating seismic data for Lithofacies Modeling: a comparison of sequential indicator simulation algorithms. *Mathematical Geology* 34 (4): 387–403. https://doi.org/10.1023/A:1015026926992.

Zehner, B., Börner, J.H., Görz, I. and Spitzer, K. (2015). Workflows for generating tetrahedral meshes for finite element simulations on complex geological structures. *Computers & Geosciences* 79: 105–117. https://doi.org/10.1016/j.cageo.2015.02.009.

14

Linkage to Process Models

Geoff Parkin[1], Elizabeth Lewis[1], Frans van Geer[2], Aris Lourens[2], Wilbert Berendrecht[2], James Howard[3], Denis Peach[4], and Nicholas Vlachopoulos[5]

[1] School of Engineering, Newcastle University, Newcastle upon Tyne NE1 7RU, UK
[2] TNO, Geological Survey of the Netherlands, 3584 CB Utrecht, The Netherlands
[3] JBA Consulting, Newcastle upon Tyne NE1 5JE, UK
[4] Retired from British Geological Survey, Keyworth, Nottingham NG12 5GG, UK
[5] Royal Military College, Kingston, ON K7K 7B4, Canada

14.1 Introduction

Three-dimensional geological models are replacing traditional printed geological maps and 2-D GIS datasets because they provide a digital and dynamic 3-D definition of the subsurface. As discussed in previous chapters, geological models efficiently convert limited direct subsurface observations to representations of the presence, characteristics, and arrangement of earth materials, including any structural discontinuities. This capability makes geological models a preferred source for documenting and archiving subsurface information and knowledge. However, as was the case with geological maps, geological models are frequently created to serve as the information source for further application-specific analyses.

Geological models can support many potential applications. Chapters 18–25 provide selected case studies organized under eight topics ranging from regulatory, planning, and management functions to scientific and engineering applications. Process models extend the dimensionality of the 3-D models beyond their spatial coordinates. Analysis of time-varying phenomena, such as modeling and prediction of the movement of fluids or the variation of pressure and temperature in the subsurface, as well as evaluating stresses and strains and groundwater conditions around mines, tunnels, and underground structures, use 4-D models consisting of three spatial dimensions plus time. Ongoing research is investigating additional dimensions such as model uncertainty (Chapter 15) and model resolution or scale (Van Oosterom and Stoter 2010). However, current practice is restricted to evaluating time-varying phenomena with 4-D models.

To date, predictive groundwater flow and contaminant transport models are the most widely applied process models that utilize 3-D geological models (Medina et al. 2017). Recently, groundwater transport modeling techniques have been adapted to assess geothermal heating and cooling capabilities (Chapter 20; Banks 2012; Fairs et al. 2015; Hecht-Méndez et al. 2010). Also the geotechnical and mining sectors are beginning to use 3-D geological models as information sources for geomechanical stability assessments of rock and earth materials surrounding underground and near-surface excavations (Barla 2016, 2017), to analyze the stability of natural and man-made slopes (Rengers et al. 2002; Stead et al. 2006), and for geohazard investigations (Rees et al. 2009; Chapter 22).

These applications use finite-difference, finite-element, or discrete-element techniques to numerically analyze time-varying flow and transport conditions or conduct kinematic analyses. Discretization of the 3-D geological model forms an appropriate grid or mesh for these numerical analyses and allows for stochastic simulations (Chapter 13); these may be used to assess uncertain property distributions in the subsurface. Sensitivity assessments of model predictions are based on repeated analyses using variable input values (D'Agnese et al. 1997, 1999; Banta et al. 2006; Doherty 2015; Doherty et al. 2018; Neuman and Wierenga 2002).

Key issues common to these modeling applications are how much complexity should be represented in process models, and how best to use diverse types of information to enhance models. Sections 14.2–14.6 of this chapter discuss the role of 3-D geological models for groundwater flow and transport predictions. Section 14.7 discusses the role of 3-D geological models when evaluating the stability of

Applied Multidimensional Geological Modeling: Informing Sustainable Human Interactions with the Shallow Subsurface, First Edition.
Edited by Alan Keith Turner, Holger Kessler, and Michiel J. van der Meulen.
© 2021 John Wiley & Sons Ltd. Published 2021 by John Wiley & Sons Ltd.

underground openings. Chapter 22 provides some case studies describing the application of 3-D geological models to geohazard investigations.

14.2 Importance of Subsurface Flow and Transport

Geological models provide critical information on how geological materials and structures may affect the storage and movement of water through the subsurface. This, in turn, has implications for the transport of dissolved substances and (increasingly) geothermal energy resources. Interactions between hydrogeological and surface hydrological conditions are important; baseflow from groundwater maintains flows in many rivers and springs, provides accessible water resources during dry seasons and important ecological habitats in rivers and wetlands. It also influences some land-climate exchange processes (Maxwell and Kollet 2008; Taylor et al. 2013).

Historically, major population centers were located on rivers, which provided the most easily accessible water resource. Although less readily accessible, groundwater resources provide long-term high-volume storage which may supplement inadequate surface water sources, especially in dry periods. Groundwater flow regimes follow hydraulic gradients which are generally topographically-controlled in natural systems (Toth 2016). In many locations, humans have modified these flow regimes over time and space by installing groundwater extraction systems, ranging from individual wells to large well fields, to satisfy agricultural, industrial, and domestic water demands.

Groundwater flows at significantly lower rates than water on the surface; a water molecule entering the ground in a recharge area may take decades or longer to emerge at a discharge zone, but water molecules in a river may reach the sea in a period of days to months at most. Variations of the properties of typical shallow subsurface materials provide for large differences in fluid storage capacities, flow rates, and hence residence times. Discontinuities, such as fissures or faults or highly permeable layers, provide pathways for much more rapid groundwater flow. These mechanisms have important implications when modeling flow and contaminant-transport processes. In a contaminated groundwater system, when conditions provide for rapid flow, some of the pollutants may quickly reach sensitive receptors, such as rivers, springs, or water abstraction boreholes. In contrast, subsurface conditions promoting very slow fluid movements result in contaminants residing in pore spaces for very long periods of time.

Groundwater flow influences the character and severity of several geohazards. A specific type of flood hazard, termed groundwater flooding, occurs when groundwater levels rise above the ground surface. This tends to occur after long periods of sustained high rainfall in low-lying areas underlain by a major unconfined aquifer, including alluvial deposits along major river valleys (Figure 14.1). In recent years, significant property damage has resulted from long-lasting and regionally extensive groundwater flooding in areas of southern England underlain by chalk aquifers. Groundwater models can provide some early warning of this hazard (Upton and Jackson 2011; McKenzie and Ward 2015).

1) Rising Water Table in an Unconfined Major Aquifer

(a) Low Water Table   (b) High Water Table   (c) Very High Water Table

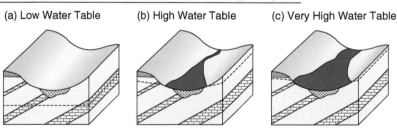

Figure 14.1 Causes of groundwater flooding in two principal settings. Source: Modified from BGS website www.bgs.ac.uk/research/groundwater/flooding.

2) Rising Water Table in a Shallow Alluvial Aquifer

(d) Low River Stage   (e) High River Stage

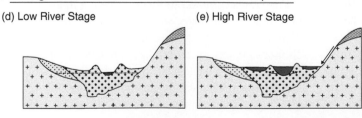

Rising pore-water pressures reduce the stability of hillslopes, especially at rock-mass discontinuities or other zones of weakness. Thus, geotechnical analyses of slope-stability require similar information on the geometry of geological structures as hydrogeological analyses, and each discipline depends on a good understanding of different physical rock properties to build appropriate process models.

14.3 Numerical Flow and Transport Modeling

The development of numerical simulation models has tended to emphasize either groundwater assessments within the field of hydrogeology or surface water assess ments within the field of hydrology (Figure 14.2). Despite groundwater and surface water being closely interconnected as parts of the hydrological cycle, this division persists with specialist journals and societies focusing on either hydrology or hydrogeology. Hydrogeology is often viewed as a specialist branch of geology, but it can also be considered a specialization within hydrology.

14.3.1 Hydrology Modeling

Hydrologists use surface water models to assess river flows in catchments, usually relying on relatively simple models of soil and groundwater processes (Parkin et al. 1996). The 2-D, or 2.5-D capabilities of typical Geographic Information System (GIS) products provide adequate spatial definition of terrain conditions and distributions of surficial materials, vegetation, and land use that are important influences on surface water supplies. GIS applications such as Arc Hydro (Maidment and Morehouse 2002) support data management for specific hydrology systems, including the

locations and properties of rivers, lakes, springs, and wells. Several widely-adopted numerical simulation models provide analysis of surface hydrology processes, including precipitation, recharge, overland flow, and channel flow (Beven et al. 1987, 1995; Ewen et al. 2000; Maxwell et al. 2016).

14.3.2 Hydrogeology Modeling

Hydrogeologists use groundwater models to analyze conditions within the saturated groundwater zone, but depend on recharge models to represent evaporation, infiltration, and runoff processes, and use surface water bodies as boundary conditions. Groundwater modeling strongly depends on 3-D geological models (Medina et al. 2017). Arc Hydro Groundwater (Strassberg et al. 2011) provides a platform for groundwater model development within a GIS environment that is suitable for layered, relatively undeformed, geological environments. Evaluation of more complex subsurface geological conditions requires true 3-D modeling, which is beyond the capability of GIS software (Turner 2006). In general, the geological framework must be transformed to define relevant hydrogeological parameters that constrain numerical flow and transport simulations. This process involves iterative progressive modifications to both the geological and the hydrogeological models (Sections 14.5.1 and 14.5.2).

14.3.3 Integrated Surface–Subsurface Modeling

Figures 14.2 and 14.3 also show that there are integrated catchment models which combine the surface water and groundwater modeling approaches. The subject of considerable research interest in recent years, these models involve the expertise of both hydrologists and hydrogeologists to provide a fully-coupled representation of both

Figure 14.2 Surface and groundwater process modeling hierarchy.

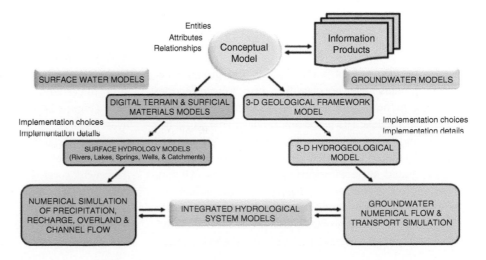

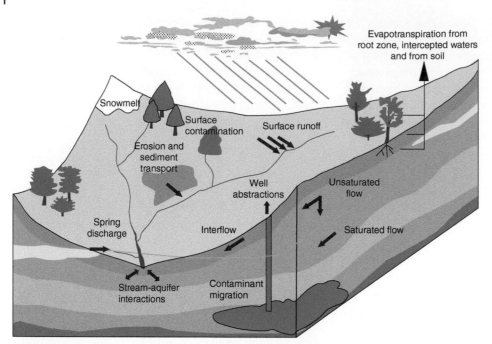

Figure 14.3 Conceptual model of hydrological processes, as represented in the SHETRAN model. Source: Newcastle University (2018).

groundwater and surface water processes and their inter-actions. Advances in computing technologies now support delivery of computing services by an online network of remote servers to store, manage, and process data. This has made it feasible to develop such complex models and use them to address large-scale problems. Two general approaches are illustrated in this chapter, each with particular advantages and disadvantages: fully-integrated modeling software with internal coupling between process components, such as MIKE SHE (DHI 2011); SHETRAN (Ewen et al. 2000; Newcastle University 2018; Figure 14.3); HydroGeoSphere (Brunner and Simmons 2012); and externally-coupled systems where different modeling software is linked through data exchanges (Mansour et al. 2013; Maxwell et al. 2016; Peach et al. 2017).

14.4 Model Classification

The broad spectrum of surface water and groundwater process models may be classified into four categories according to how they represent physical processes; (i) conceptual models; (ii) black-box models; (iii) white-box models; and (iv) gray-box models. These models are used for surface-water hydrology, groundwater hydrogeology, and integrated surface-subsurface applications. All four categories are useful; each category has unique features. Their effectiveness depends on the objectives of the study, the complexity of the problem, and the accuracy desired. These categories represent different, complementary,

levels of approximation of reality and complexity of data requirements.

Within these physical process-representation categories, individual models may be further subdivided into linear or non-linear models, time-variant or time-invariant models, lumped or distributed models, and deterministic or stochastic models (Pechlivanidis et al. 2011). However, all combinations of categories and sub-types of model are feasible or widely used.

14.4.1 Conceptual Models

Model developments for both surface and groundwater applications begin with the establishment of a conceptual model (Figure 14.2), which defines the critical entities, connections, and attributes that must be included in the simulation. The conceptual model evolves from field observations and fundamental scientific concepts; it also defines required additional data collection efforts. Hydrology evaluations are guided by the concepts of the atmospheric and surface components of the hydrological cycle (Figure 14.3). Groundwater modeling begins with a hydrogeologic conceptual model. California Department of Water Resources (2016) defines a hydrogeologic conceptual model as "*describing the geologic and hydrologic framework governing the occurrence of groundwater and its flow through and across the boundaries of a basin and the general groundwater conditions in a basin or sub-basin.*" Figure 14.4 shows an example of a conceptual hydrogeological model that was the basis for numerical groundwater

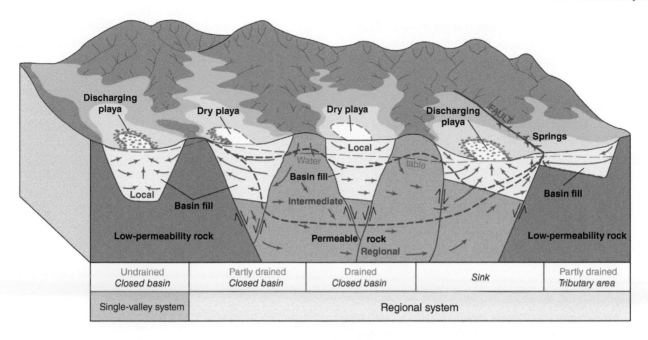

EXPLANATION

Phreatophytes	→ Groundwater flow
- - - - Approximate location of local, intermediate, and regional systems	⇌ Faults

Figure 14.4 Hydrogeological conceptual model of the Basin and Range region illustrating the structural relations between mountain blocks, valleys, and groundwater flow. (Faunt et al. 2010).

modeling in the Basin and Range desert environment of southern Nevada (Belcher and Sweetkind 2010).

14.4.2 Black-box Models

Black-box models employ empirical methods to assess both surface water and groundwater environments. They assess system inputs and outputs using mathematical relationships derived solely from measurements, with parameters that may have little direct physical significance and can be estimated only by using concurrent measurements of input and output. In many situations, empirical methods can yield accurate answers and, therefore, can be useful in decision-making. Neural network models may be considered as a type of black-box model; they can represent generalized non-parametric functions between variables, or mimic results from more complex models (Parkin et al. 2007; Rosenbaum 2001).

14.4.3 White-box Models

White-box models provide a higher level of realism because they resolve physical catchment processes and/or subsurface groundwater flow and transport processes using

differential equations based on conservation of mass, energy, and momentum. Their logical structure is similar to the real-world system, providing potential for better simulations when situations change. Examples include distributed watershed runoff models, infiltration models based on two-phase flow theory of porous media, evaporation models based on theories of turbulence and diffusion, and groundwater flow and transport models. An early example of a physically-based model is the SHE model (Abbott et al. 1986a,b), from which the SHETRAN model was further developed (Ewen et al. 2000).

14.4.4 Gray-box Models

Gray-box models (sometimes confusingly referred to as conceptual models) use a set of equations, together with logical statements, to define relationships between variables and parameters, generally with highly simplified definitions of physical laws. The observed output variable, such as discharge from a basin for a surface-water analysis, is derived by a specified function or set of functions containing measured or estimated parameters; a residual value defines the lack of fit between observed and derived output values. These models differ in the assumptions made about the selected functions and the residual values. Many models belong to this class; some evaluate processes within river catchments in simplified ways, such as lumped models, which use spatial averaging, connected reservoirs, and mathematical functions to relate storage and flux terms. Analytical element groundwater models make simplifying assumptions, such as homogeneity and uniform thickness, to spatially represent groundwater levels and flows without using detailed geological information. Gray-box models are intermediate between white-box and black-box models.

14.4.5 Applications of Models

Some of the most widely used hydrology models include the lumped gray-box models HBV (Bergström 1976, 1995) and the Stanford Watershed Model (Crawford and Linsley 1966), and distributed gray-box models such as TOPMODEL (Beven et al. 1995). The distributed gray-box category includes analytical-element models which spatially represent groundwater levels and flows with simplifying assumptions of homogeneity and layer geometry (Haitjema 1995; Strack 1989).

Widely used distributed white box models include MIKE SHE (DHI 2011), SHETRAN (Ewen et al. 2000; Newcastle University 2018; Fig. 14.3), IHDM (Beven et al. 1987), and Thales (Grayson et al. 1992). In this category, two major types can be distinguished; those requiring simulations of flow only, and those used for contaminant problems which require simulation of transport processes. For either type, a critical question is the detail and quality of geological information required to support development of the numerical models. Groundwater flow evaluations may use finite-difference models such as MODFLOW 6 (Langevin et al. 2017) or finite-element models such as ZOOMQ3D (Jackson and Spink 2004). Widely used groundwater contaminant transport models include FEFLOW (DHI-WASY 2012) and MT3D (Bedekar et al. 2016). However, integrated models such as MIKE SHE, SHETRAN, and HydroGeoSphere (Brunner and Simmons 2012) include both surface and subsurface contaminant transport models.

Hydrogeological problems resolved by flow models include large-scale water resource management, engineering construction, mine dewatering, and groundwater flooding. Flow models may be built and applied at different scales. National or regional models are used to evaluate the water resources of larger catchments or aquifers (Shepley et al. 2012); local models address specific problems with greater detail. Transport models mostly provide localized and site-specific assessment of contaminant issues (Chapter 22). Regional groundwater resource models are increasingly being used to support more localized assessments, such as studies of abstraction protection zones or wetland environmental studies.

Management of geothermal resources as a renewable energy source is a rapidly developing application of groundwater modeling. Particularly for low enthalpy environments, such as those of ground source heating and cooling systems, the equations governing groundwater transport are mathematically comparable to those of heat flow, if appropriate parameter values are used. Thus, existing numerical simulation codes that solve advection-dispersion equations can be used, within certain limitations, to represent heat transport (Banks 2012; Hecht-Méndez et al. 2010). Chapter 20 and Fairs et al. (2015) provide examples of applications for deeper high-enthalpy geothermal sources.

Popular integrated, or coupled, surface–subsurface models combine distributed gray-box and white-box models, with the choice of model components dependent on the project objectives and resources. Examples include GSFLOW (Markstrom et al. 2008) and ParFlow (Maxwell et al. 2016).

14.5 Building Hydrogeological Models Based on Geological Models

Most groundwater models are built by following a multi-step process that involves progressive refinement of a conceptual understanding of geological and hydrogeological contexts within an iterative sequence of activities. The principal stages are:

1. *Data Acquisition*: Assembling available observational data on all relevant parameters, developing a conceptual model, and undertaking additional field reconnaissance and data collection (Chapter 7).
2. *Data Management*: Development of an appropriate database to allow spatial and topical querying of the collected data, to archive the data for future use, to support statistical and quantitative assessment of the data, and to provide visualization of the data (Chapter 8).
3. *Geological Model Development*: Construction of geological models can be achieved by several methods (Chapters 9–12). The selection of the most appropriate method depends on several factors including the intended application of the model, complexity and characteristics of the geological environment being modeled, the experience of the modelers, and the institutional support provided to the model building effort.
4. *Hydrogeological Model Development*: The geological model must be translated to represent critical hydrogeologic characteristics by assigning hydraulic properties

to lithological strata or geological features such as faults and boundaries, then conversion into a numerical grid or mesh at a scale of resolution appropriate to the adequate representation of geological features and computational limitations (Chapter 13).

5. *Numerical Simulation*: Groundwater flow or contaminant transport simulations are undertaken using finite-difference or finite-element methods; the model is "calibrated" by adjustment of the hydraulic properties until some aspects of the simulated response match available observations.

6. *Model Assessment and Validation*: Evaluation of model results by comparison to geological understanding and actual observations is a key step to achieving consistent hydrogeological models. These comparisons provide information about the model capabilities, and may suggest desirable refinement in the conceptual, geological, or hydrogeological models – a "feedback" or iterative refinement. Models that pass this evaluation are considered "validated" and fit-for-purpose.

7. *Model Application*: The completed and validated model can be used to predict future scenarios and for decision-making.

14.5.1 An Early Example of Integrated Hydrogeological Modeling

Beginning in the early 1990s, a team of U.S. Geological Survey (USGS) scientists created a pioneering regional three-dimensional hydrostratigraphic model of the Death Valley regional aquifer system in southern Nevada and south-eastern California (D'Agnese et al. 1997). Developed to support the design of a proposed nuclear waste repository at Yucca Mountain, this model combined geological and hydrogeological evaluations to constrain and guide a series of increasingly sophisticated groundwater simulations. These included forward and inverse modeling techniques and evaluations of the numerical model sensitivity to assumptions or uncertainties in various controlling parameters (D'Agnese et al. 1997).

14.5.1.1 Data Acquisition and Management
The project assessed two major components of the Death Valley regional groundwater flow system – the 3-D hydrogeologic framework that established the internal controls of the regional flow system, and the surface and subsurface hydrologic conditions that defined the connections to the larger regional and global hydrologic systems and cycles. Data collected to evaluate these two components were organized into 10 principal data categories and a centralized database was used to organize and assimilate the relevant information (Figure 14.5).

While geologic maps and other spatial data sources were analyzed within a GIS, other data, such as lithologic logs and climatic data, were acquired and stored as tabular entries in the centralized database. Thus, while many of the data categories shown in Figure 14.5 involved individual databases, the centralized database contained the data required to operate the numerical groundwater flow model, and the results produced by that model.

14.5.1.2 The Model Development Framework
The hydrogeological modeling process was both integrated and iterative with numerous information flows and feedback loops. It was based on an earlier hydrogeological information flow and management concept proposed by Turner (1989) that included four interrelated fundamental modules (Figure 14.6). The original concept was adapted slightly to meet the needs of the Death Valley project. Data management was considered to be the central integrating function during the entire modeling process, and therefore the data management module is in a central position in Figure 14.6, interacting with the other modules. The integrated groundwater modeling process was considered to have two main components: the subsurface characterization module includes the building of the geological model and its subsequent transformation into a hydrogeological model. This iterative process is linked in an information and visualization loop with the data management module. Several iterations may be required to define the most probable subsurface conditions; if a unique definition is not found, then multiple alternative characterizations may be used. The data management module now held all the necessary hydrogeologic parameters for numerical groundwater simulation, and a second cycle of modeling process begins. This is shown on the right of Figure 14.6; the data management module provides input parameters required by the groundwater flow module and accepts model results for visualization and statistical analysis. The fourth module uses statistical screening techniques to evaluate the usefulness or reasonableness of the subsurface characterizations and validate the numerical model results.

14.5.2 Evolved Integrated Hydrogeological Modeling

The Death Valley integrated modeling effort was able to assess the sensitivity of the numerical model predictions to variations or uncertainties in the model input parameters. Because the centralized database contained individual parameter estimates and the system readily provided 3-D

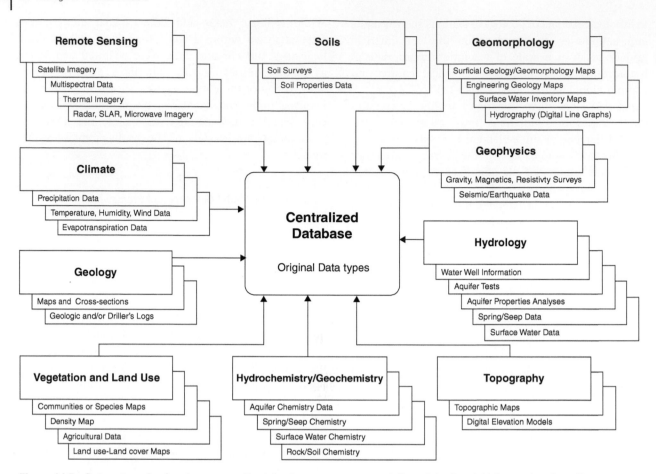

Figure 14.5 Data categories forming a centralized database to support modeling of the Death Valley regional aquifer system. (D'Agnese et al. 1997).

visualizations, it formed the basis for testing developments in parameter estimation (D'Agnese et al. 1999). The entire subsurface characterization was subsequently re-evaluated on several occasions over the following decade to reflect new data or different interpretations; these were analyzed by additional numerical modeling (Belcher and Sweetkind 2010).

The groundwater assessment techniques developed by the Death Valley project were continued with a team of USGS experts located throughout the western United States. This introduced new scientific challenges; creating robust and validated models, meeting demanding delivery schedules, avoiding inefficiencies while maintaining coordination across the multiple locations, and providing scientific documentation of all procedures. To address these challenges, structured information management approaches developed at Los Alamos National Laboratory (Nasser et al. 2003) were adapted to produce a new, more comprehensive, integrated geological data modeling, and information management system (D'Agnese and O'Brien 2003, which has five basic components (Figure 14.7).

The central technical core contains the technical tools supporting geological data management, visualization, and modeling, including both framework model development and discretization as well as numerical, predictive simulation modeling. While this technical core retains the functionality defined in the original Death Valley model (Figure 14.6), it also shows an important change. The sequence of its three components (Data Management, Framework Model Development, and Process and Predictive Model Management) has changed, with framework modeling becoming central to the process. This arrangement is followed by other reported systems (Medina et al. 2017). Framework modeling includes the creation of both geological and hydrogeological models; the next section discusses the synthesis of these two models.

The technical core is surrounded by four supporting components. The infrastructure component is required to keep the system operating smoothly. It maintains system and data security and other system operational environments. The workflow management component provides tools to

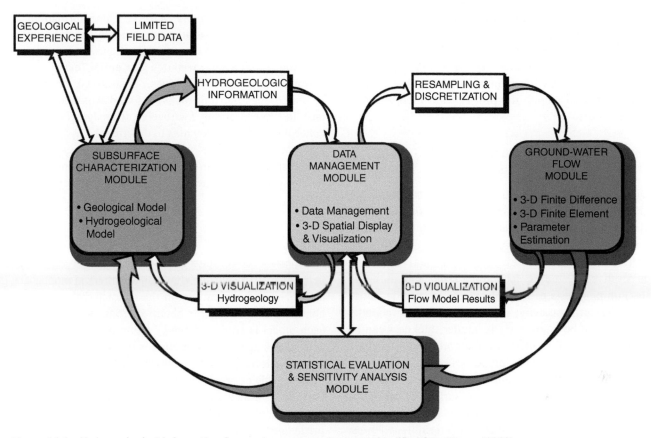

Figure 14.6 Hydrogeological information flow and management concept. Modified from Turner (1989).

Figure 14.7 Information management for integrated geological, hydrogeological, and numerical groundwater data and modeling. Modified from D'Agnese and O'Brien (2003).

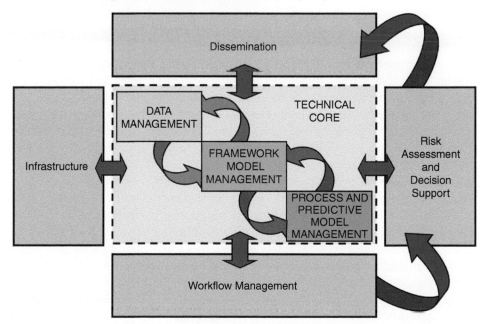

access the technical core, including user interfaces, and also links to the risk assessment and decision support component that contains software tools for risk assessment and decision making. The dissemination component manages the production and distribution of results generated by the system.

Figure 14.7 represents a high level of abstraction. Operational systems require careful system design, documentation of all components and processes, and adherence to appropriate standards. The progression from data, to information, to prediction and decision-making can occur only if all the component steps are linked through a standard data protocol; the USGS project team utilized *Extraction-Translation-Loading* (ETL) concepts to handle data transfers between processing steps (D'Agnese and O'Brien 2003). This allowed the use of data stored in older legacy systems while maintaining internal databases and coordination among the project team members and external clients.

14.5.3 Synthesizing Geological and Hydrogeological Models

Regional hydrogeological characterization studies begin with establishing the role of the geological framework in defining mappable flow units and intervening barriers to flow (aquitards). The hydraulic properties of lithofacies within each genetic unit are defined at various scales using data from core studies, laboratory tests, and pumping tests. Pumping tests can be interpreted to provide values for aquifer transmissivity; values for hydraulic conductivity can then be inferred if the aquifer thickness can be estimated. Laboratory methods and core studies provide localized values of hydraulic properties.

Many hydrogeological studies encounter layered geological units because shallow near-surface deposits are almost universally composed of layered sequences, sometimes with complex geometries and intermittent lenses, while sedimentary rocks provide many large regional ground-water basins. Geologists map these layered sequences using lithostratigraphic principles. Hydrogeologists have long noted that groundwater flow does not necessarily conform to the boundaries of recognized stratigraphic units. Two terms, *aquifer* and *hydrostratigraphic unit*, are commonly employed to subdivide the subsurface into units more relevant to groundwater hydrology. The term *aquifer* is commonly defined for water supply usage in economic terms. In many areas, an aquifer is defined by local laws and regulations, which makes it difficult to use as a technical term. Maxey (1964) introduced the concept of hydrostratigraphic units, which he defined as "*bodies of rock with considerable lateral extent that compose a geologic framework for a reasonably distinct hydrologic system.*" Seaber (1988) redefined the term as "*a body of rock distinguished and characterized by its porosity and permeability*" which he considered more consistent with established stratigraphic nomenclature and the fact that a "*hydrostratigraphic unit may occur in one or more lithostratigraphic units.*"

The synthesis of geological information for locations composed of layered sequences involves establishing an equivalence between the geological lithostratigraphic information and the desired hydrostratigraphic framework.

Figure 14.8 is an example of such a transformation; it summarizes the lithostratigraphy and hydrostratigraphy of the Cretaceous rocks forming the Edwards Aquifer, an important groundwater resource for Central Texas.

However, in some locations, unstratified crystalline igneous or metamorphic rocks are found at or near the surface. In these locations, hydrogeologic analyses utilize the concept of *hydraulic domains* to describe the hydrogeological characteristics in a form suitable for groundwater flow or transport modeling (Figure 14.9). These domains are defined primarily by the fracture patterns of the crystalline bedrock.

14.5.4 Calibration of Groundwater Models

To represent the complex groundwater systems commonly being analyzed, multiple alternative numerical models are developed and, depending on the characteristics of available data and the results obtained from the simulations, each model is either retained or eliminated from further consideration. *Model calibration* is the process of identifying and selecting sets of parameters which allow a model to be capable of providing acceptably accurate predictions to support anticipated decision-making. Demonstration that a model is capable of providing acceptably accurate predictions has been referred to as *model validation*, but validation also implies some authentication of truth or accuracy of the model. Popper (1959) argued that, since models are representations of scientific hypotheses, they cannot be proven true (or validated), only tested and invalidated. Konikow and Bredehoeft (1992) endorsed this view and recommended that the term *model validation* be avoided, although the term has been widely used for performance assessment of groundwater models developed to support decision making or regulatory processes for high-level nuclear waste repositories (Neuman and Wierenga 2002). Konikow and Bredehoeft (1992) recommended the use of alternate terms, including *model testing*, *model evaluation*, *model calibration*, *sensitivity testing*, *benchmarking*, *history matching*, and *parameter estimation*. In an engineering context, when decisions need to be made based on the best available, albeit imperfect, information, the aim of the process is to demonstrate that a model is *fit for purpose*, with clear recognition of its limitations.

Calibration of groundwater numerical simulations is accomplished by comparing observations of ground-water levels, or solute concentrations for transport models, to corresponding calculated model values. The parameter values are varied within reasonable ranges until the differences between observed and computed values are minimized. This minimization can be attempted

LITHOSTRATIGRAPHIC UNITS   HYDROSTRATIGRAPHIC UNITS

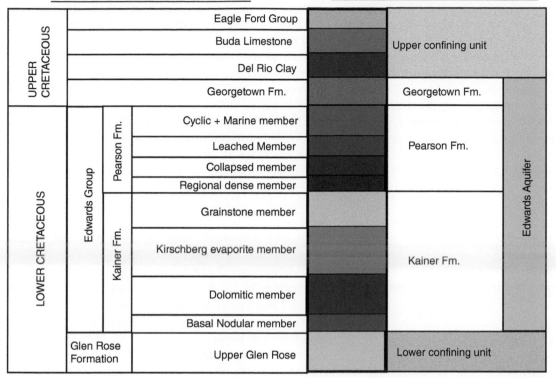

			Lithostratigraphic		Hydrostratigraphic	
UPPER CRETACEOUS			Eagle Ford Group		Upper confining unit	
			Buda Limestone			
			Del Rio Clay			
			Georgetown Fm.		Georgetown Fm.	Edwards Aquifer
LOWER CRETACEOUS	Edwards Group	Pearson Fm.	Cyclic + Marine member		Pearson Fm.	
			Leached Member			
			Collapsed member			
			Regional dense member			
		Kainer Fm.	Grainstone member		Kainer Fm.	
			Kirschberg evaporite member			
			Dolomitic member			
			Basal Nodular member			
	Glen Rose Formation		Upper Glen Rose		Lower confining unit	

Figure 14.8 Lithostratigraphy and hydrostratigraphy of the Cretaceous rocks of Central Texas. Based on descriptions by Clark et al. (2016).

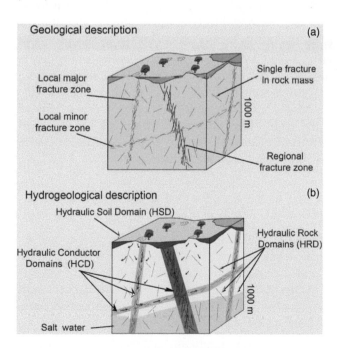

Figure 14.9 Schematic diagrams showing: (a) the geological characteristics of the crystalline bedrock and surficial deposits found at the Forsmark, Sweden, proposed spent nuclear fuel repository, and (b) the equivalent hydrogeological description of the subsurface as three hydraulic domains used to model groundwater flow and transport. Modified from Rhén et al. (2003).

through trial-and-error adjustments or by using automated *inverse modeling* or *parameter estimation* procedures (Poeter and Hill 1996, 1997). The model is considered calibrated when it reproduces some set of historical data within some subjectively acceptable level of coherence. Petroleum engineers term this approach *history matching*. Although quantitative measures are often calculated and used, there is no generally accepted quantitative standard, and acceptance is often partly based on qualitative criteria.

Sometimes, a calibrated model undergoes an additional *verification phase*; the model is used to reproduce another part of the historical record, one that was not used for the calibration. However, this does not assure the model will predict future situations, which may have very different conditions. This process is commonly used in watershed modeling but Konikow and Bredehoeft (1992) suggest it is less powerful in evaluating groundwater systems due to the very long response times for subsurface flow. *Blind validation* is an alternative approach where no calibration is carried out but reproduces more closely the way in which models are used to make predictions of future conditions (Ewen and Parkin 1996).

Calibration does not define a unique set of parameters. A poor match only suggests that errors may exist in any or all of the conceptual model, the numerical solution, or because of a poor set of parameter values. Ewen et al. (2006) classify errors in hydrologic models into model structure error; parameter error; and run-time error. The source of a poor match is not easily identifiable. A good match does not prove the validity of the model; because the solution is non-unique, the model can include compensating errors.

Poeter and Hill (1996, 1997), Hill (1998), and Hill and Tiedeman (2007) provide details of the application of automated *inverse modeling*, or *parameter estimation*, methods to calibrate groundwater models. Alternative calibration procedures based on advanced stochastic methods are discussed by Valstar et al. (2004), Carrera et al. (2005), and Hendricks Franssen and Gomez-Hernandez (2002). Neuman and Wierenga (2002) and Refsgaard et al. (2007) discuss model calibration within broader regulatory and environmental decision-making frameworks.

14.6 Alternative Approaches to Model Calibration

The model calibration process is influenced by the availability of information to define the model parameters and the potential application of the model. These influences can be clearly demonstrated by examples from the Netherlands and the United Kingdom.

Groundwater management is important in the Netherlands. The REGIS II model (Chapter 19) provides hydrogeological information for the entire country. REGIS II provides a hydrogeological framework that is entirely consistent with the lithostratigraphic classification of the national Digital Geological Model (DGM) framework (Gunnink et al. 2013). Both models are based on information provided by 16 500 boreholes; the DGM uses them to define 34 lithostratigraphic units as a stacked series of gridded surfaces, while REGIS II reclassifies each lithostratigraphic interval in each borehole into one of seven hydrologically relevant lithological classes (peat, clay, sandy-clay, fine sand, medium sand, coarse sand, and gravel), which are then interpreted hydrostratigraphically. This process subdivides the 34 DGM lithostratigraphic units into 133 hydrostratigraphic entities. REGIS II provides a spatial model by mapping the depth of the top and base, and thickness, of each of the 133 units on 100 m resolution grids. The REGIS II parameter database has additional grids to provide information on horizontal and vertical conductivity, standard deviations for these, and related properties, such as transmissivity and hydraulic resistance.

With this extensive and uniform national coverage of hydrogeological data, groundwater model calibration in the Netherlands is directed toward ensuring specific

groundwater flow models developed from these data have acceptable accuracy to support their intended use. Thus, as discussed in Sections 14.6.1 and 14.6.2, inverse modeling and parameter estimation procedures have been adapted to address the uncertainty of transmissivity estimates, a fundamental groundwater flow parameter.

In contrast, the United Kingdom has excellent geological models across different scales, but less well-developed hydrogeological parameter data. The United Kingdom has recently released a comprehensive national geological model (Waters et al. 2016) and has several higher-resolution regional geological models. The lithostratigraphic frameworks of these geological models conform to national geological classifications, so the subsurface geology is well represented. The national model was developed, in part, to support nationwide screening for a proposed nuclear waste repository (Waters et al. 2016). Many of the regional models were developed to support groundwater resource evaluations by the Environment Agency (Shepley et al. 2012). These evaluations involve integrated assessments of surface–subsurface hydrologic interactions. For example, south-eastern England is underlain by chalk aquifers and so is subject to groundwater flooding hazards.

Consequently, as discussed in Sections 14.6.4 through 14.6.6, integrated surface-subsurface modeling at the catchment scale is an important modeling activity at local, regional, and national scales (Ewen et al. 2000; Mansour et al. 2013; Lewis 2016). Section 14.6.3 discusses the calibration of such integrated catchment models, especially when they are linked with 3-D geological models to improve subsurface characterization. For United Kingdom models, calibration efforts have been influenced by the availability of relatively complex process-driven integrated surface-subsurface catchment models and 3-D geological subsurface models, but only a partially developed detailed hydrogeological parameter database.

14.6.1 Calibration Modified to Evaluate Uncertainty of Transmissivity

Figure 14.10 schematically outlines the typical groundwater modeling sequence used in the Netherlands. The geological and hydrogeological modeling steps on the left side of Figure 14.10 are undertaken within the REGIS II model (Chapter 19). REGIS II follows a four-step process to determine the model-scale values of transmissivity:

(1) Representative minimum and maximum values for each combination of geological unit and lithological class are determined from a combination of experience and published values.

(2) Measured hydraulic conductivities of small samples extracted from boreholes, supplemented by values

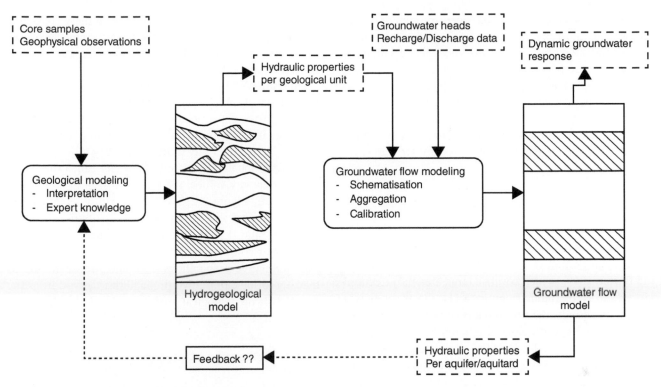

Figure 14.10 Schematic steps in groundwater modeling in the Netherlands.

derived from pump-tests and empirically derived relationships with grain-size distributions, are used to assign representative conductivity values to each lithological class within each borehole.

(3) Conductivity values are *upscaled* to produce model-scale transmissivities based on the thickness of each borehole interval and the conductivity values. A probability density function is constructed assuming a log-normal distribution with limits defined by the minimum and maximum conductivities. Representative values from this probability density function are stochastically assigned to each borehole interval and an average and standard deviation of hydraulic conductivity for each aquifer or aquitard at each borehole location is determined.

(4) A kriging interpolation uses these values to assign spatially-varying hydraulic conductivities across entire aquifers or aquitards.

As shown by the right side of Figure 14.10, groundwater flow models are developed by combining the hydrogeological property parameters provided by REGIS II with observed groundwater levels and recharge/discharge data. The groundwater modeler defines a calculation grid with a limited number of aquifers and aquitards. Apart from geological information, schematization of the groundwater flow model (grid spacing and number of model layers) depends on the hydrological conditions such as the presence of the surface water, drainage systems, and groundwater abstractions.

The observed groundwater levels reflect the hydraulic properties of the subsurface, especially the conductivity or transmissivity. A calibrated numerical flow model is achieved when the model values successfully match observed groundwater levels and recharge/discharge values. Lourens and van Geer (2016) propose the addition of a feedback loop from the calibrated groundwater flow model to the hydrogeological model (Figure 14.10). This feedback uses the prior transmissivity parameter probability density functions from the REGIS II model, updates them based on hydrological observations to produce posterior probability density functions, and then uses the differences between prior and posterior parameter estimates to suggest updates to the geological model.

The posterior parameter estimates may differ substantially from the prior estimates. According to Lourens and van Geer (2016), this feedback process allows the calibration to assess the uncertainty surrounding transmissivity estimates, and thus to determine the most likely combination of the conductivity and thickness of individual REGIS II units. They employed a two-step procedure. The first step used the REGIS II data to provide prior transmissivity probability density functions for individual REGIS II units, and then used these probability density functions as weighting mechanisms to find the posterior probability density function that defined the most likely transmissivities for each unit. The second step then found the most likely combination of conductivity and thickness for each REGIS II unit. The approach is closely related to certain optimization techniques, which have often been used to provide information on the relationship between observations and parameters through correlation matrices, but these have not yet been used routinely to reconsider geological structure.

14.6.2 AZURE Regional Groundwater Resource Model, the Netherlands

Assessment of the properties for an aquitard defined in the regional groundwater flow model AZURE (De Lange and Borren 2014), which covers a large area in the central part of the Netherlands (Figure 14.11), illustrates how a calibrated groundwater flow model can be used to improve the underlying hydrogeological schematization. AZURE's schematization, based on REGIS II, distinguishes nine aquifers separated by eight aquitards. Calibration results for aquitard number 4 were used as an example of how REGIS II can be adjusted using the feedback analysis of uncertainty described in Section 14.6.1. In the entire model area, aquitard 4 is an aggregation of several REGIS II units. In order to have a clear-cut feedback example, without deaggregation issues, two areas were selected (I and II in Figures 14.11 and 14.12) in which the aquitard

corresponds to one REGIS II unit only. In these areas, this unit (EE-kz) consists mainly of clay and sandy clay of the Eem Formation, which was deposited during the previous interglacial period.

Calibration of the AZURE groundwater model resulted in an adjusted value of the resistance of Aquitard 4 in each of its 250 m × 250 m grid cells. Following the procedure described in Lourens et al. (2015), the most likely value for the vertical conductivity of the sandy clay unit EE-kz was determined for each grid cell. These are called the posterior conductivity values, the spatial distribution of which is shown in Figure 14.13. There is a distinct difference between Areas I and II. The spatial distribution function for areas I and II were calculated separately. Figure 14.14 shows their cumulative distribution functions. The posterior cumulative probabilities have narrower distributions than the prior cumulative probabilities derived from the REGIS II hydrogeological model (the black dots in Figure 14.14).

14.6.3 Calibration of Integrated Catchment Models

The Dutch approach to calibration requires both adequate geological conceptualization and relatively detailed parameter data, and these are available at the same spatial resolution. Under these circumstances, assessment of groundwater levels and flows can lead to refinement of hydraulic properties. In the United Kingdom, there are

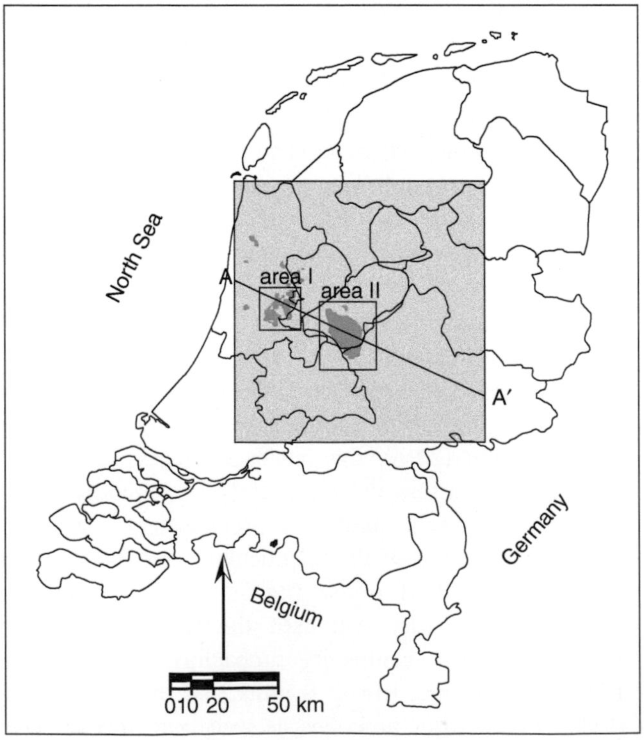

Figure 14.11 Location of the groundwater flow model Azure (gray rectangle) in the Netherlands. The yellow-color within Areas I and II defines the extent of the sandy clay unit EE-kz. The line A-A′ denotes the location of the cross-section of Figure 14.12.

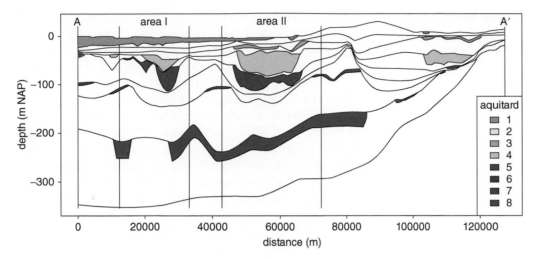

Figure 14.12 Cross-section of the groundwater flow model Azure, with all aquitards color coded. The example calculations are performed on sandy clay unit EE-kz within Aquitard 4

Figure 14.13 The most likely vertical hydraulic conductivity values of sandy clay unit EE-kz. Only conductivity values with an accompanying most likely layer thickness of more than 0.5 cm are shown.

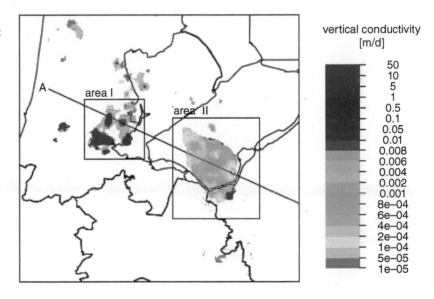

good geological framework conceptualizations, although not at a consistent level of detail across the country, and much less detailed parameter data. An alternative strategy to calibration is therefore required; one that is based on the recognition that there is inherent uncertainty in the geological structure as well as in the hydrological parameterization.

In mathematical terms, the numerical modeling of surface and subsurface hydrology conditions represents an ill-posed problem, in the sense that there are multiple possible solutions that can provide outputs, such as groundwater levels and river flows, that match observations within a certain tolerance, given available response data and hydrogeological information. This general issue has been explored within both the hydrological and the hydrogeological communities. The hydrological literature

mostly focuses on the prediction of river hydrographs. Different approaches have been proposed to determine predictive uncertainty, one of the most widely used being *Generalized Likelihood Uncertainty Estimation*, or GLUE (Beven and Binley 2014).

For hydrogeological problems, two additional issues are pertinent: spatial information products are usually required to define groundwater levels and discharges, and methods such as GLUE require many simulations to be run. This rapidly becomes unfeasible when dealing with complex 3-D hydrogeological models. While these issues have been considered in some hydrological studies (Ewen and Parkin 1996; Parkin et al. 1996), they have been addressed more thoroughly in the hydrogeological literature by parameter estimation software such as PEST (Doherty 2015; Doherty et al. 2018) and by inverse modeling techniques discussed

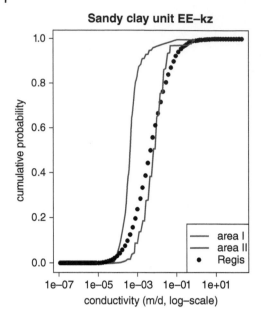

Figure 14.14 Cumulative distribution functions of the most likely vertical hydraulic conductivities of sandy clay unit EE-kz for areas I and II. For comparison, the original CDF as defined in REGIS is also depicted.

in Section 14.5.4 (Banta et al. 2006; Poeter and Hill 1996, 1997; Hill 1998; Hill and Tiedeman 2007). These methods do not address whether the underlying conceptual model may be incorrect; there is a risk that parameterization may compensate for errors in conceptual understanding.

Integrated surface–subsurface modeling at the catchment scale is an important modeling activity within the United Kingdom (Lewis 2016; Lewis et al. 2018). The UK Environment Agency has developed national and regional geological models to support groundwater resource evaluations (Shepley et al. 2012). These evaluations involve integrated assessments of surface–subsurface hydrologic interactions. For example, south-eastern England is underlain by chalk aquifers and is subject to groundwater flooding hazards. Thus, there has been increasing interest in combining the capabilities of existing integrated catchment models with the improved subsurface information provided by these geological models. The following sections provide three examples. The first is for a relatively small area subject to groundwater flooding hazard. The second assesses groundwater-surface water interactions in the Thames Basin. The third describes the use of the combined modeling approach at larger regional or national scales.

14.6.4 Chichester Integrated Flood Model

Chichester, on the southern flank of the South Downs close to the Sussex coast in the south of England (Figure 14.15),

is underlain by the Cretaceous Chalk aquifer and is subject to groundwater flooding events, as discussed in Section 14.2 and Figure 14.1. Chichester has a long history of groundwater flooding with significant flooding events in 1994, 2000–2001, 2004, 2008, and 2013–2014.

The Chichester area provided an opportunity to study the improvement in simulation accuracy of shallow groundwater levels when existing 3-D lithostratigraphic models provided information to constrain an integrated hydrological model. An existing South Downs 3-D geological model containing 10 lithostratigraphic units (Shelley and Burke 2018) provided a conceptual model of the bedrock geology that is consistent with the national UK geological model (UK3D) and corresponds to the current 1:250 000 scale geological map (Figure 14.16). A separate 3-D model of the superficial deposits near Chichester (Terrington et al. 2014) provided a good representation of near surface deposits.

Information from these existing geological models provided a readily adjusted set of conceptualizations of near-surface geological conditions within a smaller 8 km by 8 km study area centered on Chichester (Figure 14.15b). The lithostratigraphic units were assigned hydrogeological parameters required for the numerical simulations. SHETRAN, a physically-based spatially-distributed numerical modeling system (Ewen et al. 2000; Newcastle University 2018) was used to simulate the integrated surface and subsurface hydrology. The Chichester study area falls within the Environment Agency's East Hampshire and Chichester Chalk (EHCC) regional groundwater model. This allowed definition of boundary conditions for the smaller Chichester model and comparison of results from the two models. It was decided to base all simulations on the 1994 flood event data because major flood alleviation works were installed following this event and these affected later flood events in rather complex ways.

14.6.4.1 Surface Hydrology

Chichester is located on the River Lavant, which drains a predominantly rural catchment of 87.2 km² upstream of the Graylingwell gauging station on the north edge of Chichester (Figure 14.15b). Within the urban Chichester environment, the River Lavant is confined to artificial channels, having been diverted from its natural southerly route, probably during Roman times. The River Lavant Flood Alleviation Scheme follows this route to the sea at Pangham Harbour. In central Chichester, the River Lavant passes through two culverts with a maximum capacity of 4 m³ s⁻¹ before emerging at the western margin of the city and discharging into Fishbourne Channel. The River Lavant, with its source between the villages of Singleton and Charlton on the South Downs, behaves as a typical

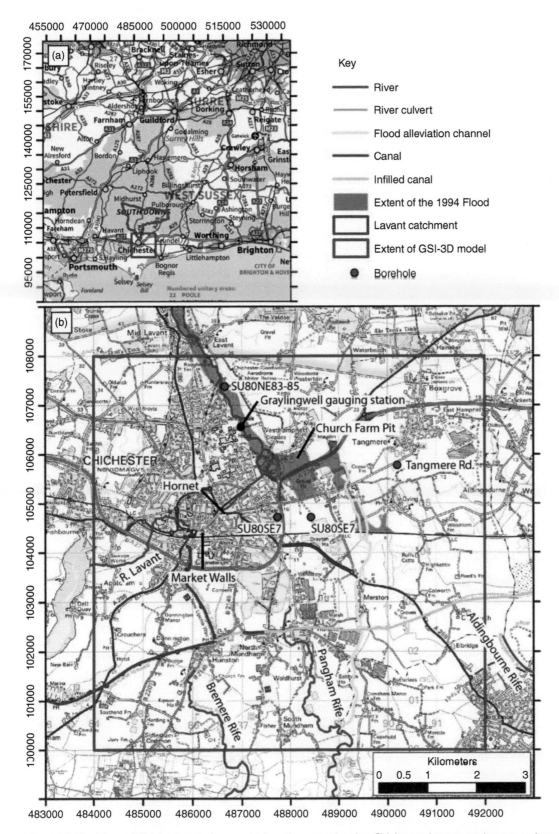

Figure 14.15 Maps of Chichester study area. (a) Location map showing Chichester, Lavant catchment, and extent of GSI3D geological model; (b) study area map showing locations of boreholes, canals, and drainage channels defined in text, and extent of 1994 flood. Contains Ordnance Survey data (© Crown copyright and database right 2018).

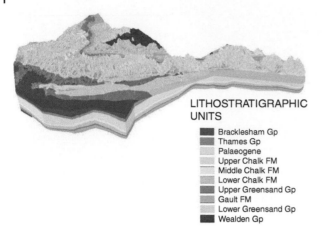

LITHOSTRATIGRAPHIC
UNITS

■ Bracklesham Gp
■ Thames Gp
Palaeogene
Upper Chalk FM
Middle Chalk FM
Lower Chalk FM
Upper Greensand Gp
Gault FM
Lower Greensand Gp
■ Wealden Gp

Figure 14.16 The South Downs 3-D geological model viewed from the south. (Shelley and Burke 2018).

baseflow-dominated chalk stream. While it is commonly completely dry in summer months, in winter its flow rate typically responds slowly to rainfall events. However, following periods of intense rainfall, the flow rates become increasingly responsive to further rainfall. There are several small streams rising on the coastal plain south of Chichester; the Aldingbourne, Pangham, and Bremere Rifes lie within the study area (Figure 14.15b).

14.6.4.2 Geology

The geology of the study area may be described with three bedrock units and seven Quaternary shallow deposits. The bedrock units are the Cretaceous Upper Chalk Formation, which forms a regional aquifer and, separated by an angular uniformity, the overlying Palaeogene Reading and London Clay formations. The Alpine Orogeny caused these formations to form gently plunging folds; the Chichester Syncline is the most significant (Figure 14.17). It contains a core of impermeable Palaeogene strata and thus exerts a strong control on groundwater flow. The Chichester Gravels, a coarse, clayey-gravel unit forming a large fan-shaped body south of the mouth of the Lavant Valley (Figure 14.17), is the most significant of the younger Quaternary age units. They are an aggregate source and form a shallow aquifer; 35 active and former gravel pits are mapped to the east and south of Chichester.

14.6.4.3 Hydrogeology

Groundwater flow in the Chalk aquifer enters the study area from the north, but the impermeable Palaeogene strata in the core of the Chichester Syncline diverts some of the flow westward toward Fishbourne springs. Underflow beneath the syncline occurs only where it is relatively shallow.

The Chichester Gravels, forming the main shallow aquifer, was the focus of the study. Gravel pits can strongly influence groundwater flow, especially when they are subsequently infilled with less permeable or impermeable materials. Many of the gravel pits near Chichester have been infilled with domestic waste, industrial waste/hardcore, silt, engineered impermeable material, and engineered permeable material. These different materials appear as infilled ground on British Geological Survey (BGS) geological maps and the 3-D geology model showed these all as infilled ground but, because accurate representation of the gravel pits was considered crucial, the hydrostratigraphy within the Chichester Gravels was expanded to include these five types of infilled material (Figure 14.18).

14.6.4.4 SHETRAN Numerical Model

The SHETRAN model used a 200 m grid, and the 3-D geological information was imported to form deep and shallow model layers. Boundary conditions for time-varying heads were imported from the regional groundwater model, and observed inflows were used for the River Lavant. Figure 14.19 shows the SHETRAN shallow model grid. Separate simulations were conducted with and without the Chalk aquifer.

Ongoing anthropogenic processes result in continual changes in ground surface elevation. This poses significant problems for the integration of historical datasets collected at different times. The Chichester model was completed in 2005 and predates the excavation and infilling of several gravel pits. It does however incorporate several gravel pits that were not present during the 1994 flood event. In any modeling study, care should be taken to ensure that ground surface and anthropogenic features are compatible with the period being modeled. To ensure the accurate topography of each pit was provided to the model, scattered data points were generated at 10 m centers across the base of every pit. These allowed each pit to be represented with a realistic steep-sided, relatively flat-floor geometry with the correct depth and infill material. Uncertainty remained concerning the distribution of made and worked ground. The Chichester geological model only included made ground at locations where it occurred on a cross-section line, but many boreholes around Chichester penetrate a layer of made ground, which suggests it is more extensively distributed.

The SHETRAN model produced a sequence of time-series evaluations of water levels in gravel pits using different assumed hydraulic properties for the pit infilling materials. At a gravel pit near the River Lavant (Location 5, Figure 14.19) and just downstream of the gravel pits (Location 4, Figure 14.19), these simulations showed that accurate prediction of interactions between surface water and groundwater levels required careful attention to

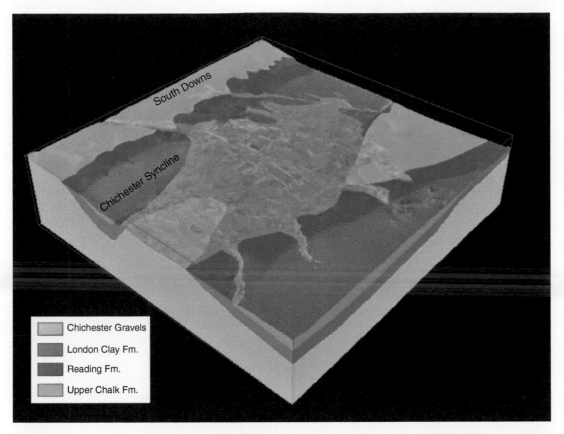

Figure 14.17 Simplified 3-D geological model of the Chichester study area showing folding of the Upper Chalk, London Clay and Reading formations, displayed at a vertical exaggeration of 10×. The extent of the Chichester Gravels is also shown.

characterization of anthropogenic features such as made ground.

Figure 14.20 compares simulated groundwater levels for July 1993 produced by SHETRAN and the regional groundwater model along two profiles shown in Figure 14.19. The two models are in close agreement for profile A, which does not intersect any gravel pits. Slight differences observed near the coast (right side of profile) are due to the coastal streams (rifes) not being included in the SHETRAN model. SHETRAN simulates more realistic, lower water levels along profile B near the gravel pits. This is due to its ability to simulate the dynamic interactions between groundwater and surface water at these locations.

14.6.5 Integrated Modeling of the Thames Basin

Management of the water resources within the River Thames Basin in southern England provides a specific example of the importance to society of integrating the data, information, and knowledge with the 3-D modeling and analysis process.

In 2011–2013, the United Kingdom experienced a very severe drought. Unusually high spring and summer rainfall averted a major crisis (Marsh et al. 2013). Although the whole country was affected, south-eastern England was hit most severely due to high water demand and abnormally lower rainfall. The situation was thought so critical that the Secretary of State for Environment, Food and Rural Affairs formed a National Drought Management Group of the main stakeholders under the leadership of the Chief Executive of the Environment Agency to answer the question: "How does the Thames Basin respond to droughts, and can abstraction be maintained under severe drought conditions?" To answer this question, several more specific questions had to be addressed, including:

1. When will London run out of water?
2. How socially acceptable is drought and the associated water restrictions to the general public?
3. When will water company groundwater sources begin to be exhausted?

These and similar questions took on an importance hitherto not appreciated by government or public. Water supply managers and the regulator had to admit that they did not know the answers to these questions with any certainty. To

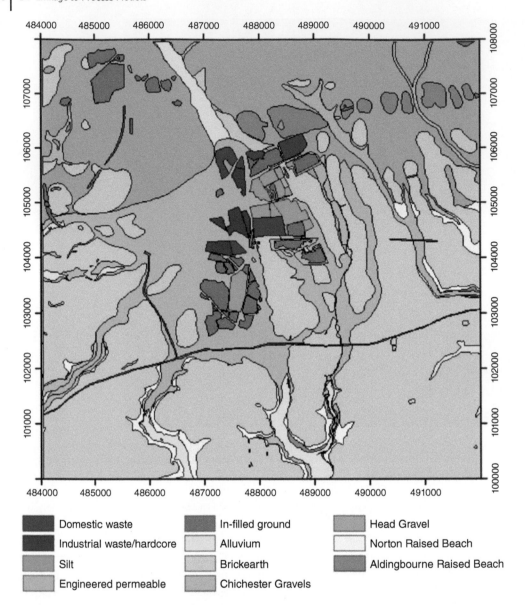

Figure 14.18 Map of revised Chichester geological model surface features, showing reclassified gravel pits.

Legend:

Domestic waste	In-filled ground	Head Gravel
Industrial waste/hardcore	Alluvium	Norton Raised Beach
Silt	Brickearth	Aldingbourne Raised Beach
Engineered permeable	Chichester Gravels	

satisfactorily answer these questions required the integration of hydrological, hydrogeological, and socio-economic information within a suite of integrated numerical models.

The Thames Basin in the south-east of England straddles two major aquifer systems and is also home to several minor aquifers. Jurassic limestones of the Cotswolds form the main aquifer in the west, while the Chalk forms the main aquifer in the central and eastern parts of the catchment. The Chalk aquifer is the most important; it overlies thick clay deposits that separate it from the Jurassic limestones. The aquifers are extensively used for public supply, private supply, and provide baseflow to the River Thames and its tributaries. Both these limestone aquifers have extreme heterogeneity and transmit water mainly through a network of fractures. Boreholes and adits have been developed to intersect flowing fractures and exploit groundwater sources. During droughts, the dewatering of these fractures causes significant reductions in pumping rates; these characteristics represent "tipping points" in the groundwater supply system.

The Chalk aquifer (Figure 14.21) has been extensively investigated over the last three decades (Wheater and Peach 2004; Peach et al. 2017). A distributed groundwater flow model had been previously developed for the Chalk aquifer (Price et al. 2000). The Jurassic limestone aquifers (Figure 14.21) are less well understood but a semi-distributed lumped parameter model had been constructed (Mansour et al. 2013).

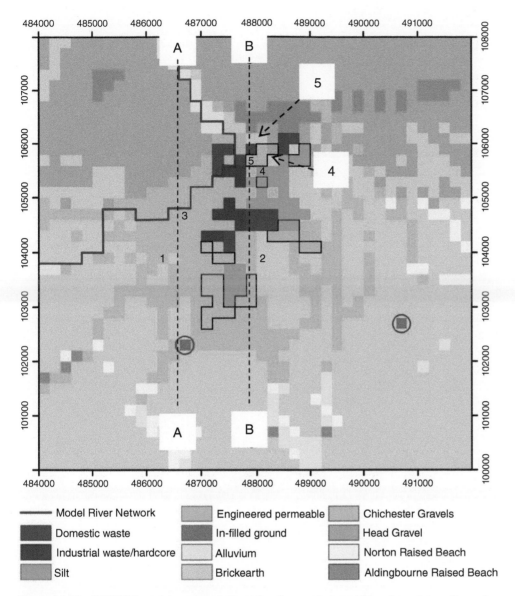

Figure 14.19 SHETRAN model representation of surface geology and River Lavant. Locations of cross-sections A and B in Figure 14.20 and locations 4 and 5 discussed in text are shown.

The Thames Integrated Model (TIM) links the detailed geology of the Thames catchment with groundwater and surface water hydrology to understand the interactions between rainfall, evaporation and runoff and thus evaluate relationships between groundwater and river flows for the River Thames and its tributaries (Figure 14.22). OpenMI, which allows different models to communicate during run-time, was used to link the fully distributed groundwater flow model of the Chalk, a simplified groundwater model of the Jurassic limestones, and a river model (Mansour et al. 2013). The relationship between river flow and magnitude of groundwater withdrawal was achieved by the inclusion of an abstraction management component that employed a simple rule to modify groundwater abstraction during runtime (Peach et al. 2017). This more holistic representation of the hydrological processes demonstrates the important role of integrated modeling in providing for better management of groundwater and surface water resources under drought conditions or future climate change.

14.6.6 National Scale Catchment Modeling in the United Kingdom

A robust, physically-based hydrological modeling system has been developed for Great Britain using the SHETRAN model (Ewen et al. 2000; Newcastle University 2018) and national datasets (Lewis 2016; Lewis et al. 2018).

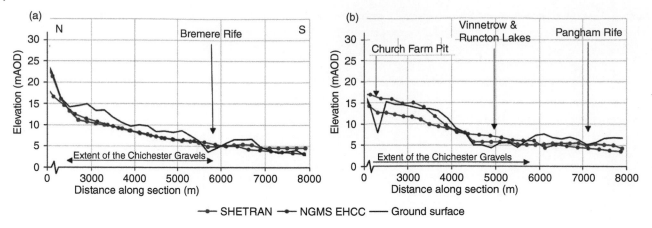

Figure 14.20 Cross-sections of simulated groundwater levels for July 1993 from SHETRAN and East Hampshire and Chichester Chalk (EHCC) models.

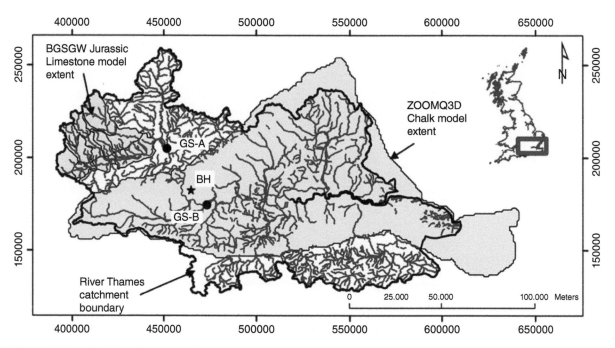

Figure 14.21 The River Thames catchment boundary, and the location and extent of models within the Thames Integrated Model composition. Modified from Peach et al. (2017).

Such a model has several advantages: (i) it is more readily applicable to ungauged catchments; (ii) because it explicitly accounts for physical processes and relationships, it may be more robust under changing climate and land cover; and (iii) a fully integrated surface and subsurface model offers a wider range of applications compared with simpler schemes. SHETRAN was used to model 306 catchments throughout Great Britain, using the most appropriate free datasets available at that time (Figure 14.23), with consistent parameterization at the national scale (each catchment was not separately calibrated), and simplified representations where more detailed data were not available, particular for aquifers

for which the active layer was assumed to be of uniform thickness.

Initially, the modeling performed well for most catchments (Lewis 2016), except for those located on the Chalk in south east England (Figure 14.24a), where a significant proportion of streamflow is derived from groundwater. SHETRAN includes full representation of coupled surface-subsurface interactions and has had previous success in modeling regions with notable groundwater interaction (Adams and Parkin 2002; Parkin et al. 2007; Koo and O'Connell 2006), so it was concluded that the issue was with the representation of aquifer geometry

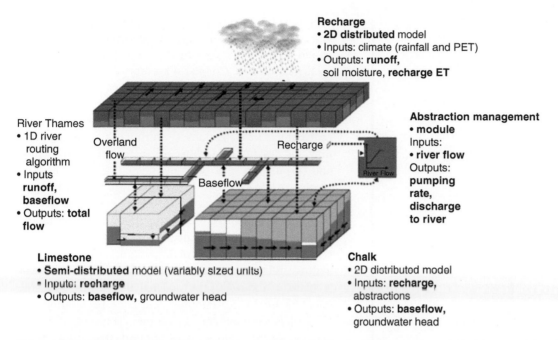

Recharge
- **2D distributed** model
- Inputs: climate (rainfall and PET)
- Outputs: **runoff,** soil moisture, **recharge ET**

River Thames
- **1D river routing algorithm**
- Inputs **runoff, baseflow**
- Outputs: **total flow**

Overland flow

Recharge

Baseflow

River Flow

Abstraction management
- **module**
Inputs:
- **river flow**
Outputs:
- **pumping rate, discharge to river**

Limestone
- **Semi-distributed** model (variably sized units)
- Inputs: **recharge**
- Outputs: **baseflow,** groundwater head

Chalk
- 2D distributed model
- Inputs: **recharge,** abstractions
- Outputs: **baseflow,** groundwater head

Figure 14.22 Model linkages and input/output parameters within the Thames Integrated Model composition. PET, potential evapotranspiration; WM, water management. Modified from Peach et al. (2017).

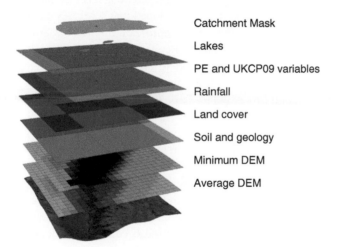

Catchment Mask

Lakes

PE and UKCP09 variables

Rainfall

Land cover

Soil and geology

Minimum DEM

Average DEM

Figure 14.23 Data layers used in the national scale SHETRAN modeling.

and hydraulic properties rather than inadequate process representation.

In 2012, the BGS released the first national-scale geological model which covered England, Wales, and Scotland. This model, known as GB3D, was revised in 2014 (Mathers et al. 2014). A new much more detailed model, which included Northern Ireland and known as UK3D, was released in 2016 (Waters et al. 2016). These models currently exist as networks of cross-sections, but continuing developments will soon produce interpolated stratigraphic surfaces.

To improve the simulation of catchments with important groundwater regimes, GB3D and UK3D were first used to redefine aquifer geometry, using the original hydraulic properties. The preliminary results showed only limited improvements in the performance of the combined model, despite a better representation of subsurface structure. Additional information on hydraulic parameters for the hydrogeological units which was available at a national scale (Allen et al. 1997) was then introduced, specifically focusing on Chalk aquifers (MacDonald and Allen 2001). The set of simulations using both improved aquifer geometry from the UK3D digital model and better information on hydraulic parameters showed some improvement (Figure 14.24b). The GB3D and UK3D models focus on bedrock geology, but river flows, groundwater recharge, and many shallow groundwater levels (affecting groundwater flooding) depend critically on models of superficial deposits, which were not readily available, or are incomplete for the whole country. Further development of national-scale hydrogeologic parameterization, including for the superficial deposits, is in progress, which could potentially improve the performance of the combined model.

The combined modeling system has great potential at national, regional, and local scales to address climate change impact assessments, land cover change studies, and integrated assessments of flooding and groundwater and

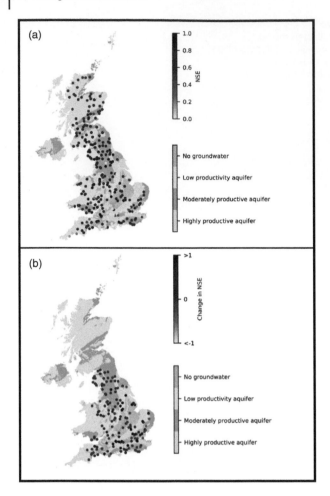

Figure 14.24 Results from national scale SHETRAN modeling: (a) Nash-Sutcliffe Efficiency (NSE) values for modeled catchments (high NSE values indicate good model performance), (b) change in NSE values for catchments with moderately or highly productive aquifers after inclusion of additional geological data.

surface water resources. With this system, it is now relatively straightforward to perform simultaneous simulations of multiple catchments including their related aquifers, or assess different management options, based on the best available hydrogeological information as represented in digital 3-D geological models.

14.7 Geotechnical Applications of Geological Models

Geotechnical design projects typically use Computer Aided Design (CAD)-based ground models, whereas most 3-D geological models are developed using procedures described in Chapters 9–12. Differences between these two approaches have, until recently, limited the widespread

adoption of 3-D geological modeling to support the design and construction of underground works. However, the situation is evolving; several recent design projects have used geological models. Chapter 4 presents the economic case for the adoption of 3-D geological models by geotechnical engineers.

Geological models have been used in three aspects of geotechnical investigations: (i) during initial project planning and design, (ii) during construction of a new facility, and (iii) as part of a back-analysis conducted following a construction problem. Veldkamp et al. (2001), Rengers et al. (2002), and Hack et al. (2006) document several relatively early uses of 3-D geological models. During the design of the second Heinenoord Tunnel, which crosses the Oude Maas channel south of Rotterdam, geological models at three different scales demonstrated how the model detail is related to the size of the modeled area and the available data (Figure 14.25). Borehole and cone penetration test (CPT) data were integrated with the 3-D geological model (Figure 14.26), deformations around the tunnel were computed by a finite element numerical model, and the resulting estimated vertical displacements (Figure 14.27) were projected onto the tunnel tube (Rengers et al. 2002).

A 3-D geological model developed using ArcGIS during the planning of the new Noord/Zuidlijn (North/South metro line) in Amsterdam (Rengers et al. 2002) was used to assess the cause of two construction incidents involving diaphragm walls at the Vijzelgracht station excavation (Bosch and Broere 2009). Barla (2017) describes additional examples of the use of geological models to investigate causes of construction problems.

Chapters 22 and 23 contain several additional examples of 3-D geological models assisting geotechnical projects. Chapter 22 provides five case studies, including the characterization of the subsurface in Christchurch, New Zealand to guide post-earthquake reconstruction, two coastal erosion assessments on the English coast, an evaluation of potentially contaminated land for an urban renewal project, and an assessment of potential groundwater pollution. Chapter 23 includes three case studies; two involving underground construction projects in London and one that involved enhancement of railway infrastructure. In these examples, 3-D geological models provided critical information required to develop and constrain various numerical models used to estimate the responses, or reactions, of rock or soil materials to changing stress or fluid flow conditions.

14.7.1 Numerical Modeling of Rock or Soil Behavior

Numerical methods were first applied to the solution of tunneling and underground excavation problems during

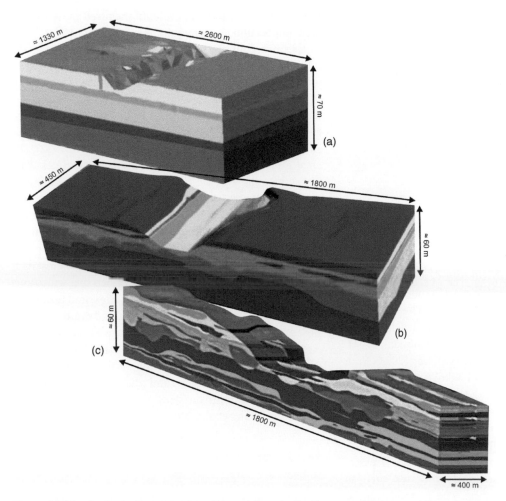

Figure 14.25 Geological models of the lithostratigraphy for the second Heinenoord Tunnel project at various levels of detail. The incision in the middle of model is the Oude Maas river. (Veldkamp et al. 2001).

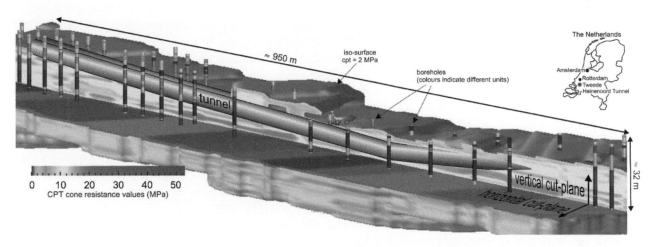

Figure 14.26 Visualization of proposed Heinenoord Tunnel alignment showing the distribution of CPT cone resistance values and boreholes showing geotechnical units. Two cut-planes show the internal modeled distribution of CPT values. Modified from Hack et al. (2006).

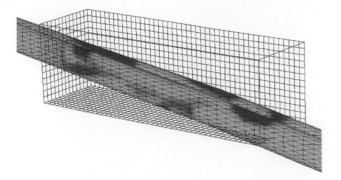

Figure 14.27 Estimated vertical displacements on the Heinenoord Tunnel projected onto the tunnel tube. Blue = small displacements; Red = large displacements. (Rengers et al. 2002).

the last half of the twentieth century (Zienkiewicz and Cheung 1967). Detailed discussion of such modeling applications is beyond the scope of this book. The interested reader is directed to the very extensive technical literature on this topic, including: Wittke (1990), Goodman (1993), Hart (1993), Hudson and Harrison (1997), Jing (2003), Pine et al. (2006), Stead et al. (2006), and Nikolić et al. (2016).

Table 14.1 summarizes the advantages and limitations of the most widely adopted alternative numerical methods for rock mechanics evaluations. Many of these techniques address complexities relating to geometry, material anisotropy, non-linear behavior, in-situ stresses, and the presence of several coupled processes, such as pore pressures and seismic loading. The methods can be classified as *Equivalent Continuum*, *Discontinuum*, and *Hybrid Continuum/Discontinuum* methods. The successful application of any of these numerical modeling methods depends on well-trained experienced users who observe good modeling practice and are aware of model/software limitations (Cundall 1971).

Continuum methods include *Finite-Difference* (FDM), *Finite-Volume* (FVM), *Finite-Element* (FEM), and *Boundary Element* (BEM) models, as well as several *Meshless Methods*. FDM and FEM are routinely used in geotechnical engineering investigations and are most appropriate for the analysis of soils or weak rocks where failure is controlled by the deformation of the intact material; in other words, where there is a continuum. The 2-D or 3-D versions of FLAC (Itasca International Inc. 2018a,b) are widely used codes that combine FDM and FVM methods. FEM allows for material heterogeneity, complex geometry and boundary conditions, and thus is ideal for simulating non-linear inelastic behavior of hard, fractured rock. PLAXIS is a widely used 2-D and 3-D FEM code. Megna et al. (2005) used FEM to evaluate stresses and displacements around normal faults. Meshless methods avoid the complexities of

mesh generation and can evaluate fractured materials but are computationally demanding (Nikolić et al. 2016).

Discontinuum methods are most useful when the stability of the rock mass is controlled by movement of joint-bounded blocks; they are routinely used by civil and mining engineers to analyze both underground structures and rock slopes. Discontinuum methods include *Distinct Element*, or DEM (Hart 1993), *Discontinuous Deformation Analysis*, or DDA (Shi and Goodman 1985; Shi 1992), *Discrete Fracture Network,* or DFN (Golder Associates 2006, 2007), and the *Particle Flow Method*, or PFC (Kulatilake et al. 2001; Itasca International Inc. 2018e). The Universal Distinct Element Code (UDEC) (Itasca International Inc. 2018c) and its three-dimensional derivative 3DEC (Itasca International Inc. 2018d) are the predominant DEM codes. Fracman, developed by Golder Associates, is the most commonly used DFN code; it has been applied to a wide range of geomechanical problems including slope and tunnel stability, predicting in situ fragmentation, and evaluating fluid flow and transport in connected fracture networks (Golder Associates 2006, 2007). Nikolić et al. (2016) briefly describe several hybrid methods that combine the capabilities and advantages of continuum and discontinuum methods. They require careful problem evaluation by experienced modelers (Zhang et al. 2000).

14.7.2 Application of Numerical Models to Evaluate Slope Stability

There are numerous published discussions of the use of numerical modeling to evaluate the stability of natural soil slopes, engineered embankments and cuttings, and rock slopes (Goodman and Kieffer 2000; Wyllie and Mah 2004; Stead et al. 2006). Kalkani and Piteau (1976) used FEM to analyze toppling of rock slopes at Hells Gate in British Columbia, Canada. Coggan et al. (2000) successfully made use of both 2-D and 3-D FDM to back-analyze failures of highly kaolinized china clay slopes. Vlachopoulos et al. (2018) utilized FEM to analyze slopes adjacent to tunnels within weak rock masses in Northern Greece.

Rengers et al. (2002) describe the use of the FEM code PLAXIS to evaluate the impact on an existing railway embankment due to construction of an adjacent 8 m (25 ft) high new embankment for a new High-Speed Railway Line between Amsterdam and Brussels. The soil below the existing embankment consisted of approximately 5 m (17 ft) of soft clay and peat. The PLAXIS analysis (Figure 14.28) indicated that the new embankment would settle approximately 1.8 m (6 ft), but the deformations to the existing tracks would be only 0.26 m (10 in.) horizontally and 0.28 m (11 in.) vertically and could be addressed by extra maintenance work on the existing track.

Table 14.1 Classification of numerical methods for rock mechanics evaluations

Method of representation	Numerical approach	References	Advantages	Limitations
Equivalent continuum methods	Finite difference method, FDM	Perrone and Kao (1975) and Itasca International Inc. (2018a,b)	Oldest and most widely used method. Good for analysis of unconsolidated materials or weak rocks where material deformation and failure, including creep deformation, are important.	Most FDM codes use a regular mesh. Not good for analysis of heterogeneous materials or irregular structures
	Finite volume method, FVM	Wheel (1996) and Fallah et al. (2000)	A later enhancement to FDM that allows use of irregular meshes. FLAC and FLAC3D are widely used codes that combine FDM and FVM	Similar to FDM, but uses irregular meshes
	Finite element method, FEM	Zienkiewicz et al. (2013) and PLAXIS by (2017, 2018)	Major alternative to FDM. Allows for material heterogeneity, complex geometry and boundary conditions. Ideal for simulating non-linear inelastic behavior of hard, fractured rock. PLAXIS is widely used 2-D and 3-D FEM code.	Cannot model crack propagation.
	Boundary element method, BEM	Brady and Braj (1978) and Crouch and Starfield (1983)	A variation on FEM that reduces the model dimensionality by one (uses 2-D model to analyze a 3-D situation). Accurately finds solution to partial differential equations.	Does not deal with heterogeneity of the material efficiently.
	Meshless methods	Zhang et al. (2000)	Avoids complexities of mesh generation. Has significant ability to evaluate fractured materials.	Computationally demanding.
Discontinuum methods	Discrete element method, DEM	Cundall (1988), Itasca International Inc. (2018c,d)	Allow for block deformation and movement of blocks relative to each other. Can model complex material and discontinuity behavior and mechanisms. Commonly used DEM codes are UDEC and 3DEC. DEM uses explicit time integration. DDA uses implicit time integration.	Need representative discontinuity geometry data on spacing, persistence, etc. Limited by availability of data on joint properties.
	Discontinuous deformation analysis, DDA	Shi and Goodman (1985) and Shi (1992)		
	Discrete fracture network method, DFN	Golder Associates (2006, 2007)	Used to evaluate fluid flow and transport in connected fracture networks. FRACMAN is commonly used code.	Requires statistically valid detailed information on 3-D orientations and characteristics of fractures.
	Particle flow method, PFM	Cundall and Strack (1979), Hart et al. (1988), Kulatilake et al. (2001), and Itasca International Inc. (2018e)	PFM models the movement and interaction of spherical particles using DEM. PFM simulates jointed rock block failures through intact material, along the joints, or through a combination of these.	Because the number of particles and the computational time increase rapidly for larger domains, PFM is only practicable for small size domains. PFM is unsuitable for modeling large rock masses.
Hybrid continuum/ discontinuum methods	Hybrid FEM/BEM	Elsworth (1986)	These methods combine the capabilities and advantages of continuum and discontinuum methods.	Require careful problem evaluation by experienced modelers.
	Hybrid DEM/BEM			
	Hybrid FEM/DEM	Mahabadi et al. (2012)		

Source: Based on discussions by Barla (2016) and Nikolić et al. (2016).

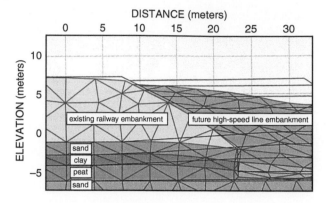

Figure 14.28 PLAXIS FEM analysis of deformations occurring on the existing railway embankment (on the left) due to the construction of the future high-speed railway embankment (on the right). Red mesh is deformed finite element mesh. Modified from Rengers et al. (2002).

DEM has been used to investigate a wide variety of rock-slope failure mechanisms, ranging from simple planar mechanisms, to complex deep-seated toppling instability and buckling (Coggan and Pine 1996; Stead et al. 2006). Adachi et al. (1991) used an early version of the 3DEC code to analyze toppling slopes along a Japanese highway. Chen and Ohnishi (1999) employed DDA to analyze a major rockfall in Japan.

14.7.3 Application of Numerical Models to Evaluate Tunnels and Underground Structures

Three broad categories of ground conditions are the chief cause of tunnel support design issues: (i) unconsolidated, usually near-surface, deposits; (ii) weak rock masses due to the inherent weakness of the rock material, intense fracturing of the rock, or a combination of these; and (iii) hard competent rock which has fractures or joints. These analyses must consider the mechanisms that are at play, whether the rock movements and failures are stress-induced or structure-induced.

Unconsolidated deposits underlie many cities. The result of anthropogenic and natural causes, these deposits are often weak and/or compressible and some are relatively permeable leading to water control issues. Many of these deposits reach considerable depths and are encountered by road or railway tunnels. Construction of tunnels or other underground structures in these deposits requires detailed assessment of their properties, including permeability and groundwater conditions. An accurate model of the subsurface can greatly assist the design of infrastructure support systems and water control options (Bakker and Bezuijen 2008). Examples of modeling applications for tunnels

and embankments in the soft ground conditions prevalent in the Netherlands have been discussed previously (Figures 14.25–14.28).

Tunnels in weak rocks encounter similar problems to those in unconsolidated deposits; these rocks readily deform around openings and require countermeasures to provide stability (Barla and Barla 2008). Due to the fact that the deformation is distributed throughout the rock mass, equivalent continuum modeling approaches are appropriate – both FDN and FEM methods are commonly used. For many situations, 2-D sections perpendicular to the tunnel alignment provide accurate assessments, since the rock characteristics are consistent along the tunnel alignment, at least within specified rock zones. While 2-D calculations have the advantage of being less computationally demanding, 3-D models are necessary in areas of complex tunnel connections, or for highly variable, anisotropic rock environments. Two-dimensional numerical models include significant limitations and assumptions of a 3-D phenomenon. Vlachopoulos and Diederichs (2009) were the first to provide design engineers with a method that provided a point of reference for the section of tunnel being analyzed with respect to the face of the tunnel – a significant consideration for analyzing tunnel support. By determining the position along the longitudinal alignment of the tunnel associated with the 2-D analysis and utilizing information on the final convergence beyond the influence of the face, it was possible to relate the displacement experienced with the final convergence. Further considerations of 2-D versus 3-D numerical analyses are investigated in depth by Vlachopoulos and Diederichs (2014).

Vlachopoulos et al. (2013) numerically evaluated the influence of selected estimates of rock strength parameters on the design of the Driskos Twin Tunnel, part of the Egnatia Highway in Northern Greece. This tunnel passes through a varied heterogeneous sequence of weak Alpine rocks (Figure 14.29). About 18 m of rock separates the twin 11 m (37 ft) wide by 9.5 m (32 ft) high horseshoe-shaped tunnels (Figure 14.30). Two numerical models were developed to assess the stability of the twin tunnels; a 3-D finite difference model using FLAC3D and a 2-D finite element model using Phase 2 from Rocscience (Figure 14.31). Vlachopoulos et al. (2013) document how the results of these numerical models compared with tunnel monitoring data of the observed interactions between the twin tunnels.

Discontinuum or hybrid continuum/discontinuum methods are used to evaluate deeper tunnels that encounter higher stress regimes in competent, but jointed, rocks (Martin 1997; Diederichs 2007; Vlachopoulos and Vazaios 2018). Vazaios et al. (2019) utilized a 2-D hybrid FEM/DEM numerical model using the Irazu code (Geomechanica Inc. 2017) to evaluate the conditions encountered at the

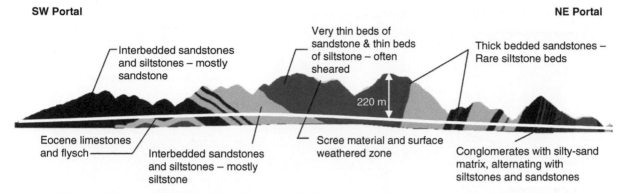

SW Portal   **NE Portal**

Interbedded sandstones and siltstones – mostly sandstone

Very thin beds of sandstone & thin beds of siltstone – often sheared

Thick bedded sandstones – Rare siltstone beds

220 m

Eocene limestones and flysch

Interbedded sandstones and siltstones – mostly siltstone

Scree material and surface weathered zone

Conglomerates with silty-sand matrix, alternating with siltstones and sandstones

Figure 14.29 Geological cross-section of Driskos Twin Tunnel alignment. White line is approximate tunnel location. Modified from Vlachopoulos et al. (2013).

Figure 14.30 Northeastern portals of Driskos Twin Tunnels during construction in 2003. (Vlachopoulos et al. 2013).

Atomic Energy of Canada Limited's (AECL) Underground Research Laboratory (URL) tunnel constructed within the Lac du Bonnet granite batholith of the Canadian Shield.

The AECL-URL Test Tunnel was excavated by mechanical means; explosives were not used to minimize the damage on the rock mass. Nevertheless, shortly after excavation of the AECL-URL was completed, the highly anisotropic stress regime produced high tangential stress concentrations at the roof and the bottom of the tunnel and the resulting spalling formed V-shaped notches, visible in Figure 14.32a (Martin. 1997; Diederichs 2007). In this case, a uniform *Excavation Damage Zone* (EDZ) did not develop but recordings of micro-seismic (MS) and acoustic emission (AE) events (Figure 14.32b) defined the existence of the EDZ beyond the failed notches. The calibrated 2-D hybrid FEM/DEM numerical model closely replicated the MS and AE event data; it predicted a combination of shear and tension fractures in the same locations as the MS and AE events (Figure 14.32c).

14.8 Discussion

Hydrogeological and geotechnical applications of spatially-detailed process models require information across the entire model domain. For hydrogeological applications, while overall rock-mass characteristics control fluid storage capacities, high-permeability flow pathways are dominant controls on dynamic behavior and interactions with surface water systems. In geotechnical problems, the geometry and properties of discontinuities control failure mechanisms. Thus, representation of discontinuities is important for both application areas, yet accurate determination of the specific geometry and properties of individual features is challenging. Observations of dynamic responses can, however, be used to infer the presence and aggregated properties of such features within a domain, providing feedback to the geological model.

The use of digital 3D geological modeling, with appropriate software for visualization and updating, allows for the development and implementation of effective workflows to combine observations of dynamic responses with geological understanding. This generally results in improved characterization of system behavior. However, incorporation of indirect information from process-response observations may lead to ambiguous results; it is often not possible to simultaneously infer quantitative values for both geological geometries and physical properties. The use of digital 3D geological models within wider frameworks of process models to infer geological structures and characterize discontinuities can benefit multiple applications and can provide a consistent conceptual and quantitative understanding of critical features that control the behavior of many dynamic processes.

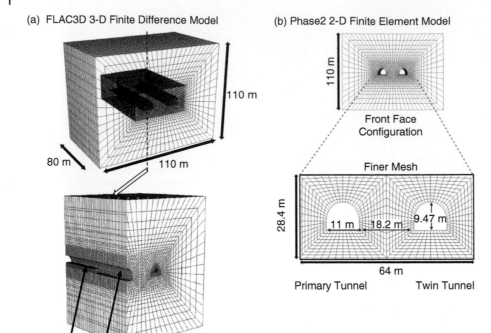

(a) FLAC3D 3-D Finite Difference Model

110 m
80 m
110 m

Bench remaining to be excavated

Top Heading and Bench excavated

(b) Phase2 2-D Finite Element Model

110 m

Front Face Configuration

Finer Mesh

28.4 m

11 m   18.2 m   9.47 m

64 m

Primary Tunnel   Twin Tunnel

Figure 14.31 Two numerical models of the Driskos Twin Tunnel. (a) 3-D finite difference model using FLAC3d code; (b) 3-D finite element model using Phase2 code. Modified from Vlachopoulos et al. (2013).

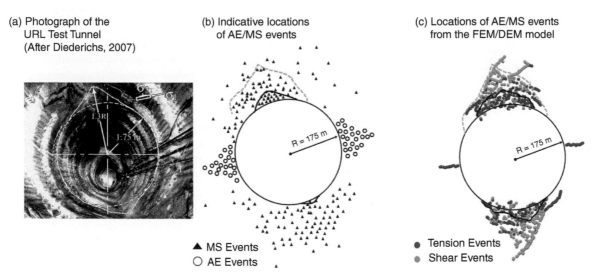

(a) Photograph of the URL Test Tunnel (After Diederichs, 2007)

(b) Indicative locations of AE/MS events

R = 175 m

▲ MS Events
○ AE Events

(c) Locations of AE/MS events from the FEM/DEM model

R = 175 m

● Tension Events
● Shear Events

Figure 14.32 Hybrid FEM/DEM modeling of tunnel in hard rock – example from the AECL-URL Test Tunnel. (a) Photograph of the tunnel showing the as-built damage profile (Source: After Diederichs 2007); (b) indicative locations of AE/MS events; (c) locations of AE/MS events from the hybrid FEM/DEM model with tension events in red and shear events in green. The highly damaged zone (black continuous line) and potential excavation damage profile (blue dashed line) are highlighted in both B and C. Modified after Vazaios et al. (2019) and Cai et al. (2001).

Acknowledgments

This chapter was developed with the cooperation and collaboration of multiple authors. Geoffrey Parkin was the lead contributor; he developed Sections 14.1–14.6, with the assistance of Elizabeth Lewis and James Howard. The topics in Sections 14.1–14.6 were enhanced by contributions from several authors. Frans van Geer, assisted by Aris Lourens and Wilbert Berendrecht, prepared Section 14.6.2 which describes the AZURE regional groundwater resource model in the Netherlands. Denis Peach provided the description of integrated modeling of the Thames Basin in Section 14.6.5. Nicholas Vlachopoulos contributed Section 14.7 which describes how geological models support geotechnical applications.

References

Abbott, M.B., Bathurst, J.C., Cunge, J.A. et al. (1986a). An introduction to the European Hydrological System – Systeme Hydrologique Europeen, "SHE", 1: history and philosophy of a physically-based distributed modelling system. *Journal of Hydrology* 87: 45–59. https://doi.org/10.1016/0022-1694(86)90114-9.

Abbott, M.B., Bathurst, J.C., Cunge, J.A. et al. (1986b). An introduction to the European Hydrological System – Systeme Hydrologique Europeen, "SHE", 2: structure of a physically-based distributed modelling system. *Journal of Hydrology* 87: 61–77. https://doi.org/10.1016/0022-1694(86)90115-0.

Adachi, T., Ohnishi, Y. and Arai, K. (1991). Investigation of toppling slope failure at route 305 in Japan. *International Journal of Rock Mechanics and Mining Sciences & Geomechanics Abstracts* 30 (3): A190–A191. https://doi.org/10.1016/0148-9062(93)93134-J.

Adams, R. and Parkin, G. (2002). Development of a coupled surface-groundwater-pipe network model for the sustainable management of karstic groundwater. *Environmental Geology* 42: 513–517. https://doi.org/10.1007/s00254-001-0513-8.

Allen, D.J., Brewerton, L.J., Coleby, L.M. et al. (1997). *The Physical Properties of Major Aquifers in England and Wales*. Keyworth, UK: British Geological Survey, Report WD/97/034. 333 pp. [Online: Available at: http://nora.nerc.ac.uk/id/eprint/13137/] (Accessed August 2020).

Bakker, K.J. and Bezuijen, A. (2008). Ten years of bored tunnels in the Netherlands: part I, geotechnical issues. In: *Geotechnical Aspects of Underground Construction in Soft* (eds. C.W.W. Ng, H.W. Huang and G.B. Liu), 243–248. Milton Park, UK: Taylor & Francis. https://doi.org/10.1201/9780203879986.

Banks, D. (2012). *An Introduction to Thermogeology: Ground Source Heating and Cooling*, 2e. Chichester, UK: Wiley Blackwell. 526 pp. https://doi.org/10.1002/9781118447512.

Banta, E.R., Poeter, E.P., Doherty, J.E. and Hill, M.C. [Eds.] (2006). *JUPITER: Joint Universal Parameter Identification and Evaluation of Reliability-An Application Programming Interface (API) for Model Analysis*. Reston, VA: U.S. Geological Survey, Techniques and Methods 6-E1. 268 pp. [Online: Available at: https://pubs.usgs.gov/tm/2006/tm6e1/] (Accessed January 2018).

Barla, G. (2016). Applications of numerical methods in tunnelling and underground excavations: recent trends. In: *Proceedings, EUROCK 2016 - the 2016 ISRM International Symposium, Cappadocia, Turkey* (eds. R. Ulusay, Ö. Aydan, H. Gerçek, et al.), 29–40. Boca Raton, FL: CRC Press.

Barla, G. (2017). Case studies of tunnel instability and interaction with the ground surface and manmade structures. In: *Proceedings, IF-CRASC 2017 Conference: Forensic Engineering in Civil, Industrial and Information Sectors*. Milano, Italy: Politecnico di Milano. 12 pp.

Barla, G. and Barla, M. (2008). Innovative tunnelling construction methods in squeezing rock. *Ingegneria Ferroviaria* 63 (12): 103–119.

Bedekar, V., Morway, E.D., Langevin, C.D. and Tonkin, M. (2016). *MT3D-USGS Version 1: A U.S. Geological Survey Release of MT3DMS Updated with New and Expanded Transport Capabilities for use with MODFLOW*. Reston, VA: U.S. Geological Survey, Techniques and Methods 6-A53. 69 pp. https://doi.org/10.3133/tm6A53.

Belcher, W.R. and Sweetkind, D.S. [Eds.] (2010). *Death Valley Regional Groundwater Flow System, Nevada and California — Hydrogeologic Framework and Transient Groundwater Flow Model*. Reston, VA: U.S. Geological Survey, Professional Paper 1711. 398 pp. https://doi.org/10.3133/pp1711.

Bergström, S. (1976). *Development and Application of a Conceptual Runoff Model for Scandinavian Catchments*. Norrköping, Sweden: SMHI, Report Hydrologi och Oceanografi Nr RHO 7 (1976). 134 pp.

Bergström, S. (1995). The HBV model. In: *Computer Models of Watershed Hydrology* (ed. V.P. Singh), 443–476. Highlands Ranch, CO: Water Resources Publications.

Beven, K. and Binley, A. (2014). GLUE: 20 years on. *Hydrological Processes* 28: 5897–5918. https://doi.org/10.1002/hyp.10082.

Beven, K.J., Calver, A. and Morris, E.M. (1987). *The Institute of Hydrology Distributed Model*. Wallingford, UK: Institute of Hydrology, Report 98. 33 pp. [Online: Available at: http://nora.nerc.ac.uk/id/eprint/5977/] (Accessed February 2018).

Beven, K.J., Lamb, R., Quinn, P. et al. (1995). TOPMODEL. In: *Computer Models of Watershed Hydrology* (ed. V.P. Singh), 627–668. Highlands Ranch, CO: Water Resources Publications.

Bosch, J.W. and Broere, W. (2009). Small incidents, big consequences: leakage of a building pit causes major settlement of adjacent historical houses: Amsterdam North-South metro line project. In: *Proceedings, ITA-AITES World Tunnel Congress: Safe Tunneling for the City and Environment. 23–28 May 2009, Budapest, Hungary*.

Budapest, Hungary: Hungarian Tunneling Association. 13p.

Brady, B.H.G. and Braj, J.W. (1978). The boundary element method for determining stress and displacements around long openings in a triaxial stress field. *International Journal of Rock Mechanics and Mining Sciences & Geomechanics Abstracts* 15: 21–28. https://doi.org/10.1016/0148-9062(78)90718-0.

Brunner, P. and Simmons, C.T. (2012). HydroGeoSphere: a fully integrated, physically based hydrological model. *Groundwater* 50 (2): 170–176. https://doi.org/10.1111/j.1745-6584.2011.00882.x.

Cai, M., Kaiser, P.K. and Martin, C.D. (2001). Quantification of rock mass damage in underground excavations from microseismic event monitoring. *International Journal of Rock Mechanics and Mining Sciences* 38 (8): 1135–1145. https://doi.org/10.1016/S1365-1609(01)00068-5.

California Department of Water Resources (2016). *Hydrogeologic Conceptual Model: Best Management Practice*. Sacramento, CA: California Department of Water Resources. 25p. [Online: Available at: http://water.ca.gov/groundwater/sgm/pdfs/BMP_HCM_Final_2016-12-23.pdf] (Accessed February 2018).

Carrera, J., Alcolea, A., Medina, A. et al. (2005). Inverse problem in hydrogeology. *Hydrogeology Journal* 13 (1): 206–222. https://doi.org/10.1007/s10040-004-0404-7.

Chen, G. and Ohnishi, Y. (1999). Slope stability analysis using discontinuous deformation analysis. In: *Proceedings, 37th U.S. Rock Mechanics Symposium, Rock Mechanics for Industry, Vail, Colorado* (eds. B. Amadei, R.L. Kranz, G.A. Scott and P.H. Smeallie), 535–541. Rotterdam, Netherlands: A.A. Balkema.

Clark, A.K., Golab, J.A. and Morris, R.R. (2016). *Geologic Framework and Hydrostratigraphy of the Edwards and Trinity Aquifers within northern Bexar and Comal Counties, Texas*. Reston, VA: U.S. Geological Survey, Scientific Investigations Map 3366. 29 pp. 1 map sheet. https://doi.org/10.3133/sim3366.

Coggan, J.S. and Pine, R.J. (1996). Application of distinct-element modelling to assess slope stability at Delabole Slate Quarry, Cornwall, England. *Transactions of the Institution of Mining and Metallurgy, Section A: Mining Industry* 105: 22–30.

Coggan, J.S., Stead, D. and Howe, J.H. (2000). Characterisation of a structurally controlled flowslide in a kaolinised granite slope. In: *Landslides in Research, Theory and Practice* (eds. E. Bromhead, N. Dixon and M.-L. Ibsen), 299–304. London, UK: Institution of Civil Engineers.

Crawford, N.H. and Linsley, R.K. (1966). *Digital Simulation in Hydrology: Stanford Watershed Model IV*. Stanford, CA: Stanford University, Technical Report 39. 210 pp.

Crouch, S.L. and Starfield, A.M. (1983). *Boundary Element Methods in Solid Mechanics: With Applications in Rock Mechanics and Geological Engineering*. London: George Allen & Unwin. 322 pp.

Cundall, P.A. (1971). A computer model for simulating progressive, large-scale movements in blocky rock systems. In: *Proceedings, Symposium on Rock Fracture of the International Society of Rock Mechanics, Nancy, France, Vol. 2* (eds. P. Blazy, C. Chambor, R. Houpert et al.), 2–8. Rotterdam, Netherlands: Balkema.

Cundall, P.A. (1988). Formulation of a three-dimensional distinct element model-part I: a scheme to detect and represent contacts in a system composed of many polyhedral blocks. *International Journal of Rock Mechanics and Mining Sciences & Geomechanics Abstracts* 25: 107–116. https://doi.org/10.1016/0148-9062(88)92293-0.

Cundall, P.A. and Strack, O.D.L. (1979). A discrete numerical model for granular assemblies. *Géotechnique* 29: 47–65. https://doi.org/10.1680/geot.1979.29.1.47.

D'Agnese, F.A. and O'Brien, G.M. (2003). Impact of geoinformatics on the emerging Geoscience Knowledge Integration Paradigm. In: *New Paradigms in Subsurface Prediction* (eds. M.S. Rosenbaum and A.K. Turner), 303–312. Berlin, Germany: Springer, Lecture Notes in Earth Sciences, vol. 99. https://doi.org/10.1007/3-540-48019-6_27.

D'Agnese, F.A., Faunt, C.C., Turner, A.K. and Hill, M.C. (1997). *Hydrogeologic Evaluation and Numerical Simulation of the Death Valley Regional Ground-Water Flow System, Nevada and California, using Geoscientific Information Systems*. Reston, VA: U.S. Geological Survey, Water Resources Investigations Report WRIR 96–4300. vii + 124 pp. https://doi.org/10.3133/wri964300.

D'Agnese, F.A., Faunt, C.C., Hill, M.C. and Turner, A.K. (1999). Death Valley regional ground-water flow model calibration using optimal parameter estimation methods and geoscientific information systems. *Advances in Water Resources* 22 (8): 777–790. https://doi.org/10.1016/S0309-1708(98)00053-0.

De Lange, W. and Borren, W. (2014). *Grondwatermodel AZURE 1.0, Actueel Instrumentarium voor de Zuiderzee Regio; Hoofdrapport: achtergronden modelbouw, stationaire kalibratie, niet-stationaire kalibratie*. Delft, Netherlands: Deltares, report 1205042-000-BGS-0067.

DHI (2011). *MIKE SHE User Manual Volume 2: Reference Guide*. Horsholm, Denmark: Danish Hydraulic Institute (DHI) 372 pp. [Online: Available at: https://manuals.mikepoweredbydhi.help/2017/MIKE_SHE] (Accessed August 2020).

DHI-WASY (2012). *FEFLOW® 6.1 Finite Element Subsurface Flow & Transport Simulation System User Manual*. Berlin, Germany: DHI-WASY GmbH. 116 pp. [Online: Available at: http://feflow.info/fileadmin/FEFLOW/content_feflow61/users_manual.pdf] (Accessed February 2018).

Diederichs, M.S. (2007). The 2003 Canadian Geotechnical Colloquium: mechanistic interpretation and practical application of damage and spalling prediction criteria for deep tunnelling. *Canadian Geotechnical Journal* 44 (9): 1082–1116. https://doi.org/10.1139/T07-033.

Doherty, J. (2015). *Calibration and Uncertainty Analysis for Complex Environmental Models*. Brisbane, Australia: Watermark Numerical Computing. 226 pp.

Doherty, J., Muffels, C., Rumbaugh, J. and Tonkin, M. (2018). *PEST: Model-Independent Parameter Estimation & Uncertainty Analysis*. Bernville, PA: Echo Valley Graphics. [Website: http://www.pesthomepage.org] (Accessed January 2018).

Elsworth, D. (1986). A hybrid boundary-element-finite-element analysis procedure for fluid flow simulation in fractured rock masses. *International Journal for Numerical and Analytical Methods in Geomechanics* 10: 569–584. https://doi.org/10.1002/nag.1610100603.

Ewen, J. and Parkin, G. (1996). Validation of catchment models for predicting land-use and climate change impacts. 1: method. *Journal of Hydrology* 175 (1–4): 583–594. https://doi.org/10.1016/S0022-1694(96)80026-6.

Ewen, J., Parkin, G. and O'Connell, P.E. (2000). SHETRAN: distributed river basin flow and transport modeling system. *Journal of Hydrologic Engineering* 5: 250–258. https://doi.org/10.1061/(ASCE)1084-0699(2000)5:3(250).

Ewen, J., O'Donnell, G.M., Burton, A. and O'Connell, P.E. (2006). Errors and uncertainty in physically-based rainfall-runoff modelling of catchment change effects. *Journal of Hydrology* 330 (3–4): 641–650. https://doi.org/10.1016/j.jhydrol.2006.04.024.

Fairs, T., Younger, P.L. and Parkin, G. (2015). Parsimonious numerical modelling of deep geothermal reservoirs. *Proceedings of the Institution of Civil Engineers* 168 (4): 218–228. http://dx.doi.org/10.1680/ener.14.00026.

Fallah, N.A., Bailey, C., Cross, M. and Taylor, G.A. (2000). Comparison of finite element and finite volume methods application in geometrically nonlinear stress analysis. *Applied Mathematical Modelling* 24 (7): 439–455. https://doi.org/10.1016/S0307-904X(99)00047-5.

Faunt, C.C., D'Agnese, F.A. and O'Brien, G.M. (2010). Chapter D: Hydrology. In: *Death Valley Regional Groundwater Flow System, Nevada and California—Hydrogeologic Framework and Transient Groundwater Flow Model* (eds. W.R. Belcher and D.S. Sweetkind), 133–160. Reston, VA: U.S. Geological Survey, Professional Paper 1711. https://doi.org/10.3133/pp1711.

Geomechanica Inc. (2017). *Irazu 2D Geomechanical Simulation Software, Version 3.0*. Toronto, ON:

Geomechanica Inc. [Website: https://www.geomechanica.com/software] (Accessed August 2018).

Golder Associates (2006). *Fractures, Fracman and Fragmentation: Applications of DFN Models to Block & Panel Caving*. Toronto, ON: Golder Associates. 8 pp.

Golder Associates (2007). *Fracman Reservoir Edition: FRED*. Toronto, ON: Golder Associates. 6 pp.

Goodman, R.E. (1993). *Engineering Geology: Rock in Engineering Construction*. New York, NY: Wiley. 432 pp.

Goodman, R.E. and Kieffer, S.D. (2000). Behavior of rock in slopes. *Journal of Geotechnical and Geoenvironmental Engineering* 126 (8): 675–684. https://doi.org/10.1061/(ASCE)1090-0241(2000)126:8(675).

Grayson, R.B., Moore, I.D. and McHahon, T.A. (1992). Physically based hydrologic modeling: 1. A terrain-based model for investigative purposes. *Water Resources Research* 28 (10): 2639–2658. https://doi.org/10.1029/92WR01258.

Gunnink, J.L., Maljers, D., van Gessel, S.F. et al. (2013). Digital geological model (DGM): a 3D raster model of the subsurface of the Netherlands. *Netherlands Journal of Geosciences* 92 (1): 33–46. https://doi.org/10.1017/S0016774600000263.

Hack, R., Orlic, B., Ozmutlu, S. et al. (2006). Three and more dimensional modelling in geo-engineering. *Bulletin of Engineering Geology and the Environment* 65: 143–153. https://doi.org/10.1007/s10064-005-0021-2.

Haitjema, H.M. (1995). *Analytic Element Modeling of Groundwater Flow*. San Diego, CA: Academic Press. 394 pp.

Hart, R.D. (1993). An introduction to distinct element modeling for rock engineering. In: *Comprehensive Rock Engineering: Principles, Practice and Projects, vol. 2: Analysis and Design Methods* (ed. J.A. Hudson), 245–261. Oxford, UK: Pergamon Press. https://doi.org/10.1016/B978-0-08-040615-2.50016-2.

Hart, R., Cundall, P.A. and Lemos, J. (1988). Formulation of a three-dimensional distinct element model part II: mechanical calculations for motion and interaction of a system composed of many polyhedral blocks. *International Journal of Rock Mechanics and Mining Sciences* 25 (3): 117–126. https://doi.org/10.1016/0148-9062(88)92294-2.

Hecht-Méndez, J., Molina-Giraldo, N., Blum, P. and Bayer, P. (2010). Evaluating MT3DMS for heat transport simulation of closed geothermal systems. *Groundwater* 48 (5): 741–756. https://doi.org/10.1111/j.1745-6584.2010.00678.x.

Hendricks Franssen, H.J.W.M. and Gómez-Hernández, J.J. (2002). 3D inverse modelling of groundwater flow at a fractured site using a stochastic continuum model with multiple statistical populations. *Stochastic Environmental Research and Risk Assessment* 16 (2): 155–174. https://doi.org/10.1007/s00477-002-0091-7.

Hill, M.C. (1998). *Methods and Guidelines for Effective Model Calibration*. Reston, VA: U.S. Geological Survey,

Water-Resources Investigations Report 98-4005. 90 pp. https://doi.org/10.3133/wri984005.

Hill, M.C. and Tiedeman, C.R. (2007). *Effective Groundwater Model Calibration: With Analysis of Data, Sensitivities, Predictions, and Uncertainty*. Hoboken, PA: Wiley. 480 pp

Hudson, J.A. and Harrison, J.P. (1997). *Engineering Rock Mechanics: An Introduction to the Principles*. Oxford, UK: Pergamon. 444 pp. https://doi.org/10.1016/B978-0-08-043864-1.X5000-9.

Itasca International Inc. (2018a). *FLAC Version 8.0: Explicit Continuum Modeling of Non-linear Material Behavior in 2D*. Minneapolis, MN: Itasca International Inc. [Online: Available at: https://www.itascacg.com/software/flac] (Accessed July 2018).

Itasca International Inc. (2018b). *FLAC3D Version 6.0: Explicit Continuum Modeling of Non-linear Material Behavior in 3D*. Minneapolis, MN: Itasca International Inc. [Online: Available at: https://www.itascacg.com/software/flac3d] (Accessed July 2018).

Itasca International Inc. (2018c). *UDEC Version 6.0 Distinct-Element Modeling of Jointed and Blocky Material in 2D*. Minneapolis, MN: Itasca International Inc. [Online: Available at: https://www.itascacg.com/software/udec] (Accessed July 2018).

Itasca International Inc. (2018d). *3DEC Version 5.2 Distinct-Element Modeling of Jointed and Blocky Material in 3D*. Minneapolis, MN: Itasca International Inc. [Online: Available at: https://www.itascacg.com/software/3dec] (Accessed July 2018).

Itasca International Inc. (2018e). *Particle Flow Code (PFC) Version 5.0*. Minneapolis, MN: Itasca International Inc. [Online: Available at: https://www.itascacg.com/software/pfc] (Accessed July 2018).

Jackson C.R. and Spink, A.E.F. (2004). *User's Manual for the Groundwater Flow Model ZOOMQ3D*. Keyworth, UK: British Geological Survey, Internal Report CR/04/140N. 107 pp. [Online: Available at: http://nora.nerc.ac.uk/id/eprint/11829/] (Accessed August 2020).

Jing, L. (2003). A review of techniques, advances and outstanding issues in numerical modelling for rock mechanics and rock engineering. *International Journal of Rock Mechanics and Mining Sciences* 40 (3): 283–353. https://doi.org/10.1016/S1365-1609(03)00013-3.

Kalkani, E.C. and Piteau, D.R. (1976). Finite element analysis of toppling failure at Hell's Gate Bluffs, British Columbia. *Bulletin of the International Association of Engineering Geology* XIII (4): 315–327. https://doi.org/10.2113/gseegeosci.xiii.4.315.

Konikow, L.F. and Bredehoeft, J.D. (1992). Ground-water models cannot be validated. *Advances in Water Resources* 15: 75–83. https://doi.org/10.1016/0309-1708(92)90033-X.

Koo, B.K. and O'Connell, P.E. (2006). An integrated modelling and multicriteria analysis approach to managing nitrate diffuse pollution: 1. Framework and methodology. *Science of the Total Environment* 259 (1–3): 1–16. https://doi.org/10.1016/j.scitotenv.2005.05.042.

Kulatilake, P.H.S.W., Malama, B. and Wang, J. (2001). Physical and particle flow modeling of jointed rock block behavior under uniaxial loading. *International Journal of Rock Mechanics and Mining Sciences* 38 (5): 641–657. https://doi.org/10.1016/S1365-1609(01)00025-9.

Langevin, C.D., Hughes, J.D., Banta, E.R., Niswonger, R.G., Panday, S and Provost, A.M. (2017). *Documentation for the MODFLOW 6 Groundwater Flow Model*. Reston, VA: U.S. Geological Survey, Techniques and Methods 6-A55. 197 pp. https://doi.org/10.3133/tm6A55.

Lewis, E.A. (2016). *A Robust Multi-Purpose Hydrological Model for Great Britain*. PhD Thesis Newcastle University. 280 pp. [Online: Available at: http://hdl.handle.net/10443/3290] (Accessed August 2020).

Lewis, E., Birkinshaw, S., Kilsby, C. and Fowler, H.J. (2018). Development of a system for automated setup of a physically-based, spatially-distributed hydrological model for catchments in Great Britain. *Environmental Modelling & Software* 108: 102–110. https://doi.org/10.1016/j.envsoft.2018.07.006.

Lourens, A., Bierkens, M.F.P. and Van Geer, F.C. (2015). Updating hydraulic properties and layer thicknesses in hydrogeological models using groundwater model calibration results. *Hydrology and Earth System Sciences Discussions* 12: 4191–4231. https://doi.org/10.5194/hessd-12-4191-2015.

Lourens, A. and van Geer, F.C. (2016). Uncertainty propagation of arbitrary probability density functions applied to upscaling of transmissivities. *Stochastic Environmental Research and Risk Assessment* 30 (1): 237–249. https://doi.org/10.1007/s00477-015-1075-8.

MacDonald, A.M. and Allen, D.J. (2001). Aquifer properties of the Chalk of England. *Quarterly Journal of Engineering Geology and Hydrogeology* 34 (4): 371–384. https://doi.org/10.1144/qjegh.34.4.371.

Mahabadi, O.K., Lisjak, A., Munjiza, A. and Grasselli, G. (2012). Y-Geo: new combined finite-discrete element numerical code for geomechanical applications. *International Journal of Geomechanics* 12: 676–688. https://doi.org/10.1061/(ASCE)GM.1943-5622.0000216.

Maidment, D.R. and Morehouse, S. (2002). *Arc Hydro: GIS for Water Resources*. Redlands, CA: ESRI Press. 203 pp.

Mansour, M., Mackay, J., Abesser, C. et al. (2013). Integrated Environmental Modeling Applied at the Basin Scale: Linking Different Types of Models using the OpenMI Standard to Improve Simulation of Groundwater Processes in the Thames Basin, UK. In: *Proceedings, MODFLOW and*

More 2013: Translating Science into Practice, Golden, Colorado, USA, 2–5 June 2013. Golden, CO: Colorado School of Mines. 5 pp. [Online: Available at: http://nora .nerc.ac.uk/id/eprint/501789/] (Accessed August 2020).

Markstrom, S.L., Niswonger, R.G., Regan, R.S. et al. (2008). *GSFLOW-Coupled Ground-water and Surface-water FLOW model based on the integration of the Precipitation-Runoff Modeling System (PRMS) and the Modular Ground-Water Flow Model (MODFLOW-2005)*. Reston, VA: U.S. Geological Survey Techniques and Methods 6-D1. 240 pp. [Online: AVailable at: https://pubs.usgs.gov/tm/tm6d1] (Accessed August 2020).

Marsh, T.J., Parry, S., Kendon, M.C. and Hannaford, J. (2013). *The 2010–12 Drought and Subsequent Extensive Flooding*. Wallingford, UK: Centre for Ecology & Hydrology. 54 pp. [Online: Available at: https://nora.nerc.ac.uk/id/eprint/ 503643/] (Accessed August 2020).

Martin, C.D. (1997). Seventeenth Canadian Geotechnical Colloquium: the effect of cohesion loss and stress path on brittle rock strength. *Canadian Geotechnical Journal* 34 (5): 698–725. https://doi.org/10.1139/t97-030.

Mathers, S.J., Terrington, R.L., Waters, C.N. and Leslie, A.G. (2014). GB3D – a framework for the bedrock geology of Great Britain. *Geoscience Data Journal* 1: 30–42. https://doi .org/10.1002/gdj3.9.

Maxey, G.B. (1964). Hydrostratigraphic units. *Journal of Hydrology* 2: 124–129. https://doi.org/10.1016/0022-1694(64)90023-X.

Maxwell, R.M. and Kollet, S.J. (2008). Interdependence of groundwater dynamics and land-energy feedbacks under climate change. *Nature Geoscience* 1 (10): 665–669. https:// doi.org/10.1038/ngeo315.

Maxwell, R.M., Kollet, S.J., Smith, D.G., et al. (2016). *ParFlow User's Manual*. Integrated GroundWater Modeling Center Report GMWI 2016-01. Golden, CO: Colorado School of Mines. 173 pp. [Online: Available at: http://inside.mines .edu/~rmaxwell/parflow.manual.2-15-16.pdf] (Accessed February 2018).

McKenzie, A.A. and Ward, R.S. (2015). *Estimating Numbers of Properties Susceptible to Groundwater Flooding in England*. Keyworth, UK: British Geological Survey, Open Report OR/15/016. iv + 8 pp. [Online: Available at: http://nora .nerc.ac.uk/id/eprint/510064/] (Accessed August 2020).

Medina, C.R., Letsinger, S.L. and Olyphant, G.A. (2017) Hydrogeologic modeling supported by geologic mapping in three dimensions: do the details really matter? In: *Quaternary Glaciation of the Great Lakes Region: Process, Landforms, Sediments, and Chronology* (eds. A.E. Kehew and B.B. Curry), 217–231. Boulder, CO: Geological Society of America, Special Paper 530. https://doi.org/10.1130/ 2017.2530(11).

Megna, A., Barba, S. and Santini, S. (2005). Normal-fault stress and displacement through finite-element analysis. *Annals of Geophysics* 48 (6): 1009–1016. https://doi.org/10 .4401/ag-3250.

Nasser, K.H., Bolivar, S., Canepa, J. and Dorries, A. (2003). New paradigms for geoscience information management. In: *New Paradigms in Subsurface Prediction* (eds. M.S. Rosenbaum and A.K. Turner), 41–58. Berlin, Germany: Springer, Lecture Notes in Earth Sciences, vol. 99. https:// doi.org/10.1007/3-540-48019-6_4.

Neuman, S.P. and Wierenga, P.J. (2002). *A Comprehensive Strategy of Hydrogeologic Modeling and Uncertainty Analysis for Nuclear Facilities and Sites*. North Bethesda, MD: US Nuclear Regulatory Commission, Report NUREG/CR-6805. 282 pp. [Online: Available at: https:// www.nrc.gov/reading rm/doc-collections/nuregs/ contract/cr6805/] (Accessed August 2020).

Newcastle University (2018). *SHETRAN Hydrological Model*. Newcastle, UK: University of Newcastle. [Website: https:// research.ncl.ac.uk/shetran/] (Accessed August 2020).

Nikolić, M., Roje-Bonacci, T. and Ibrahimbegović, A. (2016). Overview of the numerical methods for the modelling of rock mechanics problems. *Technical Gazette* 23: 627–637. https://doi.org/10.17559/TV-20140521084228.

Parkin, G., O'Donnell, G., Ewen, J. et al. (1996). Validation of catchment models for predicting land-use and climate change impacts. 2. Case study for a Mediterranean catchment. *Journal of Hydrology* 175 (1–4): 595–613. https://doi.org/10.1016/S0022-1694(96)80027-8.

Parkin, G., Birkinshaw, S.J., Younger, P.L. et al. (2007). A numerical modelling and neural network approach to estimate the impact of groundwater abstractions on river flows. *Journal of Hydrology* 339 (1–2): 15–28. https://doi .org/10.1016/j.jhydrol.2007.01.041.

Peach, D.W., Riddick, A.T., Hughes, A. et al. (2017). Model fusion at the British Geological Survey: experiences and future trends. *Geological Society of London Special Publications* 408: 7–16. https://doi.org/10.1144/SP408.13.

Pechlivanidis, I.G., Jackson, B., McIntyre, N. and Wheater, H.S. (2011). Catchment scale hydrological modelling: a review of model types, calibration approaches and uncertainty analysis methods in the context of recent developments in technology and applications. *Global NEST Journal* 13 (3): 193–214. https://doi.org/10.30955/gnj .000778.

Perrone, N. and Kao, R. (1975). A general finite difference method for arbitrary meshes. *Computers and Structures* 5: 45–58. https://doi.org/10.1016/0045-7949(75)90018-8.

Pine, R.J., Coggan, J.S., Flynn, Z.N. and Elmo, D. (2006). The development of a new numerical modelling approach for naturally fractured rock masses. *Rock Mechanics and Rock*

Engineering 39 (5): 395–419. https://doi.org/10.1007/s00603-006-0083-x.

PLAXIS bv. (2017). *Plaxis 3D 2017 Tutorial Manual*. 134 pp. [Online: Available at: https://www.plaxis.com/support/manuals/plaxis-3d-manuals] (Accessed July 2018).

PLAXIS bv. (2018). *Plaxis 2D 2018 Tutorial Manual*. 190 pp. [Online: Available at: https://www.plaxis.com/support/manuals/plaxis-2d-manuals] (Accessed July 2018).

Poeter, E.P. and Hill, M.C. (1996). Unrealistic parameter estimates in inverse modelling: a problem or a benefit for model calibration? In: *Proceedings of the ModelCARE 96 Conference, Golden, Colorado* (eds. K. Kovar and P. van der Heijde), 277–285. Wallingford, UK: International Association of Hydrological Sciences, Publication 237. [Online: Available at: http://hydrologie.org/redbooks/a237/iahs_237_0277.pdf] (Accessed August 23, 2020).

Poeter, E.P. and Hill, M.C. (1997). Inverse models: a necessary next step in ground-water modeling. *Groundwater* 35 (2): 250–260. https://doi.org/10.1111/j.1745-6584.1997.tb00082.x.

Popper, K. (1959). *The Logic of Scientific Discovery*. New York, NY: Harper & Row. 480 pp.

Price, M., Low, R.G. and McCann, C. (2000). Mechanisms of water storage and flow in the unsaturated zone of the Chalk aquifer. *Journal of Hydrology* 233 (1–4): 54–71. https://doi.org/10.1016/S0022-1694(00)00222-5.

Rees, J.G., Gibson, A.D., Harrison, M. et al. (2009). Regional modelling of geohazard change. *Engineering Geology Special Publications* 22: 49–63. https://doi.org/10.1144/EGSP22.4.

Refsgaard, J.C., Van der Sluijs, J.P., Lajer Højberg, A. and Vanrolleghem, P.A. (2007). Uncertainty in the environmental modelling process – a framework and guidance. *Environmental Modelling & Software* 22: 1543–1556. https://doi.org/10.1016/j.envsoft.2007.02.004.

Rengers, N., Hack, R., Huisman, M. et al. (2002). Information technology applied to engineering geology. In: *Proceedings, 9th congress of the International Association for Engineering Geology and the Environment: Engineering geology for developing countries, Durban, South Africa, 16–20 September 2002* (eds. J.L. van Rooy and C.A. Jermy), 121–143. Pretoria, South Africa: South African Institute of Engineering and Environmental Geologists.

Rhén, I., Follin, S. and Hermanson, J. (2003). *Hydrological Site Descriptive Model – A Strategy for its Development during Site Investigations*. Stockholm, Sweden: Swedish Nuclear Fuel and Waste Management Co, Report R-03-08. 109 pp. [Online. Available at: http://www.skb.com/publication/20393/R-03-08.pdf] (Accessed August 2020).

Rosenbaum, M.S. (2001). Neuro-fuzzy modeling in engineering geology. *Quarterly Journal of Engineering Geology and Hydrogeology* 34 (4): 415–415. https://doi.org/10.1144/qjegh.34.4.415.

Seaber, P.R. (1988). Hydrostratigraphic units. In: *Hydrogeology* (eds. W. Back, J.R. Rosenschein and P.R. Seaber), 9–14. Boulder, CO: Geological Society of America. DNAG, Geology of North America Vol. O-2. https://doi.org/10.1130/DNAG-GNA-O2.9.

Shelley, C. and Burke, H. (2018). *Model Metadata Report for the South Downs Teaching Model*. Keyworth, UK: British Geological Survey, Open Report OR/18/003. 12 pp. [Online: Available at: http://nora.nerc.ac.uk/id/eprint/519301/] (Accessed August 2020).

Shepley, M.G., Whiteman, M.I., Hulme, P.J. and Grout, M.W. [Eds.] (2012). *Groundwater Resources Modelling: A Case Study from the UK*. London, UK: Geological Society, Special Publication 364. 378 pp. https://doi.org/10.1144/SP364.0.

Shi, G. (1992). Discontinuous deformation analysis – a new numerical model for the statics and dynamics of deformable block structures. *Engineering Computations* 9: 157–168. https://doi.org/10.1108/eb023855.

Shi, G.H. and Goodman, R.E. (1985). Two dimensional discontinuous deformation analysis. *Numerical and Analytical Methods in Geomechanics* 9 (6): 541–556. https://doi.org/10.1002/nag.1610090604.

Stead, D., Eberhardt, E. and Coggan, J.S. (2006). Developments in the characterization of complex rock slope deformation and failure using numerical modelling techniques. *Engineering Geology* 83: 217–235. https://doi.org/10.1016/j.enggeo.2005.06.033.

Strack, O.D.L. (1989). *Groundwater Mechanics*. Englewood Cliffs, NJ: Prentice Hall. 732 pp.

Strassberg, G., Jones, N.L. and Maidment, D.R. (2011). *Arc Hydro Groundwater: GIS for Hydrogeology*. Redlands, CA: ESRI Press. 160 pp.

Taylor, R.G., Scanlon, B., Döll, P. et al. (2013). Ground water and climate change. *Nature Climate Change* 3 (4): 322–329. https://doi.org/10.1038/nclimate1744.

Terrington, R. L., Gow, H. and Aldiss D.T.. (2014). *Model Metadata Report for the 3D Model of the Superficial Deposits near Chichester*. Keyworth, UK: British Geological Survey, Open Report OR/14/71.

Tóth, J. (2016). The evolutionary concepts and practical utilization of the Tothian theory of regional groundwater flow. *International Journal of Earth & Environmental Sciences* 1: 111. https://doi.org/10.15344/2456-351X/2016/111.

Turner, A.K. (1989). The role of three-dimensional geographic information systems in subsurface characterization for hydrogeological applications. In: *Three Dimensional Applications in Geographic Information Systems* (ed. J. Raper), 115–127. London, UK: Taylor & Francis.

Turner, A.K. (2006). Challenges and trends for geological modelling and visualisation. *Bulletin of Engineering Geology and the Environment* 65: 109–127. https://doi.org/10.1007/s10064-005-0015-0.

Upton, K.A. and Jackson, C.R. (2011). Simulation of the spatio-temporal extent of groundwater flooding using statistical methods of hydrograph classification and lumped parameter models. *Hydrological Processes* 25: 1949–1963. https://doi.org/10.1002/hyp.7951.

Valstar, J.R., McLaughlin, D.B., te Stroet, C.B. and Van Geer, F.C. (2004). A representer-based inverse method for groundwater flow and transport applications. *Water Resources Research* 40: W05116. https://doi.org/10.1029/2003WR002922.

Van Oosterom, P. and Stoter, J. (2010). 5D data modelling: full integration of 2D/3D space, time and scale dimensions. In: *Proceedings, Geographic Information Science: 6th International Conference, GIScience 2010, Zurich, Switzerland* (eds. S.I. Fabrikant, T. Reichenbacher, M. van Kreveld and C. Schlieder), 311–324. Berlin, Germany: Springer. https://doi.org/10.1007/978-3-642-15300-6_22.

Vazaios, I., Vlachopoulos, N. and Diederichs, M.S. (2019). Mechanical analysis and interpretation of excavation damage zone formation around deep tunnels within massive rock masses using hybrid finite-discrete element approach: case of Atomic Energy of Canada Limited (AECL) Underground Research Laboratory (URL) test tunnel. *Canadian Geotechnical Journal* 56 (1): 35–59. https://doi.org/10.1139/cgj-2017-0578

Veldkamp, J.G., Bremmer, C.N., Hack, H.R.G.K. et al. (2001). Combination of 3D GIS and FEM modelling of the 2nd Heinenoord tunnel, the Netherlands. In: *Proceedings, EngGeoCity 2001 – International Symposium on Engineering Geological Problems of Urban Areas*. Beijing, China: IAEG / Bejing, China: IUGS / Ekaterinburg, Russia: UralTICIZ. 8 pp. [Online: Available at: https://www.semanticscholar.org/paper/Combination-of-3D-GIS-and-FEM-modelling-of-the-2nd-Veldkamp-Bremmer/e6a9d7fee75ed54f797b6659ccfc7fe945361d25] (Accessed August 2020).

Vlachopoulos, N. and Diederichs, M.S. (2009). Improved longitudinal displacement profiles for convergence confinement analysis of deep tunnels. *Rock Mechanics and Rock Engineering* 42 (2): 131–146. https://doi.org/10.1007/s00603-009-0176-4.

Vlachopoulos, N. and Diederichs, M.S. (2014). Appropriate uses and practical limitations of 2D numerical analysis of tunnels and tunnel support response. *Geotechnical and Geological Engineering* 32: 469–488. https://doi.org/10.1007/s10706-014-9727-x.

Vlachopoulos, N. and Vazaios, I. (2018). The numerical simulation of hard rocks for tunnelling purposes at great depths: a comparison between the hybrid FDEM method and continuous techniques. *Advances in Civil Engineering* 2018: Article ID 3868716. 18 pp. https://doi.org/10.1155/2018/3868716.

Vlachopoulos, N., Diederichs, M.S., Marinos, V. and Marinos, P. (2013). Tunnel behaviour associated with the weak Alpine rock masses of the Driskos Twin Tunnel system, Egnatia Odos Highway. *Canadian Geotechnical Journal* 50 (1): 91–120. https://doi.org/10.1139/cgj-2012-0025.

Vlachopoulos, N., Vazaios, I. and Madjdabadi, B. (2018). Investigation into the influence of excavation of twin bored tunnels within weak rock masses adjacent to slopes. *Canadian Geotechnical Journal* 55 (11): https://doi.org/10.1139/cgj-2017-0392.

Waters, C.N., Terrington, R.L., Cooper, M.R. et al. (2016). *The Construction of a Bedrock Geology Model for the UK: UK3D_v2015*. Keyworth, UK: British Geological Survey, Report OR/15/069. 22 pp. [Online: Available at: http://nora.nerc.ac.uk/id/eprint/512904/] (Accessed August 2020).

Wheater, H.S. and Peach, D.W. (2004). Developing interdisciplinary science for integrated catchment management: the UK Lowland Catchment Research (LOCAR) programme. *International Journal of Water Resources Development* 20 (3): 369–385. https://doi.org/10.1080/0790062042000248565.

Wheel, M.A. (1996). A geometrically versatile finite volume formulation for plane elastostatic stress analysis. *Journal of Strain Analysis* 31: 111–116. https://doi.org/10.1243/03093247V312111.

Wittke, W. (1990). *Rock Mechanics: Theory and Applications with Case Histories*. Berlin: Springer. 1075 pp. https://doi.org/10.1007/978-3-642-88109-1.

Wyllie, D.C. and Mah, C.W. (2004). *Rock Slope Engineering*, 4e. New York: Spon Press. 431 pp.

Zhang, X., Lu, M. and Wegner, J.L. (2000). A 2-D meshless model for jointed rock structures. *International Journal for Numerical Methods in Engineering* 47: 1649–1661. https://doi.org/10.1002/(SICI)1097-0207(20000410)47:10%3C1649::AID-NME843%3E3.0.CO;2-S.

Zienkiewicz, O.C. and Cheung, Y.K. (1967). *The Finite Element Method in Structural and Continuum Mechanics*. London, UK: McGraw-Hill. 274 pp.

Zienkiewicz, O.C., Taylor, R.L. and Zhu, J.Z. (2013). *The Finite Element Method: Its Basis and Fundamentals*, 7e. Oxford, UK: Butterworth-Heinemann. 756 pp. https://doi.org/10.1016/C2009-0-24909-9.

15

Uncertainty in 3-D Geological Models

Marco Bianchi[1], Alan Keith Turner[1,2], Murray Lark[1,3], and Gabriel Courrioux[4]

[1] *British Geological Survey, Keyworth, Nottingham NG12 5GG, UK*
[2] *Colorado School of Mines, Golden, CO 80401, USA*
[3] *School of Biosciences, University of Nottingham, Sutton Bonington, Nottingham LE12 5RD, UK*
[4] *Bureau de Recherches Géologiques et Minières (BRGM), 45060 Orléans Cedex 02, France*

15.1 Introduction

Because the subsurface is not readily accessible to direct observation, uncertainty is ubiquitous in the earth sciences. However, uncertainty should not be seen just as a limitation of our interpretations and predictions. Rather, the quantification and communication of uncertainty provides a better understanding of the system of interest. Traditional 2-D geological maps contain uncertainties which are generally indicated in a limited fashion. For example, geological boundaries are presented as continuous lines when observed in the field and as dashed lines when they are not. Many Canadian geological maps define large bedrock outcrops by brighter colors and minor outcrops with a symbol; a very useful feature when glacial or younger alluvial deposits mask bedrock exposures (Figure 15.1). A more explicit definition of the uncertainty associated with geological maps was probably considered unnecessary when they were rarely updated and geologists were the primary map users. As geological maps began to be used by a much wider range of users, and digital techniques allowed maps to be updated on an almost daily basis, the need for clearly expressed uncertainty became important (Culshaw 2005).

All 3-D geological modeling procedures generate representations of the subsurface from incomplete and uncertain data by some process of interpretation, numerical interpolation, or a combination of the two. While the geologist who produces the model may be aware of the inherent uncertainty of the available geological information, the modeling process introduces additional uncertainties; therefore, a quantification of the overall model uncertainty becomes an essential step in the modeling framework (Oreskes 2003).

Establishing the overall uncertainty of some 3-D subsurface models is a challenge; an even greater challenge is conveying that uncertainty to the end-user in a useful and understandable manner so the models can support real-world decision-making. Representing this multi-component uncertainty in a form understandable to the user community is nontrivial. Diverse approaches to the representation and interpretation of uncertainty in geological models have been proposed and tested in recent years. Wellmann and Caumon (2018) provide an extensive review of the concepts and methods that have been used to analyze, quantify, and communicate uncertainties in geological models.

15.2 Sources of Uncertainty

Geoscientists make predictions from often sparsely distributed, incomplete and low-quality data sources; the resulting models contain this uncertainty. Boreholes only sample limited volumes, seismic surveys have resolution limits, and outcrop data is obscured. Quantifying this uncertainty is important because applications of these models have important social and commercial implications.

15.2.1 Cause-Effect Analysis

A cause-and-effect diagram, known as an "Ishikawa" (after its inventor) or "fishbone" diagram (Kindlarski 1984), provides a graphical overview of the sources and causes of uncertainty in geological models. The fishbone diagram is built up hierarchically. The identified principal causes of uncertainty are placed as branches along a central axis forming the backbone (Figure 15.2). Contributing sub-causes and tertiary causes are progressively identified and added as smaller branches. Cause-and-effect diagrams

Applied Multidimensional Geological Modeling: Informing Sustainable Human Interactions with the Shallow Subsurface, First Edition.
Edited by Alan Keith Turner, Holger Kessler, and Michiel J. van der Meulen.
© 2021 John Wiley & Sons Ltd. Published 2021 by John Wiley & Sons Ltd.

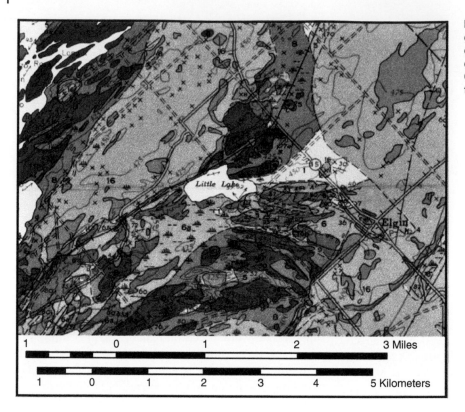

Figure 15.1 Example of a Canadian geological map showing bedrock outcrops influenced by overlying glacial deposits; a small portion of the Westport, Ontario 1:63 360 scale map. Modified from Wynne-Edwards (1967).

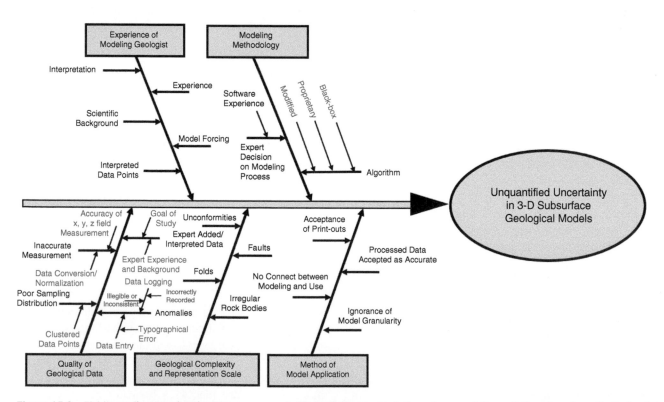

Figure 15.2 Fishbone diagram showing the causes and effects of the uncertainties when modeling a 3-D surface from limited point depth (borehole) data. Secondary causes are shown in blue; tertiary causes are shown in red. Modified from Cave and Wood (2002).

are best derived by team members working together and brainstorming. The cause-and-effect diagram in Figure 15.2 defines five main sources of uncertainty within 3-D geological framework, models, which are discussed in the following sections:

(1) The quality of geological data;
(2) The complexity of the geology being modeled and the scale at which it is conveyed;
(3) The experience of the modeling geologist;
(4) The geological modeling methodology;
(5) The application of the model output.

15.2.2 Uncertainty Source 1: Quality of Geological Data

The quality of geological data sources contributes to model uncertainty by four inter-related causes: (i) inaccurate measurement, (ii) experience in data interpretation, (iii) a poor sampling distribution, and (iv) anomalies in legacy data sources.

15.2.2.1 Inaccurate Measurement
Measurement accuracy is influenced by several factors: the method used to obtain a sample, the instrument used to measure a property, the experience of the individual(s) collecting the data, and the accuracy of the sample geo-location data. While modern survey technologies and Geographic Information Systems (GIS) tools help identify and correct geo-location errors, legacy datasets often contain unresolved positional errors. Water-well lithological logs provide notoriously inaccurate subsurface records compared with carefully monitored geological or geotechnical borings; however, they are often the most widespread existing source of shallow subsurface lithological information, and so are widely used (Chapter 7). Direct subsurface observations and boreholes are inherently more precise than geophysical sources. However, direct observations and precise measurements, which are generally defined as "hard data," are more expensive in terms of costs and time required for collection than the qualitative and interpretative information, termed "soft data," from other sources such as geophysical surveys. Therefore, hard data tend be more sparse than soft data for a given study area. The quality of data measurements is also influenced by the experience of those responsible for collecting the data. Even a qualitative, subjective, description of the inherent uncertainty of the data sources is potentially valuable to the user community.

15.2.2.2 Experience in Data Interpretation
Assessing the accuracy of geological data interpretation has been the subject of several studies. Torvela and Bond (2011) examined influences on interpretations by experts of seismic data from a folded thrust-belt site. Aitken

et al. (2008) evaluated factors influencing the uncertainty of aerial magnetic survey interpretations. Tacher et al. (2006) describe geostatistical and probabilistic modeling approaches to evaluate geological structural uncertainties in 3-D models and provided examples from the new Lötschberg railway tunnel in Switzerland. Calcagno et al. (2008) used implicit modeling procedures to define the uncertainty of geological contacts. Chapter 12 has details of this process.

Jessell et al. (2010) identified and classified information typically collected by geologists from outcrops or boreholes (Table 15.1). The uncertainties associated with this information are usually not quantified, so assessing the reliability of derived data sets becomes very difficult. Bistacchi et al. (2008) present a case study where subsurface structure was inferred by projecting observed structural trends onto cross-sections at depth. A "buffer" region may represent the uncertainty of location of geological contacts at depth resulting from the uncertainty of the orientation of the contact observed at the surface (Figure 15.3).

Jessell et al. (2010) proposed a parallel simulation approach to develop subsurface geological models using both geological and geophysical data sources. Employing the concept of multiple working hypotheses, the approach developed a series of 3-D geology models, then applied them as independent constraints on estimated geophysical rock property parameters, to produce a model that reflected both geophysical and geological information. Jessell et al. (2010) also proposed a predictive mapping approach which combines the parallel simulation approach with a sensitivity analysis of the geological observations to allow geologists to modify their field mapping and data collection strategies to produce more robust models.

Robust model predictions can be achieved by averaging the outputs from multiple models of the same system based on different interpretations and data. There are several model-averaging approaches, ranging from assigning weights to model predictions based on some metric of model performance, to more sophisticated techniques, such as Bayesian Model Averaging (BMA). Diks and Vrugt (2010) and Refsgaard et al. (2012) review the different techniques used in the context of the hydrogeological sciences.

15.2.2.3 Poor Sampling Distribution
Borehole density is frequently a proxy for model uncertainty; the model is assumed to be more reliable where the borehole density is higher. But borehole density plots only show one aspect of uncertainty. Because most density plots do not indicate the depth of the boreholes, they fail to account for the typical decrease of data density with depth and thus do not indicate total model uncertainty. For example, a 3-D model created to identify faults and displacement in the Permo-Triassic of Yorkshire and the

Table 15.1 Classes of geological information which are routinely collected to make 2-D and 3-D geological models.

Classes of geological information	Principal information sub-classes	Examples of data sources
Spatial information	Purely geometric information (x, y, z, scalar property)	Point (e.g. outcrop lithology, grain size, petrophysical properties)
		Surface (e.g. contact location and type)
		Volume (e.g. body centroid)
	Gradient information (x, y, z, vector property)	External surface and foliation orientation (e.g. orientation of bedding, fault)
		Internal surfaces and lineations (e.g. orientation of stretching lineation, cleavage)
		Derived surfaces and lineations (e.g. fold axial plane or bedding/cleavage intersection lineation)
	Topological Information (x, y, z, a vs b)	Neighborhood relationships (e.g. does unit A come into contact with unit B)
Temporal information	Absolute (x, y, z, age)	Radiometric dating (e.g. Rb/Sr)
	Relative (x, y, z, a vs b)	Stratigraphy
		Local age relationships
Hybrid information	Relative (x, y, z, fault contact)	Age relationships + contact types

Source: Adapted from Jessell et al. (2010).

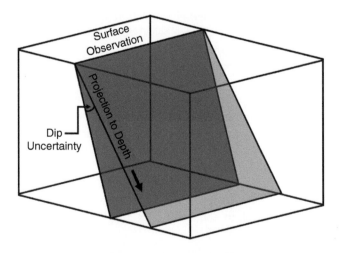

Figure 15.3 A "buffer" (cf. Bistacchi et al. 2008) i.e., a projection area at depth which honors uncertainty of dip measured at the surface.

East Midlands (Ford et al. 2006) developed a borehole density plot (Figure 15.4a) as an indicator of model uncertainty. However, it did not give any explicit information about the inherent uncertainty of fault structure locations (Figure 15.4b), and therefore had limited application.

The quality of the information provided in each borehole record influences model uncertainty. Many models must rely on existing (legacy) records (Section 15.2.2.4) with variable and low reliability. The Ontario Geological Survey (OGS) routinely provides color-coded "dot maps" associated with each stratigraphic horizon in their 3-D geological models (Box 15.1). These allow the user to qualitatively assess the uncertainty of each stratigraphic horizon, based on a visual assessment of the distribution and quality of the data points used to create it.

The Geological Survey of Canada developed a more sophisticated quantitative method of defining the uncertainty of a 3-D model of the Oak Ridges Moraine (ORM) in Ontario. The method combined distance-based qualitative ranges from different classes of boreholes and a polygon grid of outcrop locations to produce a contoured map showing data "confidence" in the range of 1 (highest confidence) to 0 (lowest confidence) (Box 15.2). A similar approach was taken by Van der Meulen et al. (2007), who derived confidence zones (high/medium/low) from the density of data used for a 3-D clay resource model, and presented it as an overlay on map visualizations of that model.

Box 15.1 Evaluating Uncertainty at the Ontario Geological Survey

Abigail Burt and Andy Bajc

Ontario Geological Survey, 933 Ramsey Lake Road, Sudbury ON P3E 6B5, Canada

The Ontario Geological Survey (OGS) has used a visual approach to describe uncertainty in its recent 3-D hydro-stratigraphic mapping projects (Bajc and Dodge 2011; Burt and Dodge 2011, 2016). This approach uses color-codes to define the reliability of the source data based on the "quality" of each data point as assessed by geologists.

The OGS modeling method is described in Chapter 19. Hydrostratigraphic horizons are interpolated from sets of 3-D points, referred to as "picks," which are manually digitized onto borehole traces. These picks are weighted as low, medium, or high according to their data source. Water-well records are generally considered to be of low quality; measured sections or continuously cored boreholes that have been logged by a geoscientist are considered to be high-quality records. When necessary, additional data points are placed to refine the geometry of the hydrostratigraphic surface.

Each interpolated hydrostratigraphic surface map is complemented by a "dot map" that indicates the location and quality of the picks used to interpolate the surface (Figure 15.1.1). In this example, from the Barrie-Oro moraine 3-D project in southern Ontario, low-quality picks are shown in blue, medium-quality picks in green and high-quality picks in yellow (Burt and Dodge 2011). Definitive picks (weighted as high during the modeling process) are from continuously cored boreholes drilled specifically for the project. Additional picks used to refine the model surface are shown by an X.

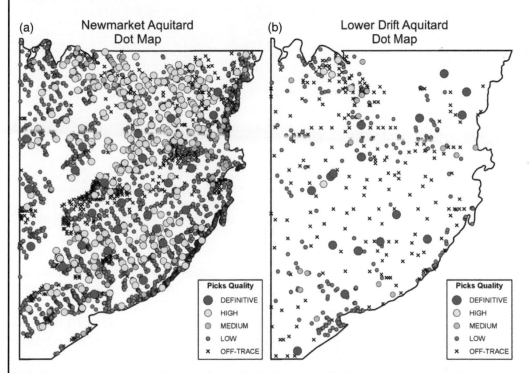

Figure 15.1.1 Examples of dot maps for two hydrostratigraphic units from the Barrie-Oro Moraine 3-D project area, southern Ontario, Canada. (a) Many high-quality picks were used to generate the Newmarket aquitard surface; thus, it can be considered a relatively reliable surface. (b) Very few picks were used to generate the Lower Drift aquitard surface, so it is considered less reliable. Source: Burt and Dodge 2011. Copyright © Queen's Printer for Ontario, 2011. Used by permission of the Ontario Geological Survey.

The dot maps provide an estimate of the reliability of the source data and give an immediate visual indication of the distribution of data across the study area. The dot map of the shallow Newmarket aquitard (Figure 15.1.1a) shows many high-quality picks; thus, the generated surface is considered reliable. In contrast, the lower drift aquitard (Figure 15.1.1b) has only limited numbers of picks, and so is a much less-reliable surface.

Box 15.2 Evaluating Uncertainty in the Oak Ridges Moraine Geological Model

C.E. Logan, H.A.J. Russell, and D.R. Sharpe
Geological Survey of Canada, Ottawa, Ontario, Canada

Evaluation of geological model uncertainty has often focused on reducing and assessing the aleatory uncertainty in interpolation. Reliable numeric geological models are based on defensible conceptual geological models with little consideration of epistemic uncertainty.

The complex aquifer system of the Oak Ridges Moraine (ORM) consists of a sequence of discontinuous strata with a prominent regional unconformity (Logan et al. 2006). A 3-D subsurface geological model, created with an innovative combined stratigraphic database-GIS approach, defines four principal Quaternary units plus Paleozoic bedrock as a succession of interpolated surfaces (Sharpe et al. 2007). The model-building process involved stratigraphically coding high-quality data to form a stratigraphic training framework, which was merged with digital elevation model (DEM)-controlled surface mapping polygons interpolated to form 50 m grids in a GIS. Additional surfaces were defined by integrating an extensive and diverse array of subsurface geological and archival datasets. Expert knowledge supplied geological context that was constrained by this conceptual stratigraphic framework allowed an expert system to automatically interpret many low-quality water well records (Figure 15.2.1).

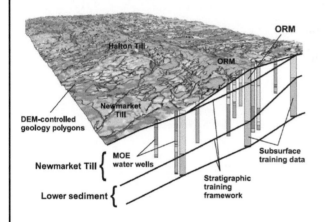

Figure 15.2.1 A schematic representation of a data-driven model with conceptual geological control. (Logan et al. 2006).

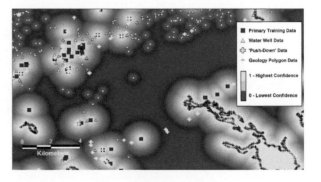

Figure 15.2.2 Sample data distribution showing types of data support and the related confidence grid map derived from combining the distance-based qualitative ranges and Boolean geology polygon grid. Source: Modified from Logan et al. (2005).

The accuracy of any grid cell value from an interpolated stratigraphic surface is directly related to the proximity and relative quality of data points. Confidence grids were produced for each stratigraphic surface to establish an estimation of potential elevation error.

The uncertainty estimation was implemented by creating a distance buffer for each data point using the same range-of-influence rules used for coding water well records (Logan et al. 2005, 2006). A separate grid was produced for training data, water well records and "push-down" points. Using grid math functions, each grid was converted into a fuzzy set of grid cell values ranging from 0 at the maximum range of influence to a maximum at the node location of either 1.0, 0.5, or 0.25 for training data, water wells, and "push-down" data points, respectively.

When used qualitatively, the confidence value "1" represents the highest confidence or good data support, and value "0" represents the lowest confidence or poor data support (Figure 15.2.2). The variance of surface elevation at confidence grid values of "0" was estimated as about 10 m, and at confidence grid values of "1" it becomes about 1 m. The confidence estimation allows the model to be easily queried at any location to provide an indication of potential accuracy. The adopted approach accounts for both aleatoric uncertainty of data support (distance and density) and partial epistemic uncertainty of contained geological knowledge.

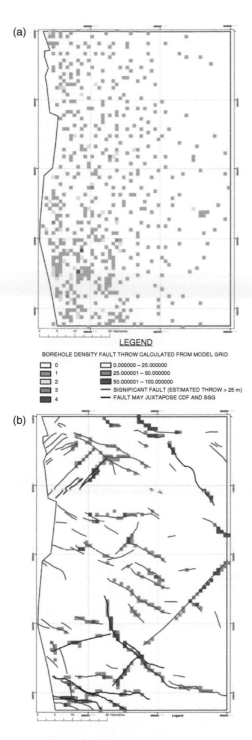

LEGEND

BOREHOLE DENSITY FAULT THROW CALCULATED FROM MODEL GRID

- 0
- 1
- 2
- 3
- 4

- 0.000000 – 25.000000
- 25.000001 – 50.000000
- 50.000001 – 100.000000
- SIGNIFICANT FAULT (ESTIMATED THROW > 25 m)
- FAULT MAY JUXTAPOSE CDF AND SSG

Figure 15.4 Qualitative model uncertainty evaluation for the Permo-Triassic of Yorkshire and the East Midlands; (a) Grid-based borehole density plot; (b) Faulting pattern – its uncertainty was not assessed by the borehole density plot. Modified from Ford et al. (2006).

15.2.2.4 Anomalies in Legacy Data Sources

Practical and economic constraints on additional field investigation or data collection cause many modeling projects to rely on existing borehole and observational data. Use of existing, or legacy, data frequently produces significant uncertainty (see Chapter 7). Older historical data may use obsolete classifications, and records may be incomplete. Some records can no longer be verified on-site due to changes in land use; this is especially common in urban areas. Transcription of old paper records to digital formats may introduce additional inadvertent errors. Although careful and extensive review of these data sources can often eliminate inaccuracies, the transcription of poor-quality records is an expensive and tedious task that is a barrier to the implementation of modeling for some projects. The Geological Survey of the Netherlands, for example, decided to digitize their paper borehole description in the late 1980s, and from the digital archive that was obtained in this way, a relational coded database was created in the late 1990s (van der Meulen et al. 2013). Digitization was a key decision that would eventually enable the data-intensive systematic 3-D mapping program that the Dutch survey runs. But this operation took two conversions – first from paper records to digital records, and then from text strings to coded information – and both involved errors and accuracies, and millions of euros have been spent to date to on dealing with these.

15.2.3 Uncertainty Source 2: Complexity of the Geological Environment

The complexity of the geological setting can also influence the uncertainty of the geological model. Greater certainty is inherent in models of deposits with more predictable geological structures such as deep marine or lake deposits, depositional environments that are physically large and have low (hydro)dynamics. Uncertainty is much greater in areas where the geological setting and units are products of relatively unpredictable, more dynamic, processes and conditions, such as fluvial or glacial deposits, faulted or folded geological units, and anthropogenic deposits, especially in urban areas (Chapter 25). In predictable environments, extrapolation between boreholes over hundreds of meters is potentially acceptable; in more complex environments, any extrapolation of borehole data may result in considerable inaccuracies (Figure 15.5). Much geological model uncertainty results from a lack of geological information, independent of scale, because sparse outcrops limit field observations and inadequate borehole or geophysical data cannot precisely define subsurface conditions.

Figure 15.5 In folded strata, such as in this cliff exposure, extrapolation between two nearby boreholes, as shown, will produce unacceptable models unless constrained by other information. (Kessler et al. 2013).

15.2.4 Uncertainty Source 3: Experience of the Modeling Geologist

To create geological models from sparse borehole datasets, geologists make interpretations about the geological characteristics of the subsurface between boreholes. This interpretation, which requires extrapolation from 1-D or 2-D information to a 3-D representation of the geology of the area of interest, also requires significant geological expertise, scientific background, and an ability to interpret geology in three dimensions. Borehole logs and geological maps produced by different geologists will reflect differences in their experience and background. Research to quantify this uncertainty was conducted by Lark et al. (2014a; see also Courrioux et al. 2015). Section 15.4.2 describes this research.

15.2.5 Uncertainty Source 4: Modeling Methodology

Model uncertainty is influenced by the choice of modeling methodology. Geological models can be developed in many ways; the two broadest categories are deterministic models and stochastic models.

Deterministic models are implemented through integration of direct and/or indirect geological observations and expert knowledge and interpretation to produce a representation of the geological setting of a certain area. Because of the substantial input of geological expertise from the modeler, deterministic models typically reflect the known geological relationships (stratigraphic, chronological, lithological, etc.) and/or established conceptualizations of the geological system. Deterministic models are further subdivided into explicit models, which can be developed by several approaches (Chapters 9–11), and implicit models (Chapter 12). For both types of models, the uncertainty

associated with this type of models is not easily quantified. Being inherently more subjective than stochastic modeling, deterministic modeling is especially reliant on the experience and expertise of the involved staff.

Stochastic models (Chapter 13) take into account the inherent randomness in the data by generating multiple, statistically equiprobable, realizations of the geological properties of the subsurface. Because these models consist of an ensemble of realizations, statistical approaches can be applied to quantify the uncertainty of the model predictions. Because these predictions rely on mathematical functions, additional input or modifications maybe required from the modeler to check the consistency of these stochastic models with geological rules (e.g. chronostratigraphic, unconformities, tectonic boundaries, etc.) and assure geological realism. In many cases, stochastic models are used to simulate smaller-scale phenomena existing within larger deterministically modeled systems, such as facies within stratigraphic units; these smaller phenomena often cannot be accurately observed or modeled using deterministic methods (see the GeoTOP model example in Chapter 11).

15.2.6 Uncertainty Source 5: Model Application

The appropriate measurement and communication of model uncertainty depends on how the model is used (Kessler et al. 2013), bearing in mind that geological model uncertainties propagate into the domain of model applications. Thus, model uncertainties affect the utility of the model to assess various important decisions such as economic or safety risks and their associated liabilities or potential precautionary measures. Geomodelers primarily see uncertainty as a measure for, or affirmation of, the intrinsic quality of their work. In fact, experienced geoscientists who produce, evaluate, or use a geological model do not rely on uncertainty information to assess the quality of a model. They have the ability to identify misrepresentations of geologic features by direct examination, and will reject geostatistically significant, but geologically improbable, features.

In contrast, non-geoscientists who use the model will have to take the model for granted; this means that quantifications of uncertainty are vital. To these model users, a 3-D model predicts the location of a condition that could be relevant to an investment decision, project design, safety measures, etc. In such contexts, the interpretation of geologic uncertainty is not always straightforward to geoscientists. For example, a large uncertainty in the location of a boundary between two units that are lithostratigraphically distinct but have comparable porosities and permeabilities could be trivial to a geohydrologist, but even a low probability of the occurrence of a minor fault that does not concern a geologist much could have huge impact on the design of a dam.

It is vital that the geological modeling methodologies and all sources of uncertainty be fully open and transparent so that the potential influence of uncertainty on the downstream application is clear. Geotechnical users who generate new data and conceptual knowledge during ground investigations require access to the original data and documentation concerning interpretations used in model development. When geological models are used to guide or constrain numerical models supporting decision-making, model users require a robust numerical assessment of uncertainty.

15.3 Alternative Approaches to Uncertainty Evaluation

The multiple sources of uncertainty within geological models can be described by various methods, which range from qualitative (or descriptive), to semi-quantitative, to quantitative (or numerical). Cave and Wood (2002) provide a comprehensive review of the approaches used in the United Kingdom.

15.3.1 Qualitative Methods

Qualitative methods emphasize three sources of model uncertainty: data quantity, data quality, and geological complexity (Table 15.2). Geological surfaces defined by high densities of high-quality observations (boreholes) in areas of low geological complexity are likely to be closer to their true locations than those surfaces defined by fewer and/or lower-quality boreholes in complex geological environments. Qualitative assessments of uncertainty in geological models are often based on maps showing the spatial variability of borehole density and/or borehole quality (Boxes 15.1 and 15.2).

Qualitative methods are both relatively easy to produce and understandable by many users, so they play a useful role in many cases. Simple graphic displays are readily understood and appreciated by many user communities, as has been the experience in Ontario with their "dot-maps" (Box 15.1).

15.3.2 Semi-Quantitative Methods

Semi-quantitative methods combine aspects of qualitative methods and more computationally-intensive quantitative methods. Semi-quantitative methods typically employ statistics to evaluate the combined uncertainties in a model produced by some or all three sources of uncertainty defined in Table 15.2. Lark et al. (2014a)

Table 15.2 Qualification of uncertainty based on borehole density, borehole quality, and geological complexity.

Model characteristic	Qualitative level of uncertainty	
	Low	High
Borehole data density	High density	Low density
Borehole quality	High quality	Low quality
Geological complexity	Low complexity	High complexity

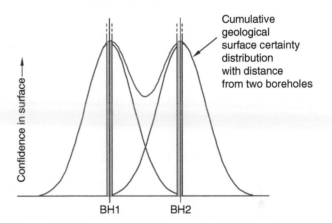

Figure 15.6 Kernel density smoothing function (red line) represents the cumulative contributions from multiple (in this case two) boreholes which have normal (Gaussian) certainty distributions with distance from the boreholes. Modified from Lelliott et al. (2009).

used semi-quantitative methods to assess qualitative uncertainty associated with the expertise of a geological modeler (Section 15.4.2.2). Lelliott et al. (2009) used a semi-quantitative method to assess the uncertainty associated with borehole density, borehole quality, and geological complexity at a small contaminated site. A kernel density smoothing procedure (Figure 15.6) provided a semi-quantitative measure of the uncertainty of surfaces defined by multiple boreholes. Lelliott et al. (2009) and Section 15.5.2.2 provide further details.

15.3.3 Quantitative Methods

Quantitative methods require the model to be sequentially modified and recalculated many times. For each calculation, a small change, representing the parameter uncertainty, is made to one or more of the input parameters. An ensemble of model runs and their outputs are stored, and statistically evaluated to provide a measure of model uncertainty. These methods are complex and may require considerable computing power. The most commonly used quantitative methods for uncertainty assessment are based on traditional statistical metrics, the confidence index, or metrics based on information entropy. Section 15.6 contains examples of these methods.

15.4 Evaluating Uncertainty of Interpretation

Much geological model uncertainty results from a lack of precise geological information which is commonly based on sparse outcrops, which limit field observations, or on inadequate borehole or geophysical data. Geological modeling (and mapping) begins by using observations to initially conceive of, and later to verify, a *conceptual geological model* (Fookes 1997; Fookes et al. 2000; Parry et al. 2014). The conceptual model encapsulates the geological evolution of the area; it thus influences the interpretation of the available observational information.

15.4.1 Uncertainty due to the Choice of Conceptual Model

Bond et al. (2007) conducted two experiments that evaluated how the choice of conceptual models influenced seismic imagery interpretations. In the first experiment, over 400 geologists interpreted a seismic profile to quantify the range in conceptual uncertainty. Results showed conceptual uncertainty has a large effect on interpreted results.

In a second experiment, Torvela & Bond (2011) provided a set of alternative conceptual models and seismic data from a folded thrust-belt site to 28 experienced geologists, and then examined variations in their interpretations. Most of the interpretations (14 out of 24) complied with all features of at least one recognized conceptual model; a sizeable minority (seven) of interpretations were influenced by one of these conceptual models but adapted to features of the data; only three interpretations were inconsistent with any of these conceptual models.

These two experiments led to the conclusion that the uncertainty in the final interpretation depends, at least in part, on the choice of the guiding conceptual model and how strongly and in what way it constrains the interpretation. Considering a number of equally plausible conceptual models in the early stages of the modeling task is good modeling practice, since making modifications of the conceptual model after the implementation of the geological

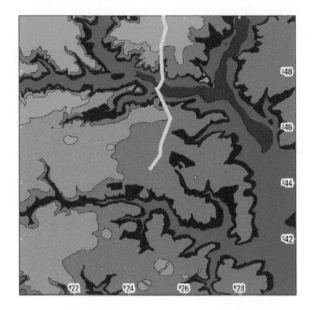

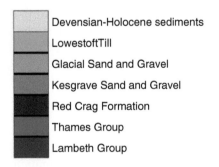

Devensian-Holocene sediments

LowestoftTill

Glacial Sand and Gravel

Kesgrave Sand and Gravel

Red Crag Formation

Thames Group

Lambeth Group

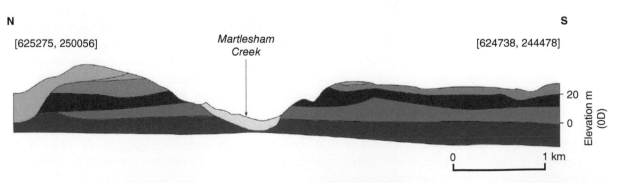

Figure 15.7 A 2-D map of the modeled units in the East Anglia study area with an example cross-section. (Lark et al. 2013).

Box 15.3 Data Resampling Techniques: The Jackknife, Cross Validation, and the Bootstrap

The term *Jackknife* was proposed by John Tukey to define one of the first resampling techniques used for variance estimation because, like a jackknife – a generic name for a Swiss army knife – the technique can be used as a "quick and dirty" replacement tool for a lot of more sophisticated and specific tools. The jackknife process generates a series of sub-samples from a dataset by systematically leaving out each observation in turn. Given a data set with n samples, the jackknife produces $n-1$ subsets. A model is calculated for each subset and the differences between observed and estimated values at each observed location are stored. These variations are aggregated to produce a Probability Density Function (PDF) for the model, which may be used as a measure of model uncertainty. Alternatively, individual PDFs can be determined for each observed location to provide an estimate of the spatial variations in uncertainty (Efron 1982; Meinrath 2000). The jackknife often works well, provided that the model under study does not change drastically with small changes in the data (Cave and Wood 2002).

The term *Cross validation* is used to define a modified jackknife procedure often employed with larger datasets. In these cases, instead of removing a single observation to produce a subset, a small fixed number of observations are chosen at random to be removed and form a subset. With this modified procedure, a smaller number of subsets are produced and analyzed than would be the case with a full jackknife procedure. Cross validation thus offers a smaller computational effort when large datasets are evaluated. The uncertainty estimates are developed with the same sequence described for the jackknife process.

The *Bootstrap* procedure uses a random sampling method to produce a series of sub-samples from a dataset (Davison and Hinkley 1997; Efron 1979; Efron and Tibshirani 1993; Wehrens et al. 2000). The random sampling method permits a selected observation to qualify for further selection; thus, a bootstrap subset may contain some observations multiple times, while other observations may be omitted. The number of observations in a subset is decided by the analyst. The uncertainty estimates are developed with the same sequence described for the jackknife process.

---

model can be time consuming and costly (Refsgaard et al. 2012).

15.4.2 Uncertainty due to Interpretation Process

Lark et al. (2013, 2014a) provide details of experiments to evaluate two potential sources of model uncertainty that are related to the interpretation of subsurface geology in 3-D framework models.

15.4.2.1 Uncertainty in Interpreting Lithostratigraphic Surfaces

Lark et al. (2013) evaluated the uncertainty of lithostratigraphic surfaces in 3-D framework models produced by five geologists. The experiment modeled a 10 km × 10 km (6.25 mi × 6.25 mi) area in southern East Anglia that is underlain by six sub-horizontal geological units (Figure 15.7). The geologists were provided with 346 borehole logs and conventional bedrock maps. The experiment used a jackknife process (Box 15.3) to evaluate the uncertainty of the interpreted model surfaces. Each geologist followed a common protocol to produce a 3-D framework model based on a unique subset of 326 boreholes. The remaining 20 boreholes had been removed from the dataset by a stratified random sampling design to form a "*validation set.*" The uncertainty was estimated by comparing the observed and estimated elevations of each geological surface at these validation boreholes.

Figure 15.8a shows the pooled results over all modeled surfaces. The straight line is the bisector; for any case where the modeled height and observed height are equal, the plotted symbol falls on this line. The dispersion about the bisector has a normal distribution (Figure 15.8b); its standard deviation provides a numerical estimate of the combined effects of all sources of uncertainty in the geometry of these models. Analysis of these data revealed no systematic bias in the models and no difference among the five geologists. The surface uncertainty was unaffected by distance to the nearest borehole or outcrop. However, there was evidence for an increase in uncertainty with depth.

15.4.2.2 Influence of Modeler Experience on Interpretation Uncertainty

Lark et al. (2014a) conducted a second experiment to assess the influence of modeler experience on the uncertainty of cross-section interpretations. A group of 28 geologists, with varying degrees of experience, were selected from attendees at a 3-D modeling conference. A set of 51 boreholes provided observations of the elevation of the base of the London Clay along a portion of a cross-section in London (Figure 15.9). Ten unique sets of modeling boreholes were prepared; each set contained five validation boreholes withdrawn at random. Geologists were randomly assigned a data set and were invited to interpret the base of the London Clay. Each participant was also asked to complete a questionnaire identifying their level of experience with 3-D geological modeling.

(a)

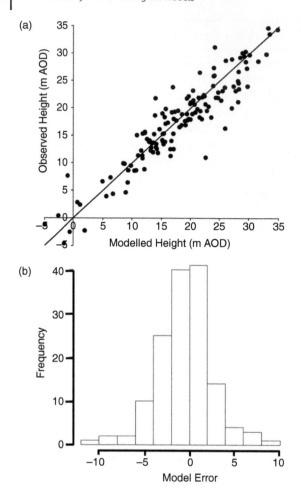

(b)

Figure 15.8 Statistical evaluation of the divergence of observed heights of geological unit bases and their modeled heights for a series of validation boreholes. (a) Scatter plots; (b) Histogram of errors. Modified from Lark et al. (2013).

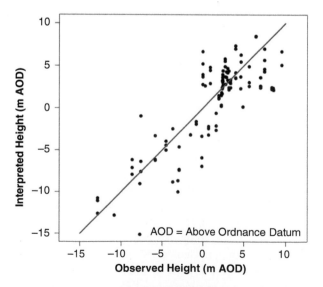

Figure 15.10 Interpreted and observed heights of the base of the London Clay at validation boreholes. Source: Reproduced from Lark et al. 2014a under CC BY 3.0 license.

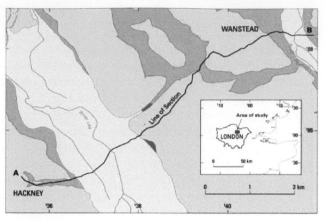

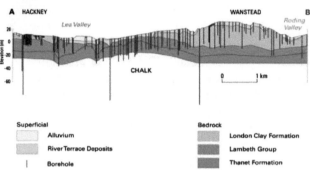

Figure 15.9 Cross-section in London interpreted by 28 geologists. Source: Reproduced from Lark et al. 2014a under CC BY 3.0 license.

Table 15.3 Decrease in model error variance with increasing geologist experience (Lark et al. 2014a).

Level of geologist experience	Between-geologist variance
None	4.44
Some (< 6 mo)	2.25
Moderate (6 mo – 2 yr)	1.32
Experienced (> 2 yr)	0.46

Figure 15.10 shows the 95% confidence limits for one interpretation of the base of the London Clay. A linear mixed statistical regression modeled the error variance as a function of potential sources of uncertainty. It revealed no difference between interpretations. However, there was a significant decrease in the error variance with increasing geologist experience (Table 15.3).

15.5 Evaluating Model Uncertainty

Examples of uncertainty evaluations are most readily described when placed in four categories: (i) the uncertainty of data sources; (ii) the uncertainty of geological models constructed by explicit modeling methods; (iii) the uncertainty of geological models constructed by implicit modeling methods; and (iv) the uncertainty of integrated modeling systems. The following sections discuss each category.

15.5.1 Uncertainty of Data Sources

Lark et al. (2014a) utilized stochastic statistical methods to further analyze the results of the experiment described in Section 15.4.2.2. Stochastic methods generated several thousand independent realizations of the lithostratigraphic surfaces. Comparisons of estimated and observed values defined interpretation errors which were used to compute a 95% confidence interval for one of the interpretations (Figure 15.11). The width of the confidence interval varies from zero (at a borehole) to about 5 m (15 ft) on the left of the section, where the boreholes are denser, to about 7.5 m (25 ft) on the right of the section, where boreholes are sparser.

Geologists at the Geological Survey of Canada used a quantitative method to define the uncertainty of a 3-D model of the ORM in Ontario. Box 15.2 provides additional details of this process, which combined distance-based ranges from different classes of boreholes with a geology map of outcrop patterns to produce a contoured "confidence map" that ranged between 1 (highest confidence) and 0 (lowest confidence).

15.5.2 Uncertainty of Explicit Models

Many 3-D geological framework models are developed using an explicit modeling approach which employs expert geological interpretation of all available data; usually, this

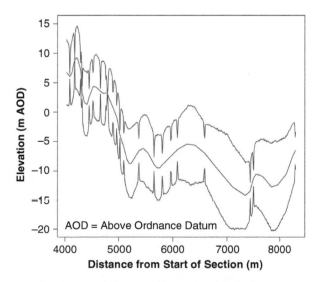

Figure 15.11 One expert interpretation of the base of the London Clay along a cross section (red line) with 95% confidence interval inferred from realizations of an error model (blue lines). Source: Reproduced from Lark et al. 2014a under CC BY 3.0 license.

consists of a combination of boreholes, geological maps, cross-sections and geophysical surveys. There are several approaches to explicit modeling (Chapters 9–11); the selected approach depends on the geological environment being modeled, the availability of existing data sources, the experience of the modelers, and the ultimate application of the model.

Because explicit geological models rely on "manual" interventions to create a model consistent with geological knowledge, the uncertainty of explicit models must be evaluated using semi-quantitative and quantitative statistical methods. These include: (i) geostatistical methods, (ii) bootstrap resampling, and (iii) cross-validation and jackknifing (Box 15.3).

15.5.2.1 Uncertainty Estimated by Geostatistical Interpolation

Blanchin and Chilès (1993) and Chilès and Delfiner (2012) describe an application of geostatistical analysis to evaluate uncertainty when planning the Channel Tunnel. The key geological issue faced by the tunnel planners was to keep the route within the relatively thin Cenomanian Chalk Marl, which was only about 30 m or 100 ft thick. This was a favorable material for the tunnel-boring machines, unlike the underlying Gault Clay and the overlying fractured and porous Gray Clay. Geostatistical (kriging) interpolation of the elevation of the upper surface of the Gault Clay, or base of the Cenomanian Chalk Marl, was based on extensive geological, geophysical, and borehole data sources. The tunnel engineers assessed the standard deviations of the kriging error for this surface along cross-sections corresponding to proposed tunnel alignments. They revised the alignment so that the route was never more than one standard deviation below the modeled surface. This corresponded to approximately a 16% probability of encountering the Gault at any point. This was regarded as an acceptable condition. In fact, during construction, the tunnel only penetrated the Gault at two locations along its route, and at both locations this was expected.

In the absence of any regional trend, ordinary kriging models the spatial distribution of a variable (assumed random) based on a model of spatial correlation represented by a semi-variogram. Any regional trend must be removed before applying the kriging interpolation approach. The most commonly employed approach is to fit an appropriate regression surface representing the trend, compute a variogram of the residuals between the data and this surface, reconstruct the spatial distribution of the residual surface using ordinary kriging, and finally combine the contributions of the trend surface and the kriged residual values. This method has been called "regression kriging," "kriging prediction with an external drift," or "universal kriging." The trend and the random residual estimates are unbiased provided that the data are unbiased (Lark

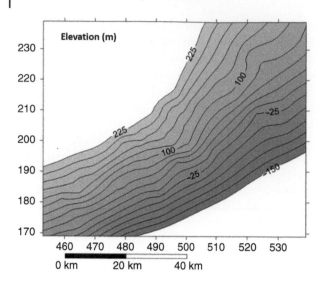

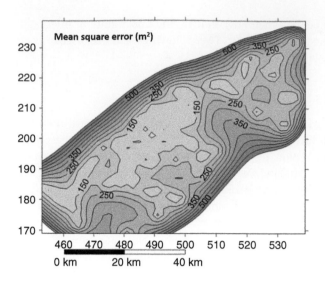

Figure 15.12 Predicted elevation (meters relative to sea level) of the sub-Upper Chalk surface. Modified from Lark and Webster (2006).

Figure 15.13 Prediction error variances of the modeled surface in Figure 15.12. Modified from Lark and Webster (2006).

and Webster 2006). Unfortunately, many geological data sources contain bias, causing the resulting prediction error variance to be underestimated. Lark and Webster (2006) provide details of an alternative method that avoids these limitations, but it is computationally more complex.

Lark and Webster (2006) re-analyzed the elevation of the Upper Chalk surface beneath the Chilterns in southern England. Because this surface has a regional dip and is gently folded, a quadratic trend surface was used to account for this regional trend before geostatistical analysis was performed on the 238 boreholes that defined the elevation of the chalk surface within an approximately 3000 km² study area. Figure 15.12 shows contours of the surface of the Upper Chalk as computed by Lark and Webster (2006). Figure 15.13, with contours of prediction error variances, demonstrates how uncertainty reflects both the model process (predictions of any trend surface are less certain near the edges of the modeled region) and the data density (kriging variance is larger where the data are sparser).

15.5.2.2 Evaluating Uncertainty by Bootstrap Resampling

Lelliott et al. (2009) used a combination of semi-quantitative methods to determine the uncertainty in a model of a small contaminated site. A kernel density smoothing procedure (Figure 15.7) was used to compute the uncertainty of surfaces defined by multiple boreholes. Three types of boreholes had been used at the site; each was assigned an appropriate maximum uncertainty level. Geological complexity was calculated from multiple realizations of each geological surface generated by "bootstrap resampling" techniques (Box 15.3). The uncertainty of the four stratigraphic model surfaces were presented as

quantitative maps showing uncertainty ranging between a minimum of 0.2 m (8 in.) and a maximum of 1.95 m (6 ft) (Figure 15.14)

15.5.2.3 Quantifying Uncertainty of Lithostratigraphic Surfaces by Cross-Validation

DGM (Digital Geological Model) is the national Dutch lithostratigraphic model of the Neogene and Quaternary down to an average depth of 500 m below the surface. It consists of a stacked sequence of 31 gridded lithostratigraphic bounding surfaces having 100 m × 100 m grid cells. DGM provides quantitative uncertainty estimates for these surfaces (Gunnink et al. 2010). Although based on a large amount of hard data (about 16 500 boreholes), development of geologically plausible lithostratigraphic surfaces required "soft" guidance from conceptual genetic models and additional geological information regarding faulting (Gunnink et al. 2010). Several of the geological units reflect regional trends; deeper marine formations typically dip to the northwest, where they are defined with very few boreholes. Some shallower units are located within bowl-shaped basins formed by glacial scour or are incised by buried tunnel valleys formed by glacial meltwaters. The borehole density is inadequate to model lithostratigraphic surfaces in these situations, so trend surfaces were used to define their anticipated shapes. Faults abruptly disrupt some units; others "pinch out" along depositional boundaries. "Steering points" were added to define these situations. The trend surfaces and steering points provided additional "soft data" observations that, in conjunction with the "hard data" of the boreholes, allowed interpolation of plausible geological units and relationships.

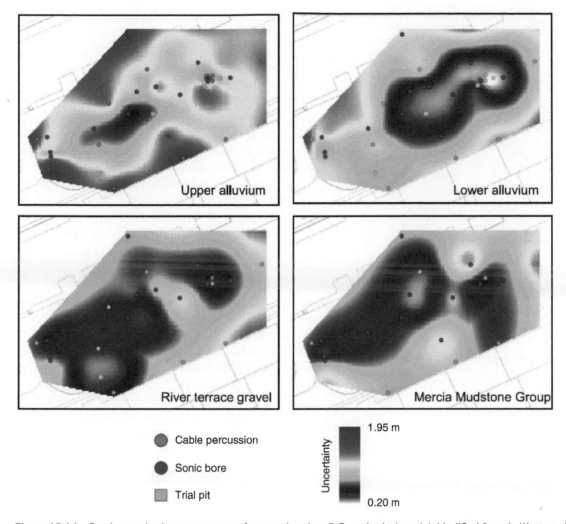

Figure 15.14 Semi-quantitative assessment of uncertainty in a 3-D geological model. Modified from Lelliott et al. (2009).

Geostatistical interpolation methods (kriging) produced the individual gridded lithostratigraphic surfaces. However, because these interpolations were based on combinations of hard and soft data points, the kriging error variance was not a reliable measure of the uncertainty of the resulting gridded surface. Thus, cross-validation (Box 15.3) was used to produce multiple elevation estimates for each surface at all borehole locations. Comparison of observed and estimated elevations produced error mean and variance statistics for each surface at all borehole locations. A moving window centered on grid nodes produced a regional variance. i.e., averaged values of the borehole statistics across the model. This was termed the *regional variance*. The size of window varied between 4 and 40 km depending on borehole density. Subsequently, the regional variance was scaled by the kriging variance to produce a local variance that accounted for clustering and distance of the boreholes; the standard deviation, rather than

this variance, provided a quantified measure of surface uncertainty (Figure 15.15).

Gunnink et al. (2010) also performed a jackknifing procedure (Box 15.3) to validate the surface uncertainty defined by the local variance. This provided an estimate of the robustness of the uncertainty estimates. For surfaces with lots of data, statistics from the jackknifing evaluation demonstrated little bias in the uncertainty estimates. However, for surfaces with sparse data, removing a considerable proportion of the boreholes led to unfavorable statistics because the amount of data removed compromised the resulting model. This approach to uncertainty quantification requires access to substantial numbers of observations.

15.5.3 Uncertainty of Implicit Models

The implicit approach to geological modeling allows construction of 3-D geological surfaces and volumes which

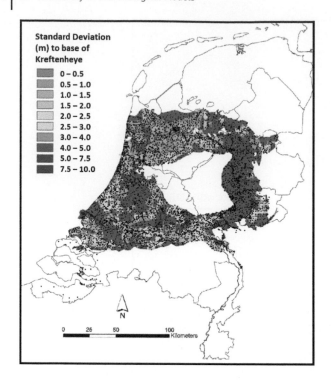

Figure 15.15 Local standard deviation estimate of the uncertainty of the base of the Kreftenheye Formation. (Gunnink et al. 2010).

honor geological observations. Lajaunie et al. (1997) recognized that often orientation observations were more frequently recorded than geological formation contacts. Thus, ignoring them represented a regrettable loss of information on geological structures. To resolve this deficiency, they proposed an implicit interpolation algorithm known as "potential-field geological modeling" which employs universal co-kriging (multivariable interpolation) to simultaneously account for: (i) scattered geological contact observations; (ii) structural orientation data observed independently of geological contacts; and (iii) fault network information (Lajaunie et al. 1997; Calcagno et al. 2008). The geostatistical formalism underlying the use of co-kriging for interpolating the potential field permits quantification of model uncertainty (Aug 2004). With the aid of co-kriging variances, it is possible to map the probability of interface locations, identifying areas of high uncertainty.

GeoModeller, discussed in Chapter 12, incorporates the implicit co-kriging potential-field approach to model complex geological structures using integrated datasets (McInerney et al. 2005, 2007). GeoModeller evaluates the data to ensure it is internally consistent, accurately represents geological concepts and observations, and is consistent with the tectonic history. These evaluations combine visual assessments and geostatistical tests of uncertainty (Aug 2004; Chilès et al. 2004). When

inconsistencies are discovered, model modifications are undertaken by employing new geological hypotheses, additional data, modifications to the fault network and other object geometries, and adjustments to the lithostratigraphic column.

Courrioux et al. (2015) discuss how, for a decade, geology students have created 3-D geological models employing the GeoModeller approach during two-week field courses in the coal basin of Alès, in southern France. The study area covers 220 km² in the southern part of the French Central Massif located between Alès and Bessèges. This area has relatively complex geological structures. Statistical analysis of the models produced during the different sessions clearly identify the different causes of uncertainty. Figure 15.16 shows the uncertainty in the interpretation of the base of the Carboniferous. Variations among the models result from a combination of incomplete field observations and different interpretations or identifications of formations, positions of faults, or natural contacts. The right-hand side of Figure 15.16 uses the spatial distribution of the standard deviation of co-kriging from multiple models to quantify the uncertainty in the horizon geometries (Courrioux et al. 2015).

These single potential field techniques have been applied to efficiently create 3-D models of several complexly deformed areas. Figure 15.17 shows one such model an area in the Ardennes; the model has been used to assess water resources.

15.5.4 Uncertainty Aspects of Integrated Multicomponent Models

Since integrated modeling involves the linking of different models, it is important to quantify the propagation of uncertainty through the "model cascade", when the outputs of one model become the inputs to another. Beven (2009, 2012, 2016) described several kinds of uncertainty encountered in these model chains. A subsequent comprehensive discussion of the uncertainty problem (Beven and Lamb 2017) clearly suggests that a precautionary approach is required when results from such complex model sequences are used for decision-making.

Pappenberger et al. (2005) assessed uncertainty propagation in a cascade of models used to evaluate rainfall-runoff and flooding predictions. Inclusion of larger numbers of models in a sequence resulted in a huge increase in the number of simulations required to assess the uncertainty. "Sub-sampling", if used intelligently, may reduce the number of simulation runs required to assess uncertainty. Beven (2007) proposes an approach to dealing with uncertainty in complex, all-encompassing models that avoids *uncertainties of model predictions and the consequential risks of potential outcomes.*

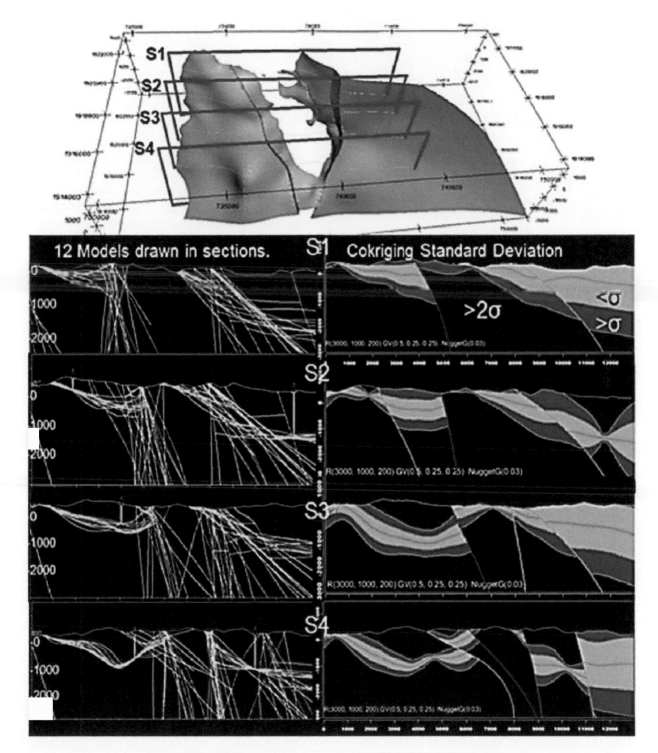

Figure 15.16 Analysis of uncertainties from models of the Alès area made during 12 successive "3-D map making" training sessions. Top: 3-D view (north at top) of carboniferous base from one of the models produced (in yellow). Bottom: Left: superimposition of 12 models of carboniferous base along four E-W cross-sections (S_1 to S_4). Right: theoretical standard deviation of co-kriging (σ) for one of the models. The dark gray area corresponds to a confidence interval of 68% ($\pm\sigma$) (Courrioux et al. 2015).

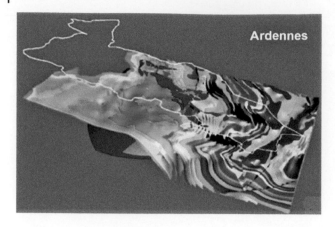

Figure 15.17 A 3-D geological model of Paleozoic horizons of the Givet area (Ardennes) obtained with a single potential field (series). Source: Courrioux et al. (2015) after Lacquement et al. (2011).

15.6 Computational Aspects of Uncertainty Evaluations

Uncertainty evaluations sometimes require longer calculation sequences or more complex analyses, chiefly because they rely on stochastic procedures that involve large numbers of model realizations (perhaps several hundreds) to develop distributions of the relative frequencies of occurrence of several statistically equiprobable outcomes. In the case of the Confidence Index, the process involves the collection of expert opinions and their subsequent evaluation.

15.6.1 Stochastic Methods

Stochastic approaches based on Monte Carlo analysis are used for modeling geological heterogeneity. They generate multiple, statistically equiprobable, realizations of the subsurface. Stochastic models provide information about the uncertainty of modeling geological structures (geometrical uncertainty) and/or the uncertainty in defining the distribution of facies within lithostratigraphic units (parameter uncertainty). Both are major sources of uncertainty in many geotechnical modeling applications, or for groundwater flow and transport simulations (Bianchi and Zheng 2016).

Geostatistical simulation can be performed using several modeling approaches (see also Chapter 13). Spatial distribution of a certain property can be modeled with traditional geostatistical simulation algorithms (Deutsch and Journel 1998), including Sequential Indictor Simulation (SISIM) for categorical variables and Sequential Gaussian Simulation (SGSIM) for continuous variables. These approaches can produce realizations of the spatial distribution of the variable of interest that honor the hard data at certain locations and a semi-variogram or covariance model representing the spatial structure of the data.

An alternative method is based on the quantification of the transition probabilities between categorical variables. These probabilities define the conditional probability that a certain category will transition into another over a certain distance from a point. They are modeled as mathematical functions (i.e. Markov chain models). When applied to simulating the distribution of geological units, the "transition probability approach" (T-PROGS) can account for geological interpretation (Carle and Fogg 1996, 1997; Carle 1999) by controlling certain properties including the proportion, the mean lengths, the connectivity, and the juxtapositional tendencies of each unit (Weissmann et al. 1999).

In recent years, another geostatistical simulation approach called "Multiple-point statistics" (MPS) has been developed to overcome the inability of traditional (two-point) geostatistical methods, such as SISIM, to reproduce complex, curvilinear, and interconnected geometries (Strebelle 2002). The key element of this approach, which can generate multiple realizations honoring both soft and hard data, is a so-called training image (TI), which acts as the representative conceptualization of geological heterogeneity for the system of interest. Statistical properties are extrapolated from the training image.

The stochastic modeling of geological structures has been of major interest to oil and gas exploration; it has been used to characterize the distribution of channels within alluvial or deltaic environments and the faulting patterns that compartmentalize reservoirs. For example, Geel and Donselaar (2007) used object-based stochastic modeling to define the tidal estuarine succession of the Holocene Holland Tidal Basin in the Netherlands. They modeled an area 5 km by 6 km to a depth of 25 m using only five borings. The tidal channels were modeled as east–west oriented, slightly sinuous, 400 m wide channels. Stochastic modeling, constrained by the borings and the characteristic size and shape of tidal channel and tidal flat sand bodies, produced a lithofacies architecture model (Figure 15.18). Different models were produced with smaller (100 m) and larger (800 m) channel sizes to test the influence of channel size on potential well productivity.

Kearsey et al. (2015) used indictor kriging and sequential indicator simulation stochastic modeling processes to generate multiple realizations of facies distributions in shallow glacial and fluvial sediments in Central Glasgow. Digitized records of classified lithology recorded in 4391 boreholes were used and the simulation was repeated 500 times. A bootstrapping method, where 50% of the input data was removed, was used to estimate model uncertainty. The accuracy of prediction was found to be generally good at depths less than 10 m. Borehole density was a limiting

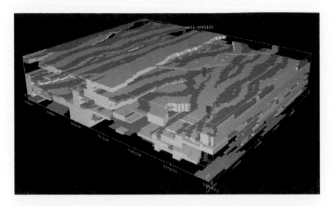

Figure 15.18 3-D view of a stochastic realization with characteristic tidal channel width of 400 m. Tidal channels in yellow, fringing tidal flats in orange, mud-dominated inter-channel deposits are transparent. The five wells on which the model is conditioned are shown. (Geel and Donselaar 2007).

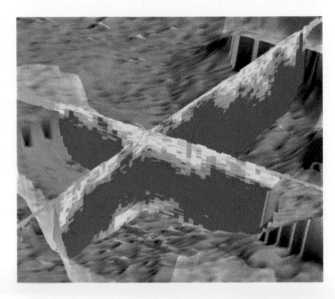

Figure 15.19 The probability that a voxel in the Walcheren member contains the tidal channel lithofacies. At the center of the channel, red color shows high probability (100%). In the upper part of the channel, green, and yellow colors reveal smaller probabilities due to expected tidal flat deposits. Lower probabilities are found at the bottom of the channel, where shells, and shell-rich sands are expected. (Stafleu et al. 2011).

factor in accurately predicting lithology; areas where additional observations would improve the prediction were highlighted.

Stafleu et al. (2011; see also Chapter 11) used an indicator kriging approach to model the distribution of lithofacies within pre-defined lithostratigraphic units in the shallow subsurface of Zeeland. The presence or absence of each lithofacies was defined by an indicator variable ("1" if present, "0" if some other lithofacies occurs). The spatial dependence of each lithofacies was then modeled individually by indicator kriging. These models were then subjected to stochastic simulation procedures to generate multiple realizations, conditioned on the observations and the model. The probability of occurrence of any lithofacies at each location was determined by examining the distributions of the lithologies in all the multiple realizations. This allows the identification, on a voxel-by-voxel basis, of the most probable lithofacies unit and also the probability of finding a particular lithology. These probabilities provide a measure of uncertainty and can be visualized (Figure 15.19).

Dell'Arciprete et al. (2012) evaluated the capability of three different geostatistical simulation approaches (SISIM, T-PROGS, and MPS) to reproduce the hydrofacies distribution within a Holocene point-bar–channel system exposed in a quarry. This comparative study concluded that each method has particular capabilities for reproducing certain characteristics of the hydrofacies distribution. While MPS simulations can efficiently reproduce the geometries of the most frequent hydrofacies, SISIM, and T-PROGS simulations can better simulate the distribution of the less abundant units.

All these approaches obtain the most accurate results when the highest-rank depositional elements are simulated independently and subsequently merged. This result highlights the fact that the application of geostatistical methods is effective only when it is possible to assume that the statistical properties of the variables of interest do not change with the domain of the simulation (i.e., they can be assumed to be stationary). In the presence of discontinuities due to tectonic or stratigraphic boundaries, or trends, corrective strategies must be employed to address non-stationarity.

15.6.2 Confidence Index

Lark et al. (2014b) proposed a Confidence Index that expresses the confidence of experts in the accuracy of a stratigraphic surface at any location in a 3-D model. The Confidence Index assesses the variation of elevation of a geological surface by evaluating its regional variability with trend surfaces and its local variability with variogram models. The opinions of experts define six confidence value parameters required to compute the Confidence Index.

Table 15.4 defines these six confidence value parameters. The "a" parameter is the confidence level at the location of a good-quality, reliably coded, logged borehole. Multiple "a" parameters may be used when boreholes have variable reliabilities. The "b" parameter is the confidence level at a point on a seismic line or other geophysical measurement. This will be less than at a borehole location because the geophysically defined contacts are less accurate. More

Table 15.4 Confidence Index parameter values of for the East Midlands study area.

Parameter	Meaning of parameter	Elicited value	Comments
a	Value of the confidence index at the site of a borehole	10	All boreholes regarded as equal in quality
b	Value of the confidence index at the site of a seismic observation	7	All seismic observations regarded as equal in quality
$c1$	Value of the confidence index at a site remote from a borehole in the absence of spatial trend	1	The lowest confidence level.
$c2$	Value of the confidence index at a site remote from a borehole in the presence of weak spatial trend	3	Weak trend when: $R^2 \leq 0.25$
$c3$	Value of the confidence index at a site remote from a borehole in the presence of moderate spatial trend	4	Moderate trend when: $0.25 < R^2 \leq 0.75$
$c4$	Value of the confidence index at a site remote from a borehole in the presence of strong spatial trend	5	Strong trend when: $R^2 > 0.75$

Source: Lark et al. (2014b).

than one "b" parameter may be necessary to represent different geophysical sources. Four "c" parameters define the minimum model confidence at locations some distance from any direct observation. The confidence in the model will be at a minimum at such locations and reflects both general and local geological variability. The most difficult situation occurs when there is no large-scale structural trend across the modeled region and the surface elevation is influenced only by fine-scale fluctuations around a constant mean depth. Such variation is not readily predictable by the modeler and is assigned confidence parameter "c1." Parameters "c2," "c3," and "c4" define the minimum model confidence in the presence of "weak," "moderate," or "strong" regional geological structures.

The coefficient of determination of the fitted trend model (R^2) allows the strength, or absence, of a trend to be defined (Table 15.4). By fitting a variogram model to the deviations from the trend surface, the rate of decay of the Confidence Index may be computed. It will range from a local maximum (defined by parameters "a" or "b") to a minimum value defined by parameter "c1" in the case of no discernable trend, or by parameters "c2," "c3," or "c4" when a trend is present.

The model surface is assessed at a series of regularly spaced (gridded) locations. At each location, the direct distances to the nearest borehole and nearest geophysical observation point are used to compute the Confidence Index. The larger of these two values is recorded. Reduced model confidence in the presence of faults is partially addressed by assuring that no fault intervenes between the location where the index is being computed and the borehole or geophysical observations used in the calculation.

Lark et al. (2014b) illustrate the use of the Confidence Index with an example from an 80 km by 50 km study area in the East Midlands of the United Kingdom. The area

includes three main geological components or structural levels: (i) gently eastward-dipping, sparsely faulted, Mesozoic strata, (ii) intensely faulted and folded Carboniferous strata containing hydrocarbon resources and source rocks, and (iii) tightly folded and faulted Lower Paleozoic and Neoproterozoic rocks. Two major unconformities separate these units: the Variscan Unconformity defines the top of the Carboniferous while the Caledonian Unconformity defines the top of the older Lower Paleozoic sequence. The 3-D model was developed using GoCAD; it contained seven structural or stratigraphic surfaces (Lark et al. 2014b). A combination of borehole logs and seismic reflection profiles were the primary information source. Due to prior exploration for hydrocarbon resources within the deep Carboniferous basin, both sources are concentrated in the north-western part of the study area (Figure 15.20).

The seismic profiles provided most of the observations used to define the regional structure with trend surfaces,

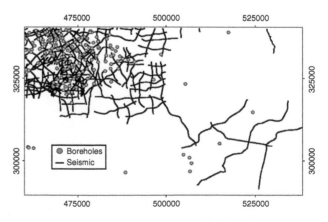

Figure 15.20 Distribution of boreholes and 2-D reflection seismic lines used to construct the East Midlands model. Modified from Lark et al. (2014b).

and for geostatistical analysis of the local variability about the trend. Table 15.4 presents the values for the six Confidence Index parameters for this study obtained from appropriate experts. A geologist with substantial modeling experience and familiarity with the target units provided values for parameters a and c. A group of three experienced geological modelers and three geophysicists provided a value for parameter b.

Calculation of the Confidence Index was implemented using ESRI's ArcGIS 10.0. Figure 15.21a illustrates the gridded surface elevations while Figure 15.21b shows the corresponding Confidence Index for the Variscan Unconformity. The Variscan Unconformity slopes from the southwest toward the northeast but is disrupted by some of the major east–west trending faults. The Confidence Index plot (Figure 15.21b) shows generally low confidence values in the south and east, but there are local large values around individual deep boreholes or clusters of boreholes. The heavily-faulted Carboniferous basin in the north-west has generally higher Confidence Index values due to the density of seismic profiles and deep boreholes; however, the influence of faults on the Confidence Index values can be seen.

Confidence indices have a clear value for summarizing expert opinion about sources of uncertainty in a model and their spatial distribution. However, they have limitations.

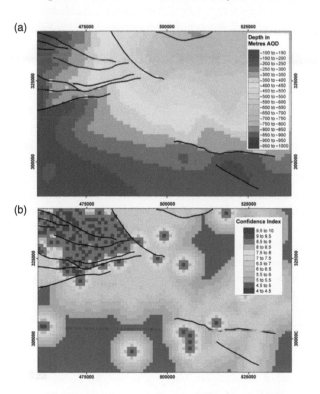

Figure 15.21 Modeled Variscan Unconformity in the East Midlands study area. (a) Elevations of the Variscan Unconformity surface, (b) corresponding Confidence Index plot. Modified from Lark et al. (2014b).

While Figure 15.21b shows higher model confidence in the north-west, it does not specify if the model uncertainty is low enough to support decision-making. However, while scientists use Confidence Index as a measure of scientific quality, it is up to the decision maker to indicate how sure he or she wants to be when making the decision at hand.

15.6.3 Information Entropy as a Measure of Prediction Uncertainty

The uncertainty of deterministic and stochastic geological models can also be evaluated using approaches based on the concept of information entropy. A thorough description of this concept can be found in the original work of Shannon (1948) and in several textbooks on Information Theory (Stone 2015). In a nutshell, for a system with a discrete number of probable outcomes, information entropy is a measure of "missing information", i.e. the amount of information required for a complete probabilistic description of the system. The concept is appealing in the context of uncertainty evaluations because it is based on a metric (the "Shannon information entropy") that is equal to zero when there is certainty about the outcome of an event and increases to a maximum value when there is the greatest uncertainty (all outcomes are equally likely). Moreover, it does not change when an additional outcome with zero probability is added.

Information entropy has been used extensively to quantify uncertainty in different fields. Wellmann and Regenauer-Lieb (2012) proposed a cell-based measure of information entropy as a method to visualize the uncertainty of a geological map. An area of interest is discretized into a number of cells using a uniform grid and the map of information entropy is defined from the probabilities of occurrence of all geological units within each cell. Areas have a low entropy where there is an abundance of hard data showing the very likely occurrence of certain units. Conversely, areas with sparse data, or located close to boundaries between the units, will have higher entropy values (indicating higher uncertainty). The approach was extended to a hypothetical 3-D geological model; this demonstrated that the information entropy approach has the advantage of combining the probabilities of occurrence of multiple geological units from multiple members.

Bianchi et al. (2015) used information entropy to evaluate and compare the uncertainty of lithological models for Glasgow and also to evaluate the impact of these uncertainties on simulations of groundwater fluxes and hydraulic heads predictions. Stochastic realizations of the distribution of four lithofacies in highly heterolithic Quaternary

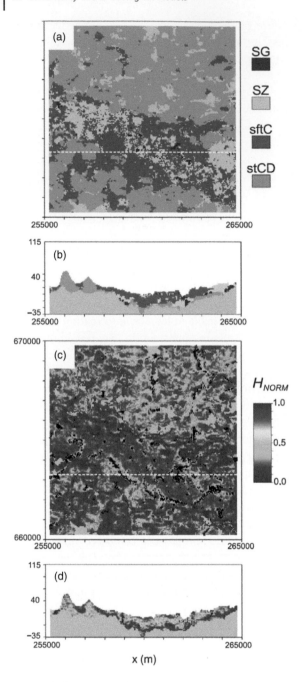

Figure 15.22 Lithofacies modeling results. (a) and (b) Most frequently occurring lithofacies in map and cross-section. (c) and (d) Corresponding normalized entropy distributions in map and cross-section. Black dots indicate borehole data locations. Light gray areas indicate where the bedrock is at the surface. White dashed lines are cross-section locations (Bianchi et al. 2015).

model was estimated using a normalized metric ranging from 0 to 1 based on the information entropy estimated using the probability of occurrence of the facies at each cell of a grid. These probabilities were estimated from the ensemble of stochastic realizations of the lithofacies distribution generated with T-PROGS (Figure 15.22c,d). The quantitative analysis of uncertainty based on this metric indicated that the incorporation of soft information from the deterministic model to complement the hard data from the boreholes resulted in a 20% reduction in prediction uncertainty. Results also showed that the information entropy is an effective metric for quantifying the uncertainty of spatially distributed stochastic geological models and for comparing different models. The Geological Survey of the Netherlands routinely computes the information entropy of its national voxel model GeoTOP and includes the parameter in the model as delivered to users (Chapter 11).

15.7 Communicating Uncertainty

Uncertainty inherent in 3-D geological models is complex; quantification of uncertainty and conveying this information in a manner that is understandable to a wide range of model users is a challenge. A key consideration is that knowledge of the general uncertainty of a model has limited scope, as the measure of uncertainty depends on how the model is used. Many models have been developed to present the geological information needed to support decisions that depend on subsurface conditions. If the information on which a decision is based is uncertain, the outcome of the decision is uncertain, and the data user is exposed to risk. Accompanying metadata and reports are an important aspect of ensuring good communication of model uncertainty.

In addition, uncertainty information must be presented in a form that can be simply understood and applied by the model users. The most comprehensive, quantitative, uncertainty analysis has little or no value if the model user cannot understand it. It is possible to compute point-wise uncertainty measures and present them as a map, but when a user interprets the map to address a particular question, the uncertainty becomes unique to the application and cannot generally be obtained from the point-wise uncertainty information. In many cases, some of the simpler qualitative uncertainty measures have the broadest acceptance. Many of these can be shown in simple graphics or maps or in the form of reports and narratives.

deposits in a 10 km × 10 km area were conditioned to both hard data from geotechnical boreholes and soft information extracted from a deterministic 3-D geological model of the area (Figure 15.22a,b). The uncertainty of the stochastic

References

Aitken, A.R.A., Betts, P.G., Schaefer, B.F. and Rye, S.E. (2008). Assessing uncertainty in the integration of aeromagnetic data and structural observations in the Deering Hills region of the Musgrave Province. *Australian Journal of Earth Sciences* 55 (8): 1127–1138. https://doi.org/10.1080/08120090802266600.

Aug C. (2004). *Modélisation Géologique 3D et Caractérisation des Incertitudes par la Méthode du Champ de Potentiel* (in French). Paris, France: École nationale supérieure des mines de Paris. 198 pp. [Online: Available at: https://pastel.archives-ouvertes.fr/pastel-00001077/] (Accessed August 19, 2020).

Bajc, A.F. and Dodge, J.E.P. (2011). *Three-dimensional mapping of surficial deposits in the Brantford–Woodstock Area, Southwestern Ontario.* Sudbury, ON: Ontario Geological Survey, Groundwater Resources Study 10. [Online: Available at: http://www.geologyontario.mndm.gov.on.ca/mndmfiles/pub/data/imaging/GRS010/GRS010.pdf] (Accessed July 15, 2020).

Beven, K. (2007). Towards environmental models of everywhere: uncertainty, data and modelling as a learning process. *Hydrology and Earth System Sciences* 11 (1): 460–467. https://doi.org/10.5194/hess-11-460-2007.

Beven, K. (2009). *Environmental Modelling: An Uncertain Future.* Abingdon, UK: Routledge. 328 pp.

Beven, K.J. (2012). How much of your error is epistemic? Lessons from Japan and Italy. *Hydrological Processes 27*: 1677–1680. https://doi.org/10.1012/hyp.9648.

Beven, K.J. (2016). Facets of uncertainty: epistemic uncertainty, non-stationarity, likelihood, hypothesis testing and communication. *Hydrological Sciences Journal* 61 (9): 1652–1665. https://doi.org/10.1080/02626667.2015.1031761.

Beven, K. and Lamb, R. (2017). The uncertainty cascade in model fusion. *Geological Society of London Special Publications* 408: 255–266. https://doi.org/10.1144/SP408.13.

Bianchi, M. and Zheng, C. (2016). A lithofacies approach for modeling non-Fickian solute transport in a heterogeneous alluvial aquifer. *Water Resources Research* 52: 552–565. https://doi.org/10.1002/2015WR018186.

Bianchi, M., Kearsey, T. and Kingdon, A. (2015). Integrating deterministic lithostratigraphic models in stochastic realizations of subsurface heterogeneity. Impact on predictions of lithology, hydraulic heads and groundwater fluxes. *Journal of Hydrology* 531 (3): 557–573. https://doi.org/10.1016/j.jhydrol.2015.10.072.

Bistacchi, A., Massironi, M., Dal Piaz, V.G. et al. (2008). 3D fold and fault reconstruction with an uncertainty model: an example from an alpine tunnel case study. *Computers & Geosciences* 34: 351–372. https://doi.org/10.1016/j.cageo.2007.04.002.

Blanchin, R. and Chilès, J.-P. (1993). The channel tunnel: geostatistical prediction of the geological conditions and its validation by the reality. *Mathematical Geology* 25 (7): 963–974. https://doi.org/10.1007/BF00891054.

Bond, C.E., Gibbs, A.D., Shipton, Z.K. and Jones, S. (2007). What do you think this is? "Conceptual uncertainty" in geoscience interpretation. *GSA Today* 17 (11): 4–10. https://doi.org/10.1130/GSAT01711A.1.

Burt, A.K. and Dodge, J.E.P. (2011). *Three-dimensional modelling of surficial deposits in the Barrie–Oro moraine area of southern Ontario.* Sudbury, ON: Ontario Geological Survey, Groundwater Resources Study 11. [Online: Available at: http://www.geologyontario.mndm.gov.on.ca/mndmfiles/pub/data/imaging/GRS011/GRS011.pdf] (Accessed August 13, 2020).

Burt, A.K. and Dodge, J.E.P. (2016). *Three-dimensional modelling of surficial deposits in the Orangeville–Fergus area of southern Ontario.* Sudbury, ON: Ontario Geological Survey, Groundwater Resources Study 15. [Online: Available at: http://www.geologyontario.mndm.gov.on.ca/mndmaccess/mndm_dir.asp?type=pub&id=GRS015] (Accessed August13, 2020).

Calcagno, P., Chilès, J.P., Courrioux, G. and Guillen, A. (2008). Geological modelling from field data and geological knowledge. Part I. Modelling method coupling 3D potential-field interpolation and geological rules. *Physics of the Earth and Planetary Interiors* 171: 147–157. https://doi.org/10.1016/j.pepi.2008.06.013.

Carle, S.F. (1999). *T-PROGS: Transition Probability Geostatistical Software (Version 2.1).* Davis, CA: University of California. 84 pp. [Online: Available at: http://gmsdocs.aquaveo.com/t-progs.pdf] (Accessed October 10, 2020).

Carle, S.F. and Fogg, G.E. (1996). Transition probability-based indicator geostatistics. *Mathematical Geology* 28: 453–476. https://doi.org/10.1007/BF02083656.

Carle, S.F. and Fogg, G.E. (1997). Modeling spatial variability with one and multidimensional continuous-lag Markov chains. *Mathematical Geology* 29: 891–918. https://doi.org/10.1023/A:1022303706942.

Cave, M.R. and Wood, B. (2002). *Approaches to the measurement of uncertainty in geoscience data modelling.* Keyworth, UK: British Geological Survey, Internal Report IR/02/068.

Chilès, J.-P. and Delfiner, P. (2012). *Geostatistics: Modeling Spatial Uncertainty, 2e.* Hoboken, NJ: Wiley. 734 pp.

Chilès, J.-P., Aug, C., Guillen, A. and Lees, T. (2004). Modelling the geometry of geological units and its uncertainty in 3D from structural data: the potential-field method. In: *Proceedings of Orebody Modelling and Strategic Mine Planning* (ed. R. Dimitrakopoulos), 313–320. [Online: Available at: http://citeseerx.ist.psu.edu/viewdoc/summary?doi=10.1.1.583.213] (Accessed August 21, 2020).

Courrioux, G., Bourgine, B., Guillen, A. et al. (2015). Comparisons from multiple realizations of a geological model: implication for uncertainty factors identification. In: *Proceedings, 17th Annual Conference of the International Association for Mathematical Geosciences – IAMG 2015* (eds. H. Schaeben, R. Tolosana Delgado, K.G. van den Boogaart and R. van den Boogaart), 59–66. Houston, TX: International Association for Mathematical Geosciences.

Culshaw, M.G. (2005). From concept towards reality: developing the attributed 3D geological model of the shallow subsurface. *Quarterly Journal of Engineering Geology and Hydrogeology* 38: 231–284. https://doi.org/10.1144/1470-9236/04-072.

Davison, A.C. and Hinkley, D.V. (1997). *Bootstrap Methods and their Applications*. Cambridge, UK: Cambridge University Press. 582 pp. https://doi.org/10.1017/CBO9780511802843.

Dell'Arciprete, D., Bersezio, R., Felletti, F. et al. (2012). Comparison of three geostatistical methods for hydrofacies simulation: a test on alluvial sediments. *Hydrogeology Journal* 20: 299–311. https://doi.org/10.1007/s10040-011-0808-0.

Deutsch, C.V. and Journel, A.G. (1998). *GSLIB Geostatistical Software Library and Users Guide*, 2e. New York: Oxford University Press. 369 pp. [Online: Available at: http://claytonvdeutsch.com/wp-content/uploads/2019/03/GSLIB-Book-Second-Edition.pdf] (Accessed October 10, 2020).

Diks, C.G.H. and Vrugt, J.A. (2010). Comparison of point forecast accuracy of model averaging methods in hydrologic applications. *Stochastic Environmental Research and Risk Assessment* 24: 809–820. https://doi.org/10.1007/s00477-010-0378-z.

Efron, B. (1979). Bootstrap methods: another look at the jackknife. *Annals of Statistics* 7: 1–26.

Efron, B. (1982). *The Jackknife, the Bootstrap and Other Resampling Plans*. Philadelphia, PA: Society for Industrial and Applied Mathematics. 92 pp. https://doi.org/10.1137/1.9781611970319.

Efron, B. and Tibshirani, R.J. (1993). *An Introduction to the Bootstrap*. New York: Chapman and Hall. 429 pp.

Fookes, P.G. (1997). Geology for engineers: the geological model, prediction and performance. *Quarterly Journal of Engineering Geology and Hydrology* 30: 293–424. https://doi.org/10.1144/GSL.QJEG.1997.030.P4.02.

Fookes, P.G., Baynes, F.J. and Hutchinson, J.N. (2000). Total geological history: a model approach to the anticipation, observation and understanding of site conditions. In: *Proceedings of the International Conference on Geotechnical and Geological Engineering; GeoEng 2000, Melbourne Australia* (eds. S.H. Chew, G.P. Karunaratne, S.A. Tan et al.), 370–460. Lancaster, PA: Technomic Publishing Co.

Ford, J., Napier, B., Cooper, A., et al. (2006). *3-D bedrock geology model of the Permo-Triassic of Yorkshire and East Midlands*. Keyworth, UK: British Geological Survey, Report CR/06/091N. 39 pp. [Online: Available at: http://nora.nerc.ac.uk/id/eprint/519850/] (Accessed October 4, 2020).

Geel, C.R. and Donselaar, M.E. (2007). Reservoir modelling of heterolithic tidal deposits: sensitivity analysis of an object-based stochastic model. *Netherlands Journal of Geosciences* 86 (4): 403–411. https://doi.org/10.1017/S0016774600023611.

Gunnink, J.L., Maljers, D. and Hummelman, J.H. (2010). Quantifying uncertainty of geological 3D layer models, constructed with a-priori geological expertise. In: *Proceedings, 14th Annual Conference of the International Association for Mathematical Geosciences – IAMG2010*. Houston, TX: International Association for Mathematical Geosciences. 13 pp.

Jessell, M.W., Ailleres, L. and de Kemp, E.A. (2010). Towards an integrated inversion of geoscientific data: what price of geology? *Tectonophysics* 490: 294–306. https://doi.org/10.1016/j.tecto.2010.05.020.

Kearsey, T., Williams, J., Finlayson, A. et al. (2015). Testing the application and limitation of stochastic simulations to predict the lithology of glacial and fluvial deposits in Central Glasgow, UK. *Engineering Geology* 187: 98–112. https://doi.org/10.1016/j.enggeo.2014.12.017.

Kessler, H., Lark, M. and Dearden, R. (2013). The challenge of capturing multi-component uncertainty in 3D models. *Geological Society of America Abstracts with Programs* 45 (7): 149. [Online: Available at: http://nora.nerc.ac.uk/id/eprint/503860/] (Accessed October 4, 2020).

Kindlarski, E. (1984). Ishikawa diagrams for problem-solving. *Quality Progress* 17: 26–30.

Lacquement, F., Courrioux, G., Ortega, C. and Thuon, Y. (2011). *Réalisation du Modèle Géologique 3D de la Pointe de Givet (08) - Étape 1 : Caractérisation des Potentialités d'Exploitation des Eaux Souterraines* (in French). Orléans, France: Bureau de Recherches Géologiques et Minières, Report BRGM/RP-60384-FR. 52 pp. [Online: Available at: https://www.brgm.fr/projet/modelisation-3d-geologie-pointe-givet] (Accessed October 4, 2020).

Lajaunie, C., Courrioux, G. and Manuel, L. (1997). Foliation fields and 3D cartography in geology: principles of a method based on potential interpolation. *Mathematical Geology* 29 (4): 571–584. https://doi.org/10.1007/BF02775087.

Lark, R.M. and Webster, R. (2006). Geostatistical mapping of geomorphic variables in the presence of trend. *Earth Surface Processes and Landforms* 31: 862–874. https://doi.org/10.1002/esp.1296.

Lark, R.M., Mathers, S.J., Thorpe, S. et al. (2013). A statistical assessment of the uncertainty in a 3-D geological framework model. *Proceedings of the Geologists' Association* 124 (6): 946–958. https://doi.org/10.1016/j.pgeola.2013.01.005.

Lark, R.M., Thorpe, S., Kessler, H. and Mathers, S.J. (2014a). Interpretative modelling of a geological cross section from boreholes: sources of uncertainty and their quantification. *Solid Earth* 5: 1189–1203. https://doi.org/10.5194/se-5-1189-2014.

Lark, R.M., Mathers, S.J., Marchant, A. and Hulbert, A. (2014b). An index to represent lateral variation of the confidence of experts in a 3-D geological model. *Proceedings of the Geologists' Association* 125: 267–278. https://doi.org/10.1016/j.pgeola.2014.05.002.

Lelliott, M.R., Cave, M.R. and Wealthall, G.P. (2009). A structured approach to the measurement of uncertainty in 3D geological models. *Quarterly Journal of Engineering Geology and Hydrogeology* 42: 95–105. https://doi.org/10.1144/1470-9236/07-081.

Logan, C., Russell, H.A.J. and Sharpe, D.R. (2005). *Regional 3-D Structural Model of the Oak Ridges Moraine and Greater Toronto Area, Southern Ontario (Version 2.1)*. Ottawa, ON: Geological Survey of Canada, Open File 5062. 27 pp. https://doi.org/10.4095/221490.

Logan, C., Russell, H.A.J., Sharpe, D.R. and Kenny, F.M. (2006). The role of expert knowledge, GIS and geospatial data management in a basin analysis, Oak Ridges Moraine, Southern Ontario. In: *GIS Applications in the Earth Sciences* (ed. J. Harris), 519–541. Saint John's, NL: Geological Association of Canada.

McInerney, P., Guillen, A., Courrioux, G. et al. (2005). Building 3D geological models directly from data? A new approach applied to Broken Hill, Australia. In: *Proceedings, Digital Mapping Techniques 2005 Workshop, Baton Rouge, Louisiana* (ed. D. Soller), 119-130. Reston, VA: U.S. Geological Survey, Open-File Report 2005–1428. [Online: Available at: https://pubs.usgs.gov/of/2005/1428/mcinerney/] (Accessed October 4, 2020).

McInerney, P., Goldberg, A., Calcagno, P. et al. (2007). Improved 3D geology modelling using an implicit function interpolator and forward modelling of potential field data. In: *Proceedings of Exploration 07: Fifth Decennial International Conference on Mineral Exploration* (ed. B. Milkereit), 919–922. Toronto, ON: Decennial Mineral Exploration Conferences.

Meinrath, G. (2000). Computer-intensive methods for uncertainty estimation in complex situations. *Chemometrics and Intelligent Laboratory Systems* 51 (2): 175–187. https://doi.org/10.1016/S0169-7439(00)00066-6.

Oreskes, N. (2003). The role of quantitative models in science. In: *Models in Ecosystem Science* (eds. C.D. Canham, J.J. Cole and W.K. Lauenroth), 13–31. Princeton, NJ: Princeton University Press.

Pappenberger, F., Beven, K., Hunter, N.M. et al. (2005). Cascading model uncertainty from medium range weather forecasting (10 days) through rainfall-runoff models to flood inundation predictions using European flood forecasting system (EFFS). *Hydrology and Earth System Sciences* 9 (4): 381–393. https://doi.org/10.5194/hess-9-381-2005.

Parry, S., Baynes, F.J., Culshaw, M.G. et al. (2014). Engineering geological models: an introduction: IAEG commission 25. *Bulletin of Engineering Geology and the Environment* 73: 689–706. https://doi.org/10.1007/s10064-014-0576-x.

Refsgaard, J.C., Christensen, S., Sonnenborg, T.O. et al. (2012). Review of strategies for handling geological uncertainty in groundwater flow and transport modeling. *Advances in Water Resources* 36: 36–50. https://doi.org/10.1016/j.advwatres.2011.04.006.

Shannon, E.C. (1948). A mathematical theory of communication. *Bell System Technical Journal* 27: 379–423.

Sharpe, D.R., Russell, H.A.J. and Logan, C. (2007). A 3-dimensional geological model of the Oak Ridges Moraine area, Ontario, Canada. *Journal of Maps* 3 (1): 239–253. https://doi.org/10.1080/jom.2007.9710842.

Stafleu, J., Maljers, D., Gunnink, J.L. et al. (2011). 3D modelling of the shallow subsurface of Zeeland, the Netherlands. *Netherlands Journal of Geosciences* 90: 293–310. https://doi.org/10.1017/S0016774600000597.

Stone, J.V. (2015). *Information Theory: A Tutorial Introduction*. Sheffield, UK: Sebtel Press. 260 pp.

Strebelle, S. (2002). Conditional Simulation of Complex Geological Structures Using Multiple-Point Statistics. *Mathematical Geology* 34 (1): 1–21. https://doi.org/10.1023/A:1014009426274.

Tacher, L., Pomian-Srzednicki, I. and Parriaux, A. (2006). Geological uncertainties associated with 3-D subsurface models. *Computers and Geosciences* 32: 212–221. https://doi.org/10.1016/j.cageo.2005.06.010.

Torvela, T. and Bond, C.E. (2011). Do experts use idealized structural models? Insights from a deepwater fold-thrust belt case study. *Journal of Structural Geology* 33: 51–58. https://doi.org/10.1016/j.jsg.2010.10.002.

Van der Meulen, M.J., Doornenbal, J.C., Gunnink, J.L. et al. (2013). 3D Geology in a 2D Country: Perspectives for Geological Surveying in the Netherlands. *Netherlands*

Journal of Geosciences 92 (4): 217–241. https://doi.org/10.1017/S0016774600000184.

Van der Meulen, M.J., Maljers, D., Van Gessel, S.F. and Gruijters, S.H.L.L. (2007). Clay resources in the Netherlands. *Netherlands Journal of Geosciences* 86 (2): 117–130. https://doi.org/10.1017/S001677460002312X.

Wehrens, R., Putter, H. and Buydens, L.M.C. (2000). The bootstrap: a tutorial. *Chemometrics and Intelligent Laboratory Systems* 54 (1): 35–52. https://doi.org/10.1016/S0169-7439(00)00102-7.

Weissmann, G.S., Carle, S.F. and Fogg, G.E. (1999). Three-dimensional hydrofacies modeling based on soil surveys and transition probability geostatistics. *Water Resources Research* 35: 1761–1770. https://doi.org/10.1029/1999WR900048.

Wellmann, F. and Caumon, G. (2018). Chapter One - 3-D structural geological models: concepts, methods, and uncertainties. *Advances in Geophysics* 59: 1–121. https://doi.org/10.1016/bs.agph.2018.09.001.

Wellmann, F. and Regenauer-Lieb, K. (2012). Uncertainties have a meaning: information entropy as a quality measure for 3-D geological models. *Tectonophysics* 526-529: 207–216. https://doi.org/10.1016/j.tecto.2011.05.001.

Wynne-Edwards, H.R. (1967) *Geology Westport, Ontario*. Ottawa, ON: Geological Survey of Canada, "A" Series Map 1182A. 1 sheet. https://doi.org/10.4095/108032.

Part III

Using and Disseminating Models

16

Emerging User Needs in Urban Planning

Miguel Pazos Otón[1], Rubén C. Lois González[1], Ignace P.A.M. van Campenhout[2], Jeroen Schokker[3], Carl Watson[4], and Michiel J. van der Meulen[3]

[1] *Department of Geography, Universidade de Santiago de Compostela, 15782 Santiago de Compostela, Spain*
[2] *Gemeentewerken Rotterdam, 3072 AP Rotterdam, The Netherlands*
[3] *TNO, Geological Survey of the Netherlands, 3584 CB, Utrecht, The Netherlands*
[4] *British Geological Survey, Keyworth, Nottingham NG12 5GG, UK*

16.1   Introduction

Three-dimensional geological models are routinely serving the hydrocarbon and mining sectors (largely outside the scope of this book), and groundwater management organizations (discussed in Chapter 14, illustrated in Chapter 19). The construction industry is increasingly using 3-D models (discussed in Chapters 3, 4, and 14, illustrated in Chapter 24), and there is an emerging interest in using 3-D subsurface information in spatial and urban planning (see Chapter 3 for legislative and policy contexts). After all, prior to developing new housing areas, business areas or infrastructures, planners will have made decisions about siting, typically with a low awareness of or regard for subsurface conditions, whether they present challenges or opportunities.

Delivery of a 3-D model for exploration purposes, resource management, or the assessment of ground conditions is a fairly well-defined transaction between specialists belonging to the same or adjoining disciplines. However, providing models that serve the needs of the urban planning community requires geomodelers to communicate effectively with people with completely different backgrounds, who cannot be expected to be able to understand or interpret basic geoscience data, evaluate the merits of alternative interpretations, or distinguish between theories and facts. Such users desire *"solutions, not data"* and *"information in understandable form"* (Turner 2003). Conversely, geomodelers typically have a limited understanding of the needs of the urban planning community, which arise in the context of spatial visions and policies, in resolving conflicting interests and competing land claims, and also communicating with the general public.

In fact, to most people, the subsurface is largely *"out of sight, out of mind"* (Van der Meulen et al. 2016). In recognition of the fact that subsurface information is underused in urban planning, the EU COST Action SUB-URBAN (2015–2017) was set up to bridge the information and communication gaps that exist between the geologists, geographers, and planners (Campbell et al. 2017). The combination of the geographic and conceptual scope, and its case-based, community-of-practice type of approach, made SUB-URBAN the first project of its kind. This chapter draws mainly from the experience gained in SUB-URBAN, and also includes an example of the required interactions between modelers and planners based on an assessment of responses to the flooding risk in New Orleans, Louisiana. Additional case material is available at www.sub-urban.eu, an online resource managed by the SUB-URBAN network that was established after the project ended.

This chapter provides geomodelers, the main audience of this book, with some basic concepts used in spatial and urban planning and discusses the information needs in this particular domain. Chapter 17 discusses communication of, or with, subsurface models, with special emphasis on novel techniques and serving new types of users.

16.2   Urban Planning in Brief

Urban planners have always focused on the surface, and have developed extensive regulations about land use, spatial zoning, and the characteristics of the built environment. Booming urbanization during the nineteenth and twentieth centuries eventually made it difficult or impossible for cities to physically accommodate the needs of their inhabitants (Knox and McArthy 2012; Jonas et al.

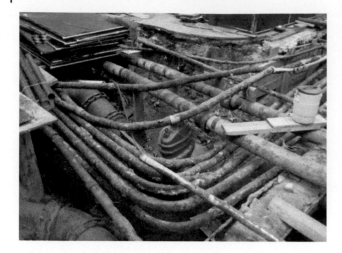

Figure 16.1 View of the extensive utilities under Fulton Street in New York City. (Environmental Protection Agency 2011)

2015). Cities started to not only build out, but also up and down. Tunnels, parking facilities, sewage pipes, electric cables, heating infrastructures, telephone cables and, more recently, internet infrastructures, were placed underground (Campbell et al. 2015). This subsurface infrastructure has not always been planned, coordinated, or administered properly; the result has produced what might be termed "underground sprawl" (Figure 16.1).

The European Regional/Spatial Planning Charter, adopted in 1983 by the Council of Europe Conference of Ministers responsible for spatial/regional planning (CEMAT), includes the following comprehensive definition of regional or spatial planning:

> *"Regional/spatial planning gives geographical expression to the economic, social, cultural and ecological policies of society. It is at the same time a scientific discipline, an administrative technique and a policy developed as an interdisciplinary and comprehensive approach directed towards a balanced regional development and physical organization of space according to an overall strategy."*

Urban planners are trained to process and analyze various types of information concerning an urban area and present it to the public and politicians to support decision making. The urban planner needs to respect national laws as well as possible local restrictions and guidelines that set the framework for the urban planning for the municipality or region (also see Chapter 3). Planning systems vary by country but are generally hierarchical. At the highest level, there will be a master plan, based on a long-term development vision, which defines the overall zonation for housing, commercial, and industrial developments as well as for infrastructure improvements throughout the city. The master plan sets boundary conditions for subordinate plans, down to the level of detailed zoning plans for individual urban sectors.

In many countries, the disregard of the urban subsurface can be attributed to a general lack of expertise in dealing with the particularities of underground planning and management. Municipal architects and city planners are, in most cases, the only professionals in charge of underground developments. Accordingly, planning documents typically do not consider the subsurface explicitly; the subsurface is considered to be either inaccessible or, at best, a featureless homogeneous space that can be dug into for any purpose. This is aggravated by the fact subsurface data, information sources, and models are often either inaccessible or uninterpretable by urban planners.

The COST SUB-URBAN Action has produced comprehensive city reports which describe city settings, economic aspects, and subsurface planning issues; working group reports which focus on various subsurface urban planning topics, data handling and 3-D modeling; and reports on short-term visits by participants which allowed for exchanges of skills and methodologies. The SUB-URBAN participants also produced scientific publications and contributions to workshops and conferences. This material can be found on the COST-SUBURBAN website: http://sub-urban.squarespace.com. The SUB-URBAN project reports include some European good-practice examples (Van der Meulen et al. 2016) and references to specific city studies (Mielby et al. 2017; see also Campbell et al. 2015; Lois-González et al. 2015; Van der Lugt et al. 2015).

Some of these city studies form the basis for case studies in this book. Chapter 25 describes how geological models have been used to map anthropogenic deposits to help the city of Bergen with groundwater management while taking into account the city's buried archeological legacy (Case Study 25.1) and to help the city of Newcastle upon Tyne to model its water system (both surface and groundwater), thus reducing the risk of groundwater flooding and soil and groundwater contamination (Case Study 25.2). Case Study 25.3 describes the technicalities and particularities of modeling anthropogenic deposits. Chapter 11 shows how a lithologically attributed voxel model was used in Tokyo to explain the variations of earthquake damage.

16.2.1 Planning Context

Urban planners ultimately support political decision making, which serves the public according to the government's political value system. Accordingly, urban planning guidance documents influence how politicians will be held accountable for their plans by their constituents; this

Table 16.1 United Nations sustainable development goals.

Goal #	Definition
1	No poverty
2	Zero hunger
3	Good health and well-being for people
4	Quality education
5	Gender equality
6	Clean water and sanitation
7	Affordable and clean energy
8	Decent work and economic growth
9	Industry, Innovation, and Infrastructure
10	Reducing inequalities
11	Sustainable cities and communities
12	Responsible consumption and production
13	Climate action
14	Life below water
15	Life on land
16	Peace, justice and strong institutions
17	Partnerships for the goals

Source: United Nations (2015).

accountability is defined in much broader terms than those of more traditional geological modeling applications.

In 2015, the United Nations General Assembly defined a collection of 17 global sustainable development goals as part of UN Resolution 70/1 – the 2030 Agenda (United Nations 2015). These UN-defined sustainable development goals (Table 16.1) have been widely adopted by many governments, irrespective of their political inclinations. Thus, urban planners often have to indicate how their plans or planning options contribute to these, or similarly formulated, goals.

Subsurface models, and geoscientists in general, can contribute to several of these UN goals, but never exclusively. For example, providing clean water involves more than finding pristine groundwater; geothermal energy is just one of several clean energy options; and CO_2 storage is just one of many actions that might be undertaken in response to climate-change regulations. So, for every solution that involves the subsurface, there are multiple alternatives that do not rely on the subsurface. Communications between modelers and urban planners play a critical role in selecting a solution from among multiple alternatives, especially with the existence of the "out of sight, out of mind" limited knowledge and experience levels of the urban planning community (Van der Meulen et al. 2016). Disregarding geology does not mean that sustainable development goals cannot be reached, but it could result in suboptimal plans,

especially where geological resources or geological hazards are present. In summary, awareness of both opportunities and challenges associated with geology must be better communicated (and accepted) if the emerging use case for geological information (and 3-D models) in urban planning is to be further developed.

16.2.2 The SUB-URBAN Toolbox

The SUB-URBAN project created a Toolbox which contains a fit-for-purpose suite of recommended methodologies, guidance in good practice, and case studies designed to enable the exchange of key subsurface data and knowledge between geoscientists and urban planners. The goal of the Toolbox is to improve urban planning and provide solutions that encourage sustainable and resilient urban development. The Toolbox consists of an online website, located at http://sub-urban.squarespace.com/toolbox-1, where this information is available to subsurface specialists, urban planners, and decision makers. Users with different backgrounds and needs require different types of access; therefore, the Toolbox provides different entry points for subsurface technical experts, urban planners, decision-makers, and policymakers.

The Dutch SUB-URBAN participants that created the toolbox used the metaphor of a bicycle to demonstrate the interactions of the Toolbox components and how the various users could navigate through the Toolbox (Figure 16.2). In this metaphor, the front wheel of the bicycle represents knowledge of urban planning while the rear wheel represents knowledge of the subsurface. Just as pedals put bicycles in motion, the metaphorical bicycle pedals convert these two knowledge sources into cooperative workflows among subsurface specialists and urban planners. Decision-makers, addressing city needs,

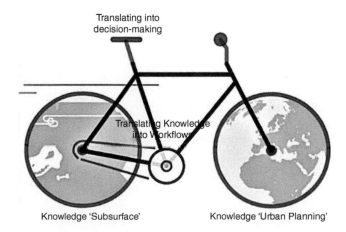

Figure 16.2 How the cost suburban toolbox works. Source: I.P.A.M. van Campenhout.

sit on the saddle. They decide when the bicycle moves and in which direction.

The Toolbox is designed to provide potential users with guidance on what might be achieved in their own organization, and how to avoid costly mistakes. While the Toolbox is expected to be most useful during the initial stages of urban planning, it includes examples of how it can be used as a training tool and it provides contact information for representatives of the SUB-URBAN network.

16.2.3 Resilience as a Key Concept

Since the early 2000s, an extended ongoing debate among much of society has resulted in a broad consensus on three facts: (i) our living environment is changing as a consequence of climate change and sea-level rise; (ii) this changing environment is partly attributable to human action; and (iii) there are risks and limitations to our ever-growing societies. Such awareness underpins all 17 UN sustainable development goals either directly or indirectly. Resilience is the leading principle that guides societal responses to global change. Resilience is defined as the ability to withstand change and keep functioning (Pelling 2003; Smith 2013). The "Risk Society" (sensu Beck 1992) is a society that has organized itself resiliently, taking a future-oriented approach, and is aware of changing conditions, pressures, and risks.

Risk is concerned with both the likelihood of a harmful event occurring and the consequences of the event. Harmful events fall, roughly, into environmental and socio-economic risk categories. Environmental risks are associated multiple naturally occurring hazards, including earthquakes, hurricanes, wildfire, severe rainfall events, and sudden collapse or subsidence in areas underlain by rocks prone to dissolution. Groundwater can dissolve limestone and, even more rapidly, gypsum to produce karst conditions – the formation of caves and sinkholes. Several cities experience severe risks from these hazards, including Paris (France), Madrid and Zaragoza (Spain), and towns peripheral to the Hartz mountains in Germany (Cooper and Gutiérrez 2013; Gutiérrez et al. 2008). The city of Ripon, in North Yorkshire, is severely impacted by the dissolution of Permian gypsum strata and has developed special planning review procedures to mitigate the hazard (Box 16.1).

Many of the socio-economic risks, ranging from planning failures to terrorist attacks, are the consequence of exposure to environmental pressure, especially if inequality is present. *Environmental pressure* is concerned with harmful or deteriorating conditions rather than single harmful events. These deteriorating conditions also fall into environmental and socio-economic categories. The environmental category includes sea-level rise, global warming, subsidence, and scarcity of natural resources; the socio-economic category includes prior and current over-population, impoverishment, and lack of access to resources.

16.3 Resilient Cities

The concept of resilience is an appropriate framework for urban planning; it stimulates the development of long-term sustainability strategies that are capable of dealing with a changing living environment and the associated risks and pressures. The Rockefeller foundation established the "100 Resilient Cities" network to promote resilience in city planning; their website (www.100resilientcities.org) is a fairly comprehensive online resource.

16.3.1 Scarcity of Space and Typical Urban Stresses

People will migrate to cities if they cannot make a living elsewhere. Proverbial "big-town problems" occur because of the large numbers of people living and working in a small, intensely used, space, where disruption, whether in infrastructure, utilities, or supplies, will impact large numbers of individuals, socially or otherwise (Pacione 2001). Urban stresses include high unemployment, inefficient public transportation, congested infrastructures, endemic violence, chronic shortages, and difficulties in sustaining cultural heritage. None of these stresses are unique to the urban environment, but they are more intense in cities than in smaller towns and villages where lower population densities promote less congestion and violence with a more relaxed lifestyle.

Except in the case of a city that is expanding into new areas, urban development is often a matter of redevelopment of built-up terrains or "brownfield" sites. With redevelopment, city planners have to locate new functions between, above, below, or at the expense of, existing functions while causing the least possible disruptions. A resilient city is primarily a system that keeps functioning under stress, including its daily stresses and those associated with its (continuous) development. Urban space is usually considered more precious than anything the subsurface has to offer. Geoscientists will rarely be asked to discover a best possible location for a certain urban development. Rather, they are typically asked to indicate whether a proposed development is feasible at a specific location.

Box 16.1 Planning for Sinkholes Caused by Gypsum Dissolution: An Example from Ripon, North Yorkshire

Anthony H. Cooper, Honorary Research Associate, British Geological Survey

The city of Ripon, North Yorkshire, is underlain by two sequences of gypsum which are 40 and 10 m thick respectively. They are sandwiched between dolomite aquifers that feed water under artesian pressure into the gypsum (Cooper et al. 2013). Because gypsum dissolves very quickly (Klimchouk 1996), a joint-controlled maze cave system is rapidly evolving in the gypsum. Sinkhole collapses, forming a pattern related to the joints (Figure 16.1.1), are common (Cooper 1986, 1998). The sinkholes can be up to 30 m across and up to 20 m deep.

Figure 16.1.1 Planning area for the city of Ripon prone to subsidence due to the dissolution of Permian gypsum, also showing the distribution of sinkholes with their year of formation. (Cooper et al. 2013)

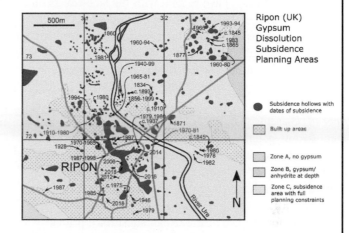

Where gypsum is near the surface it poses a geological hazard; sudden sinkhole collapse and general ground subsidence cause severe building damage and pose human safety hazards in a zone that is about 3 km wide and extends along strike to the north and south. The western limit of the subsidence-prone area is the up-dip dissolved feather edge of the lower gypsum, the down-dip limit approximates to the transition from gypsum to anhydrite which is defined by water circulation.

Local planning authorities have legally defined a special planning area with three hazard zones (Figure 16.1.1). In Zone C, with the highest hazard exposure, gypsum problems must be investigated and mitigated or avoided (Paukstys et al. 1999; Thompson et al. 1996). The completion of standard documents and a formal signing-off procedure by a competent person is required.

In 2018, as part of a NERC Urgency Grant, BGS modeled the 3-D geology of Ripon, (Figure 16.1.2). The model was developed to provide geological context for the findings of geophysical surveys undertaken to investigate the sinkhole that opened on Magdalen's Road in Ripon on November 9, 2016. By setting the sinkhole in its 3-D geological context it may be possible to improve our understanding of the processes associated with the formation of sinkholes in Ripon.

Figure 16.1.2 3-D geological model of the Ripon area looking north-west. (British Geological Survey)

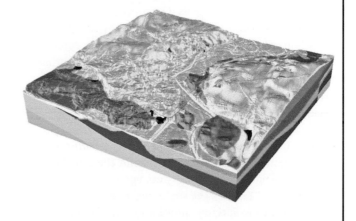

16.3.2 Geological Data and Information Needs

Traditionally, the subsurface is seen primarily as a source of energy, minerals, and water, and available subsurface knowledge is largely derived from their exploration and exploitation. Yet, while cities obviously rely on such resources, they are usually produced outside cities, and regional or national authorities are mostly responsible for their licensing and the policies that govern their operations. However, cities often have to deal with the side-effects of mining activities; if not with those of an ongoing operation, then perhaps with the after-effects of past ones. For example, about half of Glasgow is affected by disused mine workings; the mining was mostly for coal and ironstone, although some mines were underground limestone and sandstone quarries (Harrison 2012). Part of Jordanhill, one of Glasgow's most desirable neighborhoods, was identified as potentially at risk from the collapse of shallow abandoned mine workings. The risk was addressed by backfilling of the mine workings which required around six months to complete, and cost about £1 million (Harrison 2012).

Planning for the use of the urban subsurface primarily involves determining how the characteristics and conditions of the subsurface materials might affect the design or stability of proposed facilities or influence the optimal use of subsurface urban spaces. Geological surveys are the traditional custodians of subsurface data and information. However, most geological maps, the primary product of geological surveys since the nineteenth century, often do not show urban geology. Traditionally, urban areas have not been surveyed and are simply shown as "built up," but this is changing in two important ways (Huggenberger and Epting 2011; Teixeira-Guerra 2011; Forman 2014). First, geological surveys have started to work with third-party data, and are now starting to utilize, and make sense of, the vast amounts of subsurface data that are acquired in cities as part of ongoing building and construction projects (Price et al. 2010; Ford et al. 2014). Second, there is a shift from 2-D to 3-D information products which can more clearly communicate the subsurface conditions, especially to the urban planning community (Terrington et al. 2014; Price et al. 2018).

Because resilience requires anticipating changing conditions, urban geomodelers should be able to support multiple scenarios, including: "Can we keep exploiting this aquifer sustainably if recharge drops and the population grows?" or "Would slope stability change if hurricane frequency doubles?" To answer such questions geomodeling has to include time-varying "process models" and expand to 4-D assessments (Chapter 14).

16.4 Challenges to Urban Subsurface Modeling

The three main challenges affecting the use of 3-D subsurface models in support of urban planning and urban development projects are: (i) the ubiquitous presence of artificially modified ground (AMG); (ii) the scale and density of the required information; and (iii) the communication and sharing of information.

16.4.1 Modeling Artificially Modified Ground

Most urban areas are blanketed by a ubiquitous layer of AMG which has formed over centuries due to human activities. Characterization and modeling artificial deposits is still in its infancy. A limited number of studies, such as by Rosenbaum et al. (2003) and Ford et al. (2014), address the problem from a basically stratigraphic point of view. This is an important step in any form of 2-D or 3-D mapping.

AMG deposits result from anthropogenic activities, including various topographic modifications and the gradual accumulation of building debris following destruction and redevelopment of various structures. Topographic modification, including excavation of ridges or hills and filling depressions, may cause compaction and settlement of relatively porous or poorly consolidated shallow deposits due to changes in overburden loads. Urban areas are largely covered with impervious buildings and paved areas; these reduce groundwater recharge rates, which results in lowered groundwater levels in many locations. Extraction and drainage of groundwater further lowers groundwater levels. Compaction of dewatered aquifer systems and organic soils results in damaging land subsidence (Galloway et al. 1999; Nieuwenhuis and Schokking 1997). In contrast, large underground objects such as tunnels, parking facilities, sheet pile walls, or leaky sewage pipes may divert groundwater flows, or increase groundwater levels, resulting in increased risk of local or regional flooding hazards. Materials used in building construction or roadbuilding may be geochemically unstable; industrial activities can lead to subsurface pollution issues. Yet, most of the changes to the urban subsurface are not administered in such a way that they can be readily and systematically integrated in geomodeling workflows.

For the SUB-URBAN project, Schokker (2017) compared current geological modeling approaches for AMG and highlighted the importance of mapping and characterizing the anthropogenic subsurface layer to produce a useful model with four key findings:

1. The anthropogenic subsurface layer is all-inclusive and its definition and modeling should encompass everything that is present between the land surface and the

natural subsurface that is of influence to the model properties at the intended use scale of the model. This means that this layer does not only consist of reworked natural deposits and artificial deposits, but also any tunnels, cables, foundations, basements, and archeological heritage that may be present (Schokker et al. 2017).

2. The types of input data that are used to model AMG should be appropriate to the modeling purpose and the associated modeling scale, but in practice the data sources are largely determined by data availability, the time and financial constraints of the project, and the selected modeling method. The use of nationwide standardized and freely accessible datasets is preferred.

3. AMG modeling may be used to bridge the scale and information gap between above-surface information and natural subsurface information. The combination of these layers or sub-models within one model, as proposed by the GeoCIM approach (Mielby et al. 2017), requires the nesting of different model scales, both laterally and vertically.

4. The ideas and technical solutions for efficient model maintenance exist. It is very important that all parties involved acknowledge the need to keep models up to date as this preserves their value. This requires maintenance funding and a feedback loop to report model issues (Schokker et al. 2017). Model maintenance and update requires the establishment of sustainable relationships among municipal governments, research institutes, geological surveys, and the private sector.

16.4.2 Scale and Data Density

Urban planning is conducted and communicated at map scales up to 1:5000. Geological maps are typical available at scales up to 1:25 000; but such detailed geological mapping can only be undertaken in well-exposed areas. The urban environment obviously has limited outcrops and provides few possibilities for making field observations. The only viable option to get at the observation density that is necessary for urban geomodeling is to work with geotechnical borehole data. However, geotechnical data are often poorly distributed for geological modeling because they are concentrated around previous construction sites and along major transportation routes. Cone penetration tests (CPTs) offer a less expensive exploration option in locations with extensive medium or fine-grained unconsolidated surficial deposits. They are commonly acquired in the planning stage of infrastructural works. In the Netherlands, for example, CPT soundings are located about every 25 m (80 ft) along railways and motorways and throughout construction sites, so they vastly outnumber the available boreholes. Recent experimental use of Artificial Neural Network methods has successfully interpreted CPT data into facies units that define subsurface geometries and properties (Maljers et al. 2019).

16.4.3 Communicating and Sharing Information

Communicating and sharing subsurface information with urban planners is challenging. It is greatly influenced by the rapid recent adoption of Building Information Modeling (BIM) by both the construction industry and the urban regulatory authorities. Current BIM methods constrain the easy integration of geological subsurface information and models with BIM workflows. The GeoCIM concept has been proposed as a way to facilitate such integration.

16.4.3.1 Building Information Modeling (BIM)

Traditional building design has usually relied on two-dimensional technical drawings, including plans, elevations, and sections. BIM evolved as Computer Aided Design (CAD)-based tools were adapted to provide a 3-D "digital representation of physical and functional characteristics of a facility" (National Institute of Building Sciences 2014). The construction industry has increasingly used BIM for planning, designing, constructing, and operating buildings and infrastructure because it provides a platform for shared knowledge throughout the life cycle of a building. The life cycle of a facility is defined as existing from earliest conception to demolition (National Institute of Building Sciences 2014). The initial BIM has expanded by augmenting the three primary spatial dimensions (width, length, and height or depth) with the additional dimensions of time, producing 4-D BIM, and cost, producing 5-D BIM (NBS 2019). Recently, further BIM dimensions have been proposed; building environmental and sustainability analysis is included in 6-D BIM and life-cycle facility management in 7-D BIM. However, there remains some ambiguity and there are conflicting definitions about these higher-dimensioned BIM options (NBS 2019). National governments are encouraging the use of BIM, and several have already mandated its use or specified its required use in the near future.

BIM concepts, with their reliance on CAD technology, have so far mostly ignored the subsurface, in spite of the importance of the geological information concerning the subsurface to many aspects of a building life cycle. The integration of geological information and geological models into BIM workflows requires a transformation of constructed geological framework models into data structures that can be understood by BIM (Kessler et al. 2016). Several software vendors have developed solutions using proprietary formats. This constrains users because they can only integrate information through interfaces proposed

Figure 16.3 The five LODs (Levels of Detail) of CityGML 2.0. The geometric detail and the semantic complexity increase, ending with the LOD4 containing indoor features. (Biljecki et al. 2016)

Figure 16.4 BIM urban model of City of London buildings combined with part of the BGS national geological model. Source: British Geological Survey; building model courtesy of Arup.

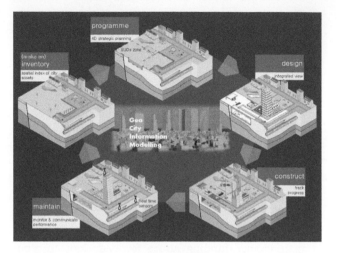

Figure 16.5 GeoCIM lifecycle. (Mielby et al. 2017)

by a single vendor's software. To increase interoperability, building Smart International (bSI) has supported the open-BIM initiative to develop software independent standards to facilitate data exchanges.

BIM represents a design as a combination of objects (or model elements), which are flexibly defined. For example objects can be solid shapes that only represent a building's outer shell, or have void spaces that allow the representation of the inside structure as well. They can be geometrically generic or very detailed. They can carry attributes such as product vendor or wall texture, as well as relationships with other objects. Thus, these objects can represent buildings and their components at differing levels of detail (LOD). The Open Geospatial Consortium (Gröger et al. 2012) defines five levels of detail (LOD0–LOD4), which are graphically represented in Figure 16.3. Biljecki et al. (2016) argue that these levels are not mutually exclusive and designed an alternative scheme with 16 levels to define model complexity. BIM is expected to become central to future building and infrastructure design and maintenance. There is an increasing interest and acceptance by city and urban planning organizations

16.4.3.2 The GeoCIM Concept

The built environment is always more precisely resolved spatially than the subsurface, especially when BIM procedures are employed. However, coarser 3-D renderings

of the built environment can be combined with subsurface models in a useful and geospatially justifiable way. Figure 16.4 shows one recent example where a BIM urban model of the buildings of the City of London have been located above a portion of the BGS national model (Waters et al. 2016). Recognizing the potential for the enhanced form of such model integration, the SUB-URBAN project proposed a Geo City Information Modeling, or GeoCIM, concept (Figure 16.5). GeoCIM expands on BIM principles by explicitly requiring the integration of the above- and below-surface data at scales appropriate to a city as part of sustainable urban planning and management. The prefix "geo" is used to mean ground or land and should be recognized as representing geology, geotechnics, geolocation, and geophysics.

GeoCIM is a process (Figure 16.6) that involves generating, sharing, integrating, and managing digital representations of physical and functional characteristics of at least the following urban environment layers:

- Surface layer: natural and man-made on-surface features.
- Anthropogenic subsurface layer: man-made ground, buried infrastructure, foundations.
- Natural subsurface layer: geological units, hazards and processes.

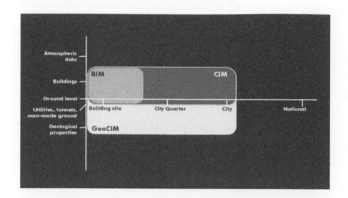

Figure 16.6 BIM, CIM, and GeoCIM relationships by geographical scale of interest (*x*-axis) and data themes above and below ground (*y*-axis). (Mielby et al. 2017)

Similar to BIM, GeoCIMs will rely on multiple software tools. Specialist data and models will be developed and exist in specialist tools while the communally relevant elements will be displayed at an appropriate scale to facilitate integration with information from other key stakeholders.

It is expected that subsurface specialists will use GeoCIMs to share crucial information about geological structures, processes, and hazards using methods that are accessible to wider audiences than traditionally. Subsurface specialists will also benefit by being able to identify the location of man-made structures that influence groundwater flow or ground stability. For urban planners, GeoCIMs will facilitate the planning process by providing a single point of information for all key 3-D and 4-D urban datasets. Once GeoCIMs incorporate environmental indicators such as depths to the water table

and air quality and soil quality data, environmental assessments can be integrated into the planning process. Decision makers will be able to visualize and quantify the impacts of high-level strategic decisions without being required to delve into the interrelated specialist domains of geoscience, construction, and city planning. GeoCIMs will also provide the general public with greater access to information about their environment. A GeoCIM developed using open web services would encourage the general public and private companies to develop new applications, thereby maximizing information re-use.

16.5 Case Example: Planning for a More Resilient New Orleans

New Orleans is located south of Lake Pontchartrain on the banks of the Mississippi River in the Mississippi River Delta, approximately 105 miles (169 km) upriver from the Gulf of Mexico (Figure 16.7). The initial settlement was on a natural levee and an area of higher ground that was slightly above mean sea level. Over time, development pressures and the advent of new technologies pushed human settlement into vulnerable areas and masked the natural processes that have traditionally sustained the landscape.

The city of New Orleans presents an unusual hydrologic situation; it is completely surrounded by higher terrain, thus rainfall will collect in the lowest sections of the city unless otherwise removed by evapotranspiration or pumping. This drainage hazard is aggravated by the fact that

Figure 16.7 Map showing principal features of the New Orleans metropolitan area. (U.S. Army Corps of Engineers 2009a)

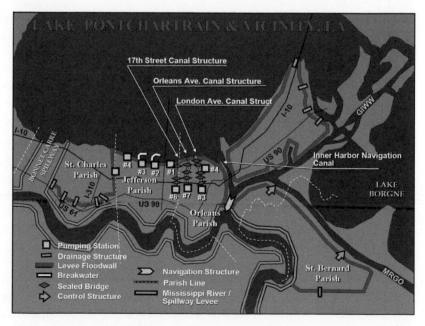

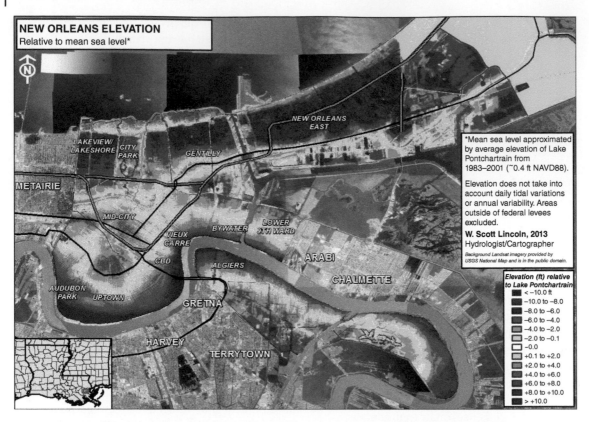

Figure 16.8 Elevation map of New Orleans. Blue/purple colors indicate elevations below the average level of Lake Pontchartrain and orange/brown colors indicate elevations above. Source: U.S. National Oceanic and Atmospheric Administration.

about 65% of New Orleans lies at or below mean sea level (Figure 16.8), as defined by the average elevation of Lake Pontchartrain during the 1983–2001 period (Schlotzhauer and Lincoln 2016). The average elevation of the city is currently between 0.30 and 0.61 m (1–2 ft) below sea level; however, some portions of the city are much lower. After the Flood Control Act of 1965, the US Army Corps of Engineers (USACE) built floodwalls and man-made levees around a much larger geographic footprint that included previous marshland and swamp. Figure 16.9 is a north–south profile across New Orleans that shows a maximum levee height of 7 m (23 ft).

It has long been recognized that New Orleans is subsiding and is therefore susceptible to catastrophic flooding. Although studies have shown the deeper geological units to be relatively stable (Dokka 2006), space-based synthetic-aperture radar measurements revealed that parts of New Orleans underwent rapid subsidence in the three years before the 2005 Hurricane Katrina (Dixon et al. 2006). Andersen et al. (2007) summarized the situation as follows:

> "*Large portions of Orleans, St. Bernard, and Jefferson parishes are currently below sea level—and continue to sink. New Orleans is built on thousands of feet of*

soft sand, silt, and clay. Subsidence, or settling of the ground surface, occurs naturally due to the consolidation and oxidation of organic soils (called 'marsh' in New Orleans) and local groundwater pumping. In the past, flooding and deposition of sediments from the Mississippi River counterbalanced the natural subsidence, leaving southeast Louisiana at or above sea level. However, due to major flood control structures being built upstream on the Mississippi River and levees being built around New Orleans, fresh layers of sediment are not replenishing the ground lost by subsidence."

Shinkle and Dokka (2004) used the precise vertical datum NAVD88 to estimate higher rates of subsidence than had previously been reported. Nienhuis et al. (2017) evaluated subsidence along the Louisiana coast and found an average subsidence rate as high as 9 mm yr^{-1}, with significantly higher rates occurring in several areas, including near New Orleans. These subsidence rates are significantly higher than had been previously reported and they concluded that subsidence is likely accelerate due to limited sediment supply and accelerated sea-level rise, as reported by Blum and Roberts (2009).

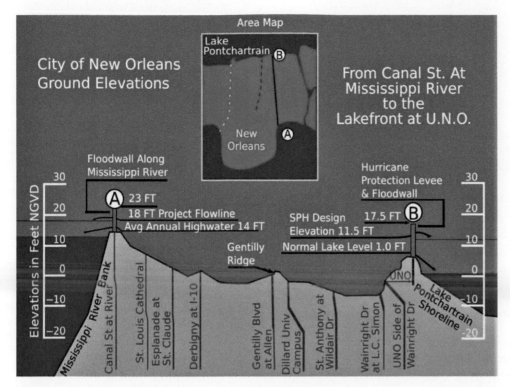

Figure 16.9 North-south topographic profile across New Orleans showing maximum levee height of 7 m (23 ft). Source: R.J. Stuurman.

16.5.1 Hurricane Katrina

New Orleans has been affected by numerous hurricanes, but the most devastating was Hurricane Katrina on August 29, 2005. Katrina made landfall south-east of New Orleans as a Category 3 storm with winds reaching speeds as high as 192 km h^{-1} (120 mph). New Orleans experienced lower velocity winds but received heavy rainfall. A tidal surge in Lake Pontchartrain caused overtopping of levees and floodwalls. Some 23 breaches in the levees and floodwalls flooded about 80% of New Orleans by August 31, 2005, with some areas under 4.6 m (15 ft) of water (Figure 16.10). The famous French Quarter and Garden District escaped flooding because they are above sea level. The flooding damaged about 70% of New Orleans' occupied housing; damage was disproportionately high in low-income neighborhoods, and many of those who lost their homes faced years of hardship. Between 80% and 90% of the residents of New Orleans were evacuated prior to Katrina; many evacuees never returned and the population of New Orleans decreased by over 50% and remains below pre-Katrina levels. However, many poorer residents, without access to personal vehicles or who were unable to receive local news bulletins, remained in the city. Many fatalities ensued; it is estimated that between 1200 and 1500 people died as a direct result of Katrina. Katrina caused an estimated $108 billion in property damage, making it the costliest storm on record in the USA.

16.5.2 Post-Katrina Investigations

The USACE immediately formed an Interagency Performance Evaluation Task Force (IPET) to evaluate the performance of the levees, to assess the causes of their failures, and to recommend implementation of measures to prevent similar disasters from recurring. Several interim draft documents were produced before the final official report was completed in 2009 (USACE 2009a). The American Society of Civil Engineers (ASCE) formed an independent Hurricane Katrina External Review Panel (Andersen et al. 2007) to conduct an in-depth review of the comprehensive work of the USACE IPET.

Although considerable international assistance was offered to the United States, political and legal barriers constrained the formal role of many international experts. However, an existing Memorandum of Agreement between the USACE and *Rijkswaterstaat* (the Dutch Directorate-General of Public Works and Water Management) facilitated technical exchanges between American and Dutch engineers. The Dutch first visited New Orleans to assist the USACE; there followed a visit by key USACE personnel to the Netherlands to meet with engineers from *Rijkswaterstaat*. In addition to participating in a

Figure 16.10 Flooding due to breach of the 17th Street Canal. Source: U.S. Army Corps of Engineers.

number of workshops and reviews, *Rijkswaterstaat* and the Netherlands Water Partnership (a Dutch consortium of government agencies, researchers, and consultants) produced a report titled "A Dutch Perspective on Coastal Louisiana: Flood Risk Reduction and Landscape Stabilization" (Dijkman 2007). A summary evaluation of the Dutch Perspective was added to the USACE IPET Final Report 2009 (USACE 2009b).

The Dijkman (2007) report provided an independent view of risk reduction and restoration issues for the Louisiana coastal area based on Dutch experience in dealing with similar issues in the Netherlands. Five alternative strategies were evaluated and ultimately a preferred strategy, termed "Protected City and Closed Soft Coast," was identified. Modeled after the flood risk reduction approach implemented by the Netherlands, the preferred strategy combines landscape stabilization with active flood reduction (Figure 16.11). It is intended to provide metropolitan New Orleans with a 1/1000 year or greater level of risk reduction, a level that the Dijkman (2007) report states is economically justifiable.

16.5.3 The Greater New Orleans Urban Water Plan

In 2010, the State of Louisiana's Office of Community Development Disaster Recovery Unit funded Greater New Orleans, Inc. (GNO, Inc.) to partner with local and international experts, neighborhood groups, and civic leaders to develop a comprehensive, integrated, and sustainable water management strategy for the Greater New Orleans region. Waggonner & Ball Architecture/Environment (New Orleans, LA) led an international team of designers, engineers, scientists, and policy experts in a multi-parish

Table 16.2 American and Dutch members of the Greater New Orleans Water Management Plan design team.

US team members	Dutch team members
ARCADIS US	BoschSlabbers Landscape Architects
Bright Moments	City of Amsterdam
CDM Smith	City of Rotterdam
Dana Brown & Associates	Deltares
Dewberry	H + N + S Landscape Architects
Eustis Engineering	Palmbout Urban Landscapes
FutureProof	Robert de Koning Landscape Architect
GCR Inc.	RoyalHaskoningDHV
LSU Coastal Sustainability Studio	Delft University of Technology
Manning Architects	
Nelson Engineers	
Sherwood Design Engineers	
Tulane University	
Waggonner & Ball Architects	

Source: Waggonner and Ball (2013a).

planning effort to produce the Greater New Orleans Urban Water Plan (Louisiana Resiliency Assistance Program 2017). The team included several Dutch experts (Table 16.2) who worked closely with local governments during plan development; however, implementation of the plan will be the responsibility of the communities.

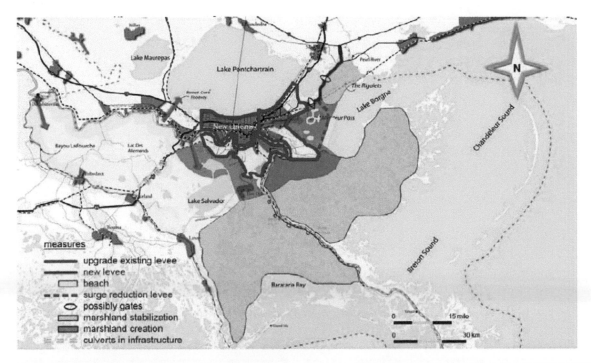

Figure 16.11 The preferred strategy of the Dutch Perspective report *"Protected City and Closed Soft Coast."* (USACE 2009b)

The Greater New Orleans Urban Water Plan provides sustainable strategies that address flooding caused by excess runoff and subsidence caused by pumping of storm water. The plan incorporates a resiliency paradigm that combines water resource management, urban design opportunities, and economic benefits of implementation or non-action (Waggonner & Ball 2013a,b,c). The plan incorporates the multiple lines of defense concept, as defined by the earlier Dijkman (2007) report of the USACE IPET Final Report (USACE 2009a). Because the traditional protection approach of levees and pumping exacerbates land subsidence, the plan promotes bio-retention and infiltration strategies to delay storm water runoff and enhancement of groundwater recharge by storing storm water in the landscape for longer periods by retrofitting canals and finding space for new canals and ponds.

The plan includes several demonstration projects that have allowed the proving of various resiliency concepts. The Mirabeau Water Garden demonstration project, started in 2015, is one of the first. It incorporates several Dutch-inspired solutions for storm water storage and filtration that uses a system of wetland terraces, rain gardens, and bioswales. A woodland provides diverse habitats for native flora and fauna (Figure 16.12). The Dutch research institute Deltares assisted with the monitoring of the subsoil and groundwater; they also evaluated the potential for water storage and the impact of reduced subsidence on the surrounding area. These activities

formed the basis for a "Soft Soils Need Solid Governance" course, conducted in New Orleans with the cooperation of the Water Institute of the Gulf, U.S. Geological Survey, and Tulane University. The Mirabeau Water Garden includes areas that can be flooded by diverted storm water as required during typical and heavier rainstorms (Figure 16.13).

16.6 Conclusions

The key message of this chapter to geological modelers is that underground planning is not merely a matter of technical optimization, communication about the geological models with the user communities in terms they understand is of crucial importance. In oil and gas exploration, which is still the best-established application of 3-D modeling, the focus is on identifying the best possible location to produce a geological resource based on geological data. In this case, communication mostly involves peer-to-peer discussions, a relatively straightforward process. In underground planning, the issues relate to weighing of diverse interests of multiple stakeholders that claim the same space. To a planner, a suboptimal solution that satisfies the largest possible group of stakeholders—including political decision makers—could be better than a winner-takes-all solution. This requires more complex communication channels, so the geological modeler cannot serve the

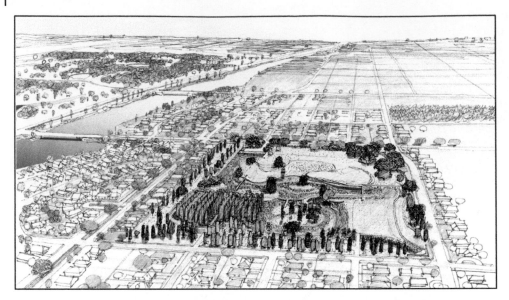

Figure 16.12   Conceptual overview of the 300 m × 300 m Mirabeau Water Garden demonstration project. Source: Waggonner & Ball Architecture/Environment.

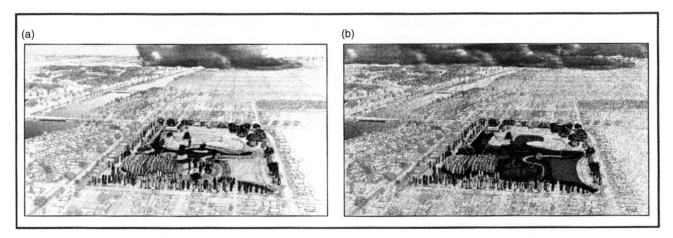

Figure 16.13   Operation of the Mirabeau Water Garden during: (A) typical rainstorm (13 inches) and (B) heavy rainstorm (10 year event). Source: Waggonner & Ball Architecture/Environment.

planning community with an exploration mindset. The successful use of a geological model must support multiple scenarios, honor considerations outside the geoscience domain, and pay special attention to model visualizations that are understandable to non-geoscience experts.

Acknowledgments

This chapter is largely based on discussions and experiences gained by the authors during their participation in the EU COST Action SUB-URBAN Project. The assistance of the many project participants in developing these topics is now acknowledged. Anthony H. Cooper, Honorary Research Associate, British Geological Survey, kindly provided the discussion of the subsidence hazard at Ripon (Box 16.1) and Hannah Gow created the 3-D model of the Ripon area. Information concerning the New Orleans post-Katrina resiliency planning (Section 16.5) was freely offered by colleagues at several Dutch organizations shown in Table 16.2. Waggonner & Ball Architecture/Environment in New Orleans agreed to our use of two conceptual illustrations of planned sustainable installations there.

References

Andersen, C.F., Battjes, J.A., Daniel, D.E. et al. (2007). *The New Orleans Hurricane Protection System: What Went Wrong and Why*. Reston, VA: American Society of Civil Engineers. 84 pp. [Online: Available at: https://ascelibrary.org/doi/pdf/10.1061/9780784408933] (Accessed October 10, 2020).

Beck, U. (1992). *Society of Risk: Towards a New Modernity*. Thousand Oaks, CA: SAGE Publishers. 272 pp.

Biljecki, F., Ledoux, H. and Stoter, J. (2016). An improved LOD specification for 3D building models. *Computers, Environment and Urban Systems* 59: 25–37. https://doi.org/10.1016/j.compenvurbsys.2016.04.005.

Blum, M.D. and Roberts, H.H. (2009). Drowning of the Mississippi Delta due to insufficient sediment supply and global sea-level rise. *Nature Geoscience* 2 (7): 488–491. https://doi.org/10.1038/ngeo553.

Campbell, S.D.G., Bonsor, H., Lawrence, D.J. et al. (2015). Datos del subsuelo y su conocimiento para las Ciudades del Mañana: Lecciones aprendidas de Glasgow y su aplicabilidad en otros lugares (in Spanish). *Ciudad y Territorio, Estudios Territoriales* XLVII (186): 745–759. [Online: Available at: https://recyt.fecyt.es/index.php/CyTET/article/view/76443/46790] (Accessed October 2020).

Campbell, S.D.G., De Beer, J., Mielby, S. et al. (2017). Transforming relationships between geoscientists and urban decision-makers: European COST Sub-Urban Action (TU1206*). Procedia Engineering* 209: 4–11. https://doi.org/10.1016/j.proeng.2017.11.124.

Cooper, A.H. (1986). Foundered strata and subsidence resulting from the dissolution of Permian gypsum in the Ripon and Bedale areas, North Yorkshire. *Geological Society Special Publications* 22: 127–139. https://doi.org/10.1144/GSL.SP.1986.022.01.11.

Cooper, A.H. (1998). Subsidence hazards caused by the dissolution of Permian gypsum in England: geology, investigation and remediation. In: *Geohazards in Engineering Geology* (eds. J.G. Maund and M. Eddleston), 265–275. Geological Society Engineering Geology Special Publication 15. https://doi.org/10.1144/GSL.ENG.1998.015.01.27.

Cooper, A.H. and Gutiérrez, F. (2013). Dealing with gypsum karst problems: hazards, environmental issues, and planning. In: *Treatise on Geomorphology, Karst Geomorphology, vol. 6*, (eds. J. Shroder and A. Frumkin), 451–462. San Diego, CA: Academic Press.

Cooper, A.H., Odling, N.E., Murphy, P.J. et al. (2013). The role of sulfate-rich springs and groundwater in the formation of sinkholes over gypsum in eastern England. In: *Sinkholes and the Engineering and Environmental Impacts of Karst, Proceedings of the 13th Multidisciplinary Conference* (eds. L. Land, D.H. Doctor and J.B. Stephenson), 141–150. Carlsbad, NM: National Cave and Karst Research Institute.

Dijkman, J. [Ed.] (2007). *A Dutch Perspective on Coastal Louisiana Flood Risk Reduction and Landscape Stabilization*. London, UK: European Research Office of the U.S. Army. 272 pp.

Dixon, T.H., Amelung, F., Ferretti, A. et al. (2006). Subsidence and flooding in New Orleans. *Nature* 441: 587–588. https://doi.org/10.1038/441587a.

Dokka, R.K. (2006). Modern-day tectonic subsidence in coastal Louisiana. *Geology* 34 (4): 281–284. https://doi.org/10.1130/G22264.1.

Environmental Protection Agency (2011). *Decision Support for Renewal of Wastewater Collection and Water Distribution Systems*. Cincinnati, OH: U.S. Environmental Protection Agency, National Risk Management Research Laboratory. 59 pp.

Ford, J.R., Price, S.J., Cooper, A.H. and Waters, C.N. (2014). An assessment of lithostratigraphy for anthropogenic deposits. *Geological Society Special Publications* 395: 55–89. https://doi.org/10.1144/SP395.12.

Forman, R.T.T. (2014). *Urban Ecology—Science of Cities*. Cambridge, UK: Cambridge University Press. 462 pp. https://doi.org/10.1017/CBO9781139030472.

Galloway, D., Jones, D.R. and Ingebritsen S.E. (1999). *Land Subsidence in the United States*. Reston, VA: U.S. Geological Survey, Circular 1182. 177 pp. https://doi.org/10.3133/cir1182.

Gröger, G., Kolbe, T.H., Nagel, C. and Häfele, K.H. (2012). *OGC City Geography Markup Language (CityGML) Encoding Standard*. Brussels, Belgium: Open Geospatial Consortium OGC 12-019. 326 pp. [Online: Available at: https://www.ogc.org/standards/citygml] (Accessed October 2020).

Gutiérrez, F., Cooper, A.H. and Johnson, K.S. (2008). Identification, prediction, and mitigation of sinkhole hazards in evaporite karst areas. *Environmental Geology* 53: 1007–1022. https://doi.org/10.1007/s00254-007-0728-4.

Harrison, J. (2012). Half of Glasgow Homes at Risk from Disused Mines. *The Herald*, October 27, 2012. [Online: Available at: https://www.heraldscotland.com/news/13078601.half-of-glasgow-homes-at-risk-from-disused-mines/] (Accessed October 2020).

Huggenberger, P. and Epting, J. [Eds.] (2011). *Urban Geology: Process-Oriented Concepts for Adaptive and Integrated Resource Management*. Basel, Switzerland: Springer AG. 216 pp.

Jonas, A.E.G., McCann, E. and Thomas, M. (2015). *Urban Geography—A Critical Introduction*. Hoboken, NJ: Wiley. 384 pp.

Kessler, H., Wood, B., Morin, G. et al. (2016). Building Information Modelling (BIM) – A route for geological models to have real world impact. In: *Three-Dimensional Geological Mapping: Workshop Extended Abstracts; Geological Society of America Annual Meeting, Baltimore, Maryland, October 31, 2015* (eds. K.E. MacCormack, L.H. Thorleifson, R.C. Berg and H.A.J. Russell), 13–18. Edmonton, Alberta, Canada: AER/AGS Special Report 101. [Online: Available at: http://nora.nerc.ac.uk/id/eprint/512137/] (Accessed October 2020).

Klimchouk, A. (1996). The dissolution and conversion of gypsum and anhydrite. *International Journal of Speleology* 25: 9–36.

Knox, P.L. and McArthy, L.M. (2012). *Urbanization. An Introduction to Urban Geography, 3e*, New York, NY: Pearson. 480 pp.

Lois-González, R.C., Pazos-Otón, M. and Van Campenhout, I.P.A.M. (2015). El subsuelo de las ciudades y áreas urbanas europeas: una propuesta general de estudio para su consideración en los documentos de planeamiento (in Spanish). *Ciudad y Territorio, Estudios Territoriales* XLVII (186): 717–731. [Online: Available at: https://recyt.fecyt.es/index.php/CyTET/article/view/76441/46788] (Accessed October 2020).

Louisiana Resiliency Assistance Program (2017). *Greater New Orleans Urban Water Plan*. Baton Rouge, LA: Louisiana State University. [Online: Available at: https://resiliency.lsu.edu/case-studies-blog/2017/11/14/greater-new-orleans-urban-water-plan] (Accessed May, 2019).

Maljers, D., Gunnink, J. and Stafleu, J. (2019). Trends and Innovation – Automated Interpretation of Boreholes and Cone Penetration Tests. *Abstracts, 5th European Meeting on 3D Geological Modelling, May 21–24, 2019, Bern, Switzerland*.

Mielby, S., Eriksson, I., Campbell, S.D.G. and Lawrence, D.J. (2017). Opening up the subsurface for the cities of tomorrow. *Procedia Engineering* 209: 12–25. https://doi.org/10.1016/j.proeng.2017.11.125.

National Institute of Building Sciences (2014). *National BIM Standard-United States*. Washington, DC: NIBS. 2 pp.

NBS (2019). *National BIM Report 2019*. Newcastle upon Tyne, UK: NBS 54 pp.

Nienhuis, J.H., Törnqvist, T.E., Jankowski, K.L. et al. (2017). A new subsidence map for coastal Louisiana. *GSA Today* 27 (9): 58–59. https://doi.org/10.1130/GSATG337GW.1.

Nieuwenhuis, H.S. and Schokking, F. (1997). Land subsidence in drained peat areas of the Province of Friesland, the Netherlands. *Quarterly Journal of Engineering Geology* 30: 37–48. https://doi.org/10.1144/GSL.QJEGH.1997.030.P1.04.

Pacione, M. (2001). *Urban Geography: A Global Perspective*. Abingdon-on-Thames, UK: Routledge. 744 pp.

Paukstys, B., Cooper, A.H. and Arustiene, J. (1999). Planning for gypsum geohazards in Lithuania and England. *Engineering Geology* 52: 93–103. https://doi.org/10.1016/S0013-7952(98)00061-1.

Pelling, M. (2003). *The Vulnerability of Cities: Natural Disaster and Social Resilience*. Abingdon-on-Thames, UK: Routledge. 224 pp.

Price, S.J., Burke, H.F., Terrington, R.L. et al. (2010). The 3D characterisation of the zone of human interaction and the sustainable use of underground space in urban and peri-urban environments: case studies from the UK. *Zeitschrift der Deutschen Gesellschaft fur Geowissenschaften* 161 (2): 219–235. https://doi.org/10.1127/1860-1804/2010/0161-0219.

Price, S.J., Terrington, R.L., Busby, J. et al. (2018). 3D ground-use optimisation for sustainable urban development planning: A case-study from Earls Court, London, UK. *Tunnelling and Underground Space Technology* 81: 144–164. https://doi.org/10.1016/j.tust.2018.06.025.

Rosenbaum, M.S., McMillan, A.A., Powell, J.H. et al. (2003). Classification of artificial (man-made) ground. *Engineering Geology* 69: 399–409. https://doi.org/10.1016/S0013-7952(02)00282-X.

Schlotzhauer, D. and Lincoln, W.S. (2016). Using New Orleans pumping data to reconcile gauge observations of isolated extreme rainfall due to Hurricane Isaac. *Journal of Hydrologic Engineering* 21 (9): 05016020. https://doi.org/10.1061/(ASCE)HE.1943-5584.0001338.

Schokker, J. (2017). *Modelling Man-Made Ground to link the Above- and Below-Ground Urban Domains*. COST TU1206 Sub-Urban STSM Report TU1206-36204. 19 pp. [Online: Available at: http://sub-urban.squarespace.com] (Accessed May, 2019)

Schokker, J., Sandersen, P.B.E., De Beer, H. et al. (2017). *3D Urban Subsurface Modelling and Visualisation – A Review of Good Practices and Techniques to Ensure Optimal Use of Geological Information in Urban Planning*. COST TU1206 Sub-Urban Report TU1206-WG2-005. 93 pp. [Online: Available at: http://sub-urban.squarespace.com] (Accessed May, 2019).

Shinkle, K. and Dokka, R.K. (2004). *Rates of Vertical Displacement at Benchmarks in the lower Mississippi Valley and the northern Gulf Coast*. Washington, DC: National Oceanic and Atmospheric Administration, Technical Report 50. 135 pp. [Online: Available at: https://www.ngs.noaa.gov/heightmod/Tech50.shtml] (Accessed October 10, 2020).

Smith, K. (2013). *Environmental Hazards: Assessing Risk and Reducing Disaster*. Abingdon-on-Thames, UK: Routledge. 478 pp.

Teixeira-Guerra, A.J. (2011). *Geomorfologia Urbana (in Portuguese)*. Rio de Janeiro: Bertrand Brasil. 280 pp.

Terrington, R.L., Thorpe, S., Burke, H. et al. (2014). *Enhanced Mapping of Artificially Modified Ground in Urban Areas; Using borehole, Map and Remotely Sensed Data*. Keyworth, UK: British Geological Survey, Open Report OR/15/010. 47 pp. [Online: Available at: http://nora.nerc.ac.uk/id/eprint/510140/] (Accessed October 20, 2020).

Thompson, A., Hine, P.D., Greig, J.R. and Peach, D.W. (1996). *Assessment of Subsidence Arising from Gypsum Dissolution (with Particular Reference to Ripon, North Yorkshire)*. East Grinstead, UK: Symonds Travers Morgan. 288 pp.

Turner, A.K. (2003). Putting the user first: implications for subsurface characterization. In: *Characterisation of the Shallow Subsurface: Implications for Urban Infrastructure and Environmental Assessment* (eds. A.K. Turner and M. Rosenbaum), 61–68. Berlin: Springer, Lecture Notes in Earth Sciences, vol 99. https://doi.org/10.1007/3-540-48019-6_5.

USACE (2009a). *Performance Evaluation of the New Orleans and Southeast Louisiana Hurricane Protection System: Volume I – Executive Summary and Overview*. New Orleans, LA: U. S. Army Corps of Engineers. 288 pp. [Online: Available at: https://biotech.law.lsu.edu/katrina/ipet/Volume%20I%20FINAL%2023Jun09%20mh.pdf] (Accessed May 2019).

USACE (2009b). *Dutch Perspective Appendix: Louisiana Coastal Protection and Restoration Final Technical Report*. New Orleans, LA: U. S. Army Corps of Engineers. 17 pp. [Online: Available at: https://www.mvn.usace.army.mil/Portals/56/docs/environmental/LaCPR/DutchPerspective.pdf] (Accessed May, 2019).

United Nations (2015). *Transforming Our World: The 2030 Agenda for Sustainable Development*. New York, NY: United Nations, Resolution A/RES/70/1. 36 pp. [Online: Available at: https://sustainabledevelopment.un.org/post2015/transformingourworld/publication] (Accessed May 2019).

Van der Lugt, P., Dick, G., Eriksson, I. and De Beer, J. (2015). Planificando la Ciudad del Mañana: reduciendo la brecha entre urbanistas y especialistas del subsuelo (in Spanish). *Ciudad y Territorio, Estudios Territoriales XLVII* (186): 731–745. [Online: Available at: https://recyt.fecyt.es/index.php/CyTET/article/view/76442/46789] (Accessed October 10, 2020).

Van der Meulen, M.J., Campbell, S.D.G., Lawrence, D.J. et al. (2016). *Out of Sight, Out of Mind? Considering the Subsurface in Urban Planning - State of the Art*. COST TU1206 Sub-Urban Report TU1206-WG1-001. 49 pp. [Online: Available at: http://sub-urban.squarespace.com] (Accessed May 2019).

Waggonner & Ball (2013a). *The Greater New Orleans Urban Water Plan: Vision*. New Orleans, LA: Waggonner & Ball Architects. 203 pp.

Waggonner & Ball (2013b). *The Greater New Orleans Urban Water Plan: Implementation*. New Orleans, LA: Waggonner & Ball Architects. 223 pp.

Waggonner & Ball (2013c). *The Greater New Orleans Urban Water Plan: Urban Design*. New Orleans, LA: Waggonner & Ball Architects. 199 pp.

Waters, C.N., Terrington, R.L., Cooper, M.R. et al. (2016). *The Construction of a Bedrock Geology Model for the UK: UK3D_v2015*. Keyworth, UK: British Geological Survey Report OR/15/069. 22 pp. [Online: Available at: http://nora.nerc.ac.uk/id/eprint/512904/1/UK3D Metadata Report_Submission copy.pdf] (Accessed February 2018).

17

Providing Model Results to Diverse User Communities

Peter Wycisk[1] and Lars Schimpf[2]

[1] Institute of Geosciences and Geography, Faculty of Geosciences, Martin-Luther University, 06120, Halle (Saale), Germany
[2] State Office for Geology and Mining (LAGB) Saxony-Anhalt, 06118, Halle (Saale), Germany

17.1   Introduction

This chapter summarizes the characteristics of tools and methodologies currently used to disseminate the outputs of geological models. Systems that are still in development or in the experimental stage are discussed in Chapter 26. New challenges have arisen as the routine use of 3-D digital geological models for multiple applications have involved new user communities that are less familiar with geological concepts. As these user communities become more digitally-enabled, the importance of providing appropriate dissemination modes and visualization options for geological models increases. Geological models may involve large data volumes; the dissemination of these models and visualization products remains constrained by the user's access to the internet and lack of appropriate hardware. This restricts the broad direct digital distribution of models and visualizations. Current technology provides some options; future increased flexibility will allow for many new dissemination options.

Libarkin and Brick (2002) introduced a very useful classification framework for visualizations of geological information which has three categories: static, animation, and interactive. Each category has different advantages and disadvantages. Libarkin and Brick (2002) concluded that the users of interactive models and visualizations were less passive and more actively engaged with the content.

Static materials predate the development of computer techniques and digital geological models; traditional geological static products include drawings, diagrams and graphs, photographs, maps, cross-sections and physical models, such as Sopwith's early wooden teaching models (Figure 17.1). While requiring computers, animations and videos can illustrate 3-D environments with ease (Piburn et al. 2002); they enhance digital representations of static 3-D visualizations or provide for time-sequence representations of geologic processes. Interactive products include both interactive visualizations and interactive models. Several technologies can produce interactive visualizations, some use widely adopted communication standards, such as 3D PDFs, while others depend on specific software products. Interactive models include both physical and computer-generated models; they allow the models to be manipulated or viewed in different ways using a variety of input-response interactions.

The increasing number and availability of scientifically accurate 3-D geological models raises questions regarding their visualization potential, best method of presentation, and necessary supporting information. As the public interest in these local and regional geological models increases, several alternative visualization techniques have been developed and tested to determine if they made the models more understandable for audiences outside of the geoscientific community (Wycisk and Schimpf 2016). Two aspects were evaluated: (i) to what extent were the models acting as "eye-catchers" that increased the attention of the viewer, and (ii) what was the appropriate method of model visualization and what additional information was needed to adequately explain the model content.

The most commonly used approach to viewing 3-D model content is through displays created by the modeling software or with the use of independent viewer systems (Berg et al. 2011; Kessler and Dearden 2014). While modern viewer systems based in web services have become much more user-friendly, both methods often remain complicated and often require the users to make investments in, and have familiarity with, appropriate computer systems. In any case, computer-based visualizations of 3-D models often create an emotional and perceptional distance between the viewer and the presented information due to the absence of a real visual 3-D perception and the lack

Applied Multidimensional Geological Modeling: Informing Sustainable Human Interactions with the Shallow Subsurface, First Edition.
Edited by Alan Keith Turner, Holger Kessler, and Michiel J. van der Meulen.

Figure 17.1 Examples of Sopwith wooden geological models illustrating geological structure. Source: The Lapworth Museum, 2019.

of interaction with the model beyond the computer-based presentation.

For communication with the general public, these limitations suggest the need for new visualization options that are independent from the direct interaction with and perception of computer-based presentations. Hidden subsurface geology is by itself a fascinating and challenging topic for most interested people, but often it is difficult for them to imagine. While the public is able to understand medical images (X-rays, CT-scans, etc.) that illustrate the human anatomy, they do not have experience with typical geological structures. The situation becomes much more complicated if the geological model includes specific effects of rock displacement, subsurface channel-fill structures, or consequences of open pit mining activities (Wycisk et al. 2009). Therefore, Wycisk and Schimpf (2016) evaluated alternative new or adaptable techniques for providing a real 3-D perspective of complex geological models in the hope of improving the communication of the content of the models to a broad spectrum of potential users.

17.2 Visualization Principles

In natural and engineering sciences, visualization provides a more direct representation of data contained in models, concepts, and data sets (Valle 2013). The most suitable visualization presents the facts in a clear manner, allowing the viewer to better identify the problem and more deeply understand the issue. Communication in this context focuses more or less strictly on the bilateral relation of the visualized scientific information and the viewer.

In geology, each conventional geological map is a 2-D representation of a conceptual 3-D model that exists in the mind of the map's creator. The map and the underlying model are interpretations derived from field observations, which form the information base. The more recent evolution of numerical subsurface 3-D geological models is

merely the latest form of such geological interpretation process. Traditional geological maps and pictures, or the alternative digital models, have to clearly show a link to the basic observations; most often an idealized illustrative picture or visualization is used for this task. However, over time, the perceived role of visualization has gradually shifted from that of a cognition amplifier of basic geological information, to that of a toolbox of software that provides pleasing images. Some scientists might begin to wrongly consider visualization to be a luxury that they can forego. But when properly used, visualization has value for both increasing scientific understanding as part of the research process, and as a communication tool for multiple audiences. Facing the increasing influence of 3-D modeling products in geosciences and the increasing demand in communication, Valle (2013) states, *"Visualization is the use of computer supported, interactive, visual representations of data to amplify cognition."* A 3-D geological model can be presented most effectively to the recipient when three well-accepted principles of scientific visualization are followed – the visualization of a data set should be: (i) expressive, (ii) effective, and (iii) adequate (Schumann and Müller 2000).

Expressivity: From the visualization point of view, pictures are considered expressive if they clearly present the information of a given dataset. Pictures follow the rule of effectiveness if they can be perceived in an intuitive way by the observer, are adequately processed, and can be produced in reasonable time and at reasonable costs. A good scientific visualization presents data in an unbiased and undistorted way (Mackinlay 1986; Ware 2013). This condition seems easy to fulfill at first glance, but when applied to 3-D models, and more specifically to statistics and interpolation of the geological data and model results, the concept becomes much more complex. In some publications, the presentation of model results and the 3-D geological model images provide unclear or even incomplete information. A standard for communicating 3-D geological model results that complies with the criteria of expressivity is a basic condition in terms of quality assurance for scientific visualization.

Effectiveness: Multiple graphic visualization options may provide suitable expressivity, but the question of their effectiveness concerns the identification of the most appropriate technique to present the results. Effectiveness involves answering the major questions about what the main predictions and objectives of the model are. Effectiveness has to be assessed in the context of the application and the intuitiveness of the visual communication method considering the capability and vision of the recipient.

Adequateness: Adequateness includes the entire economic cost of producing a specific visualization product. This aspect has to be taken into consideration, although the observed decreasing costs of many visualization techniques over the last few years have made this aspect less critical.

17.3 Dissemination of Static Visual Products

Historically, geological knowledge was transferred via paper maps; these projected the 3-D geological situation two-dimensionally. Cross-sections, and sometimes block diagrams were included on the map sheet to provide additional information about the third dimension. As more specialized user communities emerged, the generic geological map was re-evaluated according to defined rules to produce "domain maps", which were designed to answer application-specific questions such as groundwater vulnerability or foundation conditions. Depending on the complexity of the rules and the underlying geology, these maps may become extremely busy and increasingly hard to use. They cannot be changed or adapted for applications other than the original purpose.

The adoption of digital printing capabilities for both text and graphics provides major advantages for the creators and users of geological information. It provides colored illustrations without the expenses associated with traditional publications. It avoids the constraints of brevity demanded by traditional journals, allowing for a greater variety of illustrations.

While many critical evaluations of subsurface conditions require the use of 3-D geological models, many user communities, such as urban or regional planning authorities or resource management regulators, prefer 2-D domain map presentations that summarize important characteristics of the pertinent issues. These 2-D maps may be paper or digital; in the latter case they are typically requested in a standard Geographic Information Systems (GIS) format. They provide a fixed, or "static," data source that may become one component of a multiple-factor assessment.

17.4 Dissemination of Digital Geological Models or Data

The degree to which digital 3-D subsurface data are available for exchange and dissemination without cost, at low cost, for lease, or outright purchase, depends on societal expectations and economic, security, and legal considerations that vary from country to country. Digital data dissemination provides major advantages for the creators and users of geological models. However, the very large size of many of the data files defining geological models, and even their associated visualizations, are a major constraint; they are often difficult for some potential users to accept. The current internet capabilities of some users are still not entirely satisfactory, although data compression and data transmission techniques suitable for handling large files on the internet are now available. The acceptance of widespread legal sharing of large data sets depends on the universal adoption of appropriate safeguards and standards. In the interim, many national geological surveys and other public holders of geological data are making considerable volumes of digital geological data available from various websites.

17.4.1 Direct Distribution of 3-D Geological Model Data Files

For the past two decades, the most common delivery method of 3-D geological model data and information is the direct export of the geological units (or domain maps) as ArcGIS shapefiles, grids or Triangulated Irregular Surfaces (TINs) to define the extents, elevations, and thickness of geological units. Alternatively, geological models can be exported as Computer Aided Design (CAD) files that can be imported into engineering design project databases. This has served the various user communities relatively well. This process does, however, limit the user to the analysis and interrogation methods available to them within their respective systems. For example, many systems cannot generate cross-sections from a model even though this is the most common request by users. An additional user constraint is that the resolution and property attributes are fixed at time of model export, and editing of the geological model is usually not possible.

When a geological model is used as part of an infrastructure design project, or for the modeling of a process such as slope stability or groundwater flow and transport, the full value of a geological model can only be realized when all the components (or objects) of the model are available within the client's own workflow system as well as the workflows of its clients and suppliers. This can be achieved only by delivering the model and its components together with the software. This is rarely achieved, but was successfully accomplished during the design and construction of the Crossrail Farringdon station (Gakis et al. 2016).

17.4.2 Distribution of Complete Digital Models on Data Disks

Although the internet provides an efficient marketing and dissemination channel, due to practical, technical, and legal issues some models continue to be disseminated on digital media such as CDs or DVDs. The U.S. Geological Survey (USGS) released a CD-ROM containing a 3-D model of an area near San Antonio, Texas with a free version of a commercial viewer that allows the user to interrogate this single dataset (Pantea and Cole 2004). The USGS paid a fee to the software provider in order to provide this product to the public. However, while easily installed, the viewer has many functions and the user needs to read several pages of documentation to learn how to operate it. More recently, the Ontario Geological Survey has released a series of reports in Southern Ontario on DVDs that include a series of folders containing a report describing the 3-D model, along with all the data used to create the model, derivative products, and a section viewer (Burt et al. 2015).

17.4.3 Low-cost or Free Specialty "Viewer" Tools

Many bespoke and mostly free geological model viewing and analyzing systems have emerged over the years. As data and model standards are still largely absent, these systems are highly varied and sometimes closely aligned to an accompanying commercial tool. In fact, viewers belong to three classes: (i) free-standing viewers that are downloadable software tools that operate on the user's computer; (ii) web-based applications that allow the user to access and visualize 3-D data; and (iii) web-based data portals that primarily serve as data distribution hubs, but often allow the user to explore and visualize the datasets. Table 17.1 summarizes the principal currently available options in each of these three classes. Many of the viewers allow the user to rotate, zoom into, and cut through portions of the model, or interrogate the model data by producing virtual boreholes or cross-sections at locations specified by the user. These capabilities remove the many restrictions imposed by fixed views in standard publications. However, many of these viewers do not provide the flexibility to load and integrate third-party data into existing systems and workflows.

Table 17.1 is not intended to provide an exhaustive list; however, even this small number of examples illustrates the great variety of approaches that are currently in use. Some viewers are designed to visualize and analyze data on global scales; these use an Earth globe as at least the entry point for the user to navigate from. Other systems are tied to regional or local datasets. The stated purpose of all of these viewers is to allow users to easily interrogate 3-D models and visualize models created by others using sophisticated and expensive specialized software.

The first class of viewers are downloadable free-standing software products. Some are provided free of charge, while others incur considerable costs for each user. The Geological Survey of the Netherlands initially developed a compact, no-cost 3-D viewer which could be downloaded along with selected datasets of the Dutch subsurface (Van Wees et al. 2003). Currently it uses the iMOD viewer, as do several other geological survey organizations (GSOs), to distribute model data. Several examples of this class of viewer are linked to specific geological modeling software and can only display models structured in their proprietary format.

Several GSOs have experimented with using widely available web visualization tools belonging to the second class of viewers. The most commonly used option is Google Earth; ArcGIS Online is a more recently introduced option. These systems provide the user with an attractive and affordable approach to visualizing existing 3-D geological models. Although Google Earth does not allow the portrayal of information beneath the land surface, several GSOs have used it. Figure 17.2 illustrates the use of Google Earth by the British Geological Survey (BGS) to deliver a model consisting of geological cross-sections. The Ontario Geological Survey distributes its geological model data as Google Earth Keyhole Markup Language (KML) files so that users can create similar visualizations. However, these displays have limited usefulness for real decision making, because the cross-sections are not placed at the correct elevation. The more recent addition of ArcGIS Online allows users to bring in fully attributed 3-D data, and visualize it below the ground surface.

With regard to the third class of viewers, most major GSOs have developed data portals to provide online visualization of their geological models. Important contemporary European examples listed in Table 17.1 are the data portals: DINOloket (Netherlands), The Geology of Britain, Databank Ondergrond Vlaanderen (Flanders, Belgium), and the Bavarian State Geological Survey data store.

The level of usability and utility of these viewers varies considerably and, so far, there has been no systematic assessment of their performance and effectiveness' however, Bang-Kittilsen et al. (2019) report the results of a questionnaire sent to European GSOs. They conclude that further research is required if the existing viewers are to have adequate usability and utility and be capable of meeting societal demands. Further assessment should be performed by a representative group of users performing typical tasks in coordination with the GSOs.

Table 17.1 Currently available geological model viewers.

Model viewer	URL	Description	Cost
Class 1: Free-standing viewers			
iMod 4.3	https://oss.deltares.nl/web/imod	iMOD is an easy-to-use Graphical User Interface containing a version of MODFLOW capable of fast interactive 2-D- and 3-D-analysis and visualization of the subsurface.	Free
Hydro GeoAnalyst (HGA)	https://www.waterloohydrogeologic.com/product/hydro-geoanalyst	HGA is an environmental data management software that integrates analysis, visualization and reporting tools with a powerful and flexible database.	$6485.00 (Single User)
LithoFrameViewer	https://www.bgs.ac.uk/datasets/bgs-lithoframe/	A software tool to view and query complex 3D geological models constructed by the BGS. It cannot incorporate, view, model edit, modify or export other geological data.	Free
Class 2: Web-based viewers			
Google Earth	https://earth.google.com/web	Renders a 3-D representation of the Earth by superimposing satellite images, aerial photography, and GIS data onto a 3-D globe. Users can add their own data using Keyhole Markup Language (KML) – geological data is added this way. The program can also be downloaded on a smartphone or tablet.	Free
GPlates	http://portal.gplates.org/cesium/?view=Geology	GPlates is open-source application software offering both the visualization and the manipulation of plate-tectonic reconstructions and associated data through geological time. It uses a globe as basic display. Developed by an international team.	Free
ArcGIS Online	https://www.arcgis.com/home/index.html	A cloud-based mapping and analysis solution to explore and visualize 2-D and 3-D data. Operated with ArcGIS tools and technologies	Multiple level license fees
Geological Survey of Norway	http://geologi.maps.arcgis.com/home/webscene/viewer.html?webscene=80add414f9994798955ce469bbc542b3	Uses ArcGIS Online to display geological cross-section model of Norway	Free
Class 3: Data portals with visualization capability			
DINOloket	https://www.dinoloket.nl/en	Main data portal for TNO-Geological Survey of the Netherlands offers free access to subsurface data and models contained in both TNO's DINO repository and the new Key Register for the Subsurface (BRO, see chapters 2 and 4).	Free
Geology of Britain	http://mapapps.bgs.ac.uk/geologyofbritain3d	A simple viewer tool aimed at giving the general public access to geological information held by the BGS, It currently includes borehole scans, earthquake timeline, and 3-D models	Free
Databank Ondergrond Vlaanderen	https://www.dov.vlaanderen.be/portaal/	Data portal for geological and related data for Flanders. Operates with a map (GIS) interface.	Free
Bavarian State Geological Survey data store.	https://www.3dportal.lfu.bayern.de	Contains several models including the GeoMol Alpine Foreland Basins data. Uses the GST application (Le et al. 2013)	Free

Figure 17.2 Screenshot showing geological cross-sections for Northern England visualized in Google Earth. The sections form part of the UK3D national geological model. They are displayed with two times vertical exaggeration. Source: https://www.bgs.ac.uk/datasets/uk3d/.

17.4.4 GeoVisionary

In 2006, the BGS began a collaborative development project with Virtalis Ltd., a British virtual reality company, to develop software for immersive 3-D stereoscopic visualization of terabytes of high-resolution terrain, imagery, GIS, and geological model data (Napier 2011). The project quickly developed a tool that made great efficiency gains in the digital fieldwork process by creating an interactive virtual field environment where geoscientists could carry out virtual field reconnaissance before visiting the field area and review the data collected in the field on return to the office. This continues to be a key component of the BGS Digital Workflow (Jordan and Napier 2016).

Functionality was added to GeoVisionary which allowed the application to easily import and display many types of geoscientific data including boreholes, GIS data, modeled horizons, faults, cross-sections, voxel models, terrestrial laser scans, and seismic reflection data (Terrington et al. 2015). The software provides a very data-rich environment for presenting and examining geological models along with data from a wide variety of sources (Figure 17.3). One of the most powerful visualization techniques, often absent from other model viewers, is to include high-resolution models of cities, buildings, and engineering projects in the same scene as the geoscience data. This boosts communication and understanding between the geological model creators and consumers. GeoVisionary has been applied by the BGS to high speed rail link projects, coastal change, and nuclear waste repository site screening. The new UK Geo-observatories project is contributing funding to the latest GeoVisionary initiative to make scenes accessible through the internet with GVCloud.

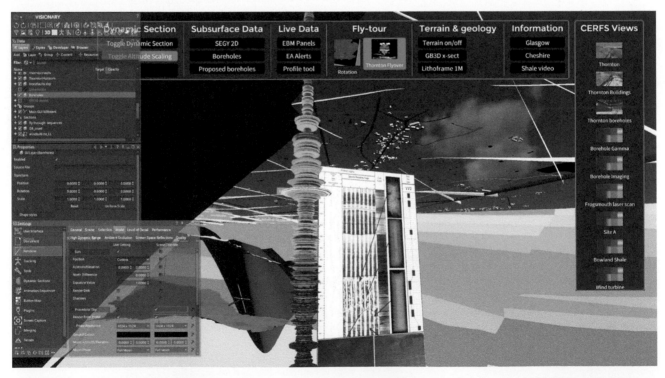

Figure 17.3 A GeoVisionary scene from the UK Geo-observatories project, showing custom GUI development and multiple data sources. In the foreground, a borehole with gamma log and imaging panel. In the background, modeled horizons, cross-sections and building models on the surface. (British Geological Survey)

17.5 Use of Animations to Explore Geological Models

Not every interested party will have access to the software used to create a geological model or, due to company software policy or limitations of available computers, have the capability to run interactive visualizations of a model. Animations provide a way to explore 3-D geological models under these constraints that is more effective than 2-D images, but less effective than well-implemented interactive model visualization tools. Because the recipient of the animation cannot alter the view perspective or the sequence of views, the effectiveness of the animation depends on the creator making visible all the important aspects of the model.

Animations can be as simple as an animated GIF file, showing a short clip of a rotating model or as comprehensive as complete presentations with text annotation and audio narration that describe the model. Animations that provide a guided tour, or fly-through, of the geological model are one of the most accessible ways to visualize 3-D models. Fly-through movies are easily uploaded to popular video sharing platforms such as YouTube. Animations are often used for educational purposes to illustrate and explain geological structures and processes.

The method of creating animations varies with the software employed. The stop-motion approach captures a series of screen-shot images with the model being moved slightly before each screen shot. The resulting sequence of images can be joined together to make a movie. Some software products, such as GeoVisionary (Jordan and Napier 2016), allow the user to capture live motion as the user navigates the model scene. Alternatively, the animation sequence can be developed using pre-defined routes and a sequence of instructions to turn-on or turn-off model features and annotations. Sometimes third-party software can be used to capture movies from the screen. There are several commercial and free applications that are commonly used by the computer gaming community.

Many sources of geological animations can be easily found with internet search engines. The British Geological Survey has produced many fly-through animations. An illustration of progress in 3-D geological modeling across Europe was requested by EuroGeoSurveys. It was shown to the European Commission in Brussels as part of a lobbying effort for the inclusion of an action focused on Geological Surveys in the Horizon 2020 program. The Gateway to the Earth uses an animated sequence of models to define the BGS science strategy. The Martin Luther University Halle Wittenberg supports a website (https://www.3d-geology.de/animated_graphics_and_videos) that contains links to a selection of animations and videos on hydrogeology and environmental geology, urban 3D geology, and mining.

17.6 Interactive Visualization of Multivariate Statistical Data

Statisticians were among the first to utilize the evolving digital graphics to dynamically visually "explore" multivariate data by using several interactive tools (Table 17.2). A notable development was the concept of a "grand tour" (Asimov 1985); this allowed the multivariate data to be "viewed from all sides" using an extension of 3-D rotation to higher dimensions. Tukey et al. (1974) provide an early

Table 17.2 Definition of commonly used interactive tools used for dynamic visual analysis of multivariate statistical data.

Interactive tool	Definition
Scaling	Maps the data to be displayed in the view window. It also is used to zoom in on crowded regions of a scatterplot, or to change the aspect ratio of a plot.
Linking	Multiple plots are linked so identifying one point in one plot will identify the same point on all other graphs. Brushing a group of points in one plot will highlight the same points in other plots. The linking can be one-to-one, or according to the values of a categorical variable in the data set.
Brushing	The mouse acts as a "paintbrush," which may be a pointer, rectangle, or polygon. As the brush moves over a point, the point will be highlighted by changing its color or glyph. Transient brushing, in which points only retain their new characteristics while they are enclosed or intersected by the brush, is most commonly used when multiple plots are visible and linked. Persistent brushing allows the identified points to retain their new appearance after the brush has been removed.
Painting	This defines a persistent brushing operation. It is used to group multiple points into clusters prior to using other operations to compare the groups,
Identification	When the cursor is brought near a point or edge in a scatterplot, or a bar in a bar chart, a label appears to identify the plot element. Sometimes called a "mouseover" this process is often linked so that the identification is shown in multiple plots.

demonstration of interactive visualization of statistical graphics during exploratory multivariate data analysis.

Cook et al. (1997) provide an early example of applying interactive visualization of geographical data; they linked the geographical display capabilities of ArcView 2.1 with a freely available multivariate statistical tool called XGobi (Swayne et al. 1998). XGobi utilized concepts developed by earlier software tools in the mid-1980s and, due to changing technology requirements, was superseded by GGobi in 1999 (Buja et al. 2003). While neither system is viable within modern computer technology, interactive exploratory data analysis of multi-dimensional data remains an important procedure that is supported by numerous commercial statistical packages.

17.7 Interactive Model Illustrations

Several digital or printing technologies allow for the production of 3-D, or pseudo 3-D images that, to a greater or lesser degree, encourage interaction with the user (viewer) of the images.

17.7.1 3-D PDFs

Because 3-D PDFs use Flash Technology, visualization of 3-D PDFs works best on PCs and laptops using a Windows operating system. Geological models can be viewed as a 3-D PDF file using the free Adobe Reader software. The BGS provides several 3-D geological models as 3-D PDFs and each is supplied with a toolbar which allows the user to change the model view by zooming in and out, rotating or panning the object; it is also possible to hide or isolate parts of the model, or make parts transparent. Figure 17.4 is a typical BGS 3-D PDF model product. The model is derived from a GOCAD model based on BGS borehole and geological mapping data. Figure 6.3 provides an additional example of BGS 3D PDF models.

17.7.2 Lenticular Printing

Lenticular printing is a well-established technique that uses lenticular plastic lenses to allow movement, change, or even a depth-effect on an image as the viewing angle changes slightly. It is used to produce smaller-format advertising documents; its use for scientific communication requires an increased illustration resolution and the

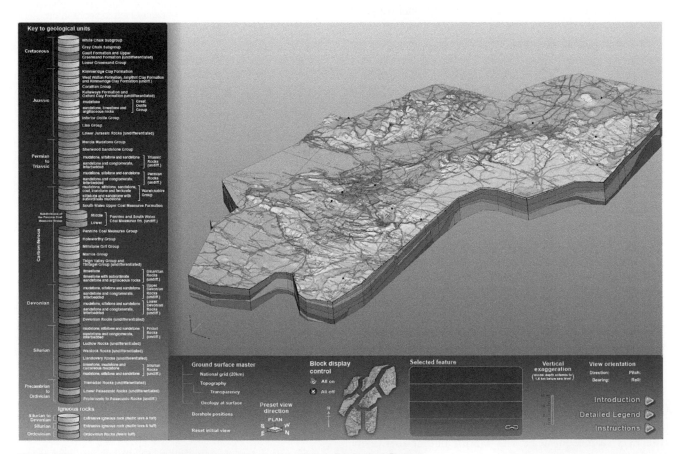

Figure 17.4   Screenshot of a typical BGS 3D PDF model. (British Geological Survey)

technical ability to produce larger printed formats (up to 2 m) at reasonable costs. Lenticular printing involves a stepwise process of creating at least two images that are combined with a lenticular lens sheet. The original images are split into vertical slices that correspond to the number of lenses and the slices are then arranged side-by-side and aligned with the lens positions. A lenticular print created from two images is called a "2-phases flip image"; as the viewing angle changes, the viewer sees first one image then the other. However, this effect also depends on the chosen size of the lenses and the size of the image itself. Larger images require a greater viewing distance. If the lenses are too small, viewers cannot easily see the original images and even very small changes of the viewing angle cause the image to change.

Figure 17.5 provides an example of a 2-phase flip. The subject is the main railway station of Halle. The first layer is an aerial photograph of the main station, and the second layer contains a horizontal section of the area's near-surface geology. The combined visualization allows the viewer to experience a view of well-known living areas and relate them to the hidden geology below the ground. This lenticular print is 60 cm × 80 cm.

Lenticular printing can also be used for 3-D depth effects. A stereoscopic effect is produced by defining the number of images and the size and number of the lenticular lenses so that very small changes of the viewing angle allows each eye to have a slightly different view. The 3-D effect is created without requiring special glasses. Figure 17.6 is an exploded view of a block model showing the Quaternary and Tertiary stratigraphy of Bitterfeld South. The 4 × 4 km modeled area has a horizontal resolution 10 × 10 m; the 3-D effect in this image required 32 individual images.

17.7.3 True Color Holograms

Although flip images can be used to create a pseudo 3-D effect, true color holograms give a much better depth perception because holograms provide motion parallax – objects nearer the viewer move more than objects further away. This is an important depth cue for the observer. True color hologram technology was pioneered by Geola Ltd., Vilnius, Lithuania. While the hologram is printed (embossed) on a flat sheet of film, it provides a very impressive 3-D colored presentation of complex structural models. The technique can produce large-scale image formats. However, producing a hologram involves pre-processing of the model data, precisely aligned cameras, and sophisticated rendering of the film. Thus, while the 3-D effect of this new technology is much more expressive, the total costs of a realization are higher than for an equivalent 3-D lenticular option. In addition, the presentation of the true

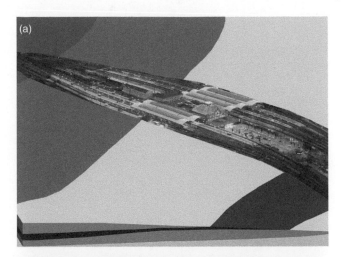

Figure 17.5 A 60 cm × 80 cm two-phase lenticular flip image of the Halle (Saale) city center that combines an aerial photograph with subsurface geological information in one picture. (a) Geology derived from the 3-D geological model of Halle (Saale) City; (b) aerial photo of the neighborhood around central station of Halle in a corrected geometric and georeferenced format. (c) Both pictures are "flipping" as the viewer's axis of view changes. (Wycisk and Schimpf 2016)

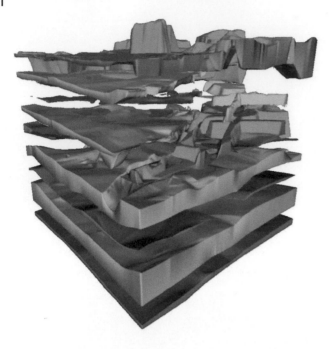

Figure 17.6 A 50 × 80 cm 3-D lenticular print illustrating the aquifer stratigraphy of Quaternary and Tertiary deposits within a 4 × 4 km area near Bitterfeld, eastern Germany. Aquifers are marked in blue while aquitards are in green. (Wycisk and Schimpf 2016)

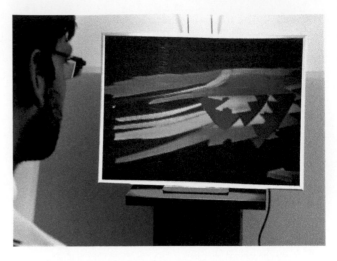

Figure 17.7 True color hologram of part of the 3-D structural model of Halle (Saale). The 30 × 40 cm image was produced by the i-Lumogram technique developed by Geola Ltd., Vilnius, Lithuania. When illuminated by an external spotlight it provides a very strong 3-D perception to the viewer. (Wycisk and Schimpf 2016)

color hologram requires a separate individual spotlight focused on the image at a 45° angle for illumination. Without this spotlight, the image appears black. Figure 17.7 shows a cut-out section of the faulted area in the northwestern part of the structural model of Halle (Saale). The 3-D effect of the hologram visualizes the inclined fault planes very clearly.

17.8 Interactive Creation or Interrogation of Digital Geological Models

The direct distribution of geological data or models (Section 17.4.2) and the use of "Viewer" software applications or web visualization tools (Section 17.4.3) provide those users able to access these options with the ability to interact with existing developed models. However, several additional technologies offer even greater levels of interactive creation or interrogation of digital geological models.

17.8.1 Minecraft Models

Several organizations with extensive digital data holdings have converted their scientific models to Minecraft worlds as an educational outreach option. Any individual with a licensed copy of Minecraft and sufficient disk space

can download these models, which closely replicate the real-world conditions, and learn about the subsurface.

In September 2013, the Ordnance Survey released a "GB Minecraft" a Minecraft world created by reprocessing digital Ordnance Survey map products. It is freely available as OS OpenData and consists of more than 83 billion blocks representing over 220 000 km^2 of mainland Great Britain and surrounding islands. Each block measures 25 m × 25 m × 12 m. Actual height data (in meters) was scaled so as to exaggerate the real-world height by approximately 2.3×; this preserves low-lying coastal features and adds interest to the landscape. About a year later, the BGS added a subsurface layer to the model providing information about shallow geological material conditions and thus created 2-D geology of the United Kingdom that can be interrogated in Minecraft. Subsequently the BGS has developed a series of 3-D geological Minecraft world models that accurately represent the geological subsurface at several locations across the country. Figure 17.8 illustrates one of these models representing the West Thurrock area in east London.

The Alberta Geological Survey has also reprocessed several of their 3-D geological models to be displayed and interrogated by Minecraft users. Figure 17.9 illustrates two versions of one of the Alberta models. When created with solid blocks (Figure 17.9a) the Minecraft player has to excavate tunnels to discover the geology. If transparent blocks are used (Figure 17.9b) the model can be "seen through" and explored in different ways.

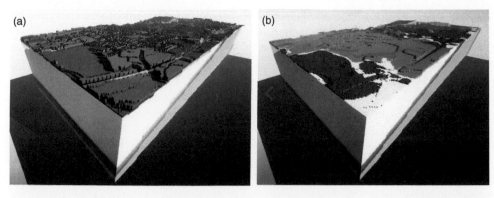

Figure 17.8 BGS Minecraft model of the West Thurrock area, east London. (a) Model with combined OS topographic and land use data combined with BGS geology information; (b) model with only the geology information. Source: British Geological Survey.

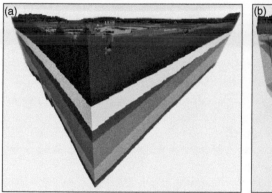

Figure 17.9 Alberta Minecraft model of the Peace River area. (a) Model built with solid blocks; (b) model built with transparent blocks. Source: Alberta Geological Survey.

17.8.2 GEOILLUSTRATOR Project Products

In January 2010, the University of Bergen (UiB) Department of Informatics and the Department of GeoSciences began a joint research project, GEOILLUSTRATOR, in cooperation with Christian Michelsen Research AS and Statoil (now Equinor, the Nowegian state energy company). This Project applied state-of-the-art computer graphics and interaction hardware to develop new and better ways to digitally model and illustrate geological data and geological concepts. GEOILLUSTRATOR developed tools to assist expert-to-expert communication, expert-to-layman communication, and interactive educational approaches for day-to-day modeling and visualization of geological data. Over the following three to four years, the GEOILLUSTRATOR Project addressed three research topics:

(1) Visualization of geology models,
(2) Visualization of ground-truth data, and
(3) Sketch-based modeling and visual hypothesis exploration.

In response to the first topic, Lidal et al. (2012) explored the challenges encountered when presenting and communicating important features in complex 3-D geological models. Cutaway visualization solutions are often used to expose important features that otherwise would be obscured; this is an important component of interactive model visualization. Five formal design principles for creating cutaway visualizations that appeared to improve user interaction and interpretation of geological models were defined and evaluated. These design principles include: (i) use an oblique viewpoint for the cutaway, (ii) use a simple cutaway shape, (iii) include familiar context, (iv) use an illumination model that effectively communicates the shape and spatial ordering inside the cutaway, and (v) take advantage of the motion parallax cue.

The second topic, visualization of ground-truth data, was addressed by Lidal's doctoral dissertation (Lidal 2013) and summarized in Lidal et al. (2013). Early stages of geological exploration typically have access to only limited ground-truth data. Model development and evaluation is thus difficult because much of the model conceptualization is based on knowledge and ideas in the mind of the geologist. The proposed solution is based on the concept of sketch-based storytelling, a method for expressing ideas and conceptual models of evolutionary processes. Geological storytelling is a graphical system for capturing and visualizing the reasoning process that leads up to the geological model, for exploring and comparing a set of such models, and for presenting and communicating the most probable ones.

A temporally ordered sequence of discrete time-event sketches, termed story nodes, describe the key steps in the process. For each story node, either a 2-D sketch is created using the "flip-over canvas" interface, or a 3-D view is directly sketched using the "box-proxy canvas" interface. The story tree, a tree-graph data structure, and visualization, for managing and accessing ensembles of alternative solutions, support the exploration of possible alternative solutions and the identification of an optimal solution. The story tree also provides an interface for navigating to specific story nodes and playing the stories as 2-D animations of either a single story or a multi-story comparison. The animations are generated by treating each story node as a key frame and then applying key-frame animation techniques to interpolate the intervening frames automatically. Automatic synthesis of illustrative 3-D animations from the 2-D sketch animations assists the presentation of optimal solutions to decision makers.

Natali (2014) and Natali et al. (2014a) addressed the sketch-based modeling aspect of the GEOILLUSTRATOR Project's third topic, while Natali et al. (2014b) considered the visual hypothesis exploration aspects. These represent extensions of the sketching and storytelling techniques described by Lidal (2013) and Lidal et al. (2013).

Natali et al. (2014a) proposed creating visual geological models interactively by sketching a sequence of stratigraphic layers representing unique erosion or depositional events. The model strata were defined by 2-D grids, which provided a compact representation of the model. Combined with an efficient rendering algorithm, this allowed rapid interactive visualization of 3-D layer-cake models. This approach is limited to 2.5-D representations, overturned or overhanging strata cannot be modeled. This was not considered to be a major concern, as the authors' primary interest was to rapidly model the evolution of rivers and deltas interactively, in order to demonstrate how they influence the character of the subsurface stratigraphy. The emphasis was on producing a user-friendly tool for the rapid creation of illustrations for communication purposes and unconstrained hypothesis testing. Therefore, the system does not include physical and geological constraints because these tended to burden the user with complex input parameters and long processing times. It is the user's responsibility to create rational simulations. It is possible to include some level of heterogeneity into the characteristics of the model units; this could allow the interactive development of 3-D models to serve as training images for Multiple-Point Statistics (MPS) simulations, as defined by Caers (2001), Strebelle (2002), and Arpat and Caers (2005).

Natali et al. (2014b) further extended the geological sketching toolbox to include the processes of faulting and compaction. The rapid model interactive visualization capabilities were then used to provide animations of fault-block displacements or compaction events. This supported visual hypothesis exploration of alternative conceptual models.

17.8.3 Visible Geology

Visible Geology is an innovative web-based application designed from a student centric perspective to support geoscience education (Cockett et al. 2016). It uses a simple and intuitive interface involving an interactive 3-D block model to visualize geologic structures and processes, thus allowing students to creatively explore concepts. It has been employed at several universities to teach geoscience concepts of relative geologic time and crosscutting relationships.

Figure 17.10 illustrates the basic functions of Visible Geology. Upon loading, the program directs the user to begin by adding a series of geologic units (Figure 17.10a,b); it then allows the user to create a series of geologic events. Geologic events implemented in Visible Geology include tilting, folding, domes, basins, unconformities, dikes, and faults. These events are additive and thus can convey the basics of structural geology. Combinations of these events can create complex geologic scenarios. Once a model is created, it can be interrogated using interactive tools to cut cross-sections or insert virtual boreholes; measurement tools allow the user to obtain distance and orientation information. A simple stereonet program that communicates with the geologic model is embedded in the software. It is also possible to explore the influence of topography using pre-made topographies or the user can sketch contour lines to create custom surfaces. The addition of topography (Figure 17.10c) allows users to explore different types of outcrop patterns found on geologic maps.

17.9 Interactive Physical Geological Models

Technological advances have made it feasible to convert digital geological data and models into physical models

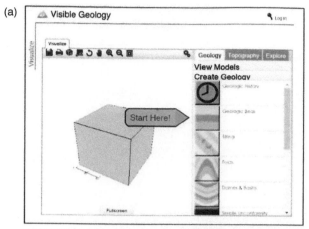

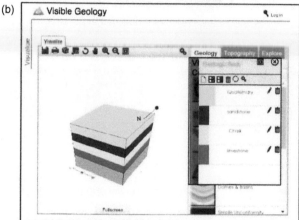

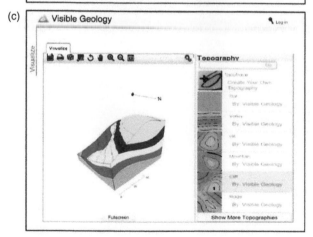

Figure 17.10 Example of visual geology: (a) introduction window, (b) creation of four flat-lying strata, (c) after assigning folding and topography to model. Source: https://app .visiblegeology.com/profilc.html.

that can be viewed, and some cases disassembled into components, to provide a "hands-on" understanding of subsurface geology. These modern equivalents of the nineteenth century wooden models shown in Figure 17.1 offer members of the general public models that have a familiar

appearance to consumer products. This helps to provide the users with a context that promotes an understanding of unfamiliar subsurface geology conditions. Three technologies have been employed to create interactive physical geological models: 3-D printing, Lego, and laser engraving of glass blocks.

17.9.1 Using 3-D Printing Technology to Create Geological Models

Three-dimensional printing, also known as additive manufacturing, refers to the various processes that sequentially deposit material in order to synthesize a three-dimensional object. Initially, 3-D printing referred to processes that sequentially deposited materials such as polymer and gypsum onto a powder bed with inkjet printer heads. Rapid development has occurred in recent years and the meaning of the term has expanded to encompass a wider variety of techniques, such as extrusion and sintering-based processes using different materials such as thermoplastics, modeling clay, ceramic materials, and metal matrix composite materials. Hull (2015) describes the invention and development of the technology.

Until quite recently, 3-D printers were quite expensive; individual researchers could not justify the expense. Today, more affordable, but quite high resolution, 3-D printers capable of producing small models (up to about 12 cm in any direction) cost a few thousand dollars. Geoscientists have started to experiment with multiple potential geoscience applications (Rosen 2014; Hasiuk 2014; Hasiuk et al. 2016). Wycisk et al. (2009) used 3-D printing methods to produce a 25×25 cm 3-D geological model that represented a 4×4 km of a lignite mine at Bitterfeld in eastern Germany.

Several geological surveys have already used 3-D printing procedures to produce physical 3-D geological models. Lin et al. (2016) provide a detailed guide to the best approach to printing small 3-D geological models from digital model files. They created models of the Mahomet aquifer (described in Chapter 5) that have proved very useful during public presentations (Figure 17.11). The Geological Survey of the Netherlands (TNO) has developed a much larger model, printed in four parts, that covers the entire country (Figure 17.12). Although expensive, this model has provided a very useful alternative public visualization of the Dutch national geological models. The Alberta Geological Survey used 3-D printing to produce several multiple-layered physical models based on its existing digital 3-D geological models (Babakhani et al. 2019; Branscombe et al. 2019). Figures 17.13 and 17.14 show these models.

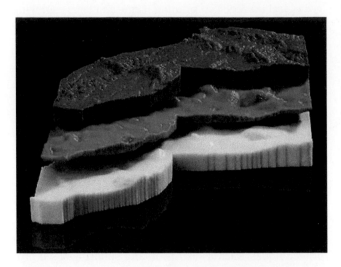

Figure 17.11 A 3-D printed model of the Mahomet aquifer in central Illinois. Lin et al. (2016). Copyright © 2016 University of Illinois Board of Trustees. Used by permission of the Illinois State Geological Survey. Photograph by Joel Dexter.

Figure 17.12 A 3-D printed model of the Netherlands is used to inform the public about the subsurface and the effects of gas extraction. Photo: TNO Geological Survey of the Netherlands.

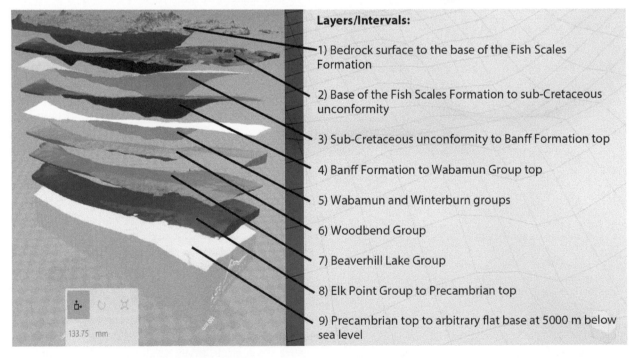

Layers/Intervals:

1) Bedrock surface to the base of the Fish Scales Formation

2) Base of the Fish Scales Formation to sub-Cretaceous unconformity

3) Sub-Cretaceous unconformity to Banff Formation top

4) Banff Formation to Wabamun Group top

5) Wabamun and Winterburn groups

6) Woodbend Group

7) Beaverhill Lake Group

8) Elk Point Group to Precambrian top

9) Precambrian top to arbitrary flat base at 5000 m below sea level

Figure 17.13 Exploded view of Province of Alberta nine-layer 3-D printed model. Source: Used by permission of the Alberta Geological Survey.

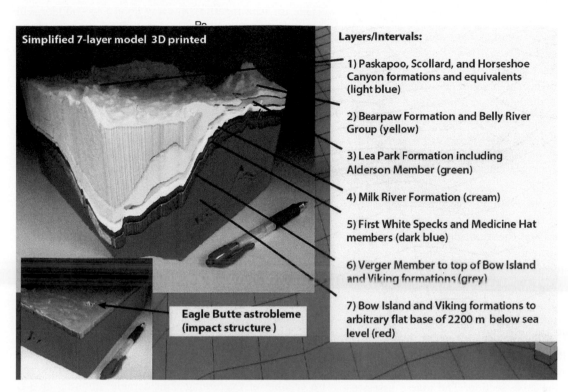

Simplified 7-layer model 3D printed

Layers/Intervals:

1) Paskapoo, Scollard, and Horseshoe Canyon formations and equivalents (light blue)

2) Bearpaw Formation and Belly River Group (yellow)

3) Lea Park Formation including Alderson Member (green)

4) Milk River Formation (cream)

5) First White Specks and Medicine Hat members (dark blue)

6) Verger Member to top of Bow Island and Viking formations (grey)

7) Bow Island and Viking formations to arbitrary flat base of 2200 m below sea level (red)

Eagle Butte astrobleme (impact structure)

Figure 17.14   South East Alberta seven-layer model produced by 3-D printing. Source: Used by permission of the Alberta Geological Survey.

LGS	Legend	Red Value	Green Value	Blue Value	Name	Lego Colour
LGS	LCK	115	255	115	Chalk	28
LGS	RTDU	246	160	89	River Terraces	191
GLT						
GLT	ALV	229	255	67	Alluvium	24
UGS	PEAT	188	130	92	Peat	192
UGS	MGR	130	130	130	Madeground	199
CK						
CK	−99	255	255	255		
CK	TPGR	254	218	174	Taplow Gravel	5
CK	WMGR	130	130	130	Worked and Made Ground	194
CK	LHGR	255	249	158	Lynch Hill Gravel	5
CK	TAB	58	13	181	Thanet Formation	23
CK	HEAD	124	64	64	Head	192
CK	LMBE	219	133	20	Lambeth Group	106
CK	HWH	255	223	242	Harwich Formation	324
CK	LC	179	156	125	London Clay	138
CK	WGR	130	130	130	Worked Ground	194
CK	BHT	255	249	158	Boyn Hill Gravel	5
LMBE	BPGR	255	249	158	Black Park Gravel	5
LMBE	CFSI	255	249	158	?	5
LMBE	HAGR	255	249	158	?	5
LMBE	CK	115	255	115	Chalk	28
LC						
LC						
LC						
LC						
LC						

Figure 17.15   Assigning Lego brick color to geological unit. Source: British Geological Survey.

17.9.2 Lego Models

The BGS has developed several 3-D geological models for educational purposes. While the BGS website has plenty of examples of digital data that the user can analyze, interrogate, and input into their own models and software, several members of the 3-D modeling team wondered if something more tangible was needed to reach young inquisitive minds. It was decided that Lego might make geological data physically real for a younger audience. Some resources were devoted to producing in Lego a cross-section from the 3-D digital London model.

Using procedures discussed in Chapter 13, existing BGS 3-D geological models can be discretized into small cubical cells or voxels and the resulting calculated model exported as a 3-D structured mesh described by a commonly used ASCII formatted file. To produce a Lego model, a voxel size of 50×50 m horizontally and 2 m vertically was chosen – when translated into Lego brick size this results in a vertical exaggeration of 25.

An Excel macro was developed to convert these voxels into the special Lego format used by Lego Digital Designer (LDD) that computes the numbers of specific Lego bricks required to create the cross-section. The macro assesses the voxel pattern and assigns a Lego brick color and style to represent the unit (Figure 17.15), calculates the location of each brick in 3-D space (Figure 17.16) and produces the appropriate codes that can be imported into the LDD. The cross-section design then can be reviewed in the LDD (Figure 17.17). A small amount of manual manipulation of the bricks is required to ensure that the finished model

is able to "lock together" (as any good Lego builder will understand) and finally a complete list of the required Lego bricks is provided to complete the order.

Once the bricks are received, the cross-section can be built following the directions provided by the LDD (Figures 17.18 and 17.19) a process that, although fun, is highly time consuming and waiting to be automated in the future. The Lego models have proved very popular and are a fantastic resource for understanding the subsurface both in terms of scale and complexity – it provides a literally hands-on approach to geology.

17.9.3 Laser-engraved 3-D Glass Models

Wycisk and Schimpf (2016) describe the development of a series of laser engraved 3-D glass representations of 3-D models. Several models (Table 17.3) were displayed at the German "Land of Ideas" national competition in 2007 and 2012 and subsequently at several other scientific exhibitions. Wycisk and Schimpf (2016) conclude that laser-engraved 3-D glass models provide an innovative visualization with considerable appeal to the public. Laser-engraved 3-D glass models produce eye-catching public views and are capable of expressing considerable amounts of 3-D geological model information (Figures 17.20 and 17.21).

The glass models are produced by using red or green lasers to transfer the information from a 3-D geological model into a polished crystal block of optical quality. The laser engraving process was developed in cooperation with Star Glas GmbH, Bünde, Germany, since 2007. Wycisk and Schimpf (2016) provide a detailed discussion of the

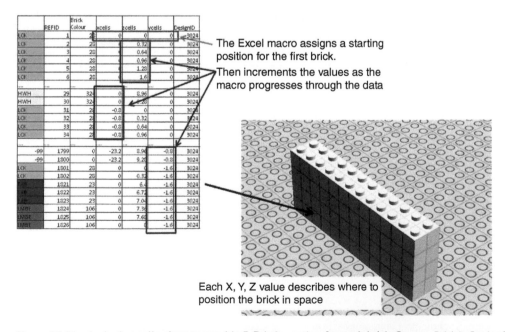

Figure 17.16 Assigning cell values to provide 3-D information for each brick. Source: British Geological Survey.

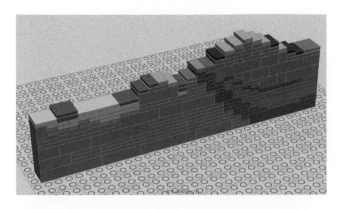

Figure 17.17 The digital cross-section imported into LDD software. Source: British Geological Survey.

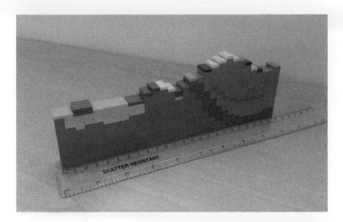

Figure 17.18 The geological cross-section modeled in Lego. Source: British Geological Survey.

Figure 17.19 A group of Lego enthusiasts building cross-sections. Source: British Geological Survey.

workflow required to convert an existing digital 3-D geological model into a laser engraved 3-D glass representation. Major challenges involved the handling of the very large datasets and development of appropriate visualization to define the original 3-D geological model information. Engraving features such as crossing fault planes potentially risks producing cracks that might damage the entire glass block. Recent advances in laser beam technology permit the representation of much more detailed structures. Laser engraving of the glass block surfaces allows for the addition of topographic or local information to the glass model. Wycisk and Schimpf created 3-D glass models based on their own modeling research and re-evaluation of existing 3-D geological model datasets in collaboration with the Federal Institute for Geosciences and Natural Resources (BGR) in Hannover, the Senate of Berlin, and the Geological Survey of Austria (GBA) in Vienna.

However, the laser engraving of glass blocks has some limitations. The presentation of very complex structural details is often infeasible, and such complex displays do not always allow a deep, insightful view of the model. Thus, the glass displays are most effective when they present only specific thematic topics based on a subset of the entire 3-D geological model data. The penetration of the laser beam is limited to about 10–12 cm; this imposes constraints on the size of the crystal blocks. The weight of the glass blocks also has to be considered. Table 17.3 provides details of the size and weight of several recent glass block models; models that are $30 \times 26 \times 12$ cm and weigh 25 kg have been used to visualize the subsurface geology of entire cities (Figure 17.22). The largest glass model produced so far is $60 \times 27 \times 12$ cm and weighs 52 kg (Figure 17.23). Smaller $10 \times 10 \times 10$ cm glass cubes that weigh only 2.6 kg have been used to display more detailed views of the subsurface or specific thematic information for selected portions of the larger regional models (Figure 17.24).

The glass models can be enhanced when illuminated with blue light or combined with other forms of presentation, such as geological maps. Combining laser-engraved glass block models with colorful, computer-based holographic animations gives the viewer a significant amount of additional information and attraction. Besides geological information, the glass models can display additional information such as a topographic surface, groundwater surfaces, boreholes, pumping wells, or mining infrastructure to show the context between the causal geology and the anthropogenic environment.

Table 17.3 Summary characteristics of geological laser-engraved glass models described by Wycisk and Schimpf (2016).

Model name	Model extent	Model content	Geological modeling software	Glass block model specifications	Reference illustration	Cooperator
Near Surface Geological 3-D Model Halle (Saale)	City Halle (Saale) 135 km², 250 m depth	*Geology:* 25 layers Focus on Tertiary and Quaternary	GSI3D	Size: 26 × 26 × 12 cm, Weight: 22 kg, Approx Model Scale: 1:60 000, exploded view.	Figure 17.20	City Halle (Saale) provided borehole data
Structural 3-D Model Halle (Saale)	City Halle (Saale) 135 km², 1000 m depth	*Geology:* 6 layers, Rotliegend to Muschelkalk	3DMOVE	Size: 30 × 26 × 12 cm, Weight: 25 kg, Approx Model Scale: 1:60 000.	Figure 17.21	Geological Survey of Saxony-Anhalt (LABG), Halle, provided borehole and seismic data
Thematic 3-D Model Halle-Neustadt	Halle-Neustadt 25 km², 1000 m depth	Cut-out of the Structural 3-D model Halle (Saale). *Geology:* 6 layers, *Environment:* groundwater level, producing wells, high flood hazard map, sewer network, and distance heating infrastructure	3DMOVE and GIS	Size: 30 × 26 × 12 cm, Weight: 25 kg, Approx Model Scale: 1:20 000	Figures 17.22 and 17.24	Halle (Saale) City Planning Dept.
3-D Model Saline, Halle (Saale)	Halle (Saale), 1 km², 1000 m depth	Cut-out of the structural 3-D model Halle (Saale) *Geology:* 6 layers, Rotliegend, Zechstein, Bunter and Muschelkalk *Environment:* two deep boreholes	3DMOVE	Size: 10 cm × 10 cm × 10 cm, Weight: 2.6 kg, Approx Model Scale: 1:10 000.	Figure 17.25	Detailed model of part of the structural 3-D model Halle (Saale) (see Figure 17.21)
Geological 3-D Model "Salt Anticline Stassfurt"	City of Stassfurt, 11 km²	*Geology:* Folded base of the potassium salt stratum "Stassfurt" (Zechstein) *Environment:* 3-D modeled mining infrastructure of the salt mine complex	openGEO	Size: 60 × 27 × 12 cm, Weight: 55 kg, Approx Model Scale: 1:7000 without exaggeration.	Figure 17.23	3-D model geology and mining infrastructure by BGR, Hannover

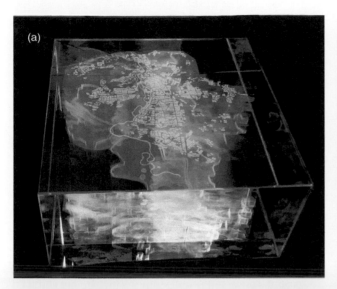

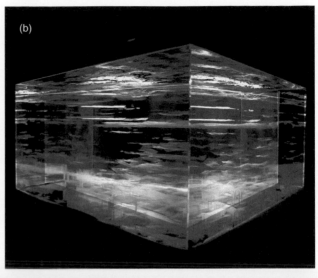

Figure 17.20 Two views of the laser-engraved glass block Near Surface Geological 3-D Model Halle (Saale) which is 26 × 26 × 12 cm. (a) Top view; (b) side view, showing the 28 individual geological layers representing pre-Tertiary, Tertiary and Quaternary units. (Wycisk and Schimpf 2016)

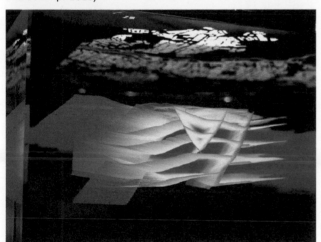

Figure 17.21 The laser-engraved glass block Structural 3-D Model of Halle (Saale) extends to a depth of about 1000 m. and represent six geological units and complex structures of the "Halle-Störung." (Wycisk and Schimpf 2016)

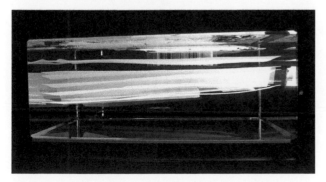

Figure 17.22 Side view of the Thematic 3-D Model Halle-Neustadt which represents part of the Structural 3-D Model Halle (Saale) shown in Figure 17.21. The underlying geology is shown in greater detail and environmental and technical city planning issues are superimposed. Approximate model scale: 1:20 000; glass block size is 30 × 26 × 12 cm. (Wycisk and Schimpf 2016)

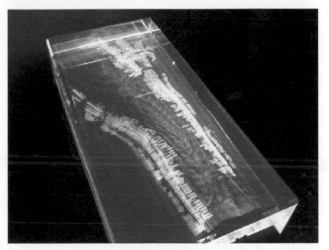

Figure 17.23 Laser-engraved 3-D model "Salt anticline Stassfurt" shows the base of the "Stassfurt" (Zechstein) potassium salt stratum and the detailed infrastructure of the mines. Approximate model scale: 1:7000; glass block is 60 × 27 × 12 cm. (Wycisk and Schimpf 2016)

However, laser-engraved glass block models often require additional information in printed or animated form to make the more or less abstract visualization more comprehensible. Figure 17.25 compares a 10 × 10 × 10 cm small glass model (Figure 17.25a) with an annotated graphical version of the same model reproduced directly from the computer (Figure 17.25b). The 3-D glass model facilitates the visual perception while the 2-D printout provides context and interpretive annotations.

Figure 17.24 More detailed view of the Thematic 3-D Model Halle-Neustadt (Figure 17.22). The stacked thematic information with relevance to city planning aspects includes: (a) groundwater table; (b) gallery of 132 production wells; (c) information of high flooding hazards; and (d) wastewater and technical infrastructure. The height of the picture is approximately 4 cm. (Wycisk and Schimpf 2016)

17.10 Conclusions

As this chapter has shown, there are many ways to disseminate geological models to end users due to the vast range of potential end uses and users and an ever-increasing range of technologies. Even during the time of compiling this chapter advances have been made, and some of these

are discussed in Chapter 26. The authors hope that this chapter provides a good overview of the existing methods and helps model producers and consumers to choose the right methodology that is fit for their purpose. Irrespective of the choice of delivery, it is apparent that the community has taken huge strides over the past decade to rise to the challenge set out by Jackson (2003) when he stated that "geologists don't listen, and the public can't read geological maps."

Acknowledgments

The authors thank Katie Whitbread at the British Geological Survey (BGS) for reviewing the entire chapter and proposing some important improvements. Several staff at the BGS and the Geological Survey of the Netherlands (TNO) contributed to the review of model "viewer" tools and Table 17.1.

While the principal authors contributed their expertise concerning several of the methods used to create interactive illustrations of models (Section 17.7) and applications of laser-engraved glass models (Section 17.9.3), many other individuals have supplied information on specific tools and methodologies currently used to disseminate the outputs of geological models.

Bruce Napier (BGS) provided information on GeoVisionary (Section 17.4.4) and the discussion on the role of animation (Section 17.5). Kelsey MacCormack at the

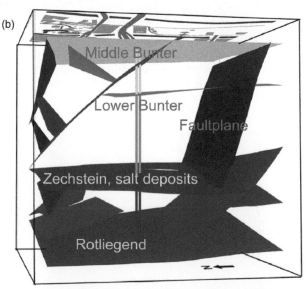

Figure 17.25 Perspective views of the detailed 10 × 10 × 10 cm cut-out model from the structural 3-D model of Halle (Saale). (a) Laser engraved glass block model (model scale approximately 1:1000) shows complex faulted geology and two deep boreholes; (b) equivalent model-based computer graphic provides the viewer with additional scientific information. (Wycisk and Schimpf 2016)

Alberta Geological Survey supplied illustrations of both Minecraft and 3-D printer generated models. The use of 3-D printing technology has been tested by several geological surveys; additional illustrations and information have been supplied by the Illinois State Geological Survey and TNO. Steve Thorpe (BGS) provided information on the creation of geological models in Minecraft and Lego (Sections 17.8.1 and 17.9.2).

References

Arpat, G. and Caers, J. (2005). A multiple-scale, pattern-based approach to sequential simulation. *Quantitative Geology and Geostatistics* 14: 255–264. https://doi.org/10.1007/978-1-4020-3610-1_26.

Asimov, D. (1985). The Grand Tour: a tool for viewing multidimensional data. *SIAM Journal of Scientific and Statistical Computing* 6 (1): 128–143. https://doi.org/10.1137/0906011.

Babakhani, M., Mckay, K. and Warren, J.E. (2019). *Southeastern Alberta 3D Geological Model: 3D Print Files (digital data, STL format)*. Edmonton, AL: AER/AGS, Digital Data 2019–0003. [Online: Available at: https://www.ags.aer.ca/publications/DIG_2019_0003.html] (Accessed April, 2019).

Bang-Kittilsen, A., Małolepszy, Z. and Kessler, H. (2019). *3D Geological Modelling: An Assessment of Web-Viewers*. Abstracts: 5th European Meeting on 3D Geological Modelling, May 21–24, 2019, Bern, Switzerland.

Berg, R.C., Mathers, S.J., Kessler, H. and D.A. Keefer [Eds.] (2011). *Synopsis of Current Three-Dimensional Geological Mapping and Modeling in Geological Survey Organizations*. Champaign, IL: Illinois State Geological Survey, Circular 578. 92p. [Online: Available at: http://library.isgs.uiuc.edu/Pubs/pdfs/circulars/c578.pdf] (Accessed April 2019).

Branscombe, P., Warren, J.E. and Mckay, K. (2019). *Simplified 3D Provincial Geological Framework Model of Alberta (version 1): 3D print files (digital data, STL format)*. Edmonton, AL: AER/AGS, Digital Data 2019–0005. [Online: Available at: https://www.ags.aer.ca/publications/DIG_2019_0005.html] (Accessed April, 2019).

Buja, A., Lang, D.T. and Swayne, D.F. (2003). GGobi: evolving from XGobi into an extensible framework for interactive data visualization. *Computational Statistics and Data Analysis* 43 (4): 423–444. https://doi.org/10.1016/S0167-9473(02)00286-4.

Burt, A., Bajc, A., Rainsford, D. et al. (2015). 3-D Geological modelling at the OGS – products and applications. In: *Three-Dimensional Geological Mapping: Workshop Extended Abstracts* (eds. K.E. MacCormack, L.H. Thorleifson, R.C. Berg and H.A.J. Russell), 89–94. Edmonton, AL: AER/AGS, Special Report 101, [Online: Available at: https://files.isgs.illinois.edu/sites/default/files/files/abstracts/14-SPE-Burt.pdf] (Accessed October 2020).

Caers, J. (2001). Geostatistical reservoir modelling using statistical pattern recognition. *Journal of Petroleum Science and Engineering* 29 (3): 177–188. https://doi.org/10.1016/S0920-4105(01)00088-2.

Cockett, R., Moran, T. and Pidlisecky, A. (2016). Visible geology: creative online tools for teaching, learning, and communicating geologic concepts. In: *Earth, Mind, and Machine: 3D Structural Interpretation*, vol. 111 (eds. R. Krantz, C.J. Ormand and B. Freeman), 53–63. Tulsa, OK: American Association of Petroleum Geologists. https://doi.org/10.1306/13561985M1113671.

Cook, D., Symanzik, J., Majure, J.J. and Cressie, N. (1997). Dynamic graphics in a GIS: more examples using linked software. *Computers & Geosciences* 23 (4): 371–385. https://doi.org/10.1016/S0098-3004(97)00015-0.

Gakis, A., Cabrero, P., Entwisle, D. and Kessler, H. (2016). 3D Geological model of the completed Farringdon underground Railway Station. In: *Crossrail Project, Infrastructure, Design and Construction*, vol. 3 (ed. M. Black), 431–446. London, UK: Thomas Telford Limited and Crossrail.

Hasiuk, F. (2014). Making things geological: 3-D printing in the geosciences. *GSA Today* 24 (8): 28–29. http://doi.org/10.1130/gsatg211gw.1.

Hasiuk, F.J., Florea, L.J. and Sukop, M.C. (2016). Three-dimensional printing: transformative technology for experimental groundwater research. *Groundwater* 54: 157–158. https://doi.org/10.1111/gwat.12394.

Hull, C.W. (2015). The birth of 3D printing. *Research-Technology Management* 58 (6): 25–30. https://doi.org/10.5437/08956308X5806067.

Jackson, I. (2003). Why geologists don't listen and the public can't read geological maps. *Geologija* 46 (2): 361–366.

Jordan, C.J. and Napier, B. (2016). Developing digital fieldwork technologies at the British Geological Survey. *Geological Society Special Publications* 436: 219–229. https://doi.org/10.1144/SP436.6.

Kessler, H. and Dearden, R. (2014). *Scoping Study for a Pan-European Geological Data Infrastructure: Technical Requirements for Serving 3D Geological Models*. Keyworth, UK: British Geological Survey, Report OR/14/072. 22 pp. [Online: Available at: http://nora.nerc.ac.uk/id/eprint/509262] (Accessed October 2020).

Le, H.H., Gabriel, P., Gietzel, J. and Schaeben, H. (2013). An object-relational Spatio-temporal geoscience data model. *Computer & Geosciences* 57: 104–115. https://doi.org/10.1016/j.cageo.2013.04.014.

Libarkin, J.C. and Brick, C. (2002). Research methodologies in science education: visualization and the geosciences. *Journal of Geoscience Education* 50 (4): 449–455. https://doi.org/10.5408/1089-9995-50.4.449.

Lidal, E.M. (2013). *Sketch-based Storytelling for Cognitive Problem Solving: Externalization, Evaluation, and Communication in Geology.* PhD thesis University of Bergen. 131 pp. [Online: Available at: http://bora.uib.no/handle/1956/7175] (Accessed October 2020).

Lidal, E.M., Hauser, H. and Viola, I. (2012). Design principles for cutaway visualization of geological models. In: *Proceedings of Spring Conference on Computer Graphics (SCCG 2012)* (ed. C. O'Sullivan), 53–60. New York, NY: Association for Computing Machinery. https://doi.org/10.1145/2448531.2448537.

Lidal, E.M., Natali, M., Patel, D. et al. (2013). Geological storytelling. *Computers and Graphics* 37 (5): 445–459. https://doi.org/10.1016/j.cag.2013.01.010.

Lin, Y.-F.F., Zhong, S. and Stumpf, A.J. (2016). *Procedure for Three-Dimensional Printing of a Digital Hydrostratigraphic Model.* Champaign, IL: Illinois State Geological Survey, Circular 593. 9 pp. [Online: Available at: http://library.isgs.illinois.edu/Pubs/pdfs/circulars/c593.pdf] (Accessed October 2020).

Mackinlay, J. (1986). Automating the design of graphical presentations of relational information. *ACM Transactions on Graphics* 5 (2): 110–141. https://doi.org/10.1145/22949.22950.

Monaghan, A.A. (2012). *Model Metadata Report and Geological Interpretation for the Clyde, Douglas Basin Model.* Keyworth, UK: British Geological Survey, Report IR/12/003. 28 pp. [Online: Available at: http://nora.nerc.ac.uk/id/eprint/19394] (Accessed October 2020).

Napier, B. (2011). GeoVisionary: Virtual Fieldwork for Real Geologists. *V1 Magazine* 9 pp. [Online: Available at: http://nora.nerc.ac.uk/id/eprint/13558] (Accessed October 2020).

Natali, M. (2014). *Sketch-based Modelling and Conceptual Visualization of Geomorphological Processes for Interactive Scientific Communication.* PhD thesis University of Bergen PhD. 94 pp. [Online: Available at: http://bora.uib.no/handle/1956/8570] (Accessed October 2020).

Natali, M., Klausen, T.G. and Patel, D. (2014a). Sketch-based modelling and visualization of geological deposition. *Computers & Geosciences* 67 (1): 40–48. https://doi.org/10.1016/j.cageo.2014.02.010.

Natali, M., Parulek, J. and Patel, D. (2014b). Rapid Modelling of Interactive Geological Illustrations with Faults and Compaction. In: *Proceedings of Spring Conference on Computer Graphics (SCCG 2014)* (ed. D. Gutierrez), 5–12. New York, NY: Association for Computing Machinery. https://doi.org/10.1145/2643188.2643201.

Pantea, M.P. and Cole, J.C. (2004). *Three-Dimensional Geologic Framework Modelling of Faulted Hydrostratigraphic Units within the Edwards Aquifer, Northern Bexar County, Texas.* Reston, VA: U.S. Geological Survey, Scientific Investigations Report 2004–5226. 10 pp. [Online: Available at: https://pubs.usgs.gov/sir/2004/5226] (Accessed October 2020).

Piburn, M.D., Reynolds, S.J., Leedy, D.E. and Mcauliffe, C.M. (2002). *The Hidden Earth: Visualization of Geologic Features and their Subsurface Geometry.* Paper presented at the Annual Meeting of the National Association for Research in Science Teaching, New Orleans, LA, April 7–10, 2002. 48 pp. [Online: Available at: https://api.semanticscholar.org/CorpusID:17686316] (Accessed October 2020).

Rosen, J. (2014). Changing the Landscape: Geoscientists Embrace 3-D Printing. *EARTH Magazine* 8: 40–47. [Online: Available at: https://www.earthmagazine.org/article/changing-landscape-geoscientists-embrace-3-d-printing] (Accessed October 2020).

Schumann, H. and Müller, W. (2000). *Visualisierung – Grundlagen und Allgemeine Methoden (in German).* Berlin: Springer. 370 pp. http://doi.org/10.1007/978-3-642-57193-0.

Strebelle, S. (2002). Conditional simulation of complex geological structures using multiple-point statistics. *Mathematical Geology* 34 (1): 1–21. https://doi.org/10.1023/A:1014009426274.

Swayne, D.F., Cook, D. and Buja, A. (1998). XGobi: interactive dynamic data visualization in the X window system. *Journal of Computational and Graphical Statistics* 7 (1): 113–130. https://doi.org/10.1080/10618600.1998.10474764.

Terrington, R., Napier, B., Buchi, A. and Procopio, P.M. (2015). Managing the mining cycle using GeoVisionary. In: *Proceedings, 5th International Symposium "Mineral Resources and Mine Development" Aachen, Germany, 27–28 May, 2015* (ed. P.N. Martens), 151–162. Aachen, Germany: RWTH Aachen University. [Online: Available at: http://nora.nerc.ac.uk/id/eprint/510702] (Accessed October 2020).

Tukey, J.W., Friedman, J.H. and Fisherkeller, M.A. (1974). *PRIM-9: An Interactive Multidimensional Data Display and Analysis System.* Alexandria, VA: American Statistical Association. [Online: Available at: http://stat-graphics.org/movies/prim9.html] (Accessed April 7, 2019).

Valle, M. (2013). Visualization: a cognition amplifier. *International Journal of Quantum Chemistry* 113: 2040–2052. https://doi.org/10.1002/qua.24480.

Van Wees, J.-D., Versseput, R., Allard, R. et al. (2003). Dissemination and visualisation of Earth system models for the Dutch subsurface. Chapter 20. In: *New Paradigms in Subsurface Prediction* (eds. M.S. Rosenbaum and A.K. Turner), 225–234. Berlin, Germany: Springer, Lecture Notes in the Earth Sciences, vol. 99. https://doi.org/10.1007/3-540-48019-6_20.

Ware, C. (2013). *Information Visualization – Perception for Design, 3e*. Amsterdam: Elsevier. 509 pp.

Wycisk, P. and Schimpf, L. (2016). Visualising 3D geological models through innovative techniques. *Zeitschrift der Deutschen Gesellschaft für Geowissenschaften (ZDGG)* 167 (4): 405–418. https://dx.doi.org/10.1127/zdgg/2016/0059.

Wycisk, P., Hubert, T., Gossel, W. and Neumann, C. (2009). High resolution 3D spatial modelling of complex geological structures for an environmental risk assessment of abundant mining and industrial megasites. *Computers & Geoscience* 35 (1): 165–182. https://doi.org/10.1016/j.cageo.2007.09.001.

Part IV

Case Studies

18

Application Theme 1 – Urban Planning

Editor's Introduction

Urban planning is typically a 2-D process; the urban subsurface is still largely "out of sight, out of mind" and not yet a daily concern of city planners and managers (Chapter 3). Yet better knowledge of the subsurface is a pre-requisite to improved planning, development, and management of resources as well as for more efficient construction of infrastructure projects. In short, the goal of urban sustainability depends on the availability of accurate subsurface information.

The designers of new urban infrastructure projects requiring approval by urban planning authorities are increasingly employing 3-D subsurface models. Chapter 23 contains three case studies that use 3-D models in infrastructure planning and design.

The application of 3-D subsurface modeling by urban planners has been limited by a lack of experience and understanding of the technology by the urban authorities. In response, the COST Action TU1206 SUB-URBAN Network was undertaken to provide detailed state-of-the-art and state-of-practice reviews on the use of 3-D information for subsurface management in several European cities (Van der Meulen et al. 2016; Mielby et al. 2017).

This chapter contains three case studies that demonstrate a variety of approaches to the incorporation of 3-D subsurface information within urban planning activities. Chapters 21 and 22 contain additional examples from Bremen, Germany; Christchurch, New Zealand; and Nantes, France.

The first case study describes an ongoing pilot study for the city of Darmstadt that employs an integrated 3-D modeling approach to combine information on underground infrastructure and building foundations with subsurface geological conditions. This approach resolved the costs of required software licenses, limited experiences with using technically challenging software, and the use of 2-D maps for most urban decision-making. Results are disseminated through desktop and web-based interfaces and as thematic 2-D maps, vertical and horizontal cross-sections, and virtual boreholes. Because project success depends on contributions from all data owners, the existing user community is being expanded and the 3-D information system further developed and established as part of the daily operations of involved institutions.

The second case study describes the long-term collaboration between the British Geological Survey (BGS) and Glasgow City Council (GCC) to provide subsurface knowledge for Glasgow development and planning processes (Whitbread et al. 2016). The BGS/GCC collaboration has specified that all new ground investigation data for Glasgow be provided in a standard format; a complementary data acquisition procedure supports this process. The Accessing Subsurface Knowledge (ASK) knowledge exchange network facilitates information distribution and has increased the private sector's awareness of the capabilities of 3-D geological models. The ASK network links a broad range of stakeholders who either use and/or generate subsurface data within any stage of the city development process (Bonsor 2017). The GCC has made the ASK data exchange procedures a contractual requirement, and other local authorities, in Scotland, Northern Ireland, England, and Wales are looking to replicate this approach.

The third case study provides an insight into the potential use of 3-D models for the urban planning required by the rapidly growing mega-cities of south-east Asia. A German-Bangladesh Technical Cooperation Project developed several 3-D geological subsurface models for Dhaka, the capital and largest city of Bangladesh, to demonstrate the value of geo-information for urban development (Ludwig et al. 2016). A combination of existing archival information and project-sponsored boreholes created 3-D geological subsurface models that provided some immediately useful geotechnical analysis products and guidance for future work.

Applied Multidimensional Geological Modeling: Informing Sustainable Human Interactions with the Shallow Subsurface, First Edition.
Edited by Alan Keith Turner, Holger Kessler, and Michiel J. van der Meulen.
© 2021 John Wiley & Sons Ltd. Published 2021 by John Wiley & Sons Ltd.

Case Study 18.1: Integrated 3-D Modeling of the Urban Underground of Darmstadt, Hesse, Germany

Rouwen Lehné, Christina Habenberger, Jacob Wächter, and Heiner Heggemann

Hessisches Landesamt für Umwelt, Naturschutz und Geologie, 65203 Wiesbaden, Germany

18.1.1 Introduction

The city of Darmstadt, with a population of approximately 150 000 (HSL 2016), occupies about 122 km² within the Frankfurt/Rhine-Main metropolitan region (Figure 18.1). Better knowledge of subsurface conditions would allow more efficient planning, development, and management of water, soil, and raw material resources. Reliable subsurface information for specific locations will permit more efficient construction. While geological 3-D subsurface modeling has recently become an economically and technically viable technology (Arndt et al. 2011; Berg et al. 2011; Lehné et al. 2013), its application for urban planning in Darmstadt has been limited by a lack of experience and understanding of the technology by the urban authorities.

Adoption of 3-D geological modeling has been hampered by the absence of standardized geological data definitions, no integration of geological and underground infrastructure information, and incompatible multiple platforms for processing geological data.

The Hessian Agency for Nature Conservation, Environment and Geology (HLNUG) developed "Project Darmstadt_3D" in 2017 to address five goals:

- Creation of a high-resolution 3-D model for the urban underground;
- Integration of infrastructure information with a geological/hydrological 3-D subsurface model;
- Development of understandable parameterized geological information that would support urban decision-making;
- Storage of outcomes in a 3-D database; and
- Dissemination of information without charges or restrictions to all relevant user groups and institutions.

18.1.2 Geological Setting

The Upper Rhine Graben (URG) is a Cenozoic sedimentary basin formed during the Alpine Orogenesis. The URG is

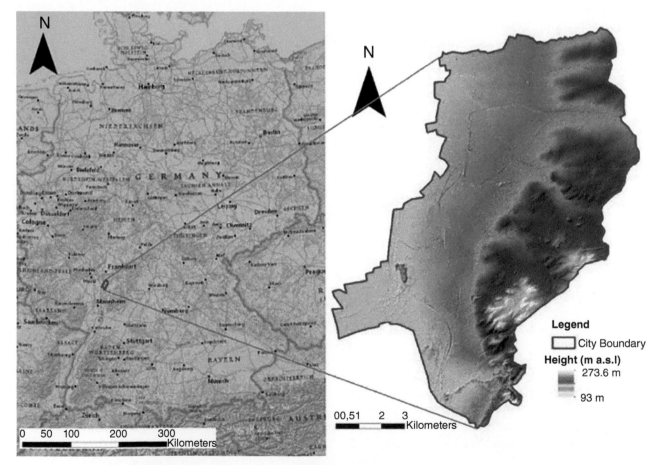

Figure 18.1 (a) Map of Germany. (b) Digital elevation model of the city of Darmstadt.

bounded by NNE–SSW trending major faults and contains many minor faults, which form distinct depositional centers within the URG (Pflug 1982; Schumacher 2002; Behrmann et al. 2005). The eastern master fault of the URG has a displacement of at least 1500 m. Geologically, the city of Darmstadt lies partly within the URG and partly on its eastern flank (Figure 18.2). Thus, the western part is flat and underlain by alluvial deposits of the rivers Rhine, Neckar and Main; while the eastern part, located on the Odenwald mountains, is dominated by a rugged topography with some steep slopes. This challenging geologic setting was a major reason for choosing Darmstadt to provide an example of the value of 3-D geological models to urban planning.

Northern portions of the URG have up to 1800 m of Tertiary sediments while Quaternary sediments reach up to 500 m thickness in the Heidelberg Basin (Ellwanger et al. 2008). Around Darmstadt the stratigraphic throw of the eastern master fault is at least 1500 m. This structure and several minor faults are believed to be recently

active (Haimberger et al. 2005; Peters and Van Balen 2007; Behrmann et al. 2005; Rózsa et al. 2005; Schlatter et al. 2005). Comparatively recent movements along the WNW–ESE trending Gräfenhäuser Fracture Zone (Figure 18.2) control the thickness of the basin fill. The Odenwald mountains forming the eastern flank of the northern URG consist of granodiorite and other crystalline rocks in the southeast and Permian sandstones, claystones, and volcanic rocks in the north (Figure 18.2). The granodiorite typically has very thick weathered zones with extensive granular disintegration. Eroded weathered rock and soil from the slopes in this hilly terrain are deposited within the URG, where they are inter-bedded with alluvial basin-fill deposits.

18.1.3 Developing the 3-D Model

The 3-D model was developed in three main stages: (i) Data acquisition and preparation, (ii) 3-D model construction, and (iii) Dissemination of model products.

Figure 18.2 Geological overview of Darmstadt.

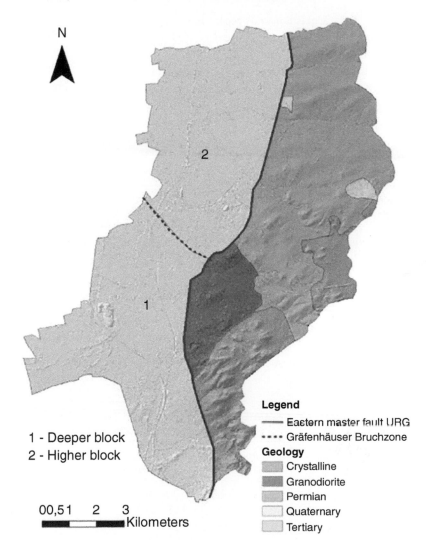

N

1 - Deeper block
2 - Higher block

0 0,5 1   2   3
████□████ Kilometers

Legend
——— Eastern master fault URG
•••• Gräfenhäuser Bruchzone
Geology
Crystalline
Granodiorite
Permian
Quaternary
Tertiary

18.1.3.1 Software Selection

Table 18.1 defines the stages of model development and lists the software products used by each stage. The data acquisition and preparation stage used four software products. GeODin Version 7.5 supported the development of a consistent borehole database. AquaInfo 11 was used to collate relevant hydrogeological information from many sources within a hydrogeological database. Building and infrastructure information was acquired from multiple sources in AutoCAD formats. ArcGIS Version 10.4 provided the basic GIS (Geographic Information Systems) capabilities required by subsequent geographical data evaluations, including data quality checking, and interpolation of hydrostratigraphic surfaces. The 3-D model construction stage used two software products: SKUA-GOCAD 14.1 (developed by Paradigm) and Leapfrog Geo Version 4.0.1 (http://aranzgeo.com). The product dissemination stage used the GST software product to store all the data in a generic 3D-database.

18.1.3.2 Data Acquisition and Preparation

Data preparation involved the review and standardization of borehole records and hydrogeological information from multiple sources, and the collection of building and infrastructure information. *Borehole records* from the HLNUG were provided in digital formats which contained HLNUG-specified layer descriptions for 1556 boreholes.

The city of Darmstadt provided an addition 336 records; these required review, standardization of descriptions, and digitization. These data were stored in a borehole database containing standardized descriptions defined by GeoDIN software, which is used by most German geological surveys. A quality-assurance review resulted in the rejection of 796 records, leaving a total of 1086 borehole records. These provide an average density of nine boreholes per square-kilometer. Most boreholes are quite shallow; 75% are limited to the upper 20 m and only 12% are deeper than 40 m.

Hydrogeological data from 35 groundwater monitoring wells operated by the Hessian geological survey and 121 groundwater gauges monitored by the city of Darmstadt were collated within a hydrogeological database using AquaInfo 11 (Table 18.1), a modular groundwater information and documentation system for the collection and evaluation of groundwater data. *Digital elevation data* were obtained from the official Hessian digital elevation model with a resolution of 1×1 m. *Building and infrastructure information* was obtained to demonstrate the potential applications of combining 3-D geological models with infrastructure data. Building foundation plans for two important buildings (the Congress Centre Darmstadtium and the Residential Palace) were provided in the AutoCAD DXF format (Congress Centre), or were digitized using

Table 18.1 Software employed by each 3-D model development stage.

Stages of 3-D model development	Software	Source reference	Application
Data acquisition and preparation	GeODin Version 7.5	Fugro Consult GmBH https://www.geodin.com/english/	Create and manage borehole database
	AquaInfo 11	GeoConcept Systeme GbR https://geoconcept-systeme.de/	Create and manage hydrogeology data
	AutoCAD	AutoDesk, Inc. https://www.autodesk.com/products/autocad/	Manage information on buildings and infrastructure
	ArcGIS version 10.4	Environmental Systems Research Institute (ESRI), Redlands, California	Conduct source data quality checks and interpolate geological surfaces
3-D model construction	SKUA-GOCAD 14.1	Paradigm, Inc. http://www.pdgm.com/products/skua-gocad	Develop 3-D model surfaces and Sgids cellular model
	Leapfrog Geo 4.0.1	ARANZ Geo, Christchurch, New Zealand http://aranzgeo.com	Evaluate and revise or update 3-D model
Dissemination of model products	GST	https://giga-infosystems.com	Create and manage database of 2-D and 3-D data for product dissemination

AutoCAD products (Residential Palace). Local infrastructure plans describing the sewage, water, and gas utility systems were provided by the city of Darmstadt and a local utility company as vector data files in either AutoCAD DXF format or ArcGIS Shapefiles. ArcGIS Version 10.4 (ESRI 2017) spatially evaluated these data sources, supported data quality reviews, and interpolated geological surfaces.

18.1.3.3 3-D Model Construction

Correlation of geologic features throughout the Darmstadt project area was very difficult, owing to the complex geological setting (HLNUG, 2007) and the heterogeneity of the available borehole records. The combined geological records provided 10 276 individual horizon identifications and over 300 petrographic descriptions. At even short distances, there are significant variations in vertical resolution, stratification details, and reliability (Figure 18.3). To overcome these issues, all lithologies were

assigned a hydraulic conductivity class, as defined by Garling and Dittrich (1979) and Krimm (2015). This reduced the complex stratigraphy to a much simpler hydrostratigraphy consisting of only seven conductivity classes, ranging from 1 (very high) to 7 (very low) (Table 18.2).

The 3-D model construction involved two software packages. SKUA-GOCAD 14.1 was used for initial 3-D model development; this included the stacking of the interpolated hydrostratigraphic surfaces, followed by discretization of the 3-D model volume into SGrids. The model was restricted to a depth of 30 m, with a voxel resolution of 25 × 25 m and a thickness of 2 m (Figure 18.4). The hydraulic conductivity was interpolated by a combination of GIS-based Empirical Bayesian Kriging (EBK) and GoCAD Discrete Smooth Interpolation (DSI). Leapfrog Geo Version 4.0.1 facilitated identification and correction of data inconsistencies and model modifications when new data became available.

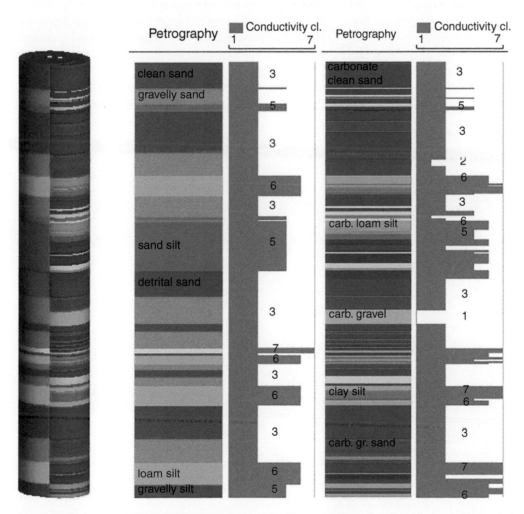

Figure 18.3 Layer descriptions of two boreholes only 10 m apart show significant petrographic variation. The assigned hydraulic conductivity classes reduce variability and support lateral correlation.

Table 18.2 Range of hydraulic conductivity k_f for loose rock according to Garling and Dittrich (1979) and hydraulic conductivity classes according to Krimm (2015).

Loose rock	Hydraulic conductivity k_f (m s^{-1})	Conductivity class
Sandy gravel	3×10^{-3} to 5×10^{-4}	1
Gravely sand	1×10^{-3} to 2×10^{-4}	2
Medium sand	4×10^{-4} to 1×10^{-4}	3
Silty sand	2×10^{-4} to 1×10^{-5}	4
Sandy silt	5×10^{-5} to 1×10^{-6}	5
Clayey silt	5×10^{-6} to 1×10^{-8}	6
Silty clay	$\sim 1 \times 10^{-8}$	7

18.1.3.4 Dissemination of Model Products

All project data and results were migrated to the generic 3D-database GST by GiGa infosystems (https://giga-infosystems.com). This provided flexible data management and product dissemination using a desktop application and a 3-D web viewer. Products can be delivered to every workplace within the HLNUG using the desktop application, which permits blending of technical information to create 2-D maps. In addition, a free 3-D data viewer, part of the generic 3-D GST-database, allows multiple external users to query and combine 2-D and 3-D information (Figure 18.5). A user management module controls user access to confidential data.

18.1.4 Model Applications

The Darmstadt pilot project demonstrated two types of model applications: geological applications and urban planning applications.

18.1.4.1 Geological Applications

While the use of just seven hydraulic conductivity classes implies a decoupling from geology, the model provided new geologic insights. Previously, inconsistencies in core descriptions prevented differentiation between alluvial valley-fill deposits and those derived from erosion of the graben shoulders. The aggregated and parameterized model clearly shows several wedge-shaped deposits with high conductivity along the eastern URG margin, formed by interbedded eroded materials from the flanking hills. These deposits become thinner with increasing distance from the graben margin (Figure 18.6). The model units can be further parameterized to define additional characteristics, such as of ease of excavation, porosity, or potential radioactivity.

Because there was insufficient geological information at an appropriate resolution, previous efforts to

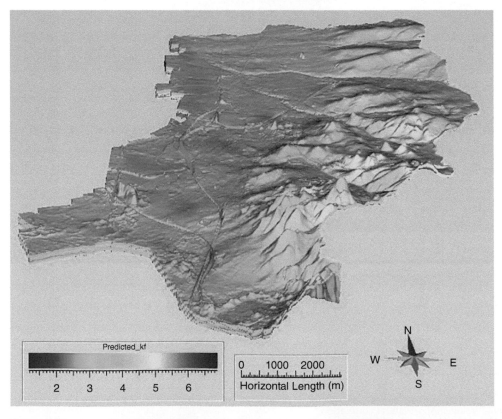

Figure 18.4 Interpolated conductivity classes for the upper 30 m, ranging from 1 (very good) to 7 (very poor). The horizontal resolution is 25 × 25 m.

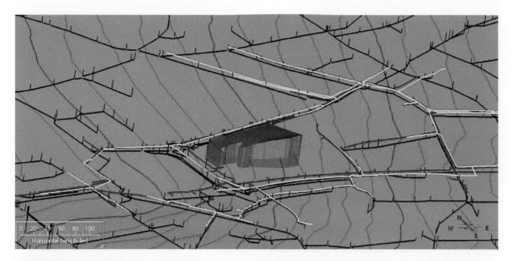

Figure 18.5 Blending of information in a 3-D space. Technical infrastructure (black lines = sewage system; yellow lines = gas supply, blue lines = water supply, gray block = building foundation) is combined with the contoured groundwater surface. The building foundation extends below the groundwater surface and affects groundwater movement; the supply systems are located above the groundwater surface and do not influence it.

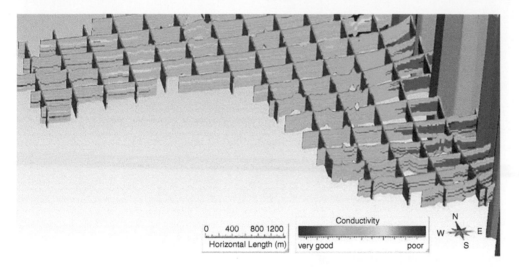

Figure 18.6 Interpolated hydraulic conductivity classes reveal wedge-shaped deposits with high conductivity along the eastern URG margin (dark blue) which can be distinguished from surrounding alluvial deposits. The brown barrier is the eastern master fault of the URG.

map groundwater elevations largely ignored geological conditions. The constructed 3-D parameterized hydro-geologic model permits the mapping of major aquifers and aquitards and predicts groundwater elevations with significantly higher reliability (Figure 18.7).

Querying the grain-size distribution in the 3-D model reveals two major zones within the URG; a northern region with mainly fine-grained sediments and a southern region with mainly coarse-grained sediments (Figure 18.8). The boundary transition zone has been interpreted as due to the "*Gräfenhäuser Fracture Zone.*" The model provides new information on this tectonic structure. The main fault trend is aligned NW–SE and consists of several segments connected by minor faults oriented perpendicular to the main fault.

18.1.4.2 Urban Planning Applications

The infiltration of clean surface water or groundwater into sewers due to leaking pipes ("*Fremdwasser*" in German) may account for up to 25% of the water running through the sewers, increasing operation and investment costs (Stolpe et al. 2016). Potential areas of interaction of groundwater with the sewage system were located by combining estimated groundwater elevation surfaces for each month during the years 2000–2015 with 3-D locations of the sewers. Figure 18.9 is a typical 2-D map product that illustrates these interactions. The interaction area varies by season. A maximum of 3.8 km² is affected in April, the month with the highest groundwater levels, and a minimum of 2.4 km² is affected in October, the month with the lowest groundwater levels.

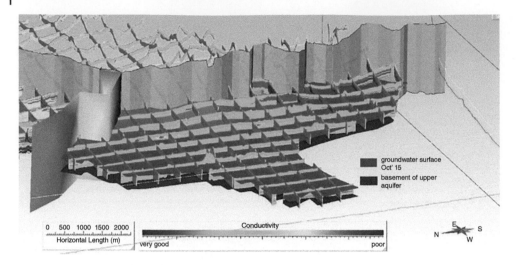

Figure 18.7 Portion of the model defined by cross-sections illustrates the October 2015 groundwater surface (blue) and the aquifer base (deep red). Brown barriers are faults.

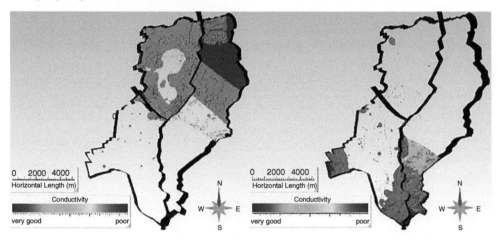

Figure 18.8 Grain-size distribution in the Darmstadt area. The northern part is dominated by fine-grained deposits (left, conductivity classes 4–7); the southern part is dominated by coarse-grained deposits (right, conductivity classes 1–3).

European legislation (guideline 2013/59/Euratom) requires national and state authorities to provide information on areas suspected of having increased radon concentrations in indoor air. Soil gas measurements, taken and interpolated without proper geological context, do not provide Darmstadt with enough information to support appropriate response action plans for building sealing, monitoring, or health check-ups. Within the URG, deposits derived from the Odenwald mountains have higher concentrations of radionuclides than deposits of the river Rhine. In eastern Darmstadt, outside the graben, volcanic rocks with relatively high concentrations of radionuclides produce soil gases with even higher radon concentrations. Detailed mapping of these locations, based on the homogeneous regions of the 3-D model, provides better demarcation of radon-hazard areas, permitting the development of more efficient response-action plans.

Building Information Modeling (BIM) provides a digital representation of physical and functional characteristics of a facility – either a building or an entire engineering project. By sharing this knowledge, BIM forms a reliable basis for management decisions during the entire life-cycle of a facility, from earliest conception to demolition. However, most existing BIM 3-D models show the ground as a grey amorphous mass – if shown at all. Thus, the BIM concept must be expanded to include the subsurface (Kessler et al. 2016).

To evaluate the interactions between building foundations and groundwater conditions, the foundation characteristics for two important buildings (the Congress Centre darmstadtium and the Residential Palace) were included in the integrated pilot model of Darmstadt (Figure 18.5). The Congress Centre foundation influences groundwater flow. However, these interactions vary significantly over time, so long-term monitoring and continuing availability of up-to-date information is important to enable evidence based decision making.

Figure 18.9 Typical 2-D Darmstadt Project map product showing locations of infiltration areas (green) and October areas of interaction between the sewers and groundwater (red).

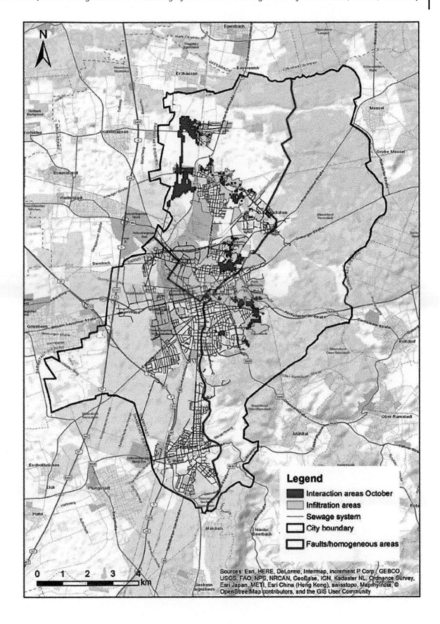

18.1.5 Conclusions

Project "Darmstadt_3D" was undertaken by the HNLUG to provide an integrated 3-D modeling approach that combined information on underground infrastructure and building foundations with subsurface geological conditions for the Darmstadt urban area. Commonly encountered barriers to 3-D modeling efforts include costs of required software licenses, limited experiences with using technically challenging software, and the fact that most urban decision-making is based on 2-D maps. The project overcame these barriers by developing an integrated hydrogeological 3-D model and storing it in a generic 3-D database that supports: (i) both geological and underground infrastructure information; (ii) dissemination of results to end users through desktop and web-based interfaces; (iii) creation of thematic 2-D maps; and (iv) urban and underground planning by providing useful tools to produce vertical and horizontal cross-sections and virtual boreholes at any specified location.

The success of the project depends on the readiness and willingness of data owners to contribute. Current contributors are: Hessian Agency for Nature Conservation, Environment and Geology (HLNUG), Technische Universität Darmstadt, Hessian Agency of Land Management and Geoinformation (HLBG), the land surveying office of the city of Darmstadt, and several private sector organizations. The pilot study ran until 2019; the existing 3-D information system is subsequently being developed further and established as part of the daily operations of the involved institutions.

Case Study 18.2: Accessing Subsurface Knowledge (ASK) Network – Improving the Use of Subsurface Information for Glasgow Urban Renewal

Hugh F. Barron[1], Helen C. Bonsor[1], S. Diarmad G. Campbell[1], and Garry Baker[2]

[1] British Geological Survey, Edinburgh, Currie EH14 4BA, UK
[2] British Geological Survey, Keyworth, Nottingham NG12 5GG, UK

18.2.1 Introduction

The city of Glasgow, situated along the banks of the River Clyde, is Scotland's largest urban development and the third largest city in the United Kingdom (Figure 18.10). In the nineteenth and early twentieth centuries, Glasgow was a major seaport and an important center of heavy industry, including steel making, chemicals, textiles, shipbuilding, and marine engineering. These industries relied on locally mined coal, ironstone, and building stone (Browne et al. 1986).

In common with many European cities, Glasgow's heavy industries declined during the latter half of the twentieth century, leaving Glasgow with a largely post-industrial landscape, a dwindling population, social deprivation, and derelict properties. This legacy included abandoned shallow mines (Figure 18.11), and many brownfield sites with contaminated and variable ground conditions (Campbell et al. 2010). Since the early 1980s, Glasgow has experienced an economic revival resulting in regeneration of the city's economy, social fabric, and environment. Redevelopment planning policy for Glasgow emphasizes sustainable development, health improvements, and social renewal (Whitbread et al. 2016). Substantial regeneration has

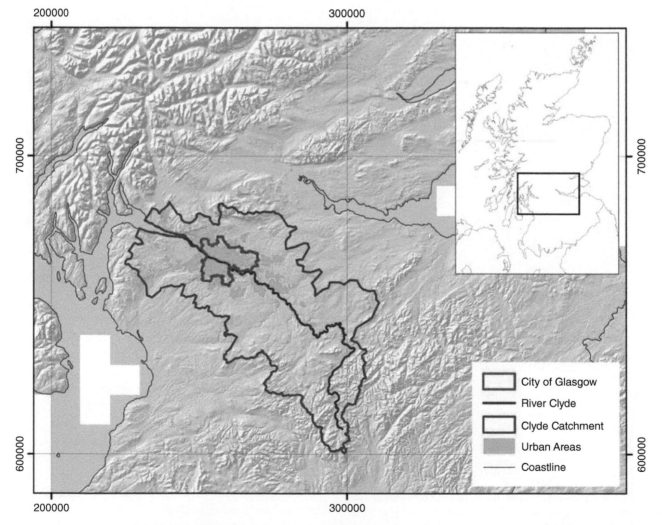

Figure 18.10 Map of central Scotland showing the location of Glasgow. The grid is British National Grid (scale in meters). The shaded terrain image is derived from the NEXTMap digital terrain model at 50 m resolution. Source: NEXTMap Britain elevation data from Intermap Technologies. The location of the main map in Scotland is shown in the inset map.

Figure 18.11 Modeled extent of abandoned mine workings under a site in eastern Glasgow.

already taken place, and will continue for decades to come, along the Clyde Corridor, including Clyde Gateway in the East End and the Clyde Waterfront area (Figure 18.12). Major new initiatives include the Sighthill Transformational Regeneration Area (TRA) in the north of the City (Figure 18.12).

Regeneration is costly. The construction industry has recognized that insufficient understanding of subsurface ground conditions results in cost over-runs, project delays, and overly conservative design (Clayton 2001; Parry 2009; Baynes 2010). Project-cost increases of 10% or more are a typical result (Paul et al. 2002; Greeman 2011). In Glasgow, the ASK Network has been developed to improve the

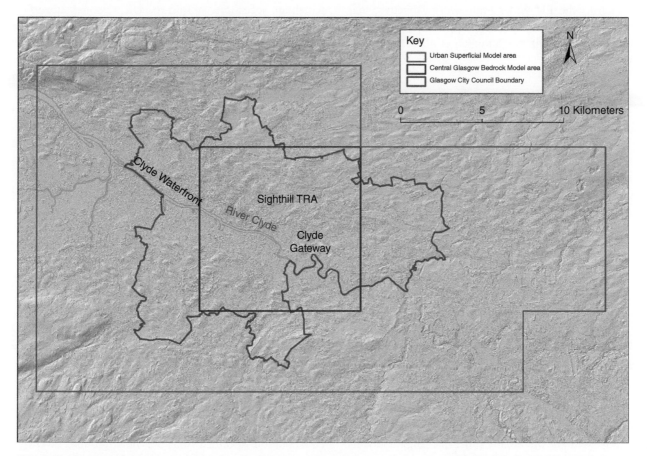

Figure 18.12 Map of Glasgow area showing the coverage of available models. Source: NEXTMap Britain elevation data from Intermap Technologies.

use of subsurface geoscience knowledge and thus reduce geotechnical risk and costs of ground engineering works (Smale 2017).

18.2.2 Urban Subsurface 3-D Modeling

Urban 3-D subsurface geological models are increasingly used to facilitate more effective use of ground investigation data and help advance understanding of subsurface conditions (Chowdhury and Flentje 2008; Royse et al. 2008; Lelliott et al. 2009; Campbell et al. 2010; Aldiss et al. 2012). European geological surveys have developed 3-D geological models on regional to local scales to underpin urban planning and sustainable development (Bridge et al. 2004; Bourgine et al. 2009; Schokker et al. 2016; Mielby et al. 2017), engineering hazard assessments (Culshaw 2005; Neumann et al. 2009), groundwater management (Lelliott et al. 2006; Carneiro and Carvalho 2010; Campbell et al. 2010; Bonsor et al. 2015) and protection of cultural heritage (De Beer et al. 2016). Three-dimensional geological models have assisted environmental regulators and local authorities to meet the requirements of recent environmental legislation, such as the EU Water Framework Directive, that demand a 3-D understanding of the geometry and properties of the main aquifers (Campbell et al. 2010; Mielby et al. 2017).

18.2.3 Subsurface Information for Glasgow

Collaboration between the BGS and GCC has extended over many years and has led to the recognition that, within the development and planning processes in Glasgow, knowledge of the subsurface environment is required to achieve effective remediation and regeneration, hazard reduction, natural resource management, and development of a sustainable economy (Whitbread et al. 2016).

The BGS has a long history in geological survey and data collection in Glasgow (and elsewhere in the UK). The BGS holds over 40 000 records of boreholes drilled within the City of Glasgow, largely from private-sector site investigations and donated voluntarily to the BGS (Whitbread et al. 2016). To improve the accessibility and utilization of these records, the BGS digitized Glasgow's extensive historical analog geoscience information (Browne and McMillan 1991; Mellon and Frize 2006). The five-year BGS Digital Geoscience Spatial Model (DGSM) project created an operational production environment within the BGS which supported the preparation, manipulation, and visualization of consistent and systematic subsurface digital models (Smith 2005). Table 18.3 summarizes the development of the 3-D geological models in the Glasgow area.

The 3-D geological models and their derivatives developed by BGS have played a central role in fostering improved communication and understanding of the geometry and properties of the very complex geological conditions in the Glasgow subsurface. Figure 18.13 shows a model of the surficial deposits, while Figure 18.14 shows bedrock structures beneath Glasgow. In addition to assisting ground investigations, the Glasgow 3-D models were designed to address specific scientific problems (Merritt et al. 2007; Whitbread et al. 2016). These include: the history of glacial oscillations; the evolution of buried valleys in and adjacent to the Clyde valley (Merritt et al. 2011); the structural evolution of the bedrock (Monaghan and Pouliquen 2009); and assessment of the thermogeological potential of mine waters and local heat-extraction and storage capacities of thick superficial deposit sequences (Monaghan et al. 2017).

18.2.4 Difficulties in Re-using Subsurface Information

However, 3-D subsurface models are only part of the solution; geologists can only collate and interpret data which are accessible. Within the UK, regional subsurface models are typically based on only a small proportion of the available data; this is a consequence of the time and cost required to extract data in non-standardized formats (Lelliott et al. 2009). For example, the BGS Glasgow urban surficial deposits model (Figures 18.12 and 18.13) was constructed from less than a third of the borehole data potentially available.

Ground investigations generate large amounts of high-quality data. Traditionally, these data were reported in printed documents or embedded in PDF digital files; both options limit the digital accessibility and efficient re-use of the information during model creation. The Association of Geotechnical and Geoenvironmental Specialists (AGS) has developed an industry-standard digital reporting format which allows for significantly faster data assimilation and accurate transfer of information between software and database systems (Association of Geotechnical and Geoenvironmental Specialists 2004).

Effective use of ground investigation data to create 3-D attributed geological models, and thus to support planning, development, and management of the subsurface, requires the data to be readily accessible in a centralized database with standardized formats (Barron et al. 2015). With the aim of establishing an unbroken (virtuous) cycle of data acquisition; data storage and data re-use, the BGS/GCC collaboration specified that all new ground investigation data for Glasgow be provided in the AGS 3.1 format (Association of Geotechnical and Geoenvironmental

Table 18.3 Development of 3-D geological models in the Glasgow area.

Date	Milestone	References
2002	BGS commenced development of pilot 3-D models of Glasgow using GSI3D® software for the surficial deposits and GOCAD® for the bedrock, using borehole records, mine plans and other geological data as part of DGSM.	Smith (2005)
2009	Initiation of multidisciplinary Clyde Urban Super Project (CUSP), bringing together 3-D subsurface model development, soil geochemistry, and monitoring and modelling of groundwater. The CUSP also served as a platform for collaborative research to tackle a range of urban subsurface environmental issues, both within BGS and between the BGS and research partners.	Whitbread et al. (2016)
2009	Report on methodology of the Clyde Regional Bedrock Model.	Monaghan and Pouliquen (2009)
2011	Production of Clyde Gateway pilot 3-D geological and groundwater models for the Clyde Gateway Urban Regeneration Company.	Merritt et al. (2011)
2012–2014	Continued development of both bedrock and surficial deposits models at various scales within the CUSP project progressively developed to produce the following: *Surficial*: Regional catchment model (>3000 km^2) (Figure 18.10) and 8 urban 'block models' (each 50–100 km^2) (Figure 18.13) developed using GSI3D™ modelling software and a deterministic methodology, in which geologist-correlated networks of cross-sections are interpolated to form lithostratigraphic surfaces using a deterministic algorithm. The superficial deposits models can be attributed with a range of geotechnical properties to predict ground conditions or model groundwater flow. *Bedrock*: Clyde catchment model (900 km^2 centred on the lower Clyde area), a higher resolution Central Glasgow model (100 km^2) (Figure 18.14), and detailed models of some areas of disused coal workings (Figure 18.11). Bedrock surfaces including stratigraphic unit bases, faults and areas of worked coal seams have been modelled from borehole records, mine plans and seismic data using GOCAD®	Whitbread et al. (2016)
2013	Release of Clyde superficial deposits and bedrock models to ASK Network members	Monaghan et al. (2013)
2015	Stochastic modelling developed to predict lithological variability between and within lithostratigraphic units directly from the borehole data.	Kearsey et al. (2015)
2017	Update of 2011 Clyde Gateway model in SKUA® software as part of UK Geoenergy Observatories, Glasgow Geothermal Energy Research Field Site project). Additional worked coal seams and mining information added to bedrock model.	Monaghan et al. (2017)

Figure 18.13 10 × 10 km surficial deposits model of Central Glasgow, looking NW, 10× vertical exaggeration.

Specialists 2004). A complementary data acquisition procedure, defined as the Glasgow SPEcification for data Capture (GSPEC), was developed to support this process. As data volumes increased, a data knowledge exchange network termed the *Accessing Subsurface Knowledge (ASK) Network* was introduced to facilitate information distribution. Subsequently, the BGS adapted the GSPEC procedures to form the *UK Data Deposit Portal* (British Geological Survey 2020a).

18.2.5 Accessing Subsurface Knowledge (ASK) Network

The BGS and GCC developed the ASK data knowledge exchange network in 2013 with the support of Grontmij

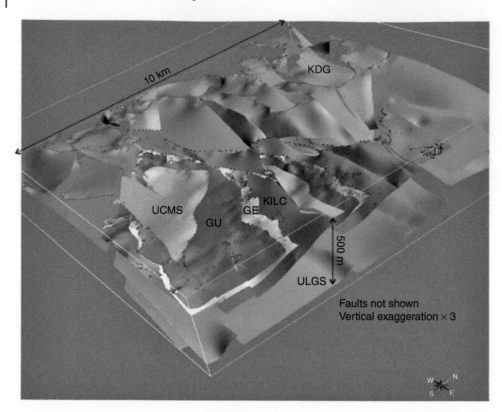

Figure 18.14 10 × 10 km bedrock model of Central Glasgow. Area of model outlined in Figure 18.12.

(now Sweco), following discussions with the private sector. The network aims to improve the exchange of information between public and private sector organizations involved in the use of ground investigation data, especially their acquisition, interpretation, reporting, and re-use (Barron et al. 2015). The ASK network links a broad range of stakeholders, including the BGS, local authorities, civil engineering and environmental consultancies, and universities, who either use and/or generate subsurface data within any stage of the city development process (Bonsor 2017). The ASK network also links to other UK initiatives and innovation hubs active within this topical area, including: Future Cities Catapult, ThinkDeep UK, the Urban Innovation Centre, GO-Science Future of Cities Foresight project, the EPSRC Urban Big Data center (Glasgow). International initiatives, such as the COST SUB-URBAN Action, Lighting Urban Community International (LUCI), and Rockefeller 100 Resilient Cities, also provide for additional collaboration and discussion (Bonsor 2017).

The ASK network originated in Glasgow to increase the private sector's awareness of the capabilities of BGS 3-D geological models. Shallow (<50 m) subsurface data from the large number of site investigations occurring in the Glasgow region were collected using the standardized AGS 3.1 digital format (Association of Geotechnical and Geoenvironmental Specialists 2004; Bonsor 2017). The ASK Network currently has 12 private-sector, seven public-sector, and four academic members. Under an innovation agreement, ASK network members can gain access to BGS 3-D geological models of the Glasgow conurbation

(Monaghan et al. 2012, 2013; Barron et al. 2015). This access has identified significant benefits, including:

- Improved ability to handle large datasets, reduced financial risk to design, and ability for rapid 3-D visualization of data. This has led to improved confidence in predicted ground conditions and attainment of optimum project designs.
- Increased certainty at an earlier stage in project planning, improved estimating and financial forecasting, and greater confidence in option selection for Glasgow infrastructure upgrades.

At its inception, the ASK Network was focused on facilitating knowledge exchange between geoscientists and geotechnical specialists involved in design and construction phases (Barron et al. 2015; Bonsor 2017). The initial phases of the Glasgow city development process remained susceptible to key knowledge gaps and utilized derived strategic knowledge of subsurface conditions in only a limited fashion. To address these issues, the ASK network is expanding to include a broader range of city stakeholders, including: local authority development planning teams and officers, national and regional planning organizations, urban designers, developers, and community organizations. Key national stakeholders, including Scottish Water and Transport Scotland, have decided to make the ASK data deposition procedures a contractual requirement of their contracts for all ground investigation work (Bonsor 2017). The GCC has also made it a contractual requirement, and other local authorities, in Scotland, Northern

Ireland, England, and Wales are looking to replicate this approach. There have been several ASK events in Scotland (Bonsor 2017).

At the heart of the ASK Network, and fundamental to its success, is a willingness on the part of all parties involved to change practice, and to share voluntarily (rather than as a consequence of any legislative requirement) data and knowledge for wider and mutual benefits. Discussions of what new knowledge could be relevant to unlocking future city development are underway amongst a broad group of actors and the BGS. One topic is how the BGS National Geoscience Data Centre (NGDC) can serve to manage information provided by diverse stakeholder sources (Bonsor 2017). Other Natural Environment Research Council components, such as the Environmental Information Data Centre (EIDC) hosted by the Centre for Ecology & Hydrology (CEH), as well as the Economic and Social Research Council Urban Big Data Centre (UBDC), are also either formal members of the ASK Network or active participants in its activities (Bonsor 2017).

18.2.6 Depositing and Accessing AGS and Geotechnical Data

A long-term, open-access, professionally managed, national data store is a critical capability for enabling public and private sectors to adopt an open flow of geotechnical data and thus receive multiple cost benefits while also meeting government initiatives. The BGS has been working to develop a web-based approach to enable increased re-use of subsurface data in the UK.

The ASK network also served as a pilot application for the increased data deposition and retrieval of data held in the BGS NGDC. The ASK network is providing a very effective forum for BGS to have ongoing conversations and iterative knowledge development with a broad range of stakeholders (Bonsor et al. 2013). The BGS has recently released a new "UK data deposit portal" (Figure 18.15) for subsurface data and information (British Geological Survey 2020a). This portal provides a platform for direct upload of data; it also provides a "single entry point" that provides access to existing national data with easy-to-use search interfaces (Figure 18.16).

The portal encourages both open data donation and compliance with industry standard formats, increasing the accessibility and future re-use of data for the user communities. Uploaded AGS files are stored in the National Geotechnical Properties Database (British Geological Survey 2020b) and complete AGS files or a single combined AGS data file can be downloaded using the search criteria and service. As part of the BIM for the Subsurface project funded by Innovate UK, BGS has collaborated with Keynetix, developers of geotechnical data management software, to provide additional web service functionality, so that the portal can be another possible data donation route for AGS project data from their latest versions of the Holebase SI software. The new, freely available BGS Groundhog software (Wood et al. 2015) enables users to consume, visualize, and model AGS data for their own projects, along with data stored in the national data store, to generate geological cross-sections that can be exported to CAD, and ultimately integrated into above- and below-ground project design

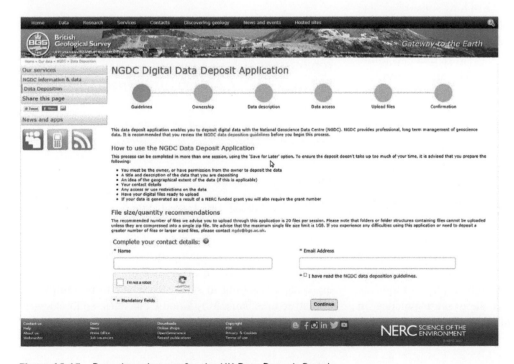

Figure 18.15 Data deposit page for the UK Data Deposit Portal.

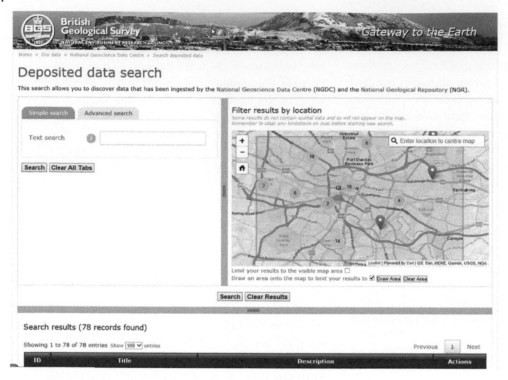

Figure 18.16 Deposited data search page for the UK Data Deposit Portal.

models. This helps the community work toward implementing Level 2 BIM, in line with the recommendations of the UK Government construction strategy.

18.2.7 Conclusions

Collaboration between the BGS, GCC, and the private sector participation in the ASK Network has demonstrated the value of introducing a step-change in how subsurface data are reported, exchanged, and managed. The public and private sectors in Glasgow are undertaking increased data utilization and sharing, and the ASK network model is under active consideration throughout the UK and internationally. ASK network members have reported significant benefits, including:

- rapid 3-D visualization of data;
- increased certainty at an earlier stage in project planning;
- improved estimating, financial forecasting, and reduced financial risk; and
- greater confidence in option selection and closer attainment of optimum project design.

The ASK network is providing a very effective forum for iterative knowledge development between BGS and an increasingly broad range of stakeholders including planners, urban designers, developers, and community organizations. Communications about best practice in the wider European community through the EU COST

SUB-URBAN Action (Watson et al. 2016) are enabling discussions of what new knowledge could be relevant to unlocking future city development. A long-term, open-access, professionally managed, national data store paves the way for public and private sectors to adopt an open flow of geotechnical data and derive multiple cost benefits, while also meeting government initiatives.

Case Study 18.3: Geological Subsurface Models for Urban Planning in Mega-Cities: An Example from Dhaka, Bangladesh

Rolf R. Ludwig and Andreas Guenther

Bundesanstalt für Geowissenschaften und Rohstoffe (BGR), 30655 Hannover, Germany

18.3.1 Introduction

Dhaka, the capital and largest city of Bangladesh, is a densely-populated megacity. Its current population of 17 million is projected to grow to 25 million by 2025. This rapid physical growth often has overlooked the importance of geological influences on economic and rational engineering design, resulting in severe damages to buildings and infrastructures caused by floods, cyclones, soil collapses, and foundation failures. In response to requests from urban planners and national planning organizations, the Geological Survey of Bangladesh (GSB) carried out extensive urban geology mapping of Dhaka (United Nations 1999; Karim 2004). Subsequently, a German-Bangladesh Technical Cooperation Project was established to provide geo-information for urban development. Within this project, 3-D geological subsurface models for Dhaka were developed (Ludwig et al. 2016).

18.3.2 Geological Setting of Dhaka

Bangladesh is located on the floodplains and delta of the Ganges, Brahmaputra, and Meghna rivers (Figure 18.17). In the Pliocene, these rivers began to transport enormous

volumes of sediment eroded from the Himalayas into the gradually subsiding Bengal Basin. The resulting landscape has low relief; most of Bangladesh is less than 12 m (40 ft) above sea level. Sedimentation rates and patterns were affected by climate fluctuations and tectonic instability within the Bengal Basin. Pleistocene glacial periods caused sea levels to drop 100 m below current levels. This resulted in cycles of valley incision followed by subsequent refilling with younger alluvial deposits. Tectonic instability produced locally-uplifted fault-bounded blocks composed of consolidated fine-grained Plio-Pleistocene terrace deposits that now are raised 1–10 m above adjacent alluvial floodplains. Dhaka was initially established at the southern tip of one of these uplifted blocks – the so-called "Madhupur

Figure 18.17 Brahmaputra/Ganges/Meghna delta system: environments of sediment deposition and main geomorphic units. (Kinniburgh and Smedley 2001).

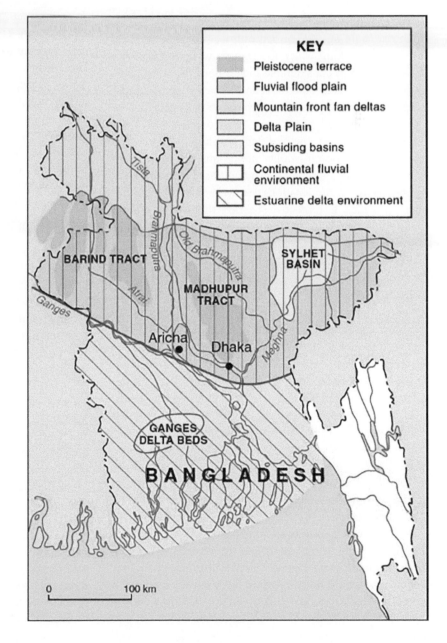

Tract." The city has since expanded onto Holocene alluvial floodplain deposits surrounding the uplifted terrains of the Madhupur Tract. Figure 18.18 shows a simplified geological map of Dhaka and the locations of the large-scale models.

The subsurface sedimentary sequence, explored up to depths of approximately 100 m, is composed of four distinct geological units (Table 18.4). The Plio-Pleistocene Dupi Tila Formation is the oldest geological unit that may be encountered by urban development. The Dupi Tila Formation comprises weathered, moderately consolidated, yellow silty fine-grained sandstones. It has no surface exposures, and fault-related block subsidence has placed it below the maximum depth of infrastructure interaction (~50 m) in some locations. However, the Dupi Tila Formation provides suitable foundation conditions for high-rise buildings and buried utilities. It is still the main source of drinking water for Dhaka; water withdrawals have resulted in considerable lowering of the water table at several locations. Potential contamination of the aquifer by percolation of polluted surface waters is a concern.

The Pleistocene Madhupur Clay overlies the Dupi Tila Formation on the uplifted Madhupur Tract. Low strength, high compressibility values, and the occurrences of swelling clay minerals cause engineering construction problems at some locations. However, the Madhupur Clay provides a protective barrier that helps control pollution of the underlying aquifers in the Dupi Tila Formation. The Madhupur Clay has been subjected to valley erosion within the uplifted Madhupur Tract and at many locations within the alluvial floodplains, due to valley incision during glacial maxima. At these locations, younger Holocene alluvium and artificial fill deposits are in direct contact with the Dupi Tila Formation, providing potential pollution pathways for pollutants to reach the aquifers.

Holocene deposits unconformably overlie the Madhupur Clay and the Dupi Tila Formation. Sandy natural levees, abandoned channel deposits, and floodplain deposits interfinger with back swamp and depression deposits in major river valleys. With highly variable strength characteristics, these deposits may extend up to depths of more than 100 m. Holocene alluvial deposits are also found within the smaller valleys dissecting the uplifted Madhupur Tract.

Artificial fill deposits are the result of three anthropogenic actions: (i) channels and depressions filled with household and industrial waste; (ii) within the Madhupur Tract, excavated hilltops redeposited in adjacent valleys to create leveled areas; and (iii) un-engineered, low density hydraulic fill consisting of fine sand with some silt and clay. The maximum observed thickness of the fill deposits is about 8 m.

18.3.3 Development of 3-D Geological Subsurface Models

The project produced three geological subsurface models. One covers the total area of the Dhaka Metropolitan City (DMC) (1529 km^2) at a regional scale of ~1:50 000 and two cover detailed pilot study areas (each less than 0.3 km^2) at a preliminary site-planning scale of ~1:10 000 (Ludwig et al. 2016). Table 18.5 summarizes the characteristics of these geological models.

Ludwig et al. (2016) provides details on the construction and characteristics of each model. The models were created using SubsurfaceViewer MX software developed by INSIGHT Geologische Softwaresysteme GmbH. This software has been used by the GSB since 2005 and follows the procedures described in Chapter 10 to produce 3-D geological models (Kessler et al. 2009; Mathers et al. 2012). Subsurface conditions were defined by combining information from a digital terrain model (DTM), existing geological maps and reports, and new project-based exploratory boreholes to produce a series of intersecting interpreted cross-sections. One advantage of using the SubsurfaceViewer MX software is that completed models may be readily used by the planning community because a free software product allows them to be viewed and analyzed.

The initial models used a stacked series of triangulated surfaces (TINs) to define the base or top of each geologic layer. Subsequently, these TIN models were converted to irregular voxel models (Figure 18.19) which represent each geologic unit by a series of small square vertical columns corresponding to the local thickness of the geologic unit. Section 18.3.4 describes the role of these voxel models for urban planning functions.

18.3.3.1 The Dhaka Metropolitan City Model (DMC Model)

The DMC model is a small-scale model covering a large area with a comparatively low density of borehole information. It consists of four major stratigraphic layers, covers an area of approximately 1500 km^2, and was developed from observations obtained from 319 boreholes developed by the project (Ludwig et al. 2016).

The model provides an overview of recent geoscientific knowledge and supports preliminary geotechnical interpretations prior to development of more specific planning products. The DMC model can supply a variety of shallow subsurface information, including: (i) distributions of different stratigraphic sequences to depths of

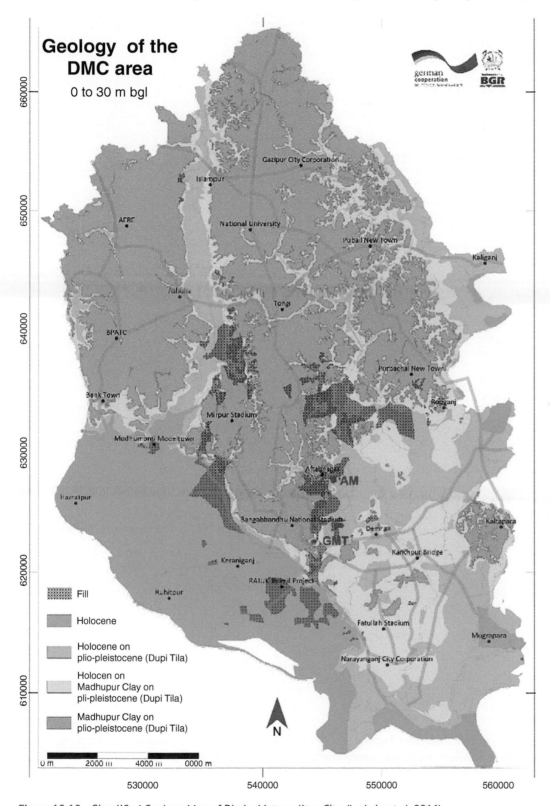

Figure 18.18 Simplified Geology Map of Dhaka Metropolitan City. (Ludwig et al. 2016)

Table 18.4 Stratigraphy of Dhaka metropolitan city.

Age	Formation	Lithology	Thickness (m)
Holocene	Artificial fill	Artificial anthropogenic deposits – three types: (1) Channels and depressions filled with household and industrial waste. (2) Excavated hilltop materials redeposited in adjacent valleys to create level areas. (3) Hydraulic fill areas containing fine sand with some silt and clay, un-engineered and not densified.	Median: 1 Maximum: ~6 (possibly up to 15 m locally)
		Local Unconformity	
	Alluvium	Natural levee and floodplain deposits: Sandy silt with varying sand and clay content, gray and friable.	Median: 24 Maximum: 59
		Back swamp and depression deposits: clay and silty clay, gray, bluish gray to dark gray	
		Local Unconformity	
Pleistocene	Madhupur clay	Red clay: Light brown to brick red and massive, with fossil wood	Median: 15 Maximum: 39
		Mottled clay: Earthy gray with patches of orange, brown color, massive containing calcareous and ferruginous nodules.	
		Unconformity	
Pliocene	Dupi tila	Sandstone: Yellow to yellowish gray, massive, cross-bedded, moistly fine to medium grained containing scattered gravel lenses, moderately consolidated.	Obs. Median: 15 Obs. Maximum: 39 Est. Total: 90+

Source: Modified after Alam (1989) with thickness information from Ludwig et al. (2016).

Table 18.5 Characteristics of the three Dhaka models.

Model name	Scale	Area (km^2)	Intended resolution	Number of layers	Number of boreholes used		
					from project	from reports	Avg. depth (m)
Dhaka Metropolitan City model (DMC)	~1 : 50 000	1529	Horiz: 50 m Vert: 2 m	4	319	0	29
Aftabnager Model (AM)	~1 : 10 000	0.18	Horiz: 2 m Vert: 1 m	8	33	0	40
Green Model Town (GMT)	~1 : 10 000	0.29	Horiz: 2 m Vert: 1 m	7	8	10	31

Source: Ludwig et al. (2016).

30 m; (ii) estimations of depth to the Dupi Tila Formation; (iii) bearing capacity for buildings; (iv) liquefaction potential; and (v) hydrogeological analyses.

The DMC model required inclusion of major faults resulting from the trans-tensional horst-and-graben tectonic regime. Ludwig et al. (2016) provides details of this modeling procedure. The traces and cross-cutting relationships of 28 major fault surfaces were mapped and elevation-registered from available sources, then verified by cross-sectional interpretations across suspected fault traces. Variable geometric projection (VGP) techniques (de Kemp 1998) were used to generate TIN surfaces defining steeply inclined normal faults with appropriate dip directions. These were then imported into the SubsurfaceViewer MX software to complete the DMC model (Figure 18.20). Figure 18.21 illustrates the completed DMC model after it was converted to an irregular voxel representation.

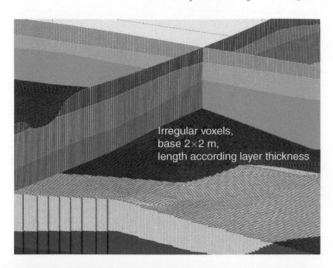

Figure 18.19 Example of a 3-D model using irregular voxels.

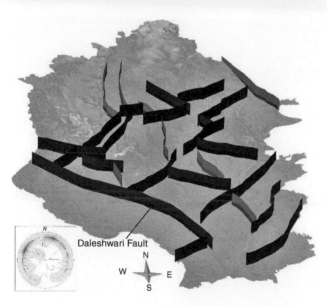

Figure 18.20 Modeled 3-D fault array. The insert is a stereonet projection (equal area, lower hemisphere) showing orientations of projected fault planes. (Ludwig et al. 2016)

18.3.3.2 The Aftabnager Model (AM Model)

The AM model was developed for a very local area (only 0.18 km²) and was based exclusively on geotechnical data collected from 33 boreholes undertaken by the project, including 842 standard penetration tests (SPT), and 561 grain-size analyses. Therefore, the source data for this model are highly reliable and consistent. The AM area is located on the floodplain deposits of a former valley of the Balu River system. The relatively thick (~40 m) Holocene sequence is built up by a series of floodplain and depression sediments, composed mainly of clayey silt, with some natural levees forming lenses of silty sand. Thus, the AM model has eight layers (units); the uppermost layer represents artificial fill and the lowest layer (layer 8) represents

the Plio-Pleistocene Dupi Tila formation. The Holocene alluvial sequence is subdivided into six units based on grain size distributions and SPT data. Figure 18.22 shows the completed AM model.

The AM model was designed to assess the data-requirements for creating subsurface models at the most detailed planning level (1:2000) within an area of considerable geological heterogeneity. As discussed in Section 18.3.4, the AM model can produce customized thematic maps of liquefaction potential, pile foundation bearing loads, thickness of landfill deposits, and depth to hard ground (the Dupi Tila Formation).

18.3.3.3 The Green Model Town Model (GMT Model)

The Green Model Town (GMT) model was developed for a local area (0.29 km²) but, in contrast to the AM model, it was based on geotechnical data extracted from archives and data from boreholes drilled by the project. Ten archival borehole records and eight project boreholes provided 372 SPT and 245 grain-size analyses. During GMT model construction, the value and reliability of existing archive borehole information was compared with information obtained from new project-financed boreholes.

The GMT area is located on the floodplain deposits of a former distributary of the Balu River system. These have similar characteristics as deposits encountered at the AM site. However, the GMT site is almost entirely covered by a layer of artificial fill consisting of 2–6 m of fine sand placed with hydraulic fill methods. Because the artificial fill is uncompacted, it has very low SPT values. The underlying relatively thick Holocene alluvial sequence is subdivided into five units. Several are discontinuous and are modeled as sand lenses. The Plio-Pleistocene Dupi Tila Formation forms the lowest model layer (layer 7). Figure 18.23 illustrates the completed GMT model as a series of cross-sections.

18.3.4 Applying Models to Urban Planning Topics

To support applications for urban planning, the three models, initially developed as a series of stacked TIN surfaces, were converted to irregular voxel models (Figure 18.19). The DMC model has horizontal voxel dimensions of 50 × 50 m; the smaller AM and GMT models have horizontal voxel dimensions of 2 × 2 m. These irregular voxel models permit the spatial distribution of physical properties defined by point borehole information to specific geological layers. This was undertaken using the "Information System Engineering Geology (ISEG)" which was originally developed by the BGR (Bundesanstalt für Geowissenschaften und Rohstoffe) in 2006 for a cooperative study in Indonesia (Günther et al. 2007). ISEG is

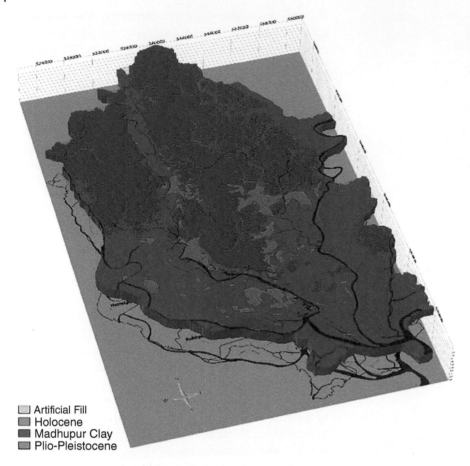

Artificial Fill
Holocene
Madhupur Clay
Plio-Pleistocene

Figure 18.21 Irregular voxel model version of completed DMC model. View is from southwest; vertical exaggeration is 30×. (Ludwig et al. 2016)

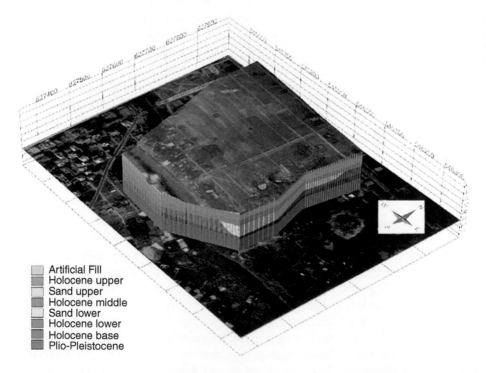

Artificial Fill
Holocene upper
Sand upper
Holocene middle
Sand lower
Holocene lower
Holocene base
Plio-Pleistocene

Figure 18.22 View of completed AM model. View is from southeast; vertical exaggeration is 2×. (Ludwig et al. 2016)

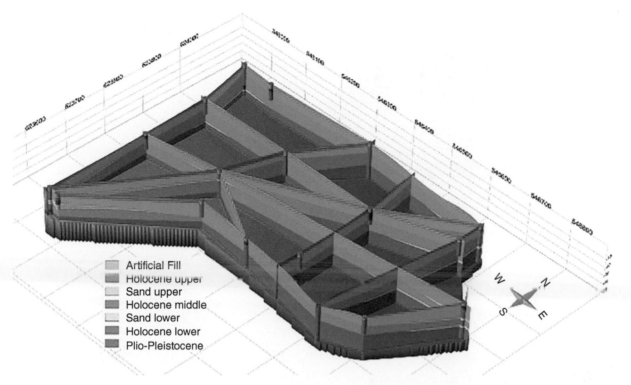

Figure 18.23 View of completed GMT model, shown as series of cross-sections. View is from southeast; vertical exaggeration is 2×. (Ludwig et al. 2016)

a multi-component expert system, incorporating a 2.5-D GIS and associated databases, which supports the collection, storage, management, analysis, and presentation of spatially-related engineering geological information (Bahls et al. 2016a).

By using the ISEG GIS client with these Dhaka 3-D models, sample properties or borehole information belonging to specific layers can be queried, depth-aggregated, interpolated, and used to populate irregular voxel models. The structure of the irregular voxel models is suitable when the physical properties of distinct geological subsurface layers have high variability laterally, but do not vary with depth within a layer. For example, borehole samples and SPT data were spatially interpolated within the AM irregular voxel model to generate customized maps or models demonstrating applications to preliminary site planning, such as the thickness of artificial fill and depth to hard ground (the Dupi Tila Formation). Figure 18.24 shows a 3-D geotechnical analytical product based on the distribution of SPT values within the AM model. Similar products were generated from the GMT irregular voxel model.

For the purposes of this project, the ISEG GIS client was linked to the DMC irregular voxel model in order to assess the distributions of nine engineering geological parameters

across the Dhaka metropolitan area for each layer within the DMC model. These parameters allowed for the calculation of regional variations of pile bearing capacities and liquefaction potentials (Figure 18.25). The results are reported in Bahls et al. (2016b).

18.3.5 Conclusions

The Bangladesh-German Technical Cooperation Project successfully created 3-D geological subsurface models of Dhaka by combining existing archival information with project-sponsored boreholes. The project produced immediately useful geotechnical analysis products and guidance for future work.

The DMC regional model is the first version of a complete and comprehensive 3-D geological subsurface model for the whole project area. It encapsulates recent geoscientific knowledge and supports preliminary geotechnical interpretations prior to development of more specific planning products. However, the DMC model will require updates with additional observational data to improve its predictive accuracy.

The Aftabnagar (AM) and GMT models show how a 3-D subsurface model based on suitable observational data can provide valuable guidance at the preliminary site planning stage. The GMT model shows that archival borehole data

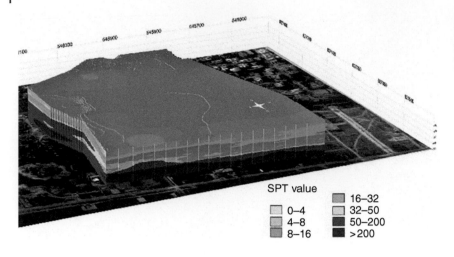

Figure 18.24 Geotechnical analytical product based on the AM model. 3-D view of heterogeneity of SPT distribution based on 3-D interpolation of SPT Nf values. View is from northeast. (Ludwig et al. 2016)

SPT value

0–4
4–8
8–16
16–32
32–50
50–200
>200

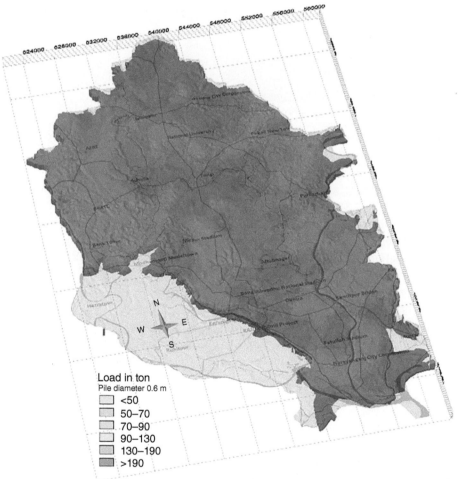

Figure 18.25 3-D view of allowable loads on 0.6 m diameter bored concrete piles on Dupi Tila Formation. View is from southwest. (Ludwig et al. 2016)

Load in ton
Pile diameter 0.6 m

<50
50–70
70–90
90–130
130–190
>190

are a valuable source of *supplemental* modeling data for depths to 20 m. Construction of the DMC regional model demonstrated that archival data were inadequate as the sole information source of an accurate 3-D subsurface geological model. Archived boreholes were not deep enough, layer descriptions were inadequate, the number of grain size analyses were insufficient, and photo-documentation of SPT samples was missing.

The three models were analyzed by the ISEG to produce geotechnical products for planning purposes. ISEG is a multi-component expert system that supports the collection, storage, management, analysis, and presentation of georeferenced engineering-geological information. Further development of these 3-D geological subsurface models and their analysis with ISEG will improve their sensitivity, reliability, and robustness.

References

Alam, M. (1989). Geology and depositional history of Cenozoic sediments of the Bengal Basin of Bangladesh. *Palaeogeography, Palaeoclimatology, Palaeoecology* 69: 125–139. https://doi.org/10.1016/0031-0182(89)90159-4.

Aldiss, D.T., Black, M.G., Entwisle, D.C. et al. (2012). Benefits of a 3D geological model for major tunnelling works: an example from Farringdon, East-Central London, UK. *Quarterly Journal of Engineering Geology and Hydrogeology* 45: 405–414. http://dx.doi.org/10.1144/qjegh2011-066.

Arndt, D., Bär, K., Fritsche, J.-G. et al. (2011). 3D structural model of the Federal State of Hesse (Germany) for geo-potential evaluation. *Zeitschrift der Deutschen Gesellschaft für Geowissenschaften* 162 (4): 353–370. https://dx.doi.org/10.1127/1860-1804/2011/0162-0353.

Association of Geotechnical and Geoenvironmental Specialists (2004). *Electronic Transfer of Geotechnical and Geoenvironmental Data (Edition 3.1)*. Bromley, UK: Association of Geotechnical and Geoenvironmental Specialists. 58 pp.

Bahls, R., Hossain, Md. A., Günther, A. (2016a). *Handbook, A Compilation of Operational Procedures, Annex 7, ISEG GIS Client Manual (Bangladesh-German Technical Cooperation Project 2013–2016: Geo Information for Urban Development, Bangladesh)*. Hannover, Germany: Bundesanstalt für Geowissenschaften und Rohstoffe. 80 pp.

Bahls, R., Günther, A., Ludwig, R.R. et al. (2016b). *Geo-information Based Land Use Suitability Map, Dhaka Metropolitan City (Report for Bangladesh-German Technical Cooperation Project 2013–2016: Geo-Information for Urban Development, Bangladesh)*. Hannover, Germany: Bundesanstalt für Geowissenschaften und Rohstoffe. 50 pp.

Barron, H.F., Bonsor, H.C., Campbell, S.D.G. et al. (2015). The ASK network: developing a virtuous cycle of subsurface data and knowledge exchange. In: *Proceedings of the XVI European Conference on Soil Mechanics and Geotechnical Engineering: Geotechnical Engineering for Infrastructure and Development* (eds. M.G. Winter, D.M. Smith, P.J.L. Eldred and D.G. Toll), 4211–4216. London, UK: ICE Publishing.

Baynes, F.J. (2010). Sources of geotechnical risk. *Quarterly Journal of Engineering Geology and Hydrogeology* 43: 321–331. https://doi.org/10.1144/1470-9236/08-003.

Berg, R.C., Mathers, S.J., Kessler, H. and Keefer, D.A. [Eds.] (2011). *Synopsis of Current Three-Dimensional Geological Mapping and Modeling in Geological Survey Organizations*. Champaign, IL: Illinois State Geological Survey, Circular 578. 104 pp. [Online: Available at: https://isgs.illinois.edu/publications/c578] (Accessed November 2020).

Behrmann, J., Ziegler, P.A., Schmid, S.M. et al. (2005). EUCOR-URGENT – upper Rhine Graben evolution and Neotectonics. *International Journal of Earth Sciences* 94 (4): 505–506. https://doi.org/10.1007/s00531-005-0513-0.

Bonsor, H.C. (2017). *NERC Briefing Note: Integrating NERC(BGS) Subsurface Environmental Research and Data to City Development Processes and Policy: Key Learning Outcomes*. Keyworth, UK: British Geological Survey, Report OR/17/005. 35 pp. [Online: Available at: http://nora.nerc.ac.uk/id/eprint/517033] (Accessed December 2017).

Bonsor, H.C., Entwisle, D.C., Watson, S. et al. (2013). Maximising Past Investment in Subsurface Data in Urban Areas for Sustainable Resource Management: A Pilot in Glasgow, UK. *Ground Engineering* 46 (2): 25–28. [Online: Available at: http://nora.nerc.ac.uk/id/eprint/502014] (Accessed November 2020).

Bonsor, H.C., Dahlqvist, P., Moosmann, L. et al. (2015). *Groundwater, Geothermal Modelling and Monitoring at City-Scales—Identifying Good Practice, and Effective Knowledge Exchange*. COST TU1206 project Sub-Urban, Report TU1206-WG2–005. 69 pp. [Online: Available at; http://sub-urban.squarespace.com] (Accessed November 2020).

Bourgine, B., Dominique, S., Marache, A. and Thierry, P. (2009). Tools and methods for constructing 3D geological models in the urban environment: the case of Bordeaux. In: *Engineering Geology of Tomorrow's Cities, Engineering Geology Special Publication, vol. 22* (eds. M.G. Culshaw, H.J. Reeves, I. Jefferson and T.W. Spink), IAEG2006 Paper 72. London, UK: Geological Society. 12 pp. https://doi.org/10.1144/EGSP22.I.

Bridge, D., Hough, E., Kessler, H. et al. (2004). Integrated modelling of geosciences information to support sustainable urban planning, greater Manchester areas, Northwest England. In: *Three-Dimensional Geologic Mapping for Groundwater Applications, Extended Abstracts of the 49th GACMAC Annual Meeting* (eds. R.C. Berg, H. Russell and L.H. Thorleifson), 16–19. Champaign, IL: Illinois State Geological Survey, Open File Series 2004–8. [Online: Available at: https://isgs.illinois.edu/publications/ofs2004–8] (Accessed November 2020).

British Geological Survey (2020a). *Geotechnical data services*. Keyworth, UK: British Geological Survey. [Web page: https://www.bgs.ac.uk/technologies/ geotechnical-data-services] (Accessed November 2020).

British Geological Survey (2020b). *National Geotechnical Properties Database*. Keyworth, UK: British Geological Survey. [Web page: https://www.bgs.ac.uk/

geological-research/science-facilities/engineering-geotechnical-capability/national-geotechnical-properties-database] (Accessed November 2020).

Browne, M.A.E. and McMillan, A.A. (1991). British geological survey thematic geology maps of quaternary deposits in Scotland. *Engineering Geology Special Publication 7*: 511–518. https://doi.org/10.1144/GSL.ENG.1991.007.01.48.

Browne, M.A.E., Forsyth, I.H. and McMillan, A.A. (1986). Glasgow, a case study in urban geology. *Journal of the Geological Society, London* 143: 509–520. https://doi.org/10.1144/gsjgs.143.3.0509.

Campbell, S.D.G., Merritt, J.E., ÓDochartaigh, B.E. et al. (2010). 3D geological models and their hydrogeological applications: supporting urban development – a case study in Glasgow-Clyde, UK. *Zeitschrift der Deutschen Gesellschaft fur Geowissenschaften* 161: 251–262. https://doi.org/10.1127/1860-1804/2010/0161-0251.

Carneiro, J. and Carvalho, J.M. (2010). Groundwater modelling as an urban planning tool: issues raised by a small scale model. *Quarterly Journal of Engineering Geology and Hydrogeology* 43: 157–170. http://doi.org/10.1144/1470-9236/08-028.

Chowdhury, R. and Flentje, P. (2008). Strategic approaches for management risk in Geomechanics. In: *Proceedings, 12th International Conference of International Association for Computer Methods and Advances in Geomechanics* (ed. M.N. Jadhav), 3031–3042. Red Hook, NY: Curran.

Clayton, C.R.I. (2001). Managing geotechnical risk: time for change? *Proceedings of the Institution of Civil Engineers, Geotechnical Engineering*, 149: 3–11. https://doi.org/10.1680/geng.2001.149.1.3.

Culshaw, M.G. (2005). From concept towards reality: developing the attributed 3D geological model of the subsurface. *Quarterly Journal of Engineering Geology and Hydrogeology* 38: 231–284. https://doi.org/10.1144/1470-9236/04-072.

De Beer, J., Boogaard, F.C., Vorenhout, M. et al. (2016). *Cultural heritage – A Review of Good Practices in Cultural Heritage Management and the Use of Subsurface Knowledge in Urban Areas*. COST TU1206 project Sub-Urban, report TU1206-WG2-009. 36 pp. [Online: Available at: http://sub-urban.squarespace.com] (Accessed November 2020).

De Kemp, E.A. (1998). Three-dimensional projection of curvilinear geological features through direction cosine interpolation of structural field observations. *Computers & Geosciences* 24 (3): 269–284. https://doi.org/10.1016/S0098-3004(97)00066-6.

Ellwanger, D., Gabriel, G., Simon, T. et al. (2008). Long sequence of quaternary rocks in the Heidelberg Basin Depocentre. *E&G Quaternary Science Journal* 57: 316–4, 337. https://doi.org/10.3285/eg.57.3-4.3.

ESRI (2017). *ArcGIS 10.4 for Desktop Quick Start Guide*. [Online: Available at: https://desktop.arcgis.com/en/arcmap/10.4/get-started/setup/arcgis-desktop-quick-start-guide.htm] (Accessed November 2020).

Garling, F. and Dittrich J. (1979). *Hydrogeologie – Gesteinsbemusterung [Hydrogeology – Rock Sampling]*. Leipzig, Germany: VEB Deutscher Verlag für Grundstoffindustrie. 54 pp.

Greeman A. (2011). A New Way of Working. *New Civil Engineer Magazine, September 2011*. [Online: Available at: https://www.nce.co.uk/features/geotechnical/a-new-way-of-working/8624001.article] (Accessed December 2017).

Günther, A., Balzer, D. and Kuhn, D. (2007). An information system engineering geology for urban spatial planning. *Geophysical Research Abstracts* 9: 06099.

Haimberger, A., Hoppe, A. and Schäfer, A. (2005). High-resolution seismic survey on the Rhine River in the northern upper Rhine Graben. *International Journal of Earth Sciences* 94 (4): 657–668. https://doi.org/10.1007/s00531-005-0514-z.

HLNUG (2007). *Geologische Übersichtskarte Hessen 1 : 300 000 [Geological Overview Map Hesse 1 : 300 000], Revised Digital Edition*. Wiesbaden: Hessisches Landesamt für Umwelt und Geologie. 1 map sheet. [Online: Available at: https://www.hlnug.de/fileadmin/dokumente/geologie/geologie/guek300.pdf] (Accessed November 2020).

HSL (2016). *Hessische Gemeindestatistik 2016 – Ausgewählte Strukturdaten aus Bevölkerung und Wirtschaft 2015 [Hesse Community Statistics 2016 – Selected Structural Data from Population and Economy]*. Wiesbaden, Germany: Hessisches Statistisches Landesamt. 247 pp. [Online: Available at: https://statistik.hessen.de/sites/statistik.hessen.de/files/HGSt_j16_3aA.pdf] (Accessed November 2020).

Karim, M.F. (2004). Geological aspects of urban planning in Bangladesh. In: *The Ground Beneath Our Feet: A Factor in Urban Planning. Atlas of Urban Geology, vol. 14* (eds. Economic and Social Commission for Asia and the Pacific), 195–207. New York, NY: United Nations.

Kearsey, T., Williams, J., Finlayson, A. et al. (2015). Testing the application and limitation of stochastic simulations to predict the lithology of glacial and fluvial deposits in Central Glasgow, UK. *Engineering Geology* 187: 98–112. https://doi.org/10.1016/j.enggeo.2014.12.017.

Kessler, H., Mathers, S.J. and Sobisch, H.-G. (2009). The capture and dissemination of integrated 3D geospatial knowledge at the British Geological Survey using GSI3D software and methodology. *Computers and Geosciences* 35 (6): 1311–1321. https://doi.org/10.1016/j.cageo.2008.04.005.

Kessler, H., Wood, B., Morin, G., Gakis, A. et al. (2016). Building Information Modelling (BIM) – A route for geological models to have real world impact. In: *Three-Dimensional Geological Mapping: Workshop Extended Abstracts* (eds. MacCormack, K.E., Thorleifson, L.H., Berg, R.C. and Russell, H.A.J.), 13–18. Edmonton, AL: AER/AGS, Special Report 101. [Online: Available at: https://www.ags.aer.ca/publications/SPE_101.html] (Accessed November 2020).

Kinniburgh, D.G. and Smedley, P.L. [Eds.] (2001). *Arsenic Contamination of Groundwater in Bangladesh. Volume 2: Final report.* Keyworth, UK: British Geological Survey, report WC/00/19. 267 pp. [Online: Available at: https://www2.bgs.ac.uk/groundwater/health/arsenic/Bangladesh/reports.html] (Accessed November 2020).

Krimm, J. (2015). *Methodischer Vergleich von 2D – und 3D-Modellierungswerkzeugen zur Interpolation von Lockergesteinsparametern in einem hochauflösenden geologischen 3D-Modell als Basis für eine numerische Grundwassersimulation – Fallbeispiel Babenhausen [Methodological Comparison of 2D and 3D-Modeling Tools to the Interpolation of Loose Rock Parameters in a High Resolution Geological 3D Model as a Basis for a Numerical Groundwater Simulation – Case Study of Babenhausen].* Master Thesis (unpublished), Technische Universität Darmstadt. 118 pp.

Lehné, R.J., Hoselmann, C., Heggemann, H. et al. (2013). Geological 3D modelling in the densely populated metropolitan area Frankfurt/Rhine-Main. *Zeitschrift der Deutschen Gesellschaft für Geowissenschaften (ZDGG)* 164 (4): 591–603. https://dx.doi.org/10.1127/1860-1804/2013/0051.

Lelliott, M.R., Bridge, D., Kessler, H. et al. (2006). The application of 3D geological modelling to recharge assessments in an urban environment. *Quarterly Journal of Engineering Geology and Hydrogeology* 39: 293–302. https://doi.org/10.1144/1470-9236/05-027.

Lelliott, M.R., Cave, M.R. and Wealthall, G.P. (2009). A structured approach to the measurement of uncertainty in 3D geological models. *Quarterly Journal of Engineering Geology and Hydrogeology* 42: 95–105. https://doi.org/10.1144/1470-9236/07-081.

Ludwig, R.R., Bahls, R., Faruqa, N. et al. (2016). *Geoscientific 3D Models of Dhaka Metropolitan City, Bangladesh (Report for Bangladesh-German Technical Cooperation Project 2013–2016: Geo-Information for Urban Development, Bangladesh).* Hannover, Germany: Bundesanstalt für Geowissenschaften und Rohstoffe. 72 pp.

Mathers, S.J., Kessler, H., Macdonald, D.M.J. et al. (2012). *The Use of Geological and Hydrogeological Models in Environmental Studies.* Keyworth, UK: British Geological Survey, Internal Report IR/10/022. 93 pp. [Online:

Available at: http://nora.nerc.ac.uk/id/eprint/20481] (Accessed November 2020).

Mellon, P. and Frize, M. (2006). A digital geotechnical data system for the City of Glasgow. *Engineering Geology of Tomorrow's Cities, Engineering Geology Special Publication, vol. 22* (eds. Culshaw, M.G. Reeves, H.J. Jefferson, I. and Spink, T.W.), IAEG2006 Paper 346. London, UK: Geological Society. 8 pp. https://doi.org/10.1144/EGSP22.I.

Merritt, J.E., Monaghan, A.A., Entwisle, D.C. et al. (2007). 3D attributed models for addressing environmental and engineering geoscience problems in areas of urban regeneration: a Case Study in Glasgow, UK. *First Break* 25: 79–84. [Online: Available at: http://nora.nerc.ac.uk/id/eprint/4130] (Accessed November 2020).

Merritt, J.E., Monaghan, A.A., Loughlin, S.C. et al. (2011). *Clyde Gateway Pilot 3D Geological and Groundwater Model.* Keyworth, UK: British Geological Survey, Report CR/09/005N. 86 pp. [Online: Available at: http://nora.nerc.ac.uk/id/eprint/21118] (Accessed November 2020).

Mielby, S. , Eriksson, I., Campbell, S.D.G. et al. (2017). *Opening Up the Subsurface for the Cities of Tomorrow: Considering Access to Subsurface Knowledge – Evaluation of Practices and Techniques.* COST TU1206 project Sub-Urban, report TU1206-WG2–001. 94 pp. [Online: Available at: http://sub-urban.squarespace.com] (Accessed November 2020).

Monaghan, A.A. and Pouliquen G. (2009). *Clyde Regional Bedrock Model (Phase 1) Methodological Report.* Keyworth, UK: British Geological Survey, Project Document IR/09/070. 14 pp. [Online: Available at: http://nora.nerc.ac.uk/id/eprint/14772] (Accessed November 2020).

Monaghan, A.A., Terrington, R.L. and Merritt, J.E. (2012). *Clyde Gateway Pilot 3D Geological Model, Version 2.* Keyworth, UK: British Geological Survey, Report CR/12/010N. 6 pp. [Online: Available at: http://nora.nerc.ac.uk/id/eprint/21118] (Accessed November 2020).

Monaghan, A.A., Arkley, S.L.B., Whitbread, K. and McCormac M. (2013). *Clyde Superficial Deposits and Bedrock Models Released to the ASK Network 2013: A Guide for Users, Version 2.* Keyworth, UK: British Geological Survey, Report OR/13/025. 32 pp. [Online: Available at: http://nora.nerc.ac.uk/id/eprint/500548] (Accessed November 2020).

Monaghan, A.A., ÓDochartaigh, B., Fordyce, F. et al. (2017). *UKGEOS: Glasgow Geothermal Energy Research Field Site (GGERFS): Initial Summary of the Geological Platform.* Keyworth, UK: British Geological Survey, Report OR/17/006. 205 pp. [Online: Available at:

http://nora.nerc.ac.uk/id/eprint/518636] (Accessed November 2020).

Neumann, D., Schönberg, G. and Strobel, G. (2009). 3D-modelling of ground conditions for the engineering geology map of the City of Magdeburg. In: *Engineering Geology of Tomorrow's Cities, Engineering Geology Special Publication, vol. 22* (eds. Culshaw, M.G. Reeves, H.J. Jefferson, I. and Spink, T.W.), IAEG2006 Paper 444. London, UK: Geological Society. 7 pp. https://doi.org/10.1144/EGSP22.I.

Parry, S. (2009). Introduction to engineering geology in geotechnical risk management. *Quarterly Journal of Engineering Geology and Hydrogeology* 42: 443–444. https://doi.org/10.1144/1470-9236/09-042.

Paul, T., Chow, F.C., Kjekstad, O. [Eds.] (2002). *Hidden Apects of Urban Planning: Surface and Underground Development* London, UK: Thomas Telford Publishing. 88 pp. https://doi.org/10.1680/haoupsaud.31012.

Peters, G. and Van Balen, R.T. (2007). Tectonic geomorphology of the northern Upper Rhine Graben, Germany. *Global and Planetary Change* 58: 310–334. https://doi.org/10.1016/j.gloplacha.2006.11.041.

Pflug, R. (1982). *Bau und Entwicklung des Oberrheingrabens [Structure and Development of the Upper Rhine Graben]*. Darmstadt, Germany: Wissenschaftliche Buchgesellschaft Darmstadt. 145 pp.

Royse, K.R., Reeves, H.J. and Gibson, A. (2008). The modelling and visualisation of digital geoscientific data as an aid to land-use planning in the urban environment, an example from the Thames gateway. *Geological Society Special Publications* 305: 89–106. https://doi.org/10.1144/SP305.10.

Rózsa, S., Heck, B., Seitz, K. et al. (2005). Determination of displacements in the Upper Rhine Graben area from GPS and leveling data. *International Journal Earth Sciences*, 94 (4): 538–549. https://doi.org/10.1007/s00531-005-0478-z.

Schlatter, A., Schneider, D., Geiger, A. and Kahle, H.-G. (2005). Recent vertical movements from precise levelling in the vicinity of the City of Basel, Switzerland. *International Journal of Earth Sciences* 94 (4): 507–514. https://doi.org/10.1007/s00531-004-0449-9.

Schokker, J., Sandersen, P.B.E., De Beer, H. et al. (2016). *3D Urban Subsurface Modelling and Visualisation – A Review of Good Practices and Techniques to ensure Optimal Use of Geological Information in Urban Planning*. COST TU1206 project Sub-Urban, report TU1206.WG2-005. 93 pp. [Online: Available at: http://sub.urban.squarespace.com] (Accessed November 2020).

Schumacher, M.E. (2002). Upper Rhine Graben: Role of Preexisting Structures during Rift Evolution. *Tectonics* 21(1): 1006. https://doi.org/10.1029/2001TC900022.

Smale, K. (2017). Sharing Geotechnical Information 'Could Cut Costs'. *New Civil Engineer 13 November 2017*. [Online: Available at: https://www.newcivilengineer.com/archive/sharing-geotechnical-information-could-cut-costs-13-11-2017 (Accessed December 2017).

Smith, I.F. [Ed.] (2005). *Digital Geoscience Spatial Model Project Final Report*. Keyworth, UK: British Geological Survey, Occasional Publication No. 9. 56 pp. [Online: Available at: http://nora.nerc.ac.uk/id/eprint/2366] (Accessed November 2020).

Stolpe, H., Borgmann, A. and Brüggemann T. (2016). Auswirkungen von Kanalabdichtungen. Teil 4 der Serie: Fremdwasser in der Kanalisation und die Ökoeffizienz der Kanalsanierung [Effects of duct sealing. Part 4 of the series: Extraneous water in sewers and the eco-efficiency of sewerage remediation]. *Wasserwirtschaft Wassertechnik* 2016(4): 35–39.

United Nations (1999). *Urban Geology of Dhaka, Bangladesh Atlas of Urban Geology, vol. 11* (eds. Economic and Social Commission for Asia and the Pacific). New York, NY: United Nations. 68 pp.

Van der Meulen, M.J., Campbell, S.D.G, Lawrence, D.J. et al. (2016). *Out of Sight, Out of Mind? Considering the Subsurface in Urban Planning – State of the Art*. COST TU1206 project Sub-Urban, Report TU1206-WG1-001. 49 pp. [Online: Available at: http://sub-urban.squarespace.com] (Accessed November 2020).

Watson, C., Jensen, N-P., Hansen, M. et al. (2016). *Data Acquisition and Management*. COST TU1206 project Sub-Urban, report TU1206-WG2–003. 68 pp. [Online: Available at: http://sub-urban.squarespace.com] (Accessed November 2020).

Whitbread, K., Dick, G. and Campbell, S.D.G. (2016). *The Subsurface and Urban Planning in the City of Glasgow*. COST TU1206 project Sub.Urban, report TU1206.WG1-005. 24 pp. [Online: Available at: http://sub-urban.squarespace.com] (Accessed November 2020).

Wood, B., Richmond, T., Richardson, J. and Howcroft, J. (2015). *BGS Groundhog ® desktop Geoscientific Information System External User Manual*. Keyworth, UK: British Geological Survey, Report OR/15/046. 99 pp. [Online: Available at: http://nora.nerc.ac.uk/id/eprint/511792] (Accessed November 2020).

19

Application Theme 2 – Groundwater Evaluations

Editor's Introduction

For several decades, evaluations of groundwater quality and availability at local, regional, or national scales have been the most important application of 3-D geological models. Geological surveys in Denmark and the Netherlands undertook major national geological modeling efforts to support hydrogeological evaluations, while numerous regional models were developed in the United Kingdom and the United States (see Chapters 5, 6, and 14). In the Great Lakes region of the USA and Canada, 3-D geological models have supported groundwater resource evaluations (Berg et al. 2016). Two reports (Berg et al. 2011; MacCormack et al. 2019) summarize the development of 3-D geological models for groundwater assessments by several geological surveys.

This chapter contains four case studies that demonstrate the application of 3-D subsurface information in developing local, regional, and national groundwater resource evaluations. Chapters 21 and 22 contain additional examples related to groundwater regulation and pollution assessment.

The first case study describes a model that supports the management of the municipal water supply for Uppsala, Sweden. The Uppsala esker is a key water source; with the growing population of Uppsala, safeguarding the esker's continued viability is a major concern. A 3-D hydrogeological model created following the procedures described in Chapter 10 guided development of a numerical groundwater flow model. Together, these models support municipal planning, assessment of groundwater vulnerability, and the need for protective measures.

The second case study summarizes the regional groundwater assessment by the Ontario Geological Survey (OGS) for the Orangeville-Fergus area of southern Ontario. This is one of a series of regional models undertaken by the OGS to provide geoscience information for the identification, protection, and sustainable use of the groundwater resources in a glaciated region where contamination poses a potential risk to groundwater sources of drinking water.

The third case study describes the construction of a geological model for Kent and Sussex Counties in the State of Delaware. A small team used a stacked-surface approach to create the model (see Chapter 9). It required a minimal investment by using traditional Geographic Information System (GIS) software. In this area, groundwater extracted from Coastal Plain aquifers is the only source of drinking water and is the dominant source of water used by agriculture and industry. By evaluating one shallow unconfined aquifer and 12 deeper confined aquifers the 3-D model and its associated database support geospatial assessment of groundwater withdrawals by water use category, geographic location, and source aquifer.

The fourth case study describes the national-scale REGIS II 3-D hydrogeological model for the Netherlands. Re-assessment of hydrogeological properties for the lithostratigraphic classifications and geological framework defined by a national geological model provides regional estimates of critical hydraulic parameters and their variability. REGIS II supports development of customized geohydrological models for specific research applications, approval of groundwater production permits, and environmental impact assessments for infrastructure construction, drinking water production, and geo-thermal energy.

Applied Multidimensional Geological Modeling: Informing Sustainable Human Interactions with the Shallow Subsurface, First Edition.
Edited by Alan Keith Turner, Holger Kessler, and Michiel J. van der Meulen.
© 2021 John Wiley & Sons Ltd. Published 2021 by John Wiley & Sons Ltd.

Case Study 19.1: Three-dimensional Geological Modeling of the Uppsala Esker to Evaluate the Supply of Municipal Water to the City of Uppsala

Eva Jirner[1], P-O. Johansson[2], and Duncan McConnachie[3]

[1] Swedish Geological Survey, 751 28 Uppsala, Sweden
[2] Artesia Grundvattenkonsult AB, 183 62 Täby, Sweden
[3] WSP Sverige AB, 121 88 Stockholm-Globen, Sweden

19.1.1 Introduction

The city of Uppsala, the fourth largest Swedish city, is located 71 km (44 mi) north of Stockholm (Figure 19.1). Uppsala Vatten (Uppsala Water & Waste Ltd.) provides water to the city of about 141 000 inhabitants by extracting groundwater from the Uppsala esker. With a growing population, safeguarding the esker's continued viability as the main water supply for Uppsala is a major concern for Uppsala Vatten.

The Uppsala Esker (an esker is a narrow, winding ridge composed of stratified sand and gravel deposited by a stream flowing in or under melting glacial ice) is a key water source. It trends north-east to south-west across the Uppland region and has a total length of 200 km (125 mi), a maximum width of 1 km (3300 ft), and a maximum height of 75 m (245 ft) (Johansson 2006). Most of the esker is covered with clay, providing some protection against contamination from adjacent land uses, but regulation of nearby activities is very important where the esker is more exposed. In 2013, Uppsala Vatten, in co-operation with the Swedish Geological Survey (SGU), initiated a strategic study of the central portion of the Uppsala Esker. The SGU developed a digital database and a 3-D geological model to investigate the geometry and stratigraphy of the

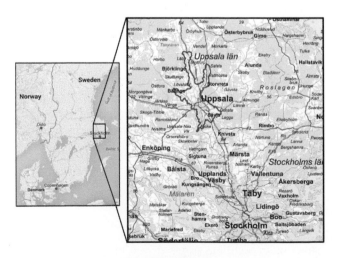

Figure 19.1   Location of the City of Uppsala, Sweden.

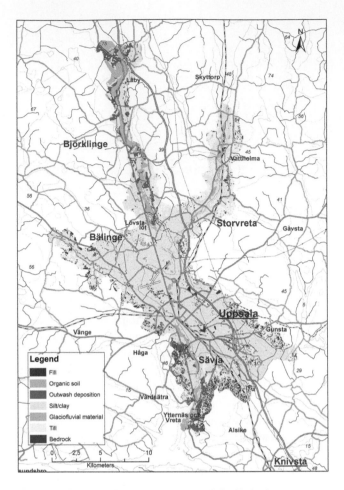

Figure 19.2   Map of the central parts of the Uppsala Esker – the study area.

Quaternary deposits of the entire groundwater catchment area. The 3-D geological model covers an area of about 300 km^2 (117 mi^2) and extends about 42 km (26 mi) in a north–south direction (Figure 19.2). Subsequently the 3-D geological model units were attributed with appropriate hydraulic properties, resulting in a conceptual 3-D hydrogeological model of the main Uppsala Esker, smaller tributary eskers, and the entire recharge area. The model revealed that the maximum thickness of the main esker to be at least 30 m (100 ft), and that it is often in direct contact with the bedrock. The tributary eskers are thinner, usually with thicknesses of 5–10 m (16–33 ft).

Groundwater conditions are controlled by a predominant flow from north to south and an annual recharge of approximately 250–300 mm/yr (10–12 in/yr) in areas where the esker outcrops. Uppsala Vatten used the conceptual 3-D hydrogeological model to develop a mathematical groundwater flow model. The process is thus similar to that described by Royse et al. (2010). The groundwater model is an important tool for evaluating the viability of current and future water supplies.

19.1.2   Development of the 3-D Geological Model

The 3-D geological model was created using *Subsurface-Viewer MX* software developed by INSIGHT Geologische Softwaresysteme GmbH. This software follows the procedures described in Chapter 10 to produce 3-D geological models (Kessler et al. 2009; Mathers et al. 2012). Subsurface conditions were defined by combining information from topographic maps, a digital terrain model (DEM), geological maps and cross-sections, exploratory boreholes, and geophysical profiles to produce a series of intersecting interpreted cross-sections.

19.1.2.1   Sources of Information

SGU had about 1000 exploratory borehole records available within the study area. Uppsala Vatten contributed a further 400 borehole records. Boring logs were manually standardized and simplified to conform to a general vertical sequence of seven material classes: bedrock, till, glaciofluvial sediment, silt and clay, outwash sand, organic soil and fill. The existing Quaternary geological map was also simplified into the same seven material classes (Figure 19.3). An existing groundwater map showed the outer limit of the aquifer and this was accepted as the limit of the glaciofluvial deposits.

In order to make the model easily manageable on desktop computers, the DEM was resampled from 2 m (7 ft) to 25 m (80 ft) resolution. The bedrock surface was modeled by combining existing information held by SGU and Uppsala Vatten, supplemented by geophysical profiles and information from older reports.

19.1.2.2   Creation of Interpreted Cross-Sections

Creation of the interpreted cross-sections within the *SubsurfaceViewer MX* software followed some general principles:

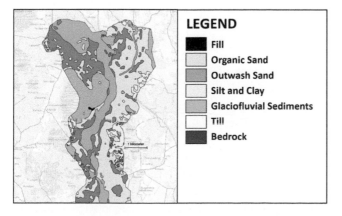

Figure 19.3   Map of the seven simplified stratigraphic units in the northern extent of the study area (Fig. 19.2).

- The geological map is always correct. If a borehole record showed a different top unit, the map definition was followed.
- The soil thickness in a borehole record takes precedence over the top-of-the-bedrock map, since not all currently available information was available when those maps were created.
- The geologist's "expert knowledge" of the esker always takes precedence.

East–west trending cross-sections were developed first, starting in the north and working southwards. The cross-sections were spaced at intervals of approximately 500 m (1650 ft). Alternate cross-sections extended across the entire model area, with intermediate shorter sections extending across the aquifer (Figure 19.4). Cross-sections were reviewed by a group of geological experts, then revised, and re-checked until they were approved. North–south cross-sections were developed next and these were reviewed to ensure aquifer continuity in the direction of groundwater flow.

19.1.2.3   Development of 3-D Volumetric Geological Model

Almost 200 approved cross-sections defined the thickness of the units. The *SubsurfaceViewer MX* software also requires the definition of the lateral extent of each unit to compute a 3-D volumetric model. The geological map defined the extents of the organic soil and fill. The extents of outwash sand and the silt and clay unit were partly based on the geological map, but were modified by geological inference to reflect their occurrence below younger units. The extent of glaciofluvial sediment was defined by a groundwater aquifer map supplemented by additional geological interpretation. The till unit was interpreted to exist below younger units throughout the area except at bedrock outcrops, or at glaciofluvial aquifer locations. Figure 19.5a shows the extent of the glaciofluvial sediments (forming the main aquifer), while Figure 19.5b shows the complete 3-D volumetric model.

19.1.3   Transfer of the Geological Model to the Mathematical Groundwater Flow Model

At the time of the project, the software selected for the groundwater modeling, *FEFLOW* (version 6.2), required layers to be continuous throughout the model area (DHI-WASY 2012). To create these continuous layers the tops of all geological units were exported as ASCII files from SubsurfaceViewer for post-processing in ArcGIS using tools in the Spatial Analyst extension (ESRI 2010; A.H. Pasanen, personal communication). In the applied methodology, fictitious layers with a minimum thickness were created where strata in the general vertical sequence

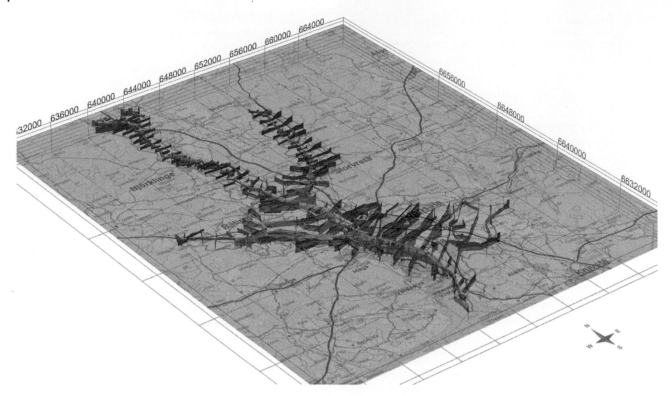

Figure 19.4 Complete set of interpreted cross-sections. Source: Cross-sections shown with 10× vertical exaggeration.

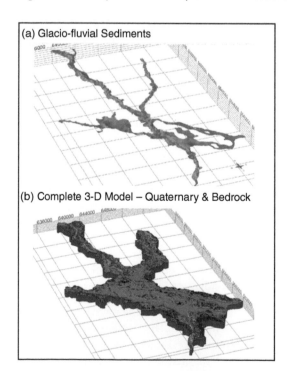

(a) Glacio-fluvial Sediments

(b) Complete 3-D Model – Quaternary & Bedrock

Figure 19.5 Two views of the completed 3-D volumetric geological model.

were missing. For instance, in areas of bedrock outcrops corrections were required so that the other six younger strata were included above the bedrock while the bedrock

was lowered by the sum of the minimum thicknesses. This was carried out working from top to bottom. In the next step the correct geological units were assigned to the fictitious layers. This task was also achieved in ArcGIS, working from the uppermost layer downwards in the geological sequence. For each layer the fictitious areas were identified and assigned the same geological unit as the first authentic layer found below. Initially, the minimum thickness of the fictitious layers was set to 0.1 m. However, these thin layers caused numerical problems, which were resolved by increasing the minimum thickness to 0.5 m. In December 2015, DHI-WASY launched *FEFLOW* Version 7.0 (DHI-WASY 2015) which supports the use of unstructured meshes and thus allows for discontinuous layers.

The elevations of the top of all layers were exported from ArcGIS to *FEFLOW* as point shapefiles, with each point representing a raster cell. In *FEFLOW*, the imported files were used to assign elevations to the nodes of the finite-element mesh by the available parameter association and interpolation facilities. The geological units of all layers were exported from ArcGIS as polygon shapefiles while their hydraulic properties (K-and S-values) were obtained from the database. The parameter association facility in *FEFLOW* was then used to assign hydraulic properties to the geological units. Figure 19.6 shows comparison

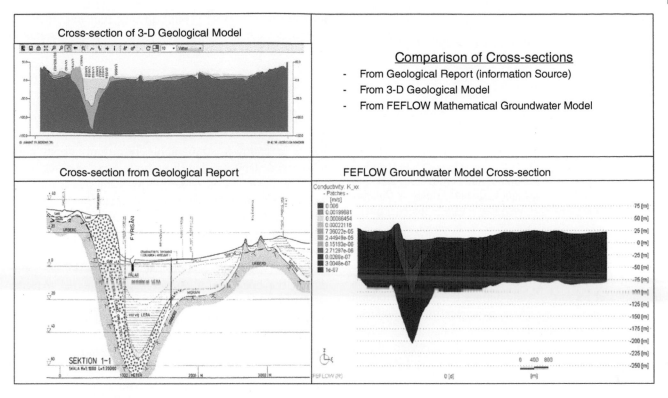

Figure 19.6 Comparison of cross-sections at three stages of model development.

cross-sections from a geological report, the 3D geological model, and *FEFLOW*.

19.1.4 Conclusions

A general experience from the modeling process is that geological expertise is essential, in addition to a well-defined working procedure and quality control at each step of the process. In particular, a sound knowledge of the geology of the model area is critical when deciding on the quantity and location of the cross-sections.

It is important that all the data used in a model are well-structured, harmonized, and have metadata. The collection of borehole data and the adaptation of the logs to the simplified general vertical geological sequence was a work-intensive part of the process.

The geological model can be used as a tool in municipal planning for assessing groundwater vulnerability and the need for protective measures in connection with different types of exploitation. A 3-D understanding of the geological conditions is, for example, important when assessing the influence on groundwater levels, flows, and quality from underground constructions; the removal or puncture of confining layers, or the accidental release of contaminants. Cross-sections and 3-D illustrations can readily be exported from the model to be used in presentations, as a basis for discussions and as a decision support.

One of the main objectives for the present study was to develop a geological model which could be used in a mathematical groundwater model requiring continuous layers. A procedure was developed in ArcGIS which enabled an efficient transfer of GRID-files of top and bottom of the geological units exported from SubsurfaceViewer to a format that could be directly imported to the groundwater model (*FEFLOW* 6.2).

Case Study 19.2: Three-dimensional Geological Modeling of the Orangeville-Fergus Area to Support Protection of Groundwater Resources

Abigail Burt

Ontario Geological Survey, Sudbury, ON P3E 6B5, Canada

19.2.1 Introduction

The OGS is undertaking regional scale 3-D hydrostratigraphic modeling of Quaternary sediments to provide geoscience information for the identification, protection, and sustainable use of the provincial groundwater resource. The 37×42 km (23×26 mi), or 1550 km^2 (600 mi^2) Orangeville-Fergus 3-D map area is located in southwestern Ontario (Figure 19.7), northwest of the city of Toronto

Figure 19.7 Location of the Orangeville-Fergus study area. (Burt and Dodge 2016). Copyright © Queen's Printer for Ontario, 2016. Used by permission of the Ontario Geological Survey.

and northeast of the city of Waterloo. Data acquisition was undertaken between 2008 and 2011, followed by modeling, generation of digital outputs and reporting. The resulting groundwater resources study defines the 3-D distribution and characteristics of surficial geological materials forming aquifers and aquitards (Burt and Dodge 2016). A free, publicly available digital release provides access to all data and interpretative products (Burt and Dodge 2016).

19.2.2 Protection of Groundwater Supplies

In May 2000, the water supply of the small community of Walkerton (Figure 19.7), became contaminated with *Escherichia coli* bacteria. Seven people died and over 2000 were sickened. A public inquiry, known as the Walkerton Commission, made recommendations for providing better protection of water supplies. Key recommendations were the development of science-based assessment reports and the implementation of watershed-based source water protection plans for all regions of the province (O'Connor 2002a; 2002b).

In response, the OGS is producing a series of robust, regional-scale 3-D maps of Quaternary deposits within areas of southern Ontario that rely on groundwater for a significant portion of their potable water supply. The importance of understanding the stratigraphic context and internal sediment characteristics of large moraine systems when developing groundwater flow models has long been recognized (Howard et al. 1995). The Orangeville-Fergus project was initiated because: (i) the Orangeville Moraine forms both a regional aquifer recharge area and the headwaters for three major watersheds, thus influencing source water protection planning for three conservation authorities; (ii) the area encompasses protected green-belt as well as non-greenbelt areas which are expected to see increased population growth; and (iii) the city of Orangeville is targeted for further population growth under the Places to Grow Act (Burt and Dodge 2016). This Act focuses on

intensification of existing built-up areas and, in particular, areas that offer municipal water and wastewater systems as a way to accommodate growth.

19.2.3 Geologic Setting

Multiple cycles of glacial advance and retreat occurred across Ontario throughout the late Pleistocene, resulting in a complex stratigraphic record of glacial and non-glacial deposits that host both local and regional aquifers and aquitards. Older tills and stratified sediments are overlain by Late Wisconsin Catfish Creek Till, an important stratigraphic marker in southwestern Ontario (Bajc and Shirota 2007). The Orangeville Moraine is composed of fine-grained glaciolacustrine sediments and diamicton which were deposited in an interlobate zone during the break-up of the ice sheet. Later ice-margin oscillations resulted in deposition of till that partially buried the Orangeville Moraine. These deposits overlie southwest-dipping Ordovician and Silurian aged bedrock formations which are separated by the prominent Niagara Escarpment (Armstrong and Dodge 2007). The eastern side of the area is characterized by a series of wide bedrock valleys that extend back from the Niagara Escarpment as shown on Figure 19.8. The valleys are incised to depths of as much as 200 m (650 ft) below the surrounding bedrock surface and are partially infilled with Quaternary sediments. They are clearly visible near the escarpment where modern streams have partially eroded the valley fill. Away from the escarpment, the valleys are narrower and are completely buried with Quaternary sediments (Burt and Dodge 2016).

19.2.4 Modeling Workflow

The Orangeville-Fergus 3-D model was developed in four stages: (i) data acquisition, compilation, and standardization; (ii) development of a conceptual geological framework; (iii) model creation; and (iv) generation of output files and products.

19.2.4.1 Data Acquisition, Compilation, and Standardization

The 3-D model is based on legacy datasets and new geophysical and geological data. Originally gathered for a variety of purposes, legacy datasets have a highly variable spatial distribution, resolution, and quality. Publicly available water well records were the primary data source (10 336 records), supplemented by geotechnical borehole logs, aggregate exploration logs, and both archived and new field descriptions. Each record was assigned a quality rating of "low," "medium," or "high" based on confidence in the source data (Burt and Dodge 2016). These ratings were

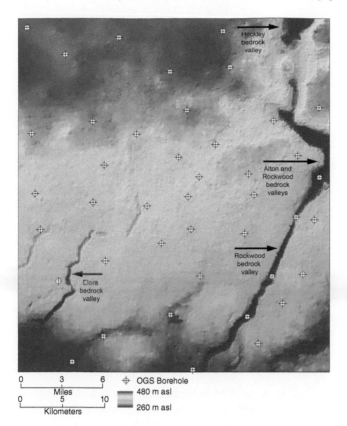

Figure 19.8 Structural contour map of bedrock surface draped over a hillshade created from the 3-D model. Note the bedrock valleys that widen toward the Niagara Escarpment on the east of the map area. (Burt and Dodge 2016). Copyright © Queen's Printer for Ontario, 2016. Used by permission of the Ontario Geological Survey.

Figure 19.9 Distribution of subsurface data points, showing the preponderance of low-quality data points (blue) and sparsity of medium- and high-quality data points (orange and green). (Burt and Dodge 2016). Copyright © Queen's Printer for Ontario, 2016. Used by permission of the Ontario Geological Survey.

used to weight the data during the interpolation process. Material descriptors were standardized into primary geological materials such as sand, silt, clay, and diamicton to allow merging of the individual datasets. The resulting master subsurface database contains 11 074 legacy records identifying 42 267 geological layers. Figure 19.9 shows the distribution of the subsurface data points, including 43 new continuously cored boreholes described in the following paragraphs. The preponderance of low-quality data points, shown in blue, is readily seen.

A high-quality surface material database was created to ensure that the 3-D model reflected the materials mapped at the ground surface. First, the provincial digital elevation model was sampled on a 100 m grid spacing, resulting in a point file with X, Y, and Z data. Next, the digital seamless surficial geology map (OGS 2010) was recoded using the hydrostratigraphic units being modeled. The re-coded map was then used to attribute the point file.

An extensive network of buried-bedrock valleys had previously been identified in southwestern Ontario (for example Gao et al. 2007); however, the sporadic distribution

and low quality of the source data prevented a precise definition of valley trends. A highly successful trial ground-based gravity survey was conducted to evaluate whether the bedrock and valley fill lithologies would allow the location of the valley network to be refined geophysically (Burt and Rainsford 2010). This survey was later expanded to a total of 54 lines with 6868 stations along 723 km of roads crossing predicted valleys (OGS 2013). Ground-based gravity survey methods are explored further in Chapter 7.

Drilling targets were selected to refine the stratigraphic relationships of tills and associated stratified sediments, establish sediment-landform associations and determine the nature of the bedrock valley fill (Burt 2008, 2009, 2011). Forty-three 8.5 cm diameter continuously cored boreholes totaling 1918.5 m (6292.7 ft) were drilled through the Quaternary sediment cover and upper 3 m (10 ft) of bedrock. The core was logged, photographed and representative samples obtained for grain size, carbonate, and heavy mineral analysis, radiocarbon dating and pollen identification. The resulting high-quality dataset allowed the low-quality datasets to be interpreted more accurately (Burt and Webb 2013).

19.2.4.2 Development of the Conceptual Geological Framework

Simultaneously, a conceptual framework subdividing the Quaternary sediment cover into 16 regional-scale hydrostratigraphic units and one undifferentiated bedrock unit was developed (Figure 19.10). These units were identified on the basis of age and the sediment characteristics resulting from deposition in different environments (Burt 2012). The conceptual geologic framework functions as a guide for the interpretation of legacy data sets.

19.2.4.3 Model Creation

The 3-D model was generated with commercially available Datamine Studio. Three tables (a location table, a formation table, and a static water depth table) were created from the master subsurface database and imported into the modeling software. The location table defines the location, elevation, drilling method, and quality rating for each borehole. The formation table contains the depths and colors of the standardized primary geological materials. These tables were used to generate borehole traces that can be viewed as 2-D cross-sections or in 3-D space. The static water depth table defines the depths of water well-screens and static water levels. Viewed as a series of flags on the borehole traces, the data aided in differentiating between aquifers.

Three-dimensional points identifying the top of a given hydrostratigraphic unit were manually digitized onto the borehole traces (Figure 19.11). The number of "picks" made on an individual borehole trace was dependent on the number of hydrostratigraphic units that could be interpreted from the primary material types. This process automatically generated a "picks table" containing the X, Y, Z location of the pick, the hydrostratigraphic unit, and the quality rating of the borehole trace. Additional picks were digitized from the borehole traces to refine the geometry of the modeled surfaces. Off-trace picks were assigned high-quality ratings because they reflect the geologist's expertise. In total, over 50 000 picks were digitized (Burt and Dodge 2016).

Once a preliminary set of picks had been made, a set of scripted routines were used to generate interpolated wireframe surfaces representing the tops of each hydrostratigraphic unit. The interpolation method used for OGS 3-D modeling is the isotropic inverse power of distance cubed (Bajc and Shirota 2007). First, the picks were validated by the modeling software to ensure that the hydrostratigraphic units were in the correct sequence. Out of order picks were flagged for correction and then ignored by the software during the current run.

In order to reduce problems with overlapping wireframe surfaces, especially in data sparse areas, hydrostratigraphic unit elevations were interpolated onto a 100 m² (330 ft²) grid of virtual boreholes. During this process, each hydrostratigraphic unit was considered individually.

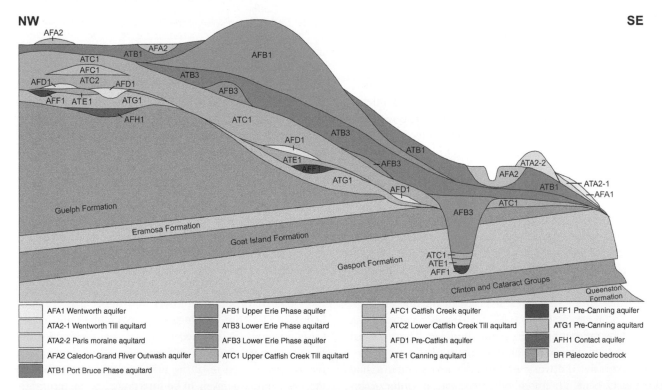

Figure 19.10 Conceptual geologic framework for the Orangeville-Fergus area. (Burt 2012). Copyright © Queen's Printer for Ontario, 2012. Used by permission of the Ontario Geological Survey.

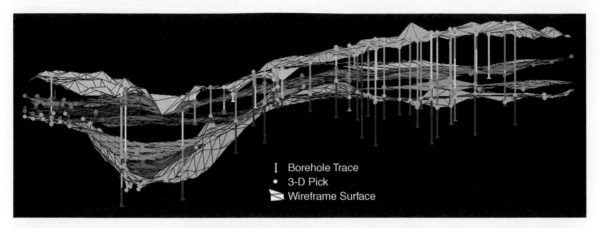

Figure 19.11 A subsection of the 3-D model showing borehole traces, 3-D "picks" of tops of hydrostratigraphic units, and interpolated wireframe surfaces. (Burt and Dodge 2016). Copyright © Queen's Printer for Ontario, 2016. Used by permission of the Ontario Geological Survey.

A search radius, defined by the geologist to reflect the perceived extent of the unit, was drawn around each virtual borehole in turn. An interpolated elevation was assigned to the virtual borehole when a minimum of one high-quality, two medium-quality, or three low-quality picks were found within the given search radius. If there were insufficient picks, the unit was considered to be absent and the elevation was automatically set as equal to the underlying unit. This process resulted in all 17 hydrostratigraphic units being represented in each virtual borehole. Finally, a continuous set of wireframes were generated from the interpolated elevations and a set of rules was applied to remove any overlaps. Pinch-outs were accommodated by draping units on top of each other. These continuous surfaces are a requirement of some groundwater flow modeling software.

A second set of wireframes were created by comparing the elevations for each triangular element. Locations where any unit had zero thickness were identified and those triangular elements were removed. This introduced holes, or pinch-outs, into the surfaces and more accurately represented the spatial extent of the units.

A 3-D block model was created by filling the space between each modeled wireframe surface with columns of blocks (Figure 19.12). The modeling software allows the planar dimensions of the columns to be defined by the user, but 100 m² (330 ft²) provided a good resolution for this 3-D model while generating output files of a reasonable size. The vertical dimension of each column was calculated automatically so that it fitted exactly between the surfaces. The resulting block model was used to calculate the volume, thickness, and surface contour of each unit.

Once a model run was complete, the results were visually compared with the borehole traces. Out-of-order picks were corrected, interpretations refined, and additional off-trace picks added. The model was then re-run. Early model runs were done on small 5 km² (3 mi²) subareas to speed up the process and then expanded to encompass the entire map area.

19.2.4.4 Generation of Model Outputs and Products

A variety of digital products were produced, including subsurface and analytical databases, a comprehensive report, cross-sections, and maps (Burt and Dodge 2016). Many of the data files were designed as inputs to other software packages, such as hydrogeological modeling or visualization software. Comma-separated value (.csv) files of the x, y, and z coordinates of each hydrostratigraphic unit on a 100 m² (330 ft²) grid were generated from the block model. Many of the graphic outputs were direct products of the modeling software, exported as pdf (portable document format) files using Adobe Acrobat. These include thickness and surface contour maps for each hydrostratigraphic unit and cross-sections selected to illustrate the stratigraphy of the units. ESRI ArcInfo was used to produce georeferenced thickness and structural contour maps viewable using the Google Earth mapping service. An interactive cross-section viewer was developed to allow generation of user-defined cross-sections. The cross-sections can be saved, then imported into the Google Earth mapping service for interactive viewing.

19.2.5 Model Application to Groundwater Protection

A key objective of the project was the identification of groundwater recharge areas and potentially vulnerable aquifers and, to this end, maps showing the depth, identification, and elevation of the first aquifer encountered were produced. Figure 19.13 shows aquifers exposed at the

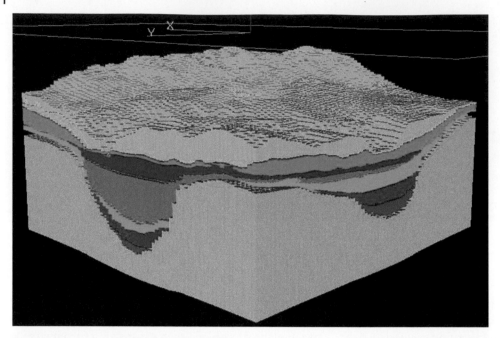

Figure 19.12 Three-dimensional block model for 5 × 5 km subarea. (Burt and Dodge 2016). Copyright © Queen's Printer for Ontario, 2016. Used by permission of the Ontario Geological Survey.

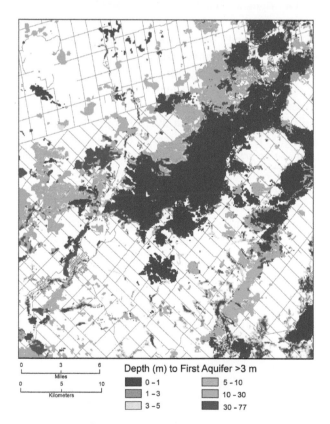

Depth (m) to First Aquifer >3 m

- 0 – 1
- 1 – 3
- 3 – 5
- 5 – 10
- 10 – 30
- 30 – 77

Figure 19.13 Example of aquifer vulnerability map. Red areas are locations with aquifers at the surface (less than 1 m cover). (Burt and Dodge 2016). Copyright © Queen's Printer for Ontario, 2016. Used by permission of the Ontario Geological Survey.

surface in red; while yellow, green, and blue colors indicate progressively thicker confining layers. This permits a quick assessment of recharge potential and vulnerability to surface contamination. Areas that have thick, laterally continuous aquifers mapped at depth (green and blue colors) are potentially valuable water sources protected from surface contamination. However, hydrogeological testing is required to establish whether these aquifers are being actively recharged.

For example, the thick, centrally located Orangeville Moraine is the largest surface aquifer in the area (Figure 19.13). Most of the moraine is high on the landscape so it will be recharged from local precipitation and, while it is highly vulnerable to contamination, much of that will be from local sources. This study demonstrates that the aquifer extends considerable distances in the subsurface and, even where deeply buried, may be susceptible to contamination that occurs on the exposed portion of the moraine. Gravel-filled channels and fine-textured sediment layers within the moraine can be expected to produce complex groundwater and contaminant flow paths.

19.2.6 Conclusions

This study produced a digital data release with many components derived from the 3-D model. The data may be viewed and used in a variety of other software products, including Google Earth mapping service, ESRI ArcInfo, and groundwater flow modeling software. The information provides guidance for the development of watershed-based source water protection plans and land use planning.

Case Study 19.3: Successful Construction of a 3-D Model with Minimal Investment: Modeling the Aquifers for Kent and Sussex Counties, State of Delaware

Peter P. McLaughlin, Jr., Jaime Tomlinson, and Amanda K. Lawson

Delaware Geological Survey, University of Delaware, Newark, DE 19716, USA

19.3.1 Introduction

Most of the State of Delaware, including all of Kent and Sussex Counties (Figure 19.14), is underlain by Atlantic Coastal Plain sediments. These consist of near-surface Quaternary and late Cenozoic deposits that are predominantly cut-and-fill alluvial and estuarine units underlain by a succession of southeast dipping Cenozoic to Cretaceous sediments. In Kent and Sussex Counties, groundwater

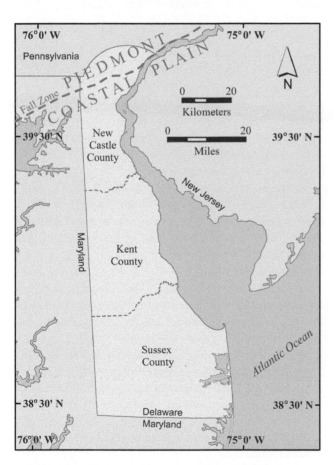

Figure 19.14 Location map identifying Kent and Sussex Counties in the State of Delaware. Source: Lillian Wang, Delaware Geological Survey.

extracted from Coastal Plain aquifers is the only source of drinking water and is the dominant source of water used by agriculture and industry (McLaughlin et al. 2016).

Over the last decade, the Delaware Geological Survey has developed and refined a 3-D subsurface geological model for Kent and Sussex Counties that defines the spatial relationships of 12 confined aquifers and one surficial unconfined aquifer (Table 19.1). The 3-D model provides essential information for assessing groundwater withdrawals and supports future groundwater management planning. This effort demonstrates the ability to construct a very useful 3-D model with limited manpower and financial investment, as the geological interpretation and modeling were done by two geologists and without the benefit of specialized modeling software.

19.3.2 Three-dimensional Model Construction

The relatively gently dipping and undeformed stratigraphy of the coastal plain sediments facilitated the creation a 3-D geological model. A series of gridded surfaces representing the tops and bottoms of each aquifer were created for the study area using ArcGIS (ESRI 2015) and Surfer (Golden Software 2015). These grids were assembled to create a "stacked-surface" 3-D geological framework model (Figure 19.15) by following the techniques discussed in Chapter 9.

One key to efficient 3-D model building was the development of a stratigraphic database. The database utilized borehole data from over 6600 sites, including geophysical borehole logs, geologist's logs, and driller's logs. The log from each borehole and well was examined and stratigraphic contact elevations ("picks") were identified. The driller's logs were the most abundant data type but commonly had quality issues and required careful evaluation, resulting in lower confidence picks that were in some cases revised or excluded from the analysis. The geophysical logs and geologist's logs were less numerous but provided higher quality information and a basis for higher-confidence aquifer picks. The resulting database contained over 14 000 stratigraphic pick records. The locations of all boreholes and wells were carefully evaluated using aerial photographs and original maps.

Gridded surfaces with a 100 m grid resolution defined the top and base of each aquifer. They were constructed from the pick elevations using the ArcMap Geostatistical Analyst Radial Basis Function. To minimize edge effects, gridding included borehole data from just outside of the study area. The initially generated grids were checked and, where necessary, corrected to ensure they did not violate three-dimensional geographic constraints. This was done by comparing each aquifer grid to (i) the digital elevation

Table 19.1 Aquifers underlying Kent and Sussex Counties, Delaware.

Age	Geologic formation	Aquifer		Usage area by county
Quaternary	Various	Mostly unconfined		Kent and Sussex
Upper Miocene-Pliocene	Beaverdam Fm.			
Upper Miocene	Bethany Fm.	(12) Pocomoke aquifer	Confined aquifers	Sussex
	Cat Hill Fm.	(11) Manokin aquifer		Sussex
Middle Miocene	Choptank Fm.	(10) Upper Choptank aquifer		Southern Kent and Northern Sussex
		(9) Middle Choptank aquifer		
		(8) Milford aquifer		
Lower Miocene	Calvert Fm.	(7) Frederica aquifer		Kent
		(6) Federalsburg aquifer		
		(5) Cheswold aquifer		
		(4) Lower Calvert aquifer		
Eocene	Piney Point Fm.	(3) Piney Point aquifer		Central and Southern Kent
	Manasquan Fm.	(2) Rancocas aquifer		Northern Kent
Paleocene	Vincentown Fm.			
Upper Cretaceous	Mt Laurel Fm.	(1) Mt. Laurel aquifer		Northern Kent

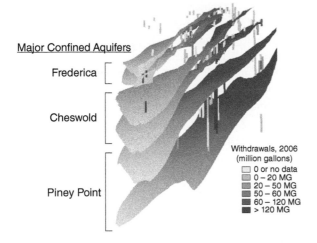

Major Confined Aquifers

Frederica

Cheswold

Piney Point

Withdrawals, 2006 (million gallons)
- ☐ 0 or no data
- ☐ 0 – 20 MG
- ☐ 20 – 50 MG
- ☐ 50 – 60 MG
- ☐ 60 – 120 MG
- ☐ > 120 MG

Figure 19.15 A 3-D image looking northwest showing the intersection of well screens of public wells, rated by pumping, mapped against the elevations of three important confined aquifers in Kent County. The complete 3-D model contains surfaces for 12 confined aquifers and the unconfined surficial aquifer. (McLaughlin et al. 2016).

model of the land surface, (ii) the immediately adjacent confined aquifer grids (first grids above and below) and (iii) the unconfined aquifer grid. The unconfined and confined aquifer grids were created using slightly different approaches.

The unconfined aquifer is made up of sands from any of several near-surface formations and occurs throughout the study area. These Pleistocene and latest Cenozoic formations all exhibit a degree of erosion at their bases, resulting in highly variable thicknesses for the unconfined aquifer. The elevation of the base of unconfined aquifer in Kent County was mapped by calculating a grid using picks from driller's logs, geophysical logs, and geologist's logs from nearly 1900 boreholes. An existing grid of the elevation of base of the unconfined aquifer for Sussex County (Andres and Klingbeil 2006) was merged with the Kent County grid to create a seamless map for the entire study area. The thickness of the unconfined aquifer was computed by subtracting the grid elevations of the base of the unconfined aquifer from the digital elevation model of the land surface. The resulting maps show the unconfined aquifer to generally be less than 100 ft thick in Kent County but to vary from a few feet to more than 200 ft thick in Sussex County.

Twelve confined aquifers occur in Kent and Sussex Counties, ranging in age from upper Miocene to Cretaceous. The confined aquifers generally have simpler and less variable stratigraphic geometries than the unconfined aquifer so could be reliably mapped using more widely spaced data points, preferentially those with higher-quality data, such as geophysical logs. Grids defining the elevations of the top and base of each confined aquifer were developed and compared to the grid of the base of the unconfined aquifer. Confined aquifer thicknesses were calculated by subtracting base-elevation grid values from top-elevation grid values for each aquifer. The youngest confined aquifers, the Pocomoke and Manokin, occur in an upper Miocene interval with fairly heterogeneous lithologies. As a result, the confining layers between the Manokin and Pocomoke aquifers and the unconfined aquifer are commonly poorly developed or absent, resulting in hydrological connections between these aquifers in many parts of Sussex County (Figure 19.16). The older confined aquifers below the Pocomoke and Manokin have simpler and less variable stratigraphic geometries. These older confined aquifers are typically tens of feet thick and occur at progressively greater depths southeastward from their northern recharge areas (Figure 19.16).

19.3.3 Model Applications

Two shallow confined aquifers are important groundwater sources in Sussex County. The lower of the pair, the Manokin aquifer, subcrops under sandy surficial formations across a wide belt of northern Sussex County and thickens to more than 100 ft toward the southeast. The upper Pocomoke aquifer subcrops in southeastern Sussex County and has a net thickness of more than 100 ft in some coastal areas. By locating the grid intersections of each confined aquifer with the base of the unconfined aquifer, "windows" between the unconfined and confined aquifer were identified (Figure 19.17). These areas are potential recharge areas to the lower confined aquifer. They also indicate potential pathways for near-surface groundwater contaminants to reach otherwise confined and protected aquifers.

The 3-D model and the database provided a means for geospatially evaluating groundwater withdrawals by water-use category, geographic location, and source aquifer. Nine categories of wells were considered, ranging

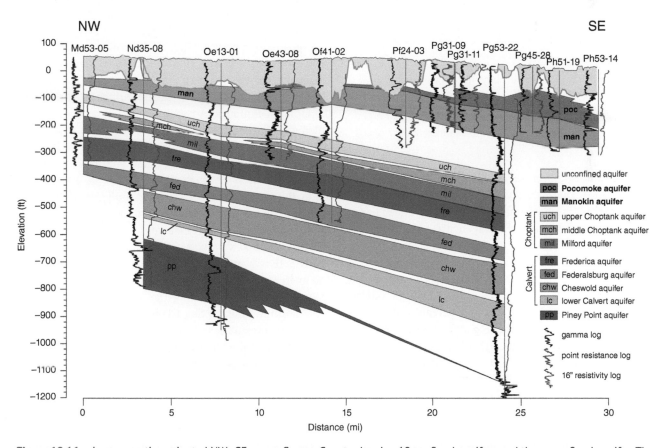

Figure 19.16 A cross-section oriented NW–SE across Sussex County showing 10 confined aquifers and the unconfined aquifer. The unconfined aquifer has direct connections to the confined Manokin and Pocomoke aquifers in this area. Figure 19.17 maps the critical Manokin subcrop areas. The Frederica, Cheswold, and Piney Point aquifers are shown in a 3-D view in Figure 19.15. (McLaughlin et al. 2016).

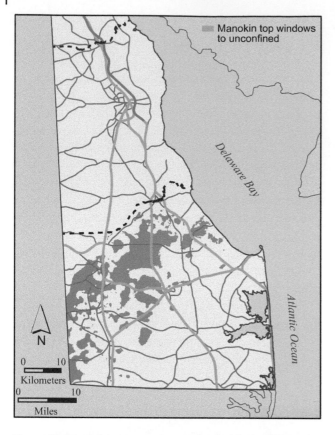

Figure 19.17 Subcrop map of the Manokin Aquifer. These areas are important for estimating recharge from the unconfined aquifer and also are areas where contaminants from the surface may be introduced to a confined aquifer. The patchiness of the subcrop extent is a reflection of the variable thickness of the near-surface formations in which the unconfined aquifer occurs. (McLaughlin et al. 2016).

from large public-community systems to individual domestic supplies (McLaughlin et al. 2016). Withdrawal volumes by the larger public and industrial users were tallied based on pumping reported to state agencies, whereas water use was estimated based on demographic, agricultural, or business data for a defined geographic area (usually census block) for the other categories. Withdrawals were assigned to exact geographic coordinates determined for public, industrial, and golf-course wells, whereas withdrawals for other categories with more poorly constrained well locations (such as agricultural irrigation and domestic use) were assigned geographically to the center point of the census block they occur within. By comparing well-screen elevations at these locations and estimated or reported withdrawal rates for the wells or defined area, it was possible to produce a regional analysis of groundwater withdrawals, both by geography and by source aquifer (McLaughlin et al. 2016).

The results of this geospatial analysis highlight the value of integrating analysis of groundwater withdrawals

with a high quality subsurface 3-D geological model. The Delaware Geological Survey and partner state agencies are currently using the subsurface geological model and the compiled database as tools for more detailed analyses of site- or problem-specific questions, including the development of numerical groundwater-flow models.

Case Study 19.4: REGIS II – A 3-D Hydrogeological Model of the Netherlands

Ronald W. Vernes, Willem Dabekaussen, Jan L. Gunnink, Ronald Harting, Eppie de Heer, Jan H. Hummelman, Armin Menkovic, Reinder N. Reindersma, and Tamara J.M. van de Ven

TNO, Geological Survey of the Netherlands, 3584 CB Utrecht, The Netherlands

19.4.1 Introduction

Until the late 1980s Dutch hydrogeological information was provided as a series of maps and reports forming the "Grondwaterkaart van Nederland" (Groundwater Map of the Netherlands). Initial attempts to convert this information to a digital map format began in the early 1990s with the goal of forming a "REgional Groundwater Information System (REGIS)." However, these attempts suffered from inconsistencies between geology and hydrogeology information sources and from limitations of early digital mapping techniques. Advances in digital technology, coupled with a new lithostratigraphic classification, led to the development of a completely revised approach to hydrogeological information management and the development of the REGIS II product, first released in 2005. REGIS II evolved from separate mapping assignments from the provincial authorities, together with the former Institute for Inland Water Management and Waste Water Treatment (RIZA) of Rijkswaterstaat (now Waterdienst) and TNO. REGIS II is a hydrogeological model describing the 3-D geometry and hydraulic properties of the subsurface of the Dutch mainland, including the IJsselmeer, the Wadden Sea, and the Wadden Sea Islands as well as Zeeland and the (former) estuaries.

REGIS II relies on the lithostratigraphic classification and geological interpretations contained within the Digital Geological Model (DGM) to define hydrogeological entities (Vernes and van Doorn 2005). REGIS II and DGM versions 2.2 are both based on the same set of around 26 500 moderately deep boreholes and both cover an average depth range to 500 m (1200 m in the center of the Roer Valley Graben). While the DGM contains 34 model units corresponding with Cretaceous, Paleogene, Neogene, and Quaternary lithostratigraphic units, hydrogeological interpretation of

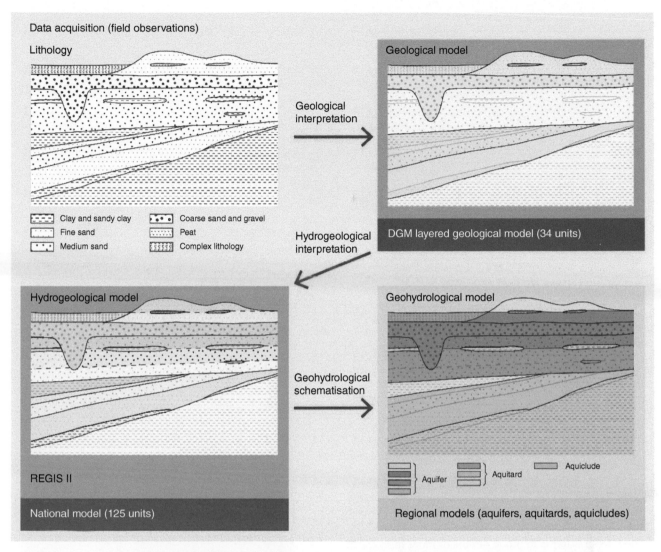

Figure 19.18 The underlying REGIS II concept; geological interpretation of field observations creates a national lithostratigraphic model (DGM); hydrogeological interpretation of the DGM model produces the REGIS II national hydrogeological model; geohydrological schematization produces regional geohydrological models for national and regional authorities defining aquifers, aquitards, and aquicludes. Modified from Vernes and van Doorn (2005).

the DGM model produced a national REGIS II hydrogeological model with 125 units (Figure 19.18). While the majority of the hydrogeological units are Neogene and Quaternary lithostratigraphic units, in South-Limburg and along the borders with Belgium and Germany, older deposits are found at the near-surface and are important for groundwater management. Further analysis within the REGIS II process assigned hydraulic property parameters to these units.

19.4.2 Hydrogeological Setting

The shallow subsurface of the Netherlands is composed almost entirely of Quaternary and Neogene sediments. These include Pliocene-Miocene marine clays and sands,

which are overlain by Pleistocene marine sediments. In most locations, these are not sources of drinking water as they mostly have very low permeability or contain brackish or saline water (Figure 19.19). However, to the south and east, these sediments are located at shallow depth, are coarser, have a higher permeability, often contain fresh water and thus are a supply of drinking water. Younger Pleistocene and Holocene continental sediments include considerable volumes of sand, interspersed with clay, peat, and lignite layers. The younger Pleistocene and Holocene units record the migrating channels of the Rhine and Maas rivers and the former Baltic river system, the interplay of alluvial deposition and marine incursions due to gradual sea level rise, and several glacial-interglacial stages. The Holocene units record various anthropogenic influences

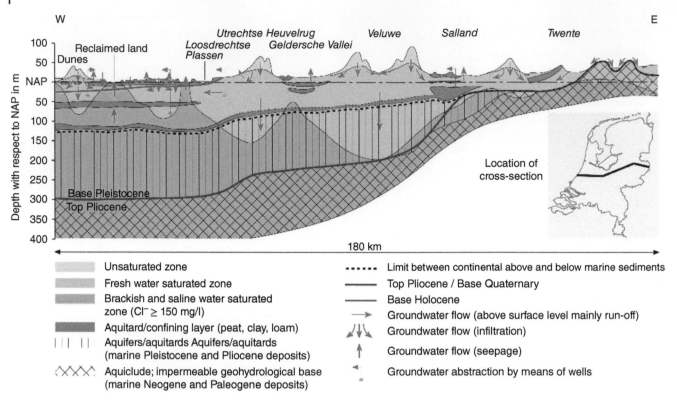

Figure 19.19 East–West cross-section across the Netherlands showing the typical groundwater conditions in the shallow subsurface. Modified from Van de Ven et al. (1986) and Dufour (2000).

on the shallowest parts of the subsurface. Drinking (fresh) groundwater supplies (defined as less than $150\,mg\,l^{-1}$ Chloride) are almost entirely found within these younger units, mostly within the upper 50–100 m; in areas like the sandy ice-pushed ridges of the Veluwe and the Roer Valley Graben, fresh water is found at depths of more than 250 m (Figure 19.19).

19.4.3 Developing the REGIS II Subsurface Models

Figure 19.20 shows the REGIS II workflow. For the initial steps, REGIS II uses the lithostratigraphic classification and geological interpretations of the DGM as a framework for distinguishing hydrogeological entities. The DGM is built as a stacked series of gridded surfaces (Gunnink et al. 2013) following procedures discussed in Chapter 9. It not only honors information from 26 500 boreholes, it also reflects knowledge of the geological evolution of the Netherlands. Shallow boreholes, those less than 10 m depth, are not used in REGIS II, except in the Province of Zeeland where a detailed hydrogeological model of the Holocene top layer was developed, and in northern Netherlands where mapping of the hydrologically important

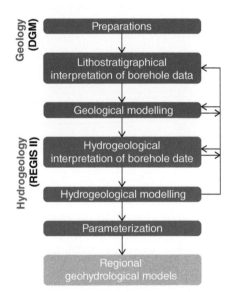

Figure 19.20 REGIS II model development workflow.

boulder clay required the use of shallow boreholes. Thus, at present, REGIS II mostly ignores the topmost 10 m. GeoTOP, a nationwide 3-D cellular geological model discussed in Chapter 11, is in the process of reaching national coverage and is planned to be attributed with geohydrological parameters. The REGIS II model creation process continues with the hydrogeological interpretation, modelling and parameterization steps shown in Fig. 19.20.

During the hydrogeological interpretation of the borehole data, each lithostratigraphic interval in a borehole record is subdivided into one or more of 14 lithological classes (lignite, peat, clay, sandy-clay, clay with unspecified sand content, fine sand, medium sand, coarse sand, sand with unspecified grainsize, glauconite sand, gravel, shells, chalk and marl, and sandstone and siltstone). Based on the hydraulic properties of the lithological classes and the thicknesses of the lithological layers in the boreholes, layers with corresponding properties are combined. In order to avoid inconsistencies and to assist easy reproduction, this aggregation process is fully automated. The results of the aggregation step are then interpreted manually. This process ultimately subdivides the 34 DGM lithostratigraphic units into 125 hydrogeological units (Figure 19.21). Based on their main lithology six types of hydrogeologic units are defined: sand, clay, peat, lignite, chalk, and complex. The overall procedure ensures that individual hydrogeological units always belong to a single geological unit.

Parameterization of the hydrogeological model is the next step. Hydraulic conductivity, in the horizontal (k_h) and vertical (k_v) directions, is a key parameter; representative minimum and maximum values for each combination of geological unit and lithological class (based on interpretations of measured hydraulic conductivities of small samples extracted from boreholes, published values, specific pumping tests, and experience) are extracted from a database. Model-scale measurements of hydraulic conductivity are developed by "upscaling" these values. The upscaling procedure takes the thickness of each borehole interval into account and uses the arithmetic mean for k_h, and the harmonic mean for k_v. By assuming a log-normal distribution, the minimum and maximum conductivities are used to construct a probability density function (pdf) from which representative values are stochastically assigned to each borehole interval. Individual borehole values are then evaluated to obtain an average and standard deviation of hydraulic conductivity for each hydrogeological unit. A kriging interpolation uses these values to calculate hydraulic conductivities across entire hydrogeological units. For hydrogeological units for which little reliable (borehole) information is available, estimates are made of the average and standard deviation of hydraulic conductivity or these values are based on empirical relationships. The resulting hydrogeological

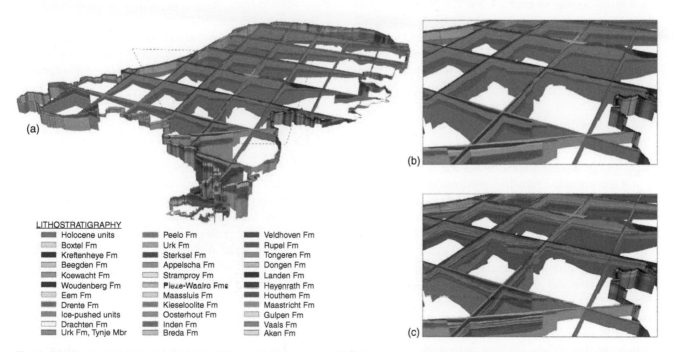

LITHOSTRATIGRAPHY

Holocene units	Peelo Fm	Veldhoven Fm
Boxtel Fm	Urk Fm	Rupel Fm
Kreftenheye Fm	Sterksel Fm	Tongeren Fm
Beegden Fm	Appelscha Fm	Dongen Fm
Koewacht Fm	Stramproy Fm	Landen Fm
Woudenberg Fm	Pieze-Waalre Fms	Heyenrath Fm
Eem Fm	Maassluis Fm	Houthem Fm
Drente Fm	Kieseloolite Fm	Maastricht Fm
Ice-pushed units	Oosterhout Fm	Gulpen Fm
Drachten Fm	Inden Fm	Vaals Fm
Urk Fm, Tynje Mbr	Breda Fm	Aken Fm

Figure 19.21 Relationship between the DGM (a and inset b) and REGIS II (inset c) representations of the shallow subsurface of the Netherlands.

Table 19.2 Green-shaded entries define the available REGIS II gridded map files defining geometry and hydraulic properties of different hydrogeological units.

Map File	Hydrogeological unit					
	Sand	Clay	Peat	Complex[1]	Chalk	Lignite
Unit Top	■	■	■	■	■	■
Unit Base	■	■	■	■	■	■
Unit Thickness	■	■	■	■	■	■
Horizontal Conductivity (k_h)	■			■	■	■
Standard Deviation of Horizontal Conductivity (σk_h)	■			■	■	■
Transmissivity (kD)	■			■	■	■
Vertical Conductivity (k_v)		■	■	■		■
Standard Deviation of Vertical Conductivity (σk_v)[2]		■	■	■		■
Hydraulic Resistance (c)		■	■	■		■

[1] With the exception of the complex Holocene deposits and complex ice-pushed ridges for which no hydraulic properties are yet available.
[2] With the exception of the boulder clay for which no standard deviation of vertical hydraulic conductivity is yet available.

model covers the entire country. The depth of the top and base, and thickness, of each of the 125 hydrogeological entities are mapped on 100 m resolution grids. Additional grids provide information on horizontal and vertical conductivity, standard deviations for these, and related properties, such as transmissivity and hydraulic resistance (Table 19.2).

19.4.4 Application of the Models

The REGIS II hydrogeological model is not only used as a standalone reference model; it is used to produce site and problem-specific geohydrological models of aquifers, aquitards, and aquicludes for input to groundwater flow models. REGIS II provides critical groundwater management information to national and regional authorities. Dutch groundwater managers accept REGIS II as the standard for acquiring conceptual insight into the hydrogeological structure and characterization of the subsurface. Provincial authorities and regional water authorities responsible for groundwater management are particularly interested in a regional version of the national hydrogeological model that defines aquifers and aquicludes that are relevant to their decision making. The national hydrogeological model is thus reclassified into a series of regional,

slightly simplified geohydrological models containing clearly defined aquifers and aquicludes (Figure 19.22).

It also provides the basis for groundwater models used for research, applications for groundwater production permits, and environmental impact assessments for infrastructure construction, drinking water production, and geo-thermal energy (Figures 19.23 and 19.24).

19.4.5 Conclusions

REGIS II provides hydrogeological information for the Netherlands that is entirely consistent with current geological concepts and information. It contains good regional estimates of the critical hydraulic parameters and their variability. It supports the development of customized framework models describing aquifers and aquitards (geohydrological models) by combining the digital map units and data from the national hydrogeological model. Geometric and parameter information is stored in grid files with 100 m resolution. These files are available free of charge from the DINOLOKET web portal (www.dinoloket .nl). The data are readily analyzed by 2-D GIS systems and can be converted into parameter arrays used by groundwater flow modeling.

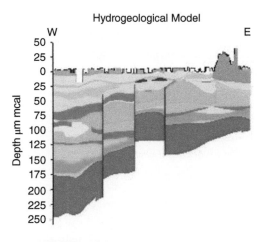

Hydrogeological Model

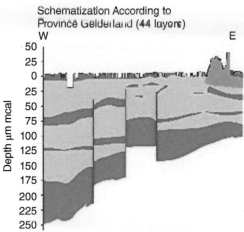

Schematization According to
Province Gelderland (44 layers)

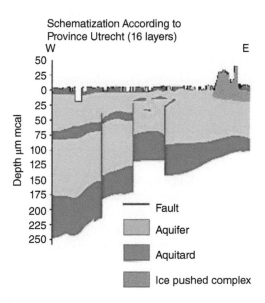

Schematization According to
Province Utrecht (16 layers)

Fault

Aquifer

Aquitard

Ice pushed complex

Figure 19.22 Conversion of the national hydrogeological model into regional geohydrological models. Modified from Vernes and van Doorn (2006).

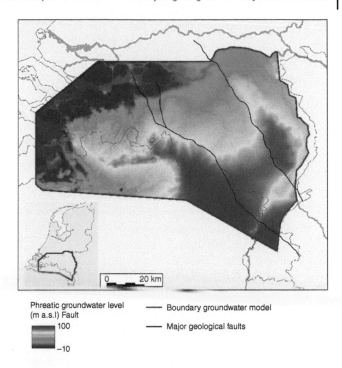

Phreatic groundwater level
(m a.s.l) Fault

100

−10

Boundary groundwater model

Major geological faults

0   20 km

Figure 19.23 A multi-regional REGIS II application, showing the computed average phreatic groundwater level in the southern part of the Netherlands and the northern part of Belgium. Modified from Verhagen et al. (2019).

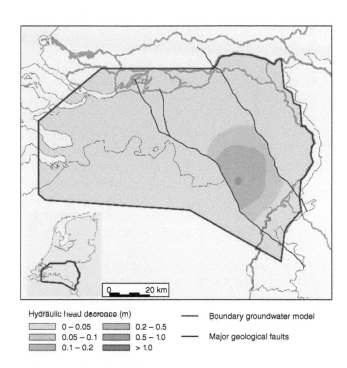

Hydraulic head decrease (m)

0 – 0.05   0.2 – 0.5
0.05 – 0.1   0.5 – 1.0
0.1 – 0.2   > 1.0

Boundary groundwater model

Major geological faults

0   20 km

Figure 19.24 Groundwater application of REGIS II showing the hydraulic head changes under the clay members of the Kieseloolite Formation as a result of groundwater abstraction for drinking water production at station Budel. Modified from Verhagen et al. (2019).

References

Andres, A.S. and Klingbeil, A.D. (2006). *Thickness and Transmissivity of the Unconfined Aquifer of Eastern Sussex County, Delaware*. Newark, DE: Delaware Geological Survey, Report of Investigations Nr. 70. 19 pp. [Online: Available at: http://udspace.udel.edu/handle/19716/2808] (Accessed November 2020).

Armstrong, D.K. and Dodge, J.E.P. (2007). *Paleozoic Geology of Southern Ontario*. Sudbury, ON: Ontario Geological Survey, Miscellaneous Release—Data 219. [Online: Available at: http://www.geologyontario.mndm.gov.on.ca/mndmfiles/pub/data/imaging/mrd219//MRD219.pdf] (Accessed November 2020).

Bajc, A.F. and Shirota, J. (2007). *Three-dimensional Mapping of Surficial Deposits in the Regional Municipality of Waterloo, Southwestern Ontario*. Sudbury, ON: Ontario Geological Survey, Groundwater Resources Study 3. 41 pp.

Berg, R.C., Mathers, S.J., Kessler, H. and Keefer, D.A. [Eds.] (2011). *Synopsis of Current Three-dimensional Geological Mapping and Modeling in Geological. Survey Organizations*. Champaign, IL: Illinois State Geological Survey, Circular 578. 92p. [Online: Available at: http://library.isgs.illinois.edu/Pubs/pdfs/circulars/c578.pdf] (Accessed December, 2020).

Berg, R.C., Brown, S.E., Thomason, J.F. et al. (2016). A multiagency and multijurisdictional approach to mapping the glacial deposits of the Great Lakes region in three dimensions. In: *Geoscience for the Public Good and Global Development: Toward a Sustainable Future* (eds. G.R. Wessel and J.K. Greenberg), 415–447. Boulder, CO: Geological Society of America, Special Paper 520. https://doi.org/10.1130/2016.2520(37).

Burt, A.K (2008). The Orangeville Moraine study: A new three-dimensional Quaternary mapping project. In: *Summary of Field Work and Other Activities 2008* (eds. C.L. Baker, E.J. Debicki, R.I. Kelly et al.), 30-1–30-9. Sudbury, ON: Ontario Geological Survey, Open File Report 6226. [Online: Available at: http://www.geologyontario.mndm.gov.on.ca/mndmaccess/mndm_dir.asp?type=pub&id=ofr6226] (Accessed November 2020).

Burt, A.K (2009). The Orangeville Moraine project: An update of field activities. In: *Summary of Field Work and Other Activities 2009* (eds. C.L. Baker, R.I. Kelly J.A. Ayer et al.), 22-1–22-3. Sudbury, ON: Ontario Geological Survey, Open File Report 6240. [Online: Available at: http://www.geologyontario.mndm.gov.on.ca/mndmaccess/mndm_dir.asp?type=pub&id=ofr6240] (Accessed November 2020).

Burt, A.K (2011). The Orangeville Moraine project: Preliminary results of drilling and section work. In: *Summary of Field Work and Other Activities 2011* (eds. R.M. Easton, O.M. Burnham, B.R. Berger et al.), 28-1–28-34.

Sudbury, ON: Ontario Geological Survey, Open File Report 6270. [Online: Available at: http://www.geologyontario.mndm.gov.on.ca/mndmaccess/mndm_dir.asp?type=pub&id=ofr6270] (Accessed November 2020).

Burt, A.K (2012). Conceptual geologic model for the Orangeville Moraine three-dimensional project. In: *Summary of Field Work and Other Activities 2012* (eds. G.P. Beakhouse, R.D. Dyer, R.M. Easton et al.), 32-1–32-6. Sudbury, ON: Ontario Geological Survey, Open File Report 6280. [Online: Available at: http://www.geologyontario.mndm.gov.on.ca/mndmaccess/mndm_dir.asp?type=pub&id=ofr6280] (Accessed November 2020).

Burt, A.K. and Dodge, J.E.P. (2016). *Three-dimensional Modelling of Surficial Deposits in the Orangeville–Fergus Area of Southern Ontario*. Sudbury, ON: Ontario Geological Survey, Groundwater Resources Study 15. [Online: Available at: http://www.geologyontario.mndm.gov.on.ca/mndmaccess/mndm_dir.asp?type=pub&id=GRS015] (Accessed November 2020).

Burt, A.K. and Rainsford, D.R.B. (2010). The Orangeville Moraine Project: Buried valley targeted gravity study. In: *Summary of Field Work and Other Activities 2010* (eds. J.A. Ayer, R.M. Easton, G.P. Beakhouse et al.), 31-1–31-6. Sudbury, ON: Ontario Geological Survey, Open File Report 6260. [Online: Available at: http://www.geologyontario.mndm.gov.on.ca/mndmaccess/mndm_dir.asp?type=pub&id=ofr6260] (Accessed November 2020).

Burt, A.K. and Webb, J.L. (2013). *Results of the 2008, 2009 and 2010 Drilling Programs in the Orangeville–Fergus Area of southwestern Ontario*. Sudbury, ON: Ontario Geological Survey, Miscellaneous Release—Data 303. [Online: Available at: http://www.geologyontario.mndm.gov.on.ca/mndmaccess/mndm_dir.asp?type=pub&id=MRD303] (Accessed November 2020).

DHI-WASY (2012). *FEFLOW 6.2 User Manual*. Berlin, Germany: DHI-WASY GmbH. 202 pp.

DHI-WASY (2015). *FEFLOW 7.0. User Guide*. Berlin, Germany: DHI-WASY GmbH. 124 pp.

Dufour, F.C. (2000). *Groundwater in the Netherlands, Facts and Figures*. Utrecht, Netherlands: Netherlands Institute of Applied Geosciences TNO. 96 pp.

ESRI (2010). *Spatial Analyst Tutorial*. Redlands, CA: ESRI. 60 pp. [Online: Available at: http://help.arcgis.com/en/arcgisdesktop/10.0/pdf/spatial-analyst-tutorial.pdf] (Accessed February 7, 2016).

ESRI (2015). *Getting to Know ArcGIS*, 4e (eds. M. Law and A. Collins). Redlands, CA: ESRI. 808 pp.

Gao, C., Shirota, J., Kelly, R.I. et al. (2007). *Bedrock Topography and Overburden Thickness Mapping, Southern Ontario*. Sudbury, ON: Ontario Geological Survey,

Miscellaneous Release—Data 207. [Online: Available at: http://www.geologyontario.mndm.gov.on.ca/ mndmaccess/mndm_dir.asp?type=pub&id=MRD207] (Accessed November 2020).

Golden Software (2015). *Surfer® User's Guide: Contouring and 3D Surface Mapping for Scientists and Engineers.* Golden, CO: Golden Software. 122 pp.

Gunnink, J.L., Maljers, D., van Gessel, S.F. et al. (2013). Digital geological model (DGM): a 3D raster model of the subsurface of the Netherlands. *Netherlands Journal of Geosciences* 92 (1): 33–46. https://doi.org/10.1017/ S0016774600000263.

Howard, K.W.F., Eyles, N., Smart, P.J. et al. (1995). The Oak Ridges Moraine of southern Ontario; a ground-water resource at risk. *Geoscience Canada, B* 22 (3): 101–120. [Online: Available at: https://journals.lib.unb.ca/index .php/GC/article/view/3878] (Accessed November 2020).

Johansson, C. E. (2006). Uppsalaåsen – A long Geosite candidate in Eastern Sweden. *ProGeo News* 2: 1-3. [Online: Available at: http://www.progeo.ngo/downloads/ PROGEO_news_2006_2.pdf] (Accessed November 2020).

Kessler, H., Mathers, S.J. and Sobisch, H.-G. (2009). The capture and dissemination of integrated 3D geospatial knowledge at the British Geological Survey using GSI3D software and methodology. *Computers & Geosciences* 35 (6): 1311–1321. https://doi.org/10.1016/j.cageo.2008.04.005.

Mathers, S.J., Kessler, H., Macdonald, D.M.J. et al. (2012). *The Use of Geological and Hydrogeological Models in Environmental Studies.* Keyworth, UK: British Geological Survey, Internal Report IR/10/022. 93 pp. [Online: Available at: http://nora.nerc.ac.uk/id/eprint/20481/] (Accessed November 2020).

MacCormack, K.E., Berg, R.C., Kessler, H. et al. [Eds.] (2019). *2019 Synopsis of Current Three-Dimensional Geological Mapping and Modelling in Geological Survey Organizations.* Edmonton, AL: AER/AGS, Special Report 112, 307 pp. [Online: Available at: https://ags.aer.ca/publication/spe- 112] (Accessed December, 2020).

McLaughlin, P.P., Tomlinson, J.L. and Lawson, A.K. (2016). Attributing groundwater withdrawals to aquifers using 3-D geological maps in Delaware, USA. In: *Three-Dimensional Geological Mapping: Workshop Extended Abstracts; Geological Society of America Annual Meeting, Baltimore, Maryland, October 31, 2015* (eds. K.E. MacCormack, L.H. Thorleifson, R.C. Berg and H.A.J. Russell), 57–62. Edmonton, AL: AER/AGS, Special Report 101 [Online: Available at: https://www.ags.aer.ca/document/SPE/SPE_ 101.pdf] (Accessed November 2020).

O'Connor, D.R. (2002a). *Report of the Walkerton Inquiry. Part 1: The Events of May 2000 and Related Issues.* Toronto, ON:

Ontario Ministry of the Attorney General. 504 pp. [Online: Available at: http://www.archives.gov.on.ca/en/e_records/ walkerton/report1/index.html] (Accessed November 2020).

O'Connor, D.R. (2002b). *Report of the Walkerton Inquiry. Part 2: A Strategy for Safe Drinking Water.* Toronto, ON: Ontario Ministry of the Attorney General. 588 pp. [Online: Available at: http://www.archives.gov.on.ca/en/e_records/ walkerton/report2/index.html] (Accessed November 2020).

Ontario Geological Survey (2010). *Surficial Geology of Southern Ontario.* Sudbury, ON: Ontario Geological Survey, Miscellaneous Release—Data 128—Revised. [Online: Available at: http://www.geologyontario.mndm.gov.on.ca/ mndmaccess/mndm_dir.asp?type=pub&id=MRD128- REV] (Accessed November 2020).

Ontario Geological Survey (2013). *Ontario Geophysical Surveys, Ground Gravity Data, Grid and Point Data (ASCII and Geosoft® formats) and Vector Data, Orangeville Area.* Sudbury, ON: Ontario Geological Survey Geophysical Data Set 1072. [Online: Available at: http://www .geologyontario.mndmf.gov.on.ca/mndmaccess/mndm_ dir.asp?type=pub&id=GDS1072] (Accessed November 2020).

Royse, K.R., Kessler, H., Robins, N.S. et al. (2010). The use of 3D geological models in the development of the conceptual groundwater model. *Zeitschrift der Deutschen Gesellschaft für Geowissenschaften* 161 (2): 237–249. https://doi.org/10 .1127/1860-1804/2010/0161-0237.

Van de Ven, G.P., Burggraaff, J.P., Huisman, P. et al. (1986). *Wetenschappelijke Atlas van Nederland, deel 15: Water* (in Dutch). The Hague, Netherlands: Staatsuitgeverij. 23 pp.

Verhagen, F., Van Steijn, T., Van der Wal, B. et al. (2019). *Update Hydrologische Gereedschapskist Noord-Brabant* (in Dutch). Amersfoort, the Netherlands: Royal HaskoningDHV. 212 pp.

Vernes, R.W. and van Doorn, Th.H.M. (2005). *Van Gidslaag naar Hydrogeologische Eenheid – Toelichting op de totstandkoming van de dataset REGIS II* (in Dutch). Utrecht, Netherlands: Netherlands Institute of Applied Geosciences TNO, Report 05-038-B. 105 pp. [Online: Available at: https://www.dinoloket.nl/sites/default/files/ file/dinoloket_toelichtingmodellen_20131210_01_rapport_ nitg_05_038_b0115_netversie.pdf] (Accessed November 2020).

Vernes, R.W. and van Doorn, Th.H.M. (2006). REGIS II, a 3D Hydrogeological Model of the Netherlands. *Geological Society of America Abstracts with Programs* 38 (7): 109.

20

Application Theme 3 – Geothermal Heating and Cooling

Editor's Introduction

Recent decades have witnessed increased interest in using renewable geothermal energy resources ranging from shallow groundwater resources to deeper hot water and hot rock reservoirs. Low enthalpy ground-source heat pump systems can provide a low-cost, low-carbon, sustainable method for heating and cooling buildings (Allen et al. 2003); these systems can be classified as either "open loop" or "closed loop" systems. Open loop systems abstract groundwater, pass it through a heat exchanger, and return it to the aquifer; they work efficiently when shallow groundwater is available but may be subject to abstraction and discharge regulations. Closed loop systems use sealed pipes; they are typically more expensive, requiring larger numbers of boreholes, but avoid most regulatory costs (Farr et al. 2017). Deeper hot water reservoirs are used to heat buildings and greenhouses and for several industrial processes. In selected locations of higher than normal heat flow, geothermal reservoirs are used to generate electricity. Some geothermal power plants directly use steam from hot water the reservoir to power a turbine, while others use the to boil a working fluid that vaporizes to power a turbine. Hot dry rock geothermal resources are not yet commercially viable; they would operate by injecting cold water down one well, circulating it through hot fractured rock, and recovering heated water from another well.

Key parameters in geothermal reservoir characterization are temperature, permeability, and thermal conductivity (Tester et al. 2006; Busby et al. 2009). In several locations, geological models have proved useful in defining the 3-D spatial variability of the parameters needed to model groundwater and heat dynamics. Recent United Kingdom examples of the role of 3-D geological models for assessing geothermal energy resources include mapping the baseline groundwater temperature, chemistry, and dimensions of the shallow unconsolidated aquifer in Cardiff (Patton et al. 2015; Boon et al. 2016; Farr et al. 2017) and the geothermal potential of flooded abandoned mine workings in Scotland and south Wales (Farr et al. 2016). Interest in better management of the shallow geothermal resource has grown in Europe too, with EU-funded projects such as MUSE (Managing Urban Shallow Geothermal Energy; Herms et al. 2019) and GeoPlasma-CE (https://portal.geoplasma-ce.eu/).

This chapter contains three case studies concerning the application of 3-D subsurface information to assess and develop shallow and deeper geothermal resources. The first case study describes the modeling used to support development of a shallow low enthalpy ground source heat pump system at Zaragosa, Spain. The 3-D model defined management policies by assessing several remediation scenarios that employed open loop groundwater heat-pumps, including thermal and geochemical impacts associated with injection of surface water under an intense exploitation regime.

The second case study describes the TransGeoTherm project which promoted and supported the development of shallow geothermal resources in the Saxon-Polish trans-boundary region and made the results of geothermal modeling and mapping available to the public. Geological, hydrogeological, and geothermal data from Polish and German sources were standardized to produce a coherent 3-D geological model which was evaluated to produce maps that illustrated the estimated geothermal conditions at four depths.

The third case study describes the modeling used to assess the deeper geothermal resources in the Upper Rhine Graben in the German State of Hesse. The combined geological-geothermal 3-D model, the first for an entire German state, quantified and analyzed the deep geothermal potential using a multi-criteria decision support system. This integration of diverse data sources produced an important new tool for the initial planning of geothermal power plants.

Applied Multidimensional Geological Modeling: Informing Sustainable Human Interactions with the Shallow Subsurface, First Edition.
Edited by Alan Keith Turner, Holger Kessler, and Michiel J. van der Meulen.

Case Study 20.1: Assessing Shallow Geothermal Resources at Zaragoza, Northeast Spain, with 3-D Geological Models

Alejandro García-Gil, Miguel Á. Marazuela, Violeta Velasco, Mar Alcaraz, Enric Vázquez-Suñé, and Albert Corbera

Institute of Environmental Assessment and Water Research (IDAEA), CSIC, Barcelona 08034, Spain

20.1.1 Introduction

During the past two decades, city managers and environmental protection agencies have paid attention to the growing utilization of ground heat pumps as an economical renewable energy resource (Lund and Boyd 2016). The use of numerical tools to analyze the environmental impacts of these systems and to support appropriate management policies is generally understood to be very important (Jaudin et al. 2013). Thus, 3-D geological modeling of the shallow subsurface has become a requirement for the quantitative management of shallow geothermal resources in urban areas.

The geometries and properties of subsurface geological entities influence heat transport processes, especially heat advection by groundwater. Therefore, an appropriate 3-D subsurface geological model provides critical guidance to the development of accurate derived hydrogeological and heat-transport models (Herbert et al. 2013; Epting et al. 2013; García-Gil et al. 2014). A 3-D geological model assisted the assessment of shallow geothermal resources in the Zaragoza urban area, northeast Spain (Figure 20.1). At this location, bedrock karst resulted in complex deformation of the overlying alluvial deposits (Acero et al. 2013; Gutiérrez et al. 2009).

A Geographic Information System (GIS) software platform was used to create a 3-D geological framework model that represents the irregular geometry of the groundwater resource. This model guided the geometries and properties incorporated into the subsequent groundwater and heat-transport numerical models.

20.1.2 Construction of the 3-D Geological Model

Over time, considerable volumes of geological, hydrogeological, hydrochemical, geophysical, and thermal data have been collected to describe the alluvial aquifer

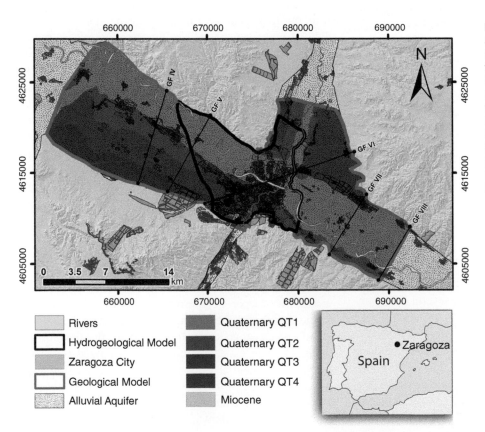

Figure 20.1 Location of the study area including the 3-D geological model and groundwater and heat transport model domains. The green line is boundary of the geological model. The black line is boundary of the hydrogeological model. Source: Geological data from Robador Moreno et al. (2011). Digital Elevation Model obtained from IGN (2012). Projection: WGS 1984 Complex UTM Zone 30 N, Datum DWGS1984.

Rivers
Hydrogeological Model
Zaragoza City
Geological Model
Alluvial Aquifer
Quaternary QT1
Quaternary QT2
Quaternary QT3
Quaternary QT4
Miocene

near Zaragoza. These data were obtained from multiple sources, including: scientific research campaigns, geotechnical engineering surveys, and aquifer monitoring networks.

All the data were loaded into a personal geodatabase within the ArcGIS environment using the HYDOR geodatabase approach (Velasc 2013). The HYDOR geodatabase allows the user to structure and store the spatio-temporal data in a clear and compact way. It was created and designed by the Hydrogeology Group (GHS), a research institute associated with both the Department of Geotechnical Engineering and Geosciences at the Universitat Politècnica de Catalunya (UPC) and the Institute of Environmental Assessment and Water Research (IDAEA) of the Spanish Council of Scientific Research (CSIC) located at Barcelona, Spain. The assessment was completed using additional tools, also developed by GHS, for analyzing, processing and interpreting data stored in the HYDOR geodatabase. These tools included:

- HEROS tools (Velasco et al. 2012), which enabled the analysis and interpretation of the geological data for creating 3-D geological models.
- HEROS 3-D tools (Alcaraz 2016), which assisted the analysis, visualization, and management of the 3-D geological models;
- QUIMET tools (Velasco et al. 2014), which were developed for analyzing and interpreting hydrochemical data; and
- HYYH tools (Criollo et al. 2016), which visualize and further interpret hydrogeological data stored in HYDOR (such as heads, pumping tests, etc.)

Using the HYDOR geodatabase approach, a 3-D geological framework model was developed to represent the alluvial aquifer of the Ebro River in the area surrounding Zaragoza city (Figure 20.1). To better analyze groundwater flow conditions, and to reduce the uncertainty of the predictions, the domain of the geological model was extended several kilometers upstream and downstream from the confluence of the respective alluvial aquifers along the rivers at Zaragoza.

In order to support the subsequent numerical modeling of groundwater flow and heat transport, the geological model described the four levels of quaternary terraces along the Ebro River. These terraces are defined as, from ancient to modern, QT4, QT2, QT3, and the current QT1 (Figure 20.1). The four terraces have distinct hydraulic behaviors and were identified by considering:

- Their surface elevation differences observed from a digital elevation model (DEM) of the area (IGN 2012),
- Their representation on a geological map of the city of Zaragoza (Robador Moreno et al. 2011), and
- Water resources study maps (MOPU 1989).

Other alluvial landforms, such as abandoned river channels and alluvial fans, were not represented due to their relatively minor spatial extents.

To reconstruct the geometry of the terraces, a total of 626 lithological descriptions were extracted from borehole observation records contained within the Water Points Inventory database (Inventario de Puntos de Agua; IPA) developed for the local water administrator – the Ebro Hydrographic Confederation (CHE 2016). The Miocene bedrock surface defines the base of the alluvial aquifer. It was defined by 174 boreholes that reach or penetrate the bedrock. As shown in Figure 20.1, six electrical resistivity profiles provided additional information to support the geometry and conditions within the Miocene bedrock (MOPU 1989). The bedrock surface is a major stratigraphic discontinuity that is affected by karst processes (Gutiérrez et al. 2007).

The boreholes were visualized and analyzed with the HEROS toolset by accessing the lithological information stored in the HYDOR database. In order to perform stratigraphic correlation between boreholes, a total of 52 cross-sections were constructed, using the HEROS toolset. These cross sections covered the whole area of interest (Figure 20.2).

The HEROS toolset also allowed interpretation of stratigraphic subunits within the terraces. Once all cross-sections were interpreted, 3-D lines representing the elevation of each stratigraphic subdivision within each terrace were interpolated to generate 3-D surfaces as "Triangular Irregular Networks" or TIN's. The bedrock surface was interpolated first. Then the terraces in contact with the bedrock were interpolated and limited with the bedrock surface. The boundaries of internal subunits within the terraces were interpolated next. Finally, the top surface of exposed terraces was defined by the DEM. This geospatial analysis process produced the 3-D geological model shown in Figure 20.3.

Our approach resulted in a 3-D geological model that represents the complexities of this particular sedimentary environment and karstic subsidence. The model defined the thickness of the urban alluvial aquifer, and therefore its hydraulic transmissivity, which is the controlling factor for heat advection.

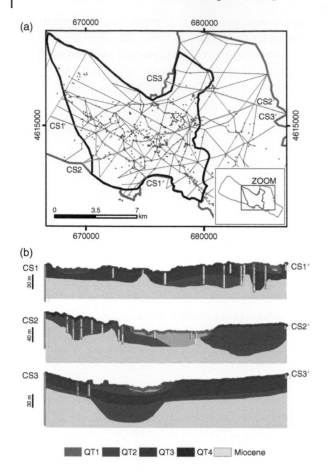

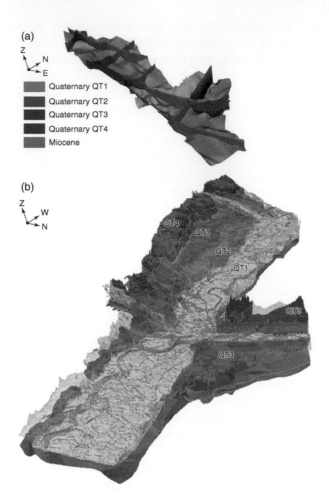

Figure 20.2 (a) Location of the geological cross-sections used to build the 3-D geological model. Points represent boreholes introduced into the HYDOR geodatabase. Points highlighted in red are boreholes that intersect with the bedrock. (b) Example of three geological cross-sections.

Figure 20.3 (a) Geological fence diagrams. (b) Representation of the 3-D geological model.

20.1.3 Three-dimensional Hydrogeological and Heat Transport Numerical Modeling

Geological processes define the spatial variability of key parameters needed to model groundwater and heat dynamics. Therefore, a more accurate reproduction of the geology will result in more accurate parameterization and predictions of the hydrogeological and thermal models (Carrera et al. 2005). In this way, geological data helps constrain the model parameters, thus reducing the difficulties associated with non-uniqueness and instability in the calibration process.

For that purpose, the entire geological model volume needs to be discretized in order to solve numerically the governing equations of interest for spatially variable media. The inverse problem is used to test all plausible proposed conceptual models against observed hydrological state variables (heads, temperature, fluxes, etc.). The inverse problem searches for the set of parameters which minimizes the differences between the observations and model predictions. This task is not trivial and poses new challenges, since the number of available observations usually cannot satisfy the number of degrees of freedom defining the problem to be solved. In addition, more than one set of parameters may lead to an acceptable solution of the forward problem, so that non-uniqueness arises.

20.1.4 Zaragoza Hydrogeological Model

At Zaragoza, the modeling process of groundwater flow and heat transport was based on available information about a real system. The data collected were analyzed, interpreted, and then used to define plausible conceptual 3-D geological models. These models provided the dimensions of subsurface geological units and reflected the conceptual interpretations of the process dynamics responsible for creating the studied subsurface volume. Hydraulic properties (transmissivity, storativity, mass flow, etc.) and thermal properties (heat capacity, thermal conductivity, porosity, etc.) depend on the characteristics of these individual geological formations, and they determine the fluid and energy dynamics of the system.

A 3-D hydrogeological model for Zaragoza was created based on the 3-D geological model of the urban alluvial aquifer. The main boundary conditions of this model are shown in Figure 20.4. The hydrogeological model included the whole metropolitan area of the city over the alluvial aquifer. first-type (or Dirichlet) boundary conditions (shown by red lines on Figure 20.4) coincide with stable-head conditions. They were used to reproduce regional groundwater inflows from the West, North, and South, and regional outflows to the East. These boundary conditions reproduce the lateral and bottom limits of the aquifer. Second-type (or Neumann) boundary conditions (shown by green lines on Figure 20.4) were used to define areas with no flow (null flux). River-aquifer relationships were implemented in the model as third-type (or Robin) boundary conditions (shown as blue lines on Figure 20.4). These allowed for the reproduction of recharge from the river during flood events and exfiltration from the aquifer to the river in times of river low-flow. Two different surface recharge rates were implemented, distinguishing different infiltration rates from less developed metropolitan areas compared with urbanized areas.

Figure 20.4 Boundary conditions of the groundwater and heat transport model. Note that the hydrogeological model limits are shown also in Figure 20.1.The 5 m pixel resolution Digital Elevation Model obtained from IGN (2012) defines the model topography.

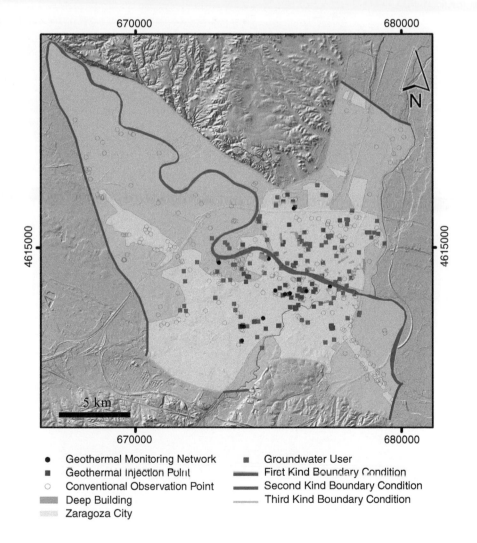

Legend:
- ● Geothermal Monitoring Network
- ■ Geothermal Injection Point
- ○ Conventional Observation Point
- ▬ Deep Building
- ▬ Zaragoza City
- ■ Groundwater User
- ▬ First Kind Boundary Condition
- ▬ Second Kind Boundary Condition
- ▬ Third Kind Boundary Condition

20.1.5 Zaragoza Thermal Plume Model

Extraction and injection wells (shown as red squares on Figure 20.4) imposed transient flow and heat source points in the model according to exploitation regimes of shallow geothermal heat pumps. The background temperature of the aquifer was imposed on the regional outflow and inflow areas. The temperature of the surface water in the rivers was considered transient and varied seasonally. Topographic surface temperatures were considered transient, and were estimated by the measured soil temperature at a depth of 20 cm. In addition, deep buildings that reach the water table (shown in yellow on Figure 20.4) were implemented in the model as heat-flow boundaries. Figure 20.4 also shows the sites used to calibrate the model; geothermal monitoring network sites are shown as black circles and conventional groundwater monitoring network sites are shown as open circles. The boundary conditions were imposed by a 3-D unstructured finite element mesh of 913 983 elements and 484 264 nodes distributed in 21 layers (Figure 20.5).

Results from the hydrogeological model produced a good fit between observed and calculated data (García-Gill et al. 2014), justifying the 3-D geological modeling effort. Figure 20.6a shows a temperature distribution map that identifies thermal plumes in the aquifer generated by groundwater heat pumps. Figure 20.6b is a cross-section of one thermal plume. This highlights the importance of vertical heat fluxes (heat conduction) and justifies the need to reproduce the geometry of the subsurface, not only of the aquifer body, but also the unsaturated zone and the bedrock.

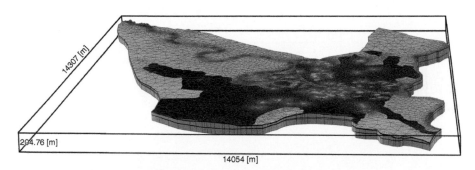

Figure 20.5 Unstructured finite element mesh of the 3-D groundwater and heat transport model.

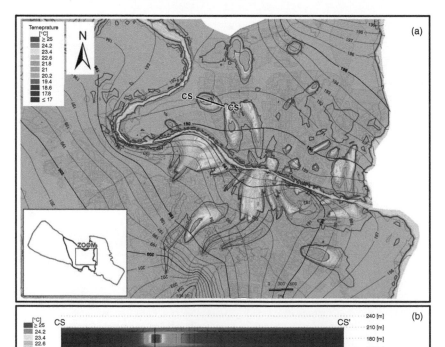

Figure 20.6 (a) Temperature distribution map showing the thermal plumes generated by groundwater heat pumps in the aquifer. (b) Cross-section of a thermal plume in the 3-D hydrogeological model. Hydraulic head contours are shown in black.

20.1.6 Conclusions

At Zaragoza, the adequate simulation of groundwater flow and heat transport processes based on a hydrogeological model proved to be decisive for developing management policies that support sustainable thermal interactions in the subsurface. Fundamental criteria for a concession approval process for new shallow geothermal exploitation have been proposed based on this model (García-Gil et al. 2015c). The hydrogeological model also allowed for the simulation of a large number of remediation scenarios, including injection of surface water under an intense exploitation regime by open groundwater heat-pumps (García-Gil et al. 2015a). Finally, thanks to the results obtained from this hydrogeological model, the geochemical impacts associated with intensive shallow geothermal resources exploitation in an urban environment have been investigated in detail (Garrido Schneider et al. 2016). Beyond support for the thermal management of this aquifer, the hydrogeological model has assisted the assessment of groundwater inundation phenomena (García-Gil et al. 2015b).

Figure 20.7 Location of the TransGeoTherm project and geological setting.

Case Study 20.2: Cross-border 3-D Models for Assessing Geothermal Resources

Sascha Görne and Ottomar Krentz

Department of Geology, Saxon State Office for Environment, Agriculture and Geology, 09599 Freiberg, Germany

20.2.1 Introduction

TransGeoTherm is a cooperative, cross-border project of the Polish Geological Institute (PGI) and the Saxon Geological Survey (LfULG). Co-financed by the European Union (EU) within the framework of the Operational Programme for Transboundary Cooperation Poland-Saxony 2007–2013, the project was active between August 2012 and December 2014. The project area is approximately 24 by 44 km in size and is located along the valley of the Neisse River, which forms the border between the German state of Saxony and Poland (Figure 20.7).

The main objective of the TransGeoTherm project was to promote and support the development of shallow geothermal resources (by definition <400 m depth) as

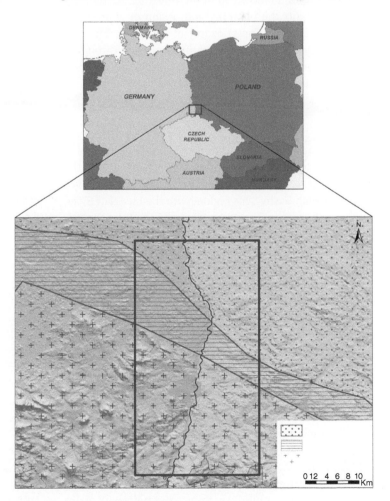

a low-emission energy alternative for the Saxon-Polish trans-boundary region, and to make the results of geothermal modeling and mapping assessment available to the public. Development of renewable energy is a priority of the EU's environmental policy (European Union 2009). Increased use of shallow geothermal systems as a replacement of the current fossil fuels, especially lignite, will reduce the emission of CO_2 and avoid pollution.

Conditions within the transboundary project area are favorable for developing and implementing geothermal energy for heating and cooling systems but no coherent transboundary information system of geothermal databases existed. The TransGeoTherm project collected and standardized specified geological, hydrogeological and geothermal data from Polish and German sources. These were used to produce a 3-D geological model; then numerical modeling methods were applied to the 3-D model to produce a series of maps, illustrating the estimated geothermal conditions at four depths (40, 70, 100, and 130 m). This was the first application of this method in the transboundary region.

The resulting geothermal maps provide information on the potential for shallow geothermal heating at any location and support the planning and design of appropriate geothermal installations (Kozdrój et al. 2014). The 3-D model and geothermal maps are disseminated through a public web portal to a large audience, especially the citizens, local authorities, and engineers of the Saxon-Polish trans-boundary area.

20.2.2 Geological Setting

The TransGeoTherm project area is located on the transition between the Bohemian Massif and the NE German-Polish Basin. This region, with the Lusatian Block in the south and the Görlitz syncline and the Intra-Sudetic Basin in the north, is characterized by late Proterozoic granitoides, Paleozoic shales and Permian and Cretaceous sediments. The geology of the project area was described in detail by Krentz et al. (2001); it is very heterogeneous and has been strongly influenced by tectonic events. Geological units older than the Tertiary may be divided into three sectors in this region. The southern part of the project area is dominated by a Proterozoic granodiorite that intrudes Proterozoic sedimentary rocks. The central part of the area in the Saxony side is composed of greatly deformed Paleozoic phyllites, shales, greywackes, siliceous shales, and marbles as well as alkaline and acidic volcanics. These extend eastward onto the Polish side as a metamorphosed sedimentary-volcanic complex. Permo-Mesozoic deposits, mostly evaporites, sandstones and marls, form the deeper subsurface creating a structural basin in the northern part of the area. Tertiary units bury these older units with about

100 m of lignite with interbedded gravel, sand, and silt. In the central part of the project area, Tertiary units are absent or are preserved only as small remnants. In the southern part of the area, the Tertiary units are up to 200–250 m thick in the Berzdorf and Radomierzyce basins. Tertiary volcanic units occur locally in the southern part of the project area.

In the northern part of the area, Quaternary deposits are 50–80 m thick and include glacial moraine deposits of Pleistocene age, along with widely distributed river and lake sediments composed of marl, gravel, sandstone, silts, and clays. The southern part of the project area has a limited distribution and thickness of glacial deposits; they are covered with eolian deposits related to the older Vistula glaciation, including dunes and loess. Recent (Holocene) sediments are found primarily along streams and rivers as clay, gravel, and sand deposits, especially along the Neisse valley.

The tectonic regime of the project area is characterized by a major northwest–southeast trending fault on the north edge of the Zgorzelec Slate Mountains that has caused the Mesozoic sedimentary rocks of the northern basin to be lowered by at least 1500 m. Its activity dates back to the early Tertiary. Other faults trending north–south or east–west result in deep (300 m) troughs filled with Tertiary units in the Berzdorf-Radomierzyce and Zittau-Turow areas.

20.2.3 Developing the 3-D Geological Subsurface Model

Model construction began with data preparation. All available Polish and German geoscientific information was collected, analyzed, and standardized. Evaluation of geological maps, printed reports and borehole records defined a set of 76 hydrogeological–geothermal (HGE) units (Małolepszy et al. 2015). Digital geological maps at different scales provided information for the 3-D model. Some problems occurred while combining these sources with the high-resolution DEM. Additional problems arose with some coordinate transformations and due to the different scales, ages, and contents of these sources (Görne and Krentz 2015). To accelerate the modeling workflow the Cenozoic and the Pre-Tertiary units were modeled separately. A consistent approach from the model base to its top would have avoided some problems experienced by the project. Due to the variety of stratigraphic units, modeling of the Quaternary turned out to be the most time-consuming step.

Seven geological cross-sections extending northeast–southwest were developed to support the 3-D modeling (Figure 20.8). Three of these sections cross the Polish-German border.

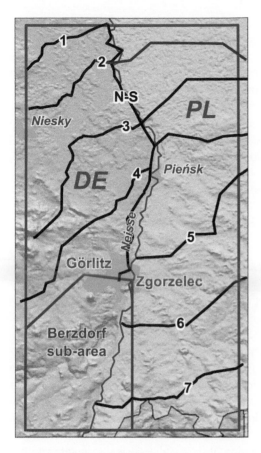

Figure 20.8 Locations of the geological cross-sections, and the extent of the Berzdorf 3-D model.

These sections are oriented approximately perpendicular to bedrock geological structures and the extensional tectonic elements of the northern basin. They include all relevant geological units and provide information about local thicknesses and characteristics of the units (Figure 20.9). An additional north–south cross-section that included information from several deep boreholes

was developed to avoid boundary effects in the transborder area.

Subsurface geology was mainly defined from 11 600 boreholes in Germany and 1250 boreholes in Poland, where most of the boreholes are not deeper than 50 m. The approximately 500 deep boreholes were important because they allowed prediction of deeper geological structures. Figure 20.10 shows the distribution of borehole depths.

The 3-D geological subsurface model was created using the SKUA-GOCAD software by Paradigm which employs an implicit approach to model construction. The model was constructed in five steps: (1) creation of horizon boundaries in 2-D; (2) definition of appropriate constraints to honor locally measured layer thicknesses and the spatial distribution of borehole data; (3) geostatistical interpolation of layer thicknesses and base heights; (4) construction of the volume model based on base layers and layer thicknesses; and (5) conversion to a regular grid with 25 m spacing.

The implicit approach to model construction (see Chapter 12) offers some assistance to the model creation process by providing computer algorithms that assist the geologist by creating model surfaces with trends, stratigraphic characteristics, and other geologically meaningful conditions. These algorithms are controlled by constraints defined by the modeler (step 2), and the surfaces are generated by geostatistical interpolation (step 3). The volume model (step 4) is defined by triangulated surfaces. Because the model results had to match the data-structure in the existing hydrogeological database, the triangulated surfaces in the geological model were converted to regular grids with a 25 m spacing. This regular grid was used to calculate the geothermal characteristics, and to compute geothermal maps showing the potential for geothermal heat extraction at several depths (Figures 20.11 and 20.13).

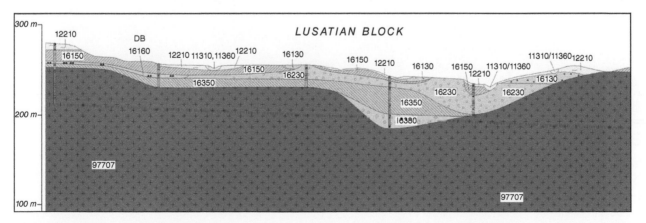

Figure 20.9 Southern portion of cross-section 4 (in Figure 20.8) showing the stratigraphic logs of boreholes and the numerical codes assigned to the different HGE (hydrogeological-geothermal units).

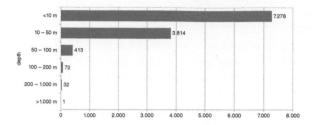

Figure 20.10 Depth and number of boreholes used for the 3-D model.

Figure 20.11 Completed 3-D model showing the geological horizons and the DEM color-coded with the thermal extraction capacity (view is from south).

The project area was split into western (German) and eastern (Polish) regions along the German-Polish border, with a buffer zone covered by both areas. The main model covered about $1000\,km^2$ and extended to a depth of 200 m below the ground level (locally down to 340 m). In addition to the main 3-D geological model, supplementary more detailed 3-D models were constructed for two former lignite deposits: the Berzdorf deposit in Saxony (Figure 20.12) and the Radomierzyce deposit in Poland (Woloszyn et al. 2015).

20.2.4 Application of the Model

The main application of the 3-D geological model was the promotion and support of the development of shallow geothermal energy resources in the Saxon-Polish transboundary region. By using the 3-D model it was possible to estimate the geothermal conditions at any location in the project area. Based on the geothermal properties of rocks and their groundwater content, a specific value of geothermal conductivity λ [W/m·K] for each rock type (layer) in every real borehole was allocated. Then a weighted mean thermal conductivity value was calculated for each borehole that belongs to a certain hydrogeological-geothermal unit (HGE) of the 3-D geological model. Next the horizontal distribution of these values was interpolated within every HGE, using the borehole data and the groundwater level. The results were converted to a series of maps and

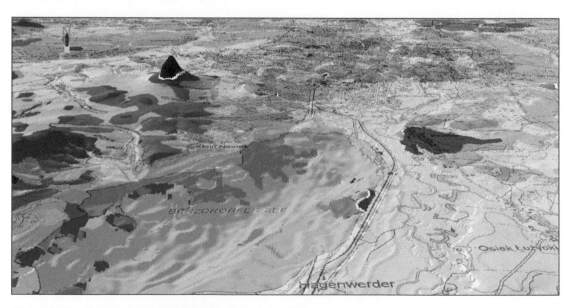

Figure 20.12 Final geological model of the Berzdorf region.

(a)

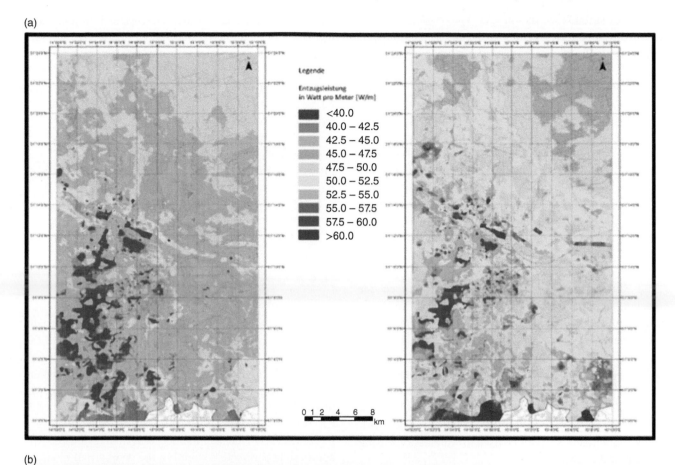

(b)

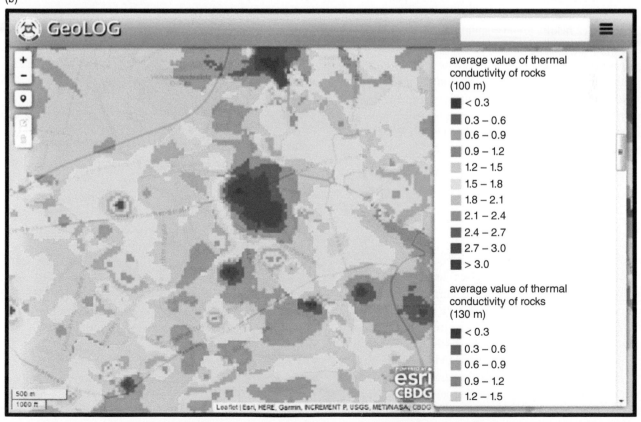

Figure 20.13 (a) Printed maps of the average geothermal extraction capacity for different operation hours (2400 and 1800 h) and probe depths (70 m and 100 m). (b) Mean geothermal conductivity map represented via an online web application (PGI-NRI 2017).

digital files suitable for use in GIS. The TransGeoTherm project developed two versions of geothermal maps: a "public version" and a "professional version".

- The "public version" maps show the spatial distribution of the *average geothermal extraction capacity* in watts per meter [W/m] for four different depth levels (40 m, 70 m, 100 m, 130 m) and two usage scenarios of annual heat pump operation: 1800 hours for heating use only, or 2400 hours for use as both heating and hot water. These maps allow evaluation and selection of depths that offer optimal sources of geothermal heat to provide a specific heat pump efficiency. This version is particularly suitable for individual investors planning the construction or modernization of houses. (Figure 20.13a)
- The "professional version" has been developed for design, architectural, engineering, drilling, and corporate offices. This version evaluates the potential of geothermal energy at any location by assessing the *mean value of the thermal conductivity of bedrock*, expressed in watts per meter-Kelvin [W/m·K]. This is again calculated for the four depth levels of 40, 70, 100, 130 m. (Figure 20.13b). These values are used for planning and designing geothermal heat pumps when combined with other parameters and can be used for calculations using specialized software.

The project results are available free-of-charge and can be used by all interested parties (TransGeoTherm 2015). The results serve as a decision-making tool for local and regional planning and development, and thus are of special interest to the local authorities, planners and the general public as well as to the business sector (for example producers and installers of heat pumps). The geothermal maps can also be used for renewable energy auditing at the local and regional scale.

20.2.5 Concluding Remarks

Based on all available data, a detailed 3-D geological model was created by the TransGeoTherm project. Some problems and challenges occurred during the process of data preparation and modeling. A lack of interoperability among the input data and data structures caused inhomogeneity in the intermediate results. With the help of a 3-D database approach, the final data could be stored consistently and made available for further projects. The derived information (3-D model and geothermal maps) is available through a public web portal, which the TransGeoTherm project has committed to maintaining for five years after the model was completed.

Case Study 20.3: Use of 3-D Models to Evaluate Deep Geothermal Potentials in Hesse, Germany

Kristian Bär[1], Dirk Arndt[2], Rouwen Lehné[3], Johann-Gerhard Fritsche[3], Matthias Kracht[3], and Ingo Sass[1]

[1] *Institute of Applied Geosciences, Technische Universität Darmstadt, 64287 Darmstadt, Germany*
[2] *Geo-Explorers AG, 4410 Liestal, Switzerland*
[2] *Hessisches Landesamt für Umwelt, Naturschutz und Geologie, 65203 Wiesbaden, Germany*

20.3.1 Introduction

The project "3D-modeling of the deep geothermal potentials of Hesse" was initiated in 2008 to systematically detect and evaluate the deep geothermal potential of the entire federal State of Hesse, not just the readily accessible hydrothermal potentials of the Upper Rhine Graben. This evaluation is based on the various rock and reservoir properties stored in the cellular 3-D geothermal model. An Analytic Hierarchy Process (AHP) that employs a multiple-criteria decision-support system evaluated the appropriate parameters stored in the geothermal model to assess the viability of different deep geothermal systems. Threshold values based on technical constraints for each parameter were defined so that the geothermal potential for each system could be identified and visualized in five classes: very high, high, medium, low, or very low potential.

Prior to this project, comprehensive datasets for deep geothermal potential evaluations in Hesse only defined the underground temperatures for the Upper Rhine Graben, a small part of the entire state (Hänel and Staroste 1988; Hurter and Hänel 2002; Hurter and Schellschmidt 2003; Schulz and GEOTIS-TEAM 2009). In addition to temperature, several additional factors influence the deep geothermal potential for open systems. The bulk permeability of the reservoir defines the achievable flow-rate of thermal water, while matrix permeability, porosity, and thermal conductivity are required to estimate the conductive and convective heat-flows within the reservoir. Information on all these had to be collected for the entire state; a large database was developed and supported 3-D models.

Assessment of deep geothermal potential requires knowledge of geological structure and geothermal properties of reservoir rocks. The project initially developed a 3-D geological model, and this in turn allowed development of a 3-D geothermal model (Sass and Götz 2012). Together, these models support comprehensive evaluations of the entire deep geothermal potential of Hesse and can display the potential for open systems, including both hydrothermal or petrothermal enhanced geothermal systems (EGS), and closed systems utilizing deep borehole heat exchangers (Bär et al. 2011).

20.3.2 Three-dimensional Geological Model

The 3-D model (Arndt 2012) was constructed using Paradigm's GOCAD software. The model area covers more than 21 000 km² and has a maximum depth of 6 km (Figure 20.14). It consists of prominent faults and six stratigraphic model units: (i) combined "Quaternary/Tertiary" unit; (ii) "Muschelkalk" (mainly limestones and marls); (iii) "Buntsandstein" (sandstones, conglomerates and pelites); (iv) "Zechstein" (limestones, dolomites and evaporites); (v) "Permocarboniferous" (sandstones, conglomerates, pelites, and volcanics); and (vi) "pre-Permian" units divided, according to Kossmat (1927), into "Mid-German Crystalline Rise" (MGCR) consisting of felsic granitoids and subsidiary metamorphic and basic intrusive rocks, and "Rheno-Hercynian and Northern Phyllite Zone" (RH and NPZ) consisting of low-grade metamorphic and volcanoclastic source rocks. The model was designed to support evaluations of deep geothermal potential.

The model is based on the Geological Survey of Hesse 1:300 000 scale geological map (HLUG 2007). This was supplemented by more than 4150 well datasets, 318 geological cross-sections with a total length of more than 3700 km, mainly from 1:25000 geologic maps, as well as numerous isopach and contour maps (Figure 20.15). After assessing over 1500 2-D seismic profiles from hydrocarbon or potassium salt exploration campaigns, 29 were chosen to support model development (Arndt et al. 2011). The distribution of existing data is variable; more data are available for the potassium mining areas, and where hydrocarbon exploration has occurred in the Upper Rhine Graben. The model has quite large areas of very low data density. Therefore, the model horizons carry an additional property defining uncertainty using five uncertainty classes, ranging from "proven" to "very uncertain". These reflect the expert knowledge of the modeler and are based on data density and anticipated geologic complexity of localities.

Faults with a vertical displacement of at least 200 m were included. Instead of the customary vertical planes, these fault zones were modeled, but with their true dip angle as observed in the field or from seismic profiles (Figure 20.14b).

20.3.3 Three-dimensional Geothermal Model

Temperature, permeability and thermal conductivity are key parameters in geothermal reservoir characterization (Tester et al. 2006). Available publications and databases contained very few locations where more than one key property was measured. According to the thermo-facies

(a)

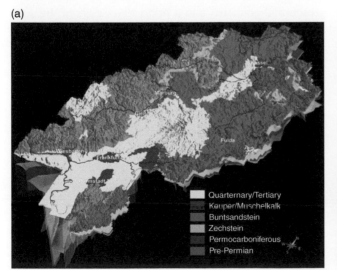

(b)

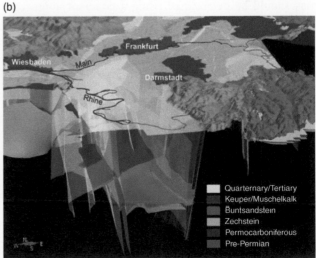

Figure 20.14 (a) Geological 3-D model of Hesse showing model units as well as major fault systems (transparent gray). The location of major cities (red) and rivers (blue) are given for orientation. (b) Detail of the geological 3-D model of the northern Upper Rhine Graben with the potential hydrothermal reservoir units Buntsandstein and Permocarboniferous bounded by the graben faults.

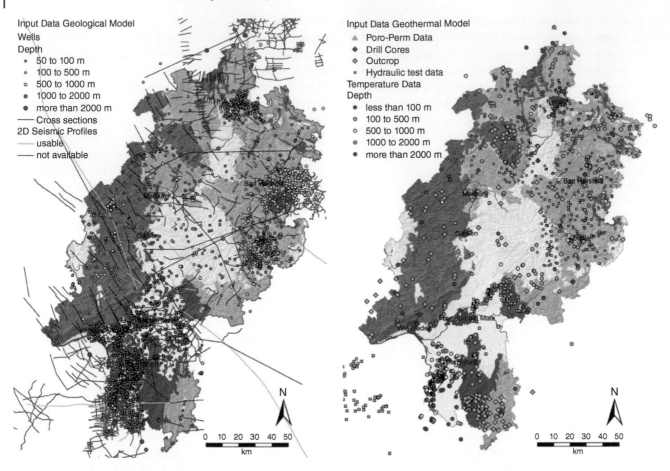

Figure 20.15 (a) Simplified geological survey map of the location of input data for the geological 3-D model including depth of the well data. Isopach or contour maps as well as existing 3-D models, which were incorporated into the model are not shown. (b) Input data used for the geothermal 3-D model showing the locations of all outcrop analog study locations conducted and all available drill cores, temperature data points, porosity-permeability datasets, and hydraulic test datasets.

concept by Sass and Götz (2012) all petrophysical properties should be determined using a coherent approach on the same set of samples for each facies type. To allow predictions of the geothermal properties, a dataset of outcrop observations at more than 600 locations, borehole data from more than 25 boreholes, core investigations of more than 500 m of cores, and hydraulic test data from more than 900 boreholes was compiled for all relevant formations (Figure 20.15b).

The subsurface temperature was modeled to a depth of 6 km using 2029 data points from the Leibniz Institute for Applied Geophysics (LIAG) and the geophysics archive of the HLNUG. Only observations at depths between 150 and 3105 m were used because shallower observations were potentially influenced by shallow thermal springs, seasonal influences, or paleo-climate signals. An empirical correlation of regional geothermal gradients and the depth of the Mohorovičić Discontinuity (Dèzes and Ziegler 2001) was used to define geothermal gradients. The gradients ranged between 24 and 40 K km^{-1} (Arndt et al. 2011).

An initial thermal model was developed using these geothermal gradients and a priori knowledge of the subsurface from the 3-D geology model. This model was compared with the observed geothermal temperatures and adjusted to match the observations as closely as possible. A final temperature model was produced where differences between predicted temperatures predicted by the model and observed values are within ±10 K. Inaccuracies occur in areas where temperature data are missing or where temperature data were measured in hydrothermal convection zones. However, this model predicted subsurface temperatures with acceptable accuracy (Figure 20.16). Subsequently, a more refined assessment of the conductive thermal field was undertaken to more accurately account for convective heat transport (Rühaak et al. 2014; Freymark et al. 2017) using a combination of conductive numerical modeling and quality-weighted interpolation.

The initial 3-D geological model used surfaces to define geological horizons and faults. The GOCAD Stratigraphic Grid (S-Grid) operation was used to create a solid cellular

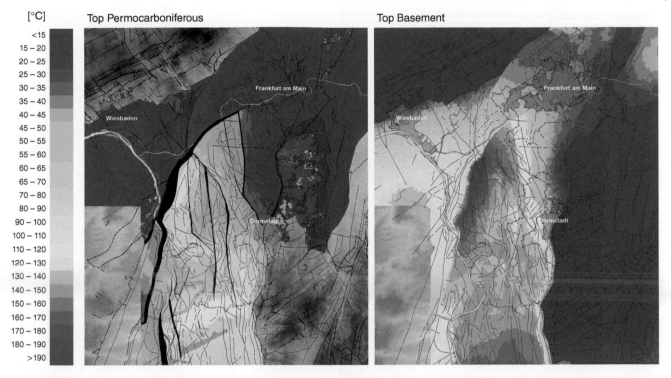

Figure 20.16 Temperature at the top of the Permocarboniferous (a) and pre-Permian basement (b) respectively close to the three biggest cities of Hesse.

3-D model whose cells could be assigned parameters defining all relevant geothermal properties (Figure 20.17). The S-Grid model is composed of conformable cells; their very small volumes allow precise definition of the geological horizons and fault surfaces (Mallet 2002).

Empirical functions were developed to define the depth and temperature dependence of the hydraulic properties of each unit by comparing outcrop observations with data from deep hydrocarbon exploration wells (Bär 2012). Thermophysical properties were defined by established functions from crustal-scale thermal models; these were adapted to represent Hesse's reservoir conditions (Zoth and Hänel 1988; Somerton 1992; Pribnow 1994; Vosteen and Schellschmidt 2003). This cellular version of the 3-D geological model was parameterized (i.e., attributed with geothermal properties) in GOCAD using these equations and the temperature model.

The resulting 3-D geothermal model contains all relevant depth-corrected and temperature-corrected geothermal properties for each geological unit, including: (i) thermal conductivity; (ii) thermal diffusivity; (iii) density; (iv) specific heat capacity; (v) porosity; (vi) matrix permeability; and (vii) bulk rock permeability (Figure 20.17). To account for fault damage zones, bulk rock permeability was gradually increased near fault systems by two orders of magnitude (Caine et al. 1996, Evans et al. 1997, Faulkner et al. 2010). The in situ stress field, as defined by the

world stress map (Heidbach et al. 2010), was not evaluated due to insufficient mechanical rock-property data. Transmissibility was calculated based on the fault corrected bulk rock permeability and the vertical thickness of the model units.

20.3.4 Quantification of Geothermal Potential

Assessment of the heat in place followed the volumetric approach of Muffler and Cataldi (1978). The extractability of heat and power production potential was assessed by considering known heat extraction rates and technical degrees of efficiency. These factors reflect reservoir temperature, effective porosity, and depth, as well as the project layout. Although they should be evaluated for specific sites, generalized values can be defined based on benchmark parameters (Jung et al. 2002). This approach was used to calculate the potential for geothermal power production for reservoir volumes with temperatures exceeding 100°C, and for direct heating applications where reservoir temperatures are between 60 and 100°C.

The deep geothermal potentials were evaluated using a multiple-criteria approach that incorporated the relevance of selected rock and reservoir properties for different geothermal systems. The most important parameters for a hydrothermal system are bulk rock permeability, transmissibility, and temperature. In contrast, the evaluation

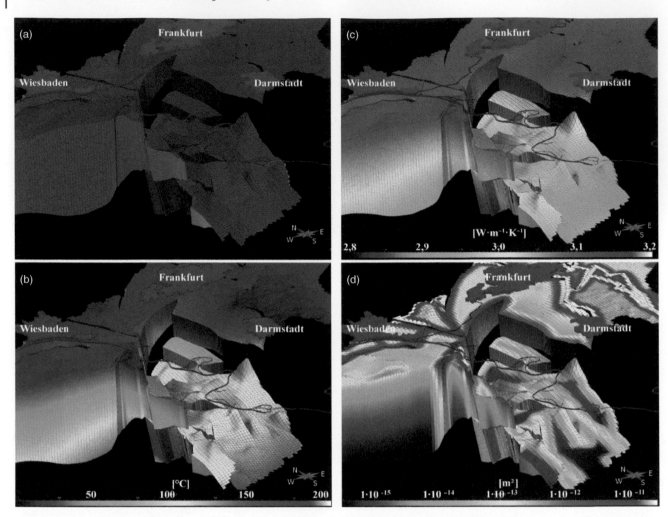

Figure 20.17 (a) S-grid of the Permocarboniferous, also showing following the properties: (b) temperature; (c) thermal conductivity; (d) and bulk permeability, including the influence of fault systems on the bulk rock permeability.

of petrothermal potential relies on temperature, and to a lesser degree on thermal conductivity, bulk rock permeability, and matrix permeability. The Analytic Hierarchy Process (AHP) introduced by Saaty (1980; 1990; 2005) provides a suitable multi-criteria decision support system for identifying the weightings to be applied to assess the relevant parameters for evaluating geothermal potential.

Arndt et al. (2011), Arndt (2012), and Bär (2012) provide details on the use of AHP within a GoCAD model. This method employs the cellular 3-D geothermal model and rapidly evaluates the relevant parameters for each geopotential system, and then identifies and visualizes different geothermal potentials using five classes, defined by benchmark parameters based on geothermal technical framework requirements Table 20.1. The five potential classes are: very high, high, medium, low, and very low.

Table 20.1 Benchmark parameters to define the degrees of efficiency for binary power plants.

Parameter[a]	Value
Minimum reservoir temperature (T_{MIN})	100°C
Maximum drilling depth (Z_{MAX})	7 km
Minimum extraction temperature (T_{MIN})	100°C
Injection temperature of thermal water (pure power production) (T_{IN})	70°C
Injection temperature of thermal water (PHC without HP) (T_{IN})	50°C
Injection temperature of thermal water (PHC with HP) (T_{IN})	30°C

a) PHC, Power-Heat Cogeneration; HP, Heat Pump.
Source: Modified after Jung et al. 2002

20.3.5 Results

High to very high deep geothermal potentials define natural reservoir conditions that are more than sufficient for economically feasible electric power production. Medium potentials define situations that may be eligible for federal geothermal R&D grants. Low to very low potentials are only capable of supporting heating of buildings. If measures to enhance reservoir properties are taken, enhanced geothermal systems (EGS) may be used in locations with low potential to generate geothermal electricity without the need for natural convective hydrothermal resources.

Medium to high hydrothermal potentials capable of supporting more than 600 TWh (2 EJ) of power production have been identified for the Permocarboniferous and the Buntsandstein successions within the northern Upper Rhine Graben and the adjacent Saar-Nahe Basin in the west (Table 20.2 and Figure 20.18). These high hydrothermal potentials are located along major faults within the graben; medium hydrothermal potentials were identified in the Permocarboniferous over almost the entire graben region at depths of more than 2000 m (Figure 20.18). Similar results were obtained for the Buntsandstein succession located further to the south and in the whole middle and southern Upper Rhine Graben. Hydrothermal potential is strongly influenced by increased transmissibility or bulk rock permeability along fault zones.

High petrothermal potentials with more than 10 000 TWh (36 EJ) of power production potential were identified for the granites, granodiorites and gneisses of the MGCR below the northern Upper Rhine Graben where temperatures exceed 150°C at depths of more than 3 km. The extensive fault damage zones in the Upper Rhine Graben may yield higher bulk rock permeability for the basement rocks than used in this geothermal model.

In other regions of Hesse, the crystalline or metamorphic bedrocks provide medium petrothermal potentials. Except for the quartzites, sandstones, and greywackes of the Taunus mountains, where about 5000 TWh (18 EJ) of power production potential are expected, the low-grade metamorphic rocks of the RH and NPZ are not well suited for petrothermal exploitation or reservoir enhancement by hydraulic fracturing, which is a prerequisite for power production by EGS technology.

20.3.6 Application in Urban Planning Processes

The hydrothermal potential of the Permo-carboniferous in the northern Upper Rhine Graben appears suitable for several large cities located adjacent to the graben

Table 20.2 Deep geothermal potential[a] of the different Hessian hydrothermal and petrothermal reservoir formations for enhanced geothermal systems (EGS)

Model units	Mean reservoir temperature (°C)	Volume (km³)	Heat in place (EJ)	Recoverable heat (EJ)	Power production potential (EJ)
RH and NPZ	110	9988	2020	42.5	4.25
	135	14 805	3830	134	15.1
	171	18 601	6370	287	35.9
RH and NPZ Total		43 394	12 220	463.5	55.2
MGCR	110	3930	959	20.1	2.01
	136	6107	1920	67.4	7.61
	197	18 934	9200	451	58.6
MGCR Total		28 971	12 080	538.5	68.2
Buntsandstein URG		40,7	13.1	2.65	0.32
Permocarboniferous URG		442	102.8	17.01	1.92

a) Geothermal potential estimates are based on technical degrees of efficiency for binary geothermal power plants (see Table 20.1).

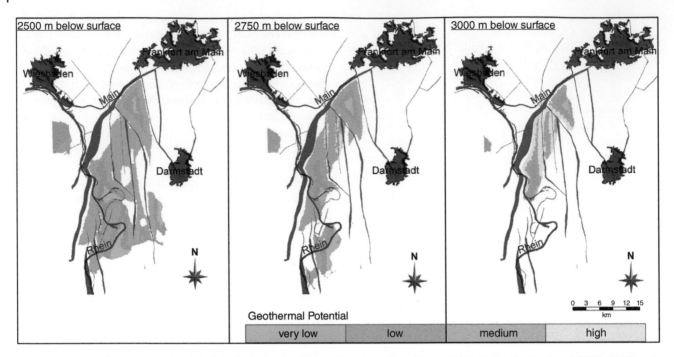

Figure 20.18 Hydrothermal potential of the Permocarboniferous in the northern Upper Rhine Graben in depths of, 2500, 2750, and 3000 m respectively. The location of major cities (red) and rivers (blue) are given for orientation. Gray shaded areas are the modeled fault interfaces within the Permocarboniferous.

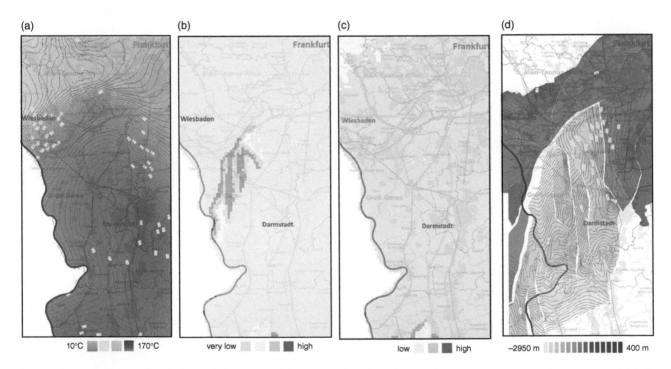

Figure 20.19 Four 2-D visualizations of the 3-D-model: (a) subsurface temperature at −3000 m depth as contours and continuous surface; (b) hydrothermal potential at −3000 m depth as potential classes; (c) petrothermal potential at −3000 m depth as potential classes; and (d) depth level of Permo-carboniferous model unit.

(Figure 20.14). These cities form a portion of the larger Frankfurt/Rhine-Main metropolitan region, one of the most dynamic and rapidly growing regions in Europe (Lehné et al. 2013). The availability of and access to geo-resources, including geothermal energy, will influence future urban development. The HLNUG has developed additional information within its spatial data infrastructure database to address conflicts over utilization of subsurface resources, and to support urban planning and decision making.

To meet the standard 2-D visualization capabilities of the existing GIS-platform and associated expert information system, the upper 4000 m of the 3-D subsurface has been subdivided into nine depth slices. Each depth slice contains information on the temperature (Figure 20.19a), hydrothermal potential (Figure 20.19b), and petrothermal potential (Figure 20.19c). Depth levels of modeled geological units are also available (Figure 20.19d). This approach maximizes the utility of the 3-D information because it recognizes (a) 3-D information can be directly accessed by a limited number of people, and (b) decision-making is still based on 2-D-maps. However, the HLNUG has introduced a 3-D database which is accessible by standard web browsers; this may bridge the gap between existing 3-D information and the requirements of many potential users of the information.

The ongoing research project "Hessen 3D 2.0" (Bär et al. 2016) focuses on medium deep geothermal potential for direct heating and seasonal heat storage, as well as the potential of EGS systems. It will employ 3-D city models to directly link geothermal potential estimates to heat demands and heat distribution infrastructures. Determination of mechanical rock properties and the influence of the orientation and magnitude of the in situ stress field are also research topics.

20.3.7 Conclusions

The 3-D geothermal model allows the quantification of the deep geothermal potential and is intended to be an instrument for public information. Because it incorporates the quantification and the analysis of the deep geothermal potentials, it is an important new tool that can be used during the initial planning for geothermal power plants. It is the first combined geological-geothermal 3-D model of an entire German federal state. The large geothermal database permits a reservoir prognosis on statistically confirmed parameters. Additionally, all thermo-physical and hydraulic parameters are depth and temperature corrected so that over- or under-estimations of the reservoir potentials are highly unlikely. The highly flexible multi-criteria approach used to evaluate geothermal potentials yields highly reproducible results. Furthermore, the model is capable of exploration risk prognosis and can also be used as a foundation for numerical reservoir models.

References

Acero, P., Gutiérrez, F., Galve, J.P. et al. (2013). Hydro-geochemical Characterization of an Evaporite Karst Area affected by Sinkholes (Ebro Valley, NE Spain). *Geologica Acta* 11: 389–407. https://doi.org/10.1344/105.000002052.

Alcaraz, M. (2016). *GIS Platform for Management of Shallow Geothermal Resources*. PhD thesis Polytechnic University of Catalonia (UPC). 248 pp. [Online: Available at: http://hdl.handle.net/2117/96311] (Accessed December 2020).

Allen, A., Milenic, D. and Sikora, P. (2003). Shallow gravel aquifers and the urban 'Heat Island' effect: a source of low enthalpy geothermal energy. *Geothermatics* 32: 569–578. https://doi.org/10.1016/S0375-6505(03)00063-4.

Arndt, D (2012). *Geologische Strukturmodellierung von Hessen zur Bestimmung von Geo-Potenzialen* (In German). PhD thesis Technical University of Darmstadt. 199 pp. [Online: Available at: https://tuprints.ulb.tu-darmstadt.de/3082] (Accessed December 2020).

Arndt, D., Bär, K., Fritsche, J.-G. et al. (2011). 3D structural model of the Federal State of Hesse (Germany) for geo-potential evaluation. *Zeitschrift der Deutschen Gesellschaft für Geowissenschaften* 162 (4): 353–370. https://doi.org/10.1127/1860-1804/2011/0162-0353.

Bär, K. (2012). *Untersuchung der tiefengeothermischen Potenziale von Hessen*. PhD thesis Technical University of Darmstadt. 265 pp. [Online: Available at: https://tuprints.ulb.tu-darmstadt.de/3067] (Accessed December 2020).

Bär, K., Arndt, D., Fritsche, J.-G. et al. (2011). 3D-Modellierung der tiefengeothermischen Potenziale von Hessen – Eingangsdaten und Potenzialausweisung (in German). *Zeitschrift der Deutschen Gesellschaft für Geowissenschaften* 162 (4): 371–388. https://doi.org/10.1127/1860-1804/2011/0162-0371.

Bär, K., Hintze, M., Welnert, S. et al. (2016). Das Verbund-projekt Hessen 3D 2.0 (in German). *Geothermische Energie* 85 (3): 24–25.

Boon, D., Farr, G., Patton, A. et al. (2016). The Contribution of Geology and Groundwater Studies to City-Scale Ground Heat Network Strategies: A Case Study from Cardiff, UK.

Geophysical Research Abstracts 18: EGU2016-4983. [Online: Available at: https://meetingorganizer.copernicus.org/EGU2016/EGU2016-4983.pdf] (Accessed December 2020).

Busby, J., Lewis, M., Reeve, H. and Lawley, R. (2009). Initial geological considerations before installing ground source heat pump systems. *Quarterly Journal of Engineering Geology and Hydrogeology* 42: 295–306. https://doi.org/10.1144/1470-9236/08-092.

Caine, J.S., Evans, J.P. and Forster, C.B. (1996). Fault zone architecture and permeability structure. *Geology* 24: 1025–1028. https://doi.org/10.1130/0091-7613(1996)024%3C1025:FZAAPS%3E2.3.CO;2.

Carrera, J., Alcolea, A., Medina, A. et al. (2005). Inverse problem in hydrogeology. *Hydrogeology Journal* 13: 206–222. https://doi.org/10.1007/s10040-004-0404-7.

CHE (2016). *Inventario de Puntos de Agua (IPA)*. Zaragoza, Spain: Ebro Hydrographic Confederation. [Website: http://www.chebro.es/contenido.visualizar.do?idContenido=14201&idMenu=3061] (Accessed December 2020).

Criollo, R., Velasco, V., Vázquez-Suñé, E. et al. (2016). An integrated GIS-based tool for aquifer test analysis. *Environmental Earth Sciences* 75: 1–11. https://doi.org/10.1007/s12665-016-5292-3.

Dèzes, P. and Ziegler, P.A. (2001). European map of the Mohorovičić discontinuity, version 1.3. In: *2nd EUCOR-URGENT Workshop: Upper Rhine Graben Evolution and Neotectonics*, Mont. Saint-Odile, France, October 7–11, 2001. France.

Epting, J., Händel, F. and Huggenberger, P. (2013). Thermal management of an unconsolidated shallow urban groundwater body. *Hydrology and Earth System Sciences* 17: 1851–1869. https://doi.org/10.5194/hess-17-1851-2013.

European Union (2009). Directive 2009/28/EC of the European Parliament and of the Council of 23 April 2009 on the promotion of the use of energy from renewable sources and amending and subsequently repealing Directives 2001/77/EC and 2003/30/EC. *Official Journal of the European Union* L140: 16–62. [Online: Available at: http://data.europa.eu/eli/dir/2009/28/oj] (Accessed December 2020).

Evans, J.P., Forster, C.B. and Goddard, J.V. (1997). Permeability of fault-related rocks and implications for hydraulic structure of fault zones. *Journal of Structural Geology* 19: 1393–1404. https://doi.org/10.1016/S0191-8141(97)00057-6.

Farr, G., Sadasivam, S.M., Watson, I.A. et al. (2016). Low enthalpy heat recovery potential from coal mine discharges in the South Wales coalfield. *International Journal of Coal Geology* 164: 92–103. https://doi.org/10.1016/j.coal.2016.05.008.

Farr, G.J., Patton, A.M., Boon, D.P. et al. (2017). Mapping shallow urban groundwater temperatures, a case study from Cardiff, UK. *Quarterly Journal of Engineering Geology and Hydrogeology* 50 (2): 187–198. https://doi.org/10.1144/qjegh2016-058.

Faulkner, D.R., Jackson, C.A.L., Lunn, R.J. et al. (2010). A review of recent developments concerning the structure, mechanics and fluid flow properties of fault zones. *Journal of Structural Geology* 32: 1557–1575. https://doi.org/10.1016/j.jsg.2010.06.009.

Freymark, J., Sippel, J., Scheck-Wenderoth, M. et al. (2017). The deep thermal field of the Upper Rhine Graben. *Tectonophysics* 694: 114–129. https://doi.org/10.1016/j.tecto.2016.11.013.

García-Gil, A., Vazquez-Sune, E., Schneider, E.G. et al. (2014). The thermal consequences of river-level variations in an urban groundwater body highly affected by groundwater heat pumps. *Science of the Total Environment* 485-486: 575–587. https://doi.org/10.1016/j.scitotenv.2014.03.123.

García-Gil, A., Vázquez-Suñé, E., Sánchez-Navarro, J.Á. et al. (2015a). The propagation of complex flood-induced head wavefronts through a heterogeneous alluvial aquifer and its applicability in groundwater flood risk management. *Journal of Hydrology* 527: 402–419. https://doi.org/10.1016/j.jhydrol.2015.05.005.

García-Gil, A., Vázquez-Suñé, E., Sánchez-Navarro, J.A. and Lázaro, J. (2015b). Recovery of energetically overexploited urban aquifers using surface water. *Journal of Hydrology* 1: 111. 531 (3): 602–611. https://doi.org/10.1016/j.jhydrol.2015.10.067.

García-Gil, A., Vázquez-Suñe, E., Schneider, E.G. et al. (2015c). Relaxation factor for geothermal use development – criteria for a more fair and sustainable geothermal use of shallow energy resources. *Geothermics* 56: 128–137. https://doi.org/10.1016/j.geothermics.2015.04.003.

Garrido Schneider, E.A., García-Gil, A., Vázquez-Suñè, E. and Sánchez-Navarro, J.Á. (2016). Geochemical impacts of groundwater heat pump systems in an urban alluvial aquifer with evaporitic bedrock. *Science of the Total Environment* 544: 354–368. https://doi.org/10.1016/j.scitotenv.2015.11.096.

Görne, S. and Krentz, O. (2015). Cross border 3D modelling – challenges and results of a transnational geothermal project. In: *Proceedings IAMG 2015, Freiberg, Germany, September 5–13, 2015* (ed. H. Schäben), 883–885. Houston, TX: International Association for Mathematical Geosciences.

Gutiérrez, F., Calaforra, J.M., Cardona, F. et al. (2007). Geological and environmental implications of the evaporite karst in Spain. *Environmental Geology* 53: 951–965. https://doi.org/10.1007/s00254-007-0721-y.

Gutiérrez, F., Galve, J.P., Lucha, P. et al. (2009). Investigation of a large collapse sinkhole affecting a multi-storey building by means of geophysics and the trenching technique (Zaragoza city, NE Spain). *Environmental Geology* 58: 1107–1122. https://doi.org/10.1007/s00254-008-1590-8.

Hänel, R. and Staroste, E. [Eds] (1988). *Atlas of Geothermal Resources in the European Community, Austria and Switzerland*. Luxembourg: Office for Official Publications of the European Communities, Publication No. EUR 11026 of the European Commission. 74 pp. 110 plates.

Heidbach, O., Tingay, M., Barth, A. et al. (2010). Global crustal stress pattern based on the world stress map database release 2008. *Tectonophysics* 482: 1–15. https://doi.org/10.1016/j.tecto.2009.07.023.

Herbert, A., Arthur, S. and Chillingworth, G. (2013). Thermal modeling of large scale exploitation of ground source energy in urban aquifers as a resource management tool. *Applied Energy* 109: 94–103. https://doi.org/10.1016/j.apenergy.2013.03.005.

Herms, I., Goetzl, G., Borovic, S. et al. (2019). MUSE – Managing Urban Shallow Geothermal Energy. A GEOERA GEO-Energy Project. In: *Proceedings, European Geothermal Congress 2019, The Hague, 11–14 June 2019*. Brussels, Belgium: European Geothermal Energy Council. 6 pp. [Online: Available at: http://europeangeothermalcongress.eu/wp-content/uploads/2019/07/41.pdf] (Accessed December 2020).

HLUG (2007). *Geologische Übersichtskarte von Hessen, 1 : 300 000 (Geological Overview Map of Hessen, 1 . 300 000)*. Wiesbaden, Germany: Hessisches Landesamt für Umwelt und Geologie. 1 map. [Online: Available at: https://www.hlnug.de/fileadmin/dokumente/geologie/geologie/guek300.pdf] (Accessed December 2020).

Hurter, S. and Hänel, R. [Eds] (2002). *Atlas of geothermal resources in the European Community, Austria and Switzerland*. Luxembourg: Office for Official Publications of the European Communities, Publication No. EUR 17811 of the European Commission. 91 pp. 89 plates. [Online: Available at: https://op.europa.eu/en/publication-detail/-/publication/9003d463-03ed-4b0e-87e8-61325a2d4456] (Accessed December 2020).

Hurter, S. and Schellschmidt, R. (2003). Atlas of geothermal resources in Europe. *Geothermics* 32 (4–6): 779–787.

IGN (2012). *Modelo Digital del Terreno (Digital Terrain Model)*. Madrid, Spain: National Geographic Institute (IGN). [Online: Available at: https://www.ign.es/web/seccion-elevaciones] (Accessed December 2020).

Jaudin, F., Angelino, L., Annunziate, E. et al. (2013). *Overview of Shallow Geothermal Legislation in Europe*. REGEOCITES EU Project D2.2: General Report of the Current Situation of the Regulative Framework for the SGE Systems. 49 pp. [Online: Available at: https://ec.europa.eu/energy/intelligent/projects/sites/iee-projects/files/projects/documents/regeocities_shallow_geothermal_legislation_in_europe_en.pdf] (Accessed December 2020).

Jung, R., Röhling, S., Ochmann, N. et al. (2002). *Abschätzung des technischen Potenzials der geothermischen Stromerzeugung und der geothermischen Kraft-Wärme-Kopplung (KWK) in Deutschland – Bericht für das Büro für Technikfolgenabschätzung beim Deutschen Bundestag* (in German). Hannover, Germany: BGR/GGA, Archiv-Nr. 122 458.

Kossmat, F. (1927). *Gliederung des varistischen Gebirgsbaues* (in German). Dresden, Germany: Sachsischen Geologischen Landesamts, Abhandlungen des Sachsischen Geologischen Landesamts, Heft 1. 39 pp.

Kozdrój, W., Kłonowski, M., Mydłowski, A. et al. (2014). 3D Geological Modelling and Geothermal Mapping – The First Results of the Transboundary Polish – Saxon Project "TransGeoTherm". *Geophysical Research Abstracts* 16: EGU2014-5132. [Online: Available at: https://meetingorganizer.copernicus.org/EGU2014/EGU2014-5132.pdf] (Accessed December 2020).

Krentz, O., Kozdrój, W. and Opletal, M. [Eds.] (2001). *Comments on the Geological Map Lausitz-Jizera-Karkonosze (without Cenozoic sediments), 1:100 000*. Warsaw, Poland: Pażstwowy Instytut Geologiczny/Praha, Czechia: Cesky Geologicky Ustav/Dresden, Germany: Landesvermessungsamt. 64 pp.

Lehné, R.J., Hoselmann, C., Heggemann, H. et al. (2013). Geological 3D modelling in the densely populated metropolitan area Frankfurt/Rhine-Main. *Zeitschrift der Deutschen Gesellschaft für Geowissenschaften* 164 (4): 591–603. https://doi.org/10.1127/1860-1804/2013/0051.

Lund, J.W. and Boyd, T.L. (2016). *Direct Utilization of Geothermal Energy 2015 Worldwide Review. Geothermics* 60: 66–93. https://doi.org/10.1016/j.geothermics.2015.11.004.

Mallet, J.-L. (2002). *Geomodeling*. New York, NY: Oxford University Press. 612 pp.

Małolepszy, Z., Dąbrowski, M., Mydłowski, A. et al. (2015). Modelling the Geological Structure of Poland – Approach, Recent Results and Roadmap. *Proceedings, Three-Dimensional Geological Mapping Workshop, 2015 Annual Meeting* (Eds. K. MacCormack, H. Thorleifson, D. Berg and H. Russell), 47–50. Edmonton, AL: AER/AGS Special Report 101. [Online: Available at: https://isgs.illinois.edu/2015-baltimore-maryland] (Accessed December 2020).

MOPU (1989). *Estudio de los recursos hidráulicos subterráneos de los acuíferos relacionados con la provincia de Zaragoza, Aluvial del río Ebro (tramo Cortes – Zaragoza) escala 1:50000* (in Spanish). Madrid, Spain: Direccion General de Obras Hidráulicas, Ministerio de Obras Publicas.

Muffler, P. and Cataldi, R. (1978). Methods for regional assessment of geothermal resources. *Geothermics* 7: 53–89. https://doi.org/10.1016/0375-6505(78)90002-0.

Patton, A.M., Farr, G.J., Boon, D.P. et al. (2015). *Shallow Groundwater Temperatures and the Urban Heat Island Effect: The First U.K. City-Wide Geothermal Map to support Development of Ground Source Heating Systems Strategy. Geophysical Research Abstracts* 17: EGU2015-3488. [Online: Available at: https://meetingorganizer.copernicus.org/ EGU2015/EGU2015-3488.pdf] (Accessed December 2020).

PGI-NRI (2017). *The Central Geological Database* [Online: Available from: http://m.bazagis.pgi.gov.pl/cbdg/#/main] (Accessed: 21 August, 2017).

Pribnow, D. (1994). *Ein Vergleich von Bestimmungsmethoden der Wärmeleitfähigkeit unter Berücksichtigung von Gesteinsgefügen und Anisotropie* (in German). Düsseldorf, Germany: VDI Verlag, Fortschrittsberichte 19 (75). 111 pp.

Robador Moreno, A., Ramajo Cordero, J., Muñoz Jiménez, A. et al. (2011). *Digital Continuous Geological Map of Spain (GEODE), 1:50.000*. Madrid, Spain: Instituto Geológico y Minero de España (IGME).

Rühaak, W., Bär, K. and Sass, I. (2014). Combining Numerical Modeling with Geostatistical Analysis for An Improved Reservoir Exploration. *Geophysical Research Abstracts* 16: EGU2014-8658. [Online: Available at: https:// meetingorganizer.copernicus.org/EGU2014/EGU2014-8658.pdf] (Accessed December 2020).

Saaty, T.L. (1980). *The Analytic Hierarchy Process: Planning, Priority Setting, Resource Allocation*. NY: McGraw-Hill. 287 pp.

Saaty, T.L. (1990). How to make a decision – the analytic hierarchy process. *European Journal of Operational Research* 48 (1): 9–26. https://doi.org/10.1016/0377-2217(90)90057-I.

Saaty, T.L. (2005). The analytic hierarchy and analytic network process for the measurement of intangible criteria and for decision-making. In: *Multiple Criteria Decision Analysis – State of the Art Surveys* (ed. S. Greco), 345–407. New York: Springer-Verlag.

Sass, I. and Götz, A.E. (2012). Geothermal reservoir characterization: a thermofacies concept. *Terra Nova* 24 (2): 142–147. https://doi.org/10.1111/j.1365-3121.2011 .01048.x.

Schulz, R. and GEOTIS-TEAM (2009). *Aufbau eines geothermischen Informationssystems für Deutschland, Endbericht* (in German). Hannover, Germany: Leibniz-Institut für Angewandte Geophysik. 114 pp.

[Online: Available at: https://www.geotis.de/homepage/sitecontent/info /publication_data/final_reports/final_reports_data /GeotIS2_Endbericht.pdf] (Accessed December 2020).

Somerton, W.H. (1992). *Thermal properties and temperature-related behavior of rock-fluid systems*. Amsterdam, Netherlands: Elsevier, Developments in Petroleum Science 37. 256 pp.

Tester, J.W., Anderson, B.J., Batchelor, A.S. et al. (2006). *The Future of Geothermal Energy: Impact of Enhanced Geothermal Systems (EGS) on the United States in the 21st Century*. Idaho Falls, ID: Idaho National Laboratory, Report INL/EXT-06-11746. 72 pp. [Online: Available at: https://inldigitallibrary.inl.gov/sites/sti/sti/3589644.pdf] (Accessed December 2020).

TransGeoTherm (2015). *TransGeoTherm: Geothermal Energy for Transboundary development of the Neisse Region – Pilot Project*. [Website: http://www.transgeotherm.eu/index.en .html] (Accessed December 2020).

Velasco, V. (2013). *GIS-based Hydrogeological Platform for Sedimentary Media*. PhD thesis Polytechnic University of Catalonia (UPC). 158 pp. [Online: Available at: https:// www.tdx.cat/handle/10803/135005] (Accessed December 2020).

Velasco, V., Gogu, R., Vazquez-Sune, E. et al. (2012). The use of GIS-based 3-D geological tools to improve hydrogeological models of sedimentary media in an urban environment. *Environmental Earth Sciences* 68: 2145–2162. https://doi.org/10.1007/s12665-012-1898-2.

Velasco, V., Tubau, I., Vázquez-Suñè, E. et al. (2014). GIS-based hydrogeochemical analysis tools (QUIMET). *Computers & Geosciences* 70: 164–180. https://doi.org/10 .1016/j.cageo.2014.04.013.

Vosteen, H.D. and Schellschmidt, R. (2003). Influence of temperature on thermal conductivity, thermal capacity and thermal diffusivity for different types of rock. *Physics and Chemistry of the Earth* 28: 499–509. https://doi.org/10 .1016/S1474-7065(03)00069-X.

Woloszyn, I., Merkel, B. and Stanek, K. (2015). 3D geological modeling of the transboundary basin Berzdof-Radomierzyce in Upper Lusatia (Germany/Poland). *Geophysical Research Abstracts* 17: EGU2015-574. [Online: Available at: https://meetingorganizer.copernicus.org/ EGU2015/EGU2015-574.pdf] (Accessed December 2020).

Zoth, G. and Hänel, R. (1988). Appendix. In: *Handbook of Terrestrial Heat-Flow Density Determination* (eds. R. Hänel, L. Rybach and I. Stegena), 449–466. Dordrecht: Kluwer Academic Publishers.

21

Application Theme 4 – Regulatory Support

Editor's Introduction

The support of regulation and management of subsurface uses is a new application of 3-D geological models, which is rapidly developing alongside their well-established application in the hydrocarbon sector. To date, the primary regulatory use of such 3-D geological models has been to support groundwater resource management (Chapters 6 and 19). More recently, 3-D models have been used for geothermal resource assessment and management; Chapter 20 provides three geothermal case studies and summarizes additional recent developments. Regulation and management issues also influence the use of models for geohazard mitigation (Chapter 22) and infrastructure (Chapter 23). Chapters 3 and 16 explore in some detail the various interactions between geological information (including models) and the urban planning issues surrounding societal desires for environmental protection, resilience, and sustainability. Chapter 14 provides additional information on the interactions between 3-D geological models and 4-D process models

Regulation of the subsurface is evolving in many countries, and the use of appropriate technologies to define and assess alternative solutions to address societal goals is increasingly becoming recognized as an inherent part of urban planning and resource management processes. Chapter 18 has three case studies of 3-D models for urban planning, including Glasgow where a long-term collaboration between the British Geological Survey and Glasgow City Council has resulted in establishment of a contractual requirement to deposit and share data in a knowledge exchange network "Accessing Subsurface Knowledge (ASK)", which provides links to a broad range of stakeholders who either use and/or generate subsurface data within any stage of the city development process (Bonsor 2017). The use of 3-D models is particularly important in the very rapidly growing cities of south-east Asia (see Case Study 18.3, which describes a model for Dhaka, Bangladesh).

The first case study describes how a 3-D geological model provided by the British Geological Survey assists the Environment Agency of England and Wales (EA) in assessing alternative sources of water supplies in the Hertfordshire and North London (HNL) area where the current groundwater withdrawals are adversely affecting the baseflow of several streams originating in the Chalk Group aquifer. The model supports the EA with appropriate responses to demands by public water suppliers and large industrial users for new licenses to abstract water from the Lower Greensand Group confined aquifer as an alternative water source to meet their business needs.

The second case study describes the regulatory uses of 3-D geological models by the Federal State of Bremen, the smallest of the three German city states, which consists of the two cities Bremen and Bremerhaven. Located in northwestern Germany, this densely populated industrial area experiences a wide range of subsurface problems related to building construction, transportation, and water supply infrastructure improvements and utilization of geothermal energy sources. These activities are influenced by contaminated soils at many locations, the risk of flooding due to storm events and, over the longer term, rising sea level and associated ground water salinization. In response to these concerns, the Geological Survey for Bremen has developed a 3-D geological model of the subsurface Quaternary and Neogene units. This model and a derived steady-state groundwater flow model support government administrative actions, planning for construction projects, and issues related to groundwater use and geothermal development. Products derived from these models are provided to public and private sectors, partially through a web-service.

Applied Multidimensional Geological Modeling: Informing Sustainable Human Interactions with the Shallow Subsurface, First Edition.
Edited by Alan Keith Turner, Holger Kessler, and Michiel J. van der Meulen.
© 2021 John Wiley & Sons Ltd. Published 2021 by John Wiley & Sons Ltd.

Case Study 21.1: The use of 3-D Models to Manage the Groundwater Resources of the Lower Greensand Confined Aquifer, Hertfordshire and North London, England

Catherine Cripps[1], Michael Kehinde[2], Melinda Lewis[1] and Marieta Garcia-Bajo[1]

[1] British Geological Survey, Keyworth, Nottingham NG12 5GG, UK
[2] Environment Agency, Welwyn Garden City AL7 1HE, UK

21.1.1 Introduction

The Hertfordshire and North London (HNL) area of England is currently water-stressed; provision of adequate water supplies while balancing ecologic and economic goals is challenging. Current groundwater withdrawals from the Chalk Group aquifer, the major groundwater source, are adversely affecting the baseflow of several streams originating in areas underlain by chalk. The Environment Agency, in its role as a public regulatory body, is implementing a comprehensive program of investigations and actions to return the HNL area to a sustainable abstraction regime. Public water suppliers and large industrial users are being compelled to look for alternative water sources to meet their business needs. The Lower Greensand Group confined aquifer, despite its higher development costs and lower yields, is becoming an increasingly important alternative source.

The Environment Agency must consider the concerns and interests of all user-groups when arriving at licensing decisions. Several water suppliers have recently requested the Environment Agency provide new licenses to abstract water from the Lower Greensand Group aquifer. Currently, the Environment Agency has no formal strategy for managing abstractions from the Lower Greensand Group aquifer; each application decision is dealt with on a case-by-case basis. Existing hydrogeological knowledge of the HNL area is relatively limited. Better knowledge would help achieve the Environment Agency goal of enhanced, sustainable development of the Lower Greensand Group aquifer. The Environment Agency commissioned the British Geological Survey (BGS) to construct a 3-D geological and hydrogeological model of the Lower Greensand Group aquifer (Figure 21.1).

21.1.2 Geological Setting

The bedrock in the HNL study area consists of a suite of Mesozoic to Cenozoic marine sedimentary rocks (Table 21.1) that dip gently to the southeast (Hopson et al. 1996, 2008). The Lower Greensand Group aquifer within the model area consists of the Woburn Sands Formation, a Cretaceous shallow marine sandstone; it is supplemented by several minor sand formations (Hopson et al. 2008). These sands are confined where relatively impermeable formations above the Lower Greensand Group form aquicludes. A varied sequence of shallow Quaternary deposits (Table 21.2) overlies a northeast–southwest trending outcrop zone of the Lower Greensand Group. This creates an important recharge zone for the Lower Greensand Group. A buried glacial channel, termed the "Hitchin Buried Valley" extends in a southerly direction in the area around Hitchin (Hopson et al. 1996). Filled with over 100 m thick glacial deposits, it extends into the underlying bedrock formations, creating an important hydrogeological feature.

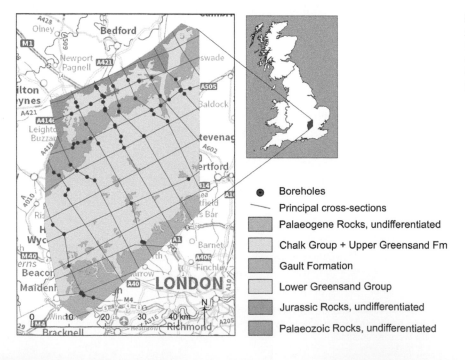

Boreholes
Principal cross-sections
Palaeogene Rocks, undifferentiated
Chalk Group + Upper Greensand Fm
Gault Formation
Lower Greensand Group
Jurassic Rocks, undifferentiated
Palaeozoic Rocks, undifferentiated

Figure 21.1 Location of the model area, showing the bedrock geology and the locations of boreholes and principal cross-sections. Source: Contains Ordnance Survey data, © Crown Copyright and database rights 2017.

Table 21.1 Bedrock stratigraphy of the Lower Greensand Group aquifer model area with hydrogeological properties.

Geologic name in model	Composition	Aquifer designation	Flow mechanism	Productivity
Palaeogene rocks, undifferentiated	Mudstone, siltstone, and minor sandstone	Unproductive	Limited groundwater	Limited groundwater
Chalk Group (including Upper Greensand Formation)	Chalk	Principal	Fracture	High
Gault formation	Mudstone	Unproductive	Limited groundwater	Limited groundwater
Lower Greensand Group	Sandstone	Principal	Intergranular	High
Jurassic rocks, undifferentiated	Mostly mudstones	Unproductive	Limited groundwater	Limited groundwater
Paleozoic rocks, undifferentiated	Crystalline basement	Secondary A	Fracture	Low

Table 21.2 Surficial deposits overlying the Lower Greensand Group aquifer outcrop area, the aquifer recharge zone. They are in stratigraphic order, youngest on top. Important hydrogeological characteristics are shown.

Geologic unit	Composition	Aquifer designation	Flow mechanism	Productivity
Peat	Peat	Unproductive	Limited groundwater	Limited groundwater
Alluvium	Clay, silt, sand, and gravel in varying proportions	Secondary A	Intergranular	Moderate
Head	Clay, silt, sand, and gravel in varying proportions	Secondary B	Intergranular	Low
River terrace deposits	Clay, silt, sand, and gravel in varying proportions	Secondary A	Intergranular	Moderate
Glaciofluvial deposits	Sand and gravels	Secondary A	Intergranular	Moderate
Till	Diamicton	Secondary B	Limited groundwater	Limited groundwater
Glaciolacustrine deposits	Clays and silts	Unproductive	Limited groundwater	Limited groundwater

21.1.3 Developing the 3-D Lower Greensand Group Subsurface Model

The 3-D geological model of the Lower Greensand Group aquifer (Figure 21.2) was developed by following the borehole and cross-section approach for model construction discussed in Chapter 10. The GSI3D software (Kessler et al. 2009) was used to develop the 3-D model. However, data preparation and model visualization utilized additional software, including Microsoft Access, Esri ArcGIS, and GroundHog desktop (Wood et al. 2015). The model was constructed with six bedrock units and 16 Quaternary surficial units (Table 21.3). Model units represent critical hydrogeological characteristics of the geologic model area; permeable bedrock units with no hydraulic connectivity to the Lower Greensand Group confined aquifer are undifferentiated in the model. The model includes Quaternary surficial deposits which overlie the Lower Greensand Group aquifer outcrop area and influence the potential for aquifer recharge.

21.1.3.1 Data Selection and Preparation

Lithological and geophysical borehole records held in the BGS database are the primary source of subsurface information. Thirty-four geophysical logs helped define geological unit boundaries. Lithologic borehole logs were of variable quality; many predated the 1950s and so contained obsolete terms or inaccurate measurements. Moreover, they were not optimally placed and many are shallow borings. There is a paucity of information in the southeastern model area. Several areas have clusters of shallow boreholes. The model is based on 203 representative boreholes selected from 20 546 boreholes within the model area. Key stratigraphic surfaces were defined using Groundhog desktop (Wood et al. 2015); this materially assisted the definition of key stratigraphic surfaces. The Ordnance Survey Terrain 50 data set (Ordnance Survey Limited 2017) defined the topography; DiGMapGB-50, a digital BGS geological map at 1:50 000 scale (British Geological Survey 2020), defined the spatial extents of geological units. Both datasets were converted to 50 m grids for use in the model.

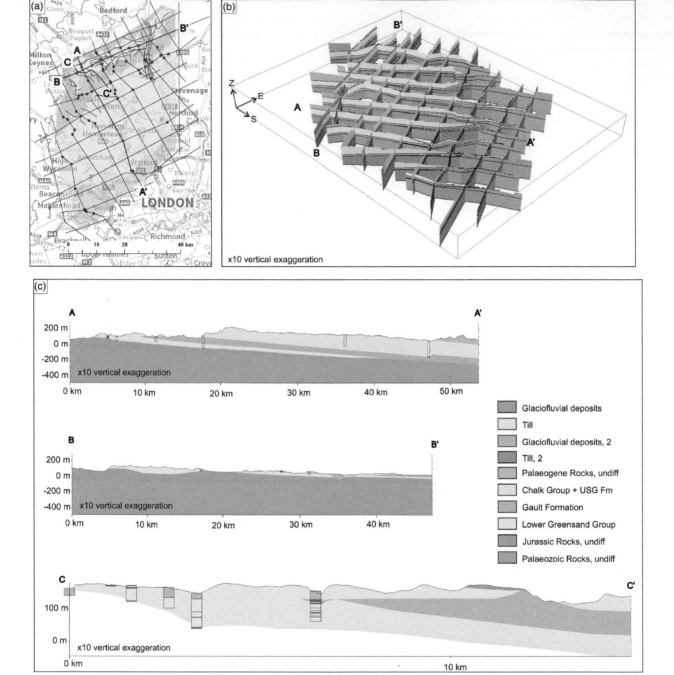

Figure 21.2 (a) Location of the geological cross-sections; (b) 3-D fence diagram of the model cross-sections; (c) example of three cross-sections, a and b being "principal" cross-sections, and c being an example of a "helper" cross-section.

GSI3D controlled the model stratigraphic relationships using a digital Generalized Vertical Section (Table 21.3).

21.1.3.2 Model Construction

The model extends from the surface to a depth of 500 m below Ordnance Datum and covers an area of 2349 km² (907 mi²). The model was constrained by 20 interpreted geological sections; 10 sections were oriented NW–SE and 10 sections NE–SW (Figure 21.2). This orientation of the sections roughly follows the regional dip and strike. Five geological cross-sections

from the nationwide geological fence diagram UK3D (Waters et al. 2015) were used to guide model development. The model initially was formed as a 3-D fence diagram (Figure 21.2b). Outcrop locations defined by DiGMapGB-50 (British Geological Survey 2020), supplemented by digitized sub-crop limits, defined the lateral extents of each geological unit. Base surfaces of alluvium and head deposits were defined by assuming a thickness for each unit within its extent. This procedure accurately represented these units while avoiding additional cross-sections.

Table 21.3 Selected entries from the Generalized Vertical Section used in the GSI3D modeling process.

Geologic name	Code used in model	Unit description	Aquifer designation	Flow mechanism and permeability
Peat	PEAT-P	Peat	Unproductive	n/a
Alluvium	ALV-XCZSV	Alluvium	Secondary A	Intergranular-moderate
Head	HEAD-XCZSV	Head; overlies terraces, cut by alluvium and sub-alluvial gravels	Secondary B	Intergranular-low
River terrace deposits	FELM-XSV	Felmersham member	Secondary A	Intergranular-moderate
	RTD1-XSV	River terrace deposits – first terrace	Secondary A	Intergranular-moderate
	RTD2-XSV	River terrace deposits – second terrace	Secondary A	Intergranular-moderate
	T1T2-XSV	River terrace deposits – first and second terrace	Secondary A	Intergranular-moderate
	RTDU-XSV	River terrace deposits – undifferentiated	Secondary A	Intergranular-moderate
Glaciofluvial deposits	GFDMP-XSV	Glaciofluvial deposits – Mid Pleistocene	Secondary A	Intergranular-moderate
Oadby member	ODT-DMTN	Oadby member	Secondary B	n/a
Glaciofluvial deposits, 2	GFDMP2-XSV	Glaciofluvial deposits – Mid Pleistocene	Secondary A	Intergranular-moderate
Till, 1	TILL1-DMTN	Till	Secondary B	n/a
Glaciofluvial deposits, 3	GFDMP3-XSV	Glaciofluvial deposits – Mid Pleistocene	Secondary A	Intergranular-moderate
Glaciolacustrine deposits	GLDMP-XSC	Glaciolacustrine deposits – Mid Pleistocene	Unproductive	n/a
Till, 2	TILL2-DMTN	Till	Unproductive	n/a
Quaternary deposits	QUU-UNKN	Quaternary deposits – undifferentiated	n/a	n/a
Palaeogene	PGU-MDSD	Palaeogene rocks – undifferentiated	Unproductive	n/a
Chalk Group and Upper Greensand formation	CK-CHLK	Chalk and upper greensand formations – undifferentiated	Principal	Fracture-high
Gault formation	GLT-MDST	Gault formation	Unproductive	n/a
Lower Greensand group	LGS-SDST	Lower Greensand formation	Principal	Intergranular-high
Jurassic rocks	JURA-MDLM	Jurassic Rocks – undifferentiated	Unproductive	n/a
Paleozoic rocks	RZRU-MDSL	Paleozoic Rocks – undifferentiated	Secondary A	Fracture-low

An iterative process was employed to convert the network of cross-sections into a completed 3-D model (Figure 21.3). Adjustments were required in areas that did not conform to expected geological conditions. In these locations, new helper cross-sections were developed to constrain geological units and produce a satisfactory model.

21.1.4 Model Products

The completed 3-D model was used to produce several digital products for the Environment Agency and the groundwater user community. Initial outputs included some 2-D derived products that could be used by standard GIS software, including:

- ASCII grids of the tops, bases, and thicknesses of the geological units;
- ESRI shapefiles of the lateral extents ("geological envelopes") of each geological unit;

- A "hydro-domain map", which provides users with a 2-D characterization of the hydrogeological properties of the intercalated surficial deposits in the recharge area (Figure 21.4). The map was created using procedures defined by Lelliott et al. (2006).

The completed 3-D model was provided to the Environment Agency accompanied by the free BGS "Lithoframe viewer" which allows viewing, investigation, and query of the 3-D geological model with a simple GIS interface (Waters et al. 2015).

Further output products were specifically related to the hydrogeology. Attributes defining the aquifer designation, the flow mechanism, and the productivity characteristics of each geologic unit (shown in Tables 21.1–21.3) allowed the model and sliced virtual cross-sections to be colored to define important geohydrological considerations (Figure 21.5). A full scientific report (Cripps et al. 2017)

(a) Looking East
x10 vertical exaggeration
Length of section closest to viewer is approx 75 km

(b) Looking North East
x10 vertical exaggeration
Length of outcrop area is approx 60 km

Figure 21.3 (a) The calculated geological model; (b) close up of the 3-D calculated surficial geology of the Lower Greensand Group outcrop/recharge zone.

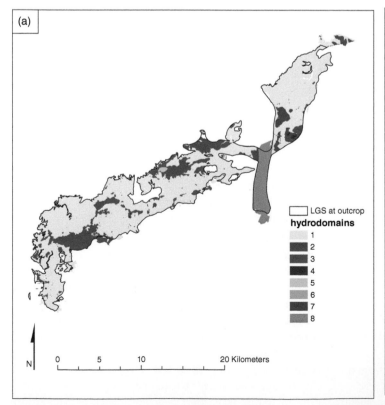

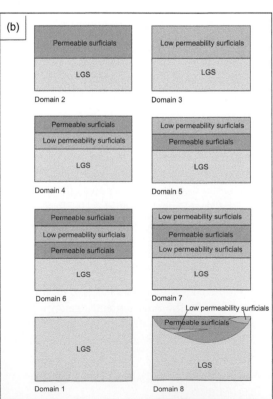

Figure 21.4 (a) "Hydro-domain map" of the Lower Greensand Group (LGS) outcrop zone; (b) "Hydro-domain" characterization.

Figure 21.5 (a) Location map identifying example cross-section outlined below; (b) example geological cross-section showing: Geology (top); Aquifer Designation (middle); and Flow Mechanism, Productivity (bottom).

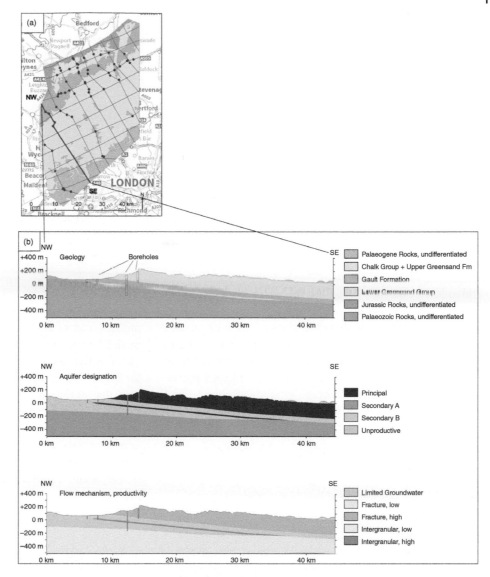

contains further hydrogeological information and provides a succinct summary of current hydrogeological knowledge for this area.

21.1.5 Applications of the 3-D Model at the Environment Agency

The Lower Greensand Group model provides a summary of the state of current geological and hydrogeological knowledge of the Lower Greensand Group in the model area. Water level data from the Environment Agency Cam-Bedford-Ouse (CBO) MODFLOW numerical groundwater model (ENTEC 2007), was added to the GSI3D geological model. This provided a crucial link in the decision-making process. It allowed the water table or potentiometric surface for each aquifer to be computed and displayed, providing for better management of the confined Lower Greensand Group aquifer as a water resource.

The Environment Agency is now embracing the use of 3-D models for improved decision-making at catchment and regional scales. The 3-D models are being used internally to educate staff and enhance their understanding of subsurface geology implications as well as for communicating with stakeholders to explain complex geological issues in a format readily understood by non-experts.

In the upper reaches of the River Chess catchment, which forms part of the regional Chalk Group aquifer in the HNL area, groundwater abstractions for public water supplies are apparently contributing to low river flows. To support hydrogeological investigation of this situation, the Environment Agency, in conjunction with both Thames Water and Affinity Water, commissioned the BGS to build an additional 3-D geological model of the River Chess catchment area (Farrant et al. 2016). The model (Figure 21.6) provided an improved understanding of the stratigraphic variations in the chalk, its spatial variability in the catchment, and how this influences aquifer response to abstraction. The model has been used to illustrate three static groundwater level scenarios: long term average, 2001

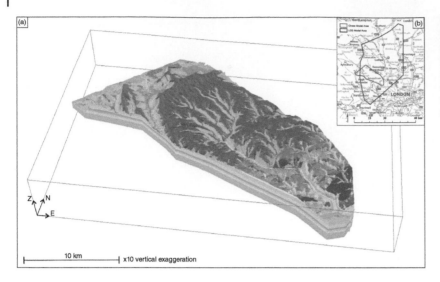

Figure 21.6 (a) Calculated model of the River Chess catchment area, showing bedrock Cretaceous strata (different green colors) and Quaternary Surficial deposits (all other colors). (Farrant et al. 2016). (b) Location map of the Chess catchment model (blue) within the Lower Greensand Group (LGS) model area (red).

flood level, and 2006 drought level. These scenarios allow individuals to relate their experiences of past extreme climatic events with potential groundwater resource use and management practices. The model has also been used to explain how groundwater resources can be protected from human activities and influences and has allowed refinement of the southwest Chilterns groundwater model and improved model calibration in the River Chess catchment.

Case Study 21.2: Regional 3-D Models of Bremen, Germany: Management Tools for Resource Administration

Björn Panteleit and Katherina Seiter

Geological Survey for Bremen (Geologischer Dienst für Bremen – GDfB), 28359 Bremen, Germany

21.2.1 Introduction

The Federal State of Bremen is the smallest of the three German city states. Located in northwestern Germany, it consists of the two cities Bremen and Bremerhaven. The city of Bremen is located along the Weser River, about 60 km inland from the port city of Bremerhaven, which is located at the mouth of the Weser River. This densely populated industrial area experiences a wide range of subsurface problems related to building construction, transportation, and water supply infrastructure improvements, and utilization of geothermal energy sources. These activities are affected by contaminated soils and waste deposits at many locations, the risk of flooding due to storm events and, longer term, a rising sea level and associated groundwater salinization.

In response to these concerns, the Geological Survey for Bremen (Geologischer Dienst für Bremen – GDfB) was commissioned to develop a 3-D geological framework model of the subsurface Quaternary units complemented by Neogene units for the Federal State of Bremen. The model was constructed in two parts (Figure 21.7). The 3-D model for the city of Bremen was completed in 2011 and contains five principal stratigraphic units; the 3-D model for Bremerhaven was completed in 2014 and contains six principal stratigraphic units.

The 3-D geological framework model and a derived steady-state groundwater flow model support government administrative actions, planning for construction projects, and issues related to groundwater use and geothermal development. Products derived from these models are provided to the public and private sectors, amongst others, through a web service.

21.2.2 Geological Setting

Bremen is located in the North German Plain; specifically, on the marshy lowlands of the Aller-Weser valley. The 3-D geological models focus on the Quaternary glacial and post-glacial Holocene deposits, but also represent Neogene units, derived from the Geotectonic Atlas (GTA) issued by the LBEG (Landesamt für Bergbau, Energie und Geologie, Federal Sate of lower Saxony). The models contain four basic stratigraphic Quaternary units and two Neogene units for Bremerhaven and one Neogene unit for Bremen (Table 21.4). The modeled units are, from top to bottom: (i) Anthropogenic/Holocene unit; (ii) Weichselian/Saalian glacial deposits; (iii) glaciolacustrine Lauenburger unit; (iv) Elsterian glacial deposits; (v) middle Miocene to Pliocene fluviatile and marine deposits; and (vi) lower Miocene marine deposits, which forms the lower boundary of the models. Glacial meltwater eroded these Neogene deposits, forming buried valleys, reaching depths of up to 360 m below sea level

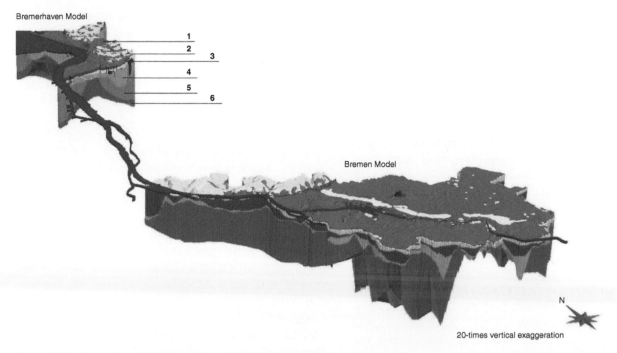

Figure 21.7 3-D geological framework model of Bremen and Bremerhaven. Units from oldest to youngest (bottom to top): 5, 6 fluviatile and marine Neogene sediments; 4 Elsterian glacial deposits; 3 Lauenburger glacio-lacustrine deposits; 2 Weichselian/Saalian glacial deposits; 1 Anthropogenic/Holocene deposits. Units 5 and 6 are combined for the Bremen model. (Seiter and Panteleit 2016).

in Bremen and 244 m below sea level for Bremerhaven, which were subsequently filled with sand, clay, and tills from three glacial cycles. These buried valleys are important groundwater sources which are partially protected from contamination from near-surface activities by the overlying predominantly clayey Holocene deposits.

The near-surface anthropogenic deposits include sands, which have been transported for construction, road building, or filling abandoned harbor basins. These are found at elevations as high as 40 m in Bremen and 26 m in Bremerhaven. Debris layers from bombed buildings are

ubiquitous. Waste deposits exist in some landfills and at a surface waste disposal site. The natural Holocene deposits include peat and organic clays in combination with fluvial and dune sands. The underlying Weichselian and Saalian glacial deposits contain gravel layers resulting from glacial valley drainage systems, the bulk of those deposits consist of sand interfingering with lenses of till. The Elsterian glacial deposits were modeled as two units: the Lauenburger unit consisting of fine glacio-lacustrine deposits, and an underlying unit composed of coarser glacial deposits, mostly sands, and occasional interfingered till deposits.

Table 21.4 Simplified Neogene to Quaternary stratigraphy of Bremen.

3-D model unit	Unit name	Description
1	Anthropogenic deposits and Holocene	Construction sand fill, building debris, waste
		Peat, organic clay, fluvial sand, dune sand, moraines
2	Weichselian/Saalian glacial deposits	Sand, interfingered till lenses, gravel layers
3	Glacio-lacustrine Lauenberger unit	Lacustrine clay and fine sand
4	Elsterian glacial deposits	Sand, basal gravel, interfingered till lenses
5 and 6[a]	Neogene sedimentary units	Fluviatile and marine fine sand and clay
	Base of model	

a) In the Bremen Model, Units 5 and 6 were combined and referred to as "Lower Miocene to Pliocene Deposits." In the Bremerhaven Model, Unit 5 referred to "Lower Miocene Deposits" and Unit 6 referred to "Middle Miocene to Pliocene Deposits."

21.2.3 Development of 3-D Subsurface Models

Over the past 150 years, various construction and exploration activities in Bremen have resulted in about 95 000 borehole records being archived by the GDfB. These boreholes range from building inspections of a few meters to deep gas exploration wells. The borehole records were digitally coded, interpreted, and stored within a GDfB borehole database. In addition to borehole locations and lithologic descriptions, many records include information on the screening depth of observation wells, geophysical parameters (gamma ray or resistance logs), groundwater chemistry data, measured water levels, and groundwater extraction rates (for production wells).

21.2.3.1 Geological Framework Model

A 3-D geological framework model was constructed using the GOCAD 3-D modeling software by Paradigm. ArcGIS software was used for data preparation and interpretation. GeODin (Fugro Consult GmbH 2016) software was used to manage and interpret the information held in the GDfB borehole database. The extracted information was imported into GOCAD.

The 3-D geological framework model was designed to support the development of a discrete-element steady-state groundwater flow model. This method of groundwater flow simulation required layers to extend continuously across the entire model domain. Thus, the model contains a base layer (base Miocene) and six continuous hydrostratigraphic units corresponding the four Quaternary and two Neogene units defined in Table 21.4.

Once the horizons defining these six hydrostratigraphic units were spatially located in the 3-D geological framework model, they were discretized into a regular 100 × 100 m grids (S-Grids in GOCAD). For further evaluation, each hydrostratigraphic unit was divided vertically into a number of layers defined according to the thickness of the unit and its heterogeneity (Table 21.5). The vertical resolution of the model varies with the thickness of each stratigraphic layer and is not constant throughout the model (Figure 21.8).

In preparation for the groundwater flow modeling, the 3-D grid model was "parameterized" – the lithologic character of each cell was assigned appropriate parameter values. Interfingering lenses within the units can now be represented by differences in their hydraulic conductivities and defined in GOCAD as "regions." This allowed inhomogeneities to be represented within specific stratigraphic units. The gridded parameter models were transferred to the USGS MODFLOW software (USGS 2005), which offers an integrated modeling environment for the simulation of groundwater flow, transport, and reactive processes.

Model construction involved a six-step process (Figure 21.9). The six steps are as follows:

- *Step 1. Spatial configuration of cross-sections*: GIS technology was used to define desirable locations for 121 geological cross-sections, 48 for the Bremerhaven area and 73 for the city of Bremen. Data queries identified about 1150 of the deepest *reference boreholes*; these defined preferred cross-section intersections. Cross-sections were oriented to represent the deep Elsterian meltwater channels. Throughout the construction process, the borehole data were quality-checked and assigned a quality rating ranging from class Q1 (the best quality data, reference boreholes) to Q5 (data unsuitable

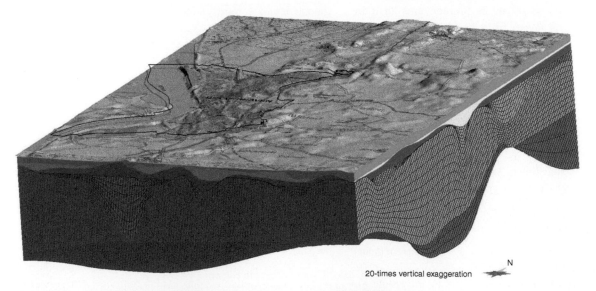

20-times vertical exaggeration

Figure 21.8 Modified example of GOCAD S-Grid. (Seiter and Panteleit 2016)

Table 21.5 Descriptions and characteristics of 3-D model horizons and component units.

Model element	Definition	Unit[a] thickness (m) Min	Max	Nr. of boreholes to horizon	Depth of horizon[b] (m) HB Min	HB Max	BHV Min	BHV Max	Nr. of of Layers per unit	Vertical Resolution of layers (m) Min	Max
Top	25 m DEM			—	−1	38	−1	26			
Unit 1	Anthropogenic/ Holocene	0.3	45		−12	31	−24	11	3	0.1	15
Horizon 1	Base Holocene			46 466							
Unit 2	Weichselian/Saalian glacial deposits	0.9	72		−41	26	−57	−1	9	0.1	7
Horizon 2	Base Saalian			3311							
Unit 3	Glacio-lacustrine Lauenberger unit	0.6	228		−177	10	−145	−3	6	0.1	38
Horizon 3	Base Lauenburger			682							
Unit 4	Elsterian glacial deposits	0.7	308		−356	−8	−244	−16	7	0.1	44
Horizon 4	Base Quaternary			820							
Unit 5[c]	middle Miocene-Pliocene deposits	1	320		−614	−9	−298	−49	10	0.1	32
Horizon 5	Base middle Miocene			no data							
Unit 6[c]	Lower Miocene	0.3	243				−373	−60	3	0.1	81
Horizon 6	Base Miocene			no data							
Neogene sedimentary units											

a) Unit Thickness derived from the rectangular model of Bremen and Bremerhaven. Borders: 3 456 200/5 870 000; 3 500 000/5 905 000, and 3 466 200/5 924 000; 3 484 000/5 949 000.

b) Calculations derived from the borders of the federal state of Bremen. HB = Bremen, BHV = Bremerhaven. Unit 5 and Unit 6 combined for Bremen.

c) Constructed with surfaces from the GTA (Baldschuhn et al. 2001).

Modified from Panteleit et al. (2013).

for modeling). Boreholes with a total depth of more than 25 m and associated geophysical logs were particularly suitable as reference data points and thus given a high-confidence classification (Q1). Gamma-ray and resistivity geophysical logs were used to determine the Neogene–Quaternary transition.

- *Step 2. Stratigraphic interpretation and construction of cross sections*: All potential reference boreholes were subjected to further stratigraphic interpretation using the GeODin software. Digital cross-sections were constructed by correlating stratigraphic boundaries identified in reference borehole logs to define the base of each hydrostratigraphic unit. Interfingering lenses had top and bottom horizons digitized. The digital cross-sections and selected borehole horizon markers were stored in a special database.

- *Step 3. Incorporation of "free wells"*: More than 34 000 deep boreholes and wells and were not used to construct the cross-sections. These wells, defined as "free wells", provided valuable additional control to the definition of

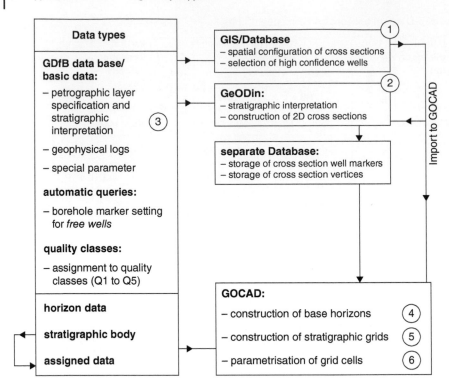

Figure 21.9 Flow diagram of stepwise model construction. Numbers correspond to steps defined in the text. Modified from Panteleit et al. (2013).

the stratigraphic surfaces. The petrographic descriptions of each layer in these "free wells" assessed and assigned quality ratings from Q2 to Q4 according to their quality. A Q4 rating was assigned to boreholes with stratigraphic horizons identified solely from geophysical logs; a Q5 rating was assigned to boreholes with information deemed unsuitable for the model construction process.

- *Step 4. Construction of stratigraphic horizons*: The reference boreholes and associated digitized horizon vertices on the cross-sections were set as "control nodes" during the construction of each horizon in GOCAD. They were fixed during the interpolation process and stratigraphic horizons were forced to pass through their locations. In contrast, interpreted horizon vertices from the cross-sections and stratigraphic positions identified in the imported free wells were used as "control points." Control points influence the geometry of the interpolated surfaces according to user-defined criteria, but the horizons do not have to absolutely match them. Based on these "control nodes" and "control points," horizons were modeled using the GOCAD DSI (Discrete Smooth Interpolation) procedure (Mallet 1992, 1997). The resulting surfaces were relatively smooth, with minimal deviations from control points. Misinterpretations of borehole stratigraphy resulted in obvious surface irregularities; when observed, the borehole interpretations were corrected and the surface recalculated. The model base horizon (top of the Neogene) had less borehole

information available, so information from regional maps was used to guide its interpolation. The ground surface forms the top of the model. A 25 m resolution Digital Elevation Model (DEM) defined the ground surface.

- *Step 5. Construction of stratigraphic grids*: The hydrostratigraphic horizons defining the model units were converted to a regular stratigraphic 100×100 m grid (GOCAD S-grid) by vertically associating the top and base horizons of each unit and dividing the distance between them into several layers according to the thickness of the unit and its heterogeneity (Figure 21.10a). Thus, the vertical resolution of the model varies, and Table 21.5 provides the range of the vertical resolution for each layer. To maintain continuous layers across the model, where a unit does not exist the unit was defined as being 0.1 m thick.

- *Step 6. Parameterization of grid cells*: In preparation for the groundwater flow modeling, each cell in the 3-D grid model was assigned appropriate parameter values based on its stratigraphic unit and property values stored for each well-log in the GDfB borehole database. This "parameterization" of the model converted lithologies to values of heat conductivity, permeability, or remediation potential (Figure 21.10b). These parameters were interpolated within the hydrogeological units from well-log observations imported into GOCAD using DSI and 3-D kriging operations supported by GOCAD.

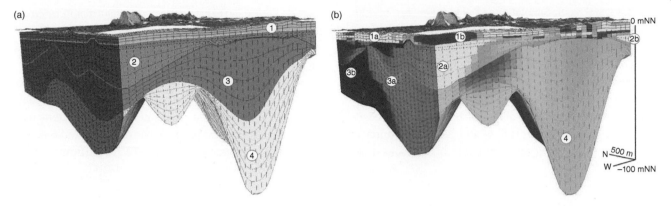

Figure 21.10 Extract of the gridded geological framework model, showing: (a) hydrostratigraphic units; (b) parameterized hydraulic permeability model with values interpolated from borehole data. (Panteleit et al. 2013)

21.2.3.2 Groundwater Flow Model

A 3-D steady-state discrete-element regional groundwater flow model was developed using the 3-D geological framework model described in the previous section. The flow modeling was accomplished using MODFLOW with different graphical user interfaces, such as PMWIN (Chiang 2005) or the open source FREEWAT platform for QGIS (Rossetto et al. 2015) and necessary pre- and post-processors. The flow model has 38 layers, corresponding to the number of layers defining the units in the geological framework model, and has the same 100×100 m grid cell size.

A gateway developed by McDiarmid (2011) allowed the rapid data transfer of parameterized GOCAD grid models to MODFLOW. Panteleit et al. (2013) provide details of the model hydraulic parameters. Each cell was assigned a hydraulic conductivity based on the parameterization derived from the lithologic borehole description (Fuchs 2010). Vertical hydraulic conductivity was calculated at nine-tenths the horizontal hydraulic conductivity by model calibration. Effective porosity, initial heads, and boundary conditions were applied to the model by analyzing external sources. Modeled groundwater levels were verified against data from about 843 observation wells and water gauges. Groundwater levels are controlled by drainage in the low-lying agricultural areas and the water level of the river Weser.

21.2.3.3 Higher-resolution Local Geological Framework Models

Several smaller areas experience severe groundwater contamination. In order to evaluate and manage pollution plumes, a series of local and higher-resolution 3-D geological framework models were developed for these areas, including Bremen (Figure 21.11). These models have in general the same hydrostratigraphic units as the Bremen model, but have a finer grid resolution, with voxels of $2.5 \times 2.5 \times 0.2$ m. They also were based on more detailed interpretations of inherent heterogeneity conditions within the hydrostratigraphic units. Local groundwater flow modeling was performed using parameterized versions of these models.

Figure 21.11 Refined 3-D geological framework models for local areas with specific groundwater contamination problems. Four local areas are shown. (Seiter and Panteleit 2016)

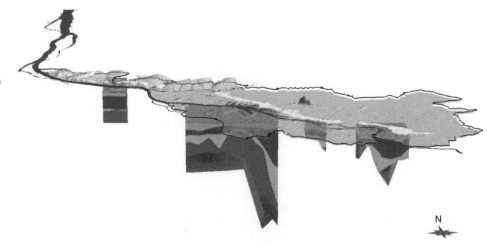

21.2.3.4 Stochastic Simulations of Heterogeneity

The flow modeling described in the previous section was not able to adequately define the transport and evolution of a contaminant plume in the heterogeneous Holocene deposits. Analysis of the extent and potential future growth of the plume required a more detailed definition of subsurface heterogeneity. Using observed boreholes, stochastic simulations (see Chapter 13) of the 3-D distribution of clay lenses produced a series of equiprobable subsurface models (Figure 21.12). In areas with insufficient or irregular distributed observations, stochastic simulations were developed from a combination of actual boreholes and randomly-placed virtual boreholes (Figure 21.12b).

These stochastic simulations of aquifer heterogeneity formed the basis for a series of parameterized grid models that permitted multiple assessments of the extent and growth of contaminant plumes. The transport and reaction model MT3D, accessed through PMWIN, performed these

(a)

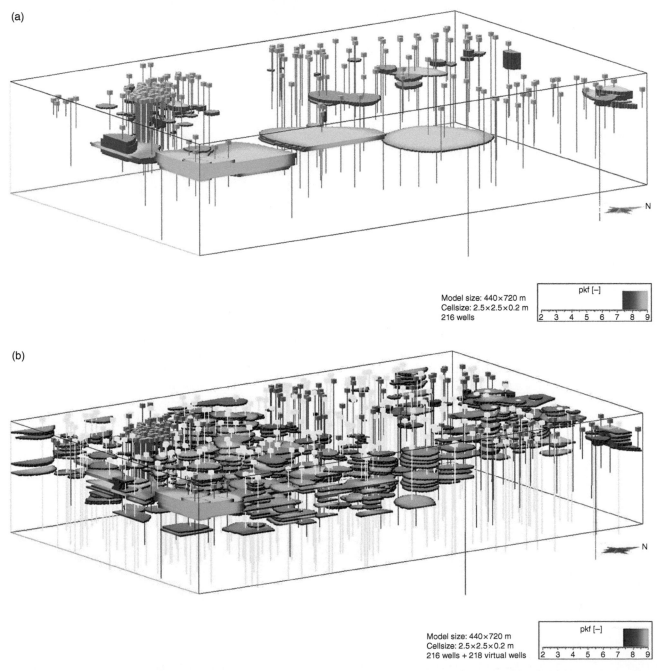

Model size: 440×720 m
Cellsize: 2.5×2.5×0.2 m
216 wells

pkf [−]
2 3 4 5 6 7 8 9

(b)

Model size: 440×720 m
Cellsize: 2.5×2.5×0.2 m
216 wells + 218 virtual wells

pkf [−]
2 3 4 5 6 7 8 9

Figure 21.12   Examples of detailed evaluations of subsurface conditions using stochastic modeling of heterogeneity: (a) model based on 216 actual wells; (b) model based on 216 actual wells and 218 virtual wells. (Seiter and Panteleit 2016)

assessments. The model solutions compare very closely to the measured data or the observed contamination plumes.

21.2.4 Application of the Models

While geology and the models are certainly 3-D, typical users prefer 2-D derivative products, generally maps, printed or online. Some applications, based on a 3-D evaluation of the appropriate model, have produced user-friendly maps with distinctive topics such as "potential for enhanced rainwater infiltration" or "thermal conductivity and possible depth limitations for borehole heat exchangers" (Figure 21.13). True 3-D applications, such as groundwater flow-model results, are designed for use by expert users; thus, these applications are supplied with restrictions to specific user groups.

Many 2-D derivative products are provided free-of-charge for use by any interested person through open-access via the internet. These products provide quick information for topics such as:

- *Potential for enhanced rain water infiltration*: To enhance groundwater recharge and minimize impacts of heavy rain events, residents are charged reduced sewage water fees if they infiltrate rain water directly onto their property (Figure 21.13a).
- *Building ground quality*: In Bremen, building site quality is often adversely impacted by compaction of peat layers or highly organic clay soils. Disposal of these soils is expensive due to oxidation of pyrite and production of acid contamination. The Holocene units in the 3-D geological framework model were analyzed to produce a 2-D map showing the thickness of organic rich sediments (Figure 21.13b).
- *Geothermal potential*: The Bremen area is increasingly using shallow geothermal systems with heat exchangers (commonly probes) down to a depth of up to 250 m

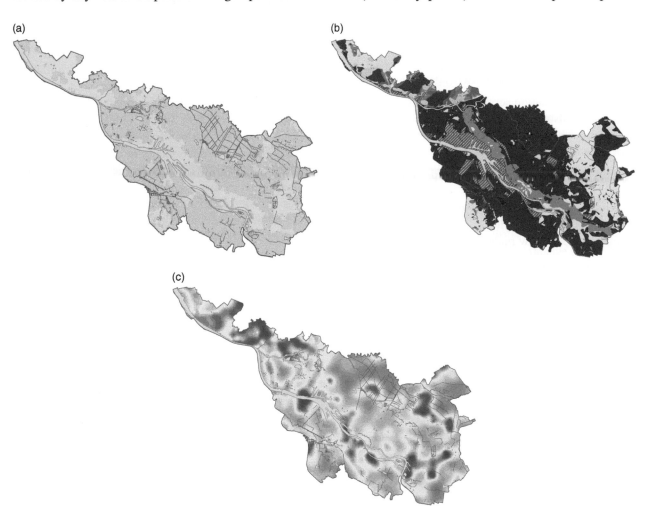

Figure 21.13 Examples of three 2-D open-access web-based application maps. (a) Potential for enhanced rain water infiltration – based on near-surface permeability and depth to water; (b) building ground quality – based on organic soils and location of marsh areas; (c) geothermal conductivity – an important parameter for assessing geothermal potential. (Seiter and Panteleit 2016)

22

Application Theme 5 – Geohazard and Environmental Risk Applications

Editor's Introduction

A wide variety of natural and human-induced geohazard assessments are being supported by use of 3-D geological models. Geohazards and risks involve time-varying phenomena; thus, such evaluations require multidimensional (4-D) analysis using a variety of numerical process models (Chapter 14). By defining the subsurface framework, 3-D geological models provide geometrical constraints to these numerical process models. The discretization and stochastic modeling methods discussed in Chapter 13 provide improved estimates of the subsurface distribution of important physical or chemical properties required by the numerical models.

Five case studies have been selected to illustrate the benefits of using 3-D geological models as part of hazard and risk investigations. Three case studies describe natural hazards: assessment of ground characteristics in Christchurch, New Zealand, a seismically active urban area; evaluating causes of coastal instability on the south coast of England; and a dynamic process-modeling approach to cliff instability on the east coast of England. Two other case studies assess environmental risks related to human action; in Nantes, France, urban redevelopment has to handle hazardous contaminated soils, while in Ljubljana, Slovenia, municipal water supplies are threatened by pollution from various industrial developments in the expanding urban area.

The first case study describes the development of 3-D geological models for Christchurch, New Zealand to guide urban reconstruction plans following major earthquakes in 2010 and 2011. Knowledge of the probable seismic response to future earthquake events identifies potentially hazardous areas. Formulation of urban redevelopment plans and regulations that identify these areas allows for appropriate cost-effective redevelopment.

The second case study explores the added value provided by 3-D modeling and visualization of typical site-investigation data. The English south coast at Barton-on-Sea consists of erodible materials; wave erosion produces slope instability. The 3-D model provided a better visualization of subsurface conditions and the role of groundwater in promoting instability.

The third case study is an example of linking 3-D geological models to multiple process-based models to provide better predictions of the rate of cliff erosion along a section of the east coast of England. This case study provides a detailed example of how the coupling a series of models produces integrated assessments of time-varying phenomena as discussed in Chapter 14.

The fourth case study describes how 3-D modeling improved the evaluation of the extent and geochemical characteristics of contaminated soils in an area of Nantes, France, that is undergoing urban redevelopment. Urban redevelopment typically involves excavation of considerable volumes of variably contaminated soil. Effective management depends on an accurate assessment of the potential quantities of contaminated material and its contamination characteristics. The modeling in Nantes employed anthropogenic deposit characterization methods more fully described in Chapter 25. Contaminated soils are commonly encountered in many urban areas; the Silvertown Tunnel case study in Chapter 23 is another example.

The fifth case study describes how a 3-D geological model of Ljubljana, Slovenia, provided better subsurface property distributions and geometrical constraints to a groundwater flow model. About 90% of Ljubljana's drinking water is provided from an unconfined shallow aquifer which is vulnerable to surface contamination. The two models form the basis of a decision support system that guides the establishment of drinking water source protection zones to safeguard water quality.

Applied Multidimensional Geological Modeling: Informing Sustainable Human Interactions with the Shallow Subsurface, First Edition.
Edited by Alan Keith Turner, Holger Kessler, and Michiel J. van der Meulen.
© 2021 John Wiley & Sons Ltd. Published 2021 by John Wiley & Sons Ltd.

Case Study 22.1: Christchurch City, New Zealand, 3-D Geological Model Contributes to Post-Earthquake Rebuilding

Mark S. Rattenbury, John G. Begg, and Katie E. Jones
GNS Science, Lower Hutt 5010, New Zealand

22.1.1 Introduction

Christchurch is the largest city on the South Island of New Zealand and is the country's second-largest urban area (Figure 22.1). A series of highly destructive earthquakes, collectively known as the 2010–2011 Canterbury Earthquake Sequence or CES (Bannister and Gledhill 2012),

began on 4 September 2010 with the M_w 7.1 Darfield Earthquake. The CES continued with an even more damaging aftershock on 22 February 2011 (the M_w 6.3 Christchurch Earthquake), followed by major aftershocks on 13 June 2011, and two on 23 December 2011. Because its hypocenter (or focus) was shallow and close to the center of Christchurch, the smaller M_w 6.3 Christchurch Earthquake produced high to extreme ground accelerations causing heavy damage to many buildings and resulting in 185 deaths and at least 6600 injuries (Begg et al. 2015b; Potter et al. 2015). There were two dominant geological manifestations of the large earthquakes in Christchurch; widespread liquefaction and differential ground settlement in the central and eastern parts of the city, particularly near river and estuary edges (Townsend et al. 2016), and

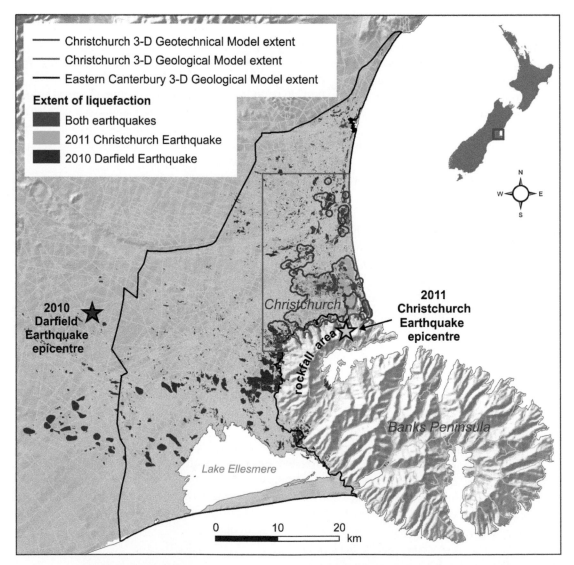

Figure 22.1   Map of the Christchurch area showing the extent of the three geological and geotechnical 3-D models and the epicenters of the 2010 Darfield and 2011 Christchurch earthquakes with associated liquefaction and rockfall. Modified from Geonet (2011), Begg et al. (2015a), Massey et al. (2014), and Townsend et al. (2016).

rockfall on the slopes of the Port Hills (Massey et al. 2014). Both caused considerable damage to buildings and infrastructure.

Prior to 2010, as the nationwide geological mapping program was being completed, GNS Science began planning a new urban geological mapping program for selected urban areas. The CES provided a powerful rationale to make Christchurch the first target for this new urban geological mapping initiative. In the Christchurch area, extensive post-earthquake geotechnical investigations produced large volumes of drill hole, geotechnical, and geophysical data. These data were acquired to help determine near-surface ground conditions and materials; this information was required for planning rebuilding and restoring services. GNS Science combined these new data with legacy information, including logs describing over 12,000 groundwater wells, to produce new geological interpretations of Christchurch and the larger eastern Canterbury area in various digital formats.

These digital data include three 3-D models (Begg et al. 2015a): (i) a regional Eastern Canterbury Geological Model ("Canterbury Model"); (ii) a higher resolution Christchurch Geological Model; and (iii) a property model derived from post-earthquake geotechnical data – the Christchurch Geotechnical Model. Sections 22.1.3 and 22.1.4 discuss the development and applications of the two Christchurch models.

22.1.2 Geological Setting

An excellent geological map of Christchurch (Brown and Weeber 1992; Brown et al. 1995) established the Quaternary stratigraphy for the area. The Christchurch 3-D geological model includes four of those geological units. From oldest to youngest, they are: (i) the Riccarton Gravel, (ii) the Avonside Member of the Christchurch Formation, (iii) the Christchurch Formation, and (iv) the Springston Formation. Table 22.1 provides brief descriptions of these units. Begg et al. (2015a) provide more detailed descriptions.

Figure 22.2 is a simplified diagrammatic east–west cross-section through the Christchurch area. The geometry of the Christchurch Formation is central to the overall geometry of the stratigraphic relationships. It forms a wedge, thick in the east and tapering to an edge in the west, with an east-dipping base and a sub-horizontal top. It underlies most of Christchurch City from the coast as far west as the western side of Hagley Park (Figure 22.2). The top of the Riccarton Gravel represents the base of materials within the Christchurch 3-D geological model. The Avonside Member of the Christchurch Formation is

Table 22.1 Late quaternary geological units in the Christchurch 3-D geological model.

Geological unit	General characteristics	Depositional environment and estimated age	Named climatic episodes
Springston Formation	Sandy gravel, gravelly sand, sand, silt, peat, clayey silt, laterally, and vertically variable. Immediately beneath the surface across the Waimakariri fan; absent where Riccarton Gravel lies at the surface, and absent in the east where Christchurch Formation is at the surface. Base down to sea level in the east.	Mostly distal alluvial and overbank deposits pro-grading across the top of the Christchurch Formation during the past 6500 years (OIS 1)	Aranui (Postglacial)
Christchurch Formation	Mainly sand and silt commonly with shells, minor peat, and thin sandy gravel lenses. Top typically close to sea level, base up to 35 m bmsl near the coast. Absent in the west.	Deposited below sea level during the past c. 9000 years (OIS 1)	
Christchurch Formation (Avonside Member)	Silt and clayey silt with minor thin sandy gravel lenses; rarely shelly. Top above sea level in the west and base up to 40 m bmsl near the coast. Absent in the west.	Deposited close to sea level in back-beach swamp environment during the past 9000 years (OIS 1)	
Riccarton gravel	Brown or blue well-graded gravel. Beneath Springston or Christchurch formations; top is typically between 10 and 40 m below the ground surface; locally at the surface where Springston and Christchurch formations are absent.	Predominantly alluvial deposits, OIS 2–4; 12–71 ka;	Otira (Last Glacial Period)

bmsl = below mean sea level.
Source: Modified from Table 4, Begg et al. 2015a.

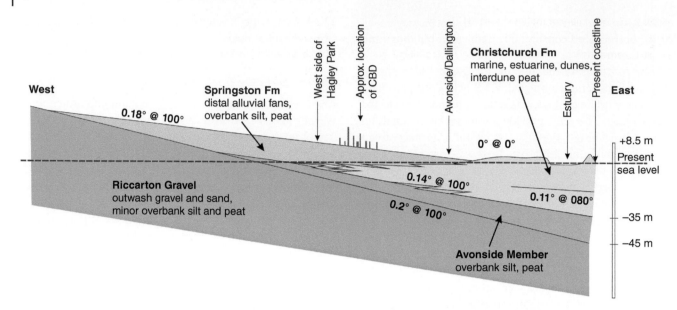

Figure 22.2 A simplified representation of the depositional relationships and structure of the defined Holocene geological units above the last glacial Riccarton Gravel along a W-E cross-section in the Christchurch area. The vertical scale indicates the maximum thickness of Holocene units near the coast (Begg et al. 2015a).

a time-transgressive unit, being older (and deeper) in the east than in the west. It does not exist in the west, in areas where the top of the Riccarton Gravel lies significantly above the current sea level. Thus, in the west, the younger Springston Formation directly overlies the Riccarton Gravel (Figure 22.2). The youngest Springston Formation contains two important components. The upper part of the Springston Formation overlies the Christchurch Formation as far east as Avonside/Dallington (Figure 22.2). The lower part of the Springston Formation probably underlies Christchurch Formation, at least locally, from the coast to the western extent of Christchurch Formation, and beyond.

22.1.3 Development of 3-D Subsurface Models

Begg et al. (2015a) describe the development and application of three separate 3-D models (Figure 22.1). Each 3-D model has its own scale and purpose. Table 22.2 summarizes the characteristics of these models. The three models are:

1. *The Eastern Canterbury Geological Model* covers an area of the Canterbury Plains. It includes all geological units within the area, although Mesozoic greywacke, Neogene sediments, and Miocene volcanic rocks are undifferentiated. It provides a wider context to the more detailed 3-D Christchurch models. The deeper Quaternary deposits are regionally significant as they host the Christchurch Artesian System, a valuable groundwater resource.

2. *The Christchurch Geological Model* provides a more detailed representation of the Late Quaternary sediment sequence beneath most of Christchurch City to a maximum depth of approximately 50 m below mean sea level. The materials at these shallower depths beneath Christchurch affect the design of foundations and near-surface utilities infrastructure.

3. *The Christchurch Geotechnical Model* covers a smaller area, mostly central and eastern Christchurch City. It is based on an interpolation of geotechnical data within localized areas where there is a sufficient density of cone penetrometer test (CPT) soundings. The model was built to corroborate, complement, and refine available geological data with more specific geotechnical information for the geological units.

These models are of two distinct types. The two geological models are "volumetric models"; they show the spatial distribution and relationships between individual geological units defined by their upper and lower bounding surfaces. In contrast, the Christchurch Geotechnical Model is a "property model" developed by interpolating geotechnical values derived from the dense network of CPT soundings and geotechnical drill holes undertaken following the CES.

22.1.3.1 Selected Modeling Software
Leapfrog Geo, developed and supplied by Seequent Limited (https://seequent.com) was selected to produce the 3-D models. Leapfrog Geo utilizes an implicit 3-D geological

Table 22.2 Summary of model characteristics.

Model characteristics	Eastern Canterbury geological model	Christchurch geological model	Christchurch geotechnical model
General location	Canterbury Plains between Amberley Beach and the Rakaia River mouth	Christchurch City urban area	Areas within Christchurch City within 500 m of CPT probe
Surface area	1836 km².	1050 km².	129.9 km²
Maximum depth	500 m bmsl	50 m bmsl	50 m bmsl
Total volume	1752.8 km³ 871.2 km³. if Banks Peninsula volcanics excluded	17.4 km³	13.1 km³ (within 10 distinct zones)
Number of geological units	12	4	Numerical property values interpolated within geological units defined in Christchurch Geological Model

bmsl = below mean sea level.

modeling methodology. As discussed in Chapter 12, implicit modeling applies rules to transform the input data and produce a model. While the model is generated by computer algorithms evaluating the observational data, the geologist creating the model provides insight as to trends, stratigraphic relationships, and other geologically meaningful conditions.

The two geological models contain a sequence of geological units; the shape and position of each unit was defined by a series of stratigraphic surfaces. The surface forming the base of the next youngest unit defined the top of each geological unit. The shape and position of each of these surfaces was influenced by surface definition constraints that helped specify its geometry. Three surface definitions were used in developing the Christchurch Geological Model:

- *deposit surface* definition allows units to conform to the principle of superposition – age relationships are defined by the user-in a "Lithology Chronology" file.
- *erosion surface* definition allows the base of a younger unit to cut downwards into an older volume and infill this space.
- *intrusion surfaces*, or even multiple intrusion surfaces, are used to model isolated units which float within another, as is the case with the two intra-Christchurch Formation units.

Users may interrogate the created models with the Leapfrog Viewer visualization software, available free of charge at https://www.seequent.com/products-solutions/leapfrog-software. The viewer enables comprehensive access to information in the models but does not have the editing or modeling capabilities of Leapfrog Geo. In addition, GNS Science distributes alternative derivative digital formats, including GOCAD Tsurf and ArcGIS shapefile

contour and grid format representations of base and top surfaces of modeled volumes and thicknesses. These derivative formats allow model data to be integrated with 2-D spatial, non-geological datasets using GIS software accessible to many potential clients who have no intention of investing in 3-D software technology.

22.1.3.2 Eastern Canterbury Geological Model

The Eastern Canterbury Geological Model sets the regional framework for the area and includes modeled geological units that do not occur near-surface in Christchurch City. The model is not discussed further here; the higher resolution Christchurch Geological and Christchurch Geotechnical models are more relevant for near-surface applications in the city area.

22.1.3.3 Christchurch Geological Model

The Christchurch Geological Model was created to provide accurate information on the near-surface geological units underlying Christchurch City (Figure 22.3). The model's topographic surface is derived from a 1 m Digital Elevation Model (DEM) constructed from a mosaic of existing Lidar point cloud datasets. The DEM was resampled within Leapfrog Geo to 100 m, corresponding to the model resolution for the contact surfaces, stratigraphic sequences, and model boundary.

The model was based on 5765 pre-existing drill hole logs located across the central Christchurch area; these were supplemented by 1328 logs from the post-earthquake drilling program. CPT soundings data supplemented the drill hole information. The combined drill hole and CPT data provided excellent sampling density across most of the model down to the top of the Riccarton Formation (Figure 22.4).

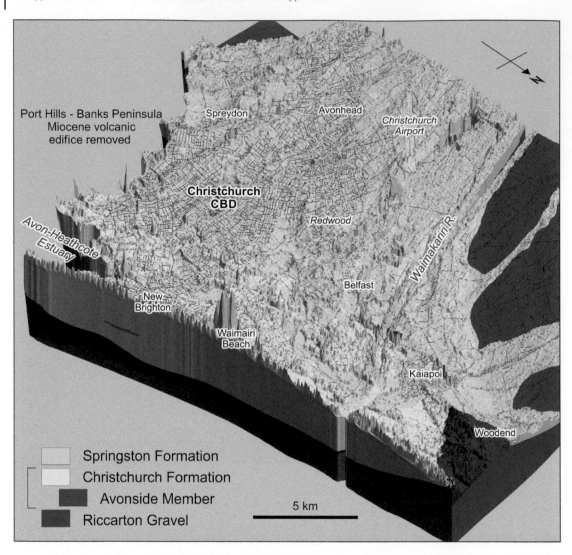

Figure 22.3 The Christchurch Geological Model, here viewed obliquely from the northeast (from 221° looking obliquely down at 41°). Riccarton Gravel is blue, Avonside Member brown, Christchurch Formation yellow and Springston Formation light green. Topo50 roads and the coastline are shown for an indication of location. The basal boundary of the model is 60 m below mean sea level. The vertical exaggeration is 100× (Begg et al. 2015a).

Prior to modeling, drill hole logs were carefully examined. Simplified lithological descriptions were viewed in 3-D to identify principal geological unit contacts. A simplified depositional structure of the model area was developed according to the conceptual model illustrated by Figure 22.2. A dense series of correlation lines defining stratigraphic unit boundaries were digitized from borehole logs shown in cross-section. These correlation lines were necessary to filter out stratigraphically anomalous outliers prior to interpolation. The Leapfrog Geo software's Radial Basis Function (RBF) algorithm interpolated these correlation lines to calculate surfaces. Table 22.3 defines the characteristics of the surfaces used to build the Christchurch Geological Model. The RBF algorithm was guided by a strongly anisotropic search

direction that mimicked the very slight southeast-dip of the stratigraphy.

22.1.3.4 Christchurch Geotechnical Model

The Christchurch Geotechnical Model is a property model based on values from about 1500 CPT soundings selected from the approximately 10,000 CPT soundings undertaken as part of the post-earthquake rebuilding effort. The soundings selected were, on average, ~127 m apart. Modern CPT methods produce highly accurate and repeatable measurements (Robertson and Cabal 2010). A CPT sounding is made using a standard-sized stainless-steel cone with three electronic sensors mounted on stainless steel rods. This is pushed vertically into the ground at a constant rate of $2 \, \text{cm} \, \text{s}^{-1}$. Measurements of:

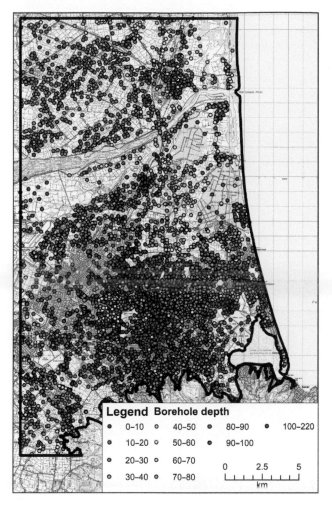

Legend Borehole depth

0–10	40–50	80–90	100–220
10–20	50–60	90–100	
20–30	60–70		
30–40	70–80		

0   2.5   5
km

Figure 22.4 The extent of the Christchurch Geological Model showing the distribution of drill hole collars with geological logs used in the modeling, color-coded according to their depth (Begg et al. 2015a).

(i) resistance on the tip of the cone (q_c), (ii) friction on the side of the cone (f_s), and (iii) dynamic pore water pressure (u) are recorded at 1 cm intervals as the probe advances.

At Christchurch, CPT soundings could only begin about 1 m below the surface at hand-dug "pre-drill pits" pits due to concerns over puncturing shallow buried infrastructure. This untested zone was defined as a separate "pre-drill volume" volume in the geotechnical model. CPT soundings are normally continued downward to "refusal," where materials are too dense to allow progress of the cone. In much of the Christchurch area, "refusal" for many CPT soundings defines the top of the Riccarton Formation. However, at some places within the Central Business District, gravels of the Springston Formation are located close to the surface. In some cases, these were drilled through before CPT soundings were run from their base downward to a second refusal level.

A variety of geotechnical derivative parameter proxies can be calculated from raw CPT data, including approximations of materials encountered (Soil Behavior Type, SBT) and many additional physical properties used by geotechnical engineers. These derivative values are calculated from data normalized for depth to compensate for overburden stress. The proprietary geotechnical software CLiq (https://geologismiki.gr/products/cliq) was used to process individual CPT soundings. Multiple derivative values, along with a CPT identifier, depth definitions, and three measured parameters (q_c, f_s, and u) were imported into a spreadsheet. Values were validated against established criteria.

The lateral extent of the Christchurch Geotechnical Model was constrained within a 500 m buffer around the CPT locations. This produced a large zone underlying

Table 22.3 Surfaces built for the Christchurch Geological Model. The top surface of the youngest volume is defined by the model topography and the model base is defined as the Riccarton Gravel to the maximum model depth about 50 m below mean sea level (Begg et al. 2015a).

Surface name	Surface type	Represents	Overlying unit	Underlying unit
Springston Formation gravel	Intrusion	intra-Springston Formation	Springston Formation	Springston Formation
Springston contact	Erosion	base of the Springston Formation	Springston Formation	Christchurch Formation
Christchurch contact	Deposit	base of the Christchurch Formation	Christchurch Formation	Avonside Member
upper intra-Christchurch silt unit	Intrusion	silt subunit of Christchurch Formation	Christchurch Formation	Christchurch Formation
lower intra-Christchurch silt unit	Intrusion	silt subunit of Christchurch Formation	Christchurch Formation	Christchurch Formation
Avonside contact	Deposit	base of the Avonside Member	Avonside Member	Riccarton Gravel
Riccarton contact	Erosion	base of the Riccarton Gravel	Riccarton Gravel	Bromley Formation

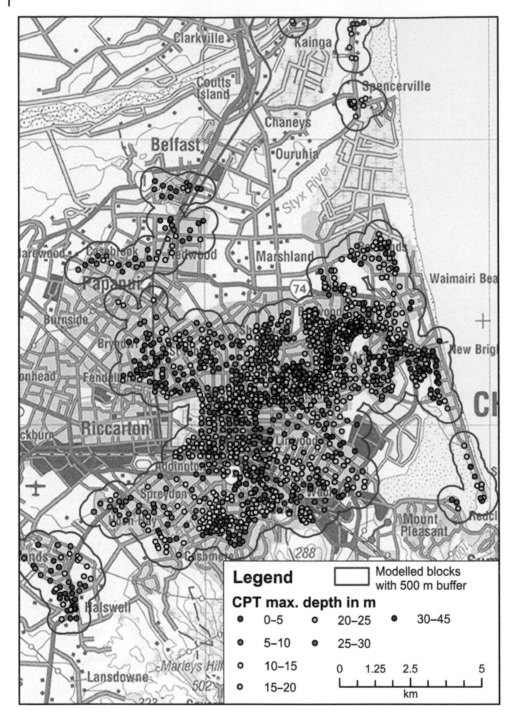

Figure 22.5 The extent of the Christchurch Geotechnical Model showing the distribution of CPT sounding sites color-coded according to their depths below surface (Begg et al. 2015a).

central Christchurch and several smaller outlying areas (Figure 22.5). Thus, the Christchurch Geotechnical Model consists of several isolated 3-D volumes. Deeper CPT refusals revealed a sub-planar surface; manual examination of borehole data defined it as the top of the Riccarton Gravel, and this became the base of the model volumes. Because no CPT values existed in the first 1 m below the

ground surface, the base of this "pre-drill zone" zone forms the top of the model volumes.

The geometry of subsurface materials within the individual defined 3-D volumes was established by viewing point data in 3-D using Leapfrog Geo. The coherence and continuity of materials and their properties was assessed by examining color-coded 3-D point displays of raw data,

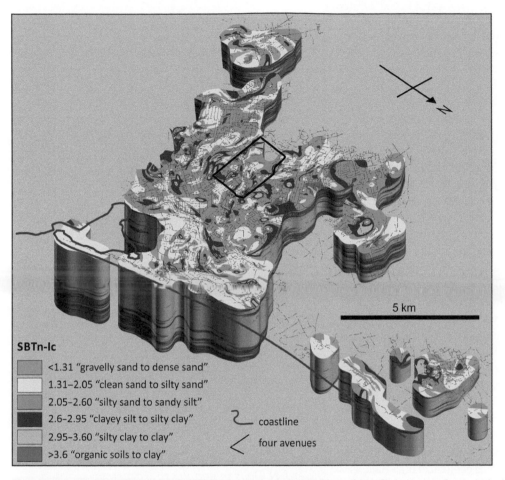

SBTn-Ic

	<1.31 "gravelly sand to dense sand"
	1.31–2.05 "clean sand to silty sand"
	2.05–2.60 "silty sand to sandy silt"
	2.6–2.95 "clayey silt to silty clay"
	2.95–3.60 "silty clay to clay"
	>3.6 "organic soils to clay"

coastline

four avenues

Figure 22.6 Oblique view of the Soil Behavior Type (Ic) distribution within Christchurch Geotechnical model, viewed from the northeast (azimuth 223°, plunge 43°). Model base is the top of the Riccarton Gravel, defined by the refusal depths of the deepest CPTs. The Topo50 coastline (blue line) and the four avenues (Fitzgerald, Bealey, Deans, and Moorhouse; heavy black lines) provide an indication of location. The vertical exaggeration is 100× (Begg et al. 2015a).

normalized raw data, and derivative values. The conceptual model of depositional geometry guided visual correlations and provided a basis for defining preferred correlation trends (Figure 22.3).

Leapfrog Geo includes geostatistical tools for 3-D spatial interpolation. Interpolation of numerical values between CPT soundings was conditioned by an oriented spheroid that reflected trends defined by the conceptual model of depositional geometry. Individual property models were developed for specific CPT raw or derivative values; each used appropriate semi-variogram sill, range, and nugget values (Begg et al. 2015a). Some derivative values have a close relationship to the measured values, and these are considered to have good reliability. Other derivative values are more distant from the raw data; these appear to have significantly less reliability, so were not modeled.

A geotechnical property model of the derivative "soil behavior type from normalized cone resistance (q_c) and side friction (f_s) values", identified by the code "Ic-SBTn",

was compared with borehole logs. For most soil types, this CPT-based property model (Figure 22.6) provided a good approximation of subsurface lithologies. Layers of loose sandy gravel and gravelly sand were not reliably identified by this CPT method, consequently their thickness was underestimated. This is important because some of the near-surface gravelly sand layers provide good foundation conditions for small light buildings and provide favorable foundation conditions for near-surface utilities.

22.1.4 Applications of 3-D Models

The Christchurch 3-D geological and geotechnical models quantify the nature and distributions of subsurface materials, allowing preliminary estimates of foundation conditions during planning and feasibility stages for future building or infrastructure development. This knowledge has facilitated optimal routing and design of subsurface utilities, enabled more accurate estimates of reinstatement costs and provide useful guidance for planning site-specific foundation testing. The models have been used to identify suitable sites for the relocation and rebuilding of structures along the Avon and Heathcote rivers.

The accuracy of the Christchurch Geotechnical Model was tested by generating synthetic CPT soundings for locations where existing CPT soundings were not used to create the model. Comparisons indicate that the synthetic CPT soundings convey the primary characteristics observed in actual CPT soundings at these locations. Synthetic CPT soundings were used only for testing the validity of the geotechnical model against new CPT data, or soundings not used in modeling. While synthetic CPTs provide approximations of soil conditions at a site, they must not be used for final site-specific design and construction purposes.

The Christchurch Geotechnical Model also makes geological sense. The distribution of materials based on the derived SBT parameter, when viewed in horizontal slices, shows evidence for river meanders, shifting coastal sand bars and marine incursions. These processes have been a feature of the area in the Holocene responding to changing sea level and varying rates of gravel aggradation.

22.1.5 Conclusions

The Christchurch 3-D models provide a comprehensive 3-D interpretation of the subsurface geological materials, as well as information on geotechnical properties. These models by themselves do not provide a detailed analysis for such issues as the extent and nature of liquefaction, or assessment of sustainable groundwater abstraction. Rather, they provide information that can contribute to future detailed studies.

Key to the 3-D geological modeling has been the development and integration of a conceptual stratigraphic and geometrical model that reflected the dynamic paleoenvironments that existed around Christchurch in the Holocene. The geological models were created by a 3-D interpolation of stratigraphic boundary data from interpreted well logs. The boundaries were filtered through manual digitization of correlation lines that removed the influence of stratigraphically anomalous outliers, and stratigraphic surfaces were generated from the correlation lines with Leapfrog Geo 3-D modeling software. The density and abundance of borehole data have justified the use of implicit interpolation modeling software, particularly as newly acquired information regularly needed to be added and the models updated. The uncertainty of such interpolation depends largely on the separation of data

points and the definitions of the boundaries of geological units (see Chapter 15).

The geological modeling has helped validate the geotechnical model derived from CPT soundings. The geotechnical model provides a prediction of what geological materials, and some of their physical properties, can be expected in the near subsurface at any location. Portrayals of the geotechnical properties of materials in 3-D are invaluable for engineering design purposes and can be used with some confidence when assessed collectively over areas that are considerably greater than the spacing of the selected CPT soundings.

These models should not be used in place of site-specific investigations. The importance of many planning, construction, or resource management activities warrants the collection and analysis of new site-specific data. Such newly acquired data will improve the validity of the existing model; the model can then be modified to conform to the new data. These models inform those who choose to use them with a "best guess" answer to geological (and geotechnical) questions.

Case Study 22.2: Evaluation of Cliff Instability at Barton-On-Sea, Hampshire, England, with 3-D Subsurface Models

Oliver J. N. Dabson and Ross J. Fitzgerald
Jacobs Engineering Group, London W6 7EF, UK

22.2.1 Introduction

Aware of the shifting technological landscape, a small earth engineering task force within Jacobs Engineering Group was able to secure funds to explore and test dedicated 3-D geological modeling software on a currently active project at Barton-on-Sea, Hampshire, on the south coast of England (Figure 22.7). Key research questions for this project were:

- How viable is the 3-D modeling process for use on consultancy projects, where time/money is often limited?
- How does 3-D modeling alter or improve our knowledge of the site?
- What additional benefits (economic, organizational or otherwise) does 3-D modeling provide to a project?

This process would identify both the perceived and the actual benefits of such software, while also highlighting areas in the modeling methodology and infrastructure

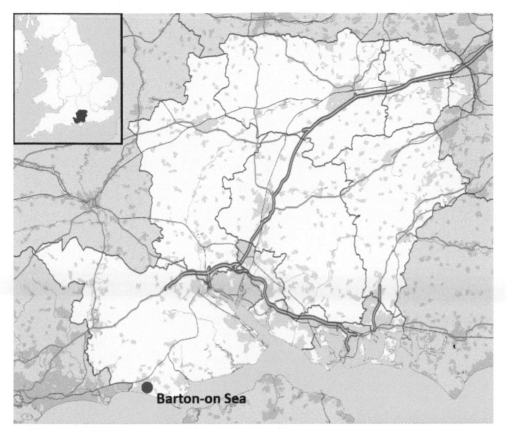

Figure 22.7 Location map of Barton-on-Sea within Hampshire. Source: Wikimedia Commons; contains Ordnance Survey data, © Crown copyright and database right.

which may need to be developed to make the tool a viable one for consultancy end use.

22.2.2 Site Description

The coastline at Barton-on-Sea exposes a sequence of sedimentary strata comprising clays, sands, and gravels of Palaeogene and Quaternary ages. These are reasonably weak; consequently, the cliffs below the town of Barton-on-Sea have been subject to rapid erosion and associated instability for decades. This has resulted in a number of ground investigations, slope remediation works and frontage protection between 1960 and 1994. However, many of these measures have reached the end of their design life and the area continues to suffer from ongoing instability.

The bedrock geology is of the Eocene Barton Group; a series of coarsening-upwards cyclothems which dip approximately 0.75° east-northeast (Hooker 1986; Bristow et al. 1991). The principal strata within these were subdivided initially on palaeontological grounds into Upper, Middle, and Lower Barton Beds by Burton (1929, 1933).

These classifications were also of use to the civil and geological engineering communities due to their distinctive lithologies (Barton and Garvey 2011). The geology of the site has also been divided, with increasing age, into the Becton Sand Formation, the Chama Sand Formation, the Barton Clay Formation and the Boscombe Sand Formation, and additionally into 16 engineering geological horizons (Table 22.4). The superficial geology consists of Pleistocene Plateau Gravels, capped by structureless, slightly gravelly, very sandy silt/slightly sandy clay (Brickearth) (West 2010). The high degree of slope instability has also resulted in colluvium over much of the site.

It is believed that the contrasts in geological material, and how these control the movement of water, are the primary controls on slope failure. There is a strong temporal relationship between 3-month antecedent rainfall and large scale instability events (Garvey 2007), suggesting that precipitation and subsequent groundwater recharge undermines stability in the Barton Group strata. However, the precise configuration of the basal shear surfaces is a subject of contention. It is generally agreed that there is a failure surface across the site at the interface between the

Table 22.4 Summary of the stratigraphic units of the Barton Group at Barton-on-Sea.

Formation		Zone	Description	Thickness (m)
Becton Sand Formation	Upper Barton Beds	K	Dense to very dense silty sand	(not exposed)
		J	Stiff dark gray (weathering gray-brown) shelly sandy silty clay	7.5
		I	Very dense dark gray (weathering yellow-white) silty shelly sand	8.0
Chama Sand Formation		H_2	Very dense blue-green (weathering orange-brown) silty shelly sand	2.4
		H_1	Stiff blue-green (weathering orange-brown) sandy silty shelly clay	3.1
		G	Very stiff/indurated gray-brown (weathering red-brown) sandy silty shelly sideritic clay/limestone	0.3
Barton Clay Formation	Middle Barton Beds	F_2	Stiff laminated dark gray-brown silty clay with sandy/shelling partings and 0.3 m gray mudstone at base	4.5
		F_1	Stiff laminated dark gray-brown silty clay with sandy/shelly partings	9.0
		E	Stiff laminated brown-gray silty clay with sandy/shelly partings and 0.2 m gray-brown siltstone at top	2.0
		D	Very stiff laminated gray-green silty variably sandy clay	7.0
		C	Very stiff laminated gray-green sandy clay with calcareous mudstone nodules at top and base (<0.3 m) and pale gray horizon near middle	3.8
	Lower Barton Beds	B	Very stiff laminated dark gray silty clay with sandy/shelly lenses	1.5
		A_3	Very stiff green/brown-gray sandy clay with regular beds of shelly sand	2.5
		A_2	Very stiff, green-gray sandy clay	(not exposed)
		A_1	Very stiff, green-gray sandy clay	(not exposed)
		A_0	Very stiff, green-gray sandy clay with basal flint pebble bed	(not exposed)
Boscombe Sand Formation			Very dense buff-brown fine- to medium-grained sands	(not exposed)

Source: After Halcrow (2011); based on information in Burton (1933); Melville and Freshney (1982); Bristow et al. (1991); and Barton and Garvey (2011).

Chama Sand Formation and Barton Clay Formation, with a shear zone at the F_1/F_2 boundary (Bromhead et al. 1991; Barton and Garvey 2011; Hosseyni et al. 2012). To the west of the site, an additional shear zone has been identified in the lower part of the D horizon in previous ground investigations (Halcrow 2011); however, there is dispute as to whether this is a single deep-seated compound slide (Barton and Garvey 2011), or a perched translational failure (zones F or H) over a lower landslide at zone D (Hosseyni et al. 2012). The design of any retaining structure will be influenced by the failure type and geometry; consequently, the application of 3-D geological modeling techniques to reconcile the surface geomorphology, the subsurface geology and the data ascertained from past ground investigations was seen as invaluable.

22.2.3 Software and Modeling Workflow

The British Geological Survey (BGS) and INSIGHT Geologische Softwaresysteme GmbH jointly had undertaken research to produce a user-friendly, knowledge-driven 3-D modeling platform that honors both borehole records and the interpretive nature of geological science. Initially known as "Geological Surveying and Investigation in 3 Dimensions (GSI3D)" and subsequently as "Subsurface-Viewer (SSV)", this SSV/GSI3D platform provides a fully digital interface to store, visualize, and interrogate ground conditions. It is based on a knowledge-driven, user-led interpretation paradigm that follows the fundaments of geological analysis (Kessler et al. 2009). Chapter 10 provides details of this approach to developing 3-D geological framework models.

Because it allowed for the synthesis of downhole factual data and user knowledge, and after investigating several existing 3-D geological modeling software packages, the Jacobs task force selected the SSV/GSI3D platform as the best method for 3-D subsurface modeling at Barton-on-Sea.

22.2.4 Results and Discussion

Figure 22.8 shows the first output of the 3-D geological model of the eastern section of the Barton-on-Sea complex. All engineering geological classifications of the Barton Group in boreholes have been incorporated into the model, and the lithostratigraphy of each unit (as described in Table 22.4) has been included as metadata.

The model provides a clear representation of the physical and functional characteristics of the subsurface, including the east-northeast dip of the geology and the distribution of non-continuous units (e.g. made ground) over 3-D space. Even where the available borehole data were poor, visualizing the ground conditions in 3-D greatly helps develop an understanding of basic geological structure. It was also possible to derive a number of secondary outputs, including "synthetic boreholes" boreholes, cross-sections, and horizontal slices; these present the modeled ground conditions at areas of interest specified by the user. In data-poor areas, these tools allow the user to use best judgment to "fill in" areas of uncertainty to improve the model, which may prove useful in attempting to reconcile shear

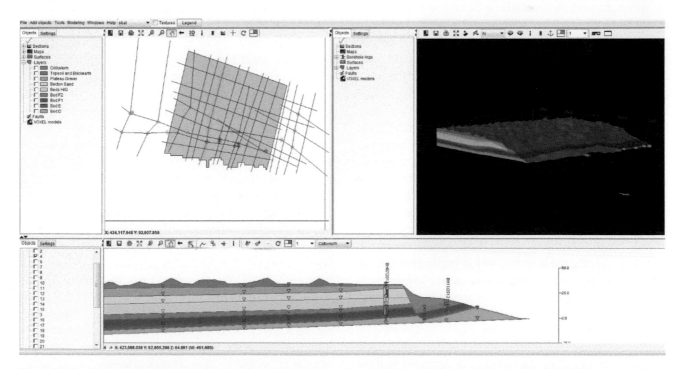

Figure 22.8 Screenshot of the modeling interface showing plan, cross-section and 3-D views of a rotational landslide modeled from borehole data at Barton-on-Sea.

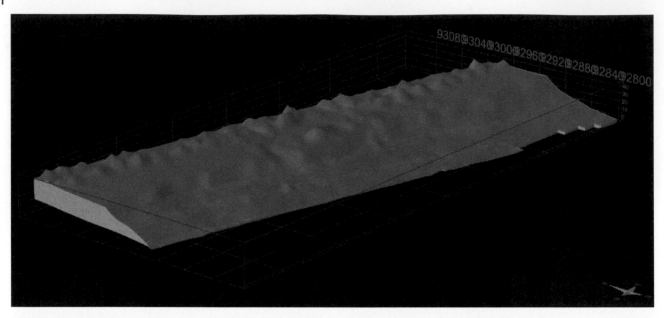

Figure 22.9 Screenshot of the 3-D model showing groundwater response horizons (in blue) at Barton-on-Sea.

zone configuration disputes found in some of the academic literature. Given the budgetary constraints of the project, no such reconciliation could be made at this stage and the shear surface configuration from Barton and Garvey (2011) was used by default. The potential to reconcile the opposing schools of thought on the landscape remains a viable use of the software.

From an applied standpoint, these tools allow for refinement of specifications for ground investigation. In light of an upcoming project to stabilize a caravan site to the west of the Barton area, Jacobs conducted a scoping exercise to determine the likely extent of ground investigation. When using "normal" practice, boreholes are positioned as a function of the current understanding of the site's geological complexity (based on a review of previous knowledge and investigations), the degree of precision required for the investigation and, most importantly, cost (Price and de Freitas 2009). These factors are normally assessed with reference to plan views of the site and individual cross-sections drawn from borehole information, which can be difficult to locate if archived poorly and are cumbersome to bring together. The process of creating the 3-D model allows the user to integrate legacy data to quickly identify locations where subsurface information is scarce, allowing the user to position future boreholes in these areas. For investigations where capital investment is limited, this allows for better quality results to be produced from fewer boreholes, resulting in better returns from investment in ground investigation.

However, geological insight is only one dimension of an investigation, and without a complete understanding of the

material characteristics of a site, it is unlikely that a project will fulfill its objectives. A review of the management history of Barton-on-Sea reveals various solutions which have had mixed effectiveness. It is likely that these result from an incomplete, as opposed to incorrect, ground model. Garvey (2007) discussed the strong influence of groundwater on slope failure; as such, modeling the distribution of groundwater response horizons across three-dimensional space would seem a useful first step in the appropriate management of drainage to improve stability. Through parameter substitution, the SSV modeling software can display these groundwater horizons by modeling groundwater response horizons as input data instead of geological groupings in a workflow improvised by Jacobs (Figure 22.9).

On a surficial level, this model produces a hydrological scenario which could be expected – the response horizons are controlled by the geology. This is apparent from the east-northeast dip of each "water layer", which is consistent with the stratigraphy. Referencing the location and elevation of the outfall of one of the recently installed drainage structures confirms that this also coincides with a "water layer", suggesting at least some value in the consideration of hydrology and hydrogeology on the 3-D platform. Further groundwater modeling can be undertaken using the 3-D model using software such as the USGS freeware MODFLOW. This would require additional parametric information and is beyond the scope of the current research project but represents an exciting prospect that will certainly be of use to the geo-engineering industry.

In addition to the geological and hydrogeological characteristics, geotechnical properties of subsurface materials are integral for many civil engineering projects, as interaction between the ground and the structure needs to be accommodated in design. Examining geotechnical logs from Barton-on-Sea ground investigations reveals the substantial variation of these properties with depth; extending this to three-dimensions is extremely challenging and any such model may suggest an unrealistic level of accuracy. Examination of available 3-D geotechnical modeling software products reveals that the approach generally used is "quasi-3D", i.e., extracting 2-D sections from the 3-D model to minimize the number of dimensions over which interpolation of a property must take place. This could conceivably be recreated using an improvised workflow created by Jacobs. Similarly, bespoke geostatistical analysis software can be used to interpolate between boreholes, and hence model the distribution of geotechnical information in three dimensions using voxel-based systems. This is an application of the software which is still being developed but looks to be an immensely powerful tool for engineering design as it adds a quantitative element to 3-D ground modeling, which previously has not been possible. It is highly likely to improve the quality of engineering design, although it is clear that further research will have to be done.

22.2.5 Conclusions

The nature and extent of the instability at Barton-on-Sea site made it one of the more challenging sites for 3-D modeling. The complexity of the geology, combined with ongoing disputes in published literature on the mechanism and location of the dominant failure horizon, have not facilitated the production of an agreed-upon ground model. This has possibly resulted in the under-design of some previous mitigation measures. Recent technological advancements offer the potential for a new way of thinking and have afforded a re-examination of the site.

Case Study 22.3: Role of 3-D Geological Models in Evaluation of Coastal Change, Trimingham, Norfolk, UK

Andrés Payo[1], Holger Kessler[1], Benjamin Wood[1], Helen Burke[1], Michael A. Ellis[1] and Alan Keith Turner[1,2]

[1] *British Geological Survey, Keyworth, Nottingham NG12 5GG, UK*
[2] *Colorado School of Mines, Golden, CO 80401, USA*

22.3.1 Introduction

Coastal erosion was widespread around the United Kingdom during the twentieth century and is expected to become even more pervasive through the twenty-first century due to sea-level rise and climate change. Erosion poses a direct hazard for coastal residents with potential risks of personal injury or loss of life and damage or complete loss of property. Failure of natural and artificial defenses during storms increases the risk of coastal flooding. Erosion and flooding resulting from a specific coastal storm may be predicted accurately, but mesoscale predictions, encompassing 10–100 km (5–50 miles) and decades to centuries, are unreliable (Payo et al. 2017). Yet these predictions are necessary to manage risks and make decisions on protective measures, or to evaluate responses to climate change impacts (Murray et al. 2013; Nicholls 2015). In response to this need, the Natural Environment Research Council (NERC) established two projects: "Integrating Coastal Sediment Systems" (iCOASST; (https://www.icoasst.net), and "Improving our understanding of processes controlling the dynamics of our coastal systems".

The 4-year iCOASST project (2012–2016) involved a consortium of UK Universities, research laboratories, the British Geological Survey (BGS), and engineering consultants, with the Environment Agency (EA) as a key embedded stakeholder. The project's task was to determine how best to predict mesoscale coastal morphological changes to guide long-term shoreline management and strategy studies (Nicholls 2012, 2015a; Van Maanen et al. 2016). This is a difficult problem because the evolution of a coastal landform, such as a beach or cliff, is influenced by interactions with adjacent coastal landforms and by human interventions. Coastal erosion is better understood and modeled than coastal recovery. The 4-year BLUEcoast project (2016–2020) is focused on both physical and biological dynamic processes and their role in coastal recovery after storm events. A principal BLUEcoast task is to produce a better representation of both transportable and source material within the coastal zone.

This case study illustrates how the sub-surface material can be represented and quantified as a bespoke-thickness model using Groundhog software (Wood et al. 2017) and how this can be used to produce better coastal evolution assessment using the Coastal Modeling Environment (CoastalME) framework (Payo et al. 2017). Geologists used Groundhog, a BGS software tool that utilizes a DTM, surface geological line work, downhole borehole information and geophysical data to produce a geological fence diagram that subsequently is used to create a 3-D geological framework model. Enhanced Groundhog capabilities supported the conversion of the 3-D geological framework model to a thickness grid model required by the numerical erosion model. CoastalME is a modeling platform that simulates decadal and longer coastal morphological changes.

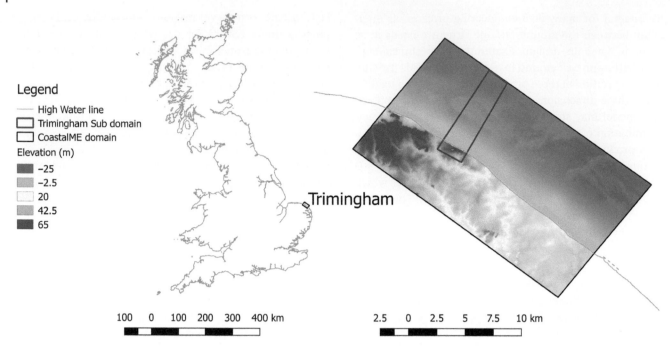

Figure 22.10 Location of Trimingham within Great Britain and 10 m resolution DTM. The Trimingham project used a larger domain (black outline) to model coastal hydrodynamics and a smaller domain (red outline) to integrate 3-D geological information with CoastalME simulation of coastal change.

Developed as a proof-of-concept in iCOASST, it is being further developed within BLUECoast.

Because the Norfolk coast is subject to severe erosion, the EA Coastal Partnerships & Strategic Overview Team for the East Anglia Area contacted the BGS in early 2017 to determine if erosion rates in that region could be more accurately assessed by combining BGS expertise in subsurface geological modeling and prior experience with the iCOASST/BLUEcoast projects. For an initial test case, a site near Trimingham was selected (Figure 22.10). The BGS organized a team of 11 staff-members with expertise in quaternary geology, landslide processes at the study area, subsurface geological modeling, and coastal morphodynamic modeling. This case study summarizes the experiences of this team in 2017 as they combined the capabilities of Groundhog and CoastalME.

22.3.2 Coastal Behavior Modeling Framework

Geomorphic coastal systems models combine understanding of hydraulics, waves, tides, sediment transport, and sediment conservation that are captured as sets of logical arguments or conceptual models, mathematical formulations, physical scaled models or statistical relationships (Payo 2017). None of these models claim to represent reality in all its complexity, but they provide a formal framework to explore, qualitatively and/or quantitatively, the behavior of coastal geomorphic systems

that are too complex to analyze through reasoning. Rather than favoring one approach over another, the prospect of integrating the different modeling approaches has been identified as a way forward to develop a system-wide capability for assessing coastal geomorphological change (Nicholls 2015b; Payo 2017). An innovative integration approach involves utilization of the essential characteristics of multiple landform-specific models using a common spatial representation within an appropriate software framework.

In the CoastalME framework, change in coastal morphology is represented by means of dynamically linked raster and geometrical objects. A grid of raster cells provides the data structure for representing quasi-3-D spatial heterogeneity and sediment conservation. Other geometrical objects (lines, areas, and volumes) that are consistent with, and derived from, the raster structure represent a library of coastal elements required by different landform-specific models. Figure 22.11 illustrates how a real coastal morphology (upper panel) is conceptualized as shoreline, shoreface profiles, and estuary elements (middle panel). All elements can share sediment among them (double-headed arrows). The shoreface comprises both consolidated and non-consolidated material that forms the cliff, shore platform, and beach respectively (bottom panel). At every time step the shoreline is delineated at the intersection of the sea level and the ground elevation. Shoreface profiles are delineated perpendicular to the shoreline.

Real world: Complex 3D-geometries

CoastalME: Simple lines and raster's sediment sharing objects

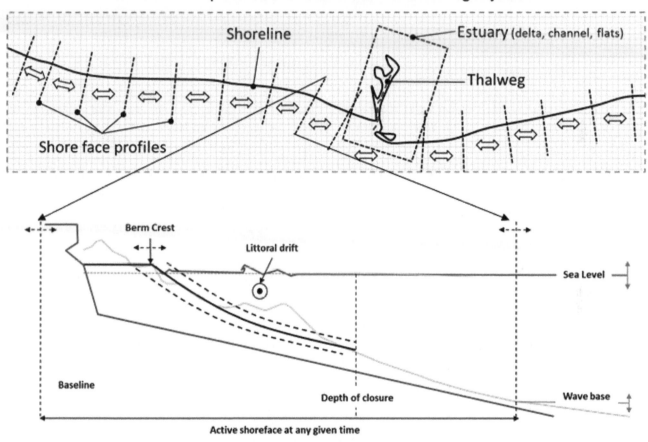

Figure 22.11 Schematic diagram of the CoastalME approach (Payo et al. 2017).

Sea level and wave energy constrain the proportion of shoreface profiles that are morphologically active at each time step. Eroded sediment from the consolidated profile is added to the drift material to advance the shoreline or is lost as suspended sediment. Gradients of the littoral drift further control the advance and retreat of the beach profile and the amount of sediment shared with nearby sections of the shoreline. Payo et al. (2017) provides additional details.

CoastalME uses a simple representation of the ground elevation and sub-surface properties (Figure 22.12), which is well aligned with our current understanding and modeling capacity of the sub-surface. Ground elevation is characterized as a set of regular square blocks. Each block has a global coordinate x, y, z. As shown by the blocks detail in Figure 22.12, each block may be composed of six different sediment fractions defined as "coarse", "sand", and "fine", with each size fraction further defined as "consolidated"

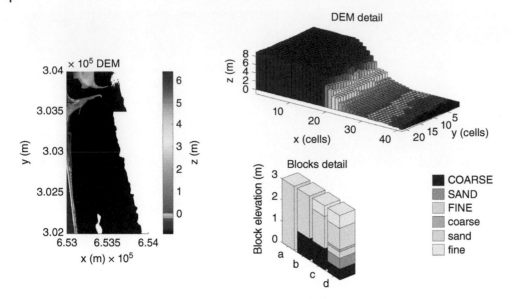

Figure 22.12 CoastalME representation of ground elevation and sub-surface at different levels of detail (Payo et al. 2017).

(capitalized) or "unconsolidated" (lower case). Block types a, b, c and d illustrate blocks having the same total elevation but with different compositions. Integration with a 3-D geological subsurface model permitted improved definition of these CoastalME grid elevations and subsurface composition. This in turn resulted in better evaluations of responses of the cliff and shore base materials to waves and currents.

22.3.3 Conditions at Trimingham

The Trimingham coast experiences periods of higher than usual rates of erosion followed by periods of relative stability. This makes prediction of erosion rates particularly difficult. Published reports suggest the cliff at Trimingham eroded between 1.5 and 2.5 m/yr in the 1966–1985 period. Evaluation of historic maps suggests 50–60 m of erosion over a period of 100 years, or between 0.5 and 0.6 m/yr. The Shoreline Management Plan Kelling Hard to Lowestoft Ness (East Anglia Coastal Group 2010) reports that erosion of 75–150 m can be expected over the next 100 years (0.75–1.5 m/yr). Littoral sediment processes on the coast around Trimingham have been studied through observations and modeling (HR Wallingford 2003a,b). These reports provide estimates of potential net longshore sediment transport, changes in beach volume and steepness, cliff recession, and sediment yields.

The geology of the region consists of a thick layer of glacial deposits unconformably overlying marl and fossiliferous limestone of the Cretaceous Chalk Group. The rockhead surface dips gently to the east and is at or below sea-level along the coast, except for a few distinctive bedrock highs; one high occurs at Trimingham. Lee et al. (2011b) provide a detailed discussion of the glacial deposits; they consist of four recognized tills separated by sequences sand, silt, and clay representing ice-marginal glaciolacustrine conditions, including large sand deltas and rhythmic, varved lake deposits.

The coastal cliffs between Trimingham and Overstrand offer a rare opportunity to examine the internal architecture these deposits (Figure 22.13). These cliffs form the eastern eroded extent of the Cromer Ridge, interpreted as a push moraine formed at the southern margins of the Middle Pleistocene ice sheet. The stratigraphy of Trimingham cliffs has presented something of a geological puzzle to many scientists due to their intense glaciotectonic deformation (Hart 1990; Hobbs et al. 2008; Lee et al. 2011b), and the frequent obscuring of exposures by large rotational landslides (Hutchinson 1976). Lee et al. (2011a) reported on studies of coastal sections at Trimingham undertaken for 10 years (1996–2006). These studies took advantage of cliff and coastal erosion which created new, often temporary, exposures.

22.3.4 Evaluation of Cliff Erosion at Trimingham

After reviewing existing information sources, the BGS team concluded that any analysis of cliff erosion at Trimingham must consider the following:

- Interactions between alongshore sediment transport and erodibility of the glaciotectonic cliff materials since these control the cliff recession rates;

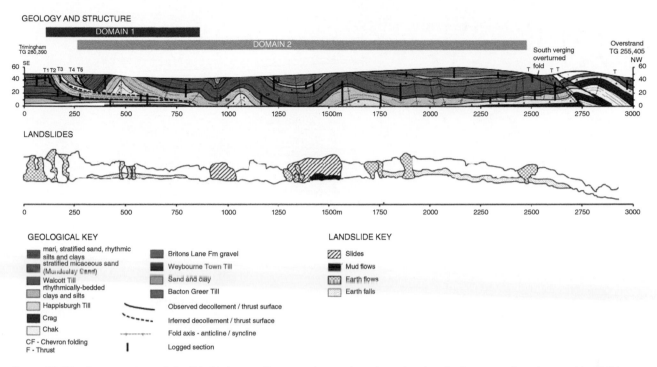

Figure 22.13 Cross-sections of the Trimingham to Overstrand coastal sections showing the interpreted geology and landslides. Source: Lee et al. (2011a); modified from Hart (1990).

- Future recession rates depend on variations in cliff height and the presence/absence and condition of artificial coastal-defense structures; and
- Small beach volumes cause highly variable observed annual recession rates, ranging from 0 to 15 m/yr between 1993 and 2001.

The limited resources available for this proof-of-concept initial project resulted in the BGS team integrating 3-D geology with CoastalME process modeling within a relatively small area at Trimingham. A larger "CoastalME domain" (entire rectangular area in Figure 22.10) was used to minimize boundary-condition influences on the CoastalME simulations of waves and alongshore currents. A much smaller domain (red outline on Figure 22.10) was used to build a 3-D subsurface geology model.

The BGS team created an integrated digital terrain model (DTM) for the larger CoastalMe domain by combining data from NextMap 5 m-resolution DTM, EA 2 m-resolution DTM of the shore areas obtained from 2016 Lidar surveys, and 10 m-resolution bathymetry data from the UK Hydrographic Office. These sources left an "information gap" along the shallow offshore; elevations in this zone were estimated by linear interpolation to produce an integrated 10 m-resolution DTM.

The 3-D geological model was constructed according to standard BGS modeling procedures (Chapter 10) using borehole logs and geological maps to create a series of interlocking interpreted cross-sections and ultimately a 3-D geological framework model (Figure 22.14). The completed model consisted of six geological units and extended from approximately 1.25 km behind the coast to 7 km offshore.

Groundhog Desktop (Wood et al. 2017) was used to manage the digital borehole and cross-section data. Its capabilities were extended to support the export of 3-D material-property distributions as raster files to define material distributions and thicknesses in a form acceptable to CoastalME (Figure 22.15).

22.3.5 Conclusions

As a test case, the BGS team used CoastalME to simulate coastal recession at Trimingham for two scenarios by changing only the subsurface composition, with all other factors unchanged. Waves were assumed constant and propagating normal to the coastline with a 1 m significant wave height and an eight second period. The simulations evaluated this wave regime for a period of 25 days. The first scenario assumed the cliff was composed of fine material that, when eroded, would be lost in suspension and not contribute to the nearshore sediment budget. The second scenario assumed a cliff made entirely of sand material that, when eroded, became part of the beach volume.

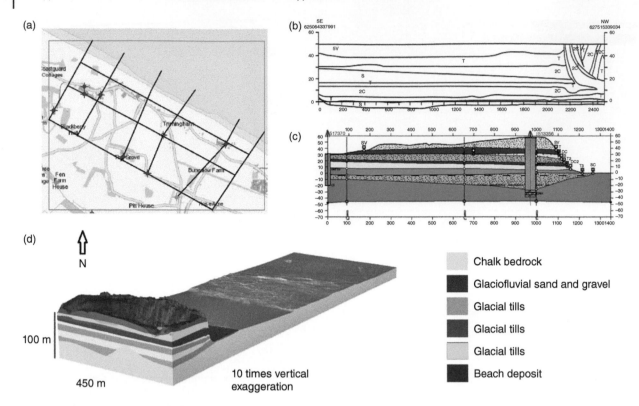

Figure 22.14 Creating the 3-D Geological Model. (a) User-defined network of interlocking cross-sections (boreholes are red dots). (b) Typical interpretive cross-section. (c) Typical digitized spatially referenced cross-section with borehole. (d) Completed 3-D geological model.

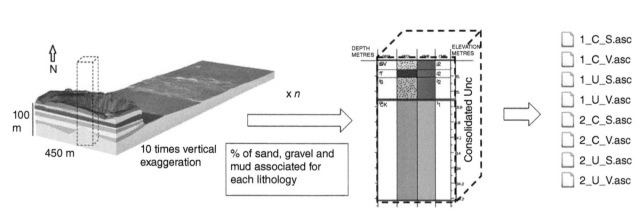

Figure 22.15 Geological model horizons are grouped into material classes, forming CoastalMe layers; these are exported as thickness rasters for CoastalME simulations.

These two simulations illustrate how different subsurface material distributions are important influences on cliff recession rates at Trimingham (Figure 22.16). The cliff recession for a fine-grained muddy cliff was up to 95 m greater than for a sandy cliff. These two scenarios indicate the sensitivity of shoreline evolution at Trimingham to cliff composition, and thus the importance of including accurate geological information in simulations.

Variations in cliff recession rates are due to influences of material contributed from the cliffs on beach volumes and how the characteristics of this material influence indirect morphodynamic conditions. The sediment yield per unit of eroded cliff is a function the cliff height and sediment characteristics. Large yields can increase beach width, resulting in an advancing shoreline. Subsequently, these wider and thicker beaches reduce the energy reaching the cliff base,

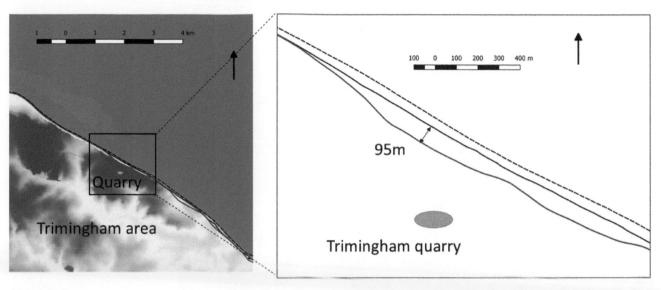

Figure 22.16 Results from two cliff erosion scenarios at Trimingham. Black line shows erosion assuming a sandy cliff; red line shows erosion assuming a fine-grained muddy cliff. Dashed line shows initial cliff location at start of simulation.

thus reducing the recession rate, and the energy reaching the shore platform, thus reducing the rate of shoreface erosion.

Case Study 22.4: Three-dimensional Geochemical Modeling to Anticipate the Management of Excavated Materials Linked to Urban Redevelopment – Example of Nantes

Cécile Le Guern, Vivien Baudouin, Baptiste Sauvaget, Maxime Delayre, and Pierre Conil

Bureau de Recherches Géologiques et Minières (BRGM), 44000 Nantes, France

22.4.1 Introduction

Historically, European cities have evolved from relatively small settlements to larger industrial communities and, most recently, to modern cities whose inhabitants value sustainability and environmental values. Urban densification (i.e., compact city planning; Jenks and Jones 2010) is a preferred solution to reduce urban sprawl while responding to two trends experienced by many cities, including Nantes: (i) deindustrialization and (ii) increased population. Urban (re)development typically involves excavation of considerable volumes of soil that may contain various levels of contamination due to former industrial activities or by the intrinsic characteristics of anthropogenic deposits. Economic and environmental issues encourage the reuse of these excavated soils because it: (i) avoids filling landfill sites

with reusable materials, (ii) preserves natural materials resources, and (iii) reduces carbon dioxide emissions by reducing transportation distances (Jeanniot et al. 2014).

Effective management of excavated soils and subsoils depends on accurate assessment of the potential quantities of contaminated materials and their contamination characteristics. Currently, such characterizations are carried out at the site scale, although large redevelopment projects typically involve much larger areas. Urban redevelopment plans for an entire city, or major districts within a city, can benefit from knowledge of historical land-use in order to define potential sources of contamination, combined with assessments of the distribution, structure, and quality of soil and subsoils. The latter aspect requires a 3-D representation of urban subsurface geology. The construction of the geological model must consider anthropogenic deposits, which often have limited lateral extents but have significant thicknesses up to 10 m (Laurent and Fondrillon 2010) in comparison with excavation depth. The usual lithostratigraphic approaches employed to describe subsurface geology for archeology evaluations (De Beer et al. 2012), or for geotechnical purposes (Royse et al. 2009), typically ignore the geochemical properties of shallow subsurface deposits.

The "Île de Nantes" is a 337 ha island located in the center of the city of Nantes in France (Figure 22.17). Originally a group of small islands located in the Loire River, the island was formed by gradual infilling to accommodate various activities, including industrial ones. The district presently contains 18,000 inhabitants, 10,000 housing

Figure 22.17 The Île de Nantes. Source: Photo: Ville de Nantes.

units and supports 16,000 jobs (Samoa 2016). The subsurface consists of mica-schist bedrock at depths of 15–20 m overlain by a sequence of alluvial deposits. Over almost the entire district, these alluvial deposits are covered by anthropogenic deposits which have accumulated gradually over the long settlement history. The recent closure of major industries and the growth of the city encouraged developers to propose to re-densify the district. The proposed 30-year redevelopment project (2000–2030) will result in about 100 ha of developable land, which would accommodate 7000 additional housing units, 30 ha of economic activities and 15 ha of urban facilities (Samoa 2016).

This redevelopment is expected to generate large quantities of excavated materials, more than 100,000 tonne/yr between 2015 and 2025. The developer needs to anticipate the costs and options for disposing or re-using these materials. Such decisions depend on their geochemical characteristics. As part of a R&D partnership with the local urban developer, the Bureau de recherches géologiques et miniéres (BRGM) developed a 3-D geochemical modeling approach to characterize the soil and subsoil geochemical quality of the Île de Nantes district. The BRGM approach involved: (i) development of a 3-D geological model of the Île de Nantes district, with special attention to anthropogenic deposits, (ii) development and validation of a typology of anthropogenic deposits according to their intrinsic contamination potential, and (iii) development of a 3-D geochemical model that allows for the assessment of the spatial distribution of contaminated materials.

22.4.2 Construction of the 3-D Geological Model

The geological model was constructed using subsurface observations and descriptions from boreholes undertaken for pollution and geotechnical investigations (Figure 22.18). Observations from 2400 boreholes were stored in a structured geodatabase created with ArcGIS 10.1 and Microsoft Access database management applications. The GDM modeling and visualization platform (Bourgine 2007, 2015; Bourgine et al. 2008; BRGM 2016) was used to create a 3-D representation of the subsurface geology of Île de Nantes, and further analysis produced estimates of the geochemical quality of the anthropogenic deposits.

The anthropogenic deposits on the Île de Nantes have a variety of sources spanning a considerable historical period. Thus, important geochemical heterogeneities were expected. The term "made ground" used here refers to the classification of anthropogenic deposits by Ford et al. (2014). As described by geologists, the anthropogenic deposits on Île de Nantes correspond to the "engineered embankment" type or the "raised fill" type (previously called "infilled ground") of the "made ground" class (Ford et al. 2010).

22.4.3 Typology of Made Ground

A typology of made grounds according to their intrinsic contamination potential was developed and validated (Le Guern et al. 2016a). The 2400 borehole descriptions defined more than 7200 lithologic layers. These were

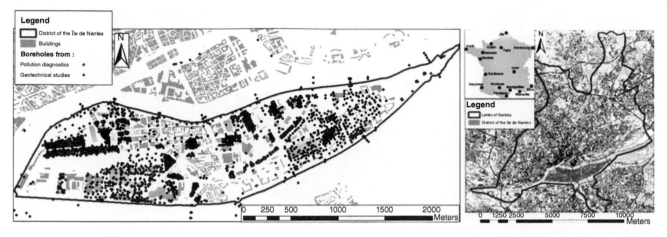

Figure 22.18 Locations of 2400 boreholes from pollution and geotechnical investigations stored in a structured ArcGIS geodatabase (Le Guern et al. 2016a).

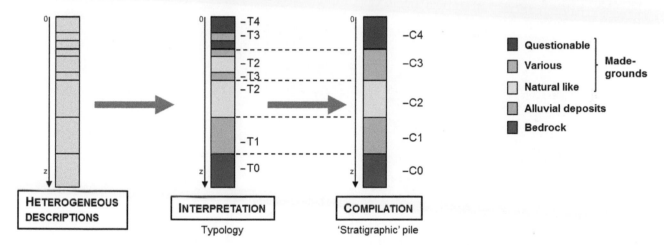

Figure 22.19 Interpreting lithology in order to define the stratigraphic classes used in the 3-D geological model (Le Guern et al. 2016a).

initially classed into four types (T1–T4) using the typology of made ground. These differentiated the potential contamination as "low" (in "natural-like" material T2), middle in "various made grounds T3", or higher in "questionable made grounds T4" (Figure 22.19). These typologies were then assessed within a simplified lithostratigraphic sequence which consisted of five strata: Bedrock (C0), Alluvium (C1), and three Made Ground strata (C2, C3, C4). The geochemical quality of made ground was defined according to classes of materials that always retained the typology with the lowest geochemical quality class, as shown in Figure 22.19.

The geometry of the anthropogenic materials on Île de Nantes was developed by analyzing the current topography of the island and historical analysis of the shape of the islands (Figure 22.20). The geometry of filling materials deposited between the former set of islands was identified by cross-referencing the contours of the islands digitized

from 15 selected old maps, dating from 1754 to the present day. The difference in altitude between the current and former topography (current digital elevation model versus reconstituted one from the 1970s) indicates the thickness of recent accumulations of made ground materials. This information was used to constrain the model.

22.4.4 Application of the 3-D Geochemical Model

The obtained 3-D representation (Figure 22.21) was used to calculate the quantities of each class of material in the top 4 m of excavated materials on the whole island, and on specific redevelopment areas (Table 22.5).

The calculations show that the island's subsoil, in the first 4 m from the surface, contains 28% of questionable materials, 13% of various made grounds, 49% of "natural-like" made grounds and 10% of alluvial deposits. In the south-western zone (the site of the next

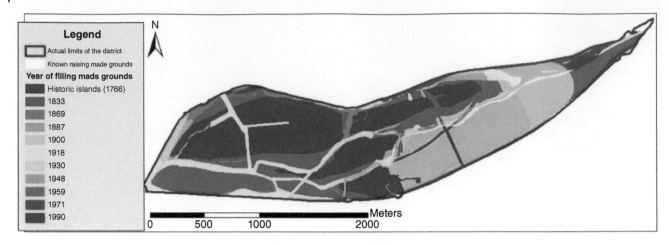

Figure 22.20 Historic infillings and accumulations of made ground materials explain the actual Île de Nantes topography (Le Guern et al. 2016b).

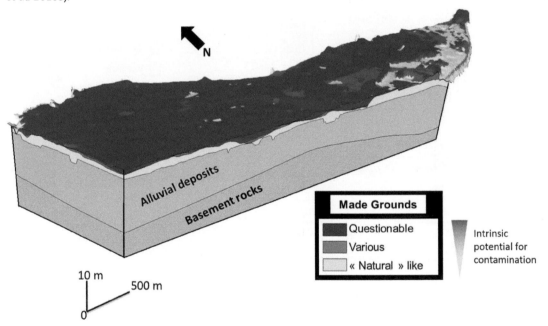

Figure 22.21 View of the 3-D geological model after assignment of the contamination potential of the made ground strata (Le Guern et al. 2016a).

Table 22.5 Calculated volumes (m³) and proportions (%) of materials for each typology within the uppermost 4 m: (a) for the entire island and (b) for a redevelopment sector in the south-western part of the island. (Le Guern et al. 2016a).

	Class of material	Whole island		South-western part	
		Volume (m³)	Proportion (%)	Volume (m³)	Proportion (%)
C4	Questionable grounds	3,596,819	28	110,429	27
C3	Various made grounds	1,746,708	13	33,245	8
C2	"Natural-like" made grounds	6,308,988	49	238,625	59
C1	Alluvial deposits	1,333,817	10	193,000	5
	Total	12,986,332		401,599	

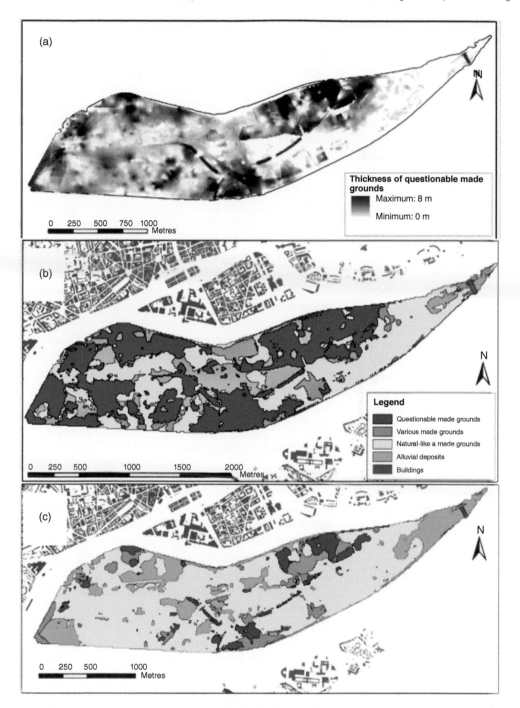

Figure 22.22 Two-dimensional representations of the distributions of subsoil classes on the Île de Nantes based on data extracted from the 3-D model: (a) thickness of questionable made grounds; (b) spatial distribution of expected materials at 1 m depth; (c) spatial distribution of expected materials at 3 m depth.

redevelopment project), the proportion of natural-like material appears higher (59%), whereas the proportions of various made grounds and alluvial deposits appear lower (8% and 5%, respectively).

Because the next development phase in the southwestern part of Île de Nantes will have to deal with less anticipated contaminated material, and this material has

good geotechnical properties, the developer will benefit from its potential for reuse and reduced management costs. However, verification of material quality is still required, as local contamination may be possible, especially in areas previously occupied by industry or service activities. To identify areas of potential contamination, a 2-D map was developed in ArcGIS showing the locations of former industrial and service activities and associated potential sources of contaminants (Le Guern et al. 2016b,c). The model developed for this aspect also serves as a decision-support tool.

The 3-D geochemical model was used to create 2-D representations of made ground thicknesses. Figure 22.22a shows one such map, the thickness of questionable made grounds. A series of horizontal slices through the 3-D model produced 2-D maps showing the expected spatial distribution of materials at various depths. Figures 22.22b and 22.22c show the anticipated contamination levels at depths of 1 and 3 m, respectively. These maps were used to prepare preliminary rules for the reuse of excavated materials on the island based on interpretations of baseline compatibility levels defined by Le Guern et al. (2016b).

22.4.5 Conclusions

The 3-D subsurface model of the Île de Nantes is a powerful tool that has improved the understanding of soil and subsoil structures and associated geochemical characteristics. Prediction of potential contamination is critically important for redevelopment. With this knowledge, projects can be adapted to minimize environmental exposure and management of excavated materials is made more cost effective. Knowledge of the distribution of anthropogenic deposits and their intrinsic potential of pollution is useful when combined with information on other potential sources of pollution such as industrial and service activities. These analyses require exploitation of historical data. Such information can also guide further field investigation that is required to guide specific redevelopment projects.

Case Study 22.5: Managing Drinking Water Supplies for Ljubjana, Slovenia with a 3-D Hydrofacies Model, Numerical Groundwater Flow and Transport Model, and Decision Support System

Mitja Janža

Geological Survey of Slovenia, 1000 Ljubljana, Slovenia

22.5.1 Introduction

Ljubljana, the capital and largest city in Slovenia, obtains its drinking water from two aquifers. The "Ljubljansko Polje" (Ljubljana Field) aquifer underlies the northern parts of the city along the Sava River while the "Ljubljansko Barje" (Ljubljana Marsh) aquifer is located to the south (Figure 22.23). The Ljubljana Field aquifer currently supplies about 90% of Ljubljana's drinking water. It is unconfined and vulnerable to surface contamination, so drinking water protection zones have been established to safeguard water quality.

For over 100 years, the water supply has not required any technical treatment, it was just occasionally chlorinated. However, a contamination event several years ago threatened the water supply and highlighted the need for better knowledge of the groundwater resource. Initial groundwater models did not accurately predict the observed path of the contaminant plume (Janža et al. 2005, 2011). Because aquifer heterogeneity controls groundwater transport processes, a detailed 3-D model of the hydrofacies distribution within the Ljubljana Field aquifer was developed, and then used to constrain a numerical simulation of groundwater flow and transport (Janža 2009).

A Decision Support System assists water managers and city officials in understanding results from this numerical modeling sequence. It combines three logically interlinked components: (i) a database of groundwater observations from the monitoring network and potential sources of pollution; (ii) the numerical groundwater flow and transport model; and (iii) a decision support system reflecting the knowledge and experiences of water managers and hydrogeologists (Janža 2015).

22.5.2 Geological Setting

Ljubljana is situated in the Ljubljana Basin, which began to subside in the Pliocene. Carboniferous and Permian rocks underlie the basin; they are exposed as a series of hills (Golovec, Grajski hrib, and Rožnik) within the city and to the south. During the Pleistocene, the Sava River transported material from alpine glaciers and filled the basin with alluvial sediments. The metropolitan area is located mostly on these sediments and partly on Holocene deposits. Two rivers flow across the Ljubljana Field aquifer. The Ljubljanica flows through the town center while the Sava, with headwaters in the Alps, flows from the northwest along the northeast side of the aquifer. While the Ljubljanica's riverbed is relatively impermeable, the Sava River flows across permeable sediments, including gravel terraces, and thus recharges the groundwater in the aquifer (Janža et al. 2016; Jamnik et al. 2003).

The Ljubljana Field aquifer is composed of permeable gravel and sand beds with lenses of conglomerate that exceed 100 m thickness in some locations (Žlebnik 1971; Janža 2009; Šram et al. 2012). The aquifer is composed of deposits formed by migrating paleo channels of the Sava River. Consequently, the internal stratigraphy is a complex pattern of flood plain sediments cut by abandoned river channels. The southwestern portion of the Ljubljana Field aquifer contains low hydraulic conductivity layers

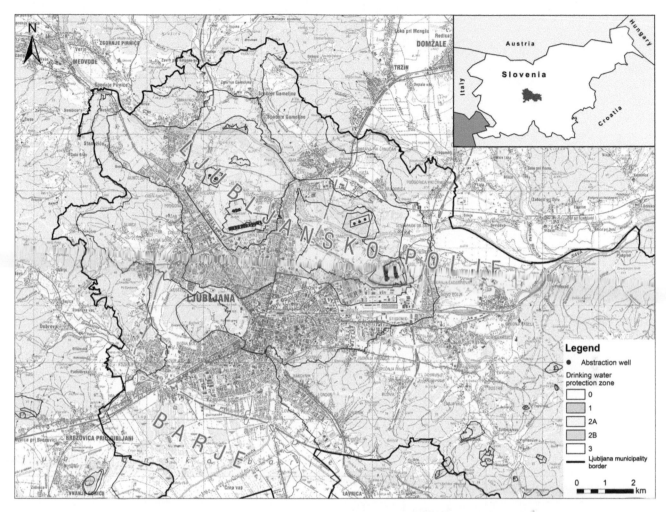

Figure 22.23 Map of the City of Ljubljana, showing "Ljubljansko Polje" (Ljubljana Field) aquifer underlying the northern part and "Ljubljansko Barje" (Ljubljana Marsh) aquifer underlying the southern part, and locations of drinking water protection zones (Janža et al. 2016).

that create local perched aquifers, mostly located in the foothills of Šišenski and Grajski hrib (Šram et al. 2012).

The Ljubljana Marsh aquifer is composed of a heterogeneous sequence of alternating fluvial and lacustrine deposits. In the northern portions of the aquifer, a low permeability layer up to 30 m thick overlies a Pleistocene gravel aquifer that is confined or semi-confined, producing artesian to sub-artesian conditions. The Ljubljana Marsh aquifer contains good-quality groundwater and is partially protected from near-surface contamination sources by the overlying clay layer.

22.5.3 Hydrogeological Model of the Ljubljana Field Aquifer

Janža et al. (2011) describe the conceptual model of the Ljubljana Field and Ljubljana Marsh aquifers originally developed by Kristensen et al. (2000). These aquifers have distinctly different hydrogeological characteristics and

only limited hydrologic connection, so they were evaluated by two separate hydrogeological models. This case study focusses on the hydrogeological modeling of the Ljubljana Field aquifer.

Descriptions from 258 borehole logs (Figure 22.24), with a total length of 6422 m, were used to evaluate the 3-D distribution of hydrofacies. These descriptions used a variety of lithological terms; they were reviewed and standardized to reflect four hydrofacies (Figure 22.25). Existing geological knowledge defined the model's lateral extent (approximately 17×10 km), boundary conditions, and base (Janža 2009).

The alluvial deposits forming the aquifer result from complex and non-uniform geological processes. Tectonic basin subsidence resulted in burial of alluvial deposits to depths of 100 m or more, with marked local variations in burial depths. Changing climate conditions within the Pleistocene produced numerous cycles of deposition and erosion, related to glacial and interglacial stages (Žlebnik

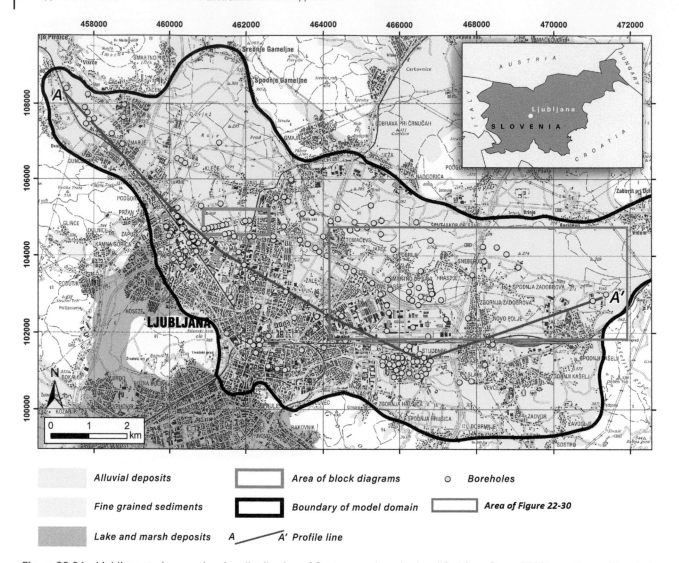

Figure 22.24 Ljubljana study area, showing distribution of Quaternary deposits (modified from Buser 2009), positions of boreholes, cross-section A–A′ (Figure 22.25), block diagrams (Figure 22.28), the pollution plume study area (Figure 22.30), and the boundary of the model domain.

1971). The subsurface distribution of hydrofacies was evaluated using stochastic modeling approaches.

22.5.3.1 Defining the Aquifer Base

The base of the Ljubljana Field aquifer is composed of very low permeability Carboniferous and Permian sedimentary rocks. They form the base of the hydrogeological model (Figure 22.26); this was defined for the initial model by Kristensen et al. (2000) and later updated with interpretation of data from new boreholes (Janža et al. 2011, 2017).

22.5.3.2 Modeling the Spatial Distribution of the Hydrofacies

The spatial distribution of the alluvial deposits forming the Ljubljana Field aquifer were modeled by a combination of

geostatistical and stochastic methods implemented within the T-PROGS (Transition PRObability GeoStatistical) software (Carle 1999). Chapter 13 discusses the T-PROGS method as well as alternative properties simulation models. The T-PROGS method involves three main steps: (i) calculation of transition probability measurements, (ii) modeling spatial variability with Markov chains, and (iii) conditional simulation.

Borehole log descriptions from 258 boreholes were standardized to four hydrofacies: "gravel *(gr)*", "silt and clay with gravel *(scg)*", "silt and clay *(sc)*" and "conglomerate *(co)*". Table 22.6 summarizes the characteristics and observed volumetric proportions of these hydrofacies. A Markov chain model generated transition rates for borehole log hydrofacies based on observations at 1-m vertical intervals (Figure 22.27); diagonal entries in the

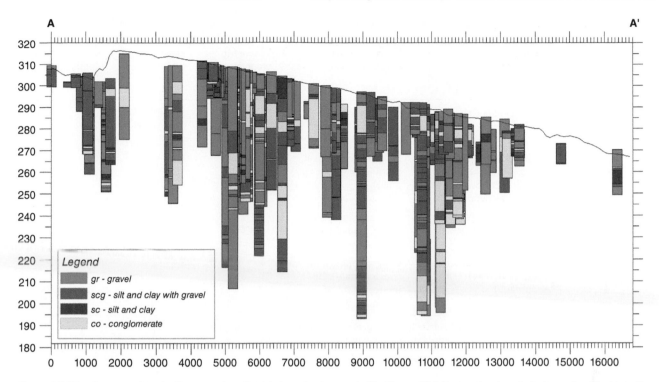

Figure 22.25 Cross-section A–A' across the alluvial deposits (presented in Figure 22.24), showing borehole logs classified into four hydrofacies groups (Janža 2009).

Figure 22.26 Interpreted base of the Ljubljana Field aquifer (Janža et al. 2017).

vertical transition rate matrix provided estimates of mean thicknesses for each hydrofacies (Table 22.6).

Transitions in the vertical direction are asymmetric: the transition rate from "hydrofacies a" to "hydrofacies b" is not the same as the rate from "hydrofacies b" to "hydrofacies a". This corresponds to field observations and so lends credence to the Markov chain model. For

example, an observed fining-upward tendency is shown by variations in transition rates between "gravel *(gr)*" and "silt and clay with gravel *(scg)*", also between "silt and clay with gravel *(scg)*" and "silt and clay *(sc)*"; these are shown in yellow boxes in Figure 22.27. Variations in transition rates between "silt and clay *(sc)*" and "conglomerate *(co)*", also between "gravel *(gr)*" and "silt and clay *(sc)*" (green boxes

Table 22.6 Attributes of the hydrofacies.

Hydrofacies	Geologic interpretation	Common driller's description	Volumetric proportion (%)	Mean thickness (m)[a]	Estimated mean lengths (m)[b]	
					Length	Width
Gravel (*gr*)	Channel deposits	Gravel, sandy gravel, pebbles, well sorted sand	45	5.5	1300	400
Silt and clay with gravel (*scg*)	Products of pedogenic processes, pebbles filled with fine grained deposits, debris flow deposits	Silty or clayey gravel, silt, and clay with gravel	36	4.5	1000	300
Silt and clay (*sc*)	Floodplain deposits, products of pedogenic processes	Silt, clay, clayey silt, poorly graded sand	5	1.4	150	100
Conglomerate (*co*)	Lithified river deposits (mainly channel sediments)	Conglomerate, conglomerate with intercalations of gravel or sand	14	3.1	800	400

a) Mean thickness determined from diagonal entries in the vertical transition rate matrix (Carle 1999).
b) Estimated mean lengths based on field observations and geological conceptual model.
Source: After Janža (2009).

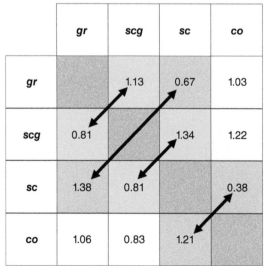

NOTES: 1) Grey boxes are unobservable transitions
2) Yellow box transition rates show upward-fining strata
3) Green box transition rates show gravel overlying clay overlying conglomerate

Figure 22.27 Transition rates in the vertical direction generated by a Markov chain model from borehole logs. Modified from Janža (2009).

in Figure 22.27) correspond to observations in the central part of the Ljubljana Field aquifer where gravel overlies a layer of clay with pebbles representing a weathered horizon of an underlying conglomerate (Žlebnik 1971).

To compute a 3-D distribution of hydrofacies, equivalent transition parameters must be defined for lateral directions. The borehole spacing prevented empirical development

of lateral Markov chain models. Incorporating conceptual geologic knowledge of depositional processes permitted estimates of volumetric portions, mean lengths, and the juxtapositional tendencies of hydrofacies in two orientations: parallel to and transverse to the depositional flow direction ("Length" and "Width" in Table 22.6). Lateral symmetry of hydrofacies was assumed; assignment of the fine-grained "silt and clay (*sc*)" hydrofacies as a background category provided computational efficiency (Carle 1999).

The 3-D Markov chain model provided input to a conditional simulation that produced a specified number of "equally likely", or stochastic, simulations of 3-D hydrofacies distributions. These are defined on a 3-D cellular grid, in this case the grid was 100 × 100 m horizontally and 1 m vertically. The process involves two stages. A sequential indicator simulation (SIS) procedure performs transition probability based co-kriging to produce stochastic realizations that both honor the field data and preserve the spatial correlation model (Carle 1999). A second simulated quenching (annealing) step produces conditional simulations that are entirely consistent with the Markov-chain model, while ensuring hydrofacies patterns closely replicate field observations (Carle 1999). Figure 22.28 shows two realizations for a small portion of the entire model; this location is outlined in Figure 22.24.

22.5.3.3 Modeling the Distribution of Perched Aquifers
Previous field investigations identified perched aquifers within the western portions of the Ljubljana Field aquifer, but their lateral extension was uncertain. Šram et al. (2012)

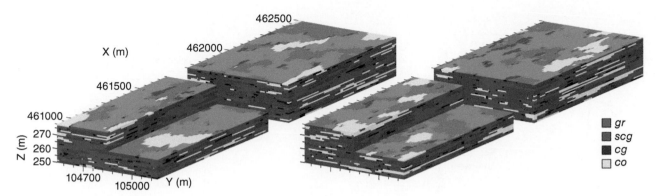

Figure 22.28 Two geostatistical realizations of the spatial distribution of hydrofacies in a portion of the alluvial deposits, outlined in Figure 22.24 (Janža 2009).

describe how stochastic modeling approaches identified the spatial distribution of low permeability clay lenses which produce these perched aquifers. Archived information provided lithological and depth information from over 1100 shallow boreholes; a database was created using the same four hydrofacies defined in Table 22.6. Hydrofacies were assigned appropriate hydraulic conductivities and JewelSuite 2011 (Baker Hughes Incorporated) was used to perform Sequential Indicator Simulation (SIS) and to produce a 3-D spatial distribution of low hydraulic conductivities. These produce perched aquifers above them; they are found at several depths in the upper 30 m around the hills (Figure 22.29).

22.5.3.4 Numerical Modeling of Groundwater Flow and Transport

Interactions between surface water and groundwater control the hydrologic regime. Thus, an integrated modeling system was required. MIKE SHE is a deterministic, physically based, distributed modeling system capable of describing flow processes within the hydrological cycle (DHI 2011b; Graham and Butts 2005). When combined with MIKE 11 (DHI 2011a), detailed evaluations of river–groundwater interactions were possible, producing an integrated modeling capability.

A numerical groundwater modeling process based on the transient and spatially distributed MIKE SHE/MIKE 11 hydrological model (Janža et al. 2011; Janža 2015) updated a model by Kristensen et al. (2000) that was developed for the water resources management of the city of Ljubljana. The model domain covers an area of 96 km² and is discretized into horizontal grid cells of size 200×200 m. Daily values of time-dependent variables (temperature, precipitation, river levels, and discharges at boundaries, groundwater level observations, abstractions) were used in the model. Calibration of the model is based on groundwater heads and Sava River water level observations.

22.5.4 Applications of the Model

The modeling effort has led to several applications. Its reliability for predicting future pollution events was tested by comparing the model predictions with a previously observed contamination event. The model was used to delineate groundwater protection areas for the city of Ljubljana (OG 2015), shown in Figure 22.23. A Decision Support System was developed to enable city officials and water managers to use the model. Recently, the model has been further developed to support the evaluation of a shallow geothermal potential for open loop ground source heat pump systems, to be used primarily for heating and cooling purposes. These activities are being performed within the GeoPLASMA-CE project (http://www.interreg-central.eu/Content.Node/GeoPLASMA-CE.html) that aims to foster the use of shallow geothermal energy (see also Chapter 20). Increased use of shallow geothermal energy supports the environmental goals of the city of Ljubljana, which are: (i) increased renewable energy share of the total energy consumption, and (ii) reduction of greenhouse gas emissions and air pollution.

22.5.4.1 Contaminant Plume Evaluations

Previous modeling efforts (Kristensen et al. 2000; Janža et al. 2005) did not accurately predict the location, extent, or growth of a Trichloroethylene (TCE) pollution plume when compared with an actual contamination event (Janža et al. 2005). The new numerical model was tested to determine if it could more accurately simulate the groundwater dynamics and transport of pollutants in the aquifer. Figure 22.30 shows four predictions of the distribution of the TCE plume. These were independently numerically modeled by MIKE SHE; each prediction is based on a different equally probable hydrofacies distribution (Janža et al. 2011). The location and range of the simulated pollution plume agreed quite well with the historical observations (Figure 22.31). This provided some assurance that the new model would

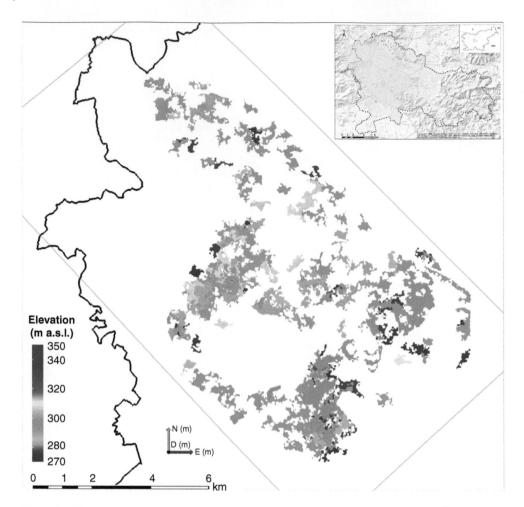

Figure 22.29 Locations with low hydraulic conductivity – potential perched aquifers (Šram 2011).

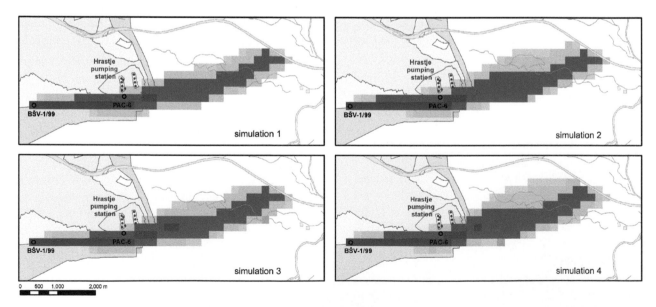

Figure 22.30 Simulated pollution spreading based on four different geostatistical realizations (area outlined in Figure 22.24).

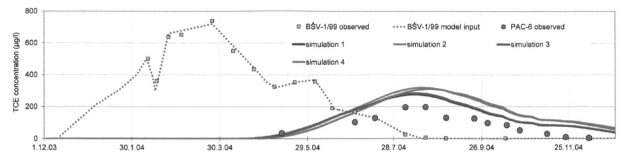

Figure 22.31 Observed and simulated concentrations of TCE in the groundwater in observation wells BŠV-1/99 and PAC-6 (shown in Figure 22.30). Simulations are based on the same pollution source input in BŠV-1/99 observation well, but different geostatistical realizations.

successfully guide emergency response and management measures for actual future contamination events.

22.5.4.2 Decision Support System

Ljubljana has developed a specific Decision Support System to suggest appropriate responses to future contamination events. The system was developed within the framework of the project INCOME by the Geological Survey of Slovenia in cooperation with several project partners and end-users (Janža 2015). A user-friendly graphical interface assists water managers in using the database, numerical modeling techniques, and expert knowledge. It provides rapid and easy access to supporting information for mitigating groundwater pollution with three logically interlinked components: the database, the hydrological model, and the decision model. The database contains data related to the groundwater resources used for drinking-water supplies to Ljubljana. It contains three types of data related to the monitoring, potential sources of pollution, and hydrogeological conditions. The hydrological model is based on the MIKE SHE/MIKE 11 modeling framework (Graham and Butts 2005; DHI 2011a,b) and enables the simulation of the propagation of pollutants in the aquifer. The decision model comprises a set of logical rules formalizing the knowledge and experiences of water managers and hydrogeologists

related to emergency activities. A wide range of possible scenarios may be analyzed to define response actions to potential groundwater pollution events.

22.5.5 Conclusions

The Decision Support System is a tool that is understandable by water managers and city officials. By reducing the quantity of input data and simplifying the use of the model, it presents outcomes: (i) anticipated travel-times for a pollution plume to reach an abstraction well, and (ii) pollutant concentrations expected in abstraction wells. While no pollution event has occurred since the construction of the DSS, the system remains active and ready for use whenever future contamination events occur.

The Decision Support System is specific to the city of Ljubljana; it utilizes local data and knowledge. However, the methodology that links logically related activities, such as detection of pollution in groundwater and simulation of pollution propagation, could be used in other locations once data sources are evaluated and 3-D models created. This upscaling of the knowledge and data exchange pathway is the next critical step for Slovenia, or for other countries where there are similar groundwater supply concerns.

References

Bannister, S. and Gledhill, K. (2012). Evolution of the 2010–2012 Canterbury earthquake sequence. *New Zealand Journal of Geology and Geophysics* 55: 295–304. https://doi.org/10.1080/00288306.2012.680475.

Barton, M.E. and Garvey, P. (2011). Reactivation of landsliding following partial cliff stabilization at Barton-on-Sea, Hampshire. *Quarterly Journal of Engineering Geology and Hydrogeology* 44: 233–248. https://doi.org/10.1144/1470-9236/09-030.

Begg, J.G., Jones, K.E. and Barrell, D.J.A. [Eds.] (2015a). *Geology and Geomorphology of Urban Christchurch and eastern Canterbury: Digital Vector Data 2015*. Lower Hutt,

New Zealand: GNS Science, Geological Map 3. 101 pp. 2 map sheets. 2 models. [DVD-ROM]

Begg, J.G., Jones, K.E., Rattenbury, M.S. et al. (2015b). A 3D geological model for Christchurch City (New Zealand): a contribution to the post-earthquake re-build. In: *Engineering Geology for Society and Territory – Volume 5* (eds. G. Lollino, G. Manconi, F. Guzzetti et al.), 881–884. Springer International Publishing. https://doi.org/10.1007/978-3-319-09048-1_171.

Bourgine, B. (2007). *Modélisation géologique 3D à l'aide du programme Multilayer (Version 3)* (in French). Orléans, France: BRGM, Report RP-53111-FR. 163 pp. [Online:

Available at: https://infoterre.brgm.fr/rapports/RP-53111-FR.pdf] (Accessed December 2020).

Bourgine, B. (2015). *MultiLayer, Manuel de référence (Version 2014)* (in French). Orléans, France: BRGM Report RP64115-FR.

Bourgine, B., Lembezat, C., Thierry, P. et al. (2008). Tools and methods for constructing 3D geological models in the urban environment: the Paris case. In: *Proceedings, GEOSTATS 2008 – VIII International Geostatistics Congress, Santiago, Chile* (eds. J.M. Ortiz and X. Emery), 951–960. [Online: Available at: https://hal-brgm.archives-ouvertes.fr/hal-00614827] (Accessed December 2020).

BRGM (2016). *GDM Suite Flyer*. Orléans, France: BRGM. 2 pp. [online: Available at: http://www.brgm.eu/sites/default/files/gdm-flyer-en.pdf] (Accessed December 2020).

Bristow, C.R., Freshney, E.C. and Penn, I.E. (1991). *Geology of the country around Bournemouth: Memoir for 1 : 50 000 geological sheet 329 (England and Wales)*. London, UK: HM Stationary Office. 116 pp. [Online: Available at: https://webapps.bgs.ac.uk/data/publications/pubs.cfc?method=viewRecord&publnId=19864774] (Accessed December 2020).

Bromhead, E.N., Chandler, M.P. and Hutchinson, J.N. (1991). The recent history and geotechnics of landslides at Gore Cliff, Isle of Wight. In: *Slope Stability Engineering: Developments and Applications* (ed. M.P. Chandler), 189–196. London, UK: Institution of Civil Engineers. https://doi.org/10.1680/ssedaa.16606.0030.

Brown, L.J. and Weeber, J.H. (1992). *Geology of the Christchurch Urban Area. Scale 1:25,000*. Lower Hutt, New Zealand: GNS Science, Geological Map 1. 104 pp. 1 map sheet. [CD-ROM].

Brown, L.J., Beetham, R.D., Paterson, B.R. and Weeber, J.H. (1995). Geology of Christchurch, New Zealand. *Environmental & Engineering Geoscience* 1 (4): 427–488. https://doi.org/10.2113/gseegeosci.I.4.427.

Burton, E. St. J. (1929). The horizons of Bryozoa (Polyzoa) in the upper Eocene beds of Hampshire. *Quarterly Journal of the Geological Society of London* 85: 223–241. https://doi.org/10.1144/GSL.JGS.1929.085.01-04.08.

Burton, E. St. J. (1933). Faunal horizons of the Barton beds of Hampshire. *Proceedings of the Geologist's Association* 44: 131–167. https://doi.org/10.1016/S0016-7878(33)80015-0.

Buser, S. (2009). *Geological Map of Slovenia 1:250,000*. Ljubljana, Slovenia: Geological Survey of Slovenia. 1 map sheet.

Carle, S.F. (1999). *T-PROGS: Transition Probability Geostatistical Software, Version 2.1*. Davis, CA: University of California. 84 pp. [Online: Available at: http://gmsdocs.aquaveo.com/t-progs.pdf] (Accessed December 2020).

De Beer, J., Price, S. and Ford, J. (2012). 3D modelling of geological and anthropogenic deposits at the World Heritage Site of Bryggen in Bergen, Norway. *Quaternary International* 251: 107–116. https://doi.org/10.1016/j.quaint.2011.06.015.

DHI (2011a). *MIKE 1D, DHI Simulation Engine for MOUSE and MIKE 11: Reference Manual*. Horsholm, Denmark: Danish Hydraulic Institute (DHI).

DHI (2011b). *MIKE SHE User Manual Volume 2: Reference Guide*. Horsholm, Denmark: Danish Hydraulic Institute (DHI).

East Anglia Coastal Group (2010). *Kelling Hard to Lowestoft Ness, Shoreline Management Plan 6*. Altrincham, UK: AECOM Ltd. 162 pp. [Online: Available at: http://www.eacg.org.uk/docs/smp6/smp/kelling%20to%20lowestoft%20ness%20smp%20-%20final.pdf] (Accessed January 2018).

Ford, J., Kessler, H., Cooper, A.H. et al. (2010). *An Enhanced Classification for Artificial Ground*. Keyworth, UK: British Geological Survey, Open Report OR/10/036. 32 pp. [Online: Available at: http://nora.nerc.ac.uk/id/eprint/509281] (Accessed December 2020).

Ford, J.R., Price, S.J., Cooper, A.H. and Waters, C.N. (2014). An assessment of lithostratigraphy for anthropogenic deposits. *Geological Society Special Publications* 395: 55–89. https://doi.org/10.1144/SP395.12.

Garvey, P.M. (2007). *A Study of the Reactivation of Landsliding at Barton-on-Sea, Hampshire, Following Stabilisation Works in the 1960s*. Unpublished MSc dissertation, University of Southampton.

Geonet (2011). GeoNet Quake Search – Geological Hazard Information for New Zealand. Lower Hutt: New Zealand: GNS Science [Website: http://quakesearch.geonet.org.nz] (Accessed Dcember 2020).

Graham, D.N. and Butts, M.B. (2005). Flexible, integrated watershed, modelling with MIKE SHE. In: *Watershed Models* (eds. V.P. Singh and D.K. Frevert), 245–272. Boca Raton: CRC Press.

Halcrow (2011). *Barton-on-Sea Cliff Instability Preliminary Study, Stage 1 Desk Study Review*. London, UK: Halcrow Group Ltd. 78 pp.

Hart, J.K. (1990). Proglacial glaciotectonic deformation and the origin of the Cromer ridge push moraine complex, North Norfolk, England. *Boreas* 19: 165–180. https://doi.org/10.1111/j.1502-3885.1990.tb00577.x.

Hobbs, P.R.N., Pennington, C.V.L., Pearson, S.G. et al. (2008). *Slope Dynamics Project Report: Norfolk Coast (2000–2006)*. Keyworth, UK: British Geological Survey, Research Report OR/08/018. 166p. [Online: Available at: http://nora.nerc.ac.uk/id/eprint/7236] (Accessed December 2020).

Hooker, J.J. (1986). Mammals of the Bartonian (middle/late Eocene) of the Hampshire Basin, southern England. *Bulletin of the British Museum (Natural History), Geology Series* 39: 191–478.

Hosseyni, S., Torii, N. and Bromhead, E.N. (2012). Discussion on 'Reactivation of landsliding following partial cliff stabilization at Barton-on-Sea, Hampshire' by M. E. Barton & P. Garvey Quarterly Journal of Engineering Geology and Hydrogeology, 44, 233–248. *Quarterly Journal of*

Engineering Geology and Hydrogeology 45: 125–128. https://doi.org/10.1144/1470-9236/11-039.

HR Wallingford (2003a). *Overstrand to Walcott Strategy Study: Cliff Processes (Part II: Technical Support Information).* Wallingford, UK: HR Wallingford, Report EX 4692. 50 pp. [Online: Available at: https://www.north-norfolk.gov.uk/media/3098/coastal_environment_013_4.pdf] (Accessed January 2018).

HR Wallingford (2003b). *Overstrand to Walcott Strategy Study: Cliff SCAPE modelling and clifftop recession analysis (Part II: Technical; Support Information).* Wallingfor, UK: HR Wallingford, Report EX 4692. 44 pp. [Online: Available at: https://www.north-norfolk.gov.uk/media/3100/coastal_environment_013_6.pdf] (Accessed January 2018).

Hutchinson, J.N. (1976). Coastal landslides in cliffs of Pleistocene deposits between Cromer and Overstrand, Norfolk, England. In: *Laurits Bjerrum Memorial Volume, Contributions to Soil Mechanics* (eds. N. Janbu, F. Asbjørn Jørstad and B. Kjærnsli), 155–182. Oslo, Norway: Norwegian Geotechnical Institute.

Jamnik, B., Zeleznik, B.B. and Urbanc, J. (2003). Diffuse pollution of water protection zones in Ljubljana, Slovenia. In: *Proceedings of the 7th IWA International Specialised Conference on Diffuse Pollution and Basin Management, Dublin, Ireland, 17–21 August 2003* (ed. M. Bruen), 7.3–7.5. [Online: Available at http://www.ucd.ie/dipcon/docs/theme07/theme07_01.PDF] (Accessed October 2017).

Janža, M. (2009). Modelling heterogeneity of Ljubljana Polje aquifer using Markov chain and geostatistics (in Slovenian with English abstract). *Geologija* 52 (2): 233–240. https://doi.org/10.5474/geologija.2009.023.

Janža, M. (2015). A decision support system for emergency response to groundwater resource pollution in an urban area (Ljubljana, Slovenia). *Environmental Earth Sciences* 73 (7): 3763–3774. https://doi.org/10.1007/s12665-014-3662-2.

Janža, M., Prestor, J., Urbanc, J. and Jamnik, B. (2005). TCE contamination plume spreading in highly productive aquifer of Ljubljansko Polje. *Geophysical Research Abstracts* 7: 09178. [Online: Available at: https://meetings.copernicus.org/abstracts/EGU05/09178/EGU05-J-09178.pdf] (Accessed December 2020).

Janža, M., Meglič, P. and Šram D. (2011). *Numerical Hydrological Modelling (Project Income, A.3.2, Final Report).* Ljubljana, Slovenia: Geological Survey of Slovenia. 66p. [Online: Available at: http://www.life-income.si/upload/fck/Image/Annex_10A_A32_Numerical_hydrogeological_modelling.pdf] (Accessed October 2017)

Janža, M., Stanič, I., Rupnik, P.J. and Bavec, M. (2016). *Ljubljana City Case Study.* COST TU1206 Sub-Urban Report TU1206-WG1-008. 14p. [Online: Available: at http://sub-urban.squarespace.com/new-index/#publications] (Accessed October 2017).

Janža, M., Lapanje, A., Šram, D. et al. (2017). Geological and geothermal conditions for assessment of shallow geothermal potential in Ljubljana area. *Geologija* 60 (2): 309–327. https://doi.org/10.5474/geologija.2017.022.

Jeanniot, E., Carreau, M., Le Guern, C. et al. (2014). *La gestion des terres excavées sur les zones d'aménagement de l'Île de Nantes.* Presented at Journées Techniques Nationales 'Reconversion des friches urbaines polluées', Paris, France, 25–26 March 2014.

Jenks, M. and Jones, C. [Eds.] (2010). *Dimensions of the Sustainable City.* Dordrecht, Netherlands: Springer. 282 pp. https://doi.org/10.1007/978-1-4020-8647-2.

Kessler, H., Mathers, S. and Sobisch, H.-G. (2009). The capture and dissemination of integrated 3D geospatial knowledge at the British Geological Survey using GSI3D software and methodology. *Computers & Geosciences* 35: 1311–1321. https://doi.org/10.1016/j.cageo.2008.04.005.

Kristensen, M., Andersson, U., Sorensen, H.R. and Refsgaard, A. (2000). *Water Resources Management Model for Ljubljansko Polje and Ljubljansko Barje – Model Report.* Horsholm, Denmark: DHI Water & Environment.

Laurent, A. and Fondrillon, M. (2010). Mesurer la ville par l'évaluation et la caractérisation du sol urbain: l'exemple de Tours (in French). *Revue archéologique du Centre de la France* 49: 307–343. [Online: Available at: https://journals.openedition.org/racf/1485] (Accessed December 2020).

Le Guern, C., Sauvaget, B., Campbell, S.D.G. and Pfleiderer, S. (2016). Sub-urban geochemistry – *A Review of Good Practice and Techniques in Sub-urban Geochemistry to Ensure Optimal Information Use in Urban Planning.* COST TU1206 Working Group 2 report; TU1206-WG2-008. 70 pp. [Online: Available at: http://sub-urban.squarespace.com] (Accessed December 2020).

Le Guern, C., Baudouin, V., Sauvaget, B. et al. (2016a). A typology of anthropogenic deposits as a tool for modelling urban subsoil geochemistry: example of the Île de Nantes (France). *Journal of Soil and Sediments* 18 (2): 373–379. https://doi.org/10.1007/s11368-016-1594-z.

Le Guern, C., Baudouin, V., Bridier, E. et al. (2016b). *Développement d'une méthodologie de gestion des terres excavées issues de l'aménagement de l'Ile de Nantes - Phase 1: Caractérisation des sols et recensement des sources de pollution potentielles* (in French). Orléans, France: BRGM Report RP-66013-FR. 128 pp. [Online: Available at: https://infoterre.brgm.fr/rapports/RP-66013-FR.pdf] (Accessed December 2020).

Lee, J.R., Pennington, C.V.L. and Hobbs, P.R.N. (2011a). Trimingham: structural architecture of the Cromer ridge push moraine complex and controls for landslide geohazards. In: *Glacitectonics Field Guide* (eds. E. Phillips, J.R. Lee and H.M. Evans), 218–227. London, UK: Quaternary Research Association. [Online: Available at: http://nora.nerc.ac.uk/id/eprint/16198] (Accessed December 2020).

Lee, J.R., Phillips, E., Evans, H.M. and Vaughan-Hirsch, D. (2011b). An introduction to the glacial geology and history of glacitectonic research in Northeast Norfolk. In: *Glacitectonics Field Guide* (eds. E. Phillips, J.R. Lee and H.M. Evans), 101–115. London, UK: Quaternary Research Association. [Online: Available at: http://nora.nerc.ac.uk/id/eprint/16086] (Accessed December 2020).

Massey, C.I., McSaveney, M.J., Taig, T. et al. (2014). Determining rockfall risk in Christchurch using rockfalls triggered by the 2010–2011 Canterbury earthquake sequence, New Zealand. *Earthquake Spectra* 30: 155–181. https://doi.org/10.1193/021413EQS026M.

Melville, R.V. and Freshney, E.C. (1982). *British Regional Geology: The Hampshire Basin and Adjoining Areas*. London: British Geological Survey HM Stationary Office. 146 pp.

Murray, A.B., Gopalakrishnan, S., McNamara, D.E. and Smith, M.D. (2013). Progress in coupling models of human and coastal landscape change. *Computers & Geosciences* 53: 30–38. https://doi.org/10.1016/j.cageo.2011.10.010.

Nicholls, R.J. (2015a). Coastal evolution and human-induced Sea-level rise: history and prognosis. In: *Proceedings of Coastal Sediments 2015* (eds. P. Wang, J.D. Rosati and J. Cheng), 1–12. Singapore : World Scientific Publishing Company. 12 pp. https://doi.org/10.1142/9789814689977_0001.

Nicholls, R.J., Bradbury, A., Burningham, H. et al. (2012). iCOASST – integrating coastal sediment systems. In: *Proceedings of the 33rd International Conference on Coastal Engineering* (eds. P. Lynett and J.M.K. Smith), 100–115. https://doi.org/10.9753/icce.v33.sediment.100.

Nicholls, R.J., French, J., Burninham, H. et al. (2015b). Improving decadal coastal geomorphic predictions: an overview of the iCOASST project. In: *Proceedings of Coastal Sediments 2015* (eds. P. Wang, J.D. Rosati and J. Cheng). Singapore: World Scientific Publishing Company. 16 pp. https://doi.org/10.1142/9789814689977_0227.

OG RS (2015). *Decree on the water protection area for the Ljubljansko Polje aquifer*. Ljubljana, Slovenia: Official Gazette, no. 43/15. 24 pp. 6 annexes.

Payo, A. (2017). Section 5.6.3 Modelling geomorphic systems: coastal. In: *Geomorphological Techniques* (eds. S.J. Cook, L.E. Clarke and J.M. Nield). London, UK: British Society for Geomorphology. 11 pp. [Online: Available at: https://www.geomorphology.org.uk/sites/default/files/chapters/5.6.3_ModellingGeomorphicSystemsCoastal_0.pdf] (Accessed December 2020).

Payo, A., Favis-Mortlock, D., Dickson, M. et al. (2017). CoastalME version 1.0: a coastal modelling environment for simulating decadal to centennial morphological changes on complex coasts. *Geoscientific Model Development* 10 (7): 2715–2740. https://doi.org/10.5194/gmd-10-2715-2017.

Potter, S.H., Becker, J.S., Johnston, D.M. and Rossiter, K.P. (2015). An overview of the impacts of the 2010–2011 Canterbury earthquakes. *International Journal of Disaster Risk Reduction* 14: 6–14. https://doi.org/10.1016/j.ijdrr.2015.01.014.

Price, D.G. and de Freitas, M.H. (2009). *Engineering Geology: Principles and Practice*. Berlin, Germany: Springer. 449 pp. https://doi.org/10.1007/978-3-540-68626-2.

Robertson, P.K. and Cabal, K.L. (2010). *Guide to Cone Penetration Testing for Geotechnical Engineering, 4e*. Signal Hill, CA: Gregg Drilling & Testing, Inc. 124 pp.

Royse, K.R., Rutter, H.K. and Entwisle, D.C. (2009). Property attribution of 3D geological models in the Thames Gateway, London: new ways of visualising geoscientific data. *Bulletin Engineering Geology & Environment* 68: 1–16. https://doi.org/10.1007/s10064-008-0171-0.

Samoa (2016). *Ile de Nantes – Fabriquer la ville autrement* (in French). Nantes, France: Amoa. [Website: http://www.iledenantes.com] (Accessed December 2020).

Šram, D. (2011). *Hydrogeology of Ljubljansko polje Perched Aquifers* (in Slovenian). BSc thesis University of Ljubljana. 83 pp.

Šram, D., Brenčič, M., Lapanje, A. and Janža, M. (2012). Perched aquifers spatial model: a case study for Ljubljansko polje (central Slovenia) (in Slovenian with English abstract). *Geologija* 55 (1): 107–116. https://doi.org/10.5474/geologija.2012.008.

Townsend, D.B., Lee, J.M., Strong, D.T. et al. (2016). Mapping surface liquefaction caused by the September 2010 and February 2011 Canterbury Earthquakes: a digital dataset. *New Zealand Journal of Geology and Geophysics* 59: 496–513. https://doi.org/10.1080/00288306.2016.1182929.

Van Maanen, B., Nicholls, R.J., French, J.R. et al. (2016). Simulating mesoscale coastal evolution for decadal coastal management: a new framework integrating multiple, complementary modelling approaches. *Geomorphology* 256: 68–80. https://doi.org/10.1016/j.geomorph.2015.10.026.

West, I. (2010). *Barton and Highcliffe, Eocene Strata: Geology of the Wessex Coast of Southern England*. Southampton, UK: Southampton University. [Website: https://wessexcoastgeology.soton.ac.uk/barton.htm] (Accessed December 2020)

Wood, B., Richmond, T., Richardson, J. and Howcroft, J. (2017). *BGS Groundhog® Desktop Geoscientific Information System v1.8.0 External User Manual*. Keyworth, UK: British Geological Survey, Internal Report OR/15/046. 178 pp. [Online: Available at: http://nora.nerc.ac.uk/id/eprint/511792] (Accessed December 2020.

Žlebnik, L. (1971). Pleistocene deposits of the Kranj, Sora and Ljubljana fields (in Slovenian with English abstract). *Geologija* 14: 5–51.

23

Application Theme 6 – Urban Infrastructure

Editor's Introduction

In recent decades, many geotechnical and civil engineering consultants and contractors have become more aware of the potential value of using 3-D geological models within the workflow of design and construction for infrastructure projects. The recent UK Government mandate requiring the use of Building Information Modeling (BIM) for all public-sector construction/design projects (HM Government 2015) has led to more systematic use of 3-D geological modeling tools, an evaluation of their capabilities, and assessment of their integration with BIM technologies (Grice and Kessler 2015). The wider application of geological models to geotechnical investigations has become a common and accepted component of geotechnical good practice.

While BIM methodologies largely depend on CAD and GIS software standards, a 3-D geological model, converted to be compatible with BIM protocols, forms a visual reference for all project data, adds strategic value to bids, and results in competitive advantages. Use of 3-D geological models also leads to efficiencies over a project lifespan. When a 3-D geological model serves as a hub for subsurface interpretation, ground investigation efficiencies result from early identification of data deficiencies and a reduced number of optimally placed new boreholes.

The first case study summarizes the experience of the London office of Dr. Sauer & Partners, an international tunneling consultancy, with the use of a 3-D model during the design and construction of the Crossrail Farringdon Station in central London. This project provided a unique, real-world opportunity to test and develop systems and databases capable of providing geological interpretations that contain valuable information for managing the design and construction process in formats that are sharable by all project partners. A commissioned geological model developed by the British Geological Survey (BGS) provided valuable initial guidance; subsequently, additional direct observations of actual ground conditions were progressively added to produce updated 3-D models. Throughout the construction phase, reference to the 3-D geological model materially reduced geotechnical risk, allowed for efficient construction, and resulted in a 4-month reduction in the planned construction time.

The second case study summarizes the experiences of TSP Projects (formally Tata Steel Projects) with the use of 3-D geological models to support the design and construction process for several recent rail infrastructure projects in the UK. Successful rail infrastructure construction activities, generally undertaken within tight schedules to minimize disruption to scheduled rail services, depend on the accurate prediction of ground conditions and geotechnical risks. The original earthworks along UK rail corridors were constructed by non-engineered methods, and many locations have been subjected to various modifications over time. The linear nature of most rail infrastructure projects, when combined with limited ground investigation opportunities, makes it is a challenge to adequately identify the significant differences in the "natural" ground conditions that are often found within a narrow rail corridor, especially at locations previously affected by Quaternary glacial and periglacial processes. Digital 3-D geological models, when developed by experienced geologists, have provided valuable information that reduces the geotechnical risk of rail infrastructure renewal.

The third case study summarizes the use of geological information within a BIM environment during the preliminary design stages of a proposed new road tunnel under the Thames in east London. Atkins, the design consultant, found increased efficiency, and thus economic benefit, from the use of a fully integrated, multidisciplinary 3-D model that combined BIM and geological data. Visualizing ground conditions in a design context resulted in a design that offered reduced project risk and project construction costs. It also provided for the rapid production of 3-D visualizations suitable for public discussions.

Applied Multidimensional Geological Modeling: Informing Sustainable Human Interactions with the Shallow Subsurface, First Edition.
Edited by Alan Keith Turner, Holger Kessler, and Michiel J. van der Meulen.
© 2021 John Wiley & Sons Ltd. Published 2021 by John Wiley & Sons Ltd.

Case Study 23.1: Design and Construction of a New Crossrail Station in London Assisted by a 3-D Ground Model

Angelos Gakis[1], Paula Cabrero[2] and David Entwisle[3]

[1] Dr. Sauer & Partners Ltd, London KT6 6QH, UK
[2] Formerly Dr. Sauer & Partners Ltd, London KT6 6QH, UK
[3] British Geological Survey, Nottingham NG12 5GG, UK

23.1.1 Introduction

Crossrail is a new underground railway system, currently under construction, which will provide a direct East-West connection through the center of London. The Crossrail route includes 21 km (13 miles) of twin-bore tunnels and eight new stations connecting with the national rail and London Underground networks. While the running tunnels were constructed using tunnel boring machines (TBMs), the stations were constructed using an open face excavation tunneling method with sprayed concrete linings (SCLs) (Spyridis et al. 2013).

The Crossrail route had to be placed around or beneath existing utilities and London Underground transit tunnels, and, in some places, within lithologically complex ground. Three-dimensional geological models were integrated within the Farringdon Station design and construction process to assess the geotechnical risks (Aldiss et al. 2009, 2012; Gakis et al. 2014; Spyridis et al. 2013).

23.1.2 Design Concerns at Farringdon Station

The new Farringdon Station will be one of London's major rail interchange stations. The station layout comprises two ticket halls, two 300 m platform tunnels, eight connecting cross passages, two concourse tunnels, two escalator inclines, plus escape and ventilation adits (Gakis et al. 2016). Existing underground infrastructure placed the planned Crossrail Farringdon Station tunnels about 30 m below the surface. The tunnels are located mostly within the Lambeth Group, which underlies the London Clay Formation, the preferred tunneling medium in London. The Lambeth Group is a challenging tunneling medium because it includes "hard grounds", water-bearing channel sands, and local gravel beds (Page and Skipper 2000). Several London tunneling projects have experienced difficulties and delays because of the variable character of the Lambeth Group (Hight et al. 2004; Newman 2009).

During the initial ground investigation at Farringdon, geological correlation of 30 boreholes revealed the expected complex lithology in the Lambeth Group but did not establish a sufficiently detailed, coherent ground model. Uncertain ground conditions included the presence of apparently randomly located water-bearing sand layers and inferred fault zones, then identified as zones of "disturbed ground" of unknown width and character. Surface deformation due to the excavation was also a concern as sensitive buildings and surface railway tracks above had to be protected. Lack of confidence in the ground model potentially required very conservative designs and construction methods, resulting in increased costs and possible delay of project completion (Gakis et al. 2016).

23.1.3 Role of a 3-D Geological Model in Station Design and Construction

In 2009, concerns about the adequacy of the ground information led Crossrail to commission the British Geological Survey (BGS) to produce a 3-D geological model of an area 850 m by 500 m containing the Farringdon station site (Aldiss et al. 2009, 2012). By modeling an area much larger than the footprint of the proposed station, it was possible to incorporate nearby geological observations and more accurately identify fault positions and character. This model was designed to guide the design of further site ground investigations and support geotechnical risk management (Figure 23.1).

The BGS built this initial 3-D model using an explicit, cross-section-based modeling methodology, described in Chapter 10. Kessler et al. (2009) describes the general workflow and methodology. Aldiss et al. (2009, 2012) provide details of the project-specific data and assumptions used to create this model. Previously, completed regional 3-D models of London (Mathers et al. 2014) aided model development. The resulting 3-D model (Figure 23.2) provided a coherent conceptual ground model of a faulted multi-layered subsurface; it defined the extent of 18 identifiable geological units and seven faults (Aldiss et al. 2009, 2012).

The modeling process and available data constrained the predicted location of each fault to an envelope 20 m wide at the level of the tunnels (Figure 23.3). These faults were not represented on published geological maps. However, the London regional 3-D geological model included the Barbican Fault; previous Crossrail desk studies identified the Smithfield Fault (Gakis et al. 2016). The model also identified sheet-like and short channel-like sand bodies in the Lambeth Group.

In April 2013, the BGS model was handed over to the contractor, a BAM-Ferrovial-Kier Joint Venture (BFK), and their specialized tunneling consultant, Dr. Sauer & Partners (DSP). It became an integral part of the site-supervision workflow (Figure 23.1). Initially, the model was updated with data from shaft excavations and additional boreholes undertaken between 2009 and 2013. Additional

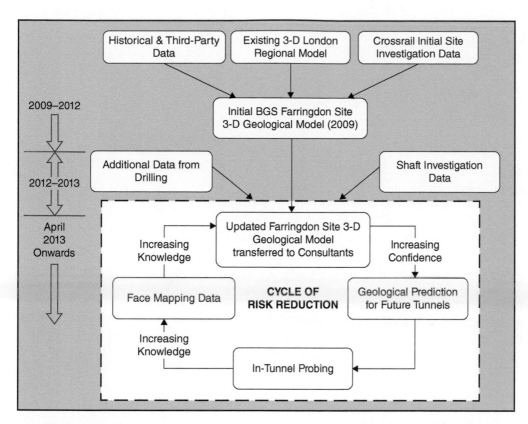

Figure 23.1 Cycle of risk reduction through the implementation of geotechnical risk management tools. Modified after Gakis et al. (2015).

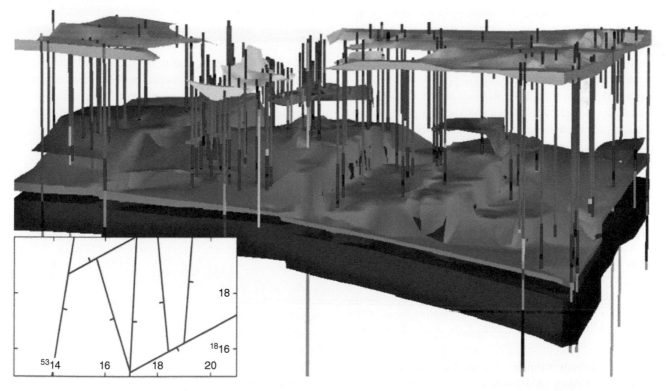

Figure 23.2 Sand and gravel (water bearing) units in the BGS Farringdon 3-D geological model, with borehole "sticks" (Aldiss et al. 2012). The inset map shows fault traces (Figure 23.3).

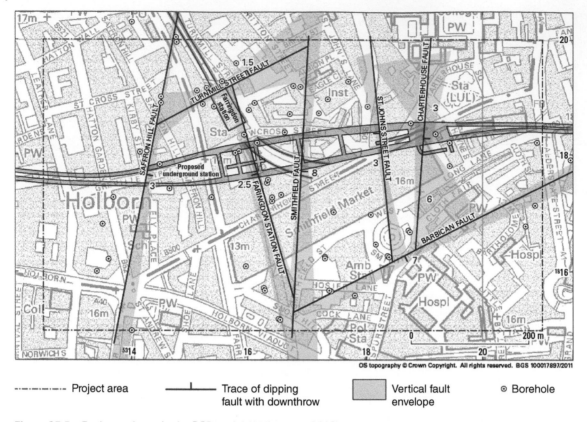

------------ Project area   ——⊥—— Trace of dipping   ▮ Vertical fault   ⊚ Borehole
                                        fault with downthrow   envelope

Figure 23.3 Fault envelopes in the BGS model (Aldiss et al. 2012).

information was provided by extending 33 boreholes, originally planned to house monitoring instrumentation, to at least 4 m below the tunnel inverts. Figure 23.4 shows the location of the additional borehole information that was included in this revised model.

From 2014 onwards, DSP used *SubsurfaceViewer MX* software developed by INSIGHT Geologische Softwaresysteme GmbH to manage the 3-D model. This provided essentially the same environment and capabilities as the original BGS modeling environment. The final modeling stages used BGS Groundhog Desktop Geoscientific Information System (Wood et al. 2015), a tool that facilitates the collation, display, filtering, and editing of a range of data relevant to subsurface interpretation and modeling.

Tunneling works to create the Farringdon Station started in May 2013. The 3-D model then became an integral part of the site supervision workflow, being updated daily with data from the tunnel face (Figure 23.1). Although the main principles and assumptions of the original BGS model were retained, additional modeling rules were adopted for inclusion of tunnel excavation data:

- In the platform tunnels, a complete face map was included for every 10 m interval.
- In the shorter cross passage tunnels, a minimum of three complete face maps were included.

- Where faults or sand lenses were encountered, complete face maps were included at every 1 m of excavation advance.
- Upon the completion of a tunnel, the continuous, geologically interpreted longitudinal section was included in the model, superseding the imported face maps.
- The inclusion of a section in the model required a subsequent update of all the intersecting sections.

Figure 23.5 shows the sections used to update the model based on the tunnel excavation data. This dynamic model became a "live" geological database with progressively increased accuracy; it provided ever more reliable geological predictions as tunnel excavations continued.

23.1.4 Applying the 3-D Model to Reduce Geotechnical Risk

Predicted fault locations in the initial 3-D model proved remarkably accurate when tunnel excavations revealed the real conditions. The model assumed faults had a dip of 70°, but a typical dip angle of 60° was encountered during the tunneling. Thus, most observed fault locations were displaced in the direction of dip from their interpreted positions. Predicted locations and extents of the

Figure 23.4 Boreholes used in the initial BGS model and additional boreholes used in the 2013 updated model (Gakis 2014).

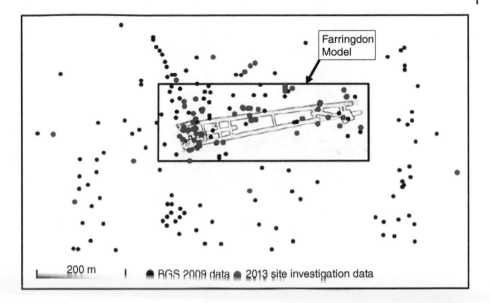

Figure 23.5 Fence diagram showing the additional sections used to update the 3-D model (Gakis et al. 2016).

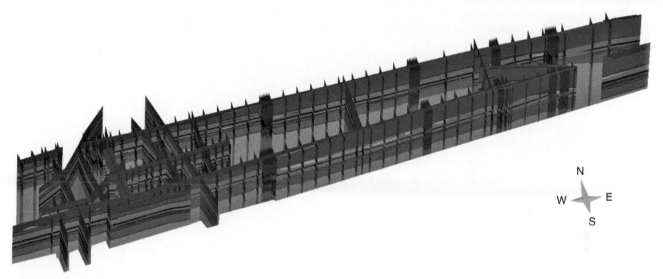

sand lenses were not always precise (Figure 23.6). Because water-bearing sand lenses imposed the highest risk on the sprayed concrete linings, they were modeled with the highest possible level of accuracy. As the station tunnels were excavated, any encountered sand lenses were carefully observed. Their geometry and continuity were updated in the model using two simple assumptions:

- The maximum thickness of a sand lens was taken to be equal to the greatest thickness observed in the excavated face.
- If the position of the edge of a sand lens was not observed, in the 3-D model the lens was extended up to the mid-distance between the last observation of the sand and the nearest reliable data point without sand.

Figure 23.6 shows sand lenses were accurately predicted in the northwest of the station, over-estimated at the east end, and partially under-estimated in the central and west parts of the station.

Many predicted geotechnical conditions applied in the design phase were very similar to conditions observed during tunnel excavation (Figure 23.7). Minor discrepancies did not adversely affect the construction (Gakis 2014; Gakis et al. 2014; Spyridis et al. 2013).

The risks to SCL tunneling posed by water-bearing sand units were compared for the design and construction phases (Figure 23.8). The construction phase used a 3-D finite-element model based on new ground investigation data and the updated 3-D geological model (Gakis et al. 2014). The risk-analysis used five risk grades ranging

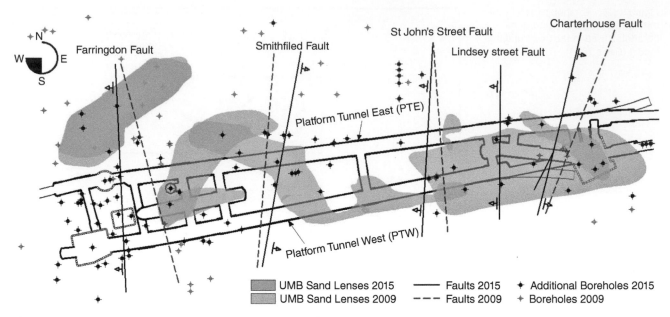

Figure 23.6 Predicted and observed locations of sand lenses and faults (Gakis et al. 2016).

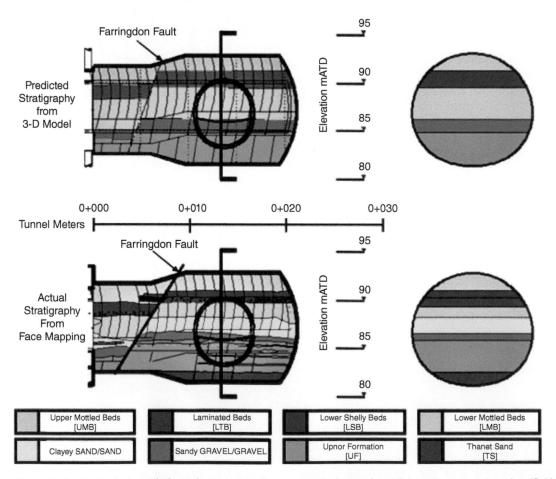

Figure 23.7 Geological prediction prior to excavation vs. actual observed conditions for a tunnel section (Gakis 2014).

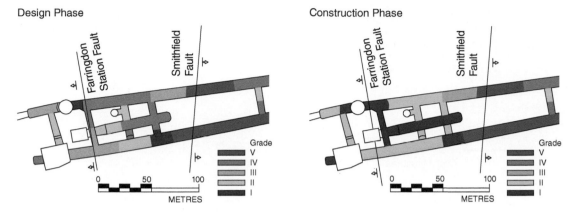

Figure 23.8 Comparison of estimated risks to SCL tunneling posed by water-charged sand units during the design and construction phases (Gakis et al. 2014).

from I (low risk) to V (high risk), and comparisons of risk estimates at the design and construction phases show the effective overall risk was reduced from Grade III during design to Grade II during construction (Gakis et al. 2014). Application of the 3-D model was responsible for much of this risk reduction.

Surface monitoring of ground subsidence due to tunnel construction could only be obtained along some Network Rail sidings, but these showed the 3-D finite-element model predictions were slightly larger than the observations. The difference may be due to the stiffness of the rails inducing a "bridging effect" and so slightly reducing the differential settlement trough (Gakis 2014).

23.1.5 Conclusions

In 2009, the BGS developed an initial 3-D geological model of Farringdon station. This model was subsequently updated by BFK/DSP between 2013 and 2015. Farringdon station was successfully designed and constructed using a 350 mm thick steel fiber reinforced SCL primary lining; this design did not require any additional reinforcement or thickening (Gakis et al. 2014). The innovative design process was materially helped by using the 3-D geological model to integrate the most recently acquired geological data. A sophisticated non-linear 3-D finite-element model accurately simulated the sequential excavation steps and the geometry of the tunnels; this assured the stability and adequacy of the SCL primary lining. Predicted in-tunnel deformations and surface settlements induced by station construction closely matched the actual monitoring results (Cabrero and Gakis 2014; Gakis 2014; Gakis et al. 2014;

Spyridis et al. 2013). Gakis et al. (2016) define several additional benefits from using a 3-D model for this project:

- A 3-D conceptual model of the geology was provided for one of the most geologically complicated station sites in Crossrail.
- Areas were identified where additional investigation was required.
- BFK/DSP were assisted in geotechnical risk mapping of the station site.
- The model became a key element of the geotechnical risk management framework as part of the site supervision workflow.
- Continuous updating with data from the tunnel excavations progressively increased the accuracy and reliability of predictions, permitted rapid examination and assessment of existing records, and supported generation of synthetic boreholes and cross-sections when required.
- The model provided geological predictions ahead of the excavations of each tunnel.
- A more efficient SCL design was achieved through enhanced geotechnical knowledge and certainty.
- A 70% reduction of in-tunnel probing, compared with the original plans, was possible.
- The model helped optimize the direction of additional in-tunnel probing and define the locations of depressurization wells.
- The use of the 3D geological model materially reduced the geotechnical risk, allowed for efficient construction, and resulted in a 4-month reduction in the planned construction time.

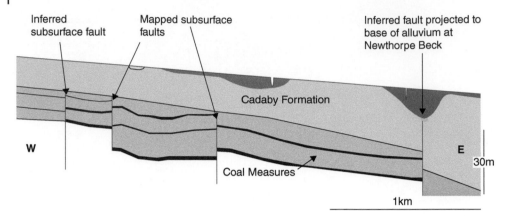

Figure 23.9 Cross-section along part of the Leeds–York CGM showing subsurface faults (Burke et al. 2015).

Case Study 23.2: Using 3-D Models to Evaluate Designs for Railway Infrastructure Renewal

Gerard McArdle

TSP Projects, York YO24 1AW, UK

23.2.1 Introduction

The North Trans-Pennine Electrification East project will electrify the 28 km of railway between Leeds and York in Northern England. TSP Projects (formally Tata Steel Projects) commissioned the British Geological Survey (BGS) to produce a detailed 3-D conceptual ground model (CGM) of the route[1]. The application of the CGM would improve the project design process by incorporating geological information into the project BIM system. Chapter 4 provides details of TSP Projects' interests in utilizing digital CGM's for UK rail infrastructure projects. During the North Trans-Pennine project, the digital CGM identified areas requiring targeted ground investigation during initial assessment of the design of electric mast foundations.

23.2.2 Model Construction

The BGS team constructed a digital CGM of the bedrock and superficial geology of the existing railway corridor (Burke et al. 2015) using GSI3D software and established workflow procedures described by Kessler et al. (2009). The CGM extends along the length of the railway but is only 80 m wide and was created to meet the specifications of TSP Projects, namely:

(1) The model extends to 30 m below track level;

---

1 Editor's note: while a "conceptual geological model" represents the first stage in geological modeling (Chapters 2, 5, 7, 9, 10, 12–15), a "conceptual ground model" is a product in its own right in engineering geology (Chapter 4; Parry et al. 2014).

(2) The modeled ground surface includes a 0.5 m deep by 3.8 m wide channel along each track alignment;

(3) The model shows all relevant geological strata and structures (faults) plus interpretations of the depth of weathering, locations of old mine workings, and karst features because these will influence mast foundation design; and

(4) Modeled outputs comply with project coordinates and project BIM and Computer Aided Design (CAD) standards.

The digital CGM was constructed using 1 : 10 000 scale digital geological map data and 102 borehole logs from the BGS national archive. All boreholes located within the modeled area were considered in the construction of the CGM, together with key boreholes that fall outside the area of study. The CGM contains 57 geological units, including 11 coal seams, and 29 mapped geological faults; most occur within the Coal Measures in the west of the model. The faults were defined as vertical planes. The bases of the affected units are stepped down across these faults to show the displacement. Mapped faults were supplemented by inferred faults where the modeling process required strata offsets (Figure 23.9).

The modeled linear route was divided into three sections to enable three geologists to construct the model simultaneously. Each modeler correlated a series of short "rung" sections, which cross the track. The spacing of the rungs is determined by the distribution of borehole data and the complexity of the geology, and thus are closer together where the geology is more complex. Every available borehole within the modeled area has been used. The rung sections were constructed first and were used to control the parallel sections. Three parallel sections were constructed, one along the center of the alignment and one spaced 36 m to either side of the center line.

Figure 23.10 illustrates a 4 km section of the central portion of the route at the eastern extent of the influence of the

Figure 23.10 A 4 km section of the central portion of the Leeds–York route. (a) is map view; (b) is isometric view of the 3-D model. Source: British Geological Survey and Ordnance Survey Data, © Crown Copyright 2015.

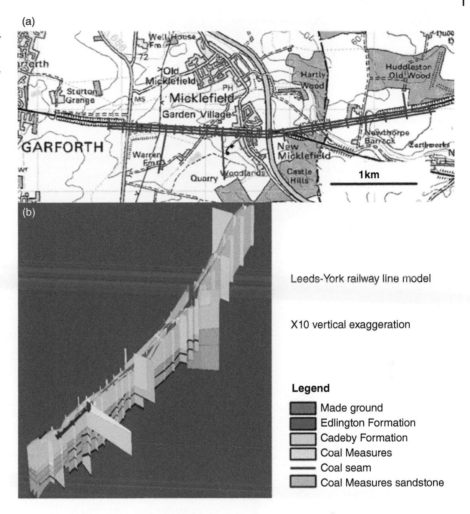

Leeds-York railway line model

X10 vertical exaggeration

Legend
- Made ground
- Edlington Formation
- Cadeby Formation
- Coal Measures
- Coal seam
- Coal Measures sandstone

Coal Measures. This section has 25 short rung sections. A few longer rung sections were extended to link to nearby borehole records.

23.2.3 Economic Benefits from Model Application

The BGS delivered the digital CGM to TSP Projects as CAD files. Thus, TSP Projects could rapidly integrate the digital CGM within their in-house BIM workflow to produce design documents (Figure 23.11).

The digital CGM provided several benefits to TSP Projects during the initial project phases. The identified benefits included:

(1) Reduced requirement to visit the site, which was 28 km long.
(2) Reduced the health and safety risks through not needing to walk alongside active railways or carry out ground investigations at night during limited periods of possession.

(3) Improved targeting of ground investigation works resulting in reduced costs.
(4) Improved initial estimates by the contractors for foundation costs; these resulted from confidence in the CGM predictions of subsurface conditions.

Subsequently, the digital CGM was used to assist the design of earthworks and foundations for bridge modifications to accommodate the installation of overhead electrification. Figure 23.12 shows a typical example of a PDF document for a bridge reconstruction site. Additional benefits resulted from the availability of high-resolution digital terrain models (DTM) derived from Lidar sources and provided by the BGS. During a geomorphological site walkover, shallow circular slip back scars were observed on the slope of an embankment. The Lidar DTM was examined within the BIM environment to confirm these slope features and to provide a better overview of their extent (Figures 23.13 and 23.14).

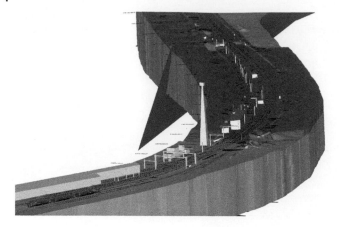

Figure 23.11 The BGS CGM integrated within the BIM workflow. At a rail maintenance facility near Leeds, the CGM defines geological conditions (sandstones and mudstones in gray and yellow and a fault in red) while the BIM shows existing and proposed overhead line electrification and building infrastructure. Source: TSP Projects.

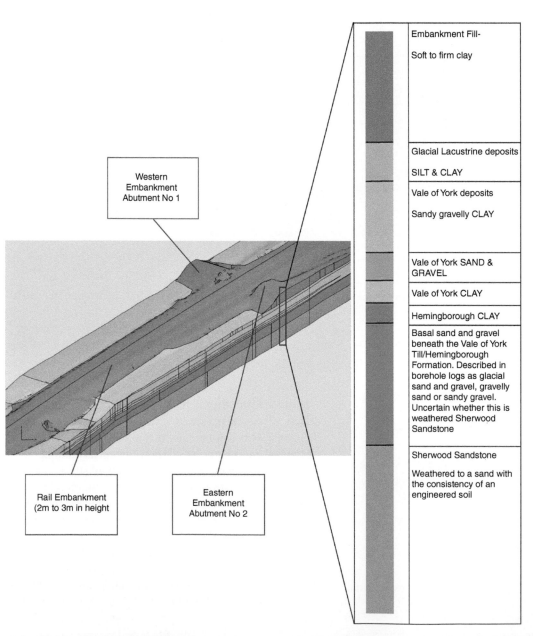

Figure 23.12 PDF-based documentation of a bridge reconstruction site based on combined terrain and digital CGM information provided by the BGS. Source: TSP Projects.

Figure 23.13 Shallow circular slip back scars on the slope of an embankment. Source: TSP Projects.

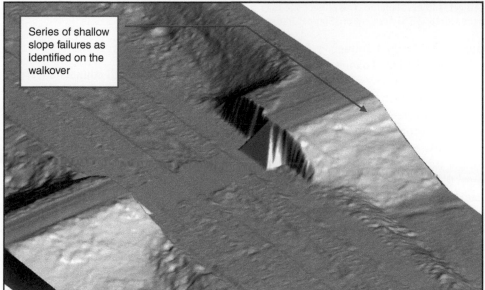

Figure 23.14 The shallow slope failures seen during a geomorphological walkover are readily visible on the Lidar digital terrain model. Source: TSP Projects.

Case Study 23.3: Use of Integrated BIM and Geological Models for the Reference Design of the Silvertown Tunnel, East London

Jerome Chamfray[1], Simon R. Miles[2] and Gary Morin[3]

[1] Jacobs, London SE1 2UP, UK
[2] Atkins, Epsom KT18 5B, UK
[3] Keynetix Ltd.,, Redditch B98 9PA, UK

23.3.1 Introduction

After many years of consideration, Transport for London, the statutory authority responsible for the transport system of Greater London, has initiated a detailed planning and design process for a new road tunnel crossing the River Thames in east London (Figure 23.15). A bored twin-tube tunnel connecting Silvertown on the north bank with the Greenwich Peninsula on the south bank (Figure 23.15) was selected as the best solution to alleviate severe traffic congestion, support economic development, and improve the resilience and reliability of the regional road transport network (Transport for London 2013, 2015a). WS Atkins plc (commonly known as Atkins) was commissioned to carry out the reference design for the Silvertown Tunnel to support the required statutory consultations.

Only a limited number of Thames river crossings for highway traffic exist in east London. There are 18 river crossings in the 29 km along the river between Vauxhall Bridge in west-central London and the western crossing by the M25 (London Orbital Motorway) near Staines, but only five river crossings in the 23 km down-river from Tower

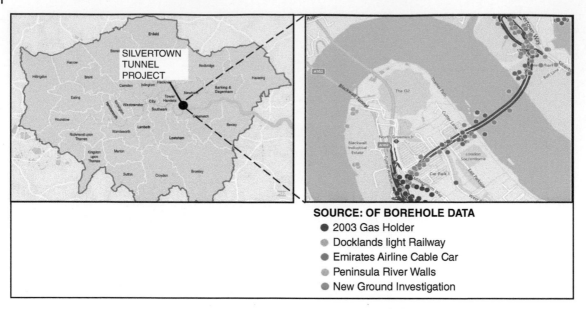

SOURCE: OF BOREHOLE DATA
● 2003 Gas Holder
● Docklands light Railway
● Emirates Airline Cable Car
● Peninsula River Walls
● New Ground Investigation

Figure 23.15 Silvertown Tunnel Project is in east London. The project site map shows the tunnel alignment and the sources and locations of the boreholes used by the HoleBASE SI Application. Source: Transport for London (2015b) and ATKINS.

Bridge to the eastern M25 crossing via the Dartford Tunnel (Transport for London 2015a). In west London, Thames road crossings are spaced on average 2 km apart, while in central London the average distance is about 1 km. In contrast, in east London the average spacing is 8 km.

This difference is largely due to the historical patterns of regional economic development. East of the Tower of London, the depth and width of the River Thames made navigation by ocean-going ships feasible; thus, industrial and port activities dominated. Limited demand for cross-river movements, combined with opposition to any low-level bridges that would impede the movement of ships, resulted in the construction of only a small number of tunnels below the river; these were supplemented by a few ferries. In contrast, in central and west London, the Thames is much narrower, there are no large ships, and residential and commercial development led to demands for convenient cross-river communications. It was relatively easy, and economically justifiable, to construct low-level bridges suitable for road vehicles, cyclists and pedestrians (Transport for London 2015a).

Over the past few decades, there has been an extensive redevelopment of east London, with former docks being transformed into a commercial hub, a leisure destination, and home to a rapidly growing population. Today, there are comparable population numbers and densities in west and east London. This redevelopment has been supported by multiple new rail crossings of the Thames. By 2020, there will be similar numbers of rail-based river crossings in east and west London. However, the provision of additional road-based river crossings in east London remains a priority.

23.3.2 Tunnel Design Challenges

The proposed Silvertown Tunnel will connect Silvertown on the north bank to the Greenwich Peninsula on the south bank (Figure 23.15). A tunnel boring machine (TBM) will excavate two 11.45 m (37.5 ft) internal diameter tunnels, each of approximately 1 km long, under the river. Cut-and-cover approaches at Greenwich and Silvertown, each of approximately 0.2 km long, will include a TBM launch chamber (Figure 23.16). Design of the Silvertown Tunnel faced several challenges related to (i) historical activities at the project location, (ii) nearby critical existing infrastructure, (iii) tunnel geometry requirements, and (iv) local geological conditions.

Both tunnel portals are in formerly heavily industrialized areas, resulting in myriad existing soil types, roads, foundations, and subsurface remnants of redundant and demolished structures. The north portal is located near the abandoned western entrance to the Royal Victoria Dock, subsequently the location of now-demolished warehouses, so that underground remnants of these and buried utilities are expected in the tunnel entrance area. Until 1987, a gas works was located near the south portal. Although the site underwent extensive remediation, contaminated soils are anticipated at this location.

Critical active infrastructure elements exist near both tunnel portals and along the tunnel alignment. Three Docklands Light Railway (DLR) stations and a DLR viaduct surround the north portal site, while the Royal Victoria Dock is located only about 100 m from the proposed north tunnel portal entrance. The south tunnel portal must connect with several critical roads, including the existing

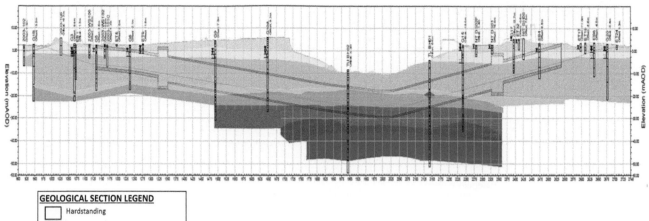

Figure 23.16 Silvertown Tunnel vertical profile with expected geological strata. Source: Atkins, by kind permission of Transport for London.

south approach to the Blackwall Tunnel, and is only a short distance south of the O2 Arena. The tunnel route closely parallels the Emirates Air Line cable car crossing.

The desired tunnel vertical profile placed the lowest point of the tunnel only slightly below the lowest point of the river channel. This produced a longitudinal gradient of 4.2% along the north tunnel section, resulting in reduced vehicle emissions and ventilation operating costs. However, it also meant that precise knowledge of the 3-D geology would help reduce geotechnical risks during construction (Figure 23.16). Faced with multiple design challenges, Atkins decided to utilize an integrated 3-D modeling approach for the preliminary design of the Silvertown Tunnel.

23.3.3 Geological Conditions at the Silvertown Crossing

Previous desk studies (Transport for London 2013) defined the local geological environment and stratigraphy (Table 23.1). Near-surface shallow deposits of "made ground" related to historic industrial uses of the site and alluvial deposits from the former marshlands bordering the River Thames underlie the Silvertown project site. These in turn overlie River Terrace Deposits composed of

silty or sandy gravels. At greater depths, a series of Eocene and Paleocene deposits include a layer of stiff London Clay above dense gravels of the Harwich Formation and dense sands of the Lambeth Group. An erosional surface separates these younger units from the much older Upper Cretaceous Chalk.

23.3.3.1 Geological Influences on Tunnel Construction

Tunnel construction along the proposed alignment must deal with very shallow overburden and geological materials with highly variable properties. Existing geological data show that the bored tunnel sections will encounter stiff London Clay overlying dense gravels of the Harwich Formation and dense sands of the Lambeth Group. The open cut, and cut and cover, sections on both sides of the river will encounter Made Ground, Alluvium, and River Terrace Deposits. Some contaminated materials are expected at the abandoned gasworks site related to its former use; asbestos was found during construction of the Emirates Air Line cable car crossing. Significant sections of the TBM alignment at each end and along the tunnel crown will run through River Terrace Deposits.

Tunnel construction and operation must also consider groundwater conditions. Previous construction in the area

Table 23.1 Regional stratigraphy of the Silvertown site.

Period	Epoch	Group	Formation
Quaternary	Holocene		Made ground, Alluvium
	Pleistocene		River Terrace Deposits
Tertiary (Paleogene)	Eocene	Thames Group	London Clay
			Harwich
	Paleocene	Lambeth Group	Woolwich, Reading, Upnor
			Thanet Sand
Cretaceous	Upper Cretaceous	White Chalk	Undivided Upper Chalk, mainly Seaford Chalk Lewes Chalk

Source: Transport for London (2013).

has encountered erratic perched water tables within the Made Ground and two water-bearing zones – typically referred to as "shallow" and "deep" aquifers (Transport for London 2013). Water-bearing silts, sands, and gravels of the Alluvium and River Terrace Deposits form the shallow aquifer. These will influence construction at both portals and are likely to be encountered along the crest of the TBM bores. The deep aquifer includes water-bearing sands and gravels of the Harwich Formation, Lambeth Group, and Thanet Sand. These are hydraulically connected to the Upper Chalk and thus are considered a major aquifer. The London Clay is relatively impermeable and separates the shallow and deep aquifers, but it is relatively thin in much of this area and its upper surface has some erosional hollows (Banks et al. 2015). One such hollow was encountered during construction of the nearby Blackwall Tunnel. Thus, there is potential for hydraulic interconnection of the two aquifers, and a requirement to control water inflows.

23.3.3.2 Geotechnical Risks

The shallowness of the alignment, compared with similar river crossing projects, presents a project risk, which will significantly influence TBM selection and specification because it restricts the range of face pressures within which the TBM can be operated safely and efficiently. The anticipated need to control groundwater inflows, combined with variable geological material properties, ranging from stiff clay to water-bearing silts, sands, and gravels and highly variable Made Ground, further raise the geotechnical risk. Thorough analysis and visualization of existing geological data within an integrated 3D model environment allowed for the development of an efficient ground investigation effort. This, in turn, allowed for rapid analysis of design alternatives and a better quantification of geotechnical risk for the Silvertown Tunnel project.

23.3.4 Creating the Integrated 3-D Model

Geological data, from historical records and concurrent ground investigations, were stored within a database that was integrated within a BIM 3-D CAD environment. This allowed for rapid combination, organization, management, and visualization of the geological data in conjunction with existing and proposed above- and below-ground structures. When new geotechnical data were added to the database, those changes were automatically reflected in the 3-D model.

23.3.4.1 Geological Data Sources

Collation of existing geological data from previous projects and from historical and public sources produced considerable information. At the Desk Study stage, information sources included: the British Geological Survey (BGS) database of historic boreholes, BGS maps and reports, a digital 3-D London regional model, and reports for the design and construction of the second bore of the Blackwall Tunnel (130 boreholes and two trial pits), London Underground's Jubilee Line Extension tunnels that cross beneath the Thames, and the Emirates Airline Cable Car crossing (36 boreholes with seven overwater, 14 trial pits, nine dynamic samples, six concrete cores, and pressuremeter tests from four boreholes). Some of these sources were a little distant from the project and so were augmented at design stage by further relevant data from the Greenwich Peninsula Decontamination (14 window samples, 10 cable percussive boreholes and five trial pits), the Docklands Light Railway City Airport Extension (17 boreholes and 13 trial pits) and the Peninsula River Walls (seven boreholes).

23.3.4.2 Managing Geotechnical Data

Much of the geological and geotechnical data were in relatively inaccessible formats, such as PDF reports and scans of old maps, plans, and site geotechnical investigations. Only some of the more recent data were available in digital form and ready for use in a comprehensive project database. Thus, Atkins utilized the digital geotechnical exchange standard developed by the Association of Geotechnical and Geoenvironmental Specialists (AGS) to make these geological and geotechnical data sources accessible within a project geotechnical database. This database was created using HoleBASE SI. By evaluating this historical data in HoleBASE SI, the number of new exploratory onsite boreholes was significantly reduced, resulting in reduced project time and cost. After completing the onsite ground investigation, new data were merged with the historical information in HoleBAS SI for engineering interpretation and stratum identification (Figure 23.17).

23.3.4.3 Modeling Infrastructure Elements

Atkins used Autodesk's AutoCAD Civil 3D (Chappell 2015) to create a 3D model of the project area showing all infrastructure elements, both above and below the surface, based on a variety of data sources. These included as-built data for the Emirates Air Line cable car foundations, and historical data for the foundations of the demolished gas works on the Greenwich Peninsula and of the warehouses and piers at the former Royal Victoria Dock entrance. The proposed tunnel alignment and other proposed structures relating to the tunnel were then added to provide a series of project design sheets (Figure 23.18).

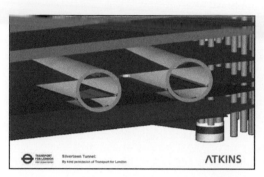

LEGEND:

Grey: Concrete, either tunnels or piles
Green: Ground Surface
Brown: base of Made ground
Mustard: base of Alluvium
Blue: base of River Terrace Deposits
Purple: base of London Clay
Pink: base of Harwich Formation
Dark grey: base of Lambeth Group – laminated beds

Figure 23.17 Three-dimensional visualization of integrated BIM infrastructure elements with the geological ground model. Source: Atkins, by kind permission of Transport for London.

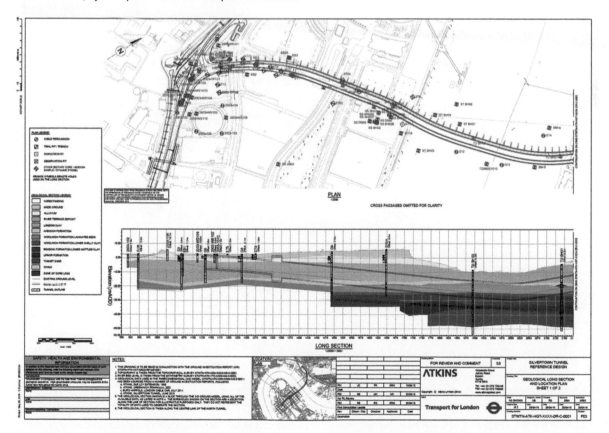

Figure 23.18 Example Design Sheet for the Southwestern part of Silvertown Tunnel illustrating the integration of BIM infrastructure elements with the geological ground model. Source: Atkins, by kind permission of Transport for London.

23.3.4.4 Integration of Geological and Infrastructure Models

The HoleBASE SI Extension for Civil 3D (Autodesk 2016) was used to integrate the geotechnical data in the Hole-BASE SI database with the Civil 3D BIM model. The geotechnical data could now be visualized in relation to the existing site and proposed design. This integration between the HoleBASE SI database and Civil 3D dramatically improved the design process; whenever new geological data were added to the HoleBASE SI database, those changes were automatically reflected in the Civil 3D BIM model. Synchronization of geotechnical data with BIM infrastructure elements facilitated the calculation of volumes of materials that required specific treatment during onsite excavations. Having all the information in a common data environment facilitated production of geological sections and other inter-disciplinary checks that maximized design efficiency and increased the level of design confidence. Atkins also used the Civil 3D project model in Autodesk Navisworks for client and partner design reviews and walkthroughs, and Autodesk 3ds Max to create high-end project renderings for Transport for London's public outreach efforts (Figure 23.19).

23.3.5 Economic Value of Integrated 3-D Model

The time saved by this synchronization allowed for more experimentation with design alternatives, resulting in refinements and improvements. Integration of geological and geotechnical data with the 3-D BIM model permitted the Atkins design team to: (i) visually understand and evaluate the design alignment, (ii) pinpoint potential construction obstructions, (iii) determine what new site investigations were needed, and (iv) easily generate earthworks quantities for project costing and risk assessment.

23.3.6 Conclusions

For the design of the Silvertown Tunnel, Atkins found increased efficiency, and thus economic benefit, from the use of a fully integrated, multidisciplinary 3-D model that combined BIM and geological data. Visualizing ground conditions in a design context resulted in a design that offered reduced project risk and project construction costs. It also provided for the rapid production of 3-D visualizations suitable for public discussions (Figure 23.19).

Figure 23.19 Example project rendering of Silvertown Tunnel produced as part of a public outreach effort. Source: Atkins, by kind permission of Transport for London.

References

Aldiss, D.T., Entwisle, D.C. and Terrington, R.L. (2009). *A 3D Geological Model for the Proposed Farringdon Underground Railway Station, central London*. Keyworth, UK: British Geological Survey, Commissioned Report CR/09/014.

Aldiss, D.T., Black, M.G., Entwisle, D.C. et al. (2012). Benefits of a 3D geological model for major tunnelling works: an example from Farringdon, east-central London, UK. *Quarterly Journal of Engineering Geology and Hydrogeology* 45 (4): 405–414. https://doi.org/10.1144/qjegh2011-066.

Banks, V.J., Bricker, S.H., Royse, K.R. and Collins, P.E.F. (2015). Anomalous buried hollows in London: development of a hazard susceptibility map. *Quarterly Journal of Engineering Geology and Hydrogeology* 48: 55–70. https://doi.org/10.1144/qjegh2014-037.

Burke, H.F., Hughes, L., Wakefield, O.J.W. et al. (2015). *A 3D Geological Model for B90745 North Trans Pennine Electrification East between Leeds and York*. Keyworth, UK: British Geological Survey, Commissioned Report CR/15/04N. 32pp. [Online: Available at: http://nora.nerc.ac.uk/id/eprint/509777] (Accessed December 2020).

Chappell, E. (2015). *AutoCAD Civil 3D 2016 Essentials: Autodesk Official Press*. Indianapolis, IN: Wiley. 416 pp.

Gakis, A. (2014). *Design of a SCL Wraparound Tunnel Utilising a 3D Geological Model for Crossrail Farringdon Station*. London, UK: British Tunnelling Society, Harding Prize Papers, 2014 Runner-Up. 21 pp. [Online: Available at: https://www.britishtunnelling.org.uk/ajax/functiongrabber.asp?loadfunction=downloadfile&f=downloads&filename=2014+angelos+gakis+design+of+an+scl+wraparound+tunnel.pdf] (Accessed December 2020).

Gakis, A., Salak, P. and John, A.S. (2014). Geotechnical risk management for sprayed concrete lining tunnels in Farringdon Crossrail station. In: *Proceedings of the World Tunnel Congress 2014 – Tunnels for a Better Life, Foz do Iguaçu, Brazil* (eds. A. Negro, M.O. Cecílio Jr. and W. Bilfinger), pap107. Châtelaine, Switzerland: ITA-AITES. 8 pp.

Gakis, A., Salak, P. and John, A.S. (2015). Innovative geotechnical risk management for SCL tunnels. *Proceedings of the Institute of Civil Engineers - Geotechnical Engineering*, 168, 385–395. https://doi.org/10.1680/jgeen.14.00070.

Gakis, A., Cabrero, P., Entwisle, D. and Kessler, H. (2016). 3D geological model of the completed Farringdon underground railway station. In: *Crossrail Project: Infrastructure Design and Construction, vol. 3*, (ed. M. Black), 431–446. London, UK: Institution of Civil Engineers.

Grice, C. and Kessler, H. (2015). *Collaborative Geotechnical BIM technologies (Presentation at the AGI BIM event 2015)*. Redditch, UK: Keynetix/Keyworth, UK: British Geological Survey, presentation. 19 pp. [Online: Available at: http://nora.nerc.ac.uk/510823] (Accessed December 2020).

Hight, D.W., Ellison, R.A. and Page, D.P. (2004). *Engineering in the Lambeth Group*. London, UK: CIRIA, PUB C583. 210 pp.

HM Government (2015). *Digital Built Britain: Level 3 Building Information Modelling – Strategic Plan*. London, UK: Department for Business, Innovation and Skills. 47 pp. [Online: Available at: https://assets.publishing.service.gov.uk/government/uploads/system/uploads/attachment_data/file/410096/bis-15-155-digital-built-britain-level-3-strategy.pdf] (Accessed December 2020).

Kessler, H., Mathers, S. and Sobisch, H.-G. (2009). The capture and dissemination of integrated 3D geospatial knowledge at the British Geological Survey using GSI3D software and methodology. *Computers and Geosciences* 35 (6): 1311–1321. https://doi.org/10.1016/j.cageo.2008.04.005.

Mathers, S.J., Burke, H.F., Terrington, R.L. et al. (2014). A geological model of London and the Thames Valley, Southeast England. *Proceedings of the Geologists' Association* 125 (4): 373–382. https://doi.org/10.1016/j.pgeola.2014.09.001.

Newman, T. (2009). The impact of adverse geological conditions on the design and construction of the Thames water ring main in greater London, UK. *Quarterly Journal of Engineering Geology and Hydrogeology* 42: 5–20. https://doi.org/10.1144/1470-9236/08-035.

Page, D.P. and Skipper, J.A.E. (2000). Lithological characteristics of the Lambeth Group. *Ground Engineering* 33: 38–43. [Online: Available at: https://www.geplus.co.uk/technical-paper/technical-paper-lithological-characteristics-of-the-lambeth-group-01-02-2000] (Accessed December 2020).

Parry, S., Baynes, F.J. and Culshaw, M.G. et al. (2014). Engineering geological models: an introduction: IAEG Commission 25. *Bulletin of Engineering Geology and the Environment* 73: 689–706. https://doi.org/10.1007/s10064-014-0576-x.

Spyridis, P., Nasekhian, A. and Skalla, G. (2013). Design of SCL structures in London. *Geomechanics and Tunnelling* 6 (1): 66–79. https://doi.org/10.1002/geot.201300005.

Transport for London (2013). *TfL River Crossings - Ground Investigation Desk Study, Preliminary Sources Study Report*. London, UK: Transport for London. 91 pp. [Online:

Available at: http://content.tfl.gov.uk/st-river-crossings-ground-investigation-desk-study-pssr.pdf] (Accessed December 2020).

Transport for London (2015a). *Silvertown Tunnel Preliminary Case for the Scheme*. London, UK: Transport for London. 327 pp.

Transport for London (2015b). *Silvertown Tunnel – Preliminary Maps, Plans and Drawings*. London, UK: Transport for London. 18 pp. [Online: Available at: http://content.tfl.gov.uk/preliminary-maps-plans-and-drawings.pdf] (Accessed December 2020).

Wood, B., Richmond, T., Richardson, J. and Howcroft, J. (2015). *BGS Groundhog® Desktop Geoscientific Information System External User Manual*. Keyworth, UK: British Geological Survey, Report OR/15/046. 99 pp. [Online: Available at: http://nora.nerc.ac.uk/id/eprint/511792] (Accessed December 2020).

24

Application Theme 7 – Building and Construction

Editor's Introduction

This chapter contains three case studies that examine the role of 3-D models in support of building and construction. Two-dimensional maps are commonly used to locate suitable sources (and volumes) of aggregates from either unconsolidated sedimentary deposits or solid rock. However, 3-D models can also provide additional valuable information.

The first case study describes a British Geological Survey (BGS) 3-D evaluation of the volume change potential of the London Clay Formation, which underlies most of the Greater London area. The London Clay contains discrete smectite horizons, which pose a significant shrink-swell hazard that can influence the development and construction of the city's transport routes and buildings. An existing London geological model defined the lithostratigraphic framework, while data from the BGS National Geotechnical Properties Database (Self et al. 2012) defined the 3-D distribution of "Volume Change Potential" (VCP). Three modeling approaches were tested and resulted in the production of the London 3-D GeoSure shrink-swell product that provides a regional susceptibility of potential shrink-swell hazard at intervals down to 20-m, using the standard GeoSure A–E range of hazard susceptibility.

Aggregate resource assessments are made at two distinct scales: *regional assessments* typically undertaken by geological survey organizations in support of minerals or spatial planning, and *local assessments* undertaken by aggregate producers prior to developing a new aggregate pit or quarry, or by building companies searching for aggregates or fill for a large project.

Aggregate resource evaluations may utilize a custom-made 3-D model but large-scale evaluations are typically based on, or derived from, existing 3-D geological models. Analyses may utilize bulk attribution of individual lithostratigraphic units within stacked-layer models (see Chapter 9), or lithological attributes stored within voxel models as discussed in Chapter 11.

The second case study describes how, in the Netherlands, 3-D geological models have assisted the regional assessments of aggregate resources, for pre-prospecting and planning purposes. While aggregate resource assessments are based on a 3-D model, 2-D maps best serve user needs. Thus, a Dutch aggregate resource information system consists of two interactive maps; one shows the "total resource" – (the cumulative thickness of the resource in place down to a chosen depth) while the other shows the "exploitable resource" (the proportion of the resource that is technically exploitable based on overburden and intercalation criteria set by the user).

The third case study describes a similar approach recently developed at the BGS to evaluate the feasibility of replacing currently-employed 2-D regional aggregate resource assessments with assessments based on a 3-D model. A prototype 3-D model of the distribution and quality of sand and gravel resources in the Thames Basin evaluated this new approach.

Case Study 24.1: Three-Dimensional Volume Change Potential Modeling in the London Clay

Lee Jones, Ricky Terrington, and Andy Hulbert
British Geological Survey, Keyworth, Nottingham NG12 5GG, UK

24.1.1 Introduction

The shrink-swell behavior of clay soils is controlled by the type and amount of clay in the soil, and by seasonal changes in the soil moisture content related to rainfall and local drainage. Variations in ground moisture cause ground movement, particularly in the upper 2 m of the ground, which may affect building foundations, pipes, or services. Shrink-swell clays may also influence the construction of deep foundations and basements, road or rail networks, and utilities. The 3-D distribution of

Applied Multidimensional Geological Modeling: Informing Sustainable Human Interactions with the Shallow Subsurface, First Edition.
Edited by Alan Keith Turner, Holger Kessler, and Michiel J. van der Meulen.

shrink-swell properties can be used to identify potential problems at the surface, in the shallow subsurface or deeper underground.

In the UK, shrink-swell damage to buildings and foundations was first recognized following the dry summer of 1947. Since then, the cost of repairing such damage has risen dramatically. After the drought of 1975–1976, insurance claims in the UK came to over £50 million. In 1991, after another drought, claims peaked at over £500 million. Shrinking and swelling of the ground (often reported as subsidence) is now one of the most damaging geohazards in Britain, costing the economy an estimated £3 billion over the past decade (Jones and Jefferson 2012). In a study of subsidence claims related to shrink-swell clays, the London Clay is described as "the most commonly encountered problem soil" (Crilly 2001). The Association of British Insurers has estimated that the average annual cost to the insurance industry of shrink-swell related subsidence is over £400 million (Driscoll and Crilly 2000), and that by 2050 this could rise to over £600 million.

24.1.2 London Clay Lithology and Shrink-Swell Potential

The London Clay Formation forms the major part of the Thames Group and underlies most of the Greater London area. It has greatly influenced the development and construction of the city's transport routes and buildings.

The presence of expansive clay minerals, such as smectite, increases the plasticity of the clay and the severity of shrink-swell phenomena. The London Clay Formation is composed dominantly of illite clay; however, discrete smectite horizons are common near the base of the London Clay in the central and eastern parts of the London Basin. Overall, the London Clay Formation is generally accepted to be highly plastic, thus posing a significant shrink-swell capability (Driscoll 1983; Cripps and Taylor 1986; Reeves et al. 2006). Detailed testing of London Clay samples from the London Basin has revealed a gradual increase in plasticity of about 30% from west to east in the London Basin (Burnett and Fookes 1974), but also showed subtle changes with depth.

Shrinkage and swelling is usually confined to the active zone (upper 1.5 m) where moisture changes are most likely to occur; this zone may be extended by the presence of tree roots (Driscoll 1983). In the London area, rising groundwater levels resulting from diminished water abstraction from the underlying Chalk aquifer will affect the London Clay. This has implications for engineering development because, if swelling occurs at deeper locations, it may cause problems for existing foundations (Lerner and Barrett 1996; Simpson et al. 1989).

24.1.3 Definition of Volume Change Potential

The Volume Change Potential (VPC) of a soil is the relative change in volume to be expected as the soil moisture content changes. In the UK, approved, established, test methods for measuring soil shrinkage and soil swell pressure are rarely undertaken during routine site investigations. Consequently, reliance is placed on estimates based on index parameters (Reeve et al. 1980; Holtz and Kovacs 1981; Oloo et al. 1987). The plasticity index (I_P) is the most widely used parameter for determining clay shrink-swell potential. However, because the plasticity index is based on remolded specimens, it cannot precisely predict the shrink-swell behavior of an in-situ soil (Jones 1999). A Modified Plasticity Index (I_P') was proposed in the Building Research Establishment Digest 240 (Building Research Establishment 1993). Because the I_P' considers the whole sample and not just the fines fraction, it gives a better indication of the "real" plasticity value. Jones and Terrington (2011) used the relationship between I_P' and VCP proposed by the Building Research Establishment (Table 24.1) and data from the British Geological Survey (BGS) National Geotechnical Properties Database (Self et al. 2012) to investigate the 3-D distribution of VCP in the London Clay.

24.1.4 Modeling the 3-D VCP of the London Clay

Prior to undertaking any 3-D modeling of London Clay VCP, Jones and Terrington (2011) undertook statistical and spatial analyses of the available data. The Geostatistical Analysis Extension in ESRI ArcMap 9.1 was used to apply an Inverse Distance Weighting (IDW) interpolation to the London Basin VCP dataset. The entire London Basin was analyzed using all available sample points; variations with depth were ignored. The resulting 2-D spatial analysis showed that the VCP tends to increase from west to east across the London Basin (Figure 24.1).

Table 24.1 Classification of volume change potential.

Classification	I_P' (%)	VCP
A	<10	Non-plastic
B	10–20	Low
C	20–40	Medium
D	40–60	High
E	>60	Very high

Source: After Building Research Establishment (1993).

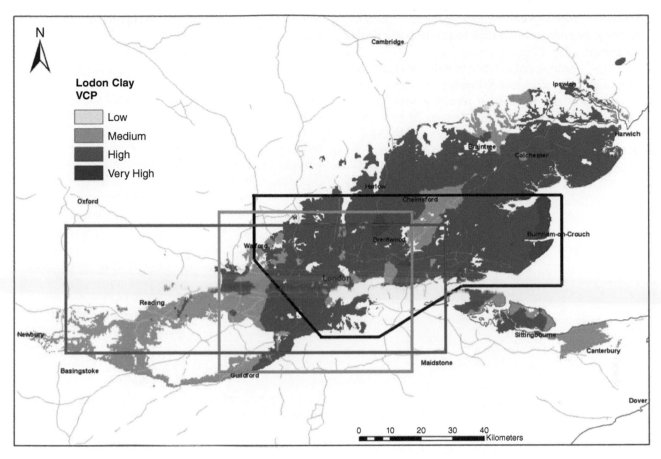

Figure 24.1   The 2-D variation in the plasticity of the London Clay across the entire London Basin. Mean I_p' values were interpolated using the IDW method. Outlined areas are the extents of the three 3-D models: black is the GoCAD S-Grid Model; green is the Facies Model, and blue is the London lithoframe and 3-D GeoSure models.

Modeling clay formations in 3-D can show trends in VCP that are ignored by 2-D, or even 2.5-D, modeling methods. The Earth Decision Suite of GoCAD and ESRI ArcGIS were used to support three different techniques for creating models to interpret and visualize the 3-D spatial VCP variations in the London Clay Formation. The models were: GoCAD S-Grid Model, Facies Model, and 3-D GeoSure Model. The available data were not ideally distributed in 3-D throughout the London Basin. Recognizing these limitations, each method was applied to slightly different sub-areas. Figure 24.1 shows the lateral extents of each model.

24.1.4.1   GoCAD S-Grid Model
The first analysis was restricted to an area covering Greater London, southern Hertfordshire and Essex, and northern Kent, defined by the black outline on Figure 24.1. The lateral extent of this 3-D model was defined by the outcrop pattern of the London Clay in this area (shown in color in Figure 24.1). The London Basin digital terrain model (DTM) defined the top of the model. The base of the model was placed 50 m below the surface defined by the DTM. Thus, the model has a uniform thickness of 50 m. The

GoCAD 3D Grid Reservoir Builder tool was used to discretize the model into 100-m cells (S-Grid) and divide the model depth into 100 uniform layers. Thus, the 3-D model volume consisted of 100-m × 100-m × 0.5-m conformable elements as illustrated in Figure 24.2.

Model elements that corresponded to the 3-D location of an observed I_p' value in the dataset were assigned that I_p' value. The remaining model elements were initialized with the overall mean I_p' value for the London Clay within the study area, determined by statistical analysis of the dataset. Then, the entire 3-D model volume was subjected to interpolation and smoothing throughout the grid, using the Discrete Smooth Interpolation (DSI) technique in GoCAD that can be applied in any dimension. Figure 24.3 illustrates the interpreted 3-D VCP distribution by showing four selected horizons between the top and base of the model.

24.1.4.2   Facies Model
Facies modeling techniques have been used by the BGS to convey lithological variation and heterogeneity of geotechnical parameters (Woods et al. 2015) and are extensively

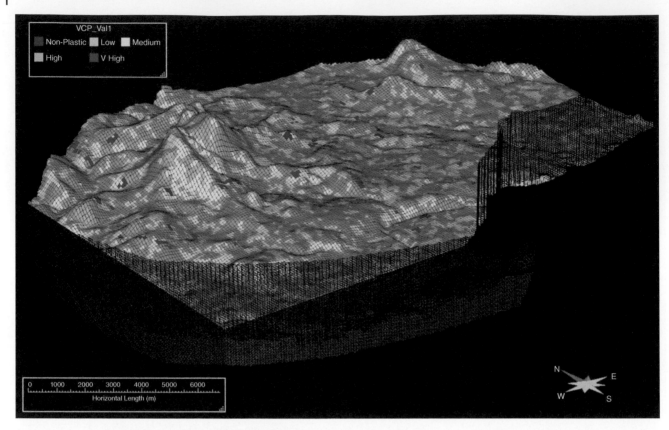

Figure 24.2 The 3-D discretized volume of the GoCAD S-Grid model.

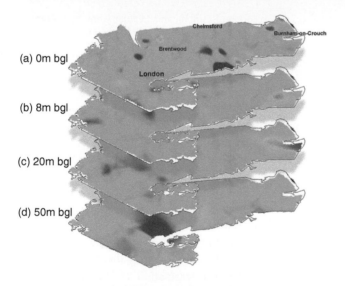

Figure 24.3 GoCAD S-Grid interpolations showing VCP values at surfaces located at 0, 8, 20, and 50-m depth. Color Code: blue = medium VCP, green = high VCP, red = very high VCP. Source: British Geological Survey.

used by other geological surveys, including TNO, Geological Survey of the Netherlands (Stafleu et al. 2012, Chapter 11).

Prior statistical and spatial analyses of the available VCP data for the London Clay by Jones and Terrington (2011)

showed a considerable range in VCP scores. However, local variations and the clustered and linear alignments of observations from road or railway development investigations obscured the 3-D characteristics of the VCP values. Therefore, a new facies model was created for the central London area shown by the green outline in Figure 24.1. About 7500 VCP observations were fairly evenly distributed across this area, making it a good candidate for testing facies modeling methods.

The SKUA-GoCAD Reservoir Properties Modeling Workflow and the Sequential Indicator Simulation tool generated the facies model. The modeling employed a regular grid with voxels approximately 100-m × 100-m × 1-m. A finer resolution, such as 50-m × 50-m × 0.5-m, would have better evaluated the data, and matched typical downhole sampling rates, but available computational capacity was incapable of modeling this area at that resolution.

Only a few simulations were run, one was chosen at random for the illustrations as a proof of concept, rather than conducting a fully stochastic simulation that could have been used to evaluate uncertainty (Kearsey et al. 2015). A single variogram was used for the London Clay data within the study area. Because 65% of the London Clay VCP values fell into the High category, this was chosen as the basis

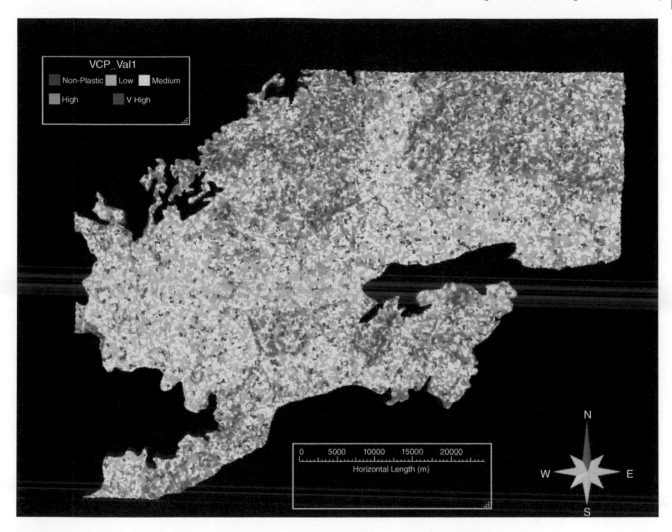

Figure 24.4 Top of the Facies London Clay Model, showing VCP distribution on ground surface.

for the variogram calculation, which was automated in the workflow.

Figure 24.4 shows the top surface of the Facies London Clay Model. The VCP values at the ground surface show no apparent lateral trends; most of the values fall within the Medium or High VCP classes, with scattered areas of Very High, Low, or even Non-Plastic classes. This agrees with the previously discussed 2-D model produced by IDW interpolation (Figure 24.1) and the simpler GoCAD S-Grid Model (Figure 24.3).

Figure 24.5 shows the base of the London Clay for this model, the I_p' sample points displayed with their VCP values, and the alignments where four vertical sections were sliced through the model to evaluate vertical variations of VCP values. These sections were aligned along concentrations of closely spaced data, usually along a linear feature such as a road. Figure 24.6 illustrates section number 1; it shows considerable local variability in the VCP values

which range from medium to high values, but there is no overall trend.

24.1.4.3 Three-Dimensional GeoSure Model

During the 1990s, BGS transformed its map output into a fully digital production system, permitting the capture of archive maps and presentation of new maps in digital formats. By 2001, a nationwide 1 : 50 000-scale digital geological map dataset was produced that included four GIS (Geographic Information System) layers: bedrock, superficial deposits, mass movement, and artificial deposits. This supported the development of additional national datasets that addressed the growing demand for geoscience information to aid land-use planning and property transactions.

The BGS GeoSure program supplied six 1 : 50 000-scale potential-hazard assessments for UK ground-stability geohazards: (i) collapsible ground, (ii) running sands,

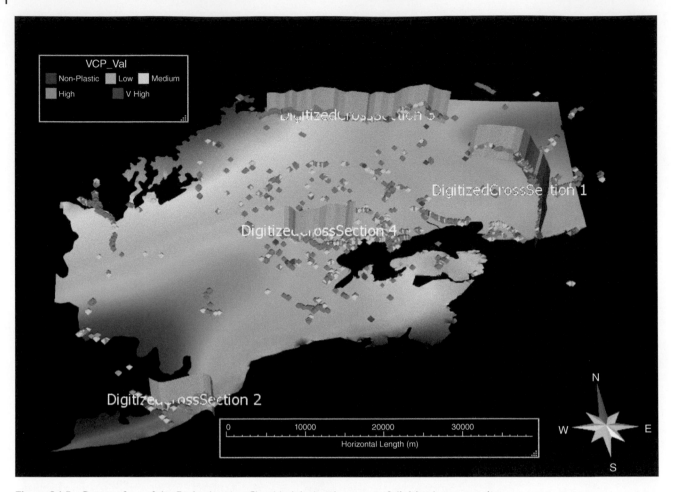

Figure 24.5   Base surface of the Facies London Clay Model, showing areas of digitized cross-sections.

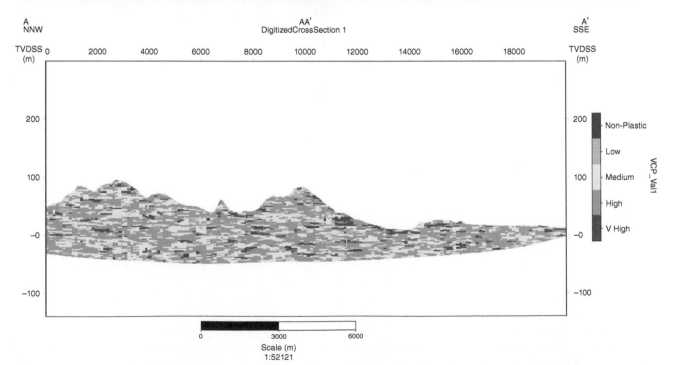

Figure 24.6   Digitized cross-section number 1 of Facies London Clay Model, showing VCP values.

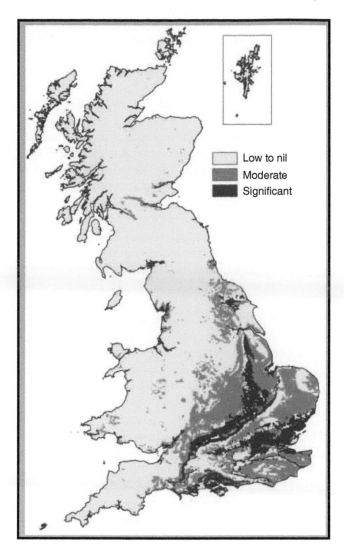

Figure 24.7 UK map of GeoSure shrink-swell geohazard. Source: British Geological Survey.

Low to nil
Moderate
Significant

(iii) compressible ground, (iv) landslides, (v) soluble rocks, and (vi) shrink-swell (Walsby 2008). These 2-D GIS GeoSure products result from deterministic assessments of appropriate causative factors. Each ground stability geohazard is classified using the same six-level rating, ranging from "NULL" through "A" (low hazard potential) to "E" (high hazard potential). Figure 24.7 illustrates the 2-D GeoSure shrink-swell geohazard within the UK. GeoSure assessments have a 25-m resolution and are typically reported as PDF files in response to individual requests (Walsby 2008).

In the London area, the severity of the shrink-swell geohazard, combined with the large population and associated construction activities, resulted in a more extensive I_P' and interpreted VCP dataset. Analysis of this dataset produced a supplemental 3-D GeoSure shrink-swell product (Jones and Hulbert 2016).

The BGS London Lithoframe geological model (Mathers et al. 2014) covers an area extending 120 km east–west and 40 km north–south (4800 km^2) located as defined by the blue outline in Figure 24.1. The model was constructed as twelve 20 × 20 km squares, arranged as six columns in the east–west direction and two rows in the north–south direction. The 3-D GeoSure Model uses the 12 subareas defined by the London Lithoframe model. Within them, the 3-D GeoSure Model contains the shrink-swell characteristics of the top 20 m of the London Clay.

ESRI ArcGIS was used to create the 3-D GeoSure shrink-swell model. The model area was defined by a regular 50 × 50 m grid; then converted to a *point feature class*, with each point located at the center of a grid cell. The Nextmap DTM (Intermap Technologies 2007) defined the ground elevation (related to Ordnance Datum) of each point. A custom Python script assessed the presence below each point of geological formations at nine specific depths (at 0, 1, 2, 3, 4, 5, 10, 15 and 20 m), and stored this information as an attribute-string attached to each point. A second custom python script converted this attributed point dataset to a series of nine gridded surfaces with 50-m × 50-m resolution, each defining the geological conditions at the specified depth.

The 3-D GeoSure shrink-swell product was created by "stacking" these gridded surfaces to produce a dataset that defined, for each 50-m cell, the geological conditions at the nine specified depths. It was then a straightforward process to convert this dataset into one that had attributes defining 3-D VCP values at specified depth intervals within this upper 20 m across the entire BGS London Lithoframe model. To ensure the compatibility of the 3-D GeoSure shrink-swell product with the existing standard 2-D GeoSure shrink-swell product, two additional attributes were added for each 50-m cell, the dominant (modal) VCP value and the range of the VCP values. The 3-D GeoSure shrink-swell product provides GIS gridded maps for each of the twelve 20-km × 20-km squares; these display the VCP values at any of the nine specified depths (Figure 24.8). Alternatively, a tabulation of the vertical distribution of VCP values for each 50-m × 50-m cell can be requested (Table 24.2).

24.1.5 Applications of the Models

Two-dimensional representations, based on statistical analyses, of the VCP in the London Clay show general trends. Three-dimensional models, such as those created by the S-Grid and facies techniques, provide a seamless interpolation and deliver a visualization of VCP that can be interpreted across a variety of depths.

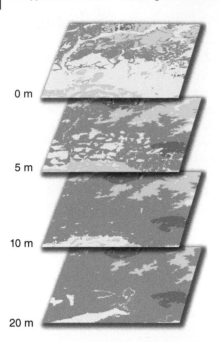

Figure 24.8 Examples of the 3-D GeoSure shrink-swell map showing distribution of VCP values at four depths within a 20 km × 20 km square. (Color code: yellow = low VCP, blue = medium VCP, green = high VCP, red = very high VCP).

Table 24.2 Example of 3-D GeoSure shrink-swell tabulation for a 50 m × 50 m cell.

Depth (m below ground level)	Formation code	Dominant VCP ranking	Range of VCP rankings
0	MGR	A	
1	ALV	C	A–C
2	ALV	C	A–C
3	LASI	B	
4	LASI	B	
5	LC	D	A–E
10	LC	D	A–E
15	LC	D	A–E
20	LC	D	A–E

Key: MGR, Made Ground; ALV, Alluvium; LASI, Langley Silt; LC, London Clay.

The GeoSure 3-D dataset provides an information resource for asset and infrastructure development and maintenance. The shrink-swell properties of the London Clay affect developers, construction companies, and local government due to increased costs of insurance, additional engineering works to stabilize land, or potential relocation of developments. Information on the 3-D distribution of shrink-swell properties permits identification of potential problems at the surface, in the shallow subsurface, or deeper underground.

24.1.6 Conclusions

The GoCAD S-Grid method allowed the $I_p{}'$ values to be examined relative to ground level. This 3-D representation permitted the evaluation of variations with depth, as well as any lateral trends that could be seen in earlier 2-D evaluations. The model showed little variation in plasticity in the uppermost part of the London Clay, which falls mainly within the Medium to High VCP class (Figure 24.3a). However, as the depth increases to 20 m, an area of Very High VCP can be seen in the east of the area (Figure 24.3c). In the west, an increasingly larger area remains at Medium VCP as the depth reaches 50 m (Figure 24.3d).

The facies model provides a better understanding of the true nature of the shrink-swell conditions in the London Clay; it provides good site-specific values and captures local variability but does not clearly define regional trends in the shrink-swell properties of any specific clay unit. More detailed site-scale models can be produced for areas where large numbers of boreholes and sample data are situated and evenly distributed.

The 3-D GeoSure shrink-swell product for London is part of the BGS GeoSure range of natural subsidence products. Based on data contained within the London Geological Model (Mathers et al. 2014), it provides a regional susceptibility of potential shrink-swell hazard, in 3-D, at intervals down to 20-m in the London and Thames Valley area using the standard GeoSure A–E range of hazard susceptibility classes.

Case Study 24.2: Dutch Experience in Aggregate Resource Modeling

MJ. van der Meulen, D. Maljers, and Jan Stafleu

TNO, Geological Survey of the Netherlands, 3584 CB Utrecht, The Netherlands

24.2.1 Introduction

"Aggregate" refers to a broad category of medium- to coarse-grained particulate material used in construction (Smith and Collis 2001). Aggregates are produced from unconsolidated sediments, crushed solid rock, and, to an increasing extent, from solid wastes and by-products, for example construction and demolition waste, blast-furnace slag and quarry spoil. They have a relatively high hydraulic conductivity and predictable, uniform load-bearing characteristics, especially when compacted. Thus, they allow for drainage and provide stable foundations for buildings, roads, and railways by helping prevent differential settling

Figure 24.9 Workflow for the assessment of aggregate resources using voxel models.

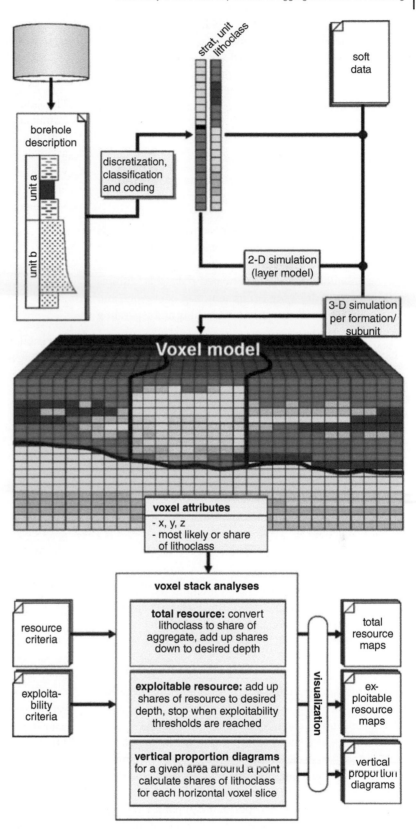

under load. Aggregates also provide bulk and strength to composite materials such as concrete and asphalt.

Available volumes of recycled aggregate materials are typically insufficient to satisfy aggregate demands, so the exploration for natural aggregate resources remains an important task that can be materially assisted with 3-D geological modeling.

24.2.1.1 Economic Considerations

Aggregates are a low-cost bulk commodity, typically supplied at low margins. Aggregate prices are heavily influenced by transport logistics and supply chain efficiency. Accordingly, aggregates are ideally sourced as closely as possible to where they are used. Aggregate production is primarily designed to generate specific grain-size mixtures from sedimentary deposits or crushed rock; this requires earth-moving or dredging equipment, belt conveyors, and machines specifically designed for crushing and separating various grain sizes.

Geological properties such as grading, roundness, and petrography of sedimentary resources, or petrography and heterogeneity of solid rock resources determine which aggregates can be produced at a specific site (Van der Meulen et al. 2005). Applications often have specified requirements, for example compressive strength for concrete aggregate, bulk density for lightweight construction, durability, and angularity to provide good traction properties for road pavement material, internal angle of friction for constructional fill and railway ballast, and permeability for fill within embankments, under roadways, or in drainage trenches.

24.2.1.2 Applications and Scope

In the Netherlands, constructional fill, land reclamations, and coastal maintenance account for the largest aggregate volumes (Van der Meulen et al. 2003; Van der Meulen 2005). These are the least demanding applications in terms of material properties. Locally available material, preferably sand, is used unless it is unsuited or inaccessible. Portland cement concrete is the second-most important use of aggregate. Aggregate grading is designed to optimize the concrete production, since it determines the quantities of water and cement required, the workability of the mortar, and the strength and durability of the concrete. Grain shape is less important; thus, provided the grain strength is satisfactory, almost any Dutch sedimentary aggregate resources are acceptable for standard concrete (Van der Meulen et al. 2005). Their only potentially deleterious component is chert in deposits of the river Meuse, which has widespread occurrences of cherty limestone in its Belgian–French catchment. Chert-bearing aggregate cannot be used for concrete that is going to be exposed to saline water, because the reaction between amorphous silica with alkaline pore water causes degradation (Ichikawa and Miura 2007). The Netherlands lacks resources for crushed aggregates, which are imported from Belgium, Norway, and the United Kingdom.

Altogether, there are fundamental differences between exploring for solid-rock and unconsolidated sedimentary aggregate resources. This case study focuses on aggregate exploration in unconsolidated sedimentary deposits, especially the production of aggregates for concrete, one of the most important and more advanced applications.

24.2.2 Three-Dimensional Modeling for Aggregate Exploration

Aggregate resource assessments are made at two distinct scales. Regional assessments are typically undertaken by geological survey organizations in support of minerals or spatial planning. Local assessments are undertaken by aggregate producers during the preparatory stages of developing an aggregate pit or quarry, or by building companies who source their own aggregates for a large project.

Regional assessments of aggregate resources can be assisted by 3-D geological models which can assist the exploration phase by addressing the following questions (Van der Meulen et al. 2005):

1. What is the grading? In other words, what is the bulk grain-size composition of the target deposit and what aggregates could be produced from it?
2. What is the cumulative resource thickness which has suitable grading?
3. What, if present, is the thickness of the overburden and intercalations? These may inhibit production, technically or economically.

Local assessments permit the calculation of the optimal proportion of various aggregates to be produced at the site under scrutiny, and thus define the economic potential of long-term supply contracts. Local assessments utilize bulk volume estimations based on a dense grid of boreholes and grain-size analyses; they do not require a spatial model. During actual production, the operator of the dredger or backhoe relies on onsite production experience rather than on some model prediction.

A regional aggregate resource evaluation can be a project in its own right (Orlić and Rösingh 1995). They can also be a based on or derived from existing 3-D geological models. If these are stacked layer models, such derivation requires applying a bulk attribution to the individual lithostratigraphic units (Chapter 9). Better results can be obtained by analyzing lithological attributes stored within voxel models as defined in Chapter 11 and by Stafleu et al. (2011, 2012). To date, the majority of the reported regional to national aggregate resource assessments based on voxel models have been conducted in the Netherlands. However, a prototype evaluation using a 3-D voxel approach has recently been undertaken for sustainable management of marine geological resources (Van Lancker et al. 2017).

24.2.3 Dutch Case Example

Maljers et al. (2015) discuss the progressive development, cost efficiency, and fitness-for-purpose of three generations of Dutch aggregate resource models.

24.2.3.1 Development History

Van der Meulen et al. (2005) produced the first national aggregate resource model for the Netherlands as a proof of concept. An improved version, which incorporated more detailed geological information and had a better resolution, was prepared for an online minerals information system (Van der Meulen et al. 2007b). Regional aggregate models routinely derived from the national 3-D mapping program GeoTOP described in Chapter 11 and by Stafleu et al. (2011, 2012) have progressively replaced this national aggregate resource model. Progress resulted from the progressive improvements in the underlying lithological geological framework models, especially the development of regional GeoTOP voxel models. The workflow for querying these models to define aggregate content remained essentially the same.

24.2.3.2 Workflow

Figure 24.9 defines the general workflow used for Dutch aggregate resource modeling. The upper part of Figure 24.9 defines the process for creating the GeoTOP voxel geological model (Chapter 11), which is not further detailed here. The lower panel shows the steps used to evaluate such a model to assess the aggregate potential, based on a relationship between lithological class and aggregate yield.

Van der Meulen et al. (2005, 2007a) defined a simple conversion process for evaluating the multiple grain-size classes required by different sand and gravel applications, arguing that more sophistication would not be supported by the non-dedicated data that are typically available for national or regional geological modeling. Table 24.3 shows this process for three typical aggregate applications. It is based on the widely used Wentworth system of grain-size classes, which in turn is translated from a Dutch sediment-sample classification system.

24.2.3.3 Visualization of Results

While aggregate resource assessments based on a 3-D model are ultimately used for site selection, resource operators are best served with 2-D prospectivity maps. Thus, the core visualization of the Dutch aggregate resource information system consists of two interactive maps; these convey the essence of the 3-D model and allow the user to observe the resource quality and exploitability criteria. One map shows the cumulative thickness of the resource in place down to a chosen depth; this is the "total resource" (Figure 24.10a). The second map shows the share of the resource that is technically exploitable, based on overburden and intercalation criteria set by the user; this is the "exploitable resource" (Figure 24.10b).

Table 24.3 The yields of concrete aggregate, masonry sand, and constructional fill expressed as percentages of sediment volume in situ, related to the commonly used Wentworth's (1922) grain size classes for clastic deposits.

Sedimentary resource in situ		Wentworth class boundary (mm)	Aggregate yield		
			Concrete aggregate (%)	Masonry sand (%)	Constructional fill (%)
Gravel					
		2.000	100		
	Very coarse			100	
		1.000			
	Coarse				
		0.500	50		100
Sand	Medium				
		0.250			
	Fine				
		0.125	0	50	
	Very fine				
		0.063			
Silt and clay				0	0

Source: Based on Van der Meulen et al. (2005).

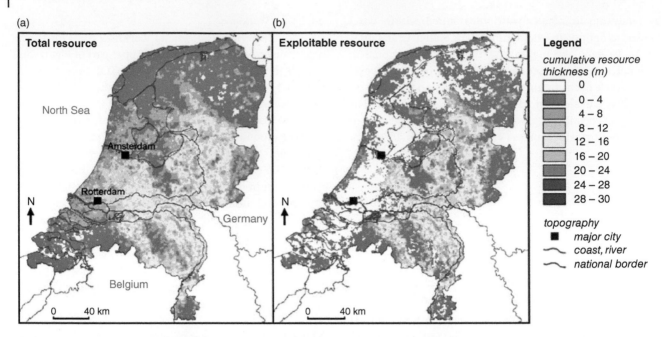

Figure 24.10 Dutch aggregate resource prospectivity maps for concrete aggregate. (a) Total resource availability; (b) Technically exploitable resource availability. Resources present in the western and northern coastal areas are unexploitable due to fact that the thickness of Holocene clayey and peaty deposits exceeds the overburden criterion (Van der Meulen et al. 2005, 2007a).

Vertical proportion diagrams allow the user to (qualitatively) understand the relationship between total and extractable volumes. Figure 24.11 shows two examples of a sandy sequence overlain by a fine-grained top layer. Figure 24.11a is a vertical proportion diagram from an (arbitrary) area where all sand is considered extractable at standard settings for the overburden criterion (i.e., with a thickness threshold of 5 m). Figure 24.11b is a diagram from an area where sand is present below approximately 10 m of clay and peat overburden which makes it unexploitable. Both diagrams were calculated for an area of 2 by 2 km by averaging multiple voxel stacks. The Dutch information system does not provide visualizations of single voxel stacks. These would approximate synthetic prospection boreholes, and so give the user a false sense of accuracy that is not supported by non-dedicated data. For the same reason, numerical model output is provided in rounded numbers only.

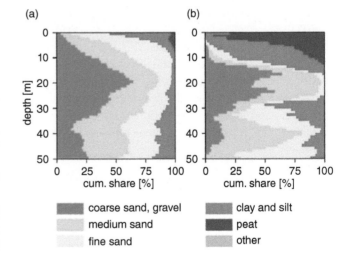

Figure 24.11 Two examples of vertical proportion diagrams. (a) Represents an area where conditions are favorable for aggregate extraction. (b) Represents an area where sand is present below approximately 10 m of clay and peat overburden which makes it unexploitable.

24.2.4 Discussion and Conclusions

The first Dutch national aggregates model was calculated as a proof-of-concept using a $1000 \times 1000 \times 1$ m voxel resolution. The second national model used a $250 \times 250 \times 1$ m resolution, and the regional models have $100 \times 100 \times 0.5$ m resolution. Each resolution improvement raised the model development costs by approximately one order of magnitude. Maljers et al. (2015) argued that this overall increase is justifiable because only the resolution and approach used by the regional models adequately met the aggregate exploration requirements. For example, the regional models can represent the floodplain architecture of the Holocene channel-belts within Rhine Meuse delta (Figure 24.12). This is significant because the channel belts have removed the thick deposits of floodplain fines that make extraction of underlying Pleistocene coarse sands uneconomic over much of the region.

The Dutch experience presents an approach that can be applied in similar geological settings (also see Case

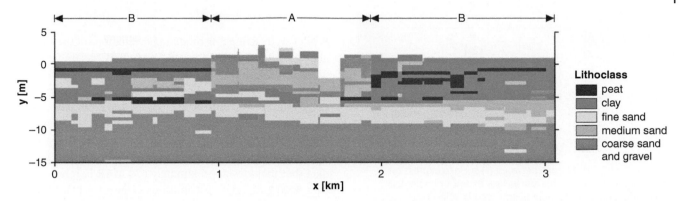

Figure 24.12 Cross-section showing how the depositional architecture of the Rhine-Meuse Delta is represented in the GeoTOP voxel model (Chapter 11), (a) Holocene channel belt sands on an Upper Pleistocene sand body: favorable conditions for sand mining. (b) Floodbasin fines on Upper Pleistocene sand body: less favorable conditions for sand mining. Modified from Maljers et al. (2015).

Study 24.3). Fairly rough modeling is sufficient to unravel general trends and to make national resource estimates. This can be accomplished with a limited investment in model development, if sufficient digital borehole data are available. Evaluations that are more detailed require careful assessment of what can or should be accomplished and the associated costs of doing so. The exponential increase in model development costs as resolution is increased suggests that model refinement should not be the primary goal. The bulk of the extra investments should be directed toward improving the underlying 3-D lithological/geological framework model.

Case Study 24.3: Modeling the Distribution and Quality of Sand and Gravel Resources in 3-D: A Case Study in the Thames Basin, UK

K. Mee, B.P. Marchant, J.M. Mankelow, and T.P. Bide

British Geological Survey, Keyworth, Nottingham, NG12 5GG, UK

24.3.1   Introduction

Maintaining adequate and reliable supplies of sand and gravel aggregates is essential to a healthy UK economy. Land-use planning decisions to safeguard these resources from sterilization by competing land-uses require accurate information on the distribution, quality, and suitability of aggregate resources (Wrighton et al. 2011).

The British Geological Survey (BGS) is the major UK national provider of spatial and statistical minerals information. Between 1968 and 1990, the BGS Industrial Minerals Assessment Unit (IMAU) produced 158 Mineral Assessment Reports (MAR's) which contain a description quantifying the resources of sand and gravel and a map at the scale of 1 : 25 000. The survey included drilling and sampling of the sand and gravel resources, logging of the borehole material and particle-size analysis to determine the proportion of gravel, sand and fines present.

Approximately 12 500 detailed borehole interpretations include more than 54 000 grading analyses collected at about 1 m intervals. The entire series of IMAU reports and maps is accessible in digital format (as .pdf files) at a BGS web portal (British Geological Survey 2019a).

Subsequently, the BGS was commissioned to produce a series of mineral resource maps defined by local planning area. When completed in early 2006, a series of 44 digitally generated maps at a scale of 1 : 100 000 covered England and parts of South Wales. In 2007, a grant from the Scottish Government Aggregates Levy Fund supported production four similar 1 : 100 000-scale digital maps for regions in Scotland. In 2009, a grant from the Welsh Government supported production six similar 1 : 100 000-scale digital maps for regions in Wales, along with an additional six digital maps delineating aggregates safeguarding areas to assist Mineral Planning Authorities (MPAs) with Local Development Plans. Six digital 1 : 100 000-scale digital mineral resource maps for Northern Ireland were developed by the BGS in co-operation with the Geological Survey of Northern Ireland. The entire series of mineral resource maps is accessible at a BGS web portal (British Geological Survey 2019b).

By combining mineral resource, environmental and other land-use information, these maps have provided baseline information to support approvals of mineral extraction operations by local authorities. The 1971–1990 systematic survey and borehole sampling program is unlikely to be repeated. However, analysis of these legacy data sources with 3-D geological modeling procedures can provide information on the properties of the mineral resources that is increasingly needed to identify and protect the most suitable sand and gravel resources.

Existing MAR maps and local planning-area mineral-resource maps are 2-D representations. They show the inferred surface expression (outcrop) of the resources. This may be too limited; any aggregate resource may extend under overlying non-aggregate deposits (overburden) due to earlier depositional or structural deformation sequences

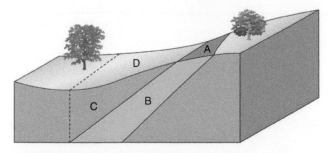

Figure 24.13 Difference between (a) surface expression (outcrop) and (b) possible subsurface extent of geological formation. (c) Shows where overburden is adjacent to the outcrop area and may conceal the full extent of the formation when viewed in 2D plan and (d) shows the area of land needing to be safeguarded for potential future extraction. Source: British Geological Survey.

(Figure 24.13). Economic and technical factors, unique to each resource and location, define the maximum depth of overburden that can be removed to profitably extract aggregate. Sand and gravel deposits often contain layers of finer-grained material that is deemed "waste." Representative volumes within a deposit can be accurately determined by modeling their 3-D distribution.

The original MAR maps were developed using criteria relevant to the existing economic conditions and limitations of extraction technology (Table 24.4). Consultation with the minerals industry revealed current extraction criteria reflecting more stringent economic and environmental considerations; these can be used to re-evaluate the original MAR data. Two different categories for deposits were identified: "Category A" deposits found within 5 m of the surface, and "Category B" deposits found within 10 m of the surface (Table 24.4).

24.3.2 Study Area

A pilot study using 3-D modeling evaluations of the quality and quantity of sand and gravel resources was undertaken in a 40 × 60 km site around Reading (Figure 24.14). This area extends along the Thames Valley from western London urban developments, including Heathrow Airport, to more rural land uses in the west; the Chiltern Hills extend NE–SW through the northwest corner of the study area. This study area has good spatial data coverage with legacy data that includes approximately 630 MAR boreholes and over 3000 grain size distributions (Figure 24.14). Mineral resource planning in the study area is divided between five county councils, three London boroughs, and six unitary authorities (Figure 24.14). Terrace deposits along the River Thames and its tributaries are the major source of sand and gravel, although some glaciofluvial deposits are also present. These deposits are underlain by various bedrock formations, including Cretaceous chalk, clay-rich Palaeogene Lambeth Group and London Clay, and the sandier Bracklesham Group (Mathers and Smith 2000).

24.3.3 Developing the 3-D Model

Selected information was extracted from the BGS Sand and Gravel Database and re-formulated in a project database (Table 24.5). Data were screened to identify and adjust erroneous or missing data, assign aggregate, overburden, or waste categories to each measured interval, and calculate the depth to the center of each borehole interval.

To permit model assessment of Category A and Category B deposits, the total thicknesses of both overburden and aggregate, and the percentage of fine material in the sand and gravel units, were calculated for the top 5 and 10 m of each borehole. The MAR grain size data were converted

Table 24.4 Extraction criteria relevant when the Mineral Assessment Reports (MAR's) were produced (old criteria), compared with current criteria developed in consultation with industry (new criteria).

Original MAR criteria	New criteria	
	Category A deposit	Category B deposit
The deposit should average at least 1 m in thickness.	The deposit should average at least 2 m thickness.	The deposit should average at least 2 m thickness.
Ratio of overburden to sand and gravel should be no more than 3:1.	Ratio of overburden to mineral should not exceed 1:1.	Ratio of overburden to mineral should not exceed 2:1.
The proportion of fines (particles passing 0.063 mm B.S. sieve) should not exceed 40%.	The proportion of fines (particles passing 0.063 mm B.S. sieve) should not exceed 20%.	The proportion of fines (particles passing 0.063 mm B.S. sieve) should not exceed 40%.
The base of the deposit should lie within 25 m of the surface, this being taken as the likely maximum working depth under most circumstances. It follows from the second criterion that boreholes are drilled no deeper than 18 m if no sand and gravel has been proved.	The deposit should lie within 5 m of the surface.	The deposit should lie within 10 m of the surface.

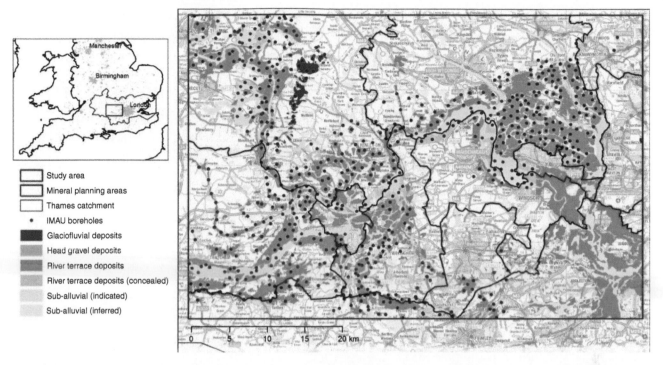

Figure 24.14 Map of study area to the west of Greater London, showing distribution of boreholes used in the model, mineral planning areas and distribution of mapped sand and gravel deposits. (Mathers and Smith 2000). © Crown copyright and database rights [2015], Ordnance Survey [100021290 EUL].

Table 24.5 Contents of the project sand and gravel table extracted from the MAR database.

	Table Headings		
Table attributes	**Location (Index2)**	**Grading Data (GDGS2)**	**Lithology (Lith12)**
	Quarter sheet	Quarter sheet	Quarter sheet
	Borehole number	Borehole number	Borehole number
	Borehole name	Base depth	Base depth
	BNG easting	Thickness	Thickness
	BNG northing	+64 mm (gravel)	Lithology
	BNG accuracy	+32 mm (gravel)	Stratigraphy
	OD level	+16 mm (gravel)	Base bed type
	OD accuracy	+8 mm (gravel)	
	Date drilled	+4 mm (coarse sand)	
		+2 mm (coarse sand)	
		+1 mm (fine sand)	
		+0.5 mm (fine sand)	
		+0.25 mm (fine sand)	
		+0.125 mm (fine sand)	
		+0.063 mm (fine sand)	
		<0.063 mm (fines)	

Shaded records in Grading Data column contain weights of separated fractions of gravel, coarse sand, fine sand, and fines obtained from analysis of a bulk sample for each borehole.

to four categories: Gravel (>4 mm), Coarse Sand (2–4 mm), Fine Sand (0.063–2 mm), and Fines (<0.063 mm). A GIS display of the cleaned and categorized data (Figure 24.15) provided a 2-D spatial visualization of the sand and gravel resources. It does not define the vertical distribution of the resources.

The data were imported into SKUA-GOCAD (Paradigm 2017) and visualized as a series of 3-D points, representing the *X*-*Y*-*Z* location of each grading analysis (Figure 24.16). The vertical grading distribution has economic consequences; two representative boreholes, shown on the right of Figure 24.16, have identical proportions of grading categories, but have very different economic values for resource extraction.

24.3.3.1 Defining the Voxel Grid

A voxel approach (Chapter 11) was used to model the aggregate resources in 3-D. Because the boreholes were on roughly 1 km spacings, and samples were collected at approximately 1 m intervals, the selected model cell size was 1000 × 1000 × 1 m. Geostatistical evaluation of the borehole observations revealed that grading predictions at distances greater than 3 km from a borehole were only marginally more accurate than the mean value. Because the Category A and Category B deposits are restricted to depths of 5 and 10 m respectively, evaluation of grid cells below these depths was unnecessary. Two grids containing only those cells located within 3 km of a borehole were developed (Figure 24.17); one extended to 5 m depth to

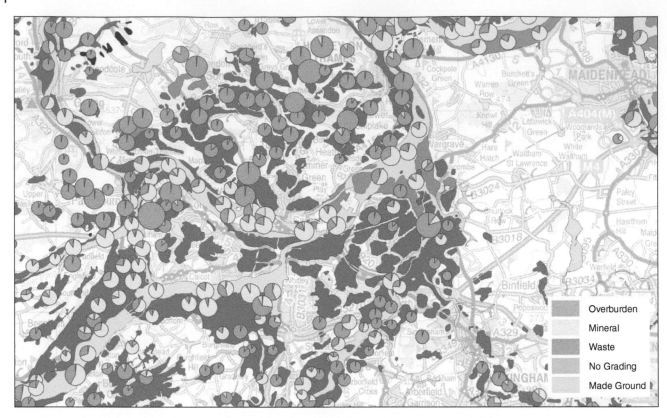

Figure 24.15 GIS assessment of sand and gravel resources (only central portion of the study area is shown). Pie charts show the relative proportions of mineral, waste, and overburden in each borehole but do not reveal their vertical distribution. Circle size is proportional to borehole depth (bigger is deeper). Source: British Geological Survey © NERC. Based on OS topography © Crown Copyright. All rights reserved. BGS 100017897/2011.

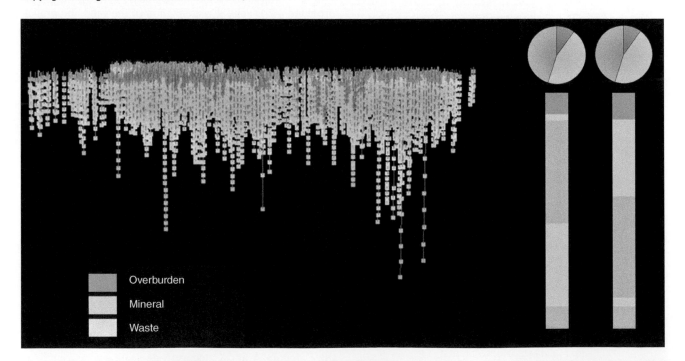

Figure 24.16 Vertical distribution of overburden, aggregate, and waste layers within boreholes in the Reading area, viewed in GOCAD (left). The two borehole logs (right) show how two locations with the same proportions of waste, aggregate, and overburden may have very different economic values for resource extraction. The borehole on the left has aggregate below thick overburden that must be removed. Source: British Geological Survey.

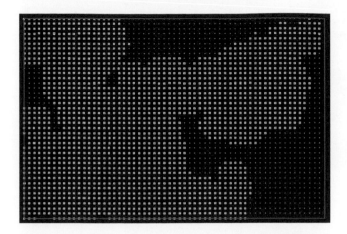

Figure 24.17 Spatial view of final voxel grid containing only cells within 3 km of a borehole. Two grids were developed; one restricted to 5 m depth and one restricted to 10 m depth.

analyze Category A deposits, and one extended to 10 m to analyze Category B deposits.

24.3.3.2 Geostatistical Computations

SKUA-GOCAD (Mallet 1992; Paradigm 2017; Chapter 12) was used to calculate the coordinates of the 3-D voxel grid and to visualize model results. The available SKUA-GOCAD geostatistical capabilities were unable to meet the project requirements, so all geostatistical modeling was coded and conducted as MATLAB scripts. These were used to interpolate the borehole mineral gradings across the 3-D grid. Variograms resulting from the geostatistical analysis were analyzed stochastically (Chapter 13) to produce multiple equally probable realizations of the spatially correlated properties at each grid node. Differences between the realizations reflect the uncertainty of the predictions at each location. The lower-upper approach to geostatistical simulation (Deutsch and Journel 1998) was used because it can quickly produce many realizations.

The aggregate resources were evaluated using two geostatistical models. The first model evaluated the depths of overburden and aggregate by assessing the observed sequence in each grid column. These evaluations defined where current MAR criteria were met for depth to aggregate deposit and overburden ratio (Table 24.4). The observed depths to the base of overburden and aggregate deposit were log-transformed prior to spatial geostatistical modeling. The material was assumed to be composed entirely of waste at locations without grading observations. The model was then used to produce 1000 simulations of the two properties at each grid location. The proportion of simulations satisfying the current MAR overburden-aggregate ratio criteria (Table 24.4) was recorded at each grid cell.

The second model evaluated the MAR grading data; it evaluated the proportions of the four grading categories present at each 3-D grid cell that met the current MAR criteria for the proportion of fines (Table 24.4). Because the

grading components must sum to 100%, standard geostatistical models might produce false correlations. The additive log-ratio transform was used to resolve this problem (Lark et al. 2015); this converted the four gradings classes into three grading ratios. Anisotropy due to differences in vertical and horizontal auto-correlations resulted in a rather complex model with 12 parameters; evaluation was by the moments estimator method (Webster and Oliver 2007). The resulting model was used to produce 1000 realizations of the transformed properties at each 3-D grid location; the simulated values were back-transformed to the original gradings. Each realization was paired with a realization of the overburden/aggregate depth model; if MAR criteria were not satisfied at a location, the material was assumed to be waste.

Geostatistical analysis of the aggregate grading information revealed a strong spatial correlation that defines coherent regional patterns in the predicted percentages of all gradings (Figure 24.19); the deposits are predominantly gravel and fine sand with little coarse sand.

24.3.4 Results

Overburden and aggregate deposit depths exhibit large variations between closely spaced boreholes and no spatial correlation between the two depths. Consequently, while regional maps revealed distinct areas of thicker overburden and shallower or deeper aggregate deposits, predicted

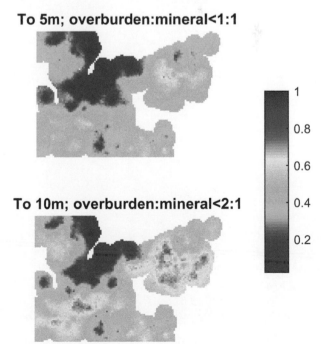

Figure 24.18 Probability of satisfying MAR criteria for Category A deposits (maximum depth of 5 m) and Category B deposits (maximum depth of 10 m).

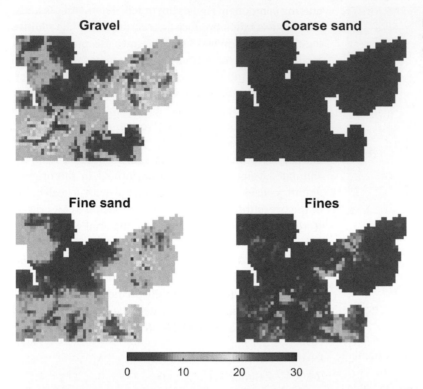

Gravel **Coarse sand**

Fine sand **Fines**

0 10 20 30

Figure 24.19 Maps of estimated percentage of gravel, coarse sands, fine sands, and fines to a depth of 5 m.

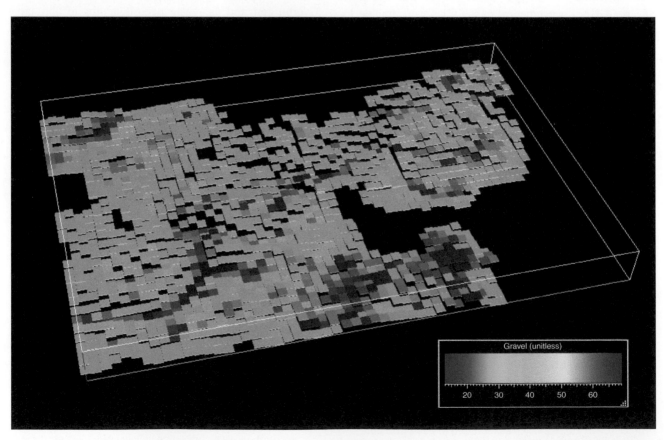

Figure 24.20 Three-dimensional display showing estimated percentage of gravel throughout the study area (vertical exaggeration is 10×).

Table 24.6 Expected volumes of four grading categories across the modeled region.

Grade	Average probability (% as decimal)	Volume in m³ (probability × grid volume)	Adjustment for presence of waste (× 0.02528)	Converted to tonnes (× 1.64)
Gravel	0.4098	3 975 299 604	1 004 955 740	1 648 127 413
Coarse sand	0.0485	470 895 736	119 042 442	195 229 605
Fine sand	0.3607	3 499 286 859	884 619 718	1 450 776 338
Fines	0.1809	1 754 993 250	443 662 294	727 606 161

values at any specific location are rather uncertain. Nevertheless, two maps were produced (Figure 24.18) that show the probability of satisfying the MAR criteria for Category A and Category B deposits across the study region.

Figure 24.20 shows the predicted 3-D distribution of the gravel grading category. Similar displays were produced for the other grading categories. Expected volumes of each mineral grading category were calculated by multiplying the average probability of the mineral grade by the volumes defined in the 3-D grid, adjusted to account for waste within the model. Table 24.6 gives the average probability of each mineral grading across the modeled region and the expected volumes in both cubic meters and tonnes.

24.3.5 Conclusions

The original MAR interpretations for the Reading area (Squirrell 1978) defined large blocks with average properties based on 2-D maps and borehole samples. This information had value to planners and decision makers, but lacked accurate volume estimations, or assessments of overburden and waste intercalations. The 3-D modeling approach addresses these limitations, plus it provides better information on the true extent and suitability of these aggregate deposits.

This new project simulated the quality of sand and gravel resources by using stochastic techniques and legacy data. Geostatistical models, when combined with stochastic simulation methods, permit quantification of uncertainty in evaluating aggregate resources. Uncertainty estimates are important for guiding land-use planning decisions. Two models produced broad-scale probability maps of aggregate viability and volumes of four aggregate grading categories. The geostatistical model assumed a single variogram throughout the study region; it ignored spatial variations among different lithologies. Lithological constraints will be considered in future work.

References

British Geological Survey (2019a). *Mineral Assessment Reports*. Keyworth, UK: British Geological Survey. [Website: https://www2.bgs.ac.uk/mineralsuk/mines/IMAU.html] (Accessed December 2020).

British Geological Survey (2019b). *Mineral Resource Maps for the UK by Planning Area*. Keyworth, UK: British Geological Survey. [Website: https://www2.bgs.ac.uk/mineralsuk/planning/resource.html] (Accessed December 2020).

Building Research Establishment (1993). *Low-Rise Buildings on Shrinkable Clay Soils: Part 1, Digest 240*. Watford, UK: Building Research Establishment. 4 pp.

Burnett, A.D. and Fookes, P.G. (1974). A regional engineering geological study of the London Clay in the London and Hampshire basins. *Quarterly Journal of Engineering Geology* 7: 257–296. https://doi.org/10.1144/GSL.QJEG.1974.007.03.02.

Crilly, M. (2001). Analysis of a database of subsidence damage. *Structural Survey* 19 (1): 7–15. https://doi.org/10.1108/02630800110384185.

Cripps, J.C. and Taylor, R.K. (1986). Engineering characteristics of British over-consolidated clays and mudrocks: I tertiary deposits. *Engineering Geology* 22: 349–376. https://doi.org/10.1016/0013-7952(86)90004-9.

Deutsch, C.V. and Journel, A.G. (1998). *GSLIB: Geostatistical Software Library and User's Guide*, 2e. New York, NY: Oxford University Press. 369 pp. [Online: Available at: http://claytonvdeutsch.com/wp-content/uploads/2019/03/GSLIB-Book-Second-Edition.pdf] (Accessed December 2020).

Driscoll, R. (1983). The influence of vegetation on the swelling and shrinking of clay soils in Britain. *Géotechnique* 38: 93–105. https://doi.org/10.1680/geot.1983.33.2.93.

Driscoll, R. and Crilly, M. (2000). *Subsidence Damage to Domestic Buildings. Lessons Learned and Questions Asked.* London, UK: Building Research Establishment. 32 pp.

Holtz, R.D. and Kovacs, W.D. (1981). *An Introduction to Geotechnical Engineering.* Englewood Cliffs, NJ: Prentice-Hall.

Ichikawa, T. and Miura, M. (2007). Modified model of alkali-silica reaction. *Cement and Concrete Research* 37: 1291–1297. https://doi.org/10.1016/j.cemconres.2007.06.008.

Intermap Technologies (2007). *NEXTMap British Digital Terrain Model Dataset Produced by Intermap.* Chilton, UK: NERC Earth Observation Data Centre. [Online: Available at: https://catalogue.ceda.ac.uk/uuid/8f6e1598372c058f07b0aeac2442366d] (Accessed December 2020).

Jones, L.D. (1999). *A Shrink/Swell Classification for UK Clay Soils.* BEng thesis Nottingham Trent University.

Jones. L. & A. Hulbert (2016). *User Guide for the 3D Shrink-Swell (GeoSure Extra) Dataset.* Keyworth, UK: British Geological Survey, Open Report OR/16/043. 14 pp. [Online: Available at: http://nora.nerc.ac.uk/id/eprint/524609] (Accessed December 2020).

Jones, L.D. and Jefferson, I. (2012). Expansive soils. In: *ICE Manual of Geotechnical Engineering. Volume 1 Geotechnical Engineering Principles, Problematic Soils and Site Investigation* (eds. J. Burland, T. Chapman, H. Skinner and M. Brown), 413–441. London: ICE Publishing.

Jones, L.D. and Terrington, R. (2011). Modelling volume change potential in the London Clay. *Quarterly Journal of Engineering Geology and Hydrogeology* 44 (1): 109–122. https://doi.org/10.1144/1470-9236/08-112.

Kearsey, T., Williams, J., Finlayson, A. et al. (2015). Testing the application and limitation of stochastic simulations to predict the lithology of glacial and fluvial deposits in Central Glasgow, UK. *Engineering Geology* 187: 98–112. https://doi.org/10.1016/j.enggeo.2014.12.017.

Lark, R.M., Marchant, B.P., Dove, D. et al. (2015). Combining observations with swath bathymetry and backscatter to map seabed sediment texture classes; the empirical best linear unbiased predictor. *Sedimentary Geology* 328: 17–32. https://doi.org/10.1016/j.sedgeo.2015.07.012.

Lerner, D.N. and Barrett, M.H. (1996). Urban groundwater issues in the United Kingdom. *Hydrogeology Journal* 4: 80–89. https://doi.org/10.1007/s100400050096.

Maljers, D., Stafleu, J., Van der Meulen, M.J. and Dambrink, R.M. (2015). Advances in constructing regional geological voxel models, illustrated by their application in aggregate resource assessments. *Netherlands Journal of Geosciences* 94: 257–270. https://doi.org/10.1017/njg.2014.46.

Mallet, J.L. (1992). GOCAD: a computer aided design program for geological applications. In: *Three-Dimensional Modelling with Geoscientific Information Systems*, NATO ASI Series, vol. 354 (ed. A.K. Turner), 123–141. Dordrecht, Netherlands: Springer. https://doi.org/10.1007/978-94-011-2556-7_11.

Mathers, S.J. and Smith, N.J.P. (2000). *Geology of the Reading district: a brief explanation of the geological map Sheet 268 Reading.* Keyworth, UK: British Geological Survey. 30 pp. [Online: Available at: http://pubs.bgs.ac.uk/publications.html?pubID=B06128] (Accessed December 2020).

Mathers, S.J., Burke, H.F., Terrington, R.L. et al. (2014). A geological model of London and the Thames Valley, Southeast England. *Proceedings of the Geologists, Association* 125: 373–382. https://doi.org/10.1016/j.pgeola.2014.09.001.

Oloo, S., Schreiner, H.D. and Burland, J.B. (1987). Identification and classification of expansive soils. In: *Proceedings of the 6th International Conference on Expansive Soils, New Delhi, India* (eds. Central Board of Irrigation and Power), 23–29. London, UK: Taylor & Francis.

Orlić, B. and Rösingh, J.W. (1995). Three-dimensional geomodelling for offshore aggregate resources assessment. *Quarterly Journal of Engineering Geology* 28: 385–391. https://doi.org/10.1144/GSL.QJEGH.1995.028.P4.08.

Paradigm Ltd. (2017). *SKUA-GOCAD: Towards High Definition Earth Modeling.* 8p. [Online: Available at: http://www.pdgm.com/resource-library/brochures/skua-gocad/skua-gocad] (Accessed December 2017).

Reeve, M.J., Hall, D.G.M. and Bullock, P. (1980). The effect of soil composition and environmental factors on the shrinkage of some clayey British soils. *Journal of Soil Science* 31: 429–442. https://doi.org/10.1111/j.1365-2389.1980.tb02092.x.

Reeves, G.M., Sims, I. and Cripps, J.C. [Eds.] (2006). *Clay Materials Used in Construction.* London, UK: Geological Society, Engineering Geology Special Publication 21. 552 pp.

Self, S., Entwisle, D. C. and Northmore, K. J. (2012). *The structure and operation of the BGS National Geotechnical Properties Database, Version 2.* Keyworth, UK: British Geological Survey, Report IR/12/056. 68 pp. [Online: Available at: http://nora.nerc.ac.uk/id/eprint/20815] (Accessed December 2020).

Simpson, B., Blower, T., Craig, R.N. and Wilkinson, W.B. (1989). *The Engineering Implications of Rising Groundwater Levels in the Deep Aquifer Beneath London.* London, UK: CIRIA, Special Publication 69. 116 pp.

Smith, M.R. and Collis, L. (2001). *Aggregates: Sand, Gravel and Crushed Rock Aggregates for Construction Purposes.* London, UK: Geological Society, Engineering Geology Special Publication 17. 360 pp. https://doi.org/10.1144/GSL.ENG.2001.017.

Squirrell, H.C. (1978). *The Sand and Gravel Resources of the Country around Sonning and Henley, Berkshire, Oxfordshire and Buckinghamshire*. Keyworth, UK: British Geological Survey, Mineral Assessment Report No. 32. 100 pp.

Stafleu, J., Maljers, D., Gunnink, J.L. et al. (2011). 3D modelling of the shallow subsurface of Zeeland, the Netherlands. *Netherlands Journal of Geosciences* 90: 293–310. https://doi.org/10.1017/S0016774600000597.

Stafleu, J., Maljers, D., Busschers, F.S., Gunnink, J.L. et al. (2012). *GeoTOP modellering* (in Dutch). Utrecht, Netherlands: TNO, Geological Survey of the Netherlands, Report TNO 2012 R10991. 216 pp.

Van der Meulen, M.J. (2005). Sustainable mineral development: possibilities and pitfalls illustrated by the rise and fall of Dutch mineral planning guidance. *Geological Society Special Publications* 250: 225–232. https://doi.org/10.1144/GSL.SP.2005.250.01.20.

Van der Meulen, M.J., Koopmans, T.P.F. and Pietersen, H.S. (2003). Construction raw materials policy and supply practices in Northwestern Europe. *Aardkundige Mededelingen* 13: 19–30.

Van der Meulen, M.J., Van Gessel, S.F. and Veldkamp, J.G. (2005). Aggregate resources in the Netherlands. *Netherlands Journal of Geosciences* 84: 379–387. https://doi.org/10.1017/S0016774600021193.

Van der Meulen, M.J., Broers, J.W., Hakstege, A.L. et al. (2007a). Surface mineral resources. In: *Geology of the Netherlands* (eds. T.E. Wong, D.A.J. Batjes and J. De Jager), 317–333. Amsterdam, Netherlands: KNAW, Royal Netherlands Academy of Arts and Sciences.

Van der Meulen, M.J., Maljers, D., Van Gessel, S.F. and Gruijters, S.H.L.L. (2007b). Clay resources in the Netherlands. *Netherlands Journal of Geosciences* 86: 117–130. https://doi.org/10.1017/S001677460002312X.

Van Lancker, V.R.M., Francken, F., Kint, L. et al. (2017). Building a 4D voxel-based decision support system for a sustainable management of marine geological resources. In: *Oceanographic and Marine Cross-Domain Data Management for Sustainable Development* (eds. P. Diviacco, A. Leadbetter and H. Glaves), 224–252. Hershey, PA: IGI Gloabal. https://doi.org/10.4018/978-1-5225-0700-0.ch010.

Walsby, J.C. (2008). GeoSure: a bridge between geology and decision-makers. *Geological Society Special Publications* 305: 81–87. https://doi.org/10.1144/SP305.9.

Webster, R. and Oliver, M.A. (2007). *Geostatistics for Environmental Scientists*, 2e. Chichester, UK: Wiley. 330 pp.

Wentworth, C.K. (1922). A scale of grade and class terms for clastic sediments. *The Journal of Geology* 30: 377–392.

Woods, M.A., Newell, A.J., Haslam, R. et al. (2015). *A Physical Property Model of the Chalk of Southern England*. Keyworth, UK: British Geological Survey, Open Report OR/15/013. 44 pp. [Online: Available at: http://nora.nerc.ac.uk/id/eprint/510117] (Accessed December 2020).

Wrighton, C.E., McEvoy, E.M. and Bust, R. (2011). *Mineral Safeguarding in England: Good Practice Advice*. Keyworth, UK: British Geological Survey, Open Report OR/11/046. 46 pp. [Online: Available at: http://nora.nerc.ac.uk/id/eprint/17446] (Accessed December 2020).

25

Application Theme 8 – Historical Preservation and Anthropogenic Deposits

Editor's Introduction

The use of 3-D models to support historical preservation and archeological investigations is a relatively recent development, one that has increasing relevance in 3-D modeling of urban areas where anthropogenic processes have significantly modified the urban subsurface. The term "Artificially Modified Ground", or AMG, is used to collectively define urban landforms, anthropogenic deposits, and excavations.

Several geological surveys have established schemas for mapping and modeling AMG based on classifications defined by landform and origin. However, they generally do not include lithological descriptions. Thus, while geological surveys provide AMG descriptions suitable for regional assessments, more detailed morphostratigraphic information derived from archeological investigations is required to characterize individual sites.

Archeologists characterize anthropogenic deposits by defining paleo-topography and paleo-geomorphological landforms, which allows them to identify likely locations of isolated archeological remains and provides a reference chronology to constrain erosional and depositional histories. Because roads, railways, or pipelines invariably cross a variety of geological units and archeological preservation environments, 3-D models have proved useful during these archeological investigations.

This chapter contains three case studies, the first two of which discuss recent examples of the application of 3-D models to historical preservation and to the assessment of the vulnerability and characteristics of anthropogenic deposits in urban areas. The third case study provides a more detailed discussion of the challenges associated with 3-D mapping and modeling of AMG.

The first case study describes the role of 3-D models in an archeological evaluation of Bryggen, the historic Hanseatic League wharf in central Bergen, Norway, where the shallow subsurface has a rich archeological record that has been preserved due to high water tables. The 1979 construction of a new hotel with underground parking in Bryggen included a drainage system enclosed by a sheet-pile wall. This lowered groundwater levels considerably in the surrounding area; oxidation of the subsurface deposits led to decomposition of organic archeological material and produced uneven ground subsidence rates of up to $8\,\mathrm{mm\,yr^{-1}}$. A 3-D model linked to a numerical groundwater model evaluated the spatial and temporal hydrogeological variations and their potential impact on archeological preservation. The model informed multiple stakeholders of these risks and helped define mitigation measures.

The second case study describes the development of a 3-D digital geological model which combines information from borehole logs with archeological evidence for the central area of the city of Newcastle upon Tyne. This allowed for reconstruction of features of the near-surface geology that control shallow groundwater movement. Portions of central Newcastle are impacted by "groundwater flooding" events; the model continues to be used to evaluate the interactions between rainfall, surface runoff, infiltration, and subsurface flows.

The third case study describes the recent approaches to classifying AMG developed by the British Geological Survey (BGS) and discusses how these have been applied to a 3-D evaluation of the veneer of artificial (anthropogenic) deposits beneath the City of London and Tower Hamlet Borough areas of central London.

Applied Multidimensional Geological Modeling: Informing Sustainable Human Interactions with the Shallow Subsurface, First Edition.
Edited by Alan Keith Turner, Holger Kessler, and Michiel J. van der Meulen.

Case Study 25.1: Evaluating Geological and Anthropogenic Deposits at the Bryggen World Heritage Site, Bergen, Norway

Johannes de Beer[1] and Jonathan R. Ford[2]

[1] *Geological Survey of Norway (NGU), 7040 Trondheim, Norway*
[2] *British Geological Survey, Nottingham NG12 5GG, UK*

25.1.1 Introduction

Bergen, Norway's second largest city, is located at the west coast of Norway; it is bounded by a fjord to the west-northwest and by mountains on the east-southeast (Figure 25.1). Bergen was founded in ad 1070 . Its natural, well-sheltered and ice-free harbor proved to be an ideal location for commanding trade along the coast. From the end of the thirteenth century Bergen was one of Northern Europe's most important ports and one of the Hanseatic League's most important towns. The settlement gradually expanded up to 140 m seawards by filling timber "boxes"

at the quay front with various materials such as wood and household rubbish. Throughout history, Bergen has experienced many disastrous fires. The current timber buildings at "Bryggen," the historic Hanseatic League wharf, were constructed after a major fire in 1702 on older sites and foundation walls. The shallow subsurface has a rich archeological record, which has been preserved due to high water tables (Christensson et al. 2004). Owing to its outstanding testimony to past traditions, Bryggen was included on UNESCO's World Heritage List in 1979.

25.1.2 Geological Setting

Bergen is located between the fjord and surrounding mountains. The relatively small city center is located on a flat valley bottom situated a few meters above sea level, with the developed areas extending up the steep hillsides. Owing to the steep topography and ample precipitation ($\sim$2250 mm yr^{-1}), there is plenty of surface water. The bedrock, composed of granitic gneisses, greenstone, phyllite, and quartzite, is exposed on the hillsides where

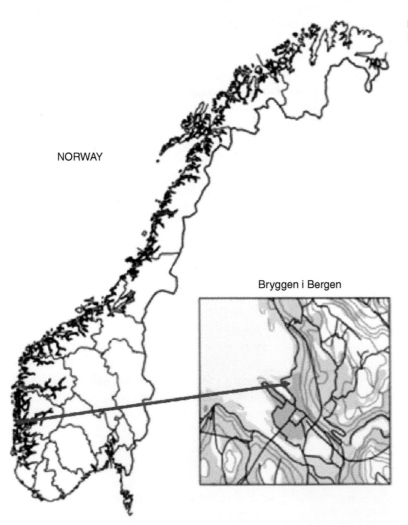

Figure 25.1 Location map of Bergen and the Bryggen area.

NORWAY

Bryggen i Bergen

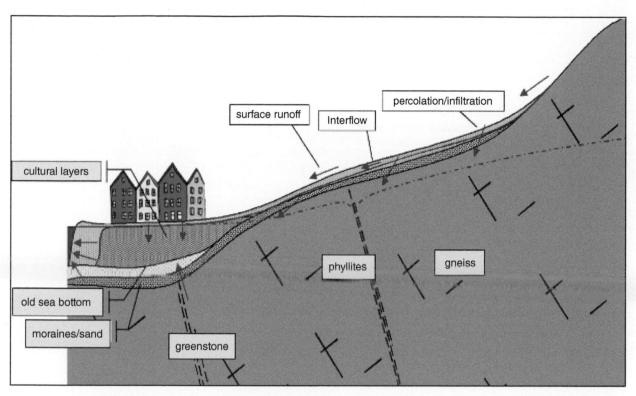

Figure 25.2 Conceptual hydrogeological model of Bryggen and the adjacent mountain slope (De Beer 2008). Section length about 150 m.

the cover is very thin. The area of the city center was below sea level during and after the last glacial maximum about 9600 years ago. Detailed ground investigations have revealed a thicker cover, consisting of glacial till, marine sand, and modern fill. The near-surface is composed of modern fill. Archeologically significant cultural deposits form a highly organic layer up to 8 m thick between the sand and the modern fill (Figure 25.2).

A system of open fractures and weak, permeable zones in the bedrock deliver groundwater to the surficial deposits under Bryggen. Local precipitation only partially infiltrates into the subsurface due to the many impervious surfaces in the urban center. Complex interactions among several factors influence the local groundwater regime, the most important of which are tidal fluctuations, urban drainage systems, natural and urban stratigraphy, and bedrock hydraulic features (Figure 25.3).

25.1.3 Bryggen's Archeological Heritage

Bryggen is one of the oldest trading ports in northern Europe; it was an important part of the Hanseatic League. The historic buildings date from shortly after the major fire of 1702 but are built upon earlier structures dating back to the eleventh century. In 1955, a large fire in the western

part of Bryggen destroyed about a third of the historic buildings. Extensive archeological excavations at this location between 1955 and 1968 (Herteig 1985) revealed cultural deposits of high archeological value sometimes exceeding 8 m in thickness, with 10 or more separate building phases on top of each other (Figure 25.4). In 1979, Bryggen was included in UNESCO's World Heritage list. All of Bryggen, from the underlying bedrock to the rooftops, including underground archeological remains plus 61 buildings, is considered a single cultural monument. Christensson et al. (2004) provide more details on the history of Bryggen.

Over the centuries, Bryggen expanded seaward up to 140 m beyond the original shoreline (Figure 25.5). Tilting of several of the famous wooden houses in Bryggen was observed for many years, but the extent of the problem has only been recognized more recently. The cultural layer deposits are highly organic, with loss on ignition values varying from 10% to 70%. Since they are below the natural water table, the organic material is saturated and protected from oxidation and decomposition (Holden et al. 2006, 2009; Marstein et al. 2007). Drainage of groundwater causes both physical compaction as well as decomposition of organic materials (including timber foundations). This results in subsidence of the deposits and the houses above (Matthiesen et al. 2006; Matthiesen 2007; de Beer 2008; de Beer and Matthiesen 2008).

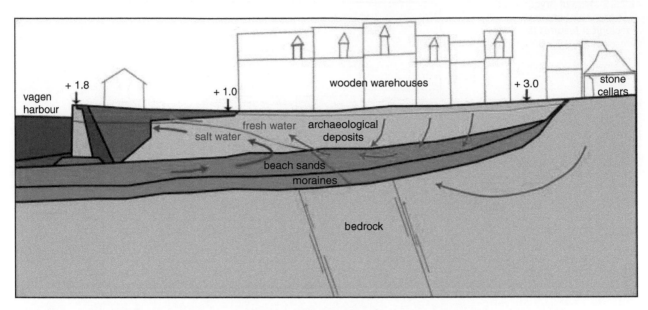

Figure 25.3 Cross-section from the harbor (left) to the rear of Bryggen (right), giving a simple picture of the subsurface layering sequence and the local water flows (De Beer and Matthiesen 2008). Section length about 150 m.

Figure 25.4 Thickness of the cultural deposits as calculated from a 3D subsurface model (De Beer et al. 2012).

Figure 25.5 Map of Bryggen and surroundings with historical hydrogeological features (De Beer et al. 2012). Area about 0.7 × 0.7 km.

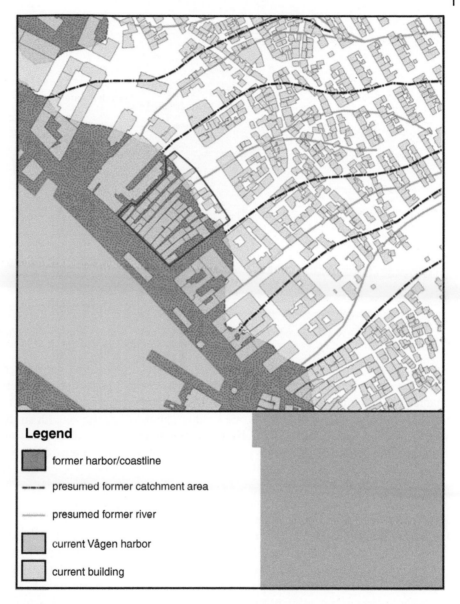

Legend

- former harbor/coastline
- presumed former catchment area
- presumed former river
- current Vågen harbor
- current building

In 1979, a hotel with underground parking was constructed west of the remaining part of Bryggen. To avoid damage to the building due to high groundwater pressures, and to protect an underground hotel parking structure, a drainage system enclosed by a sheet pile wall was constructed. This lowered groundwater levels considerably in the surrounding area, particularly at the rear of Bryggen. The oxidation of the subsurface deposits has led to decomposition of organic archeological material, which resulted in uneven subsidence rates of up to 8 mm yr^{-1}.

25.1.4 The Geological Model

The complexity of the Bryggen archeological site, as well as the large amounts of monitoring and modeling data from multidisciplinary sources that were being gathered, called for an easy-to-use 3-D visualization and subsurface framework modeling system. An interrogable 3-D subsurface model, containing multi-disciplinary archeological, geochemical, and hydrological data, was thought to provide a holistic site evaluation tool. Model outputs would enable cultural heritage authorities to better evaluate and communicate risks to the preservation of archeological records at Bryggen.

A 3-D geological framework model of the Bryggen area was constructed using the Geological Surveying and Investigation in 3 Dimensions (GSI3D) software (Mathers et al. 2011) and the modeling experience of the BGS (De Beer et al. 2012; Seither et al. 2016). The BGS has developed a classification to improve the characterization of "artificial ground" based on the genetic origin and morphology of artificial ground features (McMillan and Powell 1999;

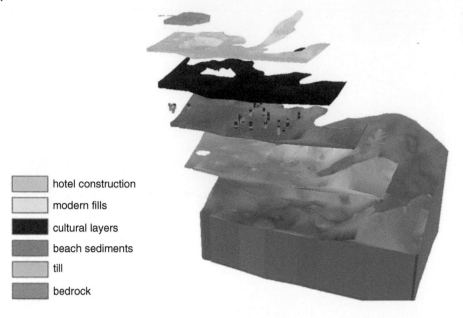

Figure 25.6 Exploded view of 3-D subsurface model covering an area of about 250 × 250 m. (De Beer et al. 2012).

hotel construction

modern fills

cultural layers

beach sediments

till

bedrock

Ford et al. 2010). As discussed in Case Study 25.3, this classification system does not define lithological or other characteristics. However, the cultural deposits and their natural or man-made surroundings are readily defined when attributes are added to the modeled stratigraphic units. Archeological drilling records provide more detailed information; this permits the subdivision of the cultural layers based on age, structure, or other characteristics. Provided that this subdivision is compatible with the stratigraphically defined geological units, cultural deposits can be integrated in the 3-D subsurface model, as shown in the "exploded model view" in Figure 25.6.

Archeological documentation typically provides information on the preservation state for the individual cultural layers at a much higher vertical resolution, typically around 0.20 m. The state of preservation generally reflects local conditions within the subsurface. These may vary with location as well as depth, so 3-D extrapolation of this descriptive attribute within any cultural layer is difficult. However, visualization of individual borehole "sticks" (profiles) in all 3-D outputs and in cross-sections provides a good method of evaluating variations in documented preservation state (Figure 25.7).

A wide range of data, such as lithology, strength or density descriptions, water level data, geotechnical, geophysical, geochemical, and archeological parameters, can be shown this way. The GSI3D software allows for additional attributes to be presented in sections and borehole logs. Bryggen ground and groundwater samples were analyzed for a range of chemical parameters that indicate preservation conditions (e.g. chloride, sulphate, oxygen, loss of ignition (LOI)). Figure 25.8 shows a single Bryggen borehole with stratigraphic subdivision, saturation conditions, state of preservation category, and chloride content. Interpreted and classified preservation conditions, describing the varying environmental conditions at the site, can also be visualized in combination with the documented state of preservation, indicating potential for archeological preservation.

Surfaces defined by raster or triangulated irregular network (TIN) files can be merged in the subsurface framework model. Figure 25.9 shows a representative annual lowest groundwater table at Bryggen. Use of contour lines and transparency allows for a clear visualization of flow direction and comparison with subsurface deposits. Cultural layers above the groundwater table are at risk of decay. By "slicing" the full 3-D subsurface framework model at the groundwater level, volumes of cultural deposits above (at risk) and below (not at risk) the groundwater level can be calculated.

The geological framework model, together with a complementary numerical groundwater model, has been used to improve volumetric and dynamic understanding of the site and its surroundings and to communicate this to relevant stakeholders. The multidisciplinary approach has led to development of sustainable mitigation solutions for long-term preservation and management of the World Heritage Site (Rytter and Schonhowd 2015).

Figure 25.7 Section view of 3-D model with borehole-sticks showing documented state of preservation (De Beer et al. 2012). Model area about 250 × 250 m.

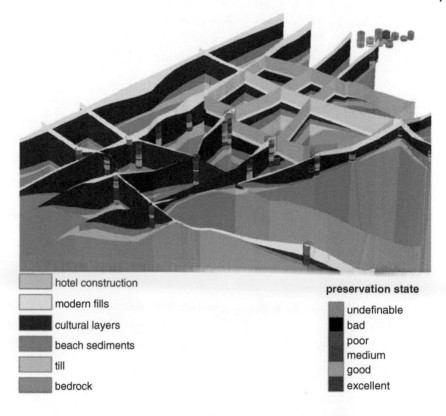

hotel construction

modern fills

cultural layers

beach sediments

till

bedrock

preservation state

undefinable

bad

poor

medium

good

excellent

25.1.5 Conclusions

Geological framework models can provide a 3-D regional environmental context for the spatial and temporal relationships between natural and anthropogenic deposits. Integration of anthropogenic deposits in geological models at different scales can provide a holistic system for the management of the subsurface and protection of archeological materials.

The Bryggen project illustrates the emerging new possibilities of combining complex model and monitoring data from different disciplines and visualizing them in three dimensions. At Bryggen, 3-D subsurface framework modeling provides insights into geological, as well as archeological and other anthropogenic deposits. It helps to explain to stakeholders, such as urban planners, heritage authorities, and the general public, how changes to natural and anthropogenic environments affect the vulnerability of the archeological deposits. The attributed 3-D model helped define a numerical groundwater model created to evaluate spatial and temporal hydrogeological variations and their potential impact on archeological preservation. Further work is needed to model the heterogeneity of subsurface properties, but an indication can be provided by visualizing selected "borehole sticks" together with the model units.

Case Study 25.2: Characterizing the Near-Surface Geology of Newcastle upon Tyne

Geoff Parkin and Elizabeth D. Hannon
School of Engineering, Newcastle University, Newcastle upon Tyne NE1 7RU, UK

25.2.1 Introduction

Newcastle upon Tyne, located in northeastern England on the River Tyne (Figure 25.10), has been an important regional city for centuries. The site has been occupied since the prehistoric period (Graves and Heslop 2013); it evolved from a military stronghold to an economic urban industrial center in the nineteenth and twentieth centuries when it was well known for its coal mining, shipbuilding, and heavy engineering industries. More recently it has become a technological center. This historical development has resulted in the uneven accumulation, throughout the city, of many centuries worth of both domestic and industrial waste, forming a highly varied sequence of anthropogenic deposits in the near surface. These deposits form a layer of artificial (or "made") ground of varying thickness, ranging from centimeters to tens of meters, composed predominantly of a mix of mine waste, ash from fires, rubble, and ship's ballast. At many locations, the initial fill was

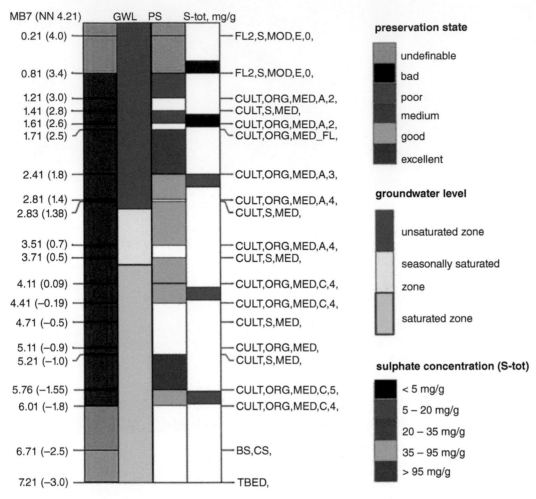

Figure 25.8 Visualization of a single borehole with categorized parameters for saturation condition, state of preservation category, and chloride content (De Beer et al. 2012).

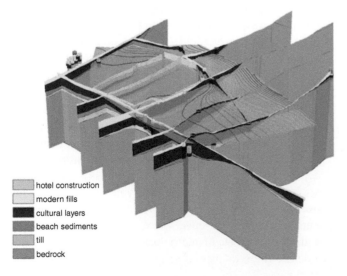

Figure 25.9 Three-dimensional section view with a representation of the annual low groundwater table (De Beer et al. 2012). Model area about 250 × 250 m.

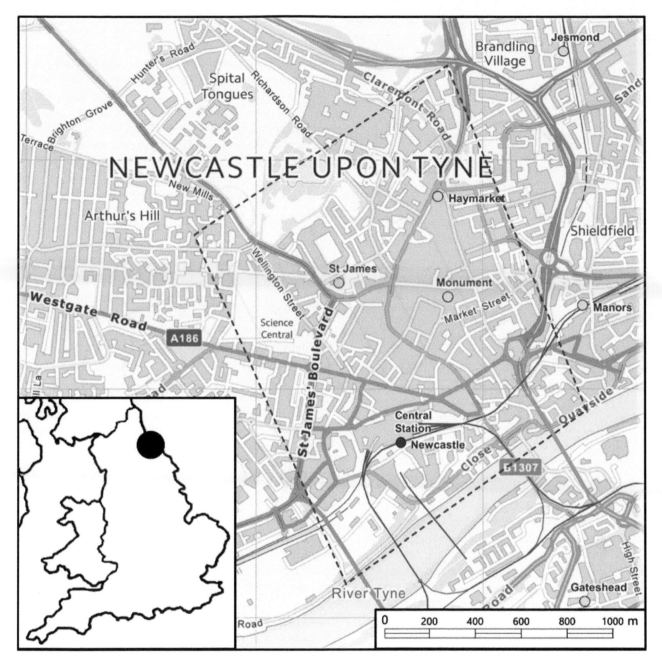

Figure 25.10 Location of Newcastle upon Tyne and extent of geological model study area.

compacted to create a level area for redevelopment; in others a capping layer of local clay soil was used to finish the site. As the city expanded, several smaller tributary streams (termed "burns" in the regional terminology) were buried in tunnels or culverts, and their valleys (termed "denes" locally) were infilled with thicker sequences of artificial ground.

Thus, the modern central area of Newcastle is built on complex anthropogenic near-surface geology. The presence of such highly heterogeneous materials in the near-surface, combined with the modification and burial

of pre-existing drainage systems, greatly influences the local groundwater hydrology. In areas of thicker artificial ground, groundwater can travel significant distances and pool in unusual places depending on the grain size and constituents of the materials. As available borehole logs often summarize these materials under generic descriptions such as "made ground" or "fill," additional detail on the presence and extent of deposits can only be inferred and evaluated by examining historical maps to define the sequence of activities conducted at specific locations.

Portions of central Newcastle are impacted by "groundwater flooding" events. Groundwater flooding is the emergence of groundwater at the ground surface away from perennial river channels and can include the rising of groundwater into man-made ground, basements, and other subsurface infrastructure. While many of these events have been documented, their scale and frequency remain uncertain as this is a more hidden problem than other sources of flooding. Substantial impacts can occur below ground without the groundwater breaching the surface, due to water inundation into underground structures and drainage pipe networks. The situation in Newcastle is further complicated by the presence of multilayered mine workings ranging from early industrial developments of open pits and adits to more recent deep mine workings. These form both potential sources of rising groundwater (groundwater levels have risen by up to 70 m over the last decade as pumping of abandoned mine workings has been reduced) and pathways to the surface for discharges. The complexity of groundwater flooding makes remediation or mitigation difficult.

To inform understanding of the urban groundwater flooding hazard in central Newcastle, a 3-D geological model was developed, based on assessment of borehole logs and a review of historical records and maps (Hannon 2015).

25.2.2 Geology of the Study Area

The bedrock geology of the region is dominated by Carboniferous Coal Measures overlying Ordovician mudstones, siltstones, and sandstones. The coal measures are split into three distinctive sections, Upper, Middle, and Lower, but only the Pennine Middle and Lower Coal Measures are found within the study area. These have been exploited over a period of several hundred years, with each colliery having its own names for the seams, resulting in complicated local nomenclature.

During the Quaternary, the North of England was inundated with numerous glacial stadials, with ice sheets flowing southwards from an ice cap over the Cheviot Hills in Northumberland, resulting in thick glacial deposits left as the final Weichselian stadial retreated. Thicknesses are typically up to 10 m but can be up to 90 m in buried valleys along the main rivers. Glacial till, mostly heavy compacted clay with a heterogeneous mix of sand, gravel, and boulders, forms the bulk of these glacial deposits, but there exist areas of glaciolacustrine and fluvioglacial deposits from ice-dammed lakes and meltwater channels. More recent deposits are found overlying the glacial units; these include river alluvium, peat, and, in urban locations, extensive areas of made ground.

25.2.3 The 3-D Geological Model

The 3-D geological model was created using Subsurface Viewer MX software developed by INSIGHT Geologische Softwaresysteme GmbH. This software was used following the procedures described in Chapter 10 to produce 3-D geological models (Kessler et al. 2009; Mathers et al. 2012). Nine geological units were initially defined, with the most recent unit (Made Ground) subsequently sub-divided into 13 classes, resulting in a total of 21 units (Table 25.1). A series of intersecting interpreted cross-sections were produced by combining information from topographic maps, a digital terrain model, published geological maps and cross-sections, and exploratory boreholes (Figure 25.11).

A paleosurface representing the pre-industrial landforms was defined in the 3-D model by removing all Made Ground layers. This revealed direct evidence of the buried valleys (often known as "hidden valleys") that existed prior to urban and industrial development. Figure 25.12 shows this paleo-surface and the underlying Quaternary geology, with inferred lines of streams ("burns") shown. The inset shows that the general stream pattern in a historical ground surface map is similar but differs in some details, which may be due to the information used as a basis for the historical map, or to lack of detail in the geological reconstruction. The comparison does, however, provide some confidence in the geological model.

25.2.4 Hydrogeological Interpretation

The 3-D geological structures combined with assessment of hydraulic parameters provides sufficient information to build flow models that can be used to investigate a range of issues. Figure 25.13 shows a simplified extract from the Newcastle 3-D model for the area around the main Newcastle University campus (see Figure 25.12). The model highlights how the existing infilled surface topography still preserves some evidence of the line of the Pandon Dene in the higher ground, but that on the flatter lower ground the valley is fully infilled and buried. However, because it still exists in the subsurface, it is likely to have a potential influence on groundwater flooding resulting from both groundwater flow and buried drainage.

Understanding of the role of Made Ground in groundwater movement requires additional information on hydraulic properties of the material. As these are highly heterogeneous, and borehole logs often provide insufficient detail, analysis of the likely provenance of materials was made based on historical maps and archives, made available for this study by Newcastle upon Tyne City Council. Construction of a stratigraphic column for sub-classification of Made

Table 25.1 Interpretation of inferred permeability for model units, based on lithologies determined from borehole logs, with subdivision of made ground lithology based on archeological reconstructions.

Model unit	Lithology	Inferred permeability
Made ground	Soil and made ground	Variable
	Tarmac and concrete	Weakly permeable
	Void	Permeable
	Fill	Variable
	Made ground – high rubble content	Permeable
	Glass works waste	Weakly permeable
	Sawmill waste	Permeable
	Pottery waste	Weakly permeable
	Made ground – rubbish and sand	Permeable
	Organic rubbish	Weakly permeable
	Ballast	Permeable
River deposits	Sand, gravel, peat	Permeable
Alluvial deposits	Silt, clay, peat	Weakly permeable
River terrace deposits	Sand, gravel, possibly with clay and silt	Permeable
Fluvioglacial deposits	Silty sand, clayey sand, and gravel	Permeable
Glaciolacustrine deposits	Laminated silts and clays	Weakly permeable
Glacial deposits	Till (boulder clay)	Weakly permeable with permeable lenses
Upper coal measures	Sandstone, coal, mudstone, siltstone	Variable – permeable
Lower coal measures	Sandstone, coal, mudstone, siltstone	Variable – permeable

Figure 25.11 Fence diagram based on borehole logs used to construct the geological model.

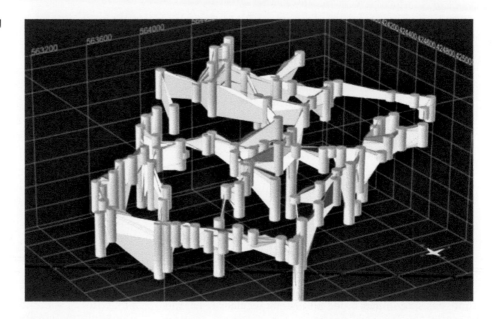

Ground (required for the 3-D geological model) proved problematic, as industrial and other activities spanned a number of centuries, with inert fill being deposited at various times. A general pattern was established; settlements from Roman times up to around the sixteenth century provided small amounts of anthropogenic materials, including organic waste, pottery, etc., while significant amounts of waste material were produced only with the

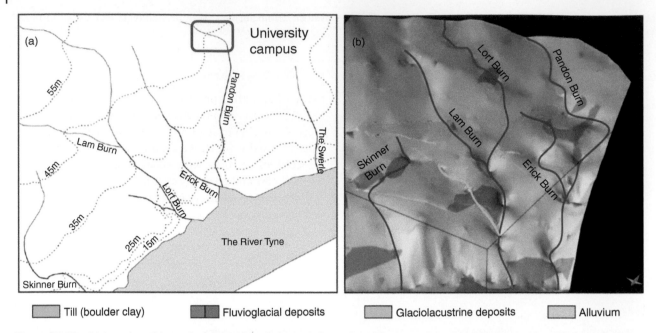

Till (boulder clay)   Fluvioglacial deposits   Glaciolacustrine deposits   Alluvium

Figure 25.12   (a) Location of burns based on historical map information; (b) perspective view of paleo-land surface extracted from digital geological model, with identifiable watercourses shown in dark blue with possible further watercourse in light blue. Steep slopes along banks of River Tyne are visible in the foreground. There is general consistency of the watercourses with previously mapped burn locations.

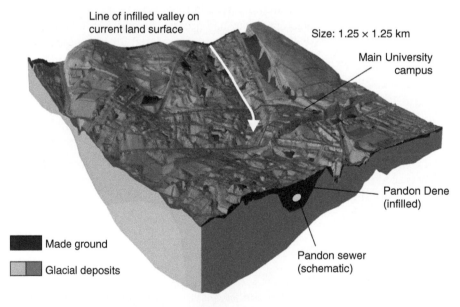

Figure 25.13   Simplified 3-D geological model of the area around Newcastle University campus (sub-divisions of made ground and glacial deposits not shown); the location of the infilled Pandon Dene is clearly visible.

onset of the industrial revolution from small industries including glassworks, lead works, and sawmills. Several classes of inert fill were defined at different times in the stratigraphic column, based on historical evidence of land development. An interpretation of the likely permeability classes for these materials is given in Table 25.1, and their general spatial distributions were inferred from the historical reconstructions together with current land cover mapping.

25.2.5 Conclusions

A 3-D digital geological model has been constructed for the central area of the city of Newcastle upon Tyne, combining information from borehole logs with archeological evidence. This use of different types of information allows a more detailed and accurate reconstruction of features of the near-surface geology that control shallow groundwater movement, including estimation of hydraulic properties, and paleo-surfaces relating to past landforms. This model continues to be used to support integrated modeling of surface water and groundwater in Newcastle upon Tyne, to better understand the interactions between rainfall, surface runoff, infiltration, and subsurface flows. This type of reconstruction using traditional geological methods with archeological interpretations can provide guidance for a range of further studies, including site investigations for urban redevelopment projects, mitigation of urban groundwater flooding hazards, locations of contaminated land, and geotechnical stability.

Acknowledgment

The authors thank Newcastle upon Tyne City Council for their co-operation and support for this study.

Case Study 25.3: Techniques and Issues Regarding the 3-D Mapping of Artificially Modified Ground

Ricky Terrington[1], Colin Waters[2], Helen Burke[1], and Jonathon R. Ford[1]

[1] *British Geological Survey, Keyworth, Nottingham NG12 5GG, UK*
[2] *Department of Geology, University of Leicester, Leicester LE4 7RH, UK*

25.3.1 Introduction

Anthropogenic processes significantly modify the urban subsurface; urban landforms, deposits, and excavations are collectively defined as Artificially Modified Ground, or AMG (Rosenbaum et al. 2003). Defining the spatial extent and properties of AMG is important because spatial variations in thickness and composition, and the potential presence of voids, affects their consolidation characteristics and creates potential hazards to future sustainable urban development and regeneration.

Mapping and modeling of AMG has been accepted as an important activity by several geological surveys. Schemas have been developed to provide additional detail about the type of AMG; these are based on classifications defined by landform and origin and generally do not include lithological descriptions. Thus, recent geological maps and databases provide only some of the pertinent information concerning ground conditions, geohazards (including potential contamination) and landscape evolution. While the AMG descriptions provided by geological survey are acceptable for regional assessments, more detailed site characterization requires additional morphostratigraphic

information derived from archeological investigation (De Beer et al. 2012).

25.3.2 Deposit Modeling by Archeologists

Carey et al. (2018) provide an extensive overview of the recent interest by archeologists in using 3-D "deposit models" to characterize anthropogenic deposits. Archeologists usually construct deposit models by combining archeological records with geotechnical data describing the thickness and geometry of sedimentary units and information derived from airborne and ground-based remote sensing surveys, academic research, maps, and online open-science resources such as the British Geological Survey (BGS) OpenGeoscience service (https://www.bgs.ac.uk/geological-data/opengeoscience/).

Archeological projects utilize the 3-D capability of deposit models for several purposes. When archeological remains are sparse or isolated, modeling key landform assemblages and their associated sediments assists the identification of likely sites. Linear construction projects for roads, railways, or pipelines are challenging for archeological investigation because they invariably cross a variety of geological units and archeological preservation environments. Deposit modeling provides a framework that defines paleotopography and geomorphological landforms. In relatively flat marshland and coastal areas, where some of the densest early human settlements were located, deposit models provide chronological frameworks to constrain erosional and depositional histories. Young et al. (2018), for example, used a combination of ArcGIS and Rockworks software to provide 3-D evaluations of the Thames floodplain paleo-landscape in support of geo-archeological investigations in the London Boroughs of Newham and Greenwich. Krawiec (2018) used similar methods to create 3-D models of archeological and environmental data gathered during a coastline realignment scheme in West Sussex.

25.3.3 Classification of AMG

McMillan and Powell (1999) provided a three-level classification of AMG (Figure 25.14); BGS maps at scales of 1 : 10 000, 1 : 25 000, and 1 : 50 000 began to show AMG according to the five classes in Level 2 (Table 25.2). However, this basic classification does not include information on the composition of AMG; for example, a waste tip cannot be distinguished from a road embankment (Figure 25.15). The BGS mapping of AMG focused primarily on mineral workings, industrial areas, and transport routes. Therefore, many BGS geological maps, such as DigMapGB-10, typically underreport the extent of AMG coverage, especially

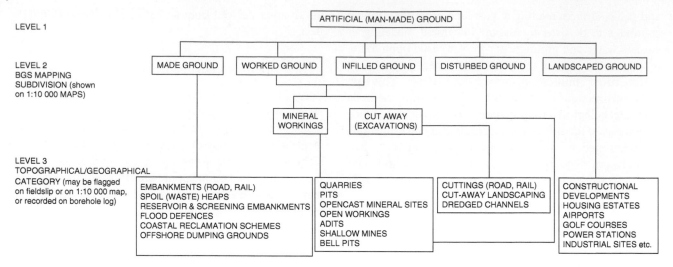

Figure 25.14 Classification of AMG proposed by McMillan and Powell (Ford et al. 2010).

Table 25.2 The five basic classes of AMG shown on BGS maps (Ford et al. 2010).

AMG class	Description
Made ground	Areas where material is known to have been placed by man on the pre-existing natural land surface (including engineered fill).
Worked ground	Areas where the pre-existing land surface is known to have been excavated by man.
Infilled ground	Areas where the pre-existing land surface is known to have been excavated (worked ground) and subsequently partially or wholly backfilled (made ground).
Disturbed ground	Areas of surface or near-surface mineral workings where ill-defined excavations (worked ground), areas of subsidence caused by the workings and spoil (made ground) are complexly associated with each other.
Landscaped ground	Areas where the pre-existing land surface has been extensively remodeled but it is impractical to delineate separate areas of made ground, worked ground, or disturbed ground.

in urban areas where most of the area is covered by a thin but continuous AMG "blanket" (Ford et al. 2014).

To resolve these limitations, Ford et al. (2010) proposed a new scheme to capture more-detailed information about AMG. The scheme expands the five classes defined by McMillan and Powell (1999) with a systematic hierarchical series of *Types* and *Units* based on the landform and origin of the deposit or excavation (Figure 25.16). This assignment of "children" to their "parents" allows numerous *Types* to be included within a *Class*, or *Units* within a *Type*. In this way, an enhanced system reflects both the long-term geological origin and the shorter-term anthropogenic characteristics of the AMG. For example, a quarry where limestone has been extracted may be mapped as *Worked Ground* at *Class* level, *Mineral Excavation* at *Type* level, and *Quarry (Hard Rock)* at *Unit* level. During surveying and map compilation, coincident areas of *Worked Ground* and *Made Ground* define *Infilled Ground* conditions. The hierarchical nature of the new scheme supports such composite definitions when detailed

information is available concerning both the "cut" and the "fill" (Figure 25.17).

Alternative methods of characterizing and classifying anthropogenic deposits generally involve combinations of lithology, landform, or soil properties. Japanese researchers use lithology and bounding surfaces to characterize anthropogenic deposits (Nirei et al. 2012). Dávid (2010) used geomorphological methods to classify quarrying and mineral extraction sites. The World Reference Base for Soils recognizes two major reference soil groups of anthropogenic soils: *Anthrosols* and *Technosols* (Rossiter 2007). Anthropogenic deposits have been mapped using *Technosol* classifications in several countries including Uruguay (Nirei et al. 2014; Mezzano and Huelmo 2011) and Lithuania (Satkūnas et al. 2011).

25.3.4 Evaluating AMG in 2-D

A more detailed understanding of AMG is possible by applying Geographic Information System (GIS) capabilities

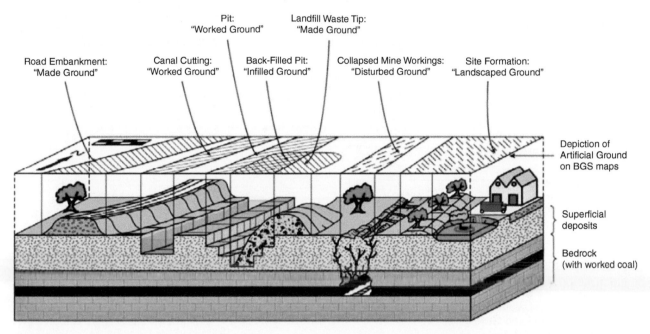

Pit:
"Worked Ground"

Landfill Waste Tip:
"Made Ground"

Road Embankment:
"Made Ground"

Canal Cutting:
"Worked Ground"

Back-Filled Pit:
"Infilled Ground"

Collapsed Mine Workings:
"Disturbed Ground"

Site Formation:
"Landscaped Ground"

Depiction of
Artificial Ground
on BGS maps

Superficial
deposits

Bedrock
(with worked coal)

Figure 25.15 Examples of diverse AMG types and their classification within the five geological map classes of AMG (Ford et al. 2010). Modified after McMillan and Powell (1999).

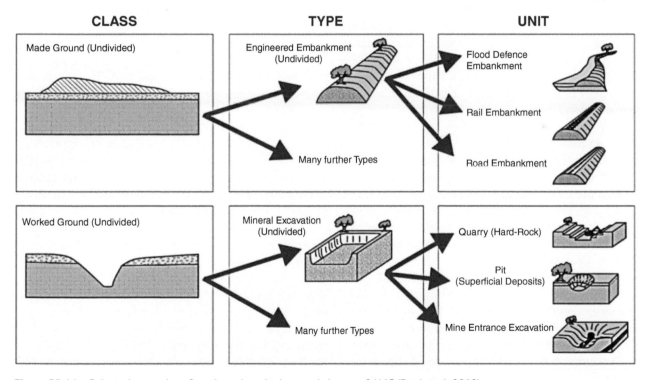

Figure 25.16 Selected examples of made and worked ground classes of AMG (Ford et al. 2010).

to supplement the AMG classes shown on digital BGS geology maps, such as DigMapGB-10, with a wider selection of data sources. However, only limited information on AMG thickness, or depth, is possible with 2-D evaluations.

When digital versions of historic Ordnance Survey topographic maps at 1 : 10 560 scale (6 in. to the mile) are geo-registered and stored in a GIS database, chronological changes in land use can be determined (Figure 25.18c). For many areas, the oldest of these maps dates from around

INFILLED GROUND EXAMPLE 1

No detail known about 'cut' but detail known about 'fill'. For example, Worked Ground (Undivided) filled with Landfill WasteTip (Domestic Waste)

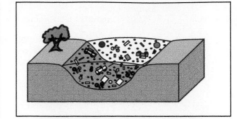

INFILLED GROUND EXAMPLE 2

Detail known about 'cut' but no detail known about 'fill'. For example, Rail Cutting filled with Made Ground (Undivided)

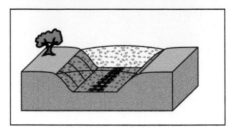

Figure 25.17 Selected examples of infilled ground class, showing combinations of made and worked ground (Ford et al. 2010).

1850, and there are updates at approximately decade intervals. Modern topographic maps show new road, motorway, or railway embankments and cuttings.

The topography layer of MasterMap produced by the Ordnance Survey (www.ordnancesurvey.co.uk/business-and-government/products/mastermap-products.html) is an excellent land use information source that is continuously updated. AMG evaluations in two relatively small central London areas combined historical maps and MasterMap information to estimate the character, extent, and thickness of anthropogenic deposits (Smith and Burke 2011; Thorpe et al. 2011). Alternative cheaper and easier to use land use information sources include Ordnance Survey open-source data, such as Vector Map, the Urban Atlas (European Commission 2012), the National Land Use Database for England (Harrison 2006), and the Eurostat Land Use/Land Cover Area Frame Statistical Survey, a similar European scheme (Eurostat 2015). Characteristic AMG landforms are readily identified with a combination of aerial photographs, topographical maps, and digital elevation models (DEMs). Figure 25.19 illustrates the value of LiDAR-derived high-resolution DEM data when evaluating AMG in an area with a complex land use history.

Many, though not all, borehole records have a date defining when the borehole was drilled; this provides key chronological data. To better define the extent and thickness of AMG deposits in the Aire Valley area southwest of Leeds, Terrington et al. (2014) assumed that the presence and thickness of AMG recorded in at least one borehole located within an Ordnance Survey MasterMap polygon identified the existence and thickness of an AMG deposit. The DigMapGB-10 AMG polygons showed 30% of the area

to be covered by AMG; the addition of the reclassified AMG polygons from the MasterMap Topography Layer increased the extent of AMG to 49.7% of the site (Figure 25.20). However, because the MasterMap polygons did not directly provide information on the process or origin of AMG features, attribution of the extracted polygons according to the BGS AMG classification scheme was very difficult.

25.3.5 Evaluating AMG in 3-D

BGS projects in key urban areas have developed new methods of employing 3-D geological models to evaluate the thickness, extent, and characteristics of AMG (Royse et al. 2009; Thorpe et al. 2011; Smith and Burke 2011; Jordan et al. 2014; Terrington et al.; 2014, 2018). High resolution Lidar accurately defines the current ground surface in many cases (Terrington et al. 2014), while older topographic maps show historical elevations. In many urban areas, predicting the thickness of AMG is problematic because urban renewal has reworked existing AMG materials to provide new building sites or transportation routes, to improve surface drainage, or for esthetic reasons. While excavation below all existing AMG is common for new large buildings, many lighter smaller and residential structures are often founded on reworked AMG.

Borehole logs are the best information source on the characteristics and thickness of AMG; thus, they form the basis of most 3-D models. Many contain a drill-date, which accurately defines the age of the AMG activity, while comparison of drill-hole collar elevations with current ground elevations provides evidence of excavation or emplacement of AMG deposits (Terrington et al. 2014).

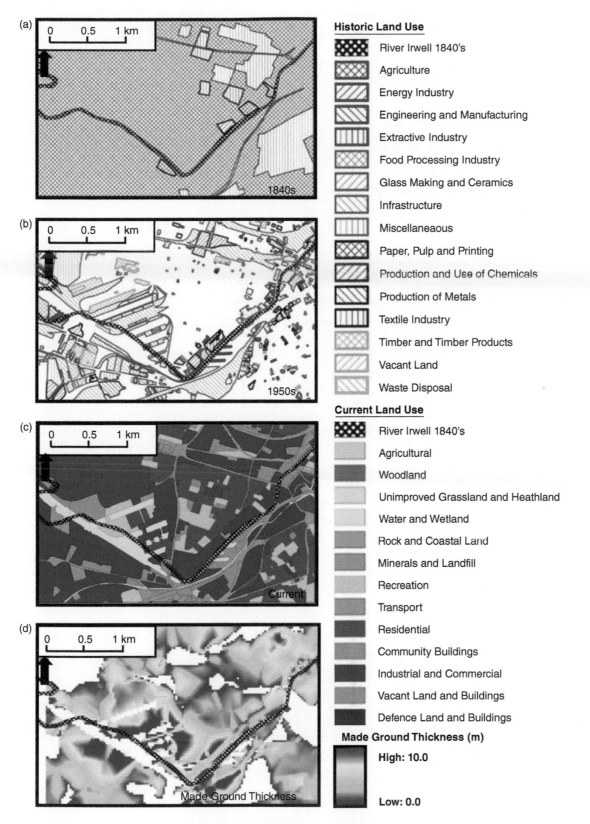

Historic Land Use

River Irwell 1840's
Agriculture
Energy Industry
Engineering and Manufacturing
Extractive Industry
Food Processing Industry
Glass Making and Ceramics
Infrastructure
Miscellaneaous
Paper, Pulp and Printing
Production and Use of Chemicals
Production of Metals
Textile Industry
Timber and Timber Products
Vacant Land
Waste Disposal

Current Land Use

River Irwell 1840's
Agricultural
Woodland
Unimproved Grassland and Heathland
Water and Wetland
Rock and Coastal Land
Minerals and Landfill
Recreation
Transport
Residential
Community Buildings
Industrial and Commercial
Vacant Land and Buildings
Defence Land and Buildings

Made Ground Thickness (m)

High: 10.0

Low: 0.0

Figure 25.18 Land use change over approximately 160 years at Trafford Park, Salford, NW England: (a) land use in the 1840s; (b) land use in the 1950s; (c) current land use; (d) estimated made ground thickness derived from a 3-D geological model. After Price et al. (2011). OS topography © Crown Copyright. All rights reserved. 100017897/2010.

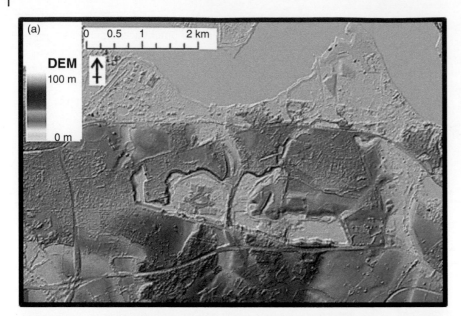

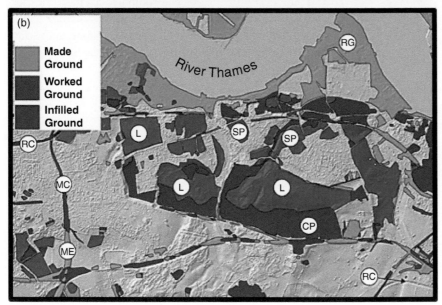

Figure 25.19 AMG mapping in the Swanscombe area, east of London using 5 m resolution digital elevation model from NextMap Lidar. (a) Digital elevation model; (b) Mapped distribution of AMG – letter codes: RC = railway cutting, MC = motorway cutting, ME = motorway embankment, L = landfill site, RG = raised ground, CP = chalk pit, SP = sand and gravel pit. After Price et al. (2011). NEXTMap Britain elevation data from Intermap Technologies. DiGMapGB50 BGS © NERC.

Price et al. (2010) used this approach with 3-D geological modeling methodologies to quantify and visualize AMG deposits and the anthropogenic landscape evolution in Manchester and Liverpool, NW England (Figure 25.18d). Made ground up to 10 m thick composed of colliery spoil, furnace waste and ash fill in the former course of the River Irwell. Equally thick deposits of similar AMG found adjacent to the Manchester Ship Canal were probably deposited during construction of the canal in between 1887 and 1894.

Table 25.3 Thickness of artificial deposits relative to first underlying natural superficial deposit.

Borough	Alluvium (m)	Terrace deposits (m)	Langley silt (m)	London clay
City of London	5.53	3.19	3.84	4.33
Tower Hamlets	3.65	2.21	2.37	N/A

(a)

(b)

(c)

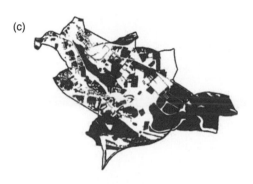

Figure 25.20 Estimated AMG coverage in the Aire Valley by: (a) DigMapGB, (b) MasterMap polygons intersected by boreholes that identified AMG, and (c) combined AMG estimate. After Terrington et al. (2014). (© Crown copyright and database rights [2015]. Ordnance Survey [100021290 EUL].

More detailed evaluations of the 3-D extents of AMG deposits require careful construction of the 3-D model. AMG is more variable than natural deposits and is often discontinuous, so normal stratigraphic rules do not apply. Careful cross-section development accompanied by appropriate surface interpolation methods are required to produce an acceptable model. Jordan et al. (2014) used 3-D modeling methods to reconstruct anthropogenic topographies, sediment thicknesses, and volumes in the Great Yarmouth area of Norfolk, UK.

25.3.6 Example of 3-D AMG Modeling

Terrington et al. (2018) applied the methods described above to evaluate the veneer of artificial (anthropogenic) deposits beneath the City of London and Tower Hamlet Borough areas of central London (Figure 25.21). Land use was defined by Urban Atlas Area polygons (Figure 25.22). Land elevation and thickness changes were mapped using a total of 3773 boreholes that had AMG recorded and coded and over 6200 boreholes that had a start height recorded (Figure 25.23). The combined average thickness of AMG for the entire area was 3.14 m (Figure 25.24). AMG forms a significant deposit in the shallow subsurface and is undoubtedly important when considering ground conditions and potential hazards.

When the AMG was assigned to each Urban Atlas Area polygon, AMG thickness could be assessed against other parameters such as building height, land-use type and age, and the underlying geology. The thickest AMG deposits occurred in the river floodplains, particularly in the alluvial deposits (Table 25.3). Much of the increased thickness of AMG is associated with the construction of river walls, sewage tunnels, bridges, and culverting of the rivers such as the Fleet and Walbrook, all predating the construction of the current buildings (Figure 25.25).

The thickest AMG deposits occur in areas underlain by Alluvium (Figure 25.26). The floodplains of the Walbrook and Fleet have been greatly altered, vaulted, and largely infilled, as has part of the channel of the River Thames along Victoria Embankment, constructed in 1864–1870. Other areas of thicker AMG are related to the disposal of bomb-damaged buildings following World War II. It was possible in many areas to distinguish pre- and post-1945 deposits (Figure 25.27).

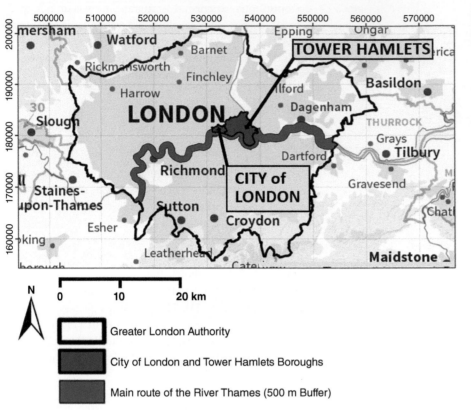

Figure 25.21 Location map of City of London and Tower Hamlets Boroughs.

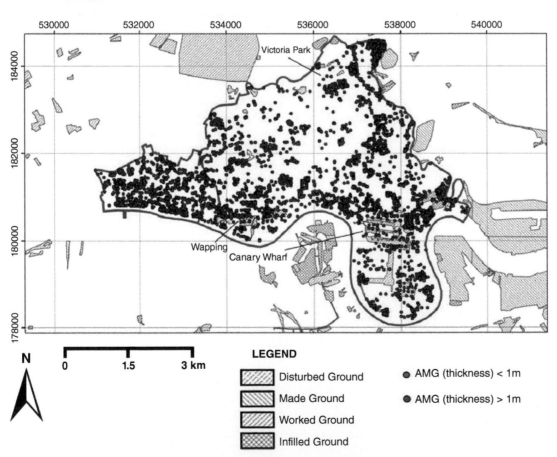

Figure 25.22 Extent of AMG based on published BGS maps (grey outline) and boreholes with more than 1 m thickness of AMG (red circles). After Terrington et al. (2018).

Figure 25.23 Distribution of main land-use categories reflecting pre- and post-war redevelopments, derived from the Urban Atlas. Significant material compositions present within boreholes are shown. After Terrington et al. (2018). (SOURCE: Urban Atlas – http://www.eea.europa.eu/legal/copyright).

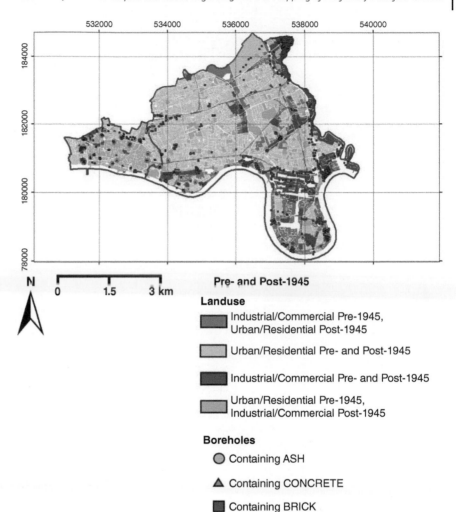

Pre- and Post-1945

Landuse

- Industrial/Commercial Pre-1945, Urban/Residential Post-1945
- Urban/Residential Pre- and Post-1945
- Industrial/Commercial Pre- and Post-1945
- Urban/Residential Pre-1945, Industrial/Commercial Post-1945

Boreholes

- ⬤ Containing ASH
- ▲ Containing CONCRETE
- ◼ Containing BRICK

Figure 25.24 Average thickness of AMG determined by comparing modern DEM with former ground level at time of site investigation. After Terrington et al. (2018).

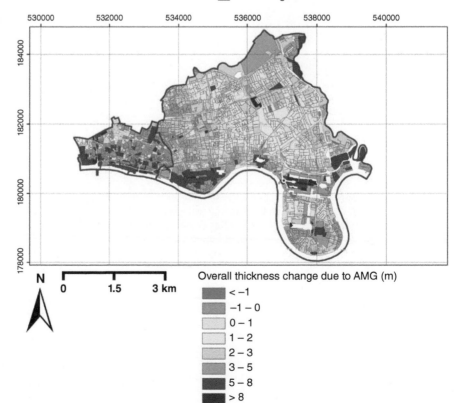

Overall thickness change due to AMG (m)

- < −1
- −1 – 0
- 0 – 1
- 1 – 2
- 2 – 3
- 3 – 5
- 5 – 8
- > 8

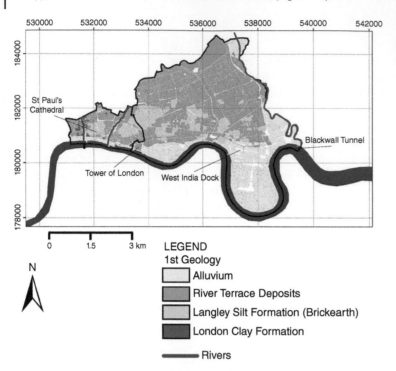

Figure 25.25 Map showing the uppermost natural geological unit present immediately beneath the AMG, sourced from the BGS DigMap 1:50 000 data. After Terrington et al. (2018).

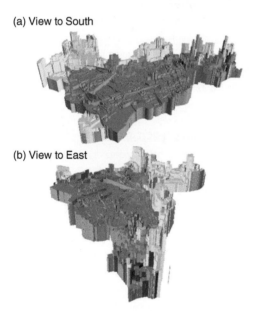

Figure 25.26 Two 3-D views of the uppermost natural geological unit present immediately beneath the AMG, illustrating the relationship between thickness of artificial deposits (shown as column height of Urban Atlas domains) and the underlying natural geology (colored as for Figure 25.25); (a) view looking toward the south, and (b) view toward the east. After Terrington et al. (2018).

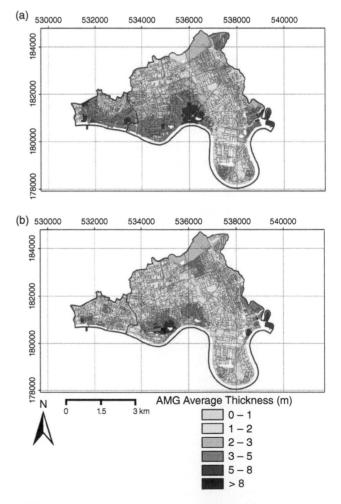

Figure 25.27 AMG thickness calculated from borehole data: (a) pre-1945; and (b) post-1945. After Terrington et al. (2018).

References

Carey, C., Howard, A.J., Knight, D. et al. (2018). *Deposit Modelling and Archaeology*. Exeter, Devon: University of Brighton, Short Run Press Ltd. 230 pp. [Online: Available at: https://www.brighton.ac.uk/_pdf/research/set-groups/deposit-modelling-and-archaeology-volume.pdf] (Accessed December 2020).

Christensson, A., Paszkowski, Z., Spriggs, J.A. and Verhoef, L. [Eds.] (2004). *Safeguarding Historic Waterfront Sites. Bryggen in Bergen as a Case Study, 1e*. Bergen, Norway: Stiftelsen Bryggen/Szczecin: Politechnika Szczecińska. 337 pp.

Dávid, L. (2010). Quarrying and other minerals. In: *Anthropogenic Geomorphology. A Guide to Man-Made Landforms* (eds. J. Szabó, L. Dávid and D. Lóczy), 113–130. Dordrecht, Netherlands: Springer. https://doi.org/10.1007/978-90-481-3058-0_9.

De Beer, J. (2008). *Statusrapport grunnvannsovervåking og hydrogeologisk modellering ved Bryggen i Bergen* (in Norwegian). Trondheim, Normway: NGU, Report 2008.069. 72 pp. [Online: Available at: https://www.ngu.no/publikasjon/statusrapport-grunnvannsovervaking-og-hydrogeologisk-modellering-ved-bryggen-i-bergen] (Accessed December 2020).

De Beer, J. and Matthiesen, H. (2008). Groundwater monitoring and modelling from an archaeological perspective: possibilities and challenges. In: *Geology for Society* (ed. T. Slagstad), 67–81. Trondheim, Norway: NGU, Geological Survey of Norway, Special Publication 11. [Online: Available at: https://www.ngu.no/en/publikasjon/geology-society] (Accessed December 2020).

De Beer, J., Price, S. and Ford, J. (2012). 3D modelling of geological and anthropogenic deposits at the World Heritage Site of Bryggen in Bergen, Norway. *Quaternary International* 251: 107–116. https://doi.org/10.1016/j.quaint.2011.06.015.

European Commission (2012). *Mapping Guide for a European Urban Atlas. The GMES "Urban Atlas" project*. Brussels: European Commission. 30 pp. [Online: Available at: http://ec.europa.eu/regional_policy/sources/tender/pdf/2012066/annexe2.pdf] (Accessed April 2018).

Eurostat (2015). *LUCAS 2015 (Land Use / Cover Area Frame Survey). Technical Reference Document C3 (Land cover & Land Use)*. Luxemburg: European Commission, Eurostat. 93 pp. [Online: Available at: https://ec.europa.eu/eurostat/documents/205002/6786255/LUCAS2015-C3-Classification-20150227.pdf] (Accessed: May 2020).

Ford, J., Kessler, H., Cooper, A.H., Price, S.J. and Humpage, A.J. (2010). *An Enhanced Classification for Artificial Ground*. Keyworth, UK: British Geological Survey, Open Report OR/10/036. 32 pp. [Online: Available at: http://nora.nerc.ac.uk/id/eprint/509281] (Accessed December 2020).

Ford, J.R., Price, S.J., Cooper, A.H. and Waters, C.N. (2014). An assessment of lithostratigraphy for anthropogenic deposits. *Geological Society Special Publications* 395: 55–89. https://doi.org/10.1144/SP395.12.

Graves, C.P. and Heslop, D. (2013). *Newcastle upon Tyne, the Eye of the North: An Archaeological Assessment*. Oxford: Oxbow Books. 304 pp.

Hannon, E. (2015). *Understanding the Influence of Near-Surface Geology on Groundwater Flooding in Urban Areas*. Unpublished MSc Dissertation, School of Civil Engineering and Geosciences, Newcastle University.

Harrison, A.R. (2006). *National Land Use Database: Land Use and Land Cover Classification: Version 4.4*. London, UK: HM Stationary Office. 79 pp. [Online: Available at: https://assets.publishing.service.gov.uk/government/uploads/system/uploads/attachment_data/file/11493/144275.pdf] (Accessed December 2020).

Herteig, A.E. (1985). *The Archaeological Excavations at Bryggen, "the German Wharf", in Bergen, 1955–68: Excavation, Stratigraphy, Chronology, Field-Documentation*. In: The Bryggen Papers, Main Series, 1 (ed. A.E. Herteig), 9–46. Oslo, Norway: Universitetsforlaget.

Holden, J., West, L.J., Howard, A.J. et al. (2006). Hydrological controls of in situ preservation of waterlogged archaeological deposits. *Earth-Science Reviews* 78: 59–83. https://doi.org/10.1016/j.earscirev.2006.03.006.

Holden, J., Howard, A.J., West, L.J. et al. (2009). A critical review of hydrological data collection for assessing preservation risk for urban waterlogged archaeology: a case study from the City of York, UK. *Journal of Environmental Management* 90: 3197–3204. https://doi.org/10.1016/j.jenvman.2009.04.015.

Jordan, H., Hamilton, K., Lawley, R. and Price, S.J. (2014). Anthropogenic contribution to the geological and geomorphological record: a case study from Great Yarmouth, Norfolk, UK. *Geomorphology* 253: 534–546. https://doi.org/10.1016/j.geomorph.2014.07.008.

Kessler, H., Mathers, S.J. and Sobisch, H.-G. (2009). The capture and dissemination of integrated 3D geospatial knowledge at the British Geological Survey using GSI3D software and methodology. *Computers and Geosciences* 35 (6): 1311–1321. https://doi.org/10.1016/j.cageo.2008.04.005.

Krawiec, K. (2018). Medmerry managed realignment scheme, West Sussex: a Holocene deposit model of a coastal environment. In: *Deposit Modelling and Archaeology* (eds. C. Carey, A.J. Howard, D. Knight, et al.), 149–158. Exeter, Devon: University of Brighton, Short Run Press Ltd. [Online: Available at: https://www.brighton.ac.uk/_pdf/research/set-groups/deposit-modelling-and-archaeology-volume.pdf] (Accessed December 2020).

Marstein, N., Paludan-Müller, C., Loska, A. et al. (2007). *The Monitoring Manual. Procedures & Guidelines for the Monitoring, Recording and Preservation/Management of Urban Archaeological Deposits*. Oslo, Norway:

Riksantikvaren/NIKU. 213 pp. [Online: Available at: http://hdl.handle.net/11250/175912] (Accessed December 2020).

Mathers, S.J., Wood, B. and Kessler, H. (2011). *GSI3D 2011 Software Manual and Methodology.* Keyworth, UK: British Geological Survey, Report OR/11/020. 152 pp. [Online: Available at: http://nora.nerc.ac.uk/id/eprint/13841] (Accessed December 2020).

Mathers, S.J., Kessler, H., Macdonald, D.M.J. et al. (2012). *The Use of Geological and Hydrogeological Models in Environmental Studies.* Keyworth, UK: British Geological Survey, Internal Report IR/10/022. 93 pp. [Online: Available at: http://nora.nerc.ac.uk/id/eprint/20481] (Accessed December 2020).

Matthiesen, H. (2007). Preservation conditions of urban deposits studied through detailed chemical analysis of ground water from Bryggen in Bergen. *Journal of Archaeological Science* 35: 1378–1388. https://doi.org/10.1016/j.jas.2007.10.003.

Matthiesen, H., Dunlop, R., Jensen, J.A. et al. (2006). Monitoring of Preservation Conditions and Evaluation of Decay Rates of Urban Deposits – Results from the First Five Years of Monitoring at Bryggen in Bergen. In: *Conference Proceedings, Preserving Archaeological Remains in situ 3* (eds. H. Kars and R.M. van Heeringen), 163–174. Amsterdam, Netherlands: VU University, Geoarchaeological and Bioarchaeological Studies, vol. 10.

McMillan, A.A. and Powell, J.H. (1999). *BGS Rock Classification Scheme Volume 4: Classification of Artificial (Man-made) Ground and Natural Superficial Deposits- Applications to Geological Maps and Datasets in the UK.* Keyworth, UK: British Geological Survey, Research Report RR/99/04. 65 pp. [Online: Available at: http://nora.nerc.ac.uk/id/eprint/3228] (Accessed December 2020).

Mezzano, A. and Huelmo, S. (2011). Occurrences of technosols in Montevideo City, Uruguay. *Geologija* 53 (4): 187–191.

Nirei, H., Furuno, K., Osamu, K. et al. (2012). Classification of man-made strata for assessment of geopollution. *Episodes* 35: 333–336. https://doi.org/10.18814/epiiugs/2012/v35i2/004.

Nirei, H., Mezzano, A., Satkūnas, J. et al. (2014). Environmental problems associated with man-made strata and their potential management. *Episodes* 37 (1): 33–40. https://doi.org/10.18814/epiiugs/2014/v37i1/004.

Price, S.J., Burke, H.F., Terrington, R.L. et al. (2010). The 3D characterisation of the zone of human interaction and the sustainable use of underground space in urban and peri-urban environments: case studies from the UK. *Zeitschrift der Deutschen Gesellschaft für Geowissenschaften* 161 (2): 219–235. https://doi.org/10.1127/1860-1804/2010/0161-0219.

Price, S.J., Ford, J.R., Cooper, A.H. and Neal, C. (2011). Humans as major geological and geomorphological agents in the anthropocene: the significance of artificial ground in Great Britain. *Philosophical Transactions of the Royal Society A* 369 (1938): 1056–1084. https://doi.org/10.1098/rsta.2010.0296.

Rosenbaum, M.S., McMillan, A.A., Powell, J.H. et al. (2003). Classification of artificial (man-made) ground. *Engineering Geology* 69: 399–409. https://doi.org/10.1016/S0013-7952(02)00282-X.

Rossiter, D.G. (2007). Classification of urban and industrial soils in the world reference base for soil resources. *Journal of Soils and Sediments* 7: 96–100. https://doi.org/10.1065/jss2007.02.208.

Royse, K.R., Rutter, H.K. and Entwisle, D.C. (2009). Property attribution of 3D geological models in the Thames Gateway, London: new ways of visualising geoscientific data. *Bulletin of Engineering Geology and the Environment* 68: 1–16. https://doi.org/10.1007/s10064-008-0171-0.

Rytter, J. and Schonhowd, I. [Eds.] (2015). *Monitoring. Mitigation. Management. The Groundwater Project – Safeguarding the World Heritage Site of Bryggen in Bergen.* Oslo: Riksantikvaren. 213 pp. [Online: Available at: http://hdl.handle.net/11250/300104] (Accessed December 2020).

Satkūnas, J., Gregorauskienė, V., Kanopienė, R. et al. (2011). Man-made formations and geopollution: state of knowledge in Lithuania. *Geologija* 53 (1): 36–44.

Seither, A., Ganerød, G.V., de Beer, J. et al. (2016). *Bergen.* COST TU1206 Sub-Urban Project, Report TU1206-WG1-003. 49 pp. [Online: Available at: http://sub-urban.squarespace.com/new-index] (Accessed December 2020).

Smith, H. and H.F. Burke (2011). *The Anthropogenic Land Use History and Artificial Ground of Rotherhithe.* Keyworth, UK: British Geological Survey, Report IR/11/041. 14 pp. [Online: Available at: http://nora.nerc.ac.uk/id/eprint/510140] (Accessed December 2020).

Terrington, R.L., Thorpe, S., Burke, H. et al. (2014). *Enhanced Mapping of Artificially Modified Ground in Urban Areas; Using Borehole, Map and 6 Sensed Data.* Keyworth, UK: British Geological Survey, Open Report OR/15/010. 47 pp. [Online: Available at: http://nora.nerc.ac.uk/id/eprint/510140] (Accessed December 2020).

Terrington, R.L., Silva, É.C.N., Waters, C.N. et al. (2018). Quantifying anthropogenic modification of the shallow geosphere in Central London, UK. *Geomorphology* 319: 15–34. https://doi.org/10.1016/j.geomorph.2018.07.005.

Thorpe, S., Burke, H.F. and Terrington, R.L. (2011). *The Anthropogenic Land Use History and Artificial Ground of the River Fleet.* Keyworth, UK: British Geological Survey, Internal Report IR/11/042. 10 pp.

Young, D., Batchelor, R. and Green, C. (2018). Deposit modelling in the lower Thames Valley (East London): correlating the sedimentary sequence with archaeological and palaeoenvironmental evidence of prehistoric human activity. In: *Deposit Modelling and Archaeology* (eds. C. Carey, A.J. Howard, D. Knight, et al.), 121–135. Exeter, Devon: University of Brighton, Short Run Press Ltd. [Online: Available at: https://www.brighton.ac.uk/_pdf/research/set-groups/deposit-modelling-and-archaeology-volume.pdf] (Accessed December 2020).

Part V

Future Possibilities and Challenges

26

Anticipated Technological Advances

Matthew Lato[1], Robin Harrap[2], and Kelsey MacCormack[3]

[1] *BGC Engineering Inc., Ottawa, ON K1S 2L1, Canada*
[2] *Queen's University, Department of Geological Sciences and Geological Engineering, Kingston, ON K7L 3N6, Canada*
[3] *Alberta Geological Society, Alberta Energy Regulator, Edmonton, AL T6B 2X3, Canada*

26.1 Looking Forward

Technological developments are driven by interactions among scientific advances, insights into manufacturing techniques, combinations of previously separate tools or approaches, recognition of market opportunities, and social acceptance – especially the enthusiastic uptake of a "shiny new thing." The transition from idea to realization, uptake, and obsolescence can take several decades or longer. Some wonderful technical innovations fail because of social or business forces; one example is the case of Betamax versus VHS video cassettes. However, some mediocre technical innovations or outright throwbacks succeed because of social movements; the re-emergence of vinyl records in recent years is one example.

Humans resist significant change that upsets their normal routine. In order for substantive technological advancement to occur, the new approach must be an order of magnitude better, and adoption must be either less painful than continuing to use what is known and comfortable or be mandated by an outside agency or Government regulation. Several automobile safety features, including seat belts and airbags, are examples of government regulation influencing adoption of new technologies. Given how unpredictable human nature is, Denis Gabor famously said: "*Futures cannot be predicted, but they can be created*".

Geological modeling and visualization occupy a relatively small market segment within the overall field of computing hardware and software. Geological modeling requirements are typically satisfied by adapting tools and techniques developed for other market segments. These offer mature and well-tested capabilities. The alternative is to use specialized products developed by small enterprises that are aimed directly at the geoscience market. While these are developed specifically to address geological modeling requirements, the limited financial and technical support available to their developers may result in less robust products. Consequently, many widely adopted technical solutions use customized components and technologies that modify procedures and processes developed for larger markets. For example, while specialized scanners are used to evaluate rock faces in geoengineering, most geospatial software used to evaluate the data defining those rock faces was developed for far larger markets connected with the construction and industrial manufacturing industries. New techniques from these and other areas are often adopted into geoengineering. These have dramatically changed our ability to model and visualize geological environments and their adoption has triggered unanticipated secondary changes. For example, adopting GPS did more than make determining location more efficient. It also allowed for entirely new types of analysis. Yet while GPS is very well suited for geoscience applications, we were not the primary market for the technology. It was adopted rather than designed for use by geoprofessionals.

Advances in computational processing power and visualization capacity have led directly to the development of sophisticated modeling and visualization approaches, but the availability of these tools, and especially their outputs, have also contributed toward rising societal expectations. What was previously state-of-the-art is now *expected* for smaller, often low-budget, projects. Many projects are now competing not only with past successes, but also with expectations that arise from computer games and movie special effects. These technologies are now mainstream and set a high bar for any new state-of-the-art products, especially with regards to visualization.

It is inevitable that future geological modeling and visualization tools will be more sophisticated and will

Applied Multidimensional Geological Modeling: Informing Sustainable Human Interactions with the Shallow Subsurface, First Edition.
Edited by Alan Keith Turner, Holger Kessler, and Michiel J. van der Meulen.

be capable of supporting a larger range of applications. While current impediments to modeling and visualization of complex geological environments are very likely to be reduced given the pace of new technology developments, geological needs and priorities will not directly drive these improvements. For example, some geological modeling and visualization needs are similar to the underlying requirements of a much larger domain – video gaming – and so the utilization of existing "video game engines" can potentially enhance simulations of some geological processes. But it would be foolish to suppose that geoengineering applications will drive the development of new game technology directly. Because many, if not most, advances will come from "outside", they will need to be recognized as relevant, adapted for use, and then proven using realistic test cases before they are accepted by the geoscientific community.

New geoscience projects now are expected to incorporate complex factors, such as site-specific and often data-limited environmental protection plans and sustainable development, which requires an assessment of the suitability of designs for modeled climate change scenarios. Ever more complex models and techniques are needed to support quantitative assessments of subsurface conditions; these will be aided in part by hardware and software advances. While modeling is currently often limited by computational power, visualization capacity, and challenging user requirements, the final and overriding constraint is the insufficiency of adequate site-specific data to constrain the models and their visualization (Oreskes et al. 1994).

With these general principles in mind, this chapter briefly explores several areas where emerging technological advancements are influencing geological modeling and geological model visualization.

26.2 General Technological Trends

The following twenty technological trends are believed to be key influences that are likely to affect geological modeling and visualization in the near to mid-term future. Important implications of each technological trend are shown as parenthetical comments.

1. Continually increasing processing power in desktop and mobile computers and in devices incorporating computers, such as scanners and sensors. (This is a consequence of Moore's Law.)
2. Increased connectivity through 5G wireless and wider availability of wireless service in remote areas. (Will provide fast access everywhere.)
3. Increased use of cloud computing and centralized data services. (Will result in fewer data silos.)
4. Increased memory and storage in all devices. (Especially important for sensors lacking fast connectivity.)
5. Increased battery efficiency and use of solar and other power generation modules. (Will result in greater use of sensors in remote field locations.)
6. Decreased power use in all devices. (Will enable longer lasting and smaller field devices.)
7. Significantly increased graphics processing, and in particular the ability to use graphics units as general-purpose processors to solve challenging computational problems. (Faster processing times reduce the response times for complex computational tasks.)
8. Emerging cloud-based graphical processing capabilities. (An important aspect of on-demand supercomputing.)
9. Increased use of reconfigurable processors that incorporate Field Programmable Gate Arrays (FPGAs), which will allow mass customization of devices. (Expanded use of customizable high-performance sensors.)
10. Increased sophistication, decreased size, and decreased cost of drones. (Increased ability to acquire just-in-time 3-D data.)
11. Increased accuracy and dramatically decreased cost of short-range scanners such as Lidar and specialized video capture, driven largely by drone technology. (Availability of more sophisticated sensors on small devices.)
12. Emergence of multispectral Lidar. (Provides better discrimination of geological materials.)
13. Emergence of third-generation radar satellite missions with L-band mapping technology. (Provides significantly better vegetation penetration.)
14. Novel low cost, long life, and self-networking sensor nodes for data collection. (Increased options for providing in-situ data from sites of interest.)
15. Convergence of many different software environments, such as Computer Aided Design (CAD), Geographic Information System (GIS), Building Information Management (BIM), and modeling tools, to support new more interoperable software suites that combine new and old forms of static and real-time data. This will be largely driven by urban information, management, and security systems. (Will increase data inter-operability and further reduce software silos.)
16. Increased sophistication and ease of use of video game engines as platforms for geospatial simulation

and construction of interfaces. (Easier modeling simulations and outreach products.)

17. Artificial Intelligence (AI) pattern recognition: increased sophistication of deep learning AI methods will allow autonomous use of sensor network data, navigation of drones, and more sophisticated supervisory control and data acquisition (SCADA) implementations. (Development of decision support aids to deal with data volume and 24/7 monitoring.)

18. Development of "helper AI" tools ("AI guides") as components of computer interfaces to provide intelligent assistance during all stages of geological modeling and visualization. (More sophisticated modes in tools and better help systems.)

19. Expanded use of network-centric computing at all scales, from sensor to supercomputer, that supports software used by scientists, engineers, and clients. (Increased and improved shared access to large project datasets.)

20. Development of immersive interfaces, including Augmented Reality (AR) and Virtual Reality (VR), which are well suited to some aspects of geological visualization. (Increased insight into complex 3-D and 4-D data and better outreach products.)

Many of these trends can be simply grouped under the themes *"computers get cheaper and faster"*, and *"mobile data gets better and easier to acquire"*, but others offer real opportunities in geoengineering that may influence the near future processes. Some may be a surprise to some readers, and other readers may have additions to the list, since any list of this type is doomed to be somewhat myopic. A few of these represent areas that have not yet been fully realized in the geoscience community yet seem to offer something between likely and inevitable.

26.3 Current Successes and Conundrums

Technology advancements, large and small, have shaped the geoscience community and numerous recently developed technologies are considered essential in daily activities. For example, a geologist does not initiate a field survey without a handheld GPS, satellite communication device, digital camera, and tablet for note taking; yet none of these were likely a part of a field kit at the turn of the millennium. Technology has similarly revolutionized all aspects of mineral exploration. For example, multi- and hyperspectral images from satellites orbiting 400 km above the Earth are being processed with machine learning

methods to better understand the spatial distribution of mineralogy and constrain geological models. Likewise, radar, Lidar, and photogrammetry are beginning to resolve deformation on and below the surface of the Earth at below the millimeter level and thus to provide new insight to active processes at a range of scales.

On the other hand, the current state-of-practice in using software to manage spatial data is less encouraging. Spatial software, including CAD and GIS, were developed by user communities requiring tools that met their needs in the early days of scientific computing. The complexity of their interfaces, their workflows, and their interrelationships with other tools (especially with regards to data standards) were based on the expectation of their use by expert users in a relatively data-poor environment and working over relatively long timeframes. When the use of GIS in cartography first appeared, simple spatial analysis and map production was miraculous because manual methods were the only alternative. These tools took months to years to master and required dedicated hardware on expensive workstations. Networking was in its infancy, and data were often quite literally moved around in boxes full of tapes, hard disks, CDs and, eventually, DVDs.

The existence of multiple tools developed for and by diverse user communities, poorly interlocking workflows, and overly complex interfaces still plagues efforts to accomplish many relatively simple tasks, especially for those recently introduced to these tools, since there is very little consistency between different organizations. A single urban geology project is currently likely to need to use CAD, GIS, BIM, 3-D visualization, and sensor-specific data capture-process-transform tools (for GPS, Lidar, photogrammetry, etc.) to complete the various study stages. There may be one *actual* building on a project site, but a half-dozen representations may be required to analyze it with the different tools. This not only results in an inefficient project workflow; it also strongly impedes the development of new tools. Software tools clearly need to converge from the dozen or so that might be used on a typical project to, at most, a couple. Our ability to incorporate new ideas into revised workflows is at risk. Interoperability needs to be dramatically improved where multiple tools are required.

Given how complex our software environments already are, the idea of taking on new tools incorporating new approaches is met by resistance. Adoption of new tools requires several questions to be considered. Where would they be hosted? Would they be new applications? What data formats do they support, or worse, require? How to train potential project staff, convince them of the value of the new tools, and maintain their expertise to use these tools, which will likely change with yearly updates?

Many tools support customization using visual tools, scripting languages, and full-on development, but few geoscience experts have the time or skill to use these, and most projects don't warrant a software development team. Custom tools that *are* developed are often kept inside one organization; a notable exception is the Groundhog desktop application developed by the British Geological Survey (Wood et al. 2015). Four potentially high-impact approaches that may resolve these issues are:

- *Changes to how we approach both data ownership and data collection.* Crowdsourcing of data collection reduces costs and increases diversity of available data. Governmental-industry-citizen partnerships over data are likely to improve the trust in decision-making.
- *Changes to how we balance data presentation and simulation in project planning and presentation.* While some areas of geotechnical engineering make heavy use of simulation (rock mechanics, for example), other areas see very little use. Simulation-based planning tools, especially those that allow communities to examine choices and outcomes, improve trust. Tools such as SimCity (a game) and CommunityViz (an urban planning tool) can lead to different perspectives on urban infrastructure planning, development, and outcomes, especially over contentious issues.
- *Recognition that mobile users with, essentially, supercomputers containing high resolution cameras in their pockets are everywhere.* Mobile information collection, and especially mobile onsite visualization, are dramatically increasing. Public data such as photographs can be used to build 3-D models of complex environments. Onsite visualization by communities, ranging from simple "hold the phone up and compare" to augmented reality techniques, allow design tradeoffs to be recognized early in projects. High-speed communication networks make phones and tablets extensions of a sophisticated computing infrastructure that can support high-performance computing, very large data stores, and linkages to diverse communication platforms. Geoscientists have not, as a community, thought deeply about how smart mobile devices will influence their activities, let alone their impacts on the larger community of clients and affected citizens.
- *Recognition of the potential for the use of advanced visualization technologies for individual or group work.* Virtual reality has made a comeback in recent years, and augmented reality has made an even larger splash. Group-based visualization, utilizing

visualization-equipped meeting rooms, has been available for some time but not seen widespread use. Is this because none of these has real value, or because the time was not right, or because the hardware was not ready to handle realistic geological datasets, or because they are, in fact, not that useful? In other words, is the relatively poor uptake of advanced visualization tools a tool issue, an adoption issue, or a gap in fit-for-purpose?

26.4 Three Technology Cases in Detail

An exhaustive examination of each of the potentially high-impact approaches discussed above is beyond the scope of this chapter, but the following three sub-sections provide examples of where the advances are happening, why they are happening, and their likely outcomes. Of course, the availability of a set of techniques is no guarantee of their adoption.

26.4.1 Using Game Engines to Simulate Rockfalls

Linear infrastructure such as rail lines and roadways, as well as open pit mining environments, are often at risk from rockfall. Addressing such problems requires the utilization of a considerable variety of data, including: 3-D spatial data to understand the slope and its rock structure, knowledge of rock properties, simulation tools to examine trajectories of moving particles, incident databases to calibrate and validate tools, and geotechnical insights about processes. Thus, the mitigation of rockfall risks involves adapting analytical tools developed by "outside" sources, incorporation of new data and user interfaces, and adoption of new modeling procedures. Ultimately a well-executed study might lead to: (i) improved protection of infrastructure, (ii) modified operating rules and guidelines for site investigations, and (iii) better recognition of the severity of direct and indirect hazards.

There are several tools currently in use to simulate rockfalls, including 2-D profile-based tools and 3-D slope-based tools (Turner and Schuster 2012). What they have in common is that they are purpose built, rely on custom codes for both simulation and visualization, and are variably limited in how flexible they are to new data types, new processing ideas, and new output types. Most are essentially as-is tools.

At Queen's University in Kingston, Ontario, a group of researchers has been working on the problem of infrastructure protection for over a decade. Embracing the idea that approaches from other, emerging, disciplines might address some of the limits on current simulation techniques, they undertook a research project on the use of customizable game engines as a platform for rockfall simulation. Two MSc theses (Ondercin 2016; Sala 2018) have been completed to date.

The initial project (Ondercin 2016) implemented rockfall modeling using the Unity game engine to evaluate slope and geological data from the White Canyon railway corridor in southern British Columbia, Canada as a test area. "Unity" is a low-to-no-cost game engine with extensive data import, modeling, animation, and physics-simulation capability developed by a large team of programmers and tested against many projects ranging from mobile games through complex desktop games. The included PhysX physics engine allows realistic physical simulation to control how game objects (such as rocks, buildings, and the slope itself) behave and change as they interact (Figure 26.1). The resulting rockfall simulation code is viewable, modifiable, and shareable. Ondercin (2016) demonstrated the general applicability of the method to 3-D rockfall simulation, and further demonstrated straightforward workflows that directly connected the Unity rockfall model to other visualization tools.

The second research project, by Sala (2018), focused on realistic handling of rockfall creation, fragmentation, and accumulation and furthermore validated the tool against a number of known rockfalls, including some where rocks were instrumented and dropped down a slope (Figure 26.2).

These projects show that the Unity-based simulation approach gives results that are at least as accurate as any existing, bespoke rockfall simulation tool. The new approach also has the advantage that it is open, extendable, and furthermore relies on a codebase that has seen professional development and scrutiny at the highest level. Data import and export is also straightforward. The Unity environment has a number of environmental simulation tools – built, of course, to make realistic games – that might also find a place in future simulation tools such as weather

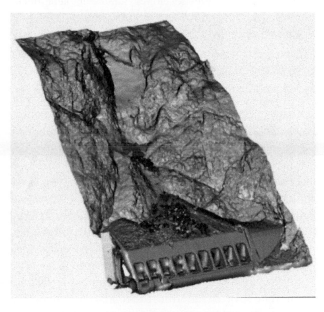

Figure 26.2 One-thousand fragment rockfall simulation using discrete and variable fragment shapes. Total volume is 100 m³. Modified from Sala et al. (2019).

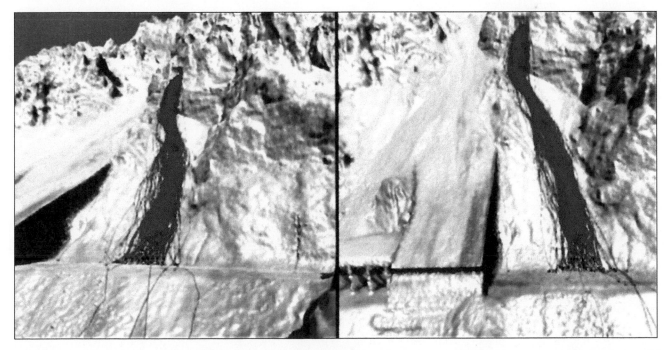

Figure 26.1 Two screenshots of the same rockfall simulation. The slope is captured by terrestrial Lidar scanning. This model has 287 blocks of cubic shape and 0.064 m³ size. Modified after Ondercin (2016).

and lighting systems, particle systems, simulations of agents such as vehicles and pedestrians, support for augmented and virtual reality, and more. Furthermore, the most recent Unity release supports multiple physics engines and much higher volumes of particles in a simulation; these capabilities will be capable of simulating many engineering design and remediation phenomena. While the underlying technology is very much a mass-market product, the Unity-based simulation approach is intended for expert use.

26.4.2 BGC Case Study – 3-D Visualization and Stakeholder Communication

Communicating 3-D problems that change over time due to physical, chemical, and environmental processes is challenging. Reliance on 2-D maps and drawings creates barriers to understanding. If projects have not been effectively communicated, their chance for success is significantly lower. Dating back to the early 1990s, several computer visualization systems were developed to bring 3-D models to life and allow people to visualize the data in 3-D environments. The first widespread commercial applications of virtual reality systems emerged in the 1990s (Lanier 1992). These systems enabled people to visualize an alternate environment from the comfort of their present location. The initial virtual reality systems were slow, complex and expensive; these characteristics hindered their adoption with both scientists and the general public.

The evolution of 3-D data visualization technologies was originally slow, but between 2010 and 2018 such systems have undergone significant advancement. The advancements are due to increases in computational power and investment by video game companies in real-time 3-D graphics. These same technologies are being utilized by geologists and engineers for visualization of 3-D data. The three most commonly deployed data visualization techniques include virtual, augmented, and mixed reality (Figure 26.3).

The need for 3-D geological modeling and for communication of information in 3-D is clear. Projects around the world are increasing in complexity, regulations are becoming stricter, and being able to demonstrate sustainable project plans to stakeholders is increasing. There is a growing need for technologies that allow stakeholders to view a project environment before, during and after construction in 3-D. True 3-D visualization tools foster better personal connections through shared experiences and facilitate discussions that were not possible previously.

Virtual reality allows a single user to immerse themselves in a virtual 3-D world completely constrained by 3-D data by using a headset typically connected to a computer. Augmented reality allows a user to visualize a 3-D scene or object superimposed in their physical environment using a 2-D screen, which enables the user to physically move around the superimposed object. Mixed reality, the most sophisticated technology, merges the physical and virtual world to overlay virtual objects in true 3-D space by

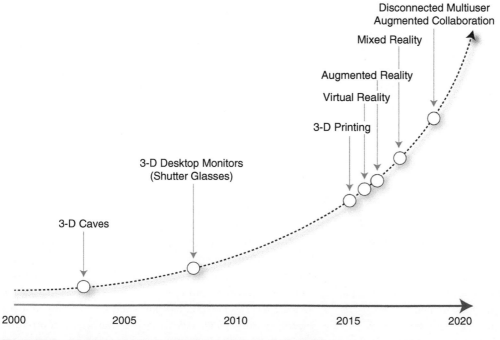

Figure 26.3 Evolution of 3-D visualization techniques (Lato 2018).

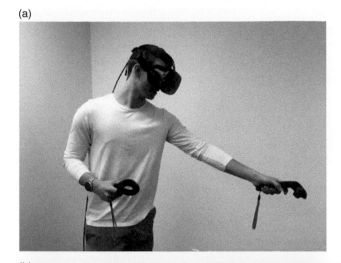

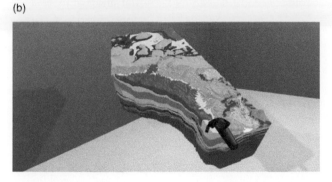

Figure 26.4 Two aspects of currently available virtual reality: (a) user employing VR headset to interactively visualize a geological model; (b) illustration of the virtual reality experience. Source: Courtesy of Alberta Geological Survey.

projecting the information through self-contained wearable holographic computers. These enable multiple users to visualize and interact with data in 3D, be disconnected from computers (no tether), while having the ability to see their actual physical environment (Figure 26.4). Mixed Reality allows geoprofessionals to combine data defining surface and subsurface features with engineering designs, while its real-time computer graphics can present complex models, regardless of background knowledge. It enables experts and community members to walk around and interact with a virtual version of a project from any angle or point in time.

Figure 26.5 illustrates one of the first uses of mixed reality for multi-stakeholder engagement in a public setting. It was decided to use artificial reality for this project to facilitate the understanding of a complex mine remediation plan of a contaminated site in Northern Canada. Use of 2-D maps and sections was judged as incapable of adequately communicating the complexities of this project to the public. The artificial reality component of the

project involved people with specialization in video game development, mobile computing, networking, geomatics, storytelling, alongside civil and geological engineers converting the various datasets from their native formats to artificial reality. The use of the HoloLens on this project provided positive outcomes that resulted from community members developing a better understanding of the complexities of the problem and the proposed solution.

The future of 3-D geological modeling will most certainly include tools to visualize data in 3-D. However, which technology will ultimately be embraced by geoprofessionals is yet to be determined. Technologies such as artificial reality that enable people to be immersed in a true 3-D environment are emerging and successes to date show a very promising trend for the future. As more successes are achieved, the use of artificial, virtual, and mixed reality technologies for understanding and communicating applied earth science projects will become the norm and will no longer be considered novel.

The reality is that geoscientists are slow to adopt new methods. Discussions at a meeting of European Geological Surveys, in Bern, Switzerland, in May 2019, addressed the problems associated with the disparate capabilities of various organizations to harness the rapidly evolving modeling technologies. Two particularly troubling trends were identified: (i) there is a danger of technological advances being too fast and laggards being unable to catch up with the leaders, and (ii) the critical baseline tasks such as data management and digitization are often ignored and are frequently underfunded. In a highly related area, Pavlis and Mason (2017) illustrate that 3-D geological field mapping is seeing significant adoption, but that this has taken decades, and was highly resisted by those familiar, comfortable, and committed to traditional mapping methods. Only the confluence of 3-D data collection, 3-D Visualization, efficient tablet-based data capture, and 3-D modeling tipped the balance in favor of new methods.

26.4.3 Using 3-D Geological Models to Enhance Decision-making and Stakeholder Communication

Many government geoscience agencies are responsible for making decisions at a regional or jurisdictional scale, and thus often require the use and implementation of extremely large coverage 3-D geological models, which can be extremely time consuming to build and typically require extensive datasets. However, building large national or jurisdictional-scale models has become a reality due to recent technological developments that allow computers and modified software programs to access cloud computing

Figure 26.5 HoloLens model of the Giant Mine project used at a community meeting. Source: Courtesy of BGC Engineering.

resources; these developments have made it possible for users to leverage additional computational power.

Building 3-D geological models to support decision-making and facilitate stakeholder engagement and communication provides significant value. These models provide the foundation for integrating a wide range of geospatial data types, such as urban infrastructure, surface water and groundwater resources, mineral resources, hydrocarbon resources, environmental conditions, and soil geochemistry. The ability to integrate geospatial information within a consistent and reliable 3-D context is critical to achieving a better understanding of the integrated relationships between surface and subsurface resources, geological features, and environmental conditions (MacCormack 2018). In turn, this integration supports decision-making based on a holistic characterization and understanding of the surface and subsurface geology, resources, and environment. The result is a more informed populace and, ultimately, better decisions.

Using 3-D models based on real data rather than conceptual drawings is essential to building trust and confidence with stakeholders. Such models demonstrate an agency's ability to understand the geological setting and make

holistic decisions based on the best available information. Because these models facilitate transparent communication of complex geological environments using tangible graphics and visualizations that are easy to understand, they are ideal tools for providing enhanced stakeholder communications and for geoscience education. Many geoscience organizations have been leveraging 3-D geological models to develop applications which are understandable by stakeholders and the general public. Augmented reality and virtual reality applications, interactive Minecraft models, and 3-D physical models produced by the recently developed 3-D printers can engage stakeholders of various ages and levels of expertise on geoscience subject matter (Figure 26.6).

26.4.4 Discussion of the Technology Cases

The future of geological modeling procedures and the future daily routines of geoscientists will undoubtedly differ from current experiences. The most significant drivers for this change will be cloud computing, which enables artificial intelligence and offers increased computational capacity, access to large volumes of diverse observational

(a)

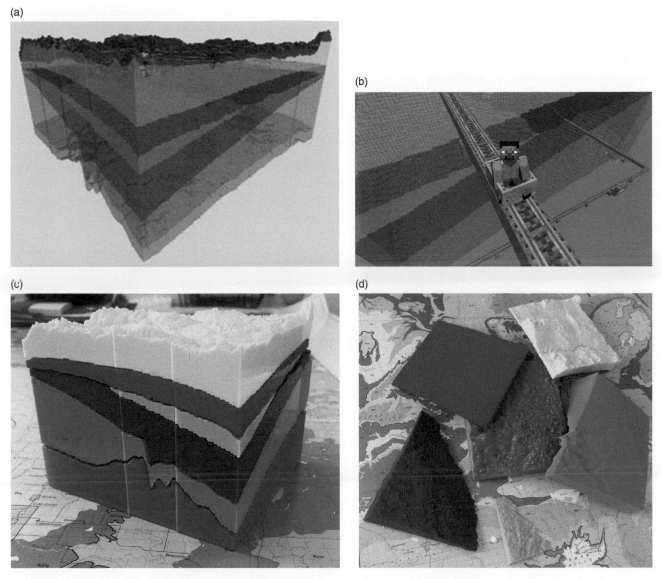

(b)

(c)

(d)

Figure 26.6 Examples of a 3-D geological model that are suitable for public participation and outreach: (a) conversion into an interactive Minecraft model; (b) Minecraft character on track provides users with a narrated tour of the model; (c) stacked 3-D print of Minecraft geological model shows stratigraphic relationships; (d) 3-D print model can be taken apart to allow users to explore the geometry of the individual geological units. Source: Courtesy of Alberta Geological Survey.

and real-time data, 5G mobile connectivity, and data visualization. How these technologies manifest as tools to support geoscientists is yet to be determined. It is unclear if specialty tools will be created for geoscientists or if leadership and adoption will emerge from other industries. All three case studies just presented illustrate the power of emerging technologies that offer revolutionary steps beyond existing standards-of-practice. Each example is a standalone, it is not a platform and does not reduce the complex workflow faced by a geoscientist. Workflows need to become simpler if significant advancement and widespread adoption of these new technologies is to be achieved.

26.5 Future Operational Considerations

Although it is always worthwhile to balance obvious trends against wild speculation, there are several areas where advancements with high impact to geological modeling activities appear likely to occur in the near future. Advanced information technologies will increasingly support and integrate future geological modeling and visualization procedures. Faster computers with better graphics, more integrated software tools, and "helper AI" will all make the use of sophisticated techniques much easier. This will promote the routine application of increasingly complex geological models to more projects and provide decisions that satisfy user needs and expectations. What, then, are the corresponding challenges and needs from within the specific community of geoscientists?

Major future operational considerations facing geological modeling projects undertaken by industry, government, and academia include the visualization of uncertainty, design of easily comprehended visualizations and user interfaces, and efficient methods for Internet-based dissemination.

Challenges remain in producing models and visualizations that clearly incorporate and display uncertainties in data or interpretation. Implicit realism can produce misplaced trust. Modern computer graphics are passing the point where a casual examination can distinguish real from artificial, and the expectation that seeing is believing is no longer valid to anyone who has played a modern game or seen a recent movie. Under these conditions, several critical conceptual and procedural questions arise. How do users identify realism? Does the model visualization indicate accuracy, or does it attempt to mislead? If we use advanced visualization methods to show uncertainty, will trust increase or disappear? Given the intimate links between geological knowledge, modeling results, and uncertainty defined by Oreskes et al. (1994), can the geoscience community provide benefits to clients by showing them uncertainty and the resulting high economic costs?

Design of suitable user interfaces is critical when responding to these questions. Expert users want to enter requests rapidly and precisely; they expect shortcuts. In contrast, novice users need assistance in understanding the options, perhaps by online help or tutorials. The novice user would be mystified by the "expert" interface, while the expert will become exasperated with the "novice" interface. This suggests a "family" of distinctive delivery products will be needed, each targeted to a distinct class of users. Sophisticated users must have capabilities to review and reprocess original data. In contrast, the general-public users usually want an answer, not the original data, which they probably cannot process and may not understand. While these multi-level interfaces could be designed based on existing CAD or GIS product interfaces, successful implementation depends on interfaces designed to use familiar interaction modes. Such interfaces are more likely to achieve traction than those that are unfamiliar. Therefore, future interfaces must consider the role of mobile phones or tablets, which are the most widely adopted method of information exchange and interaction. Furthermore, any information interfaces must support experimentation. Novices will click on things to see what happens, and they need to be able to "undo." Much research is needed on interfaces;

the arrival of entirely new devices, such as voice activated searches and head-mounted computers, highlights the dramatic changes in the technologies people use to interact with information. "Hey phone, check status of sensor array 4 and report back" was science fiction just a decade ago; now it is an emerging reality for many users.

26.5.1 Holistic Decision-making

Making decisions using 3-D models will be soon become the new norm; it will become an expectation of clients, decision-makers, and stakeholders. Recent scientific advances have demonstrated the close interconnections and relationships among geologic entities, resources, and the environment. Because a decision regarding one resource will likely result in impacts on other resources, decisions made regarding surface resources, subsurface resources, and the environment should be undertaken within a holistic geospatial environment that allows geoscientists, decision-makers, and stakeholders to visualize the relevant information and data within an integrated 3-D context (MacCormack 2016). The following case study from the Alberta Geological Survey demonstrates the value of integrating geospatial information within a 3-D geological model. The model was used to support science- and evidence-based decision-making, as well as facilitating communication of the subsurface conditions and resulting decisions to stakeholders (MacCormack 2016, 2018).

26.5.2 Alberta Geological Survey Case Study – Application of 3-D Models

The Alberta Geological Survey was asked to assist with an investigation into increased odors and emissions related to the production of heavy oil near the town of Peace River, Alberta. A 23-layer geological model (Figure 26.7a) was constructed to provide a more accurate representation of the geospatial data; this improved the Alberta Geological Survey staff's understanding of the subsurface conditions and supported a holistic understanding of the geological setting, integrated resources, and local environment.

The 3-D model was integrated with a wide range of geospatial data types defining roads, infrastructure, water wells, oil and gas wells, oil geochemistry data, and resource plays (Figure 26.7b). This provided a basis for assessing the spatial correlations between the various data elements and the occurrence of odors (Anderson et al. 2015). The ability to visualize the data in 3-D allowed the team to rapidly and easily determine that the wells reported as having the strongest odors were all produced from a similar zone in the subsurface (Figure 26.7c; Anderson et al. 2015).

(a)

(b)

(c)

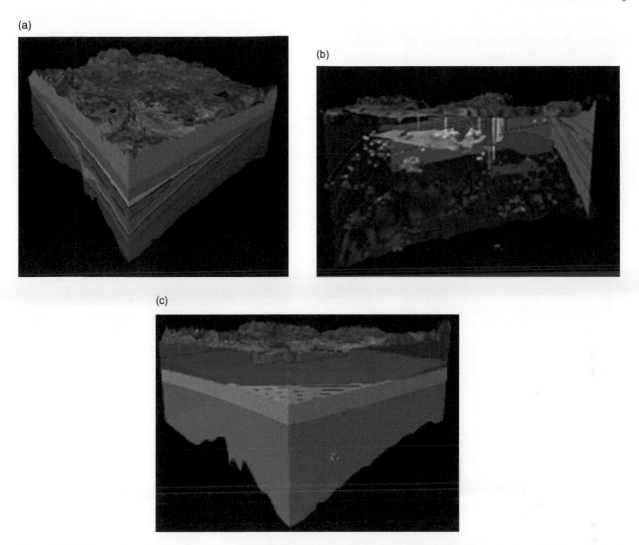

Figure 26.7 Example application of Alberta Geological Survey 3-D models: (a) 23 layer 3-D geological model; (b) multiple types of geospatial data are integrated within a single, holistic 3-D context; (c) the resulting 3-D zones showing the probability of future odors and emissions from heavy oil produced in this area. Source: Alberta Geological Survey.

Showing this information and data to decision-makers in a 3-D context ensured that everyone had a consistent understanding of the subsurface conditions and allowed them to more efficiently decide about future development within this region (MacCormack, 2016, 2018). The information provided to decision-makers would have looked quite different if the team had not been able to show it in 3-D. For example, all the highly volatile hydrocarbon data appear to be clustered inside the polygons in the 2-D view (Figure 26.8a). However, the holistic 3-D view (Figure 26.8b) revealed that the data are actually more laterally distributed. Therefore, the 3-D geological model and the ability to visualize the oil geochemistry data in 3-D space ensured that decision-makers were provided with the most accurate information possible. This 3-D model was also used to facilitate communication of the decisions

regarding mitigation of the odors and emissions associated with production in this region to stakeholders living in the area (Figure 26.8c). The 3-D visualization allowed them to easily communicate about the subsurface conditions, and what was being done to mitigate future production related odors.

26.6 Economic and Legal Issues

Societal decisions concerning funding and intellectual property rights are perhaps the most critical controls on the future of geological modeling and visualization. Subsurface characterization is a potentially critical part of resource assessment and planning, as well as site investigation. Society has typically under-funded these activities, except in the case of unusual or highly visible projects

(a)

(b)

(c)

Figure 26.8 Example model information provided to decision-makers: (a) 2-D representation of oil geochemistry data making it appear that the higher density oil (highlighted by the yellow box) occur within a specific location; (b) the same data visualized within a 3D context; (c) the higher density oil likely covers a much larger area within a 3-D zone (for which the base is represented by the green line). Source: Alberta Geological Survey.

or concerns. Thus, until very recently, the application of 3-D geological models and visualizations was restricted to a relatively few large, long-term, "important" projects. As technological advances reduce the cost of performing geological modeling and visualization, these methods will become more widely applied to a greater variety of studies.

However, several important issues involving questions concerning who is to pay, and how much, are unresolved. Society demands careful assessment of scarce resources, such as groundwater, and sustainability in development plans will encourage the wider use of these technologies. Failures of corporations to protect citizens are becoming a major driver for information sharing. For example,

the Investor Mining and Tailings Safety Initiative led by the Church of England and the Swedish Council of Ethics, which represents investors having over 10.3 trillion (USD) under management, has requested the executives of 683 extractive (mining) companies to publicly disclose information on all their tailings storage facilities.

26.7 Conclusions

Continued evolution and improvement of the existing methods to convert data to information, information to knowledge, and knowledge to action will create new pathways to a better understanding of geological systems. Future developments will see better integration between data within and external to the geosciences. Better access to data through open source or citizen science portals will feed AI systems and allow development of models that can learn from data across the world. The examples presented in this chapter represent vignettes of what is emerging in the field of 3-D geomodeling, but they are only scratching the surface of what is possible. As more prototype applications are developed, consolidation will occur, and these advanced bespoke solutions will become standard. As stated by William Gibson *"The future is already here, it's just not very evenly distributed"*.

References

Anderson, S.D.A., Filewich, C., Lyster S. and MacCormack, K.E. (2015). *Investigation of Odours and Emissions from Heavy Oil and Bitumen in the Peace River Oil Sands Area: 3-D Geological Modelling and Petroleum Geochemistry*. Edmonton, AL: AER/AGS, Open File Report 2015-07, 162 pp. [Online: Available at: https://ags.aer.ca/publication/ofr-2015-07] (Accessed December 2020).

Lanier, J. (1992). Virtual reality: the promise of the future. *Interactive Learning International* 8 (4): 275–279.

Lato, M. (2018). *Leveraging three-dimensional remote sensing in geotechnical engineering (Invited lecture, 2018 Canadian Geotechnical Society Colloquium, GeoEdmonton 2018)*. Ottawa, ON: BGC Engineering. 65 pp. [Online: Available at: http://v-g-s.squarespace.com/s/CGS_Colloq_2018_local-chapters.pdf] (Accessed December 2020).

MacCormack, K.E. (2016). *Three-Dimensional Geological Mapping: Workshop Extended Abstracts* (eds. K.E. MacCormack, L.H. Thorleifson, R.C. Berg and H.A.J. Russell), 95–99. Edmonton, AL: AER/AGS, Special Report 101. [Online: Available at: https://www.ags.aer.ca/publications/SPE_101.html] (Accessed November 2020).

MacCormack, K.E. (2018). Developing a 3D geological framework program at the Alberta Geological Survey; optimizing the integration of geologists, geomodellers, and geostatisticians to build multi-disciplinary, multi-scalar, geostatistical 3D geological models of Alberta. In: *Three-Dimensional Geological Mapping and Modeling: Workshop Extended Abstracts* (eds. R.C. Berg, K.E. MacCormack, H.A.J. Russell and L.H. Thorleifson), 64–67. Champaign, IL: Illinois State Geological Survey, Open File Series 2018-1. [Online: Available at: https://isgs.illinois.edu/2018-vancouver-british-columbia] (Accessed December 2020).

Ondercin, M. (2016). *An Exploration of Rockfall Modelling Through Game Engines*. Unpublished MSc thesis, Queen's University at Kingston. 200 pp.

Oreskes, N., Shrader-Frechette, K. and Belitz, K. (1994). Verification, validation, and confirmation of numerical models in the earth sciences. *Science* 263 (5147): 641–646. https://doi.org/10.1126/science.263.5147.641.

Pavlis, T.L. and Mason, K.A. (2017). The new world of 3D geologic mapping. *GSA Today* 27 (9): 4–10. https://doi.org/10.1130/GSATG313A.1.

Sala, Z. (2018). *Game-Engine Based Rockfall Modelling: Testing and Application of a New Rockfall Simulation Tool*. Unpublished MSc Thesis, Queen's University at Kingston. 206 pp.

Sala, Z., Hutchinson, D.J. and Harrap, R. (2019). Simulation of fragmental rockfalls detected using terrestrial laser scans from rock slopes in South-Central British Columbia, Canada. *Natural Hazards and Earth System Sciences* 19: 2385–2404. https://doi.org/10.5194/nhess-19-2385-2019.

Turner, A.K. and Schuster, R.L. [Eds.] (2012). *Rockfall: Characterization and Control*. Washington, DC: Transportation Research Board, National Academy Press. 658 pp.

Wood, B., Richmond, T., Richardson, J. and Howcroft, J. (2015). *BGS Groundhog® desktop Geoscientific Information System external user manual*. Keyworth, UK: British Geological Survey, Report OR/15/046. 99 pp. [Online: Available at: http://nora.nerc.ac.uk/id/eprint/511792] (Accessed December 2020).

Index

Note: Page references in *italics* refer to Figures; those in **bold** refer to Tables